Springer Praxis Books

Environmental Sciences

More information about this series at http://www.springer.com/series/4109

Donald Rapp

Assessing Climate Change

Temperatures, Solar Radiation and Heat Balance

Third Edition

Springer

PRAXIS

Donald Rapp
South Pasadena, CA
USA

Published in association with Praxis Publishing Chichester, UK

ISBN 978-3-319-33099-0 ISBN 978-3-319-00455-6 (eBook)
DOI 10.1007/978-3-319-00455-6

Springer Cham Heidelberg New York Dordrecht London

Printed on acid-free paper

Springer is part of Springer Science+Business Media (www.springer.com)

Contents

Figures

Tables

Acronyms and abbreviations

ABL	Atmospheric Boundary Layer
ACRIM	Active Cavity Radiometer Irradiance Monitor
AE	Auroral Electrojet (index)
AIT	*An Inconvenient Truth*
AOGCM	Atmosphere–Ocean General Circulation Model
BC	Black Carbon
BP	Before the Present
C&L	Chylek and Lohmann (2008)
CCN	Cloud Condensation Nuclei
CCPI	Climate Change Performance Index
CET	Central England Temperature
CFI	Comprehensive Flare Index
CFR	Climate Field Reconstruction
CME	Coronal Mass Ejection
CNES	Centre National d'Etudes Spatiales
CO2e	Equivalent CO_2 concentration to produce the equivalent effect of all greenhouse gases
CPS	Composite Plus Scaling
CQSM	Constant Quiet Sun Model
DIC	Dissolved Inorganic Carbon
DTR	Diurnal Temperature Range
EA	East Antarctica
EAIS	East Antarctica Ice Sheet
ECS	Esper, Cook, and Schweingruber (2002)
ELA	Equilibrium-Line Altitude
ENSO	El Nino/Southern Oscillation
EOS	Earth Observing System
EPA	Environmental Protection Agency
ERB	Earth Radiation Budget
ERBS/ERBE	Earth Radiation Budget Satellite/Earth Radiation Budget Experiment

EOF	Empirical Orthogonal Function
EPICA	European Project for Ice Coring in Antarctica
GCM	Global Climate Model
GDP	Gross Domestic Product
GHG	GreenHouse Gas
GICC	Glacial–Interglacial CO_2 Cycles
GIS	Greenland Ice Sheet
GISS	Goddard Institute for Space Studies (NASA)
GMST	Global Mean Surface Temperature
GNP	Gross Domestic Product
GRACE	Gravity Recovery And Climate Experiment
GSL	Global Sea Level
GSLR	Global Sea Level Rise
GSN	Group Sunspot Number
GST	Ground Surface Temperature
H&A	Hargreaves and Annan (2009)
HadCM3	Hadley Climate Model 3
HFC	HydroFluoroCarbon
IJC	International Journal of Climatology
IMF	Interplanetary Magnetic Field
IPCC	Inter-government Panel on Climate Change
IR	InfraRed
JGR	Journal of Geophysical Research
LFO	Low Frequency Oscillation
LGM	Last Glacial Maximum
LIA	*Little Ice Age*
LULC	Land Use/Land Clearing
M&M	McIntyre and McKitrick
M&W	McShane and Wyner (2010)
MBH	Mann, Bradley, and Hughes
MDI	Michelson Doppler Imager
MIROC3.2	Model for Interdisciplinary Research on Climate
MM	Maunder Minimum
MOC	Meridional Overturning Circulation (Atlantic)
MPH	Mobile Polar High
MSU	Microwave Sounding Unit
MWP	Medieval Warm Period
NADW	North Atlantic Deep Water
NAO	North Atlantic Oscillation
NCEP–NCAR	National Centers for Environmental Protection/National Center for Atmospheric Research
NDVI	Normalized Difference Vegetation Index
NH	Northern Hemisphere
NOAA	National Oceanic and Atmospheric Administration
NSF	National Science Foundation

OHCA	Ocean Heat Content Anomaly
OLR	Outgoing Long-wave Radiation
OPEC	Organization of Petroleum Exporting Countries
PC	Principal Component; Politically Correct
PCA	Principal Component Analysis
PDI	Power Dissipation Index
PFC	PerFluorohydroCarbon
ppmv	parts per million by volume
RE	Radiative Effectiveness
RSL	Relative Sea Level
SAT	Surface Air Temperature
SD	Standard Deviation
SEAS	NOAA's XBT program
SETI	Search for Extraterrestrial Intelligence
SH	Southern Hemisphere
SMAX	Sunspot MAXimum
SMIN	Sunspot MINimum
SMM	Solar Maximum Mission
SN	Sunspot Number
SO	Southern Oscillation
SOHO	SOlar Heliospheric Observer
SOI	Southern Oscillation Index
SORCE	SOlar Radiation and Climate Experiment
SST	Sea-Surface Temperature
TAV	Tropical Atlantic Variability
TIM	Total Irradiance Monitor
TMN	Temperature Measurement Network
TOA	Top Of Atmosphere
TOPEX	Poseidon Ocean Topography Experiment
TSI	Total Solar Irradiance
UAH LT	University of Alabama in Huntsville Lower Troposphere
UARS	Upper Atmosphere Research Satellite
UHI	Urban Heat Island
UNFCCC	U.N. Framework Convention on Climate Change
U/P	Umbra/Penumbra
USHCN	U.S. Historical Climate Network
UT	Upper Troposphere
VEI	Volcano Explosivity Index
VIRGO	Variability of solar IRradiance and Gravity Oscillations
WA	West Antarctica
WAIS	West Antarctica Ice Sheet
XBT	eXpendable BathyThermographs
YBP	years before the present

Preface

THE GLOBAL-WARMING DEBATE

The Earth has gone through incredibly wide climate changes over hundreds of millions of years. A number of factors contributed to this, including a gradually strengthening Sun, drift of continents, incidence of volcanism, etc. But it is widely believed that the Earth's thermostat was mainly controlled by prevailing CO_2 concentrations in the tug of war between CO_2 emissions and CO_2 burial. Thus, the "accepted paradigm" is that, over geological time, variability of CO_2 was the main controller of the Earth's climate (Foster *et al.*, 2009). For example, 500 million years ago, the CO_2 concentration was much higher (possibly 20 times) than it is today and the Earth was very balmy. There is little doubt that, if we again increased CO_2 to 20 times the pre-industrial value, the Earth would become considerably more tropical. The question before us now is how the Earth's climate will respond to a much smaller increase in CO_2: a mere doubling from 280 ppm to 560 ppm? There is little doubt that some temperature rise will result from this. Alarmists believe that the temperature rise will be sufficiently great to cause great harm to humanity. Skeptics think we can take in stride the relatively modest temperature rise that ensues. In between, there are so-called "lukewarmers" who are unsure. Perhaps the biggest worry is that, if the alarmists are right, we would require immediate draconian reductions in carbon emissions in which we would ramp down use of fossil fuels by 80% (or more) in a few decades. With world population growing and future energy demand projected sharply upward, how would we provide the world with needed energy while eliminating fossil fuels? There does not seem to be an answer.

Global-warming alarmists (aka "warmists") believe that human production of greenhouse gases, particularly carbon dioxide, with its concomitant water vapor feedback mechanism, has begun to add to the natural greenhouse effect, thereby raising global temperatures inordinately during the 20th century, with predictions of further increases in the 21st century that they claim would be catastrophic.

James E. Hansen, a leading spokesman for the alarmists, said that "Ignoring the climate problem at this time, for even another decade, would serve to lock in future catastrophic climatic change and impacts that will unfold during the remainder of this century and beyond". The Earth "is close to dangerous climate change, to tipping points of the system with the potential for irreversible deleterious effects. ... The planet is on the verge of dramatic climate change". We "are forced to find a way to limit atmospheric CO_2 more stringently than has generally been assumed. ... We cannot shrink from our moral responsibilities ... to preserve the planet for future generations".

Al Gore's film *An Inconvenient Truth* has carried the message to many millions of people. This has spawned a growing world movement that is seeking controls on greenhouse gas emissions. Because such controls would have serious economic consequences and, furthermore, attempts to apply controls have been unbalanced relative to developed countries vs. developing countries, there has been strong resistance to such moves by skeptics.

The majority of research climatologists at universities are alarmists to some degree. In today's political environment, it is necessary to be an alarmist (or at least to appear to be one) in order to obtain research funding. Some university and government agency alarmists have banded together in an informal association to (1) exert pressure on journal editors to prevent contrary views from being published, (2) to very quickly publish rebuttals to those few contrarian papers that slip through their net, and (3) immediately produce nasty, vicious attacks on contrarian papers on their blogs. These alarmists pompously refer to their interpretations as "*climate science*", implying that any contrary views are not *climate science*. Yet, most of the work published by these university climatologists is based on sparse, noisy, short-term data and their results typically are based on drawing a dollar's worth of conclusions from a penny's worth of data.

Skeptics span a wide range of viewpoints ranging from uncertainty regarding how much of recent global warming was primarily induced by rising CO_2 levels, to outright denial that there is any connection between rising CO_2 and global warming. Skeptics maintain blogs and circulate reports, but have only occasionally penetrated the peer-reviewed scientific literature that is dominated by alarmist publications. Alarmists provide the impression of scientific integrity through peer-reviewed publications, while skeptics often (but certainly not always) lack the credentials of alarmists. The alarmist position is often "sold" on the basis of the number and importance of climatologists and institutions that subscribe to this persuasion. But the important thing is data, not credentials and number of adherents.

Both sides have argued like trial lawyers with a case to be made, by craftily selecting ("cherry picking") bits and pieces of data to support their preconceived viewpoints. Fact, supposition, speculation, and pseudo-science have been mixed together in a brew that is confusing and difficult to resolve. I recently checked *amazon.com* books to see what's available in the way of climate books. There seemed to be about 100 new books in the last two years. All of those written by establishment climatologists present the alarmist view. All of those written by anti-establishment writers present the skeptical view. Both sides are absolutely sure they are right. Neither side has any doubt that I can discern. One thing the world does not need is another high-level book arguing with assurance on one side of the question or the other, with little actual data content, manipulated to achieve an apparent conclusion based on specious arguments and cunningly selected references.

In this book, I have investigated a large body of technical data relevant to global climate change, approaching each element with necessary (but hopefully neutral) scientific skepticism. As Einstein said: "The goal is to be as simple as possible, but not simpler." Thus, by necessity, this book is quite technical, but hopefully still quite readable.

The essential questions are:

(1) How well has the world monitored near-surface temperatures of the 30% land and 70% ocean areas on the Earth during the past 120 years or more, and how well can we characterize the changes in climate over that time span?
(2) What are the utility and significance of a single global average temperature?
(3) How has the Earth's climate varied over the past Ice Ages, the Holocene, the last millennium, and the past century, and what can we infer about "natural" variability of the climate prior to industrialization by humankind?
(4) How reliable are proxies for historical temperatures? What do we really know about past temperature variations? Is the *hockey stick* version of millennium temperatures credible, in which temperatures were relatively flat for 2,000 years prior to a sudden rise in the 20th century?
(5) How does the current global-warming trend compare with past fluctuations in the Earth's climate, and what is the likelihood that the warming trend we are experiencing now is primarily just another in a series of natural climate fluctuations as opposed to a direct result of human production of greenhouse gases?
(6) How credible are the global climate models that claim that greenhouse gases produced most of the temperature rise of the 20th century, and forecast much greater impacts in the 21st century?
(7) How good were the "good old days"? Was the climate of the *Little Ice Age* ideal, should we abhor warming from that baseline, and do we want to return to the climate of the early 19th century?
(8) How will limits on fossil energy supplies constrain future CO_2 production and climate change, even if the climate models are accurate?
(9) How can the world provide itself with energy needed for a burgeoning population that will demand more and more energy in the future, considering the finite limits on fossil fuel resources?

According to Beckman and Mahoney (1998):

> "The vested interests on both sides of the argument between the 'greenhouse' party and the 'solar warming' party are obvious. Scientifically, the meteorologists, climatologists, and atmospheric physicists, who were responsible for 'discovering' the human contribution to the terrestrial greenhouse effect, have been the most consistent champions of its importance, while the solar physics community, and especially those interested in solar–terrestrial relations, have

> increasingly stressed the possible importance of the long-term variations of the solar constant as the chief cause of climate change. Both communities tend to take the change for granted, and to neglect any purely statistical or chaotic effects which could lead to excursions of the Earth's surface temperature during periods of a couple of decades, without requiring a secular change either in the solar constant or in atmospheric transparency. In addition, the debate is conditioned by more powerful vested interest groups. The oil industry in all its guises would obviously like to believe, and would like the public to believe, that greenhouse warming has been greatly exaggerated, and exploits any genuine scientific differences to undermine the credibility of the climatologists."

Unfortunately, many global-warming alarmists have weakened their cases by building them around models and analyses of dubious veracity and, in the case of the infamous *hockey stick* temperature profile, mathematically incorrect manipulations of past temperature data from proxies. From this, they have concluded improperly that the late 20th century is far warmer than any time in the distant past, and made other elaborate claims regarding recent warming trends and dire predictions for the near future that are unsupportable by the evidence. Furthermore, the network for monitoring the Earth's temperatures is inadequate to precisely characterize the trends in climate for the past 100 years, and the utility of a single global average temperature in defining the state of the climate is limited.

Projections for the 21st century are typically far out of line with realistic expectations. The credibility attributed to global climate models belies their inherent fragility. This has provided the skeptics with plenty of ammunition with which to debunk these exaggerated claims. On the other hand, most of the skeptics made up their minds *a priori* that global warming in the 21st century due to CO_2 emissions is not a potential problem, and their arguments are often vague and hardly convincing.

A major problem in discussing climate change is that we lack a time period that we can objectively define as a base for comparison. As Anon. (N) emphasized, temperatures near the end of the 20th century were generally higher than those of the preceding four centuries. Taken at face value, this seems to imply that the preceding temperatures were normal, while the relatively higher temperatures at the end of the 20th century are comparatively abnormal. However, the preceding four centuries extended across the *Little Ice Age*, and therefore one might state the proposition differently: temperatures during the preceding four centuries were colder than they were at the end of the 20th century. Stated this way, the abnormality is attributed to the *Little Ice Age*. Perhaps the most accurate statement is that there is no normal climate, and the climate of the Earth has always varied widely, and continues to do so to this day. As Balling, Vose, and Weber (1998) said:

> "... it is entirely possible that the warming in the record of the past century has been caused by an unusually cool period 100 years ago as opposed to an unusually warm period in recent decades."

By carefully sifting through the evidence, we find that there are no ironclad answers to major questions on global climate change. Our temperature data for the past are fragmentary and sparse, both spatially and temporally. Urban heat islands and land clearing have affected measured temperatures. Past variations in solar irradiance can only be estimated with speculations. Proxies used to estimate the temperature history of the Earth over past millennia are noisy and inconsistent, leaving us with uncertain indications of the past. Climate models do not deal realistically with water vapor, aerosols, and clouds, resulting in wide variations from model to model.

The thesis of this book is that our data and models are presently inadequate to reach credible conclusions regarding how much global warming is likely to take place in the 21st century. We have emerged from the *Little Ice Age* in the latter half of the 19th century and the Earth has warmed, but the connection to greenhouse gases is uncertain. Estimates for the future depend on climate models that cannot yet properly account for changes in humidity, aerosols, and cloud cover as greenhouse gas concentrations increase.

Scientists (and the public) abhor a vacuum. They can't seem to shrug their shoulders and admit that we just don't know the answers to important questions. They introduce explanations, however speculative. Thus, we are besieged with models purporting to describe the past millennium's history of the Earth's climate, and making firm predictions about the future, none of which stands up to detailed review. The *alarmist cabal* that controls the paleo-climatological literature has a political agenda to promote public concern about greenhouse gases and, in many cases they have lost objectivity. We have ended up with two opposing camps: the alarmists and the skeptics, each 100% convinced they are right, and each firmly for or against a global-warming catastrophe, each seemingly more concerned with furthering their agendas than with discovering truth. What seems to be missing in all this is a little bit of humility. Speculation is rife. Data are lacking. Use of complex models obscures the fact that the data are limited and, in many cases, untrustworthy. Meanwhile, there is much to be gained on a personal basis from the international preoccupation with climate change. Grants and contracts are awarded, often to the most strident voices supporting the orthodoxy. Blogs have evolved like religious factions, swearing allegiance to their orthodoxies, and twisting the facts to fit their preconceived notions. Any moron can have his say—and does.

The world will face a crisis sometime around or after 2030. But that crisis will not be calamitous global warming. While global warming might become a significant problem late in the 21st century, the greater, and nearer-term crisis will be that, with oil, gas, and coal production going at full bore, the world will not able to supply the energy that is demanded by a growing world population intent on using energy at higher rates. This could lead to significantly higher energy costs, resulting in worldwide economic recession or depression. However, on the positive side, it will provide great incentive to develop renewable energy that will then become more competitive. Whether renewable energy can be developed and expanded rapidly enough to stave off economic collapse remains to be seen. The same people who are of the climate alarmist persuasion also seem to be unduly optimistic about the prospects for renewable energy to supplant fossil fuels.

SUMMARY

There is a good deal more that we don't understand about climate variability than we do understand. As Wunsch (1999) emphasized:

> "Sometimes there is no alternative to uncertainty except to await the arrival of more and better data."

The warming of the 20th century represents emergence from the *Little Ice Age*. The quantitative role of greenhouse gases in this warming remains uncertain. We need better data and better models; there are too many speculations that have hardened into beliefs.

Scientists have studied past climates using proxies. Proxies are residual data from processes that occurred in the past, when the processes were dependent on local temperatures at the times they took place, and the evidence is preserved in the present in an accessible form.

Chapter 1 deals with long-term climate change.

Chapter 2 examines what we have learned from proxies about major climate changes that have occurred over the past two millennia. A detailed study of proxy evidence for climate fluctuations in the last millennium or two has led to a controversy. Some climatologists claim that temperature variability in the last millennium or two was small prior to the 20th century, leading to a *hockey stick* graph of temperature vs. time with an unprecedented sudden rise in the 20th century. Alternatively, there is considerable evidence that temperatures varied significantly during the past few millennia, with a warm period about 1,200 years ago and a cold period from about 1600 to about 1850—the so-called *Little Ice Age*. This controversy is examined in considerable detail. Is the temperature rise of the 20th century unique, and must it be attributed mainly to increased CO_2?

Chapter 3 analyzes the measurements of Earth surface temperature that were made in the past century or so and discusses the limitations of the measurement network. Temperature data are reviewed in considerable detail. The primary trend during the 20th century has been upward from the base of the *Little Ice Age*, but this temperature rise has not been uniform, either geographically or in time. The limitations of attempting to describe the Earth with a single global average temperature are emphasized.

Chapter 4 provides a discussion of generic global scares. It also discusses subjective science: the science of inference when we cannot go back in time to validate our hypotheses of what might have taken place in the past. It then goes on to discuss the alarmist movement and the ways that climatologists think.

Chapter 5 provides an in-depth review of solar irradiance: historical observations, recent measurements, theories, and models, and use of proxies to estimate past irradiance. Reliable measurements only exist for the past 30 years. There remains considerable uncertainty as to how solar irradiance varies over time periods of centuries. There are a number of models that attempt to estimate solar intensity in the past, but each of these is based on assumptions that are impossible to validate.

In Chapter 6, the Earth's heat balance and the greenhouse effect are discussed. Recent estimates of energy transfer within the Earth system are reviewed. The Earth's current energy balance is discussed with emphasis on the heat content of the oceans—which represent most of the heat capacity of the Earth's surface. Climate variability due to volcanoes and El Niños are reviewed. The concept of climate sensitivity is introduced. Finally, the relationship of CO_2 concentration to climate over 500 million years is reviewed.

Chapter 7 deals with anthropogenic influences on climate change. While most studies have focused exclusively on the putative effects of greenhouse gases, other important effects include the deposition of Arctic black carbon snow and ice, land modification, and generation of sulfate aerosols. The major challenge of the 21st century is not global warming, but rather providing the people of the world with energy. Problems due to climate change must be discussed as a subsidiary element of energy production and consumption.

Chapter 8 deals with the potential impacts of future global warming by contrasting the views of alarmists and skeptics. While alarmists have exaggerated many risks, a significant concern is the potential future rise in sea level. Other threats of global warming have been exaggerated by alarmists.

Chapter 9 deals with climate change and public policy. Governments are rushing to judgment on the putative role of CO_2 and enacting policies that will cost trillions, and may induce economic disaster. Economists have invaded the field of climate change and as with all aspects of climate change, there are opposite viewpoints expressed with great certainty based on flimsy foundations.

Chapter 10 provides a summary and conclusions. The Earth's climate is very complex. There is more that we don't know than we do know about climate change. Past temperatures; long-term variation of solar irradiance; variations in ocean circulation; reaction of cloud cover, humidity, and aerosols to rising greenhouse gas concentrations; heat exchange between the Earth and the atmosphere all remain (like most analyses of climate) speculative, conjectural, and unproven. Nevertheless, the predictions of alarmists can be treated as worst-case possibilities and these are a cause for concern. The bigger problem, however, is an increasing population with growing need for energy.

Appendix I reviews the widely viewed Al Gore film *An Inconvenient Truth*, for which he received a Nobel Prize. This glossy, glib presentation has little actual content, and much of it is wrong. Appendix II provides a model for warming of the oceans as greenhouse gas concentrations increase. While some websites claim that increased greenhouse gas concentrations do not warm the ocean, we show that in fact they do.

In this Third Edition, I have added more than 285 new references, over 180 manuscript pages of new text, and more than 80 new figures. Particularly notable are the following additions:

- Additional data and discussion related to the Medieval Warm Period and the Little Ice Age.
- Information and discussion of so-called "climategate" revelations

- Further revelations on manipulations of proxy data and the "*hockey stick*"
- Expanded coverage of the connection of climate to El Niños
- Updated data on global and regional temperatures
- Data and discussion of weather extremes
- Expanded data and discussion of mountain glacier retreat
- Greatly expanded treatment of Earth heat balance
- Greatly expanded treatment of global warming due to CO_2
- Expanded discussion of humidity and cloudiness effects
- Inclusion of Spencer and Eschenbach models
- Extensive discussion of estimates of climate sensitivity from paleo geological data
- Updates on heat content of the oceans
- Greatly expanded treatment of climate forcings
- Expanded discussion of models for future emissions
- Expanded discussion of the role of black carbon
- Greatly expanded treatment of sea level change
- Greatly expanded treatment of sea ice extent
- Comparison of Holocene to previous deglaciations
- Expanded discussion of "Stern Report" and other economic analysis of remediation of climate change
- Discussion of ocean warming due to increased greenhouse gas concentrations

1

Long-term climate change

1.1 WEATHER AND CLIMATE

The surface of the Earth is a three-dimensional system that is dominated by water that can exist in three phases: solid, liquid, and vapor. At any point in the evolution of the Earth's climate, randomly varying factors such as winds, ocean currents, air masses, cloud formation, volcanic eruption, the spinning of the Earth, and other factors produce daily, monthly, and yearly fluctuations in the weather in various regions. If we average out these fluctuations over the whole Earth over a period of time, we can attempt to attribute a climate to the Earth over that time period. There is no universally accepted procedure for doing this.

First of all, it is not clear *a priori* how long a duration is required to define a climate. Some books claim that an average of weather over 30 years is adequate, but this is rather arbitrary. There is some evidence that there might be multi-decade cycles within a basically stable climate, so it would be safer to use a century as the duration to define a climate. The yearly variations in weather are typically so large that it is very difficult to define a long-term average with only a few decades of weather data.

Lovejoy (2013) argued that the long-term average of weather is what he calls "macroweather". He concluded: "True climate processes only emerge from macroweather at even longer times, and this thanks to new slow internal climate processes coupled with external forcings." While there is no specific timescale for identifying climate, the implication is that it is usually longer than a century.

Secondly, it is not clear *a priori* how to define the climate. Is it the average temperature of the air near the Earth's surface for all locations on the Earth? What about tropospheric temperatures? And what about the oceans? How does the temperature profile of the oceans vs. depth enter into climate? Other factors that might enter into defining the climate include precipitation patterns, temperature extremes, degree of glaciation, and prevalence of intense storms.

It is fairly common for climatologists to define climate as the average temperature of the Earth near the surface. This introduces various problems. One problem is inadequate coverage by measurement stations. Another is the question of

D. Rapp, *Assessing Climate Change*, Springer Praxis Books,
DOI: 10.1007/978-3-319-00455-6_1

how to average in ocean temperatures, noting that oceans cover about 70% of the Earth's surface.

1.2 TIMESCALES TO CHARACTERIZE CLIMATE

Climatologists have attempted to characterize the Earth's climate over many time periods, as discussed below.

Temperatures over the last decade: The Argo system to monitor world ocean temperatures was installed.
Temperatures over the past three decades: Use of satellite observations to infer global tropospheric temperatures.
Temperatures over the past ~120 years: Several groups have combined results from a worldwide network of land measurement stations plus sea-surface temperatures (SSTs) from ships to infer global average temperatures since about 1880 or 1890.
Temperatures over the past ~210 years: The so-called "BEST" group led by Richard Muller combined results from a worldwide network of land measurement stations to infer global average land temperatures since about 1800.
Temperatures over the past ~2,000 years: A number of groups have attempted to use "proxies"[1] to estimate historical temperatures at various locations, and combine these into an estimate of global average temperatures over two millennia. Unfortunately, the number of proxies drops sharply as one goes back in time, and the spatial coverage leaves much to be desired. The veracity of many proxies is also suspect.
Temperatures over the past ~400,000 years: Ice cores drilled into the Greenland ice sheet reveal $^{18}O/^{16}O$ ratios that can be used to infer temperatures near Greenland dating back as far as 400,000 years.
Temperatures over the past ~800,000 years: Ice cores drilled into the Antarctica ice sheet reveal $^{18}O/^{16}O$ ratios that can be used to infer temperatures near Antarctica dating back as far as 800,000 years.
Temperatures over millions of years: $^{18}O/^{16}O$ ratios in ocean sediments have been used to estimate global temperatures over millions of years.
Temperatures over hundreds of millions of years: Geological evidence in terms of fossil flora and fauna indicate the levels of global climate dating back as far as 500 million years or more. The data suggest that the Earth has gone though rather wide swings in climate over hundreds of millions of years, as shown in Figure 1.1.

[1] In the context of historical temperatures, proxies are residual data from processes that occurred in the past, when the processes were dependent on local temperatures at the times they took place, and the evidence is preserved in the present in an accessible form. In all cases, extraction of implied past temperature data from confounding influences requires considerable analysis and manipulation. Some prominent proxies include: tree rings (width, density, stable isotope composition), ice cores (oxygen isotope ratios, gas content in bubbles), ocean sediments (isotope ratios), pollen, boreholes, and corals.

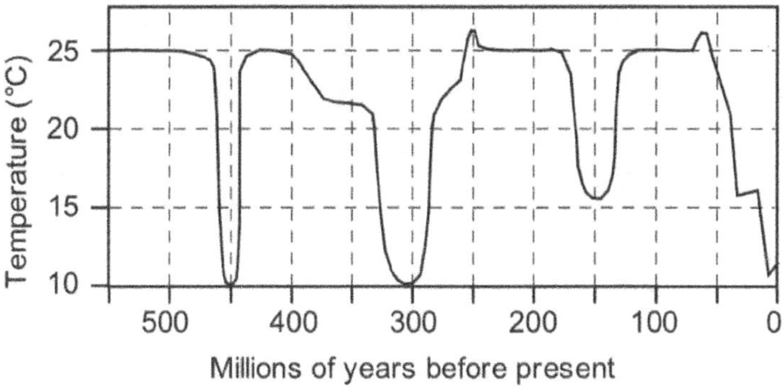

Figure 1.1. Estimate of global average temperature over the past ~500 million years. (adapted from Scotese, 2002).

1.3 FACTORS THAT AFFECT LONG-TERM CLIMATE

The Earth's climate is basically determined by an energy balance between the radiant flux emitted by the Sun, the fraction of this flux that is absorbed by the Earth system (with the remainder reflected back into space), and the temperature profile of the Earth from the surface to the stratosphere that determines the rate of radiant heat loss from the Earth to space.

1.3.1 Variability of solar luminosity

Physical models of the Sun indicate that four billion years ago, the innate solar luminosity was only about 70% of what it is today. As time progressed, the luminosity increased. Five-hundred million years ago, the solar luminosity was about 6% lower than today. Since then, it has increased slowly. It is interesting to ask how the Earth has managed to remain a habitable planet with globally distributed liquid water during the past 4 billion years and why it did not turn either into an icy planet. This question is known as the "The early faint Sun paradox". Some believe that higher levels of greenhouse gases from active volcanism kept the Earth warm despite the much lower solar irradiance. This is discussed further in Section 5.1.4.

The rate of change of the fusion process in the Sun is extremely small (10^{-4} W/m^2) per 1,000 years. Therefore, the variability potential of the fusion process is completely negligible in regard to timescales relevant to current climate change (tens to hundreds of years). On the other hand, the appearance of the Sun changes over tens to hundreds of years and there is an ~11-year solar cycle with associated small changes in luminosity. Does the solar luminosity vary enough to affect climate over decades and centuries? We have data on solar luminosity over the past 30 years. However, there are some problems aligning the data from several instruments with different calibration procedures, so there is some uncertainty in the results. The data seem to suggest that variability over this 30-year period cannot account for observed changes in climate. In addition, a number of models have been developed to estimate solar luminosity over the past few centuries. Some models indicate that variability of luminosity was small over the past few hundred years, while others indicate that variability of luminosity was sufficient to impact the climate.[2] There is no definitive answer to this question, but it seems unlikely that changes in solar luminosity had a major effect on climate over the past several centuries. The data and models are discussed in Chapter 5.

1.3.2 Variations in reflected solar flux

While variations in solar luminosity over decades and centuries are probably small, variations in the fraction of incident solar flux that is reflected by the Earth into space (the "albedo") can be significant. The energy flux absorbed by the Earth system is (1—albedo) times the solar flux impinging on the Earth. Since the solar flux impinging on equatorial and lower mid-latitude regions is the major solar input to the Earth, it is the albedo in these regions that is of the greatest importance. The albedo depends primarily on the area and location of landmasses, the amount and location of glaciation on Earth, and the distribution and thickness of clouds.

In general, water has a low albedo, land has a higher albedo, and snow and ice have very high albedos. Fresh snow and ice have albedos as high as 0.9 while dirty snow and ice and sea ice might have albedos as low as 0.3 to 0.4. Land encompasses a wide range of albedos from 0.15 to 0.45 with an overall average of roughly 0.25. Oceans (which presently cover 70% of the Earth) have an average albedo of less than 0.1 when the Sun is not at a large zenith angle. Clouds determine the fraction of incident solar flux that reaches the surface. Thick clouds have an albedo of 0.6 to 0.9 and thin clouds have an albedo of 0.3 to 0.5. An estimate of the average solar flux over the Earth and the absorption and reflection by the atmosphere and surface for the period 2000–2010 is given in Figure 1.2. In this model, 22% of the incident solar flux is absorbed by the atmosphere, 48.5% is absorbed at the surface (predominantly by oceans), and 28.7% is reflected back into space. However, other assessments suggest that the albedo of the Earth varies around 31%.

The net albedo of the Earth depends on a global average of all land and ocean areas, but clouds add significantly to the global average albedo.

Since the general acceptance of the continental drift theory, it has been widely surmised that changing continental geometries likely contributed to long-term climate change. However, it is not exactly clear how this occurred.

> "Continents are important to climate for three main reasons: They are a platform upon which polar glaciers can form; They are the primary sites of the silicate weathering reaction that governs atmospheric CO_2, and the amount of weathering is strongly affected by the continental configuration; They affect the geometry of ocean basins, and hence the ability of oceans to transport heat from one latitude to another" (Pierrehumbert, 2009).

[2] This conforms to the first two laws of climatology. Law #1: If you have two climatologists, they both differ. Law #2: They are both wrong.

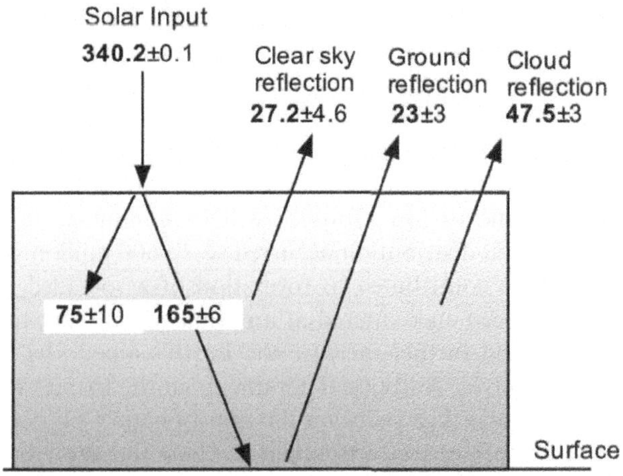

Figure 1.2. Solar power input to the Earth (W/m^2) for the decade 2000–2010 (Stephens *et al.*, 2012).

Dietz and Holden (1970) provided a good description of the continental drift process. Campbell pointed out:

> "... at certain times in the past, the equatorial ocean currents were able to circulate the Earth. This allowed more warming because of a much higher rate of ocean re-circulation and currents diverging from the equator to the north and south would be warmer. Equatorial flows at other times have been blocked resulting in higher latitude currents forming circumpolar currents. This isolates the polar continents and causes polar temperatures to drop."

Frakes and Kemp (1972), Sellers and Meadows (1975), Kennett (1977), Ravelo *et al.* (2004), and many others have discussed the effects of continental drift on climate.

Several studies have been carried out using climate models to estimate the effect of hypothetical landmass distributions on global climate. However, none of these is entirely convincing. Nevertheless, without using climate models, we can draw a few conclusions in this regard.

During epochs when there are landmasses at high latitudes or at the poles, the potential for glaciation of polar areas increases significantly since snow falling on land can accumulate. This can increase the overall average albedo, leading to further cooling. Thus, landmasses at high latitudes are widely believed to be conducive to colder climates. Conversely, when most of the continents are at tropical or mid-latitudes, accumulation of snow and ice will be constrained and the albedo of the Earth will be smaller. This will induce warming, and the lack of polar ice indicates that the oceans will be higher. Thus, continental margins will be flooded, and the area of exposed continents will decrease (i.e. land area is converted to water area). This will decrease the global albedo further, producing more warming. Hence, the occurrence of landmasses at polar or moderate latitudes, promotes global cooling or warming, respectively.

While landmasses in polar areas are ideal sites for ice sheet formation, the total heat balance of the Earth is determined by how much solar energy gets absorbed. Since the preponderance of solar energy input to the Earth is in the tropics (where solar energy per unit area is a maximum, and surface area per unit latitude is also a maximum), absorption of solar energy in the tropics is of paramount importance. Since land has a much higher albedo than water, an unusual preponderance of landmasses within middle to low latitudes will be conducive to global glaciation. Indeed, such a continental distribution occurred some 600 million years befire oresebt (MYBP), and may have contributed to formation of a so-called "snowball Earth". This situation has not been encountered at any time subsequent to that period. Any resultant glaciation would further increase the Earth's albedo by lowering sea level, exposing continental shelves. Additional continents in the tropics would also increase the silicate weathering rate, thus reducing the atmospheric CO_2 concentration. It was suggested that these combined effects might lead to the growth of large ice caps, nucleated on islands or continents bordering the polar seas (Kirschvink, 2002).

Burrett (1982) carried out paleocontinental reconstructions for the period 570 to 200 MYBP and made rough estimates of land distribution amongst deserts and forest, in order to estimate the albedo of the Earth. His goal was to test how geographical placement of land and overall global albedo affected paleoclimates. He did not find any obvious correlations of land placement and albedo with the onset of glaciation and suggested that the issues are more complex. It seems likely that merely having a landmass at one pole without major barriers to overall ocean flow may not lead to glaciation.

Another very important factor in determining the global climate is the network of pathways for ocean currents to transport heat. When the polar areas are openly exposed to ocean currents from equatorial zones, heat is efficiently transported to polar areas, thus reducing glaciation, raising the oceans, and warming the planet. When the polar areas are thermally isolated from equatorial zones, they are more likely to freeze over, thus cooling the planet. The presence of a wide network of mid-latitude landmasses can obstruct transport of heat to polar areas.

Warm, wet landmasses located in the tropics enhance uptake of CO_2 by silicate rock weathering. An additional factor is the placement of the landmasses. If they are conjoined, the humidity in the interior is likely to be low, thus reducing CO_2 uptake by the land. Conversely, if the land is distributed as separate bodies with close access to moisture from nearby oceans, CO_2 uptake by tropical landmasses is enhanced. Thus, the CO_2 concentration in the atmosphere can undergo wide variations over geologic time as continental drift rearranges the continents. This will affect global climate via the greenhouse effect.

In one study, two idealized continental geometries based on present-day total land area were analyzed with a climate model: (1) a tropical land belt 17°N to 17°S and (2) polar land caps from 90° to 45°N and S (Barron, 1984). The polar land cap model was subdivided into two subordinate cases, one of which was unconstrained, and the other had imposed a thin permanent snow cover from 70° to 90°, N and S. The resultant temperature profiles vs. latitude are shown in Figure 1.3. Note that the modeled profile for polar landmass with snow is similar to that which exists today.

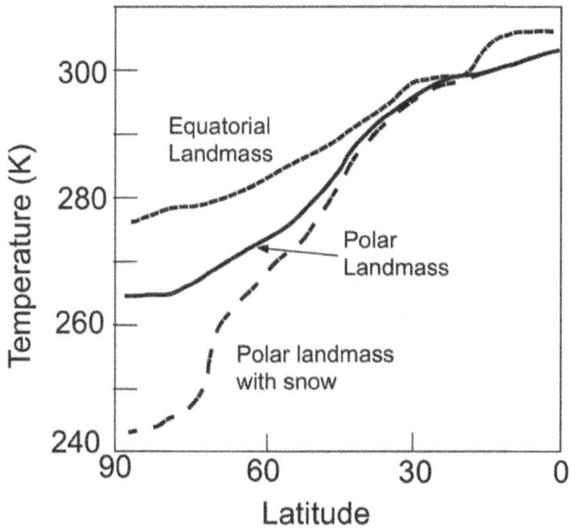

Figure 1.3. Dependence of temperature on latitude for three hypothetical distributions of landmass.

Smith and Pickering (2003) proposed what they called a "unifying explanation for the four major icehouses during the past ~620 million years". All four icehouses developed when a large continent lay within or less than 1,000 km from one or both geographic poles but there have been periods when a polar continent such as Antarctica has not been glaciated. Thus, a polar or sub-polar position for a continent appears to be a necessary (but not sufficient) condition for widespread glaciation. High topography has also been invoked as an important factor. Other important factors for establishment of continent-wide ice sheets are the opening of high-latitude gateways and the closing of subtropical gateways. However, whether the changes in circulation lead to increased snow and ice accumulation in high-latitude regions depends in part on the strength of the contemporaneous circumpolar circulation. The problem is complex and requires numerical modeling. The authors believed that astronomical factors and other processes became significant in driving glacial-interglacial events only after the continental configurations gateways and associated ocean gyres were established.

Gerhard and Harrison (2001) suggested:

> "The primary driving force behind [long-term] climate cycling is tectonic, specifically by controlling distribution of landmasses on the Earth's surface, which in turn controls the geometry of ocean currents and thus the transfer of heat around the Earth. When equatorial ocean currents exist, the Earth tends to be in a greenhouse state. In contrast, when continents exist in positions that impede or block significant equatorial currents, the Earth tends to be in the icehouse condition. Transitions between the two states are slow but may be punctuated by rapid shifts."

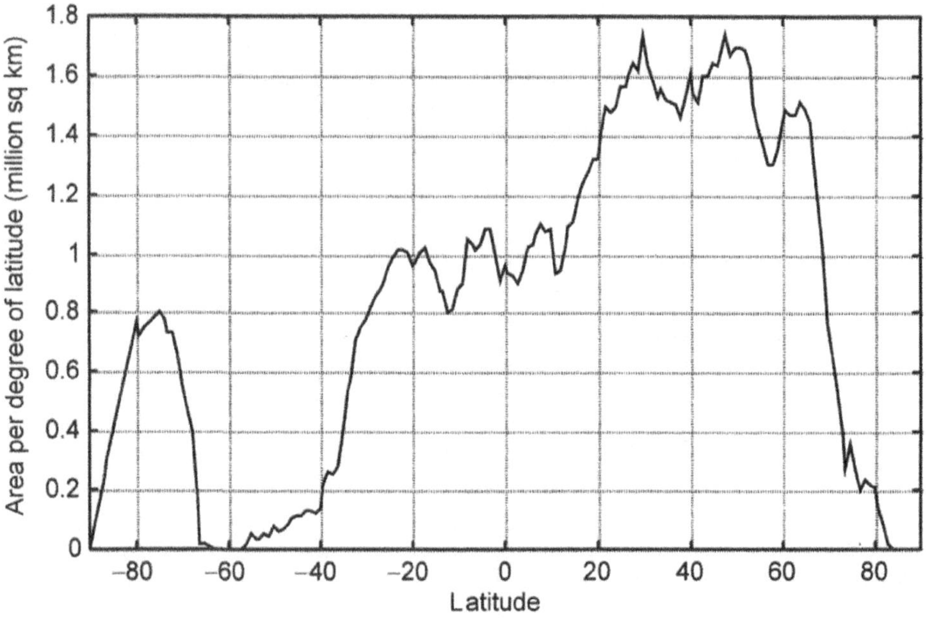

Figure 1.4. Land area vs. latitude on the Earth. Two-thirds of the land mass is north of the equator (M&M, p. 189 by permission of Praxis Publishing).

The present distribution of land as a function of latitude is shown in Figure 1.4. Two-thirds of the land area occurs in the Northern Hemisphere (NH). This undoubtedly explains why ice sheets form primarily in the NH during Ice Ages. The presence of land is necessary to allow the accumulation of ice. In addition, land responds more readily to seasonal changes than do the oceans. This might suggest that the level of summer solar flux would control the ability of ice fields to expand or contract.

As M&M said:

> "The fundamental reason for this is lack of convection. In the oceans, heat can convect between depths as well as horizontally, but on land, heat must diffuse, and that is a much slower process. Seasonal changes rarely penetrate more than about two meters. Ocean water mixes readily; cooling water contracts and sinks to the bottom until the temperature of the entire depth of water drops to 4°C. The uppermost 50 meters or so of the ocean, called the *mixed layer*, is thoroughly mixed by wind and waves. The importance of land in the formation of large glaciers is illustrated by the presence of glaciers in Greenland and Antarctica. These are the only landmasses that extend close to the poles, and they are the only ones with extensive glaciers remaining from the Ice Ages. In Greenland, the glaciers extend southward almost 30 degrees from the North Pole; in Antarctica, they extend about 20 degrees from the South Pole. Large areas in Canada and Russia that are closer to the North Pole than southern

Greenland, but don't reach as close as northern Greenland, have no glaciation. This suggests that *polar roots* on land are necessary and, if they exist, then the glaciers can extend much further from the polar regions. Although sea ice covers the Arctic Ocean, it does not appear able to provide the same kind of roots that are provided by land."

Once the glaciers begin to form, the increased albedo from their surfaces may provide a positive feedback to enhance their formation, although precipitation is also needed, and it is not obvious that increased cold alone will lead to increased ice.

The modeled history of the Earth's climate and its relationship to arrangement of landmasses is described in considerable detail by Christopher R. Scotese on his website: *http://www.scotese.com*. He provides maps of the landmasses roughly every 20 to 40 million years, starting around 540 million years ago, with brief descriptions of the prevailing climates. However, it is not always apparent why climates changed dramatically while the continents hardly changed over the same interval. For example, Scotese's maps for 480 million years ago and 440 million years ago are quite similar, yet he says that mild climates covered most of the globe 480 million years ago while a "South Polar Ice Cap covered much of Africa and South America" 440 million years ago.

Geological evidence suggests that during the Cambrian Period (about 570 to 510 million years ago), the Earth was a hothouse with essentially no polar or high-altitude glaciers. During this period, most of the continents as we know them today were either under water or part of the so-called Gondwana "supercontinent" located deeply into the Southern Hemisphere. The oceans were all interconnected, bringing warm water to polar areas. About 85% or more of the Earth's surface was covered with water (compared to approximately 70% at present) and there was a lack of significant topographic relief. Chemical weathering was at a minimum because the landmasses were minimal and they were all conjoined. Since the major portion of incident solar flux is in the tropics, and the equatorial regions were almost entirely ocean, much less solar flux was reflected to space compared to today.

1.3.3 Variations in the Earth's orbit

It is well established that over the past three million years, the Earth has undergone a long series of Ice Ages interspersed with interglacial warm periods. The Ice Ages were not continuous and were heavily interpolated by rather sudden large swings in temperature, up and down. Nevertheless, aside from short-term fluctuations, the data display quasi-periodic variations that span multiple tens of thousands of years.

Quasi-periodic variations in the Earth's orbital parameters change the summer peak solar intensity at higher latitudes with periods of multiple tens of thousands of years. The fact that peak summer solar inputs to high latitudes and data on past climate variations both show quasi-periodic variations with comparable time spans, suggests that the two may be inherently coupled. Spectral analysis supports this viewpoint to some extent. It has been widely presumed that this variability has a significant effect on the ability of surface and sea ice at higher northern latitudes to withstand the onslaught of summer heat. It has been theorized that during time periods when the peak summer solar intensity at higher northern latitudes is lower than average, the lower solar input may "trigger" feedback processes that lead to spreading of ice cover and the start of Ice Ages. Conversely, time periods with high peak solar intensity at higher northern latitudes may trigger feedback processes that cause melting, leading to deglaciation. Thus, it is generally accepted that variability of the Earth's orbit about the Sun is a primary factor in determining the timing of Ice Age–interglacial cycles. The data and models for Ice Ages are discussed in detail by Rapp (2012).

1.3.4 Effect of greenhouse gases on gadiant emission by the Earth

In Sections 1.3.1 to 1.3.3, we discussed variability of the solar energy input to the Earth. This solar energy input warms the Earth. As the Earth warms, it emits a greater amount of radiant energy. The climate of the Earth is determined by the temperature profile the Earth must reach in order to radiate enough energy flux to space to reach an energy balance between the solar input and the rate of energy loss. Here, we are concerned with the rate of energy loss.

Greenhouse gases in the atmosphere absorb some of the radiation emitted by the Earth's surface, and re-radiate in all directions. Since some of this re-radiation is downward, the net effect is to inhibit heat loss from the surface to the upper reaches of the atmosphere.

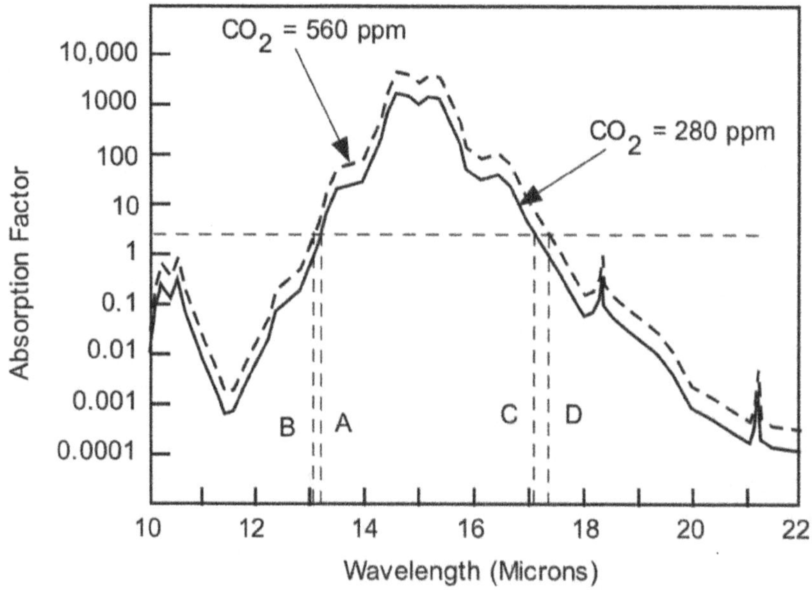

Figure 1.5. Absorption factor (absorptivity concentration integrated over vertical path through atmosphere) for CO_2 vs. wavelength. The horizontal dashed line corresponds to an absorption factor of 3 (essentially complete absorption along a vertical path through the atmosphere).

A simplistic explanation for why increased CO_2 concentration can lead to increased temperature of the Earth is illustrated in Figure 1.5. Carbon dioxide absorbs outgoing infrared (IR) radiation emitted by the Earth primarily in the wavelength band between 13 and 17 mm. The absorption of any wavelength in the atmosphere is dependent on the integral of the absorptivity times the concentration on a vertical path through the atmosphere. This integral is called the absorption factor. Since the absolute amount of absorption depends on the exponential function of the absorption factor, an absorption factor of 3 corresponds to roughly 99% of complete absorption. As Figure 1.5 shows, at the pre-industrial level of 280 ppm of CO_2, the entire absorption band from 13 to 17 mm is fully saturated. Adding more CO_2 to the atmosphere does not increase the absorption significantly within the saturated region. Only at the "wings" of the absorption band is there any significant increase in absorption by the atmosphere when the CO_2 concentration is increased. Thus, with a CO_2 concentration of 280 ppm, absorption is saturated between vertical dashed lines **A** and **C** in Figure 1.5.

If, in the future, the CO_2 concentration is doubled (from the pre-industrial value) to 560 ppm, the absorption curve moves up by a factor of 2 on the vertical log scale, and the saturated region expands to the region between vertical lines **B** and **D**, producing a net heating effect. However, the additional heating effect (from absorption in regions between **A** and **B**, and between **C** and **D**) is much smaller than the original heating effect in going from 0 ppm to 280 ppm (region between **A** and **C**). Thus we see that, as more and more CO_2 is added to the atmosphere, the heating effect decreases per unit amount of CO_2 added.

Lindzen (2007) described the effect of additional absorption by CO_2 and other greenhouse gases in the atmosphere. He pointed out that the simplistic description of the greenhouse effect is that greenhouse gases and clouds "inhibit cooling of the Earth by thermal radiation, and serves as a blanket which causes the Earth to be warmer than it otherwise would be". Lindzen objected to this description because, as he said:

> "... the surface of the Earth does not cool primarily by thermal radiation .... There is so much greenhouse opacity immediately above the ground that the surface cannot effectively cool by the emission of thermal radiation. Instead, heat is carried away from the surface by fluid motions ranging from the cumulonimbus towers of the tropics to the weather and planetary scale waves of the extra-tropics. These motions carry the heat upward and poleward to [altitudes] where it is possible for thermal radiation emitted from these levels to escape to space."

The lower atmosphere is relatively opaque to IR radiation but as the altitude increases, the density decreases and transmission of IR improves. Lindzen defined an altitude sufficiently high that the optical depth for IR is about 1 so that transmission of IR is attenuated roughly as e^{-1}. This is a region of the atmosphere that can radiate energy from the Earth to space. He provided a description of the effect of adding greenhouse gases to the atmosphere as shown in Figure 1.6.

The temperature of the atmosphere decreases with altitude to some level known

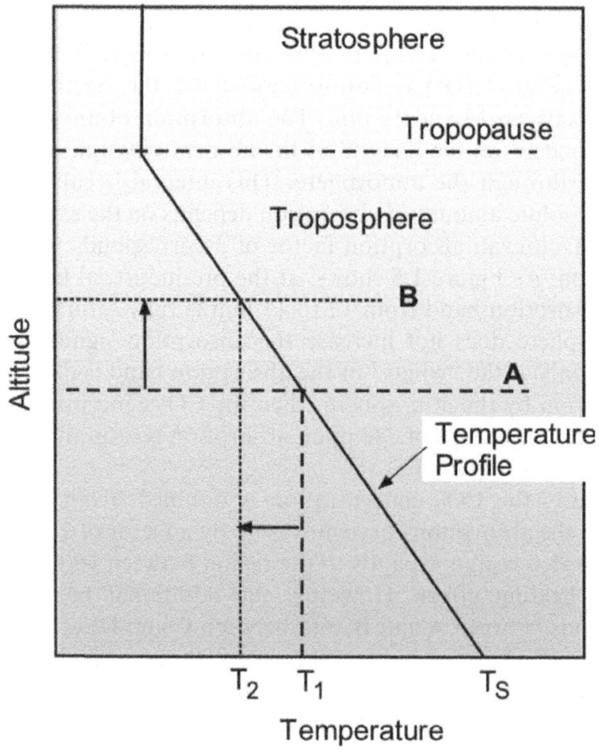

Figure 1.6. Lindzen's picture of how the greenhouse effect works (part one).

as the tropopause. The height of the tropopause varies from about 16 km in the tropics to about 12 km at 30° latitude, and to about 8 km in polar regions. Before adding greenhouse gases to the atmosphere, the altitude at which the optical depth for IR is around 1 is denoted **A** in Figure 1.6. When greenhouse gases are added to the atmosphere, the level at which the optical depth is around 1 is raised in altitude due to the additional absorption by the greenhouse gas. As Lindzen pointed out, "because the temperature of the atmosphere decreases with altitude, the new characteristic altitude for emission is colder than the previous altitude". The new altitude for emission is **B**, and $T_1 > T_2$. Since IR emission is proportional to the fourth power of the temperature, emission from altitude **B** is reduced compared to what it was at altitude **A**. Thus, the Earth cannot cool as effectively. The Earth will therefore warm until the temperature at altitude **B** approaches the original temperature. This is shown in Figure 1.7. The original temperature at the surface is T_S, leading to the vertical temperature profile shown as the slanted dotted line. Before the temperature rises, the temperature at the altitude for emission is T_2 at point 2. After the surface temperature rises to $T_{S'}$, the new vertical temperature profile is the slanted solid line and the temperature at altitude **B** rises back to T_1. The Earth is now back in thermal balance.

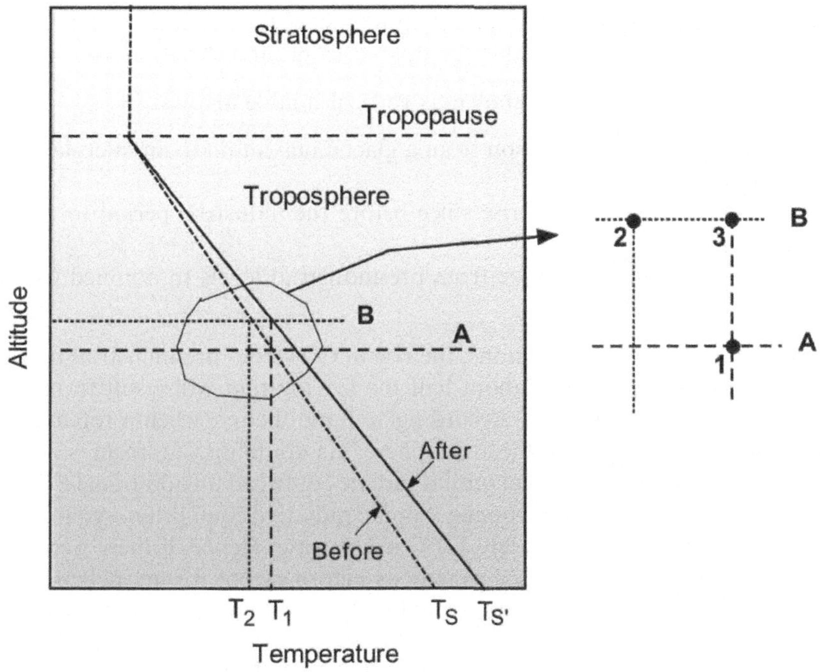

Figure 1.7. Lindzen's picture of how the greenhouse effect works (part two).

Lindzen also pointed out that climate models indicate that when the Earth warms due to the greenhouse effect, warming is strongly peaked in the tropical troposphere near the altitude where the optical depth $\tau \sim 1$. He concluded:

> "Roughly speaking, the warming at $\tau = 1$ in the tropics is about two to three times larger than near the surface regardless of the sensitivity of the particular [climate] model. This is, in fact, the signature (or fingerprint) of greenhouse warming. Stated somewhat differently, if we observe warming in the tropical upper troposphere, then the greenhouse contribution to warming at the surface should be between less than half and one third the warming seen in the upper troposphere."

Climatologists describe the effect of a greenhouse gas such as CO_2 in absorbing outgoing IR radiation as exerting a "forcing" at the top of the atmosphere. This forcing is a hypothetical heat flow downward measured in W/m^2. Hansen *et al.* (2000) have estimated the forcing on the climate due to changes in CO_2 concentration as shown in Figure 1.8.

Estimating the quantitative shape of the curve in Figure 1.8 is of great importance for understanding the effect of future CO_2 emissions on future climate change. In Figure 1.8, vertical lines represent:

A = typical CO_2 at glacial maximum in an Ice Age;
B = typical CO_2 during an interglacial period between Ice Ages;

C = current CO_2 level due to human impact on environment; and
D = CO_2 level after it doubles compared to pre-industrial levels.

The estimated forcing are shown as vertical double arrows:

F1 = forcing in the transition from a glacial maximum to an interglacial period ($\sim$3 W/m^2);
F2 = forcing due to CO_2 rise since before the industrial period to the present ($\sim$2 W/m^2);
F3 = forcing due to change from pre-industrial levels to doubled CO_2 ($\sim$3.7 W/m^2).

According to these calculations, the rise in CO_2 from pre-industrial times to the present has already produced about half the forcing that will result from doubling CO_2 from pre-industrial times. According to basic theory, when a forcing of $\sim$3.7 W/m^2 is applied to the top of the atmosphere (via doubling CO_2 from $\sim$280 ppm to $\sim$560 ppm), the Earth will warm until it radiates outward an additional 3.7 W/m^2 to compensate for the downward forcing. Simple radiative equilibrium requires that the Earth will warm up approximately 1.2°C in this case. Hence, if there were no other changes, and the Earth remained exactly as before except for an increase in global average temperature, doubling CO_2 would produce a global average temperature rise of about 1.2°C.

However, if CO_2 rises from $\sim$280 ppm to $\sim$560 ppm, various secondary consequences would undoubtedly occur. Some of these could change the heat

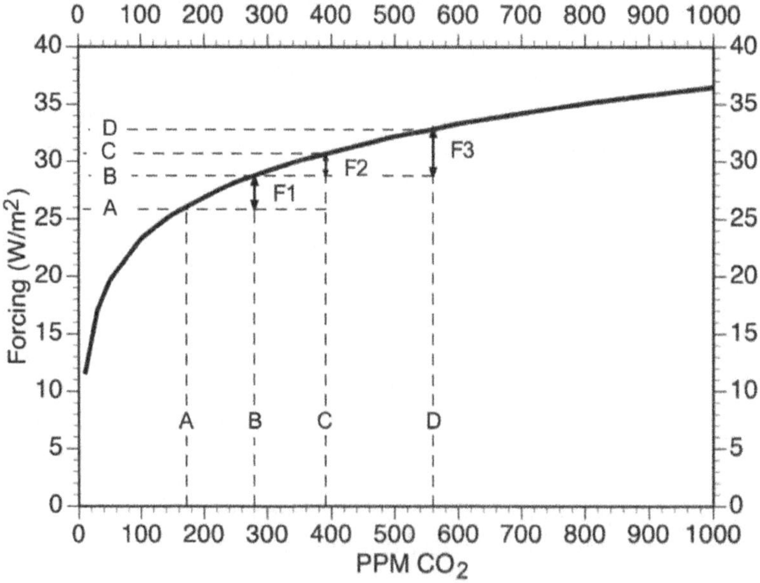

Figure 1.8. Estimated forcing of the climate due to changes in CO_2 concentration (Hansen, *et al.*, 2000).

balance of the Earth, producing additional temperature changes. These factors are called *feedbacks*. One feedback factor is that as the Earth warms, glaciers and ice sheets tend to retreat. Since these regions reflect incoming sunlight, the net effect will be additional solar energy absorbed by the Earth, resulting in additional warming. Another feedback factor is due to the fact that, as the oceans warm, they will tend to evaporate more water vapor, and water vapor (like CO_2) is a greenhouse gas that can absorb IR radiation emitted by the Earth. Some climate modelers have assumed that as the Earth warms due to doubling of CO_2, the whole Earth will experience a uniform increase in water vapor content in the atmosphere, resulting in a significant amplification of the original warming effect of CO_2. Estimates of the combined effect of warming due to a doubling of CO_2 plus the effects of feedbacks range from a temperature rise of 2°C to 9°C. However, there remains great uncertainty in these estimates. While many estimates have been made, the consensus value often used is ~3°C. Like the porridge in *The Three Bears*, this value is just right—not so great as to lack credibility, and not so small as to seem benign. Unfortunately, all of the estimates made to date by various procedures lack adequate data and require considerable speculation. The various estimates of increased humidity suffer from the facts that much of the increase in water evaporation will occur in the tropics where (i) the air is typically heavily laden with humidity and the water absorption bands are already saturated, and (ii) much of the heat transfer from the surface to the atmosphere is via upward cumulus convection, not by radiation. The only regions where increased humidity would provide a strong positive feedback would be desert areas, but if the Earth warms, these areas might become drier, not wetter. Another important factor is the effect of clouds. Several climatologists of the alarmist persuasion have written papers claiming that the effect of rising temperature due to increased CO_2 will be to reduce cloudiness, producing even more warming of the Earth by allowing more solar energy to reach the surface. However, there are many good reasons to believe that increasing evaporation will produce more (not less) cloud cover and will act contrary to the warming due to CO_2. Our knowledge of the Earth's cloudiness is very limited and this issue remains not understood to any reasonable degree.

The bottom line is that climatologists of the alarmist persuasion are convinced that the full effect of doubling CO_2 from ~280 ppm to ~560 ppm will be a global temperature rise of around 3°C or more, whereas an honest examination of their analyses shows that these projections are quite specious and unreliable. Some warming due to CO_2 from ~280 ppm to ~560 ppm is expected but it is not yet clear whether the net feedback is positive or negative.

The descriptions given above refer to a relatively modest variation in the CO_2 concentration (280 ppm → 560 ppm) that is likely to occur in the 21st century. Over very long periods of time (tens or hundreds of millions of years), it is likely that CO_2 concentrations varied over much wider ranges. While the curve in Figure 1.8 flattens out somewhat as the CO_2 concentration rises beyond 1,000 ppm, nevertheless it is still rising. It is widely believed that very high concentrations of CO_2 (thousands of ppm) produce significant warming of the Earth. A number of investigations have attempted to find a quantitative correlation of climate with CO_2 concentration over ancient time periods. These are discussed in Section 1.4.

1.4 ANCIENT CLIMATES AND CO_2 CONCENTRATION

1.4.1 Summary

Some climatologists have sought to estimate the dependence of the climate on CO_2 concentration by analyzing paleoclimatic data on climate and CO_2 concentration, with the intent of using the climate sensitivity derived from this to estimate the global average temperature increase if the CO_2 concentration doubles from the pre-industrial value of about 280 ppm. The widely held view amongst geologists and climatologists alike, is that the primary cause of long-term climate change is variability of CO_2 concentration due to imbalances between CO_2 degassing at spreading centers and the conversion of atmospheric CO_2 to mineral carbon through long-term silicate weathering and oceanic carbonate formation. The argument goes (more or less): *If it wasn't CO_2, what else could it have been?* Foster *et al.* (2009) described this as the "accepted paradigm" that requires CO_2 to vary in unison with global temperature. So, paleoclimatologists have been trying for decades to establish a relationship between climate and CO_2 concentration over many millions of years. The more audacious of these have attempted to establish a quantitative relationship between climate and CO_2 concentration in order to try to estimate the Earth system climate sensitivity. Unfortunately, the proxy data for CO_2 over many millions of years are very widely scattered and the results are equivocal. There is a general tendency for warmer climates to be associated with higher CO_2 concentrations, but this mainly relates to very large temperature excursions, and, even so, there are many exceptions. There is no direct one-to-one correspondence between CO_2 and climate. Evidently, other factors than CO_2 must also influence the climate. In his *Perspective* article, Ruddiman (2010) emphasized that there is no present explanation for the fact that there was no significant drop in CO_2 concentration over the past 22 million years while the climate cooled substantially. Nevertheless, it is noteworthy that many paleoclimatologists are so convinced from the start that CO_2 is the main driver of long-term climate change, that, even with noisy data, they claim support for their theory. Royer (2010) began his commentary with the unwarranted statement:

> "Global temperatures have co-varied with atmospheric carbon dioxide (CO_2) over the last 450 million years of Earth's history."

It is noteworthy that prior to 2004, a number of climatologists pointed out discrepancies between the geological records of climate and CO_2 over 500 million years, whereas after 2004, most published papers emphasized the correlation of climate and CO_2 over that period. It is not clear whether new data makes the difference, or whether it is necessary to support the orthodoxy to obtain research funding. Our conclusion here is that CO_2 is probably an important factor in long-term climate change, but other factors are also influential such as the placement of the continents on Earth, the functionality of ocean currents, worldwide distribution of clouds and aerosols, the past history of the climate, the orientation of the Earth's orbit relative to the Sun, the luminosity of the Sun, the presence of aerosols in the atmosphere, volcanic action, land clearing, biological evolution, etc. Hence, there is probably no single curve relating global average temperature to CO_2 concentration, but rather, a set of curves that depend on the above factors.

There are good estimates available of the global average temperature and the CO_2 concentration at the Last Glacial Maximum (LGM) 20,000 years ago, and, if these data are compared with values in the pre-industrial era, one can thereby estimate the sensitivity of the climate to CO_2 concentration over the range ~180 ppm to ~280 ppm. Using this estimated climate sensitivity, one can then estimate the global average temperature rise in going from 280 ppm to 560 ppm. The various investigators have come up with a range of projections. It is noteworthy that this range of estimates for the real-world ΔT due to doubling CO_2 from 280 ppm to 560 ppm is from ~1°C to ~3°C. However, these estimates do not take into account possible differences in humidity and cloudiness. Similar analyses have been made for much longer time periods (up to 540 million years ago) but the data are very noisy and the results are of uncertain veracity. Our conclusion here is that CO_2 is probably one of several major factors in long-term climate change, but other factors such as the placement of the continents on Earth, the functionality of ocean currents, the past history of the climate, the orientation of the Earth's orbit relative to the Sun, the luminosity of the Sun, the presence of aerosols in the atmosphere, volcanic action, land clearing, biological evolution, etc., exist. Hence, there is probably no single curve relating global average temperature to CO_2 concentration, but rather, a set of curves that depend on the above factors.

1.4.2 Introduction

While the principal current interest in climatology is in the putative climate change induced in the 21st century by increasing the CO_2 concentration from 280 ppm to 560 ppm, multiple efforts by many climatologists cannot seem to overcome the uncertainties inherent in performing this analysis, and their models still lack credibility and consistency. It has therefore occurred to a number of climatologists that perhaps by studying the past (tens of thousands of years ago to hundreds of millions of years ago) and finding relationships between CO_2 and climate during those periods, we might be able to obtain real-world data on how climate and CO_2 are connected. This real-world data will presumably have built into them all the secondary processes that take place. For example, we have very good data on CO_2 concentration during the LGM, some 20,000 years ago, so that is one important historical point for further study. In addition, there are also a variety of estimates of CO_2 concentration that go back as far as 500 million years, but unfortunately such data are very scattered and are of uncertain reliability. But climatologists are a sturdy lot, and they are willing to derive a dollar's worth of conclusions from a penny's worth of data.

What we seek is a relationship between CO_2 concentration and the Earth's climate over long geological periods during which the CO_2 concentration varied over a wide range. There is evidence that the CO_2 concentration may have been well over 20,000 ppm in the distant past, and it has been as low as ~180 ppm only 20,000 years ago. It would be very nice if there were a single curve relating the global

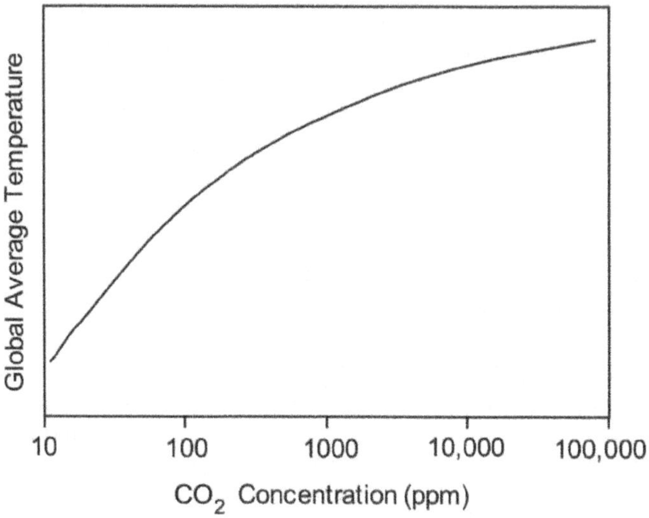

Figure 1.9. Hypothetical single curve relating T_G to CO_2 concentration.

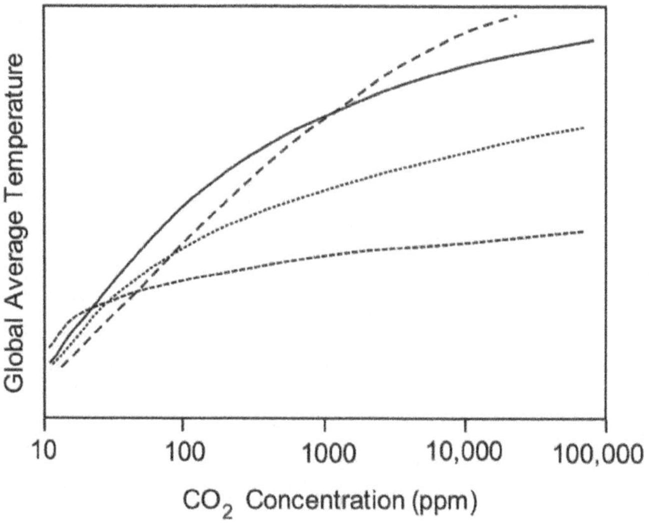

Figure 1.10. Hypothetical curves relating T_G to CO_2 concentration.

average temperature (T_G) to CO_2 concentration such as that shown in Figure 1.9. In that case, if we could find several points on the curve, we could attempt to map out a good portion of the curve.

However, over long time periods, the variation of T_G with CO_2 concentration depends on various factors such as the placement of the continents on the Earth, the functionality of ocean currents, the past history of the climate, the orientation of the

Earth's orbit relative to the Sun, the luminosity of the Sun, the presence of aerosols in the atmosphere, volcanic action, land clearing, biological evolution, etc. Hence, there is probably no single curve relating T_G to CO_2 concentration, but rather, a set of curves that depend on the above factors (see Figure 1.10).

Over the past few million years, the Earth has vacillated between alternating Ice Ages and interglacial periods, probably driven by variations in the Earth's orbit about the Sun, accompanied by variations in CO_2 concentration and other global and regional changes. Over the past ~10,000 years (the *Holocene*), the Earth has been in a relatively quiescent interglacial period with only small variations in climate. There is a tendency amongst the public, and even amongst many climatologists, to downplay variations within this relatively stable period, and consider the climate during this limited period to be relatively constant and "normal". There is some evidence that the CO_2 concentration remained in the range 270 ppm to 290 ppm during the Holocene, and variations in T_G for the most part were probably less than ± 1°C. With the advent of the industrial age, power plants, cement production, and other industrial activity spewed out more CO_2 than the Earth could absorb, and the CO_2 concentration rose steadily in the 20th century, recently reaching about 395 ppm. Depending on future world consumption of fossil fuels and energy policy, the CO_2 concentration appears poised to grow in the 21st century, and may reach some level in the range ~500 ppm to ~800 ppm by the year 2100. There is little doubt that this will produce further warming of the climate, but how much? Climatologists have selected a benchmark of a doubling of the pre-industrial level (280 ppm to ~560 ppm) as the standard basis for estimating the future rise in T_G. The holy grail of climatology is thus to seek an estimate of how T_G varies with CO_2 concentration over the range 280 ppm to 560 ppm. To put this in perspective, we show this range in Figure 1.11. Not only do we not know *a priori* which curve applies to our present situation, but the vertical slice of greatest interest is a very narrow one in the total scheme of things.

Climatologists have mainly concentrated on the realm of CO_2 concentration between 280 ppm and 560 ppm, with some concern for higher concentrations up to ~900 ppm. In this regard, we can magnify the gray slice from Figure 1.11, and combine this with known values of T_G over the past ~120 years, as shown in Figure 1.12. Curves 1 to 4 show various estimates of the temperature rise that will be induced by further increases in CO_2 concentration.

The estimates of temperature rise due to increasing CO_2 concentration shown in Figure 1.12 ultimately depend on the following analysis. If we start with the pre-industrial climate (T_G ~14.3°C and CO_2 ~280 ppm) and allow the CO_2 concentration to increase with no other changes to the Earth, we can calculate the additional absorption of upward IR radiation emitted by the Earth due to increased CO_2 concentration, and thereby estimate from the laws of radiant heat transfer, how much the Earth needs to warm in order to establish a new equilibrium whereby incident heat input from the Sun is counterbalanced by heat radiated by the Earth to space. This calculation has been carried out, and it is widely accepted that if feedbacks are neglected, an increase in CO_2 concentration from 280 ppm to 560 ppm would increase T_G by roughly 1.2°C. This is represented by curve 1 in Figure 1.12. As the Earth warms due to increasing CO_2, various secondary processes take place.

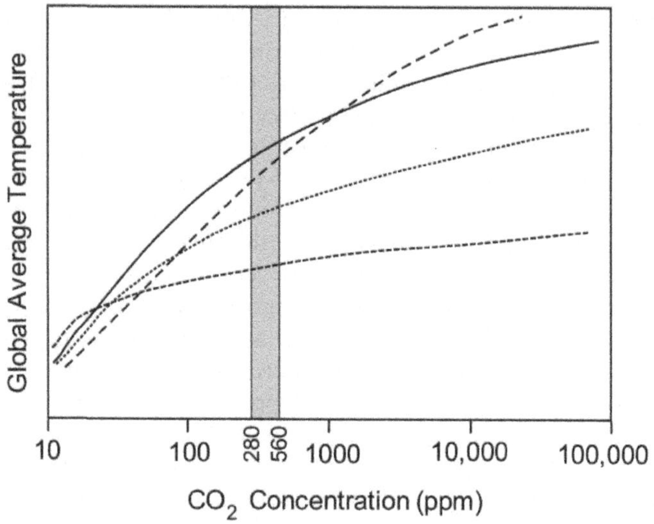

Figure 1.11. Range of CO_2 concentration (280 ppm to 560 ppm) for 21st-century climate change (gray slice).

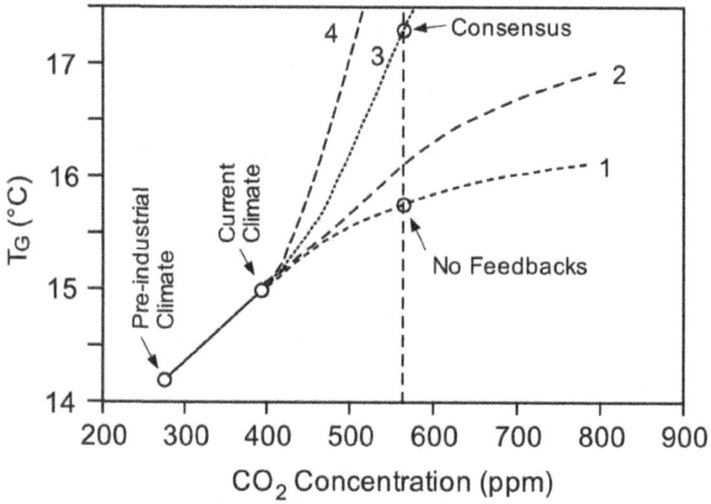

Figure 1.12. Variation of T_G with CO_2 concentration.

Additional evaporation from bodies of water will tend to increase the humidity of the atmosphere, ice sheets (with their high reflectivity of sunlight) will shrink, cloudiness might increase or decrease in some regions, aerosol concentrations might vary in unknown ways, precipitation patterns will vary, ocean currents may change somewhat, and other changes may take place in the Earth's biota. These so-called feedback processes will also affect T_G. There is considerable uncertainty in the magnitude (and, in the case of cloudiness, even the sign) of such feedback effects, and as a result, there have been many diverse estimates of T_G due to a doubling of CO_2 when feedbacks are included. Some of these are shown in Figure 1.12 as curves 2, 3, and 4. Curve 3 represents a rough estimate that has been adopted as a consensus by climatologists of the alarmist persuasion. Some skeptical climatologists believe that the truth lies nearer to curves 1 and 2.

It is not our purpose in this chapter to review the many estimates of warming due to an increase in CO_2 concentration from 280 ppm to 560 ppm. Instead, we are concerned here with the broader picture as shown in Figure 1.11. Can we somehow estimate the shape of the broad curve of T_G vs. CO_2 concentration over a very wide range of CO_2 concentration, and thereby find where this curve passes through the gray region in Figure 1.11? Therefore, our intent is to review the literature on the broad dependence of T_G on CO_2 concentration over a very wide range of CO_2 concentrations over various geologic time spans.

A few comments of caution need to be made at this point. One comment is that if feedbacks prove to be as amplifying as the consensus would advocate, this would indicate that the effect of CO_2 on climate is greatly changed by secondary factors, and the use of data and models from hundreds of millions of years ago may produce misleading results, since we may currently be on a very different curve in Figure 1.11 than they were a few hundred million years ago. Of particular importance in this regard is the arrangement of the continents. In addition, the Sun was some 6% reduced in intensity 500 million years ago. These, and other changes, add uncertainty as to whether such data can properly be extrapolated to the 21st century. Another comment is that proxies for T_G and CO_2 concentration over geologic time spans are not likely to be very reliable, and these should be critically reviewed before relying on them. As Zeebe (2011) said:

> "By studying the relationship between greenhouse gas forcings and global temperature changes during past climate episodes, palaeoclimatology currently has a unique opportunity to fundamentally contribute to understanding climate sensitivity. At present, one of the standard tools for estimating climate sensitivity is the use of numerical climate models. Unfortunately, model-derived climate sensitivities are subject to large uncertainties ... Studying past climates to estimate climate sensitivity inarguably has one great advantage over theoretical computer models: it is based on actual data. Unfortunately, palaeo data-derived climate sensitivities have large uncertainties as well. Errors can arise from issues such as dating, alteration of the climate signal after deposition, insufficient spatial and/or temporal coverage, and various uncertainties associated with the proxies for environmental variables such as temperature and past atmospheric CO_2 concentrations."

1.4.3. The transition from the LGM to the pre-industrial era

The current holy grail of climatology is to seek an estimate of how much the global average temperature will increase if the CO_2 concentration doubles from the pre-industrial value of 280 ppm to 560 ppm. Attempts to estimate this directly are difficult due to uncertainties in secondary factors that accompany warming from increased CO_2 (changes in humidity, cloud cover, winds, ocean currents, glaciers, ice sheets, etc.). Some climatologists have sought to estimate the dependence of the climate on CO_2 concentration by analyzing paleoclimatic data on climate and CO_2 concentration, with the intent of using the climate sensitivity derived from this to estimate the global average temperature increase if the CO_2 concentration doubles from the pre-industrial value of about 280 ppm.

Over the past couple of million years in which we have had alternating Ice Ages and interglacials, there is good evidence that the trigger to set the cycles in motion is solar input to higher northern latitudes. But what does it mean to set the cycle in motion? It means that an albedo effect begins in the $\sim$60°N latitude range as snow and ice accumulate, causing a regional cooling. As the ice sheet builds, and sea ice expands, and the ocean drops, other albedo effects occur and a greater regional cooling takes place. Dust is stirred up and this further amplifies the cooling. Then, as the cooling spreads, the CO_2 concentration in the atmosphere decreases, producing additional negative forcing worldwide, which lowers the worldwide temperature. However, changes in humidity and cloudiness are unknown and may be very large factors. At the height of the LGM some 20,000 years ago, the negative forcing produced a global average temperature decrease of roughly 4.5°C. Hansen and Sato (2011), Chylek and Lohmann (2008), and Kohler *et al.* (2009) all independently estimated the forcing at the LGM. The contribution of the diminution of CO_2 at the LGM to the total cooling was estimated by these studies to be in the range 16% to 33%. While it seems likely that solar input to higher latitudes triggered the cycles, the variability of CO_2 concentration played a part in determining the extremity of the temperature cycle that resulted from this trigger. The changes in CO_2 concentration between glacial maxima and interglacials ($\sim$180 ppm to $\sim$280 ppm) are well documented in ice core records, although no one seems to have a satisfactory explanation for why the CO_2 concentration changed this much (simple solubility in the oceans does not suffice). However, the estimates of forcings, particularly due to dust, vary considerably from investigator to investigator and it is difficult to pin down the climate sensitivity to CO_2 change. There are good estimates available of the global average temperature and the CO_2 concentration at the LGM 20,000 years ago, and if these data are compared with values in the pre-industrial era (a few hundred years ago), one can thereby estimate the sensitivity of the climate to CO_2 concentration over the range $\sim$180 ppm to $\sim$280 ppm. Using this estimated climate sensitivity, one can then estimate the global average temperature rise in going from 280 ppm to 560 ppm. The various investigators have come up with a range of projections. It is noteworthy that this range of estimates for the real-world ΔT_G due to doubling CO_2 from 280 ppm to 560 ppm is from $\sim$1°C to $\sim$3°C. However, these estimates do not take into account possible differences in humidity and cloudiness.

About 20,000 years ago, the most recent Ice Age was at its maximum extent with gigantic ice sheets in the higher latitudes of the NH. There is reliable evidence from ice cores that the CO_2 concentration at that time was roughly 170–180 ppm. The first question is what was T_G at the LGM.

Dwyer *et al.* (1995) utilized the ratio of magnesium to calcium (Mg/Ca) in fossil ostracodes from Deep Sea Drilling Project Site 607 in the deep North Atlantic to infer that the change in bottom water temperature changed by ~4.5°C in going from the LGM to pre-industrial times.

According to Leroux (2005), the difference in temperature between an Ice Age and an interglacial was about 10°C in the Antarctic and about 6°C globally.

Taylor *et al.* (2001) carried out an analysis in which they took into account the reduced CO_2 concentration and the extended ice sheets of the LGM in climate models to estimate the amount of cooling at the LGM compared to pre-industrial times. Using six different climate models, they obtained values of 3.5, 3.7, 3.8, 4.4, 5.2, and 5.9, for an average of 4.4°C.

Crucifix (2006) provided a less optimistic view of the precision to which this is known: "The global temperature change is therefore is estimated to be comprised between 3°C and 9°C with 95% confidence." He also estimated that the tropical ocean sea-surface temperature decreased between 1.7°C and 2.7°C, and Antarctic surface air temperature decreased by 7°C to 11°C at the LGM compared to pre-industrial times.

Shakun and Carlson (2010) carried out an extensive review of the LGM–interglacial transition. They found, as expected, that the ΔT in this transition varied with latitude as shown in Figure 1.13. Their estimate of ΔT_G (the temperature at the LGM minus the pre-industrial temperature) was –4.5°C. If we couple the temperature during the LGM (14.3°C—4.5°C = 9.8°C) with an estimated CO_2 concentration of 170–180 ppm, we can plot a point representing the LGM, as shown in Figure 1.14.

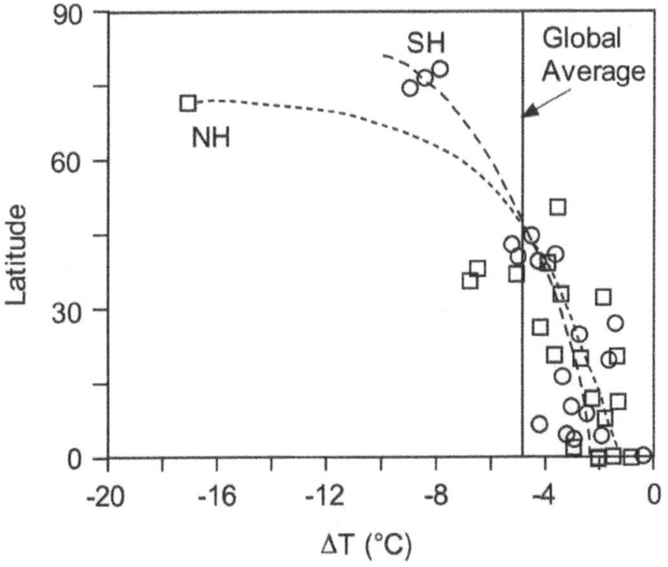

Figure 1.13. Variation of ΔT_G with latitude (adapted from Shakun and Carlson, 2010).

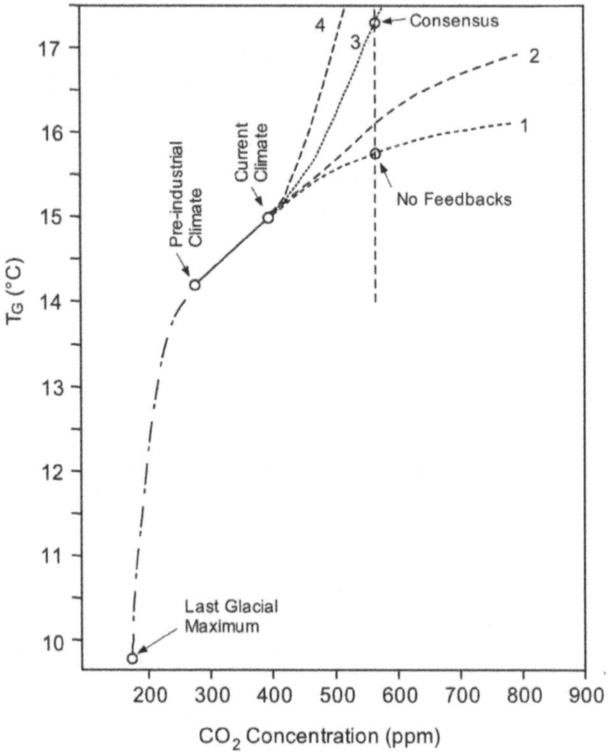

Figure 1.14. Inclusion of LGM point in relationship between CO_2 concentration and T_G.

The reason why the LGM point lies so low is because important changes took place on the surface of the Earth as the ice sheets expanded. These changes go beyond the purely spectroscopic effect of less absorption of IR by CO_2 in the atmosphere as the CO_2 concentration was lowered to below 200 ppm. The growth of large ice sheets from recent pre-industrial times to the LGM resulted in an increase in the Earth's albedo across the ice sheets, as well as for mountain glaciers. In addition, the drop in sea level moved shorelines outward, converting ocean to land, thereby further increasing the Earth's albedo. As the climate got colder, biomass and vegetation grew less abundantly, increasing the Earth's albedo still further. Undoubtedly, there were other effects as well (humidity, cloudiness, dust, etc.).

Hansen and Sato (2011) carried out an analysis in which they attempted to utilize the data relating the LGM to pre-industrial times as a basis for estimating the temperature rise in going from pre-industrial times to a doubling of the CO_2 concentration (from 280 ppm to 560 ppm). According to Hansen and Sato (2011), based on Hansen *et al.* (2008), the transition between the LGM and pre-industrial times can be characterized by two major sources of forcing:

- changes in concentration of greenhouse gases (about 75% due to CO_2);
- surface changes: (ice sheet area, vegetation distribution, shoreline movements).

Their estimates for greenhouse gas forcings were 2.25 W/m^2 for CO_2 (185 ppm → 275 ppm), 0.43 W/m^2 for CH_4 (350 ppb → 675 ppb) and 0.32 W/m^2 for N_2O (200 ppb → 270 ppb) for a total greenhouse gas forcing of 3.0 W/m^2. They also estimated the forcing due to surface changes to be 3.5 W/m^2, but this estimate appears to be rather approximate. Nevertheless, they argued that a total negative forcing of 6.5 W/m^2 would bring about the LGM–pre-industrial transition. They assumed that the ΔT_G associated with this transition was 5°C. In that case, the Earth's climate sensitivity would be:

$$\lambda = \Delta T_G/(\text{Forcing}) = 5.0/6.5 \sim 0.75^\circ\text{C}/(\text{per W/m}^2).$$

There are several problems with this calculation.

One problem is that the estimate of the forcing due to greenhouse gases appears to be a bit low. According to Hansen's papers, the forcing due to various levels of CO_2 (without feedbacks) are as shown in Figure 1.8. Thus, the forcing due to CO_2 in the LGM to pre-industrial transition is not 2.25 W/m^2 (as claimed), but 3 W/m^2, and the total forcing due to all greenhouse gases is not 3.0 W/m^2 (as claimed), but 3.7 W/m^2.

Another problem is that Hansen and Sato used $\Delta T_G = 5.0$°C, whereas 4.5°C appears to be a better choice.

In addition, Hansen and Sato did not appear to adequately consider the forcing due to high dust levels in the atmosphere during the LGM. One estimate is that dust would produce a forcing of about 1 W/m^2 (Crucifix, 2006). It is also noteworthy that Bielefeld (1997) estimated that at the height of the last Ice Age (18,000 years before present (YBP)) global radiation absorption was lower by 7%–10% than it is today. That would indicate a negative forcing of 24 W/m^2 to 34 W/m^2 that is far greater than other factors. Another concern is that no consideration was taken of possible changes in humidity or cloudiness and these are likely to be of significant magnitude.

If we modify Hansen and Sato's estimate by taking the forcing as 8.2 W/m^2 (instead of 6.5 W/m^2), and if we choose $\Delta T_G = 4.5$°C instead of 5.0°C, we obtain:

$$\lambda = \Delta T_G/(\text{Forcing}) = 4.5/8.2 \sim 0.55^\circ\text{C}/(\text{per W/m}^2).$$

Hansen and Sato used their value for the climate sensitivity (0.75 °C/(per W/m^2)) in conjunction with the forcing due to doubling of the CO_2 concentration (from 280 ppm to 560 ppm): 3.7 W/m^2 to obtain $\Delta T \sim 3.0$°C for a doubling of the CO_2 concentration. With the lower value of λ, one would obtain 2.2°C.

Crucifix (2006) provided alternate estimates of the forcings:

- change in sea level and vegetation changes ($\sim$4 W/m^2);
- reduction of greenhouse gas concentrations ($\sim$2.85 W/m^2);
- other forcings, difficult to quantify, such as increased dust concentration ($\sim$1 W/m^2).

This sums to 7.85 W/m^2, but Crucifix added: "There is also a small contribution due to the surface being, on average, more elevated than today" that might bring the total close to the value 8.2 W/m^2 previously estimated. Crucifix felt that the value of λ could not be pinned down well, primarily because of uncertainty in ΔT_G. He attempted to use climate models to bridge this gap, but concluded that "the ratio

Table 1.1. Parameters used by C&L.

	CO_2 (ppm)	*CH_4 (ppb)*	*ΔT (°C)*	*Forcing via GHG (W/m^2)*
42K yrs ago	209	548		
LGM	182	340		
Pre-industrial	285	667		
42K → LGM			2.16±0.23	0.93
LGM → pre-industrial			4.6±0.5	2.67

between LGM and CO_2 feedback factors cannot be accurately estimated from current state-of-the-art coupled models".

Chylek and Lohmann (2008) carried out an independent estimate of climate sensitivity by comparing the LGM to pre-industrial times. They asserted that "One of the uncertainties in the radiative forcing calculation during the LGM to the Holocene transition is the radiative forcing due to increased aerosol optical depth during the peak of the last ice age". In their analysis, they used the LGM to pre-industrial transition and the cooling period between the warm period around 42,000 years before present and the LGM to deduce the change in aerosol radiative forcing and to estimate climate sensitivity. It was assumed that the climate sensitivity was the same for both periods.

Chylek and Lohmann (2008) (C&L) utilized data from the Vostok ice core for transitions between two time periods:

(i) warm period around 42,000 years ago → LGM (about 20,000 years ago);
(ii) LGM → pre-industrial period (about 200 years ago).

Smoothed data from the Vostok ice core used by C&L are shown in Figure 1.15.

Based on these data, C&L estimated temperature differences and forcing due to greenhouse gases as shown in Table 1.1.

The radiative forcing due to the surface albedo changes (extent of ice sheets, sea ice and snow cover, exposure of a new land in a low-sea-level state, change in surface characteristics, and vegetation cover) for the LGM → pre-industrial transition was estimated to be roughly 3.5 W/m^2, but C&L used a range of values from 3.0 W/m^2 to 4.0 W/m^2.

C&L pointed out that the dust measurements in the Vostok ice core suggested that aerosol concentration differences from 42K to the LGM were about 53/58 as great as aerosol differences from pre-industrial time to the LGM. However, they were not able to attribute forcings to these changes *a priori*. They assumed that the forcing due to aerosols were 58X and 53X for the LGM → pre-industrial, and 42K years ago → LGM transitions, respectively, but X could not be specified *a priori*. In order to estimate the forcing due to aerosols, C&L carried out a comparison of the two transitions, assuming that the relation between ΔT and total forcing was the same for both transitions. Thus, they put:

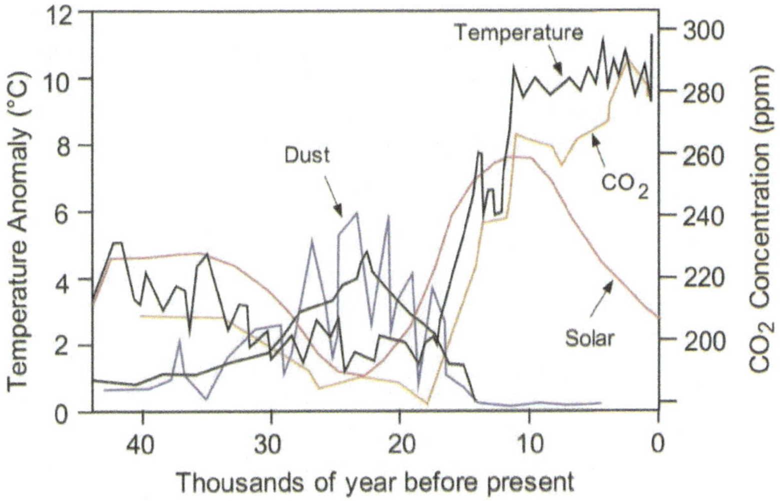

Figure 1.15. Smoothed data from Vostok ice core. The dust and solar scales are arbitrary. The solar curve represents midsummer solar intensity at 65°N (adapted from C&L).

$$\frac{DT_1}{F_{GHG1} + F_{Alb1} + F_{Aer1}} = \frac{DT_2}{F_{GHG2} + F_{Alb2} + F_{Aer2}}$$

in which transition (1) refers to LGM → pre-industrial, and transition (2) refers to 42K years ago → LGM. Their estimates for ΔT_1 and ΔT_2 and F_{GHG1} and F_{GHG2} are given in Table 1.1. Their estimate for F_{Alb1} was 3.5 W/m^2, but they did not seem to specify what they used for F_{Alb2}. If F_{Alb2} is known, and setting $F_{Aer1} = 58X$ and $F_{Aer2} = 53X$, the above equation provides a means to estimate X. C&L reported that their best estimate for X was 0.056 W/m^2. Working backwards, we may surmise that they must have used $F_{Alb2} = 1.58$ W/m^2. Using the above value for X, they estimated the total forcing for the LGM → pre-industrial to be $2.67 + 3.5 + 58 \times 0.056 = 9.4$ W/m^2, and with $\Delta T_1 = 4.6$°C, the climate sensitivity is $\lambda = 4.6/9.4 \sim 0.5$°C/(per W/m^2). This implies that when CO_2 goes from ~280 ppm to ~560 ppm, the expected temperature rise is $0.5 \times 3.7 \sim 1.8$°C. C&L also examined prior glacial to interglacial transitions and, from this, estimated slightly higher values for λ. However, they pointed out:

> "At this time it is not clear whether these higher sensitivities, compared to the climate sensitivity deduced from the LGM to Holocene transition, really reflect higher climate sensitivity at the time of the considered climate transitions or whether they are artifacts due to imperfect ice core data and uncertainties in the used approximations."

The main difference between the calculations of C&L and Hansen and Sato are the much higher values of aerosol forcing used by C&L. As in the case of Hansen and Sato, C&L did not consider changes in humidity or cloudiness.

Hargreaves and Annan (2008) (H&A) wrote a commentary on the paper by

C&L. They pointed out (properly) that the data in Figure 1.15 are vacillating, and depending on exactly how one chooses the data points, one can derive different results. They provided two examples. In their first example, they chose to read the temperature curves such that $\Delta T_2 \sim 0.8°C$ instead of the value 2.16°C used by C&L.[3] In this case, however, the dust forcing turns out to be negative—the implication is that it was less dusty at the LGM. Hargreaves and Annan (2008) seem to imply that this is an equally good interpretation of the Vostok data. However, there are two things wrong with this. One is the choice of temperatures by H&A does not fit the data well in Figure 1.13. But, more importantly, the end result of a negative dust forcing at the LGM is contrary to our physical understanding and suggests that the data chosen by H&A cannot be correct. In their second example, H&A claimed that they arrived at a dust forcing of 0.9 ± 1.2 W/m^2, as compared to the estimate by C&L of $58 \times 0.056 = 3.25$ W/m^2. This led to an estimate of $\Delta T_G \sim 2.5 \pm 0.7°C$ for a doubling of CO_2 from 280 ppm to 560 ppm. However, H&A did not specify which temperatures they used in this calculation, so it is impossible to reproduce what they did. H&A then extrapolated beyond science by asserting that an estimate $\Delta T_G \sim 2.5 \pm 0.7°C$ for a doubling of CO_2 from 280 ppm to 560 ppm "does not pose any significant challenge to the widely held view that climate sensitivity is likely to lie in the range 2–4.5°C [3.25 ± 1.25]". H&A evidently desired to derive as high a climate sensitivity as they could from glacial–interglacial transitions, and the best they could do was $2.5 \pm 0.7°C$, which they say does not pose a challenge to $3.25 \pm 1.25°C$. If we consider that C&L (known skeptics) derived a value of $\Delta T_G = 1.8°C$ for a doubling of CO_2 from 280 ppm to 560 ppm, and H&A (defenders of the orthodoxy[4]) derived 2.5°C, it seems likely that perhaps the most credible value from this type of analysis might be near 2.1°C.

None of these calculations takes into account potential changes in humidity and cloudiness during these transitions, which are likely to be as large as or larger than the forcings that were included.

Kohler *et al.* (2009) also performed an estimate of climate sensitivity based on glacial–interglacial cycles. They said: "Although water vapor is the most important GHG, the following compilation does not consider any changes in water vapor in the past due to missing constraints on its variability." In other words, they more or less said: *Water vapor may be the biggest factor, but since we have no data on it, we will neglect it!* Some of the data used by Kohler *et al.* (2009) are compared with data used by C&L and Hansen and Sato (2011) in Table 1.2.

If one were to simplistically take the result of Kohler *et al.* (2009) that a forcing of 12.43 W/m^2 produces a ΔT_G of 5.8°C, one might conclude that their estimate of climate sensitivity is $\lambda = 5.8/12.43 = 0.47$°C/(per W/m^2) that agrees with the result of C&L, although the data are different in both cases. It appears likely that Kohler *et*

[3] They did not actually provide the number 0.8°C but they did provide a graph and that was the value I read from their graph.

[4] One can discern the attitude of these authors toward the orthodoxy regarding ΔT for doubling of CO_2 from their other publications. Furthermore, in the cited reference, H&A emphasize that the estimates by the Inter-governmental Panel on Climate Change (IPCC) are inviolable.

Table 1.2. Parameters for analyzing LGM—pre-industrial transitions. Forcings are in W/m^2. Blank elements are not available. Elements with dashes represent items that were not included.

	Chylek & Lohmann (2008)	*Kohler et al. (2009)*	*Hansen and Sato (2011)*
CO_2 forcing	2.4	2.1	2.25
CH_4 forcing	0.27	0.4	0.43
N_2O forcing	–	0.3	0.32
Total GHG forcing	2.67	2.8	3.0
Land cryosphere		4.54	
Land ice		3.17	
Sea ice		0.55	
Snow cover		0.82	
Sea ice		2.13	
Sea ice—north		0.42	
Sea ice—south		1.71	
Vegetation		1.09	
Total albedo	3.5	7.76	3.5
Dust/aerosols	3.2	1.88	
Water vapor, lapse rate and clouds	–	–	–
Total forcing	9.4	12.43	6.5
ΔT_G (°C)	4.6	5.8[(5)]	5.0

al. made the most detailed analysis of the forcings, and it seems likely that their estimate of the total forcing (12.4 W/m^2) is likely to be the most reliable. However, great uncertainty remains regarding ΔT_G. If ΔT_G is as small as that estimated by C&L and Shakun and Carlson (2010), at 4.5°C, the implied climate sensitivity would be $\lambda = 4.5/12.43 = 0.36°C/(\text{per } W/m^2)$. This in turn would suggest a $\Delta T_G \sim 1.3°C$ for doubling CO_2. However, Kohler *et al.* somehow arrived at a figure of 2.4°C by arguments that are difficult for this writer to comprehend.

In summary, we have the results shown in Figure 1.16. All of the estimates for the ΔT_G for doubling CO_2 are lower than the consensus value of 3.0°C based on climate models, except for the unmodified result of Hansen and Sato (2011). None of the estimates exceeds the consensus value.

All of these calculations suffer from a lack of understanding of changes in lapse rate, cloudiness and humidity in the LGM → pre-industrial transition.

It is noteworthy that Hansen and Sato (2011), which came along three years after Chylek and Lohmann (2008) and two years after Kohler *et al.* (2009), did not refer to these prior papers that essentially carried out the same calculation but with different parameters, and, in the case of Kohler *et al.*, with much greater detail.

[5] Kohler *et al.* (2009) emphasized at considerable length that although reasonable estimates can be made for the ΔT at Antarctica, the value of ΔT_G is far more elusive. They suggested that 5.8°C was perhaps one of the better estimates but emphasized that ΔT_G is not well pinned down.

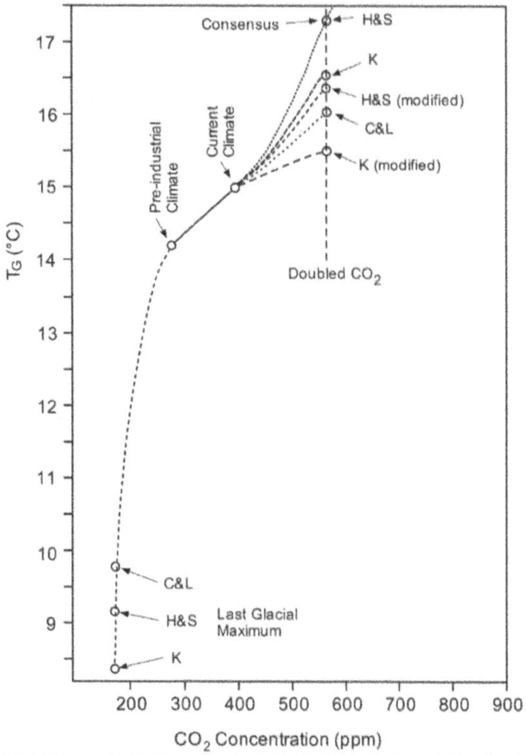

Figure 1.16. Summary of estimates from LGM → pre-industrial period transitions. C&L refers to Chylek and Lohmann (2008). H&S refers to Hansen and Sato (2011). K refers to Kohler *et al.* (2009). Modified values are produced herein as described in the text.

1.4.4 Paleo climates and CO_2

Rapp (2012) provides an extensive review of the relationship between ancient climates and CO_2 concentration. In this section, only a few excerpts from this review are given.

The use of proxies and climate models to infer relationships between climate and CO_2 concentration has been carried out by a number of investigators over various timescales ranging up to hundreds of millions of years. In general, the results require distant extrapolations from short, recent calibration periods. Typically, there is much disagreement between different data sets, and considerable scatter within any particular set of data.

The early Pliocene: 3 to 5 million years ago

There is also some evidence that CO_2 concentrations were between 365 ppm and 415 ppm about 4.5 million years ago when temperatures were 3°C to 4°C warmer than pre-industrial values. If these estimates are correct, CO_2 concentrations were comparable to those of today, yet the Earth was considerably warmer. Alarmists such as Pagani *et al.* (2010), who assume that CO_2 is the primary forcing for climate change, conclude that the longer-term Earth system climate sensitivity is much higher than the fast feedback climate sensitivity. Those who are devoted to the religion of CO_2 as the sole arbiter of climate (e.g. Pagani *et al.* (2010)) attribute the warming of the Pliocene entirely to CO_2 and will therefore conclude that if we wait long enough at a fixed CO_2 concentration of ~395 ppm, the Earth will slowly approach Pliocene conditions, and that if we hold CO_2 at 560 ppm and wait long enough, T_G may rise by as much as 9.6°C. Other models suggest that CO_2 is only one of several factors that affect climate and these estimates are exaggerated.

The past ~20 million years

Pearson and Palmer (2000) described "the boron-isotope ($\delta^{11}B$) approach to pCO_2 estimation that relies on the fact that a rise in the atmospheric concentration will cause more CO_2 to be dissolved in the surface ocean, causing a reduction in its pH". They used Benthic ($\delta^{18}O$) measurements as an indicator of climate. They produced a chart with ($\delta^{18}O$) and CO_2 concentration plotted for the past 25 million years. Any putative relationship between CO_2 and climate is difficult to discern.

van de Wal *et al.* (2011) provided a review article on CO_2 and climate over the past 20 million years. Kohler (2011) reviewed this work and carried out his own analysis. There is a wide diversity in estimates of CO_2 concentration over this time period. Neglecting the huge amount of scatter in the data by various investigators, and focusing on only one set of data, they reached the tenuous conclusion that

$$(\Delta T_G) \sim 0.05\ (CO_2\text{—}300\ \text{ppm})$$

in which (ΔT_G) is the change in global average temperature when the CO_2 concentration varies from a baseline of 300 ppm. This would indicate that if CO_2 were to double to 600 ppm, (ΔT_G) would be 15°C. In going from the pre-industrial level of ~280 ppm to the present level of ~395 ppm, (ΔT_G) would be $0.05 \times 115 = 5.8$°C. There are three possibilities: (i) one possibility is that if we hold CO_2 at 395 ppm and wait long enough, (ΔT_G) will approach 5.8°C; (ii) the second possibility is that the climate is determined by factors other than CO_2; (iii) the third possibility is that the data in these studies are inaccurate. This writer leans to the second and third possibilities. It seems likely that more things than CO_2 in Heaven and Earth are dreamt of in the philosophy of paleoclimatologists.

Foster *et al.* (2009) showed that while the period from 25 to 5 million years ago was "a period of relative warmth" and only Antarctica was glaciated, "paradoxically" CO_2 concentrations were comparable to "pre-industrial values or even lower". Foster *et al.* (2009) pointed out that the "accepted paradigm"[6] requires CO_2 to vary in unison with global temperature. However, they emphasized: "Recon-

[6] The "accepted paradigm" is an almost religious belief that only CO_2 concentrations control climate change, and paleoclimatologists often interpret data and models with considerable bias toward that belief.

structing the concentration of atmospheric CO_2 beyond the reach of the Quaternary ice cores is, however, a notoriously difficult task. Nonetheless there is a growing consensus that pCO_2 did decline over the Cenozoic, but not exactly sympathetically with climate as the paradigm suggests."

Tripati *et al.* (2009) had a goal "to test the hypothesis that CO_2 and climate were closely coupled across … major transitions". They used boron/calcium ratios in foraminifera to estimate pCO_2 during major climate transitions of the past 20 million years. They concluded:

> "These results show that changes in pCO_2 and climate have been coupled during major glacial transitions of the past 20 myr, … supporting the hypothesis that greenhouse gas forcing was an important modulator of climate over this interval via direct and indirect effects."

However, they also said:

> "During the Middle Miocene, when temperatures were ~3° to 6°C warmer and sea level was 25 to 40 meters higher than at present, pCO_2 appears to have been similar to modern levels."

One is left with this inference: assuming the results of Tripati are accurate, there appears to be a general tendency for pCO_2 to be higher during warmer periods, but, as Foster *et al.* (2009) said, the variation is "not exactly sympathetically with climate as the paradigm suggests".

Initiation of Antarctic glaciation 34–33 million years ago

Liu *et al.* (2009) said:

> "About 34 million years ago, Earth's climate shifted from a relatively ice-free world to one with glacial conditions on Antarctica characterized by substantial ice sheets.… The abrupt shift to glacial conditions [occurred] in ~300,000 years.… Proposed causes for this fundamental change in Earth's climate state include changes in ocean circulation due to the opening of Southern Ocean gateways, a decrease in atmospheric CO_2, and a minimum in solar insolation."

Liu *et al.* (2009) reported SST changes, which were determined from the alkenone unsaturation index and the tetrather index from 11 globally dispersed ocean localities. They estimated Benthic cooling of 3°C to 5°C during the transition at 33.7 Ma.

Pearson *et al.* (2009) compared CO_2 concentration with climate as expressed in Benthic ($\delta^{18}O$) measurements as shown in Figure 1.17.

Pearson *et al.* (2009) interpreted their results to:

> "… strongly suggest that the primary cause [for the transition to Antarctic glaciation] was a diminishing greenhouse effect.… Ours is the first proxy-based study to confirm a substantial pCO_2 decline during the climate transition. We also find a sharp pCO_2 increase after maximum ice growth as the global carbon cycle adjusted to the presence of a large ice cap and there was a nonlinear

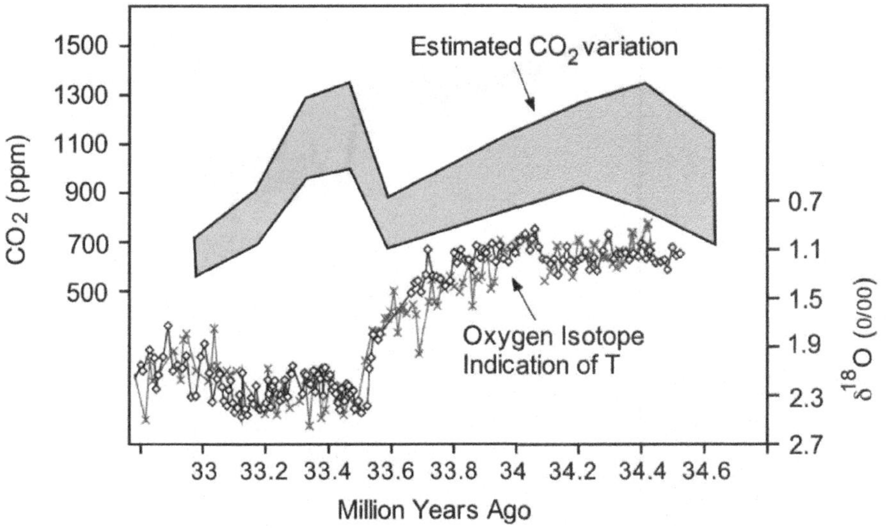

Figure 1.17. Comparison of CO_2 and temperature proxy across the Eocene–Oligocene boundary (adapted from Pearson *et al.*, 2009).

> hysteresis effect as the ice cap withstood this transient pCO_2 rise. This study reaffirms the links between cryosphere development and atmospheric carbon dioxide levels at the largest and most important climatic tipping point of the last 65 million years."

They also suggested that the threshold for initiation of Antarctic glaciation is in the range 700 ppm to 850 ppm.

These conclusions seem to be influenced by adherence to the "accepted paradigm" that CO_2 is the main factor in climate change. While there was indeed a moderate decrease in CO_2 as the Earth approached the Eocene–Oligocene boundary, the so-called "hysteresis effect" cannot be brushed away so easily. CO_2 levels popped up to well above the threshold while temperatures remained low. While CO_2 is clearly a factor in climate change, once again the comment by Foster *et al.* (2009) that CO_2 variations are "not exactly sympathetically with climate as the paradigm suggests".

Peters *et al.* (2010) used "an unusually well exposed coastal incised river–valley complex in the Western Desert of Egypt to show that eustatic sea level fell and then rose by ~40 m, 2 million years prior to establishment of a permanent Antarctic Ice Sheet".

They concluded:

> "This fall in sea level is associated with a positive oxygen isotope excursion that records buildup of an Antarctic Ice Sheet with a volume ~70% of the present-day East Antarctic Ice Sheet. Both the sea-level fall and subsequent rise were coincident with a transient oscillation in atmospheric CO_2 concentration

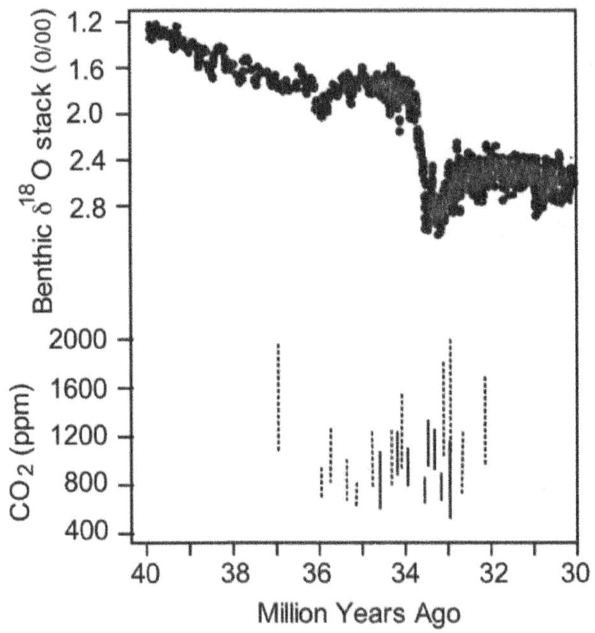

Figure 1.18. Comparison of CO_2 and temperature proxy across the Eocene–Oligocene boundary (adapted from Peters *et al.*, 2010).

down to ~750 ppm, which climate models indicate may be a threshold for Southern Hemisphere glaciation. Because many of the carbon emission scenarios for the coming century predict that atmospheric CO_2 will rise above this same 750 ppm threshold, our results suggest that global climate could transition to a state not unlike the Late Eocene, when a large permanent Antarctic Ice Sheet was not sustainable."

The result presented by Peters *et al.* (2010) is shown in Figure 1.18. How in the world they reached their detailed conclusions from this mess of CO_2 data is beyond the ability of this writer to comprehend. As is the case in most paleoclimatological studies, they drew a dollar's worth of conclusions from a penny's worth of data.

Peak warming around 40 million years ago

There is good evidence of a so-called Middle Eocene Climatic Optimum (MECO) as a short-term warming event that represents an abrupt reversal in a period of long-term cooling. The event was centered on 40 million years ago with a duration of about half a million years. While it is generally proposed that a sudden release of CO_2 caused this event, there are no credible data on CO_2 to validate this hypothesis.

60 to 40 million years ago

Pearson and Palmer (2000) compared estimated CO_2 concentration with Benthic

($\delta^{18}O$) measurements as an indicator of climate over the past 60 million years. They showed that there is a general tendency for higher CO_2 concentrations to be associated with higher T_G, although direct one-to-one correspondence is lacking.

Others have determined that CO_2 concentrations were relatively high about 50 million years ago. For example, Lowenstein and Demicco (2006) estimated that CO_2 was ~1,000 ppm to 3,000 ppm about 50 million years ago. Pagani *et al.* (2005) pointed out that "the relation between the partial pressure of atmospheric carbon dioxide (pCO_2) and Paleogene climate is poorly resolved".

Edwards *et al.* (2010) provided a result that is similar in some ways to that of Pearson and Palmer (2000) in that the warm period from 60 to 40 million years ago is associated with generally higher values of the CO_2 concentration. However, their data show very considerable scatter, and, furthermore, there isn't much variation in CO_2 while temperatures changed considerably over the past 20 million years. These results seem to suggest that, on balance, the warmest climates are associated with higher CO_2 concentrations, but the wide scatter in estimates of CO_2 concentration preclude detailed comparisons between CO_2 and climate.

One interesting event during this era was the so-called Paleocene–Eocene Thermal Maximum (PETM) that occurred about 55 million years ago. There is good evidence that T_G rose by at least several degrees (some estimates range from 4°C to 9°C) in as little as 10 to 30 thousand years. It is widely believed that this could only result from a sudden massive input of greenhouse gases. However, Zeebe (2011) said:

> "... estimated the size of the PETM carbon input based on sediment records of deep-sea carbonate dissolution and showed that the subsequent rise in atmospheric CO_2 alone was insufficient to explain the full amplitude of global warming. We concluded that in addition to direct CO_2 forcing, other processes must have caused a portion of the PETM warming.... Our study showed that there were processes in addition to CO_2 forcing that caused part of the warming, not that CO_2 was irrelevant. The processes are as yet unidentified—some may have operated independently, others as a response or feedback to the CO_2 release."

Cui *et al.* (2011) discussed the transient global warming event known as the PETM that occurred about 55.9 million years ago. "The warming was accompanied by a rapid shift in the isotopic signature of sedimentary carbonates, suggesting that the event was triggered by a massive release of carbon to the ocean–atmosphere system." They claimed "the source, rate of emission and total amount of carbon involved remain poorly constrained".

In contrast to Zeebe's indication of uncertainty regarding the PETM, Kump (2011) asserted that he understands the whole process. The initial release of CO_2 provided warming that added CH_4 to amplify the effects of CO_2. In fact, Kump (2011) provided a detailed description of the Earth during the PETM. Most of this seems to be subjective cloth woven from invisible thread. The methane hydrate hypothesis was discussed by Higgins and Schrag (2006), who concluded that analysis of the PETM leads to "a high climate sensitivity". Pagani *et al.* (2006) concluded that "the PETM either resulted from an enormous input of CO_2 that currently defies a mechanistic explanation, or climate sensitivity to CO_2 was extremely high".

As Royer *et al.* (2011) pointed out, "the PETM is considered a paleo-analog of present day climate change in terms of rate and magnitude of carbon release", although, as Kump (2011) emphasized, the annual release of carbon during the PETM was far less than today's, but it was sustained over a much longer time.

100 to 300 million years ago

Royer *et al.* (2011) compared crude estimates of CO_2 concentration to estimates of Benthic $\delta^{18}O$ and SSTs over the time range from 125 million years ago to 50 million years ago. The CO_2 and SST data show considerable scatter. There was a warm period at around 55 million years ago but it does not seem to have been accompanied by higher CO_2. In this regard, Royer *et al.* (2011) chose to ignore multiple CO_2 measurements near 55 million years ago that were low, and instead, accepted one outlier measurement that was four times higher. From this, they derived a high sensitivity of T_G to CO_2 concentration. This result does not seem credible to this writer.

Climate sensitivity based on CO_2 and climate in the Phanerozoic Eon

During the Phanerozoic Eon (the past ~540 million years), the Earth experienced significant changes. These included redistribution of continents via continental drift, the emergence of vascular plants driving up oxygen content in the atmosphere, changing CO_2 concentrations (as high as 20 times current levels at some periods), and many other changes, as discussed by Berner (2004). One particular time period, the so-called Permo-Carboniferous period between about 330 and 280 million years ago, was marked by extensive world glaciation, low CO_2 levels, and high oxygen content (30% to 35%). In addition, the brightness of the Sun increased with time across this eon.

As with almost every area of climatology, the data on the Phanerozoic climate and CO_2 concentrations are sparse and noisy, and the interpretation of the data in terms of climatological parameters requires complex models and a number of assumptions. Various investigators have arrived at different interpretations regarding the connection between CO_2 concentrations and climate change during the Phanerozoic Eon. Some concluded that changing CO_2 concentrations was the main factor producing long-term climate change. Others claim that the effect of CO_2 was secondary and galactic cosmic ray variability was the important factor. Rapp (2012) provides a detailed review.

The evidence suggests that the Earth was relatively warm from about 550 to 400 million years ago, although the temperature may have varied considerably within that time frame. Around 400 million years ago, the Earth began to cool, and the cooling bottomed out with extensive glaciation for about 40 to 50 million years approximately 330 million years ago. Subsequently, the Earth warmed again, and finally cooled again during the most recent 100 million years. Variations of the climate on shorter timescales within that general scope are subject to considerable uncertainty. The data and models for CO_2 suggest that R(CO_2) was very large prior to about 300 million years ago, peaking about 550 million years ago. R(CO_2) declined slowly from 550 million years ago, and very rapidly from 400 to 350 million years ago. The cold period centered around 300 million years ago was associated with very low values of R(CO_2). Temperature rose significantly after about 280 million years ago but CO_2 rose only very moderately. Over the last 100 million years, temperatures and R(CO_2) both declined. There seems to be very little doubt that major changes in the Earth's climate are at least sometimes associated with large changes in R(CO_2).

Boucota and Gray (2001) provided a very lengthy, detailed review of Phanerozoic climatic models and the relationship to the CO_2 content of the atmosphere. They concluded:

> "... considerable disparity exists between the curves generated by the varied models.... The wide disparities between the various published curves suggest that the presently published models are inadequate. Considerable disparity also exists between all the models and the geological climatic evidence indicating changing climatic gradients through the Phanerozoic. This indicates, based on present knowledge of climates of the geological past, that there is no simple straightforward relation between levels of atmospheric CO_2, as estimated by the various modelers and changes in the global climatic gradient."

Nevertheless, the widely held view amongst geologists and climatologists alike, is that the primary cause of long-term climate changes was variability of CO_2 concentration due to long-term imbalances between CO_2 degassing at spreading centers and the conversion of atmospheric CO_2 to mineral carbon through long-term silicate weathering and oceanic carbonate formation. As we pointed out previously, the argument goes (more or less): "If it wasn't CO_2, what else could it have been?" Foster *et al.* (2009) described this as the "accepted paradigm" that requires CO_2 to vary in unison with global temperature. Thus, paleoclimatologists have been trying for decades to establish a relationship between climate and CO_2 concentration over many millions of years. There is some evidence that over many millions of years, higher CO_2 concentrations are often, but not always, associated with warmer climates. However, there is a great deal of scatter in the CO_2 proxy data, and this relationship is difficult to pin down quantitatively. Royer (2010) began his commentary with the statement:

> "Global temperatures have covaried with atmospheric carbon dioxide (CO_2) over the last 450 million years of Earth's history. Critically, ancient greenhouse periods provide some of the most pertinent information for anticipating how the Earth will respond to the current anthropogenic loading of greenhouse gases. Paleo-CO_2 can be inferred either by proxy or by the modeling of the long-term carbon cycle. For much of the geologic past, estimates of CO_2 are consistent across methods."

This seems to be a rather optimistic view, considering the data from his paper are widely scattered. Royer's point (I think) is that throughout this period, SST was at least several degrees warmer than today, and even though there is much scatter in the CO_2 estimates, the general level of CO_2 concentration was much higher than today. This argument seems to make some sense from 120 million years ago to ~90 million years ago. Yet, there are difficulties from 70 million years ago to 50 million years ago when SST remained high, yet the CO_2 concentration appears to have been much lower. In any event, the extreme scatter in his data does not convey confidence that any firm conclusions can be drawn.

There is a general tendency over the years for temperatures and CO_2 concentrations to vary in unison, although there is no discernable proportionality, and there are several significant exceptions. For example, Uriarte (2009) pointed out:

> "The major paradox of this Ordovician (450 to 430 million years ago) glaciation is that CO_2 concentrations throughout the whole period remained well above modern-day levels. Indeed, according to some analyses of the carbon content of rocks dating from this time, they were up to 16 times higher. It therefore seems likely that it was geographical factors, rather than the chemical composition of the air, which played a key role in triggering that glacial period."

Our conclusion here is that CO_2 is probably one of several major factors in long-term climate change, but there are other factors, such as the placement of the continents on the Earth, the functionality of ocean currents, the past history of the climate, the orientation of the Earth's orbit relative to the Sun, the luminosity of the Sun, the presence of aerosols in the atmosphere, volcanic action, land clearing, biological evolution, etc. Hence, there is probably no single curve relating global average temperature to CO_2 concentration, but rather, a set of curves that depend on the above factors.

2

Temperatures during the past few millennia

A recurring theme in most of Isaac Asimov's science-fiction novels is the search for historical roots in the fog that obscures the past. These books have been very successful and are very readable, partly because they are well conceived and well written, but also because they evoke an empathetic response from our natural desire to understand our origins and roots. Similarly, the history of climatic variations leaves behind a fog that is difficult to penetrate. Many incredibly ingenious proxy methods have been devised to peer into the past. However, none is entirely satisfactory, and many uncertainties remain. In addition to proxy data, there are many anecdotal accounts in historical records that indirectly infer information about past climate (e.g., extent of glacier expansion and contraction in the Swiss Alps, paintings showing skaters on lakes that presently don't freeze, etc.). All of these have been utilized in the imperfect attempt to estimate past climates.

Today, we are concerned about potential climate changes that might occur in the future. To seek background data, we search for variability in past climates. However, random changes in weather from year to year and decade to decade are so large as to mask the underlying trends in climate change (if there is such a thing as climate). Climatologists have employed sophisticated statistical techniques to identify underlying trends, combining large numbers of proxies. The confounding thing is that the more proxies that are used, the more one averages out local and temporal variability by mixing the noise of many proxies with the signal of a few good ones. The inevitable result is a relatively flat temperature in the past, and, when this is combined with an exaggerated profile of recently measured rising temperatures, one obtains the "*hockey stick*" profile.

2.1 USE OF PROXIES TO ESTIMATE HISTORICAL TEMPERATURES

In the context of historical temperatures, proxies are residual data from processes that occurred in the past, when they were dependent on local temperatures at the times they took place, and the evidence is preserved in the present in an accessible form. In all cases, extraction of implied past temperature data from confounding influences requires considerable analysis and manipulation. As a result, the

D. Rapp, *Assessing Climate Change*, Springer Praxis Books,
DOI: 10.1007/978-3-319-00455-6_2

credibility and reliability of such proxy data vary widely from data set to data set, as well as in the eye of the beholder. For example, there are trees that are several thousand years old. The growth (width and density) of tree rings depends on the temperature prevailing during the growth period. By examining old tree rings corresponding to historical times, one can infer past temperatures. However, tree growth is also affected by other factors (water availability, humidity, wind, cloudiness, CO_2 content in the atmosphere, nutrients, etc.). These add noise to the temperature signal. Hence, it is not a simple matter to extract accurate historical temperature data from tree rings (or other proxies, for that matter). Some prominent proxies include: tree rings (width, density, stable isotope composition), ice cores (oxygen isotope ratios, gas content in bubbles), ocean sediments (isotope ratios), pollen, boreholes and corals.

2.2 PROXIES AND CLIMATE

2.2.1 Processing proxy data

The common approach to historical climate reconstruction from proxies is to establish a relationship between actual temperature measurements and the variability of the proxy over this recent period of overlap (calibration period). This provides a transfer function that enables the proxies to be used to infer the past climate in historical times when proxy data are available but direct temperature measurements are not. Ideally this works out as shown in Figure 2.1. Actually, one never obtains nice smooth curves as shown. Real data have many wild oscillations about the trend line. The fit of the proxy data to the measured data almost always leaves a great deal to be desired.

During the calibration period, measurements of the temperature are available for the locality where the proxy is located. A mathematical connection is made between

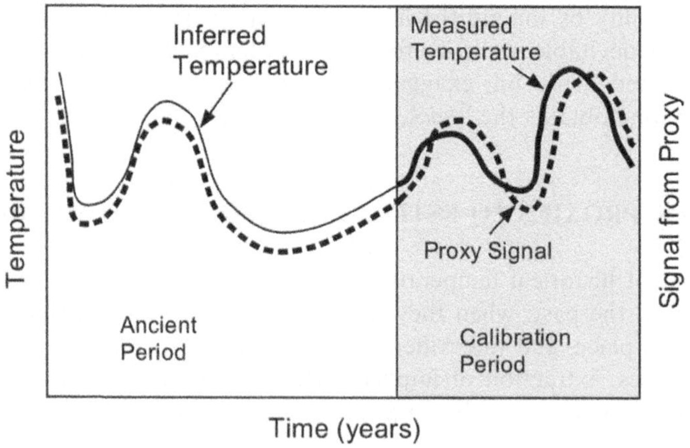

Figure 2.1. Concept of calibration period for a proxy.

the proxy signal and the temperature during this period. The proxy signal extends back in time prior to the calibration period. It is assumed (without proof) that the same relationship between proxy and local temperature holds in the past, and from this, past temperatures are estimated prior to the calibration period.

However, in many cases, the "fit" between the proxy and the measured temperature during the calibration period is not nearly as good as that depicted in Figure 2.1. There is usually a wide scatter in the data points and the putative relationship between proxy and temperature is not always clear. This is because many factors other than temperature often affect the proxy signal. Furthermore, in almost all published papers, the details of the proxy–temperature relationship during the calibration period are either not presented, or are so complex as to defy simple evaluation.

Over the past several decades, quite a number of scientists have analyzed data from one or several proxies and derived estimates of climate history in various localities and regions. Since much of the available data are from the northern hemisphere (NH), a good deal of this analysis pertains mainly to the NH.

2.2.2 Challenges in using proxies

Ogilvie and Jonsson (2001) noted that essentially all current calibrations of proxies to large-scale instrumental measurements have been made over periods of rising temperature. They raised the concern that a different calibration response might arise when the procedure is extended to an untested climate regime associated with a persistent cooling phase. They also raised other issues as well.

Jones *et al.* (1998) presented an extensive review of proxies. The spatial and timescale constraints on proxies were described in some detail. It was concluded that each reconstruction

> "... is probably limited in its ability to reproduce past temperature variations faithfully on the longest of timescales. This limitation varies from proxy to proxy and it is virtually impossible to quantify the degree to which this has occurred because instrumental series are not long enough."

It was found that, for most proxies, the correlation with instrumental data during the calibration period was not as good as reported in the journal articles.

2.2.3 Combining multiple proxies

Most proxies provide estimates of ancient temperatures at specific localities or regions. Climatologists are interested in the overall global climate, which necessitates developing a synoptic view of temperatures around the globe. The challenge is to find ways to combine temperature estimates from multiple proxies at various locations, over variable time periods.

Starting as early as the late 1970s and following through the 1980s and 1990s, culminating in a pair of very influential papers in 1998 and 1999 (Mann *et al.*, 1998, 1999), and continuing to this day, a loosely allied cadre of scientists has attempted to statistically combine large numbers of proxies (indeed, all the proxies that were available) into a reconstruction of global (or at least NH) average temperatures for the past millennium or two. They have assembled as many as 1,000 proxies into a database. These proxies include a variety of geographical locations, ranges of time, and degrees of credibility. A major question is how should these proxies be combined? Measurements at different locations, particularly different latitudes, will have different absolute temperatures and different temperature trends with time. Another major question is how to assign weights to various proxies based on their degree of credibility. Equally challenging is the question of how (or whether) to include documentary information that is typically discontinuous and often anecdotal in nature. These scientists typically employed sophisticated statistical approaches for combining proxy data sets into reconstructions of past NH or global average temperatures. The underlying basis for these approaches is the assumption that each proxy supplies an estimated temperature as a function of time (T_E) that contains a temperature signal (T_S) plus noise (T_N):

$$T_E = T_S + T_N.$$

It is further assumed that, if one utilizes a collection of proxies, the signals T_S will have similar trends for the various proxies while the trends for the noise T_N will vary randomly from proxy to proxy, sometimes plus and sometimes minus. Hence, if one adds up a number of proxies, the signals will tend to reinforce while the noise will tend to cancel out, leading to an estimate of T_S with less noise. Since the noise is typically quite large, sophisticated statistical mathematical schemes have been utilized to extract the signal according to this hypothesis.

Scientists who process proxies with complex statistical procedures have produced a steady stream of journal articles that justify their procedures (e.g., Rutherford *et al.*, 2005). These articles seem to rarely show the actual original proxies during the calibration period, but only the result of feeding them into their analysis machines. The end result is typically a relatively flat meandering curve of temperature over the past millennium with a sudden rise in the 20th century (the so-called "hockey stick"). In the latest in this series of self-justifying reports, Jones *et al.* (2009) provided a very lengthy review of the use of proxies to unravel the climate of the past millennium. The review covered (1) high-resolution proxy disciplines (trees, corals, ice cores, and documentary evidence), (2) various approaches for combining multiple climate proxy records to provide estimates of past hemispheric climates, and (3) use of climate model simulations of the past millennium. The end result for each proxy is a wiggly line representing a plot of temperature vs. year for a location. They then faced the problem of combining a large number of wiggly lines into a regional or global climate representation. The major stumbling blocks in combining multiple wiggly lines are (1) the spatial and temporal diversity and sparseness of the data, (2) the fact that most wiggly lines are heavily laden with local variations, and (3) large areas of the globe are underrepresented in the database. As a result, when large numbers of independent wiggly lines from various regions and time periods are simply summed and averaged democratically, the result tends to average out differences due to the wide variety of phases and amplitudes of the wiggles. Jones *et al.* (2009) describe a number of sophisticated approaches for "reconstructing the underlying spatial patterns of past surface temperature changes at global scales" and "assimilating proxy records into reconstructions of the underlying spatial patterns of past climate change". Basically, this means extracting T_S from $T_S + T_N$. However, as several critics have shown, the resultant regional or global climate reconstruction does indeed depend upon the method used for reconstruction, and, depending on the method used, almost any result can be obtained. Furthermore, the net result of combining many wiggly lines with variable amplitude and phase tends to be a relatively flat profile. When this flat profile for the past millennium is combined with a measured upward trend of temperature vs. time in the 20th century, the inevitable result is a hockey stick type of figure (relatively flat for the prior 1,000 years with a sudden upturn in the 20th century). It is also remarkable that Jones *et al.* (2009) made no mention of major criticisms of methods used to combine multiple wiggly lines but simply pretended that such criticisms do not exist. While it is true that most of these criticisms do not appear in the published literature, the reason for this is that it is difficult to get climate papers published that do not support the alarmist position.

2.3 THE LITTLE ICE AGE AND THE MEDIEVAL WARM PERIOD

2.3.1 Proxy evidence for the LIA and the MWP

Historic proxy studies have distinguished two periods of particular interest in the past millennium. One is the putative *Medieval Warm Period* (MWP) centered near 850–1050, which was supposedly an unusually mild climate. The other was the so-called *Little Ice Age* (LIA) from perhaps 1600 to about 1850 (depending on location) when temperatures were unusually cold. Apparently, the Earth was not uniformly warm during the MWP or uniformly cold during the LIA at all locations at all times. The same is true for the warming of the 20th century in which a third of the measurement stations report that their regions cooled during the 20th century while the other two-thirds warmed (Muller *et al.*, 2011a). There has not been any period over the past 2,000 years in which all regions of the Earth warmed or cooled in lock-step.

A number of independent proxy studies show evidence of distinct MWP and LIA. For example, the GISP2 ice core record shows evidence of a MWP and a LIA (Rapp, 2012). Thorsteinsson showed evidence for the MWP and LIA in the Camp Century ice core. Dansgaard (2005) claimed that the MWP and the LIA "were recognizable" and "stand out clearly" in the Camp Century ice core. He also presented the data shown in Figure 2.2. More recently, Vinther *et al.* (2010) revisited the matter of using Greenland ice cores to infer temperature variations over the past few thousand years. (This paper was partly based upon the Ph. D. dissertation "Greenland and North Atlantic climatic conditions during the Holocene—as seen in high resolution stable isotope data from Greenland ice cores" by Bo Møllesøe Vinther at the University of Copenhagen in 2006):

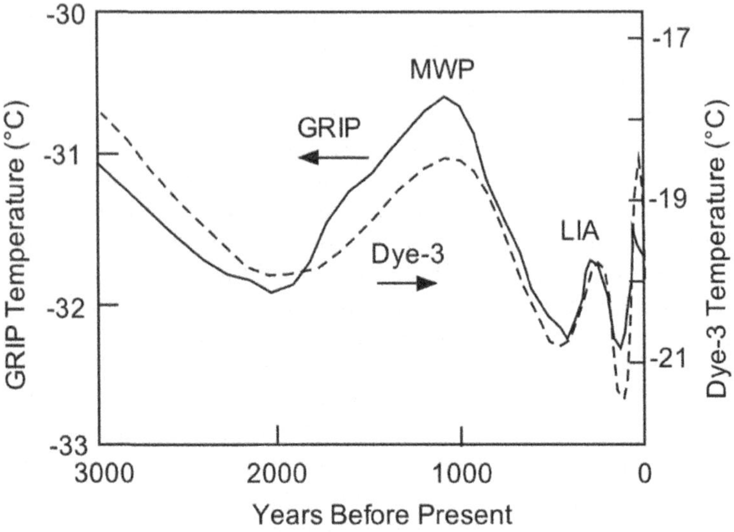

Figure 2.2. Ice core records showing LIA and MWP (Dahl-Jensen *et al.*, 1998).

"These authors worked with 20 ice core records from 14 different sites, all of which stretched at least 200 years back in time, as well as near-surface air temperature data from 13 locations along the southern and western coasts of Greenland that covered approximately the same time interval (1784–2005), plus a similar temperature dataset from northwest Iceland (said by them to be employed—in order to have some data indicative of climate east of the Greenland ice sheet)" (NIPCC, 2011).

Vinther *et al.* (2010) proceeded to demonstrate that "Greenland winter temperatures are much more variable than summer temperatures and thus dominate the annual average variability". So, they utilized winter ice core measurements of $\delta^{18}O$ at three sites on the Greenland ice sheet to examine the variability of climate from year 600 to year 2000. They found that the winter climate was highly variable with rather wild swings and, even with 50-year smoothing, the oscillations were large. Nevertheless, temperatures from about years 800 to 1000 were comparable to those of today, and temperatures from about 1400 to the late 19th century were demonstrably lower. This provides further support for the existence of a MWP and a LIA.

Richey *et al.* (2007) used Mg/Ca analyses in the white variety of the planktic foraminifera delta, which were obtained from the northern Gulf of Mexico as a measure of historical sea-surface temperatures. The results are shown in Figure 2.3.

Loehle (2007) produced a reconstruction of past temperatures that avoided the use of tree-ring proxies. His results are shown as Figure 2.4. The MWP and LIA are clearly delineated and the MWP is indicated to be warmer than the present.

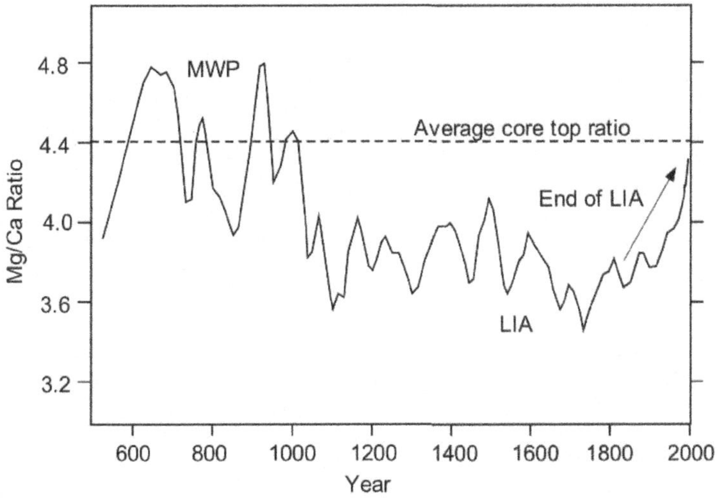

Figure 2.3. Mg/Ca analyses in the white variety of the planktic foraminifera delta, which were obtained from the northern Gulf of Mexico as a measure of sea-surface temperatures (Richey *et al.*, 2007).

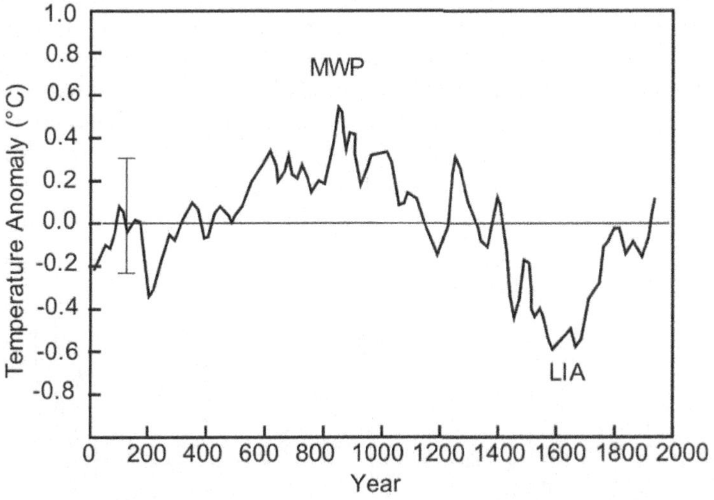

Figure 2.4. Mean temperature anomaly for 18 non-tree-ring proxy series (Loehle, 2007).

Yang *et al.* (2009) updated previous results for arid central Asia with new data over the last 2,000 years. They reported:

> "The most striking features are the existence of the *Medieval Warm Period* (MWP) and the *Little Ice Age* (LIA). The MWP was recorded in the 9–12th centuries and was accompanied by an anomalously dry climate, whereas the LIA extended from the 15–18th centuries and was accompanied by pluvial conditions."

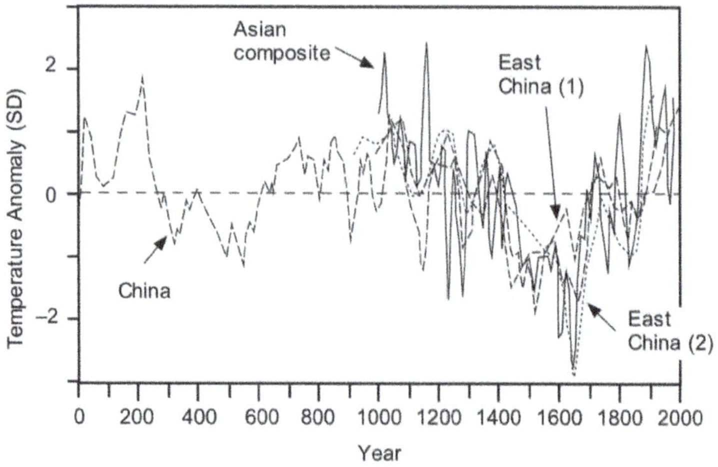

Figure 2.5. Estimates of historical temperatures in Asia (adapted from Yang *et al.*, 2009).

Their result is shown in Figure 2.5.

Esper *et al.* (2012a, b) developed a 2,000-year summer temperature reconstruction based on 587 high-precision maximum latewood density (MXD) series from northern Scandinavia. The record was "developed over three years using living and sub-fossil pine (*Pinus sylvestris*) trees from 14 lakes and 3 lakeshore sites at latitudes > 65°N, making it not only longer but also much better replicated than any existing MXD time series". The reconstruction was calibrated against regional June–July–August (JJA) temperature (1876–2006) and spanned from 138 BC to AD 2006. The calibration curve is shown in Figure 2.6. It is to the credit of this team that, unlike most paleoclimatologists, they provided revealing data on the calibration period.

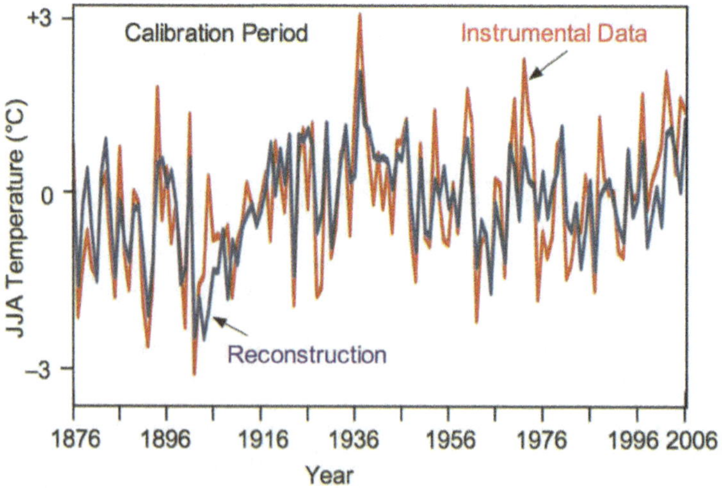

Figure 2.6. Calibration curve for northern Scandinavian tree-ring data (Esper *et al.*, 2012a, b)).

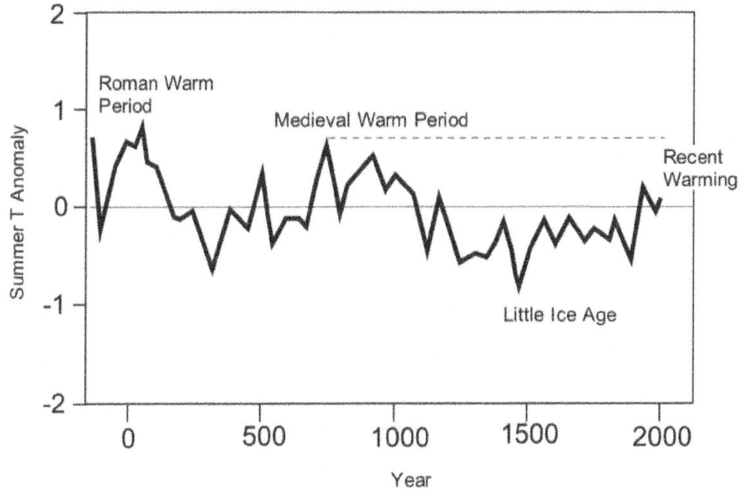

Figure 2.7. Reconstruction of 2,000-year history of Northern Scandinavian temperatures from tree ring data (Esper *et al.*, 2012a, b).

The resultant estimate of JJA temperatures for the past 2,000 years showed considerable noise from year to year with variations of $\pm 1°C$ being typical. Using a 100-year filter, they obtained the curve shown in Figure 2.7. These results indicate that past temperatures during Roman times and during the MWP were higher than the present, and there is a recognizable LIA. Esper *et al.* (2012a, b) chose to fit a straight line to the data, thus showing a fairly constant millennial-scale cooling trend. However, this interpretation is affected by the relatively high temperatures that occurred at the starting point in Roman times. Had the starting point been, say, AD 400, the trend would have been oscillatory with a positive lobe from AD 400 to AD 1200 and a negative lobe from AD 1200 to AD 1900. The interpretation of a fairly constant millennial-scale cooling trend might suggest a sudden change in the 20th century due to human influence, whereas the cyclical interpretation might suggest a new positive lobe beginning near AD 2000 (see Figure 2.8).

Idso (2008) provided extensive evidence in favor of the existence of the MWP and the LIA. In addition, prior to the MWP, there were numerous temperature excursions comparable with or greater than current global warming.

Shindell (2007) studied paleoclimatic data from a number of sources and concluded:

> "Historical data spanning the past millennium show substantial variations in aridity in the dry bands of the subtropics ... Palaeoclimatic records from a variety of sources and subtropical locations suggest that the MWP was generally marked by drier conditions, including prolonged droughts, which became less prevalent during the LIA. These records are supported by additional sediment and lake level records, including some showing wetter conditions near the equator, as well as fire residue and cultural records."

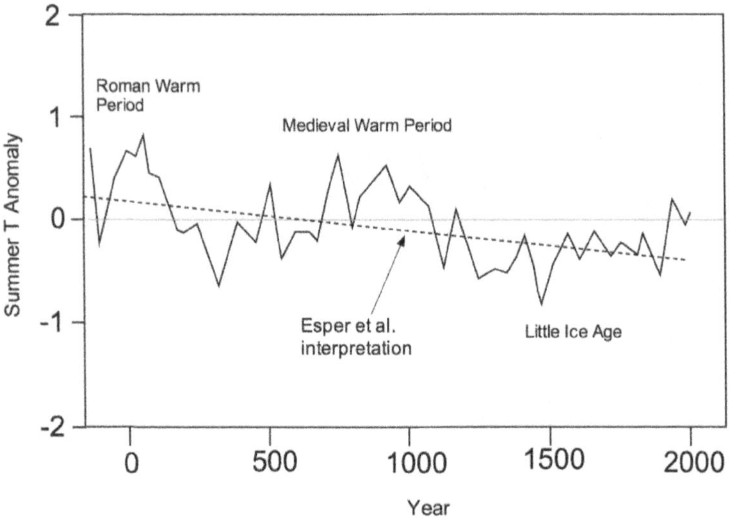

Figure 2.8. Reconstruction of 2,000-year history of Northern Scandinavian temperatures from tree-ring data with linear interpretation by Esper *et al.* (2012a, b).

Weckström *et al.* (2006) studied temperature patterns over the past eight centuries in Northern Fennoscandia inferred from sedimentary diatoms. Although temperature is not necessarily regarded as the strongest environmental variable affecting the distribution of diatoms, it was claimed that temperature has been shown to be a relatively powerful variable in this study area. The results show a pronounced MWP prior to 1200 and a wide flat bottom temperature from 1650 to 1900, representing the LIA. The temperature rise after 1900 is significant, but does not reach the temperatures of the MWP. Tree-ring data do not show this behavior.

Rørvik *et al.* (2009) examined sediment cores from a Norwegian fiord to infer temperature changes over the past 1,000 years. They found that "The periods from c. AD 500 to 790 and c. AD 1500 to 1940, stand out as cold periods".

Kobashi *et al.* (2010) derived Greenland temperatures over the past 1,000 years using nitrogen and argon isotope data, rather than oxygen isotope data. Their procedure is complex and appears to involve a number of assumptions. Nevertheless, they addressed these issues in considerable detail. Their conclusion was: "The data show clear evidence of the *Medieval Warm Period* and *Little Ice Age* in agreement with documentary evidence."

Barclay *et al.* (2009) presented data on the Tebenkof glacier in Alaska that clearly show the existence of a MWP and a LIA.

Cronin *et al.* (2010) used Mg/Ca ratios from ostracodes and oxygen isotopes from benthic foraminifera as proxies for temperature and precipitation-driven estuarine hydrography to prepare a 2,400-year paleoclimate reconstruction from Chesapeake Bay. This was compared to other paleoclimate records in the North Atlantic region to evaluate climate variability during the MWP and LIA. To the credit of these authors, they showed a detailed comparison of their proxy with temperature records (many paleoclimatologists hide these data). However, the comparison was very unimpressive to this writer, and the value of their proxy is highly debatable. Nevertheless, using several sediment cores, they consistently found warm temperatures between years 600 and 1000, and colder temperatures from 1300 to 1600. In some cases, temperatures from 600 to 1000 were warmer than today and, in some, today's temperatures were warmer. As they showed, other proxy-based estimates at other NH locations produced variable results. The accuracy of these proxies is uncertain.

Lüdecke (2011) published a new reconstruction of temperatures over the past two millennia. Two long-range annual records were utilized: "a stack of tree rings and further biological proxies" (AD 0–1979) and a record of stalagmites from the Spannagel Cave (AD -90 to AD 1935). The variance in the cave data was seven times greater than that of the proxy data. However, both indicated a clear MWP and LIA. Neither indicated warming today as being greater than that of the MWP.

Additional insights may be gained from proxies at Antarctica. Hall *et al.* (2010) obtained moss, peat, and reworked shells from Antarctica ice that indicated that the Antarctic Peninsula had retreated to or beyond present levels 700–1000 YBP and earlier, suggesting additional evidence for a MWP.

The website: *http://co2science.org/data/mwp/mwpp.php* provides an extensive list of peer-reviewed scientific journal articles pertaining to the MWP, and provides brief summaries of the findings of each paper. The locations of these studies are plotted on an interactive map of the globe. Studies are categorized by the degree to which they support the notion of a MWP as well as by geographical location (Africa, Antarctica, Asia, Australia/New Zealand, Europe, North America, NH, Oceans, and South America). An amazingly large number of studies exist. The majority of studies find good evidence for a pronounced MWP although a few studies find no evidence that the MWP was as warm as present-day temperatures. Many of these studies also provide evidence on the LIA. In addition, the studies by Idso, Carter and Singer (2011) and Idso (2008) provide extensive evidence in favor of the existence of significant MWP and LIA.

2.3.2 Anecdotal inferences on the MWP and the LIA

Fagan (2000) provided many anecdotal descriptions of the climate during the MWP and the LIA. How accurate or general these may be remains an open question.

Fagan described the mild period prior to the LIA. The overseas conquests of the Norse from about 800–1200 took place in "unusually mild and stable weather". He went on at length to describe the reduced pack ice around Iceland when "winter and summer temperatures were usually higher than today". By contrast, during the great cold of 1350–1380, sea ice came so close to land that Greenland polar bears came ashore on Iceland. He said: "For five centuries [9th through 13th centuries], Europe basked in warm, settled weather ... Compared with what was to follow, these centuries were a climatic golden age."

Fagan said: "In the 13th century, Greenland and Iceland experienced increasing cold. Sea ice spread southward around Greenland and in the northern Atlantic, creating difficulties for Norse ships ..." Icelandic recollections of the LIA are typified by: "In the extreme winter of 1695, ice blocked the entire coast in January and stayed all summer."

He also described changes in weather patterns other than temperature: drought and excessive rainfall. There was a deluge seven weeks after Easter in 1315. Rain fell essentially continuously from May through August, followed by a cold September, resulting in widespread famine and dislocation.

Fagan said:

> "The *Medieval Warm Period* saw long successions of warm, settled summers. Then, starting around 1310, and continuing for about five and a half centuries, the climate became more unpredictable, cooler, occasionally stormy, and subject to sporadic extremes—the LIA. Anecdotal accounts of the cold during the *Little Ice Age* abound in paintings of the period, showing skaters on frozen lakes and streams that do not presently freeze.
>
> "Between 1680 and 1730, the coldest cycle of the *Little Ice Age*, temperatures plummeted, the growing season in England was about five weeks shorter than it was during the twentieth century's warmest decades. The number of days each winter with snow on the ground in Britain and the Netherlands rose to between twenty and thirty, as opposed to two to ten days through most of the twentieth century. The winter of 1683–84 was so cold that the ground froze to a depth of more than a meter in parts of southwestern England and belts of sea ice appeared along the coasts of southeastern England and northern France. The ice lay thirty to forty kilometers offshore along parts of the Dutch coast. Many harbors were so choked with ice that shipping halted throughout the North Sea. Conditions around Iceland were now exceptionally severe. Sea ice often blocked the Denmark Strait throughout the summer. In 1695, ice surrounded the entire coast of Iceland for much of the year, halting all ship traffic. The inshore cod fishery failed completely, partly because the fish may have moved offshore into slightly warmer water, but also because of the islanders' primitive fishing technology and open boats. On several occasions between 1695 and 1728, inhabitants of the Orkney Islands off northern Scotland were startled to see an Inuit in his kayak paddling off their coasts. On one memorable occasion, a kayaker came as far south as the River Don near Aberdeen. These solitary Arctic hunters had probably spent weeks marooned on large ice floes. As late as 1756, sea ice surrounded much of Iceland for as many as thirty weeks a year ... The cold polar water spread southward toward the British Isles. The cod fishery off the Faeroe Islands failed completely, as the sea surface temperature of the surrounding ocean became 5°C cooler than today. Just as it had in the 1580s, a steep thermal gradient developed between latitudes 50° and 61–65° north, which fostered occasional cyclonic windstorms far stronger than those experienced in northern Europe today. The effects of colder *Little Ice Age* climate were felt over enormous areas, not only of Europe but the world." (Fagan, 2000).

There are anecdotal indications that the so-called *Medieval Warm Period* (MWP) from about 800 to perhaps 1200 was warmer than any period that followed it (e.g., grapes suitable for wine-making were reportedly grown in England, and the tree line in Scandinavia was 100–200 m higher than it is at present (Crowley and Lowery, 2000), although some believe that the late 20th century was warmer than the MWP. The degree of warmth in the MWP remains uncertain. There is considerably better anecdotal evidence that the so-called *Little Ice Age* (LIA) that followed the MWP (approximately 1400–1850) was considerably colder than the MWP, although there are uncertainties as to how consistently cold and how widespread the LIA was. The warmth of the MWP has been estimated by two means: (1) global climate models based on rather uncertain forcings, or (2) proxy analysis based on relatively few proxies available of uncertain veracity.

However, Crowley (2002) showed that historical written reports cannot always be taken at face value. An example is cited regarding the often-misused freezing of the River Thames. Between 1408 and 1914, the Thames in London froze over 22 times. Century counts are: 1400s (two times), 1500s (five), 1600s (nine), 1700s (five), and 1800s (one). Taken at face value, this would seem to imply that the 1600s and 1700s were coldest. However, there were a number of modifications to this bridge, including reductions in the number of piers of London Bridge in 1756 that reduced ponding effects. Replacement of the bridge between 1825 and 1835 widened the piers further and removed the small weir, enabling the tide to encroach farther upstream. No complete freezes have occurred since then. Changes to the river and its channel are important factors that must be considered alongside cooler winters as causes of freeze-over. For example, in the winter of 1962–1963, the third coldest since 1659 in the Central England Temperature (CET) record, the river only froze upstream of the modern tidal limit (20 km upstream of central London). In the two coldest CET winters in 1683–1684 and 1739–1740, the Thames froze, but it also froze during nine other winters between 1659 and 1820. Painter *et al.*, (2013) argue that the decline of the LIA was likely driven by black carbon deposits on alpine glaciers from expanding industrialization of Western Europe in the mid-to-late 19th century.

2.3.3 Challenges to the notion of the LIA and MWP

In the past decade, some climatologists emphasized the lack of uniformity in evidence for the MWP and the LIA, and thereby attacked the very notions of the MWP and the LIA, arguing either that they were regional, minor, and variable, or in some cases they were claimed to be non-existent.

However, as Soon and Baliunas (2003a, b) pointed out:

> "The term *Medieval Warm Period* has been the subject of considerable controversy. Its nature and even its existence has been questioned ... as has that of the *Little Ice Age* ... They were not periods of unbroken cold and warmth respectively. Climate varied on small scales both spatially and temporally, as it has also in the twentieth century. Nevertheless, climatic conditions were such during the *Little Ice Age* that mass balances were sufficiently predominant for the glaciers to remain enlarged, although their fronts oscillated. Similarly during the *Medieval Warm Period* climatic conditions caused mass balances to be

negative, and volumes of glaciers to be reduced, so that they retracted substantially, though their fronts no doubt fluctuated, as they have been observed to do during the warming of the twentieth century."

One can deal with the MWP and the LIA in various ways. Grove (2001) reduced the difficulty in deciphering the nature of the MWP and the LIA by not directly referring to climate, but rather by limiting the definition to the extent to which glaciers extended globally and remained primarily enlarged or primarily retracted, while their fronts fluctuated about these forward or backward positions.

Various proxy studies have derived different time periods, and different spatial extents of the putative LIA. Ogilvie and Jonsson (2001) pointed out "the difficulty in defining exactly the onset of the LIA". However, it is widely agreed that the end of the LIA was either late in the 19th century or early in the 20th century. They emphasized the many discrepancies and lack of coherence in data purporting to define the temporal extent of the LIA (and the MWP as well). However, it is not clear how much of the variability in the observed timing of the LIA is due to true variability and how much is due to difficulty in interpreting proxy data. Ogilvie and Jonsson (2001) said:

> "This lack of agreement could be due in part to uneven distribution and character of the evidence available, to the dating techniques used, and their resolution, and possibly due to differing degrees of effort devoted to unraveling glacial history."

There is no precise onset and there is no way to define the LIA except as a multi-century period when temperatures were predominantly (but certainly not exclusively) relatively low compared with the eras that preceded and followed the LIA.

It is understood that neither the LIA nor the MWP were periods of unbroken cold and warmth, respectively. Climate varied on small scales both spatially and temporally, as it has also in the 20th century. Nevertheless, climatic conditions may have been such during the LIA that mass balances were sufficient for the glaciers to remain predominantly enlarged, although their fronts oscillated. Similarly, during the MWP, climatic conditions would have caused the volumes of glaciers to be reduced, so that they retracted substantially, although their fronts no doubt fluctuated, as they have been observed to do during the warming of the 20th century.

Soon and Baliunas (2003a, b) carried out a detailed study of a number of proxies that indicated that the LIA and the MWP existed as a distinguishable climatic anomaly in almost all regions of the world that were assessed. Furthermore, they concluded that most of the proxy records do not suggest the 20th century to be the warmest or the most extreme.

Soon and Baliunas (2003a, b) provided a very long table listing the various proxies used in the study. For each proxy, they provided the spatial extent, latitude, and longitude (where applicable), type of proxy, reference, and qualitative evaluations of whether MWP and LIA trends were discernible. These included 14 worldwide proxies, and > 100 proxies that are regional or local. The results were provided in several figures accompanied by a lengthy discussion (about 25 pages) of detailed information regarding specific proxy results that led to these figures. These results indicated that:

- The proxy data suggest that the LIA existed as a distinguishable climatic anomaly in almost all regions of the world that were assessed. Only two records, did not exhibit any persistent or unusual climatic change over this period.
- The MWP is a distinguishable climatic anomaly with only two unambiguous negative results.
- Most of the proxy records do not suggest the 20th century to be the warmest or the most extreme in its local representations. There are only three unambiguous findings favoring the 20th century as the warmest of the last 1,000 years. An interesting feature is that the warmest or most extreme climatic anomalies in the proxy indicators often occurred in the early to mid-20th century, rather than throughout the century.

The reaction of the *paleoclimatic cabal* was quick and forceful. In a series of emails that were not revealed until they were hacked in November 2011, *cabal* members strategized to repair the damage from the Soon and Baliunas paper. In one email from Malcolm Hughes to a dozen members of the *cabal*, he cautioned that "an appeal to the National Academy of Sciences (NAS) could be counterproductive—remember the poor treatment of high-res paleo in the NAS report requested by the White House the other year". (Note that he is apparently referring to the Wegman Report that was sanctioned by the NAS.) Michael Mann, in his usual arrogant stance, referred to "two awful papers written by those clowns" and, yet, what could be more awful than Mann's publications? Mann also referred to their paper as "an assault on the science of climate change"—which is exactly what his papers constitute. The outcome of this *cabalistic* exchange was the publication by Mann and Jones (2003) that has little technical content and represents mainly an affirmation of the faith of the *cabal* in its orthodoxy. Many websites lit up with the news: "*Leading Climate Scientists Reaffirm View that Late 20th Century Warming Was Unusual and Resulted From Human Activity*."[1]

This report was authored by the founding members of the *cabal*: Michael Mann, Caspar Ammann, Kevin Trenberth, Raymond Bradley, Keith Briffa, Philip Jones, Tim Osborn, Tom Crowley, Malcolm Hughes, Michael Oppenheimer, Jonathan Overpeck, Scott Rutherford, and Tom Wigley. The report was not made public but was available only to journalists. Dozens of websites blared this headline but few details were revealed. The point was: 13 *cabalists* can hardly be wrong—or can they?

Crowley and Lowery (2000) argued that anecdotal reports as well as studies of individual records from MWP suggest that the present warmth of the 20th century is not unusual and therefore cannot be taken as an indication of forced climate change from greenhouse gas emissions. But Crowley and Lowery then asked the question: "Were all of these changes synchronous, with hemispheric amplitudes comparable to

[1] (e.g.: *http://www.agu.org/news/press/pr_archives/2003/prrl0319.html*).

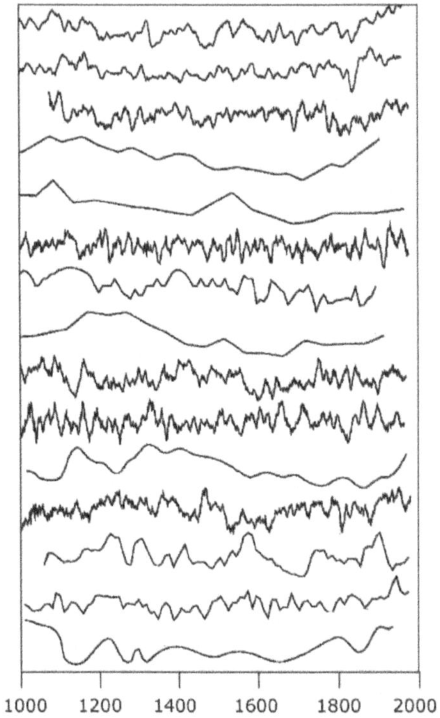

Figure 2.9. Fifteen individual proxies from various locations (adapted from Crowley and Lowery, 2000). Vertical scales are temperature anomalies.

or warmer than present?" However, this question seems to imply that present warming is spatially universal and synchronous—which it is not. (Indeed, the BEST study found that a third of reporting land sites experienced a decrease in temperature during the 20th century while the global average temperature advanced by roughly 0.6°C.) Crowley and Lowery revisited the controversy regarding the existence of the putative MWP by carrying out another proxy analysis, incorporating additional time series not used in previous hemispheric compilations. The 15 proxies used in the study are shown in Figure 2.9. It is to the credit of Crowley and Lowery that the individual proxies are shown; this is not often the case when assemblages of multiple proxies are analyzed. They combined the various proxies to obtain Figure 2.10. However, the process used to combine proxies is not clear to this writer.

Based on this result, Crowley and Lowery (2000) reached the rather incredible conclusions:

"Despite clear evidence for medieval warmth greater than present in some individual records, the new hemispheric composite supports the principal conclusion of earlier hemispheric reconstructions and, furthermore, indicates that maximum medieval warmth was restricted to two–three 20–30 year intervals, with composite values during these times being only comparable to the mid-20th century warm time interval. Failure to substantiate hemispheric warmth greater than the present consistently occurs in composites because there are significant offsets in timing of warmth in different regions; ignoring these offsets can lead to serious errors concerning inferences about the magnitude of Medieval warmth and its relevance to interpretation of late 20th century warming."

"Because of uncertainties in the proxy-instrumental temperature calibration, it is still difficult to unequivocably [*sic*] assert that the late 20th century warming is significantly greater than the peak warmth of the *Medieval Warm Period.* But there is even less justification to assert the opposite—it is not possible to make a robust statement that the *Medieval Warm Period* was warmer than the last two decades."

These conclusions seem far-fetched to this writer because:

- The huge variation from proxy to proxy suggests that these divergences may not represent true differences in climate from place to place, but rather, noise and error in the proxies themselves.
- Taking averages of highly divergent individual proxies must tend to average out variations, and add uncertainty and large error bars to the resultant average.
- The belief that the averages shown in Figure 2.9 can be trusted to the extent that differences between 20-year periods can be affirmed seems unwarranted by the lack of precision and consistency of the underlying data.
- The insistence that the MWP must perforce involve uniformly high temperatures at all locations for the entire period is unreasonable. The issue is not whether such uniform warmth occurred, but rather, allowing for spatial and temporal variations, the preponderance of evidence favors relative warmth compared with other eras. Furthermore, as we have pointed out, even the current warming in the 20th century is far from uniform, spatially and temporally.

McIntyre (2007) examined the results of Crowley and Lowery (2000) in some detail. McIntyre prepared a new figure, similar to Figure 2.10, except that it showed contributions from each of the 15 individual proxies with color-coding. McIntyre pointed out:

> "A distinctive 'hockey-stick' shape can be discerned in the 4 lowest records. Indeed, whatever '*hockey stick*iness' exists in Crowley and Lowery (2000) is entirely due to these 4 series, which consist of 2 bristlecone pine series, Briffa's Polar Urals series and Thompson's Dunde series ... The bristlecone pine series are prominent in the MBH99 reconstruction and the Polar Urals series in the Jones *et al.* (1998) series. Both series have problems [as discussed elsewhere in McIntyre, 2007]."

McIntyre (2007) then modified the color-coded version of Figure 2.10 by omitting four suspect proxy series. Instead of deleting the Sargasso Sea and Central Michigan proxies, both of which are claimed by McIntyre to be well linked to

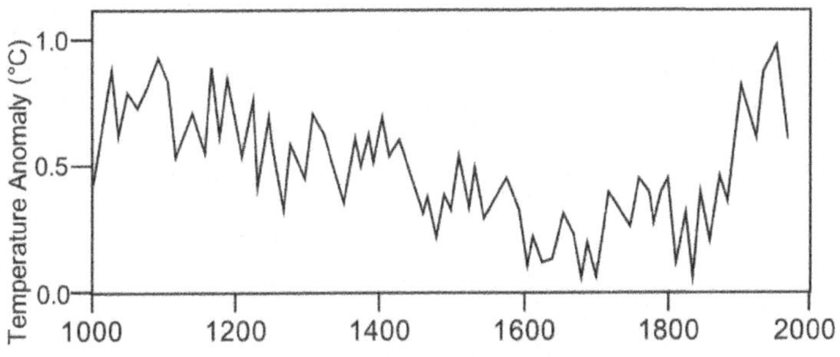

Figure 2.10. Derived NH temperature anomalies (adapted from Crowley and Lowery, 2000).

temperature, the two bristlecone pine series were excluded (as not being good temperature proxies) and the first century of the Polar Urals series was excluded on quality control grounds. Without the contribution of the bristlecones and Polar Urals, the MWP peak is comparable with the 20th-century peak.

Hegerl *et al.* (2007) added to the work of Crowley and Lowery (2000) using "updated records, a modified reconstruction method, and a new calibration technique". The stated goal of the study (as evidenced by the title of the paper) was to detect human influence on the climate:

> "The reconstruction consisted of three individual segments. A baseline reconstruction used 12 decadal records and covered the period to 1505. One longer, less densely sampled land temperature reconstruction ... was based on seven records back to A.D. 946, and [the third] consisted of five records back to A.D. 558."
>
> As is usual in such studies, they did not show the actual comparisons between proxies and temperatures during the calibration period so there is little basis to judge their adequacy *a priori*. Nevertheless, Hegerl *et al.* (2007) arrived at a slightly modified hockey stick result. Their hockey stick had a very minor MWP and a ~0.5°C LIA. They used Jones' "trick" of tacking on the recent instrumental record to the proxy results. However, they show an instrumental record that gained almost 1.5°C since 1900, that is far out of line with other estimates, and provides a very exaggerated view of the rise in temperature in the 20th century. It defies the logical mind to imagine how 12 or 7 or 5 proxy records, each of dubious credibility, centered in Europe, and with no representation from the Southern Hemisphere or the 70% of the Earth covered by oceans, could adequately define the global climate over 1,500 years."

What is missing from the proxy analysis of Hegerl *et al.* (2007) (as well as every other published proxy analysis that I am aware of) is a presentation of the comparison of each proxy at each location with the temperature as measured at that location during the calibration period (as well as after the calibration period). The variations from proxy to proxy are enormous. It seems likely that these vastly different patterns have little to do with temperature. But, if they do properly represent temperature at each location, and the temperature patterns vary by that much, we might ask how many proxies (locations) are needed to approximate a global average temperature? Certainly, use of only 15 proxies appears on the face of it to be grossly and incredibly inadequate.

2.4 GLOBAL AND HEMISPHERIC AVERAGE TEMPERATURES IN RECENT MILLENNIA

2.4.1 The "MBH" model

Realizing that there exists a number of local, regional, and hemispheric proxies, with variable spatial and temporal extent, Mann, Bradley, and Hughes (1998, 1999) attempted a comprehensive analysis of the history of global average temperatures using a multi-proxy network consisting of "widely distributed high-quality annual-resolution proxy climate indicators, individually collected and formerly analyzed by many paleoclimatic researchers". The network included annual-resolution dendroclimatic ice-core, ice-melt, and long historical records previously assembled, combined with other coral, ice-core, dendroclimatic, and long instrumental records. This was intended to integrate as many proxy sources as possible into a single comprehensive view of how a single global average temperature (or NH average temperature) varied over the past millennium. A number of subsequent related studies have also been published by the same group, as well as other allied groups. The final result is a reconstruction of a single NH or global average temperature over the past one or two millennia with a so-called *hockey stick* structure: a rather flat profile for most of the millennium, prior to the 20th century, with a significant rise in the 20th century.

Mann, Bradley, and Hughes (1998) has been referred to as "MBH" after the names of the three authors of the principal paper. Subsequently, Mann, Bradley, and Hughes (1999) extended the period of analysis from 1400 back to 1000, and Mann and Jones (2003) added an additional millennium back to 200. Mann *et al.* (2008) updated previous results. There are also a number of other relevant papers by other investigators.

MBH is a compact paper, full of jargon, and difficult to follow. However, this is a characteristic shared by many papers that deal with large data sets for historic Earth temperatures. Wegman, Scott, and Said (2006) said:

> "The papers of Mann *et al.* in themselves are written in a confusing manner, making it difficult for the reader to discern the actual methodology and what uncertainty is actually associated with these reconstructions. Vague terms such as 'moderate certainty' give no guidance to the reader as to how such conclusions should be weighed. While the works do have supplementary websites, they rely heavily on the reader's ability to piece together the work and methodology from raw data. This is especially unsettling when the findings of these works are said to have global impact, yet only a small population could truly understand them."

Wegman, Scott, and Said (2006) also said: "The description of the work in Mann, Bradley, and Hughes (1998) is both somewhat obscure and as others have noted, incomplete."

The reference period for calibration of proxies with actual temperature data was 1902–1980. The various proxies were more numerous in recent times and far less

Table 2.1. Number of proxies vs. earliest date according to Mann, Bradley, and Hughes (1998, 1999).

Earliest date	*Number of proxies*
1000	12
1400	22
1450	24
1600	57
1700	74
1763	93
1820	112
1854	219
1902	1,082

numerous in the more distant past. The number of proxies vs. earliest date is shown in Table 2.1. This array of proxies is vastly inadequate to represent the global climate. Of the dozen proxies that extended from 1000 to 1400, four were ice cores from a single small ice cap in Peru, and three were tree rings from the southwest U.S. How could that possibly lead to a global climate? These facts were well hidden in propaganda for the hockey stick, like a pea under walnut shells.

Each of the proxy data sets had variable geographical distribution. The task was to combine these into a uniform function that best expresses the putative single global average temperature over a long time span. The process used for data reduction is too complex to discuss in any detail here. As is usual in such studies, these references worked with variances from the mean, rather than actual temperatures. However, a crucial factor was that they chose the mean during the calibration period (1902–1980) rather than the mean for the entire data set. As we shall see, this had major repercussions regarding the form and credibility of the result. Another key aspect was the use of principal components analysis (PCA) to identify the primary trends in data containing scatter and noise.

The final result from Mann, Bradley, and Hughes (1998) is shown in Figures 2.11 and 2.12. Note that the mean is the mean for the calibration period 1902–1980, and therefore most of the data (1400–1920) lie below the mean. We will have more to say about this in the following sections. These *hockey stick* figures were published in subsequent papers with a relatively flat profile prior to 1900 and a sudden rise after 1900. Note that the "*X*" at the far right of Figure 2.12 is meant to be the current temperature. As shown, it is 1.1°C higher than the 1895 temperature, whereas it is widely believed that this temperature differential is more like 0.7°C. This exaggerated the shape of the hockey stick.

Mann and Jones (2003) extended the work of Mann, Bradley, and Hughes (1998, 1999) back to year 200. Their result is similar to that shown in Figure 2.12 with the addition of essentially no change in temperature from year 200 to year 1000. Taken at face value, these figures would suggest: (1) there was no MWP, (2) there was a very minor LIA, (3) Earth temperatures have been remarkably stable for 2,000 years, and (4) the only significant change in Earth temperature took place in the

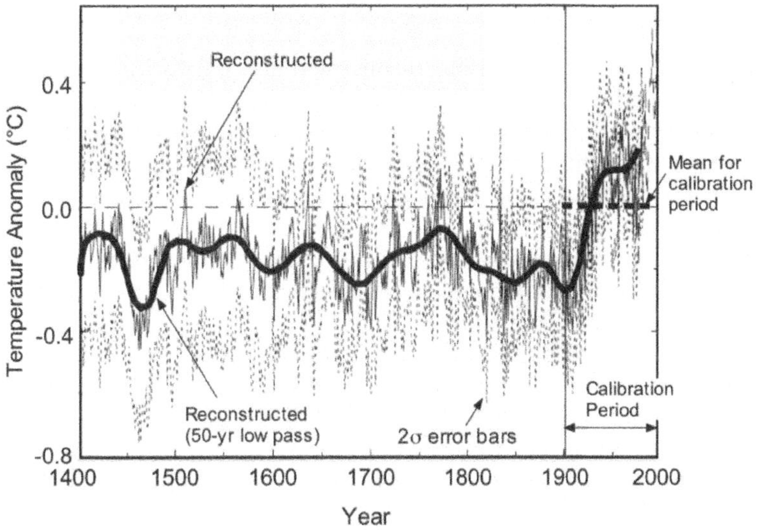

Figure 2.11. Reconstructed temperatures since 1400 (Mann *et al.*, 1998). Note that the mean is for 1902–1980. Also note that the 2σ error bars are so wide that they could hide almost any imaginable temperature curve.

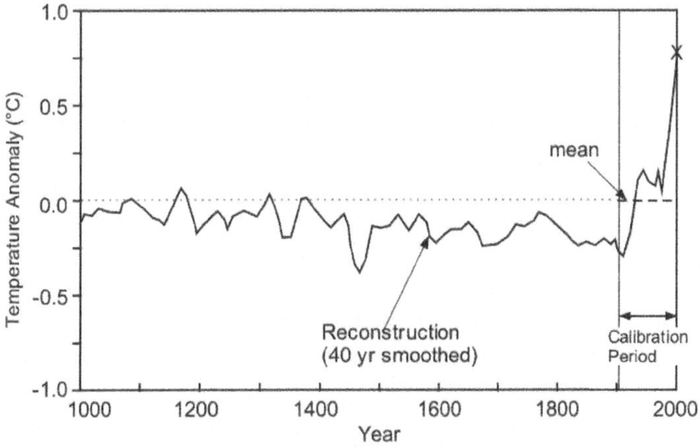

Figure 2.12. Temperature anomaly vs. year since AD 1000 (adapted from Mann *et al.*, 1999). The *X* at the far right is their estimate for 1998. The measured temperature rise from 1895 to 2008 is around 0.7°C, whereas Point *X* suggests a rise of 1.1°C. Note that the mean is for 1902–1980.

20th century with a sudden and decisive sharp rise after 1900. However, as we previously mentioned, MBH chose the mean for the calibration period (1902–1980) rather than the mean for the entire data set. As we shall see, this had major repercussions regarding the form and credibility of the result.

2.4.2 Other related models

A number of other groups published reconstructions of historical temperatures using similar methods. For example, Jones, Osborn, and Briffa (2001) obtained similar results using similar data-processing schemes. It is particularly noteworthy that Jones, Osborn, and Briffa (2001) decided not to show some of the proxy data late in the 20th century because it ticked sharply downward and conflicted with the desire to emphasize recent global warming (as we will discuss in a later section). Esper *et al.* (2005) discussed differences between various reconstructions based primarily on tree rings and presented a comparison. There is considerable variation in amplitude of the MWP and the LIA from study to study.

Moberg *et al.* (2005) indicated that, although differences in the amplitude of centennial temperature variability have been discussed in the literature, the picture with relatively small variability prior to the 20th century (i.e., the *hockey stick*) "is arguably best known by a wider audience. One reason for this is the prominent role that the multi-proxy reconstruction by MBH had in the latest IPCC report and in public media". However, they went on to point out that recent findings suggest that considerable underestimation of centennial NH temperature variability may result when regression-based methods (like those used by MBH) are applied to noisy proxy data with insufficient spatial representation. Moberg *et al.* (2005) also referred to well-documented difficulties in reliably reproducing multi-centennial temperature variability based on tree-ring proxies. Note the emphasis on data that "are best known" due to promotion by the IPCC and the media. von Storch *et al.* (2004) used a coupled atmosphere–ocean model simulation of the past 1,000 years to test empirical reconstructions of historical temperatures, specifically those of MBH. They found that centennial variability of the NH temperature is underestimated by the MBH regression-based methods. Their results also suggest that actual centennial variability may have been at least twice as large as the variability obtained in the MBH studies. Juckes *et al.* (2006, 2007) provided an extensive survey of a number of recent temperature reconstructions based on proxies. Juckes *et al.* (2006) presented a number of graphs of reconstructions of historical temperatures. However, all of these were based on MBH-type models in which the mean was chosen only for the calibration period (20th century) and, as a result, almost all of the temperature data for the past millennium (except for the 20th century) lie below the mean.

A comparison of the results of Mann *et al.* (2008) with results of other models shows wide variation from model to model (see Figure 2.13).

2.4.3 Fallacies in reconstruction of millennial temperatures

As we have pointed out, generally, the published papers on reconstruction of millennial temperatures tend to be very terse and full of jargon. The MBH papers are particularly bad in this respect. These papers present their results in small graphs with poor resolution but provide little insight into the calibration periods of specific proxies. Comparisons of proxies with temperatures during the calibration

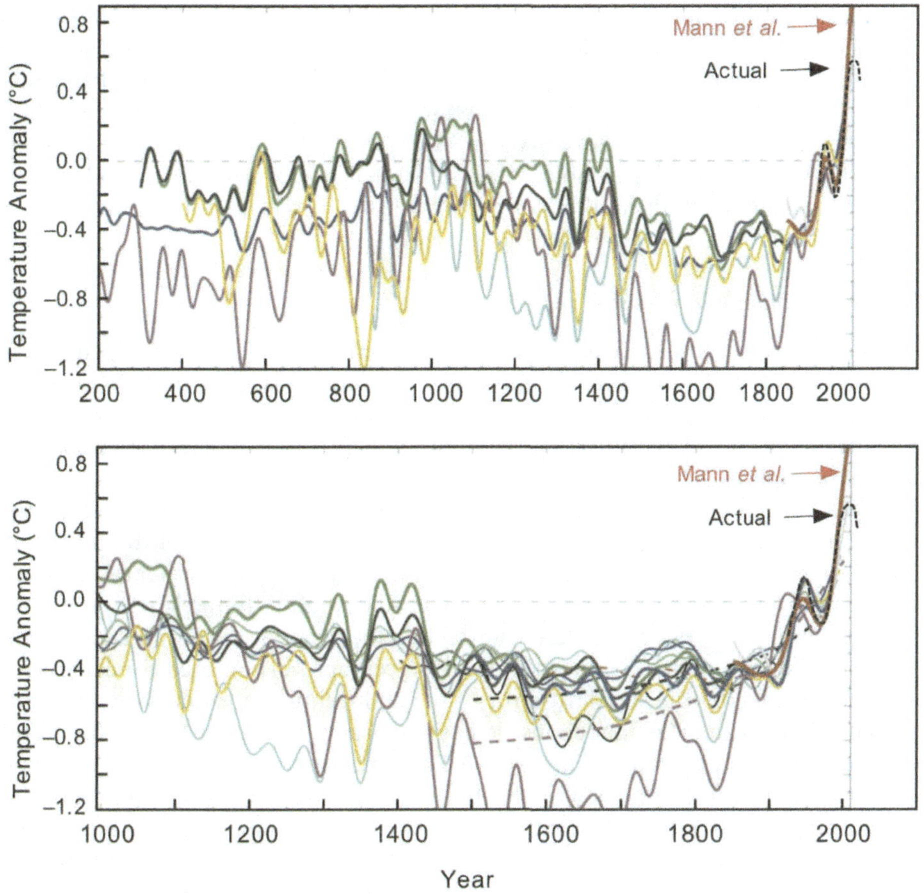

Figure 2.13. Estimates of historical temperatures by Mann *et al.* (2008) and others. Mann *et al.* (2008) exaggerated the temperature rise at the far right. The actual temperature change is shown as a black dashed line.

period are rarely provided and probably for good reason; the comparison is likely to be poor.

The only way to really understand what was done is to go back to their original data and follow the original procedures. As more and more of these reconstructions appeared in the literature with their typical hockey stick results, McKitrick (2005a) and McIntyre and McKitrick (2005, 2006, 2007) took it upon themselves to review these reconstructions by working with the original data in detail. These original studies by McIntyre and McKitrick continue to this day in the form of sporadic entries on the blog: *climateaudit.org*. The first obstacle they ran into was obtaining the data from the authors. Publications in journals are highly compressed and do not provide adequate means for others to reproduce the claimed results of the paper. Some journals require that authors archive the detailed data for access by the public but this is rarely enforced. Sensing that McIntyre and McKitrick (M&M) were antithetical to the hockey stick results, authors of papers on reconstruction of millennial temperatures resisted providing M&M with data and script from their work. Evidently, they were defensive about their work and did not cooperate in allowing their work to be checked. When M&M utilized the *Freedom of Information Act* (FOIA) in an attempt to obtain data generated by government-funded work in the U.S. and England, the authors of papers enlisted help from politicians to pervert and circumvent the FOIA on specious grounds.

2.4.3.1 The fallacy of choosing the wrong mean

After much perseverance, M&M succeeded to a considerable degree in penetrating the MBH data and procedure. In the course of doing this, they uncovered several major errors in the MBH approach. The principal problem was summarized by McKitrick (2005a). A paraphrased rendition of some of his remarks is given in the next paragraph.

In a conventional PCA, the temperature data are standardized by subtracting the mean of entire data set and dividing by the standard deviation of the entire data set. This re-centers and re-scales all the data to a mean of zero and measures deviations from the mean in units of the standard deviation. In the MBH program, a scaling was applied, but rather than subtracting the mean of the entire data set over all years, they subtracted the mean of the 20th-century portion used for calibration, and then divided by the standard error of the 20th-century portion. While this may appear at first glance as innocuous, it has important consequences for the results derived from this procedure. The overwhelming majority of individual proxy series do not have the form of hockey sticks, but appear as random noise, and, since they don't change much in the 20th century, this procedure did not make much difference for them. For these proxies, the mean of the calibration period is roughly the same as the mean of the whole series (as is the standard error) so either way of standardizing yields more or less the same result. But a few of the proxy series trend upward in the 20th century. For these, the MBH method has a huge effect. Since the mean of the 20th century portion is higher than the mean of the whole series, subtracting the 20th-century mean de-centers the series, shifting it off a zero mean. This, in turn, inflates the deviations from the mean of these series with increases in the 20th-century. PCA algorithms inflate the weights of proxies with the highest deviations. If one proxy series in the group has a relatively high level of deviation from the mean, its weight in the PC1 gets inflated. The MBH algorithm did just this. The PCA procedure would, in effect, sift through a data set and identify series with a 20th-century uptrend, and then load almost all the weight onto these series. In effect, it data-mines for hockey stick trends.

Consider the hypothetical set of proxies shown in Figure 2.14. If PCA based on the mean for the calibration period is used for the flat proxies, no problem arises. However, when the mean for the calibration period is used for the proxy with an uptrend during the calibration period, the majority of deviations from the mean (*b*) will be much greater than deviations calculated from the mean for the entire data set

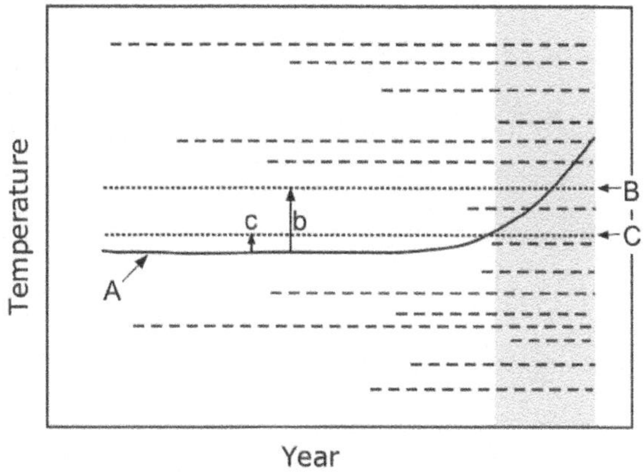

Figure 2.14. Hypothetical set of proxies where all proxies except one are flat and only one rises. The calibration period is shown as the gray rectangle. Flat proxies are shown as horizontal dashed lines. The one proxy (*A*) with variance is shown as a solid curve. Dotted line (*B*) is the mean for proxy (*A*) over the calibration period, while dotted line (*C*) is the mean for proxy (*A*) over the entire series. Deviations of proxy (*A*) from the two means are shown as (*b*) and (*c*).

(*c*). The point here is that, for the hypothetical set of proxies shown in Figure 2.14, PCA based on the mean for the calibration period will produce a "trend" similar to the one proxy with a slope and will essentially zero out the contribution of all the horizontal proxies to the estimated trend. Yet the preponderance of evidence is that the trend representing the overwhelming majority of data is actually horizontal. In cases such as that shown in Figure 2.14, PCA emphasizes the proxy with the greatest trend but it is not representative at all of the whole data set.

Figure 2.15 provides an example of the data-mining effect. It shows 2 of the 90 full-length series in the MBH database. The top panel is a tree-ring chronology from a stand of bristlecone pines at Sheep Mountain, California. The bottom panel is a tree-ring chronology from Mayberry Slough, Arkansas. In the bottom panel, the mean over the last 80 years is roughly equal to the mean for the previous 500 years, but, in the top panel, the post-1900 mean is above that for the pre-1900 portion. The MBH algorithm attributes 390 times as much weight to the top series as it does to the bottom series in the first principal component (PC1).

As it turns out, of 1,082 proxies used by MBH, only a handful exhibit the form shown in the upper panel of Figure 2.15, and all of these are tree-ring proxies that probably suffer from the potential CO_2 fertilization problem in the 20th century.

To test the MBH data-mining algorithm, M&M ran an experiment in which they input only trendless random red noise, simulating the data one would obtain from trees in a climate that is only subject to random fluctuations with no warming trend. In 10,000 repetitions, they found that a conventional PC algorithm (using the mean for the entire data set) almost never yielded a hockey stick-shaped PC1, but the MBH algorithm using the mean for only the calibration period yielded a pronounced hockey stick-shaped PC1 more than 99% of the time. The MBH algorithm efficiently

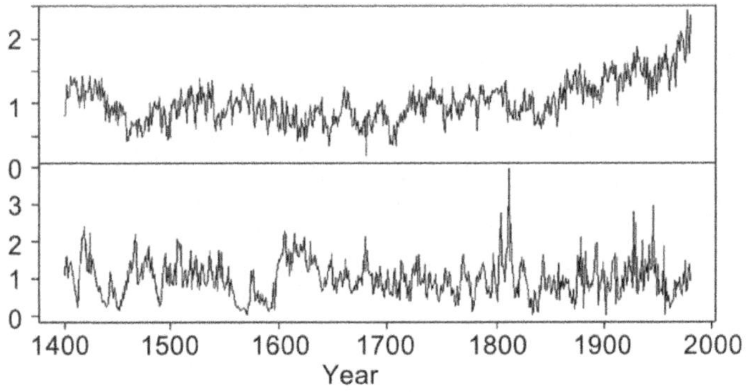

Figure 2.15. Two tree-ring temperature anomaly series from the MBH data set. Top: Sheep Mountain, CA. Bottom: Mayberry Slough, Arkansas (*climateaudit.org*).

looks for those kinds of series and flags them for maximum weighting. It concludes that a hockey stick is the dominant pattern even when pure noise is the input!

M&M extended their study in two ways. First, they showed that the MBH data-mining procedure did not just pull out a random group of proxies—it pulled out an eccentric group of bristlecone pine chronologies. These trees (the Sheep Mountain series in Figure 2.15 is an example) all turned out to exhibit a 20th-century growth spurt that has not been fully explained, but is likely to be at least partly due to CO_2 fertilization and is known not to be a temperature signal since it does not match nearby temperature records. The original authors who made the measurements (and others) have stressed that these series do not constitute proper climate proxies. So, M&M examined the consequences to the MBH results if these 20 bristlecone pine proxies were excluded. The result showed no hockey stick at all. Without these proxies with their rising shapes in the 20th century to mine for, the MBH method generates a result just like that from a conventional PC algorithm, and shows the dominant pattern is not hockey stick-shaped at all. In other words, without the bristlecone pines, even the flawed procedure of MBH would not have had a hockey stick shape.

Since the MBH papers purported to be the ultimate reference for estimation of the Earth's climate for one or two millennia, these results are likely to play a pivotal role in influencing the understanding of climatology. As such, the data and methodologies used by MBH should be readily available for replication and evaluation by others. McKitrick (2005b) provided an informal review of the way that he and McIntyre became interested in the Mann hockey stick, and their efforts to replicate the MBH results in order to understand the basis for their findings. In his discourse, McKitrick (2005b) described the difficulties in obtaining the required information, and the obstructive attitude of the MBH team. It became evident that the MBH team had manipulated the data unwittingly to greatly amplify the weight assigned to the few proxies with hockey stick form. It also became evident that the peer-review process did not penetrate into the MBH papers, and probably operates at a rather superficial level in most cases. M&M were thwarted by the journal *Nature* in their attempts to publish their criticisms and *Nature* crassly and cynically bowed to pressure from the *paleoclimatic cabal* and allowed the misleading publications of MBH to stand unchallenged.

A.W. Montford wrote a book that provides a very detailed history, background, and review of the entire hockey stick saga and the ensuing "climategate" revelations (Montford, 2010). Montford goes into considerable detail on the specifics of the proxy data and how they were processed, as well as the work by M&M in unraveling what MBH actually did, and the errors and misconceptions in the MBH analysis. He describes the resistance put up by the *paleoclimatic cabal* and the culpability of journals in shielding them from justified criticism. His penetration into the whole grisly mess is far deeper than I have attempted in this book. As the story unfolds, Montford shows that these paleoclimatologists, with their lousy data, and worse methods of processing the lousy data, have hoodwinked the science community and the world at large into believing their results.

In a more recent paper, Mann *et al.* (2008) updated their previous work by including additional proxies of various types. The spatial distribution of these was heavily concentrated in the U.S. and Europe (about 85%) with very few in the rest of the world (about 15%). As is usual in papers authored by Mann and co-workers, the paper is difficult to decipher. Oceans, which cover 70% of the Earth, were claimed to be included in some of the studies but it is not clear how ocean temperatures from a thousand years ago were obtained and averaged over all the oceans—if indeed that is what was done. It is difficult to understand how they incorporated ocean data such as they may be from the terse description given in the paper. Of the 1,209 proxies utilized, 59 extended back 1,000 years and 25 extended back 2,000 years from the present. The mean duration of a proxy was about 270 years. Some of their reconstructions utilized all proxies, and some were restricted to a subset of proxies that passed "a screening process for a local surface temperature signal. The screening process required a statistically significant correlation with local instrumental surface temperature data during the calibration interval". The period 1850–1995 was used for calibration.

When a typical set of proxies from various regions is compared, the differences between proxies are huge compared with the similarities (assuming that similarities exist at all). The proxies used by Mann *et al.* (2008) were no exception. Figure 2.16 shows some of the proxies that they utilized. Evidently, the variations from proxy to proxy outweigh any consistent signal that may underlie these time series. Hence, this set of time series represents a data set with low signal-to-noise ratio, and simply adding up these proxies is bound to produce little more than noise. Nevertheless, Mann *et al.* (2008) remained undaunted. They applied a variety of sophisticated statistical methods in an attempt to unravel a signal from the noise. However, as the saying goes, it is difficult to "convert a sow's ear into a silk purse", or, in a more modern vernacular, it may be a case of "garbage in–garbage out". As Burger and Cubasch (2005) showed, almost any desired result can be obtained from interpreting this very noisy data, depending on how they are processed.

It is noteworthy that the temperature anomalies shown in Figures 2.11 and 2.12 are essentially all negative prior to about 1950, which proves that the mean that they

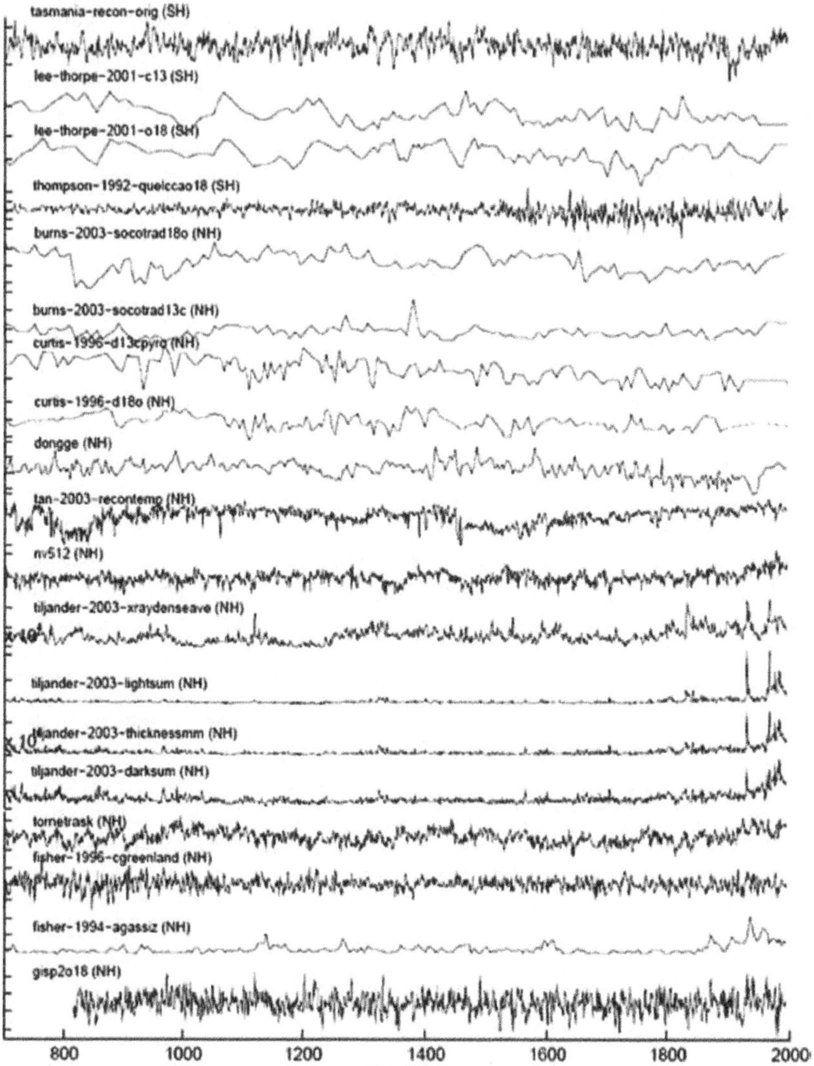

Figure 2.16. Some of the proxies used by Mann *et al.* (2008).

used to calculate anomalies was based on the calibration period, not the full data set (temperatures were higher on average during the calibration period). This casts doubt on the entire statistical procedure. In this connection, no mention was made of the criticisms of the methodology made by M&M, Wegman, and others, nor were any of their concerns addressed. Like the Wizard of Oz when exposed, they seemed to say: "pay no attention to that man behind the screen".

Clearly, the results of MBH and related papers are faulty. Juckes *et al.* (2006) attempted to deal with the criticisms of M&M (2003) claiming: "... the deficiencies in the description of the data used and possible irregularities in the data itself. These issues have been largely resolved in [Mann *et al.*, 2004]". However, aside from deficiencies in the data, which Mann, Bradley, and Hughes (2004) did not resolve except to obfuscate the matter, the critical issue of using the wrong mean, resulting in mining for *hockey stick* results was not even mentioned. Juckes *et al.* (2006) is just another in a series of papers using flawed statistics in a feeble effort to provide a pseudo-basis for their contention that we are in a state of unprecedented runaway global warming. There is no great point in reproducing their graphs here since they all display *hockey sticks* of one form or another.

It is noteworthy that Richard Muller said:

> "... carbon dioxide from burning of fossil fuels will prove to be the greatest pollutant of human history. It is likely to have severe and detrimental effects on global climate. I would love to believe that the results of Mann *et al.* are correct, and that the last few years have been the warmest in a millennium." (*http://muller.lbl.gov/TRessays/32-Global_Warming_Bombshell.htm*)

Hence, he was far from being a confirmed skeptic. Nevertheless, he went on to roundly criticize the methods of Mann *et al.* and concluded that the hockey stick is actually an "artifact of poor mathematics" and said: "A phony hockey stick is more dangerous than a broken one—if we know it is broken".

It is also worthwhile to review the response of *Nature* magazine to M&M when they attempted to publish their critique of Mann *et al.* (1998). On January 2004, M&M submitted their critique as a letter to *Nature*. One referee provided a favorable review. The other offered some confusion emphasizing the complexity of the details, but said: "In general terms I found the criticisms raised by M&M worthy of being taken seriously. They have made an in depth analysis of the MBH reconstructions and they have found several technical errors that are only partially addressed in the reply by Mann *et al.*" *Nature* issued a "favorable revise and resubmit" to which M&M responded in March 2004 with a revised manuscript. *Nature* then asked M&M to reduce the manuscript to 800 words. This was difficult, but was achieved and reduced manuscript was submitted in April 2004. In August 2004, *Nature* declined to publish the article that now (for reasons unexplained) needed to be reduced to 500 words. The main reason given was that the matters involved were "too technical" for a science journal. In other words, in a matter concerning the legitimacy and validity of the most widely accepted model of the Earth's climate over the past millennium, *Nature* decided that they would allow the erroneous publication by Mann *et al.* to stand unchallenged because the issues involved were too complicated. The ironic thing was that the *paleoclimatic cabal* could then claim that the criticisms of M&M could not be taken seriously because they were not published in a peer-reviewed journal.

Esper, Cook, and Schweingruber (2002a) started out by repeating the mantra of the global-warming alarmists:

> "... the MBH reconstruction indicates that the 20th century warming is abrupt and truly exceptional. It shows an almost linear temperature decrease

from the year 1000 to the late 19th century, followed by a dramatic and unprecedented temperature increase to the present time. The magnitude of warmth indicated in the MBH reconstruction for the MWP, 1000–1300 is uniformly less than that for most of the 20th century."

However, Esper, Cook, and Schweingruber (2002a) (ECS) admitted: "the MBH reconstruction has been criticized for its lack of a clear MWP." It was admitted that critics doubt that tree-ring records can preserve long-term, multi-centennial temperature trends. However, as usual in papers written by members of the *paleoclimatic cabal*, no mention is made of the devastating criticism of the MBH reconstruction made by McIntyre (2007). ECS then went on to present a defense of tree-ring reconstructions using centuries-long ring-width trends in 1,205 radial tree-ring series from 14 high-elevation and middle to high-latitude sites distributed over a large part of the NH extra-tropics. While ECS intended to support Mann, Bradley, and Hughes (1999), the large differences between their results and those of MBH lead this writer to the opposite conclusion. This raises the question whether any reconstruction based on proxies is credible. Furthermore, the anomalies in the result of ECS are mostly negative, suggesting that the mean used for data processing was not the mean for the entire time period, but only for the calibration period.

Nevertheless, based on their result, ECS reached the following conclusions:

(1) Multi-centennial temperature variability in long tree-ring records can be preserved if the appropriate tree-ring data and proper methods of analysis are used.
(2) The MWP appears to be more temporally variable than the warming trend of the last century and may have begun in the early 900s.
(3) The warmest period covers the interval 950–1045, with the peak occurring around 990.
(4) Past comparisons of the MWP with the 20th-century warming back to the year 1000 may not have included all of the MWP and, perhaps, not even its warmest interval.

McIntyre (2007) examined the data in ECS in considerable detail and wrote at length on their analysis. The issues are intricate and detailed and beyond the scope of the present write-up. McIntyre commented on the difficulty in obtaining the original data: "It's obviously been pulling teeth to get data from Esper. After only two years of trying, I've recently obtained all but one site chronology ... and gobbledy-gook about methodology." Using the 13 site chronologies that he had available, McIntyre plotted the individual proxies as shown in Figure 2.17. This is sometimes referred to as a "spaghetti chart".

McIntyre pointed out that only 2 of the 13 series have strongly elevated closing values. They both entail foxtail pines (interbreeding cousins of bristlecone pines) both from sites very close to Sheep Mountain, California. He cast considerable doubt on the validity of these two proxy sites. McIntyre then went on to present individual plots for each proxy, and perform a simple average. These results show that the proxies vary widely, and cast doubt on the consistency and credibility of the

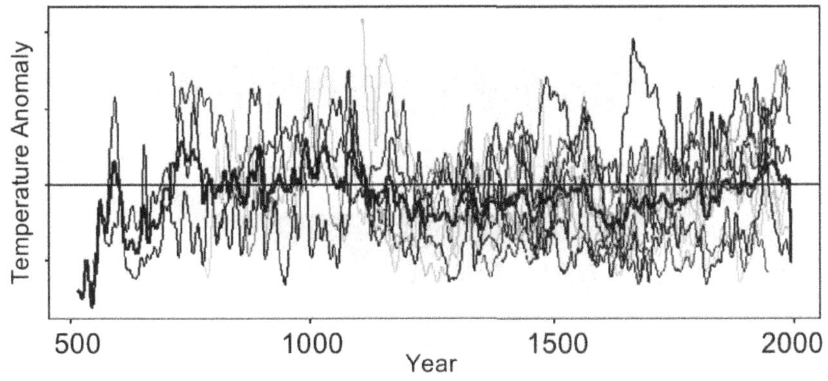

Figure 2.17. "Spaghetti chart" of individual proxies (except Mongolia) (adapted from McIntyre, 2007).

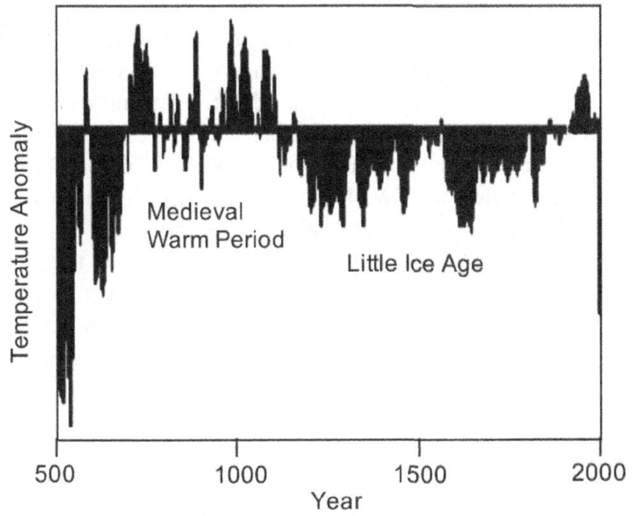

Figure 2.18. Simple average of proxy data from Esper *et al.* (2002) (adapted from McIntyre, 2007).

various proxies. While ECS provides us with assurance that "multi-centennial temperature variability in long tree-ring records can be preserved if the appropriate tree-ring data and proper methods of analysis are used", Figure 2.17 suggests otherwise.

McIntyre (2007) presented a simple average of all the proxies as shown in Figure 2.18. The result suggests an MWP and an LIA. Nevertheless, it seems evident that ECS used statistical data manipulation that unreasonably and illegitimately overemphasized the weighting of the two suspect proxies with high closing values.

Some *paleoclimatic cabalists* have argued that the MBH procedure is supported by other, more recent studies that also lead to a hockey stick. However, amazingly

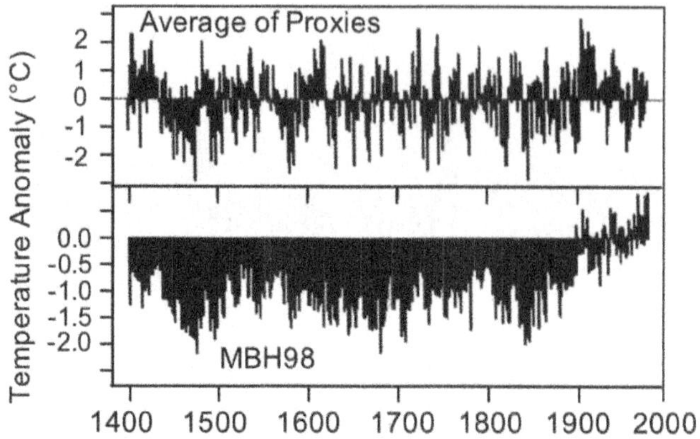

Figure 2.19. Temperature anomalies from MBH. The lower graph is the result of MBH98, while the upper graph is a simple average of their proxies (Montford, 2010).

enough, none of these further studies paid any attention at all to the M&M criticisms of the MBH procedure, and they blithely went ahead and used the mean of only the calibration period, making the same mistake as MBH, over and over again. Proof of this assertion is the fact that, in all of these reconstructions, the temperature anomaly remains starkly negative at all times prior to the calibration period. Montford (2010) supplied Figure 2.19 attributed to M&M. The lower graph is the result of MBH98 while the upper graph is a simple average of their proxies.

Paleoclimatologists who use PCA to extract a "trend" from the very noisy data seem to lose sight of the fact that, in cases where only a few proxies show a trend, PCA will weight these heavily to the exclusion of large numbers of proxies that do not show a trend. While Figure 2.14 is a rather extreme case, it illustrates a point. The object should not be to extract a "trend" but rather to represent the information contained in the entire data set. The average of all proxies in Figure 2.14 will be a line that is horizontal across most of the pre-calibration period and has a very slight rise through the calibration period due to the one proxy that rises. PCA will simply select the one proxy with a trend and ignore the data in all the horizontal proxies. In that respect, PCA is a useless, misleading method. But the reality is that the true trend is not the one PCA picks out, but rather, the average of all the data.

2.4.3.2 Hiding the decline

Tree-ring proxies are important in attempting to discern historical temperatures over the past millennium or two because they sometimes date back 2,000 years or more. Hence, tree-ring proxies are prominent in the MBH and other related reconstructions of global temperatures over one or two millennia. However, as we pointed out previously, tree growth is also affected by other factors (water availability, humidity, wind, cloudiness, CO_2 content in the atmosphere, nutrients, etc.) that add noise to the temperature signal. Hence, it is not a simple matter to extract accurate historical temperature data from tree rings (or other proxies, for that matter). There is ample evidence that the climatologists who have developed models for global temperatures over the past millennium or two have had a vested interest in proving that rising temperatures in the 20th century are unique and unprecedented, thus suggesting that natural causes cannot account for this change, and it must be attributed to growth of greenhouse gas concentrations. We can speculate on their motives. One might be a true idealistic desire to save the world from what they believe is an impending catastrophe due to global warming. Another might be the crass fact that funding for climate research will be proportional to the degree of catastrophe that is predicted. Whatever their motives, unfortunately, the behavior of tree-ring proxies was not supportive of this belief. Tree-ring proxies showed aberrations at various times, but the most serious problem was that tree-ring data typically show a downward trend in the late 20th century, while measured global temperatures were rising. The goal of the alarmists was to preserve the hockey stick, which they felt was necessary to show that rising greenhouse gases in the 20th century produced continuously rising temperatures. The "solution" to the problem of this divergence was "the trick" of not showing the down-trending proxy data in the late 20th century, and replacing it with measured temperatures that were rising.

> "I've just completed Mike's Nature trick [Michel Mann's publication in Nature where he replaced tree-ring proxy data with actual data because the tree-ring data went in the 'wrong' direction] of adding the real temperatures to each series for the last 20 years (i.e. from 1981 onwards) and from 1961 for Keith's to hide the decline." (Excerpt from email by Phil Jones)

It is particularly revealing to note some results of Briffa *et al.* (2001). Figure 2.20 shows eight different reconstructions using various procedures with one preferred reconstruction. Note that all reconstructions decline in the second half of the 20th century while measured temperatures rise. Briffa *et al.* (2001) also compared their results with reconstructions by others as shown in Figure 2.21. The divergence is readily seen and Jones's "trick" produces the hockey stick.

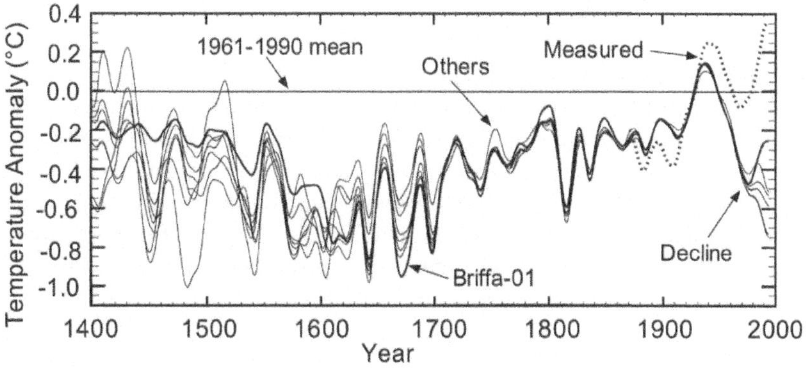

Figure 2.20. Eight reconstructions of historical northern non-tropical summer temperatures using various procedures. The heavy line (Briffa-01) is "preferred" (adapted from Briffa *et al.*, 2001).

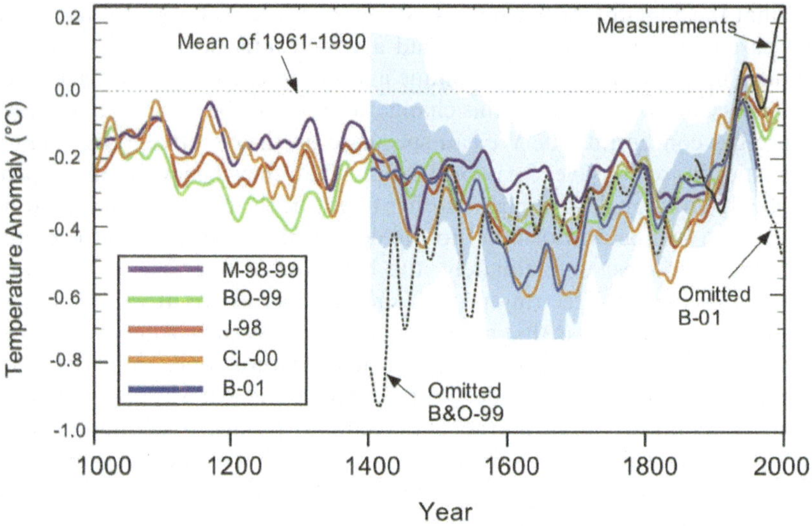

Figure 2.21. Comparison of reconstructions. M-98-99 = Mann *et al.* (1998, 1999). BO-99 = Briffa and Osborne (1999). J-98 = Jones *et al.* (1998). CL-00 = Crowley and Lowery (2000). B-01 = Briffa *et al.* (2001) (adapted from Briffa *et al.*, 2001).

Hegerl *et al.* (2007) also used Jones's "trick" of tacking on the recent instrumental record to the proxy results. Mann, Bradley, and Hughes (1998) and Mann, Bradley, and Hughes (1999) also cleverly substituted the measured temperatures for the modeled temperatures (Jones's "trick") to exaggerate the rise in the 20th century and thus accentuate the *hockey stick.*

The ultimate test for reliability of proxies is how well they track temperatures. Of all the many papers on proxies that I have reviewed, very few if any have provided such data in any detail. Briffa *et al.* (1998) is an exception. They compared tree-ring proxies with temperatures at many sites in the NH from 1880 to 1990. They said:

> "When averaged over large areas of northern America and Eurasia, tree-ring density series display a strong coherence with summer temperature measurements averaged over the same areas, demonstrating the ability of this proxy to portray mean temperature changes over sub-continents and even the whole Northern Hemisphere. During the second half of the twentieth century, the decadal-scale trends in wood density and summer temperatures have increasingly diverged as wood density has progressively fallen. The cause of this increasing insensitivity of wood density to temperature changes is not known ..."

Although Briffa *et al.* (1998) pointed out the discrepancy between tree-ring data and temperature after 1950, their assessment that proxies tracked temperatures prior to 1950 may be somewhat optimistic.

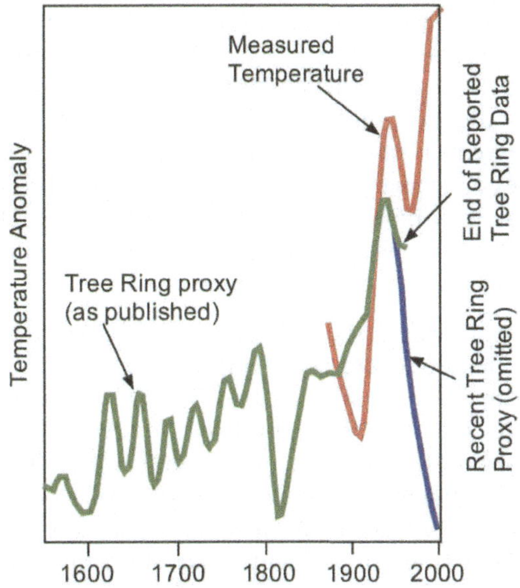

Figure 2.22. "Trick" of replacing tree-ring proxy by measured temperature (McIntyre, *http://climateaudit.org/2011/03/17/hide-the-decline-sciencemag/#more-13285*).

McIntyre[2] analyzed the data from several hockey stick reports (e.g. Briffa and Osborn, 1999). This is known as the "hide the decline" syndrome. Figure 2.22 shows the tree-ring proxy (combination of green and blue curves). However, when Briffa and Osborn were confronted with late-20th-century tree-ring data indicating a sharp drop in temperature, they craftily omitted the more recent data (blue curve) from their paper and ended the tree-ring data with the green curve. They then replaced the more recent tree-ring data by the measured curve (red curve).

Since the publication of these data in 1998, a number of additional papers have appeared dealing in one way or another with tree-ring proxies.

Jacoby *et al.* (2000) said:

> "Data from annual tree-ring widths are used to reconstruct May–September mean temperatures for the past four centuries. These warm-season temperatures correlate with annual temperatures and indicate unusual warming in the 20th century. However, there is a loss of thermal response in ring widths since about 1970."

Thus they admit to a divergence problem after 1970. However, when one examines their data prior to 1970, the correlation of tree-ring data with temperature even prior to 1970 is not impressive to this writer. D'Arrigo *et al.* (2006a) came to the defense of tree-ring proxies. They began their paper with the usual orthodoxy that "recent warming in the Northern Hemisphere appears to have been unprecedented over the past

[2] *http://climateaudit.org/2011/03/21/hide-the-decline-the-other-deletion/*.

millennium and that this warming is most likely a result of the anthropogenic release of greenhouse gases into the atmosphere"—which has not much to do with the reliability of tree-ring proxies. They mention that D'Arrigo *et al.* (2006b) used "simple averaging of tree-ring records (after accounting for differences in mean and variance over time), followed by linear regression". Simple averaging is a step in the right direction, but how good are the proxies as temperature indicators? As is usual in almost all papers on proxies, the data for the calibration period were not shown. They do allude to the divergence between tree-ring proxies and temperature reported by Briffa *et al.* (1998) and others cited in their paper, where they said:

> "Theories for the cause (s) of this observed divergence, which may vary from site to site, include decreased temperature sensitivity due to warmer temperatures, drought stress, increased winter snowmelt and ozone effects. This divergence needs to be considered to avoid bias in dendroclimatic reconstructions; however it is not present everywhere. For example, temperature-sensitive elevational treeline sites in Mongolia and the European Alps exhibit dramatic growth increases in recent decades. Greater attention to site selection (e.g. avoidance of drought-prone sites) and careful comparison of adjacent sites with regards to their ecological characteristics can help circumvent this problem. [It has been] demonstrated that the divergence appears to be limited to the recent period (after ~1950) and to trees from some northern locations (at some sites within ~55-70°N), and that there is no evidence for a comparable divergence prior to this time (e.g. during the *Medieval Warm Period*). These observations suggest a unique, anthropogenic cause for the recent divergence and argue very strongly that tree-ring temperature reconstructions for the past millennium should not be called into question based on these recent observations."

One problem with site selection is that, if one is attempting to estimate a global average temperature, one needs all the sites one can find. If only a few sites provide reliable data, how can one derive global or even hemispheric temperatures in the past? The claim that "there is no evidence for a comparable divergence prior to this time (e.g. during the *Medieval Warm Period*)" is sheer nonsense because there are no measured temperature data for that period and hence there is no way to ascertain whether such a divergence exists. D'Arrigo *et al.* (2006a) closed their paper with further homage to the orthodoxy of "unusual recent anthropogenic warming on a hemispheric to global scale" but their defense of tree-ring proxies falls flat.

To hide the divergence between proxies and reality, MBH terminated their calibration phase in 1980 even though more recent data were available. (Unfortunately, the proxies went down while measured temperatures went up after 1980.)

Wilmking and Singh (2008) discussed the "divergence effect" between measured temperatures and tree-ring proxies in the second half of the 20th century and pointed out that this "seriously questions the validity of tree-ring based climate reconstructions, since it seems to violate the assumption of a stable response of trees to changing climate over time". In their study, they claimed to have

> "… eliminated the 'divergence effect' in northern Alaska by careful selection of individual trees with consistently significant positive relationships with climate (17% of sample) and successfully attempted a divergence-free climate reconstruction using this subset."

However, they did admit:

> "The majority of trees (83%) did not adhere to the uniformitarian principle as usually applied in dendroclimatology. Our results thus support the notion that factors acting on an individual tree basis are the primary causes for the 'divergence effect' (at least in northern Alaska)."

Unfortunately, even the small subset of 17% of trees that are claimed to show good consistency with temperatures over the last century provide somewhat doubtful consistency. The diagram provided by Wilmking and Singh (2008) in their Figure 2 is a tiny little diagram that compresses the excursions between the temperature and tree-ring curves. Nevertheless, accepting the claim that 17% of the trees show good correlation with temperature for the sake of argument, the question arises as to whether it makes sense to select a subset of trees that happen to fit the temperature curve, and use these for estimating temperatures 1,000 years or more ago. Apparently, Wilmking and Singh suggest that there occurs a "mixture of trees with stable and non-stable climate growth relationships" and the ones with stable relationships provide a basis for estimating past climates. However, it may be equally likely that all the tree-ring records are randomized by other variables than temperature, and, by happenstance, about 17% of the records happen to have correlation coefficients with temperature that satisfy the criterion adopted by Wilmking and Singh (which is not impressive to this writer). There is then no great reason to believe that even these 17% of trees would remain as accurate temperature indicators over much longer periods. If you have a theory and desire to test it against data, you can cherry pick a small fraction of the data that seems to agree with the theory, but that is not science.

2.4.3.3 Sparse data set

Aside from all the other problems in reconstruction of millennial temperatures, the data set from which the analyses were conducted was very sparse. When MBH09 extended the time scale of MBH98 back from year 1400 to 1000, they depended on just 13 proxy series. Four were ice cores from a single small ice cap in Peru, and three were derived from southwestern U.S. tree rings. How could one possibly claim to have estimated global temperatures from such a sparse data set?

2.4.3.4 Lack of uniqueness

Burger and Cubasch (2005) is a difficult paper to read. It is full of jargon and uses a number of acronyms that are not defined. It is likely that only a reader who is intimately connected to statistical processing of long-term historical climate data could follow this paper in detail. Nevertheless, it appears to be an important paper and must be considered here. Burger and Cubasch (2005) examined the mathematical procedure used by MBH for the NH temperature reconstruction and noted that there were six key junctures where a fork in the road occurred, and MBH had to choose one or the other pathway for the ensuing computations. Since any one choice of path at one juncture could be combined with any other choice of path at another juncture, and there are six junctures, there must be a total of $2^6 = 64$ possible pathways to carry out the entire calculation.

We will not describe all of the junctures and choices here, but it is important to mention that one of the junctures was the choice of alternatives: (1) the MBH calculation of temperature anomalies based on the mean over the calibration period vs. (2) calculation of the mean for the entire time span of the data. As M&M have shown, use of only the calibration period mines for *hockey stick* results.

The various pathways can be described by means of six-digit binary numbers. The MBH method is described by one of these 64 binary numbers. Burger and Cubasch (2005) said: "No *a priori*, purely theoretical argument allows us to select one out of the 64 as being the 'true' reconstruction." Burger and Cubasch (2005) also argued that the alternate paths at each juncture are "a priori sound, with numerous applications elsewhere, and can hardly be dismissed purely on theoretical grounds". However, in regard to the juncture where one chooses the mean as a basis for calculating anomalies, it appears from the results of M&M that use of the mean for only the 20th century is fundamentally wrong *a priori*, and therefore the assertion by Burger and Cubasch (2005) is incorrect in this specific instance. The choice of the time period for the mean is not one of reasonable alternatives, but rather a choice of right vs. wrong.

The results of Burger and Cubasch (2005) are shown in Figure 2.23. Unfortunately, there is no key given to identify which of the 64 pathways correspond

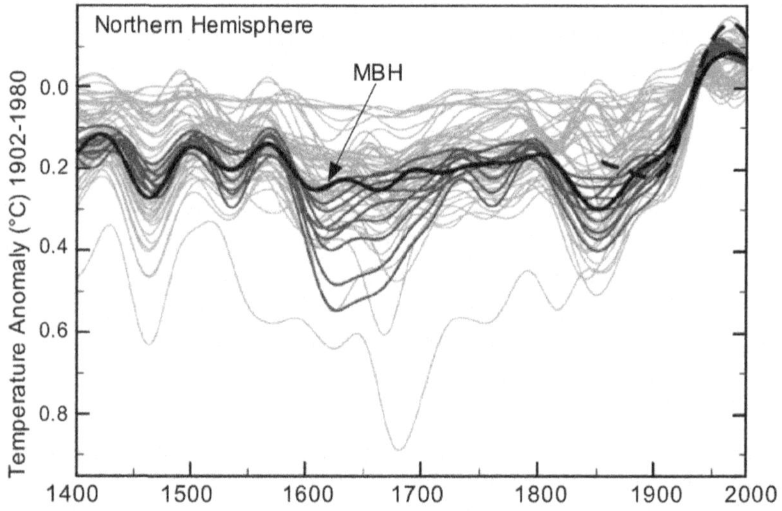

Figure 2.23. Temperature reconstructions (adapted from Burger and Cubasch, 2005).

to the various temperature reconstructions. A worrisome feature is that the zero line appears to be the mean of the 20th-century data and most anomalies are negative. This leads one to wonder whether they actually did calculations with the mean for the whole time period, and, if they did, why are the anomalies (for non-MBH options) not centered vertically on the zero line? Burger and Cubasch (2005) concluded:

> "Any robust, regression-based method of deriving past climatic variations from proxies is therefore inherently trapped by variations seen at the training stage, that is, in the instrumental period. The more one leaves that scale and the farther the estimated regression laws are extrapolated the less robust the method is. The described error growth is particularly critical for parameter-intensive, multi-proxy climate field reconstructions of the MBH type. Here, for example, co-linearity and over-fitting induce considerable error already in the estimation phase."

It appears that almost any result can follow from an integration of proxies, and no proxy result can be trusted.

2.4.3.5 Other criticisms

Zorita and von Storch (2005) examined methods used to construct historical temperatures. Unfortunately, it was difficult for this writer to follow all the arguments and jargon in the paper. However, it is apparent that Zorita and von Storch (2005) found significant problems with the methods used by MBH to infer long-term historical temperature records. Zorita and von Storch (2005) started with a model that the authors developed for a 1,000-year record of global temperatures called ECHO-G. The methods used by MBH were tested by following the MBH procedure, in which all data were considered to be deviations from the 1900–1980 mean value, even though this period is not representative for the temperature history of the past millennium. They followed the procedure of MBH but utilized different levels of noise as inputs. Inevitably, the methods of MBH led to a so-called *hockey stick* type of figure and grossly underestimated the variability of temperature in the past. Mann *et al.* (2007) argued against the results of Zorita and von Storch (2005), and Zorita *et al.* (2007) rebutted this criticism.

Esper *et al.* (2005a, b) concluded:

> "Our understanding of the shape of long-term fluctuations is better than commonly perceived, but the absolute amplitude of temperature variations is poorly understood ... Overall, amplitude discrepancies are of the order of the total variability estimated over the past millennium ..."

Even this conclusion seems optimistic.

Typically, when a set of proxies from various regions is compared, the differences between proxies are huge compared with the similarities (assuming that similarities exist at all). The proxies used by Mann *et al.* (2008) were no exception. Figure 2.16 shows some of the proxies that they utilized. Evidently, the variations from proxy to proxy outweigh any consistent signal that may underlie these time series. Hence, this set of time series represents a data set with low signal-to-noise ratio, and simply adding up these proxies is bound to produce little more than noise. Nevertheless, Mann *et al.* (2008) remained undaunted. They applied a variety of sophisticated statistical methods in an attempt to unravel a signal from the noise. As shown in Figure 2.23, almost any desired result can be obtained from interpreting this very noisy data, depending on how it is processed. Figure 2.21 shows results compiled by Mann *et al.* (2008) from their calculations, along with other estimates of historical temperatures. Several aspects of this figure are worth noting:

(1) There is considerable divergence between various estimates.
(2) None of the estimates shows a strong MWP. This may be due to a scarcity of data dating back 1,100 years.
(3) Most of the estimates show a rather weak LIA, probably due to averaging too many proxies.
(4) The steep upward curve at the far right, stated to be the measured temperature, is upwardly exaggerated.

In some of their results, Mann *et al.* (2008) provided estimates of the uncertainties in their results as a standard deviation envelope around the linear curve of temperature vs. time. In their Figure S11, they indicate that when alternate calibration periods are included, the total standard deviation of a temperature estimate for any year is typically about $\pm 0.6°C$, which suggests that any variations smaller than 0.6°C are uncertain by a significant amount. Since the uncertainties in calculations other than those of Mann *et al.* (2008) are likely to be at least as great, it must be concluded that, when an uncertainty of at least $\pm 0.6°C$ is imposed on the spread of estimates from various models, the resultant envelope would be large enough to hide almost any historical temperature profile that one desires.

It is noteworthy that the temperature anomalies shown in Figure 2.21 are essentially all negative prior to about 1950, which proves that the mean that they used to calculate anomalies was based on the calibration period, not the full data set (temperatures were higher on average during the calibration period). This casts doubt on the entire statistical procedure. In this connection, no mention was made of the criticisms of the methodology made by M&M, Wegman and others, nor were any of their concerns addressed.

There seems to be an inverse relationship working here: the more proxies that are included, the lower is the signal-to-noise ratio in the product. As Mann and associates relentlessly add new proxies, the quality of the result diminishes.

2.4.3.6 The Wegman Report

A team led by Professor Edward J. Wegman performed an independent examination of the *hockey stick* controversy. They produced a lengthy report, full of details. According to Wegman, Scott, and Said (2006):

> "The controversy of Mann's methods lies in that the proxies are centered on the mean of the period 1902–1995, rather than on the whole time period ...

Principal component methods are normally structured so that each of the proxy data series are centered on their respective means and appropriately scaled. The first principal component attempts to discover the composite series that explains the maximum amount of variance. The second principal component is another composite series that is uncorrelated with the first and that seeks to explain as much of the remaining variance as possible. The third, fourth, and so on follow in a similar way. In the MBH approach the authors make a simple seemingly innocuous and somewhat obscure calibration assumption. Because the instrumental temperature records are only available for a limited window, they use instrumental temperature data from 1902–1995 to calibrate the proxy data set. This would seem reasonable except for the fact that temperatures were rising during this period, so that centering on this period has the effect of making the mean value for any proxy series exhibiting the same increasing trend to be de-centered low. Because the proxy series exhibiting the rising trend are de-centered, their calculated variance will be larger than their normal variance when calculated based on centered data, and hence they will tend to be selected preferentially as the first principal component. Thus, in effect, any proxy series that exhibits a rising trend in the calibration period will be preferentially added to the first principal component."

Wegman, Scott, and Said (2006) went on to say:

"The centering of the proxy series is a critical factor in using principal components methodology properly. It is not clear that the MBH Team even realized that their methodology was faulty at the time of writing the MBH paper. The net effect of the de-centering is to preferentially choose the so-called *hockey stick* shapes. While this error would have been less critical had the paper been overlooked like many academic papers, the fact that their paper fit some policy agendas has greatly enhanced their paper's visibility. Specifically, global warming and its potentially negative consequences have been central concerns of both governments and individuals. The *hockey stick* reconstruction of temperature graphic dramatically illustrated the global-warming issue and was adopted by the IPCC and many governments as the poster graphic. The graphic's prominence together with the fact that it is based on incorrect use of PCA puts Dr. Mann and his co-authors in a difficult face-saving position. We have been to Michael Mann's University of Virginia website and downloaded the materials there. Unfortunately, we did not find adequate material to reproduce the MBH materials."

Wegman, Scott, and Said (2006) performed a calculation similar to that of M&M by comparing the results of an analysis of the North American tree network PC1 using the MBH data with centering based either on the calibration period mean or the mean for the whole time span of the data set. The result is shown in Figure 2.24. In addition to the *hockey stick* shape of the upper panel, it is worth noting that the lower panel exhibits considerably more variability. PCA seeks to identify the largest contributor to the variance. The MBH offset of the mean value shifts the main variance from the bulk of the data set to the 20th-century data.

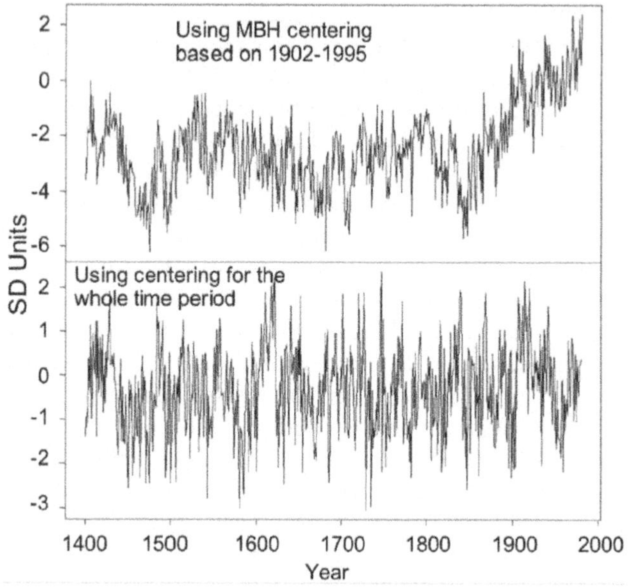

Figure 2.24. Comparison of rework of the North American tree network PC1 using MBH centering vs. using centering across the whole time span of the data set. The *hockey stick* is shown to be an artifice (adapted from Wegman *et al.*, 2006).

The findings of Wegman, Scott, and Said (2006) are quite lengthy and only a very brief summary is given here:

(1) In general, they found the writing of MBH somewhat obscure and incomplete. (This writer found the same.)
(2) In general, they found the criticisms by M&M to be valid and their arguments to be compelling.
(3) Use of the temperature profile in the 1902–1995 time span for centering leads to misuse of the PCA. However, the narrative in MBH on the surface sounds entirely reasonable, and could easily be missed by someone who is not extensively trained in statistical methodology.
(4) The cryptic nature of some of the MBH narratives requires that outsiders would have to make guesses at the precise nature of the procedures being used.
(5) Much of the discussion on the *hockey stick* issue has taken place on competing web blogs. Web blogs are not an appropriate way to conduct science and thus the blogs give credence to the fact that these global-warming issues have migrated from the realm of rational scientific discourse. Unfortunately, the factions involved have become highly and passionately polarized.
(6) Generally speaking, the paleoclimatology community has not recognized the validity of the M&M papers and has tended to dismiss their results as being developed by biased amateurs. The paleoclimatology community seems to be tightly coupled, and has rallied around the MBH position.

(7) The widely quoted assessments that the decade of the 1990s was the hottest decade in a millennium and that 1998 was the hottest year in a millennium cannot be supported by a proper rendition of the MBH analysis ... The paucity of data in the more remote past makes the hottest-in-a-millennium claims essentially unverifiable.
(8) Use of bristlecone pine proxies are inappropriate because they were probably CO_2-fertilized. It is not surprising therefore that this important proxy in MBH yields a temperature curve that is highly correlated with atmospheric CO_2. There are clearly confounding factors for using tree rings as temperature signals.
(9) There are other detailed statistical problems with the MBH treatment that require specialized knowledge to understand.

McIntyre revisited the Wegman Report in May 2011. He pointed out that the combination of criticisms of the MBH hockey stick by the *climateaudit.org* website and the Wegman Report generated a great deal of controversy on the blogs. The amazing thing is that, as McIntyre put it:

> "Rather than conceding even seemingly indisputable points, Mann and his associates contested every single issue—even the seemingly indisputable and elementary observation that Mann's principal components mined datasets for hockey stick shaped data. To this date [May 23, 2011], neither Mann nor any of his associates has conceded the point."

McIntyre also discussed the question: "... given the defects of Mann's principal components, how did the methodology pass peer review and then remain unchallenged by specialists in the field?" He stated further:

> "The Wegman Report hypothesized that this failure was due to the inter-connectedness of climate scientists through co-authorship and, in particular, by the extent of Mann's network of co-authorship, a level of inter-connectedness that the Wegman Report seemed to think as not existing in their own field. Wegman speculated that members of Mann's closest circle ('clique' in network terminology) reviewed papers of other members of the clique, resulting in non-independent and weak peer review, which, in turn, had resulted in the failure to identify the incorrectness of Mann's principal components in both the original article and subsequently."

It is noteworthy that not only did the *paleoclimatic cabal* ignore the criticisms of its hockey stick and blithely continue to promulgate their incorrect results, but they also went on the attack to attempt to besmirch the criticizers on a personal level. The mechanism for doing this was to accuse Wegman (and myself) of plagiarism in our writings. This seems to have originated on the *deepclimate.org* website where the scurrilous person hiding behind a cloak of anonymity known as "DC" apparently stayed up nights comparing word for word the criticisms of MBH with references to identify passages that were copied or slightly modified without attribution. This was further propagated by a report written by John Mashey. DC and Mashey succeeded in their goal, which was to divert attention from the technical errors in the hockey stick, and focus attention on the issue of trumped-up claims of plagiarism. There have been thousands of nonsensical blog entries about plagiarism in connection with the Wegman Report, whereas the perpetration of false science by the *paleoclimatic cabal* has mainly been ignored.

Bouville (2008) wrote a treatise on plagiarism. He said:

> "... even though ... copying other people's intellectual contribution is wrong, they do not apply to the copying of words. Copying a few sentences that contain no original idea (e.g. in the introduction) is of marginal importance compared to stealing the ideas of others. The two must be clearly distinguished, and the 'plagiarism' label should not be used for deeds that are very different in nature and importance."

The point is that plagiarism is only a serious malpractice when an intellectual concept is stolen for personal gain. When background material is presented without attribution, that is an inadvertency or an indiscretion, but not a crime. The thrust of the Wegman Report was twofold: (1) the hockey stick was based on bad science, and (2) collusion between members of the *paleoclimatic cabal* allowed the hockey stick to get repeatedly published despite the errors in the methods used. There was no plagiarism in these elements of the Report. Unfortunately, some of the introductory and background material was not given proper attribution. In the spirit of Bouville's paper, big deal.

It is interesting that both DC and Mashey accused me of plagiarism as well in this book. The second edition of my book: *Assessing Climate Change* contains 1,348 specific citations to references giving credit to authors for their work. It also includes 411 specific quotations of authors with their own words included in quote signs in my book. It is possible that, in a few places, I may have slipped up and used words from a paper and forgotten to give attribution. That was not, is not, and cannot be plagiarism. It was simply the very small fraction of inadvertent cases where I failed to give credit. Anyone with any sense can immediately see that, with 1,348 citations of references and 411 direct quotes, I could hardly have "intentionally, knowingly and recklessly" failed to reference the authors in a few cases. What did I have to gain? Why would I reference 1,348 but not the other 10 or 20? Ridiculous!

One of the henchmen involved in the attempt cast aspersions on me was a person named Samuel Cohen. Unaccountably, he sent an email to me out of the blue in October 2011:

> Dear Dr. Rapp, I am unclear about your affiliation with the University of Southern California. It is listed on your online CV, but no trace can be found on the USC website outside of an old seminar you gave at the behest of that demented crank George Olah. Can you clarify you status as a Research Professor there?
> Sincerely, Sam Cohen
> (By the way, Olah is a Nobel Laureate.)

The real meaning of this was that Mr. Cohen was under the impression that I was on the faculty at USC and he desired to poison that appointment by accusing me of plagiarism. But he was (and remains) confused on my relationship to USC. As it turns out, I did have a one-year interim appointment in 2010 but that had lapsed by the time Mr. Cohen went on the attack. After I left USC, Mr. Cohen sent a scurrilous complaint to USC accusing me of plagiarism but it had no effect; I was already long gone. As he said in an email:

> "I assume you have seen the detailed complaint made to USC?"

I heard from Mr. Cohen again on 2/22/2012. He said:

> Dear Dr. Rapp,
> How's your good buddy Edward Wegman doing lately? Heard anything about the bullshit plagiarism charges that were brought against him? I guess we will now see how the NIH and Department of Defense handle sanctions. Could be fun to watch. They certainly will act faster than GMU. Talk to Ed lately?
> Sam

I tried to explain to him the difference between academic sloppiness in not making an attribution on well-known and widely accepted introductory material vs. stealing someone's intellectual innovation for personal gain. I quoted Bouville (2008), who wrote a treatise on plagiarism. The point is that plagiarism is only a serious malpractice when an intellectual concept is stolen for personal gain. When background material is presented without attribution, that is an inadvertency or an indiscretion, but not a crime. The thrust of the Wegman Report was twofold: (1) the hockey stick was based on bad science, and (2) collusion between members of the *paleoclimatic cabal* allowed the hockey stick to get repeatedly published despite the errors in the methods used. There was no plagiarism in these elements of the Report.

To Mr. Cohen, there is no distinction between failing to give attribution on some quotes vs. stealing ideas for personal gain. In fact, if you used one word from another publication, that would be plagiarism in his view. I asked Mr. Cohen what he ever accomplished in his life and he said:

> "Besides turning Wegman in to the NIH for plagiarism for the CSDA article, a few other things such as becoming a full professor at a top-tier research university, that sort of stuff."

He seems proud to proclaim that he "turned in" Ed Wegman. Evidently, he tried to bestow the same favor on me. When I asked him if he had no shame (à la McCarthy trials), he responded:

> "I feel no shame whatsoever. I did what I consider right, and my actions are supported by the vast majority of the academic community. My actions have been lauded by many, many people. The McCarthy allusion is false."

The orthodoxy seems to have enlisted Mr. Cohen as their hatchet man, using accusations of plagiarism as their weapon. And, while these charges prevail, the real message that Wegman was technically correct that the hockey stick is bogus gets lost. Meanwhile, Cohen admitted:

> "I don't know Wegman and I have no real understanding of his work."

Cohen claims to be " an expert in academic and research misconduct".
Despite his admitted lack of knowledge, Mr. Cohen went on to proclaim:

> "The so-called hockey stick has been validated many times, most recently by the group at UC Berkeley (*http://berkeleyearth.org/*). They were experts, they were skeptical, they found that the paleoclimate data supported global warming, they were honest in both their approach and their conclusions (as so many denier are not), so I trust they did the work correctly and I trust their conclusions."

The Berkeley group never dealt with long-term climate change. They never dealt with the hockey stick. They only dealt with measured temperatures over the past couple of hundred years. One-third of their sites reported cooling, not warming. The *climateaudit.org* site has clearly demonstrated that the hockey stick is *phonus balonus*. Mr. Cohen also said:

> "The hockey stick model is a fact. No one has ever demonstrated that it is an inaccurate representation of global temperatures, as far as I know. If you have citations to the primary literature demonstrating this, please send them to me."

Now it matters little what Sam Cohen thinks about the hockey stick but this exchange of emails provides some insight into the sinister thinking and actions of the *deepclimate.org* mentality. These people are stupid and dangerous. What is most sinister about Sammy is that he seems to have taken delight in disparaging professional scientists who committed minor indiscretions and wringing his hands in joy, relishing his attempts to destroy their reputations and their careers. As he said: "It could be fun to watch" (the downfall of Ed Wegman). This man is a cruel, vicious, nasty hangman. The *deepclimate.org* crowd has enlisted him to damage the reputations of those who disagree with them, while never refuting the arguments of those who disagree with them. Sammy is certain the hockey stick is correct yet he admits he does not know squat about it. What makes Sammy run?

We have previously reviewed the response of *Nature* magazine to M&M when they attempted to publish their critique of Mann *et al.* (1998). As we said, "in a matter concerning the legitimacy and validity of the most widely accepted model of the Earth's climate over the past millennium, *Nature* decided that they would allow the erroneous publication by Mann *et al.* to stand unchallenged because the issues involved 'were too complicated' ". There is at least the appearance (and more likely the reality) that the *paleoclimatic cabal* was in complete control of the situation.

2.4.3.7 The paleoclimatic cabal

There is strong evidence that the climatologists who earn their living by reconstructing paleoclimates over the past few millennia are in frequent communication with one another and are mutually supportive of their various efforts. There is nothing fundamentally wrong with a collegial relationship between scientists in a field and this can, in principle, be a very positive thing. However, in the case of paleoclimatology, the relationship fostered very belligerent, malicious, insidious, unprofessional behavior by the principals. For several years, the various paleoclimatologists published their reconstructions and acted as reviewers for one another's manuscripts. All was well. The hockey stick was widely accepted and became one of the pivotal supporting foundations of global-warming alarmism. Then, in around 2005, the *climateaudit.org* website began reviewing these studies in great detail and found that they were all flawed due to (1) use of a mean for only the calibration period, (2) hiding the decline and using Jones's "trick" of substituting measured temperatures for proxies, as well as (3) various other problems discussed on *climateaudit.org*. This threatened to undermine years of work upon which the paleoclimatologists' reputations were based. Instead of admitting their errors and fixing them, they dug in and became defensive (and indeed paranoid) at first, and then went on the offense against their critics. The most defensive of them all was Michael Mann. Even a *cabalist* in an email cautioned: "but he would probably go ballistic" in regard to any criticism of his work.

Ball (2007) was critical of how Phil Jones (Head, Climate Research Group, East Anglia University) came up with his estimate of uncertainty in his temperature reconstructions and wrote to Jones asking for an explanation. Ball (2007) claimed that Jones replied in an email:

> "We have 25 or so years invested in the work. Why should I make the data available to you, when your aim is to try and find something wrong with it?"

When a world-leading climatologist is more concerned with protecting his *turf* than finding *truth*, things have taken a very bad turn. Jones also said in an email to Michael E. Mann, professor of climatology, Penn State University:

> "And don't leave stuff lying around on anonymous download sites—you never know who is trawling them. McIntyre and McKitrick have been after the *Climatic Research Unit* ... data for years. If they ever hear there is a *Freedom of Information Act* now in the United Kingdom, I think I'll delete the file rather than send it to anyone."

These scientists would rather destroy data than allow others to check up on them. In 2009, 2010, and 2011, extensive sets of emails between principal figures in the *paleoclimatological cabal* were made public (by unknown, but clearly illegal means). These emails revealed a deeply imbedded agreement amongst these climatologists to promulgate their orthodoxy that the Earth's climate has hardly wavered over the past 2,000 years, and that CO_2 was the principal cause of unprecedented global warming in the 20th century. The collection of emails is now referred to as "*climategate*". As Mosher and Fuller (2010) pointed out, they:

> "... ruthlessly suppressed dissent by insuring that contrary papers were never published and that editors who didn't follow their party line were forced out of their position. When *Freedom of Information* requests threatened to reveal their misbehavior, the emails showed them actively conspiring to delete emails to frustrate legitimate requests for information. Worst of all, one

scientist threatened to delete climate data rather than turn it over, and that data is still missing."

The defensive posture of the *cabal* seems to have been to disclose nothing, prevent others from delving into their work, totally ignore criticisms, and continue publishing bad science, acting as reviewers for one another's manuscripts. Amazing at it seems, even to this day, I am not aware of any principal paleoclimatologist responding to or even admitting that a criticism was made of their use of the mean for only the calibration period. McIntyre revisited the Wegman Report in May 2011. He pointed out that the combination of criticisms of the MBH hockey stick by the *climateaudit.org* website and the Wegman Report generated a great deal of controversy on the blogs. The incredible thing is that, as McIntyre put it:

> "Rather than conceding even seemingly indisputable points, Mann and his associates contested every single issue—even the seemingly indisputable and elementary observation that Mann's principal components mined datasets for hockey stick shaped data. To this date [May 23, 2011], neither Mann nor any of his associates has conceded the point."

Dr. Phil Jones, head of the Hadley CRU in 2009, said that the U.S. Department of Energy was funding his data collection—and that officials there agreed that he should not have to release the data. In a 2009 email, he said:

> "Work on the land station data has been funded by the U.S. Dept of Energy, and I have their agreement that the data needn't be passed on. I got this [agreement] in 2007."

Two months later, Jones said that the information "has to be well hidden. I've discussed this with the main funder (U.S. Dept of Energy) in the past and they are happy about not releasing the original station data". Evidently, the U.S. Department of Energy is in cahoots with the *cabal* to evade the requirements of the FOIA as well as the basic tenets of science. It should be emphasized that this is not a case of intellectual property produced by some brilliant new concept. It merely represents collecting climate data and storing them in columns. If the U.S. Department of Energy paid for it, it should be in the public domain.

The *cabal* also went on the offense. Since the principal figures in the *paleoclimatological cabal* were widely published, they tended to be chosen as reviewers for new manuscripts submitted to journals. They were able to act in collusion to prevent contrary papers from being accepted for publication and put pressure on editors who did not cooperate. The *cabal* refereed one another's papers submitted to journals, communicated improperly in a mutual back-scratch environment subverting the peer-review process, pressured journal editors not to publish papers contrary to the orthodoxy, conspired to write rebuttals to any papers that did slip through their barrier to publication of contrary views, and conspired to act in partnership to disparage and ridicule anyone with contrary findings. A particularly egregious episode in the shenanigans of the *cabal* is documented in great detail at the Bishop Hill website.[3] The challenge to the hockey stick by McIntyre needed to be rebutted in time for inclusion in the 2007 IPCC Report. Two Mann associates, Caspar Amman and Eugene Wahl, were chosen to do this. However, they missed the IPCC deadline and their papers were originally rejected by journals (for good reason—they are misguided). However, the *cabal* managed to circumvent the IPCC deadline and manipulate the journals to their advantage.

In response to the many charges of malfeasance by the *cabal* members that appeared on the blogs, several reviews of the *climategate* activities were carried out by vested interests (e.g. the so-called "Russell Report", *www.cce-review.org/*). These generally provided a whitewash that was only to be expected, considering who were on the review boards.

Kevin Trenberth, Senior Scientist as the National Center for Atmospheric Research has emerged as a defender of the *cabal*.[4]

He did admit to "lack of openness in sharing data and violations of the Freedom of Information Act" but he pointed out that five investigations failed to find any of the alleged misconduct. Unfortunately, these five investigations were conducted by friends of the *cabal*. He also asserted that "scientists would not make up stuff that could be disproven by others!" but the nature of paleoclimatic data is that they are not susceptible to proof, disproof, verification, or validation, and hence are a very safe field to work in. He cited an excerpt from a Phil Jones email:

> "I can't see either of these papers being in the next IPCC report. Kevin and I will keep them out somehow—even if we have to redefine what the peer-review literature is!"

and implied that this was Jones's invention and he (Trenberth) had nothing to do with this. Whether this is true or not, this excerpt reveals the intellectual environment of the climate *cabal*. However, I have to agree with one slide in Trenberth's presentation that says the Internet is "An open sewer of untreated, unfiltered information and the American public is incapable of deciphering between facts, fiction and opinion".

In late 2011, additional emails between *cabalists* were hacked. Some of these exposed some chinks in the armor. Self-doubt began to creep in. Tim Osborne was quoted as saying: "Also, we set all post-1960 values to missing in the MXD data set (due to decline), and the method will infill these, estimating them from the real temperatures—another way of 'correcting' for the decline, though may be not defensible!" Richard Alley (who is not really a *cabalist*, but a fellow traveler in alarmism) was quoted as saying:

> "Unless the 'divergence problem' can be confidently ascribed to some cause that was not active a millennium ago, then the comparison between tree rings from a millennium ago and instrumental records from the last decades does not seem to be justified, and the confidence level in the anomalous nature of the

[3] *http://bishophill.squarespace.com/blog/2008/8/11/caspar-and-the-jesus-paper.html.*
[4] *www.cgd.ucar.edu/cas/Trenberth/Presentations/ClimategateS.pdf.*

> recent warmth is lowered. I think the best way to sum up all of this is: Where does all this lead us? It is very likely that the NH mean temperature has shown much larger past variability than caught by previous reconstructions. We cannot from these reconstructions conclude that the previous 50-year period has been unique in the context of the last 500-1000 years. *Of course we all know that the IPCC reports differently.*" (emphasis added)

Another hacked email was reported to have said:

> "I am afraid the Mike [Mann] and Phil [Jones] are too personally invested in things now (i.e. the 2003 GRL paper that is probably the worst paper Phil has ever been involved in—Bradley hates it as well), but I am willing to offer to include them if they can contribute without just defending their past work—this is the key to having anyone involved. Be honest. Lay it all out on the table and don't start by assuming that ANY reconstruction is better than any other." (Email refers to Mann and Jones, 2003).

Jonathan Overpeck was quoted as saying:

> "... what Mike Mann continually fails to understand, and no amount of references will solve, is that there is practically no reliable tropical data for most of the time period, and without knowing the tropical sensitivity, we have no way of knowing how cold (or warm) the globe actually got.... Unsatisfying, perhaps, since people will want to know whether 1200 AD was warmer than today, but if the data doesn't exist, the question can't yet be answered. A good topic for needed future work."

Tim Osborne was quoted as saying: "Also we have applied a completely artificial adjustment to the data after 1960, so they look closer to observed temperatures than the tree-ring data actually were."

It seems to me to be unfortunate that, since publishing that excellent article in 2003, Dr. Soon has published some articles and made a number of public presentations expounding a very extreme skeptical point of view. He has been quoted as saying:

> "Most of the weather and climate variations we observed are essentially related to the sun and the changing seasons—not by CO_2 radiative forcing and feedback. The climate system is constantly readjusting naturally in a large way—more than we would ever see from CO_2. The CO_2 kick [impact of CO_2 emissions] is extremely small compared to what is happening in a natural way. Within the framework of a proper study of the sun–climate connection, you don't need CO_2 to explain anything."

I personally think he is completely misguided in this viewpoint. But his views are his right and privilege. It is unfortunate that organizations allied with the *cabal* went on campaign to besmirch Soon's reputation and the Internet is full of accusations and attacks. In fact, if you dial "willie soon climate change" into Google, you obtain mainly derogatory claims (apparently planted by Greenpeace) saying, for example,

"Climate skeptic Willie Soon received $1M from oil companies, papers show ...".[5] What these websites don't reveal is that *cabal* members have received many tens of millions of dollars to fund their alarmist research.

2.4.3.8 The climate alarmism cabal

Initially, the *cabal* consisted of paleoclimatologists. However, as time progressed, the *paleoclimatic cabal* was joined by other climatologists not necessarily involved in reconstruction of past climates, who had vested interests in climate alarmism, and viewed attacks on the *paleoclimatic cabal* as destructive to their cause. So the *paleoclimatic cabal* expanded to become the *climate alarmism cabal* dedicated to propagating the orthodoxy of alarmism. A new set of emails within *climategate* appeared in 2011. An exchange of emails between members of the *climate alarmism cabal* was revealed. Members appear to include (amongst others): Tom Wigley, Jonathan Overpeck, Caspar M. Ammann, Raymond Bradley, Keith Briffa, Tom Crowley, Malcolm Hughes, Phil Jones, Tim Osborn, Kevin Trenberth, Ben Santer, Steve Schneider, Malcolm Hughes, Michael E. Mann, Andrew Dessler, and Michael Oppenheimer (see *http://junkscience.com/2011/11/27/climategate-2-0-mann-suggests-harvard-take-action-against-soon-baliunas/*). The goals of the *climate alarmism cabal* seem to be to prevent contrary papers from getting published, to harass editors that pass contrary papers, to immediately combat any contrary papers or influential blog entries with counter papers and blog entries, and, unfortunately, in some cases it appears that attacks of a more personal nature might be considered. They have pompously and arrogantly claimed that their interpretations are *climate science* while work by other climatologists reaching different conclusions is something other than *climate science*. We see evidence of this in many publications and press releases. In particular, in regard to the effect of clouds, Dessler (2011) said: "In recent papers, Lindzen and Choi (2011), and Spencer and Braswell (2011) have argued that ... clouds are the cause of, and not a feedback on, changes in surface temperature. If this claim is correct, then significant revisions to *climate science* may be required." In other words, he regards "*climate science*" as that which the orthodoxy subscribes to. It is not his interpretation of climate science—it *IS CLIMATE SCIENCE*!

Another bizarre aspect of Dessler's publication was discussed by Pielke, Sr. (*http://pielkeclimatesci.wordpress.com/2011/09/06/comments-on-the-dessler-2011-grl-paper-cloud-variations-and-the-earths-energy-budget/*). He said: "Dessler's paper was received 11 August 2011 and accepted 29 August 2011. This is some type of record ... and indicates that the paper was fast-tracked. This is certainly unusual ...", to say the least. He went on to say:

> "It is not clear whether the Editor of *GRL* included Roy Spencer as one of the referees, [and if they did not] they were derelict in their responsibilities. Dessler's paper should have been submitted to *Remote Sensing* as a Comment [on Spencer's paper]. Then Roy Spencer would submit a Reply."

[5] *E.g., http://www.examiner.com/seminole-county-environmental-news-in-orlando/harvard-astro-physicist-dismisses-agw-theory-challenges-peers-to-take-back-climate-science#ixzz1fDDWbbz1.*

We are now witnessing a phenomenon in climatology publications that is occurring repeatedly. The climatology orthodoxy seems to have united into an informal association dedicated to (1) prevent contrary analyses and interpretations from being published, and (2) to quickly respond to those few contrarian publications that slip through their net with vitriolic attacks on the paper on orthodoxy blogs, and in the literature via rapid rebuttal publications such as that of Dessler (2011). It seems evident that many editors are in cahoots with the orthodoxy; certainly the editor of *GRL* is, and the editor of *Remote Sensing* who let Spencer and Bradwell's paper through the net, suddenly resigned for unclear reasons.

One topic that gets *cabal* members upset is the claim by some contrarians that persistent El Niños since 1976 were dominant in causing warming in the NH in the latter part of the 20th century. If this were true, it would suggest that the role of CO_2 in climate change may be far less than the orthodoxy believes. Thus, when the article by McLean *et al.* (2009) appeared in the literature suggesting an important role for El Niños as a dominant cause of warming in the NH in the latter part of the 20th century, it produced great animosity and consternation amongst the members of the *climate alarmism cabal.* McLean *et al.* (2009) concluded that the El Niño index

> "... is a dominant and consistent influence on mean global temperature. Shifts in temperature are consistent with shifts in the [El Niño index] that occur about 7 months earlier. The relationship weakens or breaks down at times of volcanic eruption in the tropics ... Since the mid-1990s, little volcanic activity has been observed in the tropics and global average temperatures have risen and fallen in close accord with the [El Niño index] of 7 months earlier. Finally, this study has shown that natural climate forcing associated with ENSO is a major contributor to variability and perhaps recent trends in global temperature, a relationship that is not included in current global climate models."

According to their estimates, changes in the El Niño index could account for about 70% of the variance in the global tropospheric temperature over the past 50 years. This paper was reviewed and accepted by three independent referees. One referee commented in part: "I found the paper to be well-organized, well-written, and clear on the importance of the research ... The findings are likely to be of interest to a wide variety of readers." A second referee commented in part: "This very clear and well-written manuscript is an analysis of the relationship between MSU-derived and radiosonde-based tropospheric temperature variability and the Southern Oscillation, as modified by major tropical volcanic eruptions."

After McLean *et al.* (2009) was published, a flurry of emails was exchanged between *cabal* members, strategizing on how to carry out damage control for their orthodoxy by preparing a rebuttal. Soon afterwards, a group of *cabalists* (Grant Foster, James Annan, Phil Jones, Michael Mann, Jim Renwick, Jim Salinger, Gavin Schmidt, and Kevin Trenberth) decided to prepare a rebuttal, and, to ensure speedy publication, they pressured the editor of the JGR and suggested the following persons as possible reviewers for their submitted critique: Ben Santer, Dave Thompson, Dave Easterling, Tom Peterson, Neville Nicholls, and David Parker (with Tom Wigley, Tom Karl, and Mike Wallace also mentioned). All of these were professionally associated in some way to the Foster *et al.* group and are thought to be members of the *climate alarmism cabal*. Phil Jones commented: "All of them know the sorts of things to say—about our comment and the 'awful original', without any prompting." (They all subscribe to the same orthodoxy.) McLean *et al.* describe the whole sordid story at *scienceandpublicpolicy.org/originals/censorship_* at_agu.html.

In their rush to rebut the original McLean article, the *climate alarmism cabal* posted their rebuttal on a website, in violation of *Journal of Geophysical Research* (JGR) rules. The results of McLean *et al.* (2009) would seem to be a major stumbling block for alarmists who attribute most of the warming of the 20th century to greenhouse gases. It is therefore not surprising that the alarmists struck back with members of the *cabal* publishing Foster *et al.* (2010), that claimed that the results of McLean *et al.* (2009) "are seriously in error" and concluded "In fact, the general rise in temperatures over the 2nd half of the 20th century is very likely predominantly due to anthropogenic emissions of greenhouse gases". Foster *et al.* (2010) fell back on climate models that attribute only 15–30% of temperature variation in the 20th century to variability of the El Niño index. Foster *et al.* (2010) constituted a rather vicious criticism of McLean *et al.* (2009), but JGR refused to publish McLean's response. Evidently, the JGR is acting in collusion with the *alarmist cabal*, and probably regrets that McLean *et al.* (2009) "slipped through". McLean (2010) provides all the details.

McLean *et al.* attempted to rebut the criticism by Foster *et al.* (2010), but the JGR refused to publish it. Their rebuttal, which will be referred to as "McLean2010", is available at: *icecap.us/images/uploads/McLeanetalSPPIpaper2Z-March24.pdf* and *http://scienceandpublicpolicy.org/originals/censorship_at_agu.html.*

There are several important aspects of this episode that require further elaboration. These include (1) technical aspects, (2) attitudes and collusion amongst *cabal* alarmists, and (3) collusion of the JGR with *cabal* alarmists.

In regard to technical aspects, the issue revolves about methods used for filtering in statistical processing of data. Foster *et al.* (2010) appear to have made some valid criticisms of specific details, but these do not negate the strong correlation of the El Niño index with climate change. Perhaps the contribution of the El Niño index is less than the 70% claimed by McLean *et al.* (2009), but clearly the El Niño index is an important factor in climate change. It seems doubtful that climatology has sufficient data and analytical insight to pin down its quantitative share in influencing climate change. A number of authors, even members of the *alarmist cabal*, have admitted that climate models do not adequately account for El Niño effects. McLean (2010) presented excerpts from the *climategate* emails that clearly show that the *alarmist cabal* regarded McLean *et al.* (2009) as a threat to their orthodoxy, and they colluded together to disparage McLean *et al.* (2009). The *cabal* regards itself as a police force to eradicate any contrary evidence or analysis that would refute their emphasis on greenhouse gases.

McLean *et al.* submitted a response to the published comment by Foster *et al.* but the JGR sent their response to the Foster *cabal* for review—like asking the fox to guard the henhouse. Needless to say, the McLean response was rejected and never published by JGR, although it appears at *http://scienceandpublicpolicy.org/*. One does not need to be an expert on statistics of large data sets to see that persistent El Niños since about 1976 have contributed significantly to warming in the NH (see Figures 3.33 and 3.37 to 3.39). The science of climatology is not capable of assigning accurate estimates of the percentage contribution of El Niños to total warming. Skeptics suggest perhaps 70%; alarmists suggest less than 30%.

A rather parallel situation occurred in regard to the paper by Douglass *et al.* (2007) that examined measured tropospheric temperature trends and compared them with "Climate of the 20th Century" model simulations. They concluded that observed temperature trends were in significant disagreement with model predictions in most of the tropical troposphere. These conclusions contrasted strongly with those of recent publications by *cabalists*. It has been claimed that a major problem for climate models is the disparity between the temperature trends observed at the Earth's surface and the much smaller trends observed in the lower troposphere that is just the opposite of what global climate models (GCM) predict. (Figure 3.34 shows that the forcing due to doubling CO_2 from the pre-industrial value is much higher at the tropopause than at the Earth's surface). Douglass *et al.* (2007) compared tropical temperature trends with climate model predictions for temperatures in the so-called "characteristic emission layer" (CEL) (2–6 km altitude) where the role of water vapor is most important. Over the period from 1979 through 2004, the models predicted a rising temperature trend of roughly 0.2°C to 0.3°C per decade, whereas satellite temperature measurements indicate essentially no increase below 10 km altitude, and a negative trend above 10 km. This was cited as evidence of the inadequacy of current climate models.

It has come to pass that a few determined skeptics (Douglass, Lindzen, McLean, Spencer, McIntyre, *et al.*) continue to publish contrarian papers (in those rare cases where the *cabal* does not succeed in censoring publication), and immediately thereafter, a flurry of emails is exchanged between *cabal* members (Mann, Jones, Schmidt, Trenberth, *et al.*) castigating the skeptics, and strategizing to achieve damage control to protect their orthodoxy that rising CO_2 is essentially the sole cause of global warming. The most pugnacious and aggressive of these is Michael Mann. It is ironic that his own research, responsible for the *hockey stick*, is far less believable than the works of those he would criticize.

After publication of Douglass *et al.* (2007), the *cabal* came forth with Santer *et al.* (2008) as a rebuttal. This paper begins with the sentence: "There is now compelling scientific evidence that human activities have influenced global climate over the past century" which, aside from the fact that the statement is not true, reveals the orthodoxy to which the authors subscribe religiously. The details of the statistical processing of large data sets are complex. The issue is whether tropical tropospheric temperatures have risen more than surface temperatures as climate models would predict for the effect of greenhouse gases on climate. Douglass *et al.* (2007) concluded that models and data disagreed to "a statistically significant extent". Santer *et al.* (2008) claimed to achieve a "partial resolution of the long-standing 'differential warming' problem", although they also said:

"We may never completely reconcile the divergent observational estimates of temperature changes in the tropical troposphere. We lack the unimpeachable observational records necessary for this task. The large structural uncertainties in observations hamper our ability to determine how well models simulate the tropospheric temperature changes that actually occurred over the satellite era. A truly definitive answer to this question may be difficult to obtain."

Yet, this did not prevent Santer *et al.* from producing a so-called "Fact Sheet" that said "We've gone a long way towards such a reconciliation" (between climate models and tropical tropospheric temperatures).[6]

In 2009, McIntyre[7] pointed out that, when the data used by Santer *et al.* (2008) that ended in 1999 are extended through 2008, the discrepancy reported by Douglass remains, and "the claim by Santer *et al.* (2008) to have achieved a 'partial resolution' of the discrepancy between observations and the model ensemble mean trend is unwarranted". McIntyre also noted the difficulty in obtaining data from Santer *et al.*, and indicated that the *International Journal of Climatology* (IJC) was stalling in responding to him. It appears that this article will never pass through the *cabal*'s lock on the IJC, and McIntyre had to be content with merely archiving his article (*http://arxiv.org/abs/0908.2196*). Yet, alarmists continue to refer to Santer *et al.* (2008) as evidence that climate models have been adequately tested.

Douglass and Christy[8] presented evidence for their claim that Ben Santer, Phil Jones, Timothy Osborn, Tom Wigley, and 13 other *climate alarmism cabal* members apparently conspired to compromise the peer-review process, with the willing cooperation of the editor of the IJC, Glenn McGregor. This evidence involved dozens of emails over nearly a year, suggesting "(a) unusual cooperation between authors and editor, (b) misstatement of known facts, (c) character assassination, (d) avoidance of traditional scientific give-and-take, (e) using confidential information, (f) misrepresentation (or misunderstanding) of the scientific question posed by Douglass *et al.* (2007), (g) withholding data, and more." Douglass and Christy provide the entire sordid story; there is no need to reproduce the details here.

An example of the need by the *climate alarmism cabal* to respond to challenges by contrarians is the paper by Santer *et al.* (2011). This paper was concerned that measurements indicated that tropospheric temperatures had not risen since 1998 despite continued increases in CO_2 concentration. The paper had 17 authors in an expression of support by the *cabal*, although it is difficult to figure out what contributions (if any) were made by the various authors. The listing of these authors seems to be more a political statement than a scientific statement. The goal was to produce an analysis that concludes that temporary periods with no temperature gain may be viewed as a temporary fluctuation superimposed on an ever-present

[6] *https://publicaffairs.llnl.gov/news/news.../NR-08-10-05-factsheet.pdf.*

[7] *http://climateaudit.org/2009/01/27/submited-article-on-tropical-troposphere-trends/* and *http://climateaudit.org/2009/04/14/tropical-troposphere-march-2009/*

[8] *http://www.americanthinker.com/2009/12/a_climatology_conspiracy.html*

underlying upward trend due to rising greenhouse gas concentrations. The logic of the paper is quite shaky, however, as we discuss in Section 3.4.3.2.

Lindzen and Choi (2009) examined data on the outgoing radiation budget from the Earth Radiation Budget Experiment (ERBE) in the tropics in an attempt to determine whether observations of the Earth's radiation imbalance can be used to infer feedbacks and climate sensitivity. From this, they concluded that the climate sensitivity is considerably lower than the values predicted by climate models. Later, Lindzen and Choi admitted: "This work was subject to significant criticism by Trenberth *et al.* (2009), much of which was appropriate." As a result, they wrote a revised paper (Lindzen and Choi, 2011) that was "an expansion of the earlier paper in which the various criticisms are addressed and corrected. . . ." As might be expected, Lindzen and Choi (2011) found that feedbacks were primarily negative, resulting in relatively low climate sensitivity. This is contrary to the alarmist position that feedbacks are positive leading to higher climate sensitivity (and therefore produce a greater increase in global temperature as greenhouse gas concentrations increase). The manuscript by Lindzen and Choi (2011) was rejected by the *Proceedings of the National Academy of Sciences* (PNAS). The revelation of the reviewers and their comments led to a very extensive series of blog entries at *climateaudit.org*. In the course of these blog entries, we find (along with the usual trivia) several nuggets of information worth mentioning. Lindzen is a member of the NAS and it is very rare that a paper submitted by a member would be rejected (96% are accepted). In a highly unusual move, the PNAS rejected Lindzen's suggestion for reviewers, and instead chose reviewers who were obviously antagonistic to Lindzen's viewpoint. The reviews of this paper were incredibly detailed and penetrating. It seems likely that papers expressing the alarmist agenda glide through the review process with little friction and no depth of review. One blog contributor was a reviewer for the paper by Wahl and Ammann (2007). His review was discarded by the *Journal of Climate* because it was not in conformity with the alarmist agenda. It appears that most of the papers in climatology are based on inadequate data: lacking in spatial and temporal coverage. The sophisticated data processing used to cover this up, whether filtering, smoothing, use of principal components, or otherwise, hides the fact that the foundations are typically very weak. Had other landmark papers in climatology that are repeatedly referred to in biblical tones been given the same kind of penetrating review as Lindzen's manuscript, they would also have been rejected. Indeed, most of the literature in climatology would have to be cleared out. Finally, the Lindzen and Choi paper was published in the *Asia-Pacific Journal of Atmospheric Science*.

The entire set of pirated emails in *climategate* provides strong evidence that there is indeed a *climate alarmism cabal*, including both paleoclimatologists who seek to show that the Earth's climate has been relatively constant for thousands of years prior to the 20th century, climate modelers who seek to use climate models to infer that most of the 20th century warming was due to greenhouse gas buildup, and the members of this *cabal* are dedicated to their preconceived beliefs, conspire with one another to prevent publication of dissenting views, conspire with one another to oppose and rebut dissenting papers that slip through their net of referees for major journals, and make frequent alarmist press releases in a losing effort to win over the public. It seems possible that the motivation for all this is to create a climate of fear so that governments will exponentially increase funding for climate research; in that respect, they have been very successful.

2.4.4 Proxy analysis

Esper, Cook, and Schweingruber (2002a) started out by repeating the mantra of the global-warming alarmists:

> "... the MBH reconstruction indicates that the 20th century warming is abrupt and truly exceptional. It shows an almost linear temperature decrease from the year 1000 to the late 19th century, followed by a dramatic and unprecedented temperature increase to the present time. The magnitude of warmth indicated in the MBH reconstruction for the MWP, 1000–1300 is uniformly less than that for most of the 20th century."

However, Esper, Cook, and Schweingruber (2002a) admitted that "the MBH reconstruction has been criticized for its lack of a clear MWP". It was admitted that critics doubt that tree-ring records can preserve long-term, multi-centennial temperature trends. However, as usual in papers written by the *paleoclimatic cabal*, no mention is made of the devastating criticism of the MBH reconstruction made by McIntyre (2007).

Esper, Cook, and Schweingruber (2002a) then went on to present a defense of tree-ring reconstructions using centuries-long ring-width trends in 1,205 radial tree-ring series from 14 high-elevation and middle- to high-latitude sites distributed over a large part of the NH extra-tropics. Their final result is shown in Figure 2.25.

Esper, Cook, and Schweingruber (2002a) (Figure 2.26) concluded that MBH did not miss an MWP, but rather, "it had a reduced expression of the LIA". They took comfort from the fact that some of the short-term "bumps" in the RCS and MBH curves were synchronous but there are significant differences between the two reconstructions.

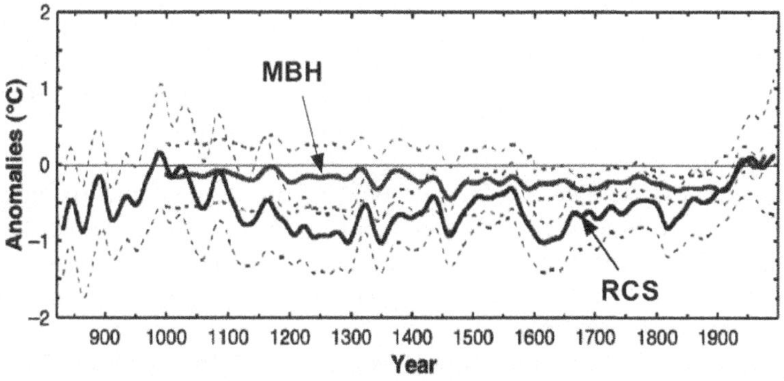

Figure 2.25. Comparison of temperature reconstructions by Esper *et al.* (2002) ("RCS") with that of MBH (1999) (adapted from Esper *et al.*, 2002).

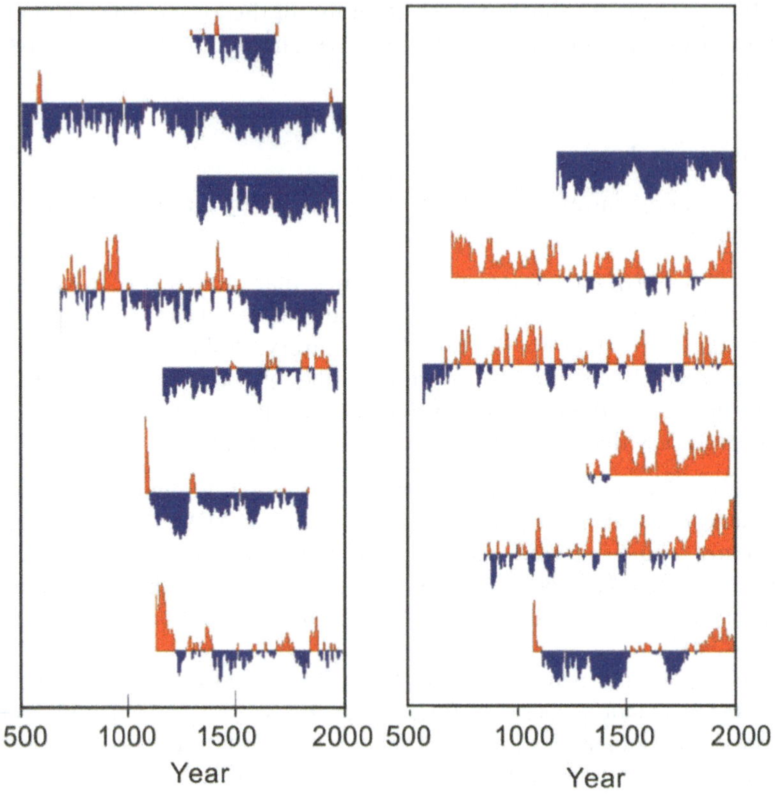

Figure 2.26. Rendition of individual proxies from Esper *et al.* (2002) as adapted by McIntyre (2007).

While Esper, Cook, and Schweingruber (2002a) intended to support Mann, Bradley, and Hughes (1999), the large differences between their results lead this writer to the opposite conclusion. This raises the question of whether any reconstruction based on proxies is credible. Furthermore, note that the anomalies in Figure 2.25 are mostly negative, suggesting that the mean used for data processing was not the mean for the entire time period and, as we have shown, this introduces significant doubt about the veracity of the results.

Finally, in Figure 2.27, McIntyre (2007) presented a simple average of all the proxies. The result suggests an MWP and an LIA. Nevertheless, it seems evident that Esper, Cook, and Schweingruber (2002a) used statistical data manipulation that unreasonably and illegitimately overemphasized the weighting of the two suspect proxies with high closing values.

McShane and Wyner (2010) (M&W) analyzed the methods of combining proxies into estimates of global average temperature from the point of view of mathematical statisticians. In discussing a typical hockey stick graph, M&W said:

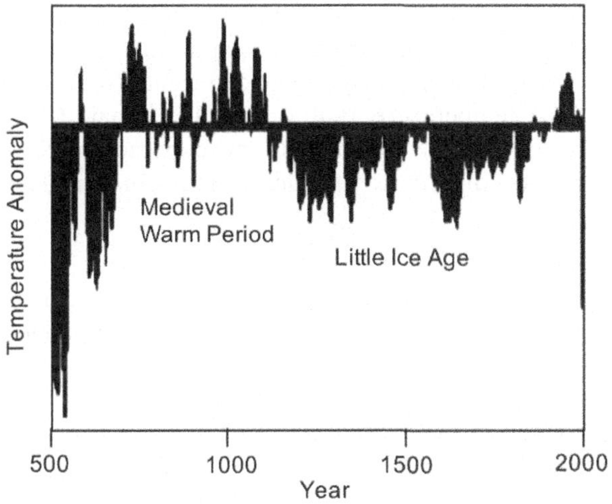

Figure 2.27. Simple average of proxy data from Esper *et al.* (2002) (adapted from McIntyre, 2007).

> "... the [proxy result] closely matches the [measured temperatures] from 1850 AD to 1998 AD because it has been calibrated to the instrumental period which has served as training data. This sets up the erroneous visual expectation that the reconstructions are more accurate than they really are. A careful viewer would know to temper such expectations by paying close attention to the reconstruction error bars given by the wide gray regions. However, even these are misleading because these are, in fact, point-wise confidence intervals and not confidence curves for the entire sample path of surface temperature. Furthermore, the grey regions themselves fail to account for model uncertainty."

What this paragraph means is that the hockey stick is a clever piece of public relations but it does not have the valid implications that a naive reader might suppose. The apparent agreement between the model and the data in the period after 1850 was forced by tuning the model to the data. The estimated implicit uncertainty in the data is optimistic, but, even so, is so wide as to encompass almost any past temperature profile.

In addition, it is noteworthy that M&W did not discuss the quality of the data *per se*, or comment on possible wide divergence in reliability of different proxies.

M&W went on to say:

> "... the task [of analyzing past proxy data to infer global average temperature history] is highly statistical [and] extremely difficult. The data is spatially and temporally auto-correlated. It is massively incomplete. It is not easily or accurately modeled by simple autoregressive processes. The signal is very weak and the number of covariates greatly outnumbers the number of independent observations of instrumental temperature."

This paragraph attests to the fact that estimating the global average temperature over an ~1,000-year period requires detailed spatial and temporal data, and the proxy data are very sparse. It is not easy to detect a signal from the noise and there are factors other than temperature that affect the signal. Climatologists and statisticians have used sophisticated methods to try to extract trends from this sparse mess of crude data, but it is difficult to produce a silk purse from a sow's ear. M&W said:

> "McIntyre and McKitrick (2003, 2005) (M&M) [reviewed] the original Mann *et al.* (1998) study [and showed] that it (i) used only one principal component of the proxy record and (ii) calculated the principal components in a 'skew'-centered fashion such that they were centered by the mean of the proxy data over the instrumental period (instead of the more standard technique of centering by the mean of the entire data record). Given that the proxy series is itself auto-correlated, this scaling has the effect of producing a first principal component that is hockey-stick shaped and, thus, hockey stick shaped temperature reconstructions. That is, the very method used in Mann *et al.* (1998) guarantees the shape of [the hockey stick figure]. M&M made a further contribution by applying the Mann *et al.* (1998) reconstruction methodology to principal components computed in the standard fashion. The resulting reconstruction showed a rise in temperature in the medieval period, thus eliminating the hockey stick shape.".

Mann and hs colleagues vigorously responded to M&M to justify the hockey stick (Mann *et al.*, 2004). According to M&W:

> "Mann *et al.* (2004) argued that one should not limit oneself to a single principal component as in Mann *et al.* (1998), but, rather, one should select the number of retained principal components through cross-validation on two blocks of held-out instrumental temperature records (i.e., the first fifty years of the instrumental period and the last fifty years). When this procedure is followed, four principal components are retained, and the hockey stick re-emerges even when the PCs are calculated in the standard fashion."
>
> "The furor reached such a level that Congress took up the matter in 2006. The Chairman of the Subcommittee on Oversight and Investigations formed an ad hoc committee of statisticians to review the findings of M&M. Their Congressional report (Wegman *et al.*, 2006) confirmed M&M's finding regarding skew-centered principal.... In his Congressional testimony (Wegman *et al.*, 2006), committee chair Edward Wegman excoriated Mann *et al.* (2004)'s use of additional principal components beyond the first after it was shown that their method led to spurious results: 'In the MBH original, the hockey stick emerged in PC1 from the bristlecone/foxtail pines. If one centers the data properly the hockey stick does not emerge until PC4. Thus, a substantial change in strategy is required in the MBH reconstruction in order to achieve the hockey stick, a strategy that was specifically eschewed in MBH ... a cardinal rule of statistical inference is that the method of analysis must be decided before

looking at the data. The rules and strategy of analysis cannot be changed in order to obtain the desired result. Such a strategy carries no statistical integrity and cannot be used as a basis for drawing sound inferential conclusions'."

There are two points here. One is that, by implication, MBH changed the rules of the game to achieve a desired result in an unscientific, unprofessional manner. The other is that, by doing this, the hockey stick shape was relegated to the fourth level of PCA, a very weak inference of low credibility. Michael Mann, in his rebuttal testimony before Congress, continued to stubbornly defend his turf, and why shouldn't he, since his whole scientific reputation depended on it? He claimed that many subsequent peer-reviewed studies confirmed his findings, but all of these studies used variants of his approach and were peer-reviewed by members of the *paleoclimatic cabal*. In particular, they all used skew-centering, as evidenced by the fact that the data were not spread roughly equally across the zero line. Furthermore, as climategate has revealed, peer review in paleoclimatology is mainly a mutual affirmation by members of the club. This was the backdrop for the study by M&W to reexamine the issues involved in reconstructing past global climates from proxy data.

M&W began with the statement: "We are not interested at this stage in engaging the issues of data quality. To wit, henceforth and for the remainder of the paper, we work entirely with the data from Mann *et al.* (2008)." M&W then went on to support this database, saying:

> "We *assume* that the data selection, collection, and processing performed by climate scientists meets the standards of their discipline. Without taking a position on these data quality issues, we thus take the dataset as given. We further make the assumptions of linearity and stationarity of the relationship between temperature and proxies, an assumption employed throughout the climate science literature...." (Italics added).

However, the validation of these assumptions in the literature leaves much to be desired. The precision of various proxies in representing temperature, and only temperature, remains a mystery unstated. One thing is certain: not all proxies are equally credible. M&W were concerned with the statistical processing of data, assuming the data were adequate. More likely, the data were poor in quality, so that even the best statistical analysis would result in GIGO.[9] Each proxy provides an estimate (often a rough estimate) of temperature for one locality. Temperatures are known to vary widely with locality. As M&W demonstrate, the various proxies in different localities vary widely in form. It challenges one's imagination to think that these wildly different proxies can provide much useful information about global average temperature. M&W began with a discussion of the over-determined system in which there are many more proxies than years in the instrumental temperature record. If, for example, one takes a particular year during the instrumental record, say 1950, there might be, say, 200 proxies for temperature that year. But there is only

[9] GIGO = garbage in; garbage out.

one global average temperature. In attempting to calibrate these 200 proxies against the single global average temperature, there is no unique solution to this over-determined problem. Certainly, one cannot hope to set all the proxies at all locations to the global average temperature. In some cases, one can segregate proxies that compare better with historically measured temperatures at their localities, but the connection to the global average temperature remains weak. Another approach involves dividing the calibration period (1850–1998) into segments, calibrating proxies over one segment, and testing the result over the other segments. While sophisticated analysts such as Mann *et al.* and M&W have used various procedures to try to work around this problem, the simple truth is that there is no solution.

The bases of all these manipulations by Mann *et al.* are the assumptions that each proxy measures a local or regional temperature, that each such local or regional temperature is the sum of a global average temperature influence term and a regional bias term, that the regional bias terms are either positive or negative, and that, when averaged over many proxies, they sum up to zero. Hence, adding up many proxies produces a measure of the global average temperature according to this theory. Unfortunately, these assumptions do not seem to have been validated. Even a casual examination of the actual proxies shows that they vary wildly from one to another. What seems to happen is that even if these assumptions were true (a consummation devoutly to be wished), the regional bias terms are orders of magnitude greater than the global average term, and the global average term is buried in an ocean of noise generated by the regional bias terms. Mann and company have attempted to use complex methods to extract the signal from the noise but, as M&W and many commentators have shown, the process suffers from many ills, not least of which is the imperfection of the proxies themselves. I would quote Carl Wunsch:

> "Sometimes there is no alternative to uncertainty except to await the arrival of more and better data." (Wunsch, 1999).

As Steve McIntyre[10] pointed out:

> "The fundamental problem in paleoclimate is not the need for some novel multivariate method, but better proxies and reconciliation of discordant existing 'proxies'.... Team reconstructions use highly stereotyped proxies over and over again in different guises...."

McIntyre also pointed out that M&W unwittingly adopted "the Mann *et al.* 2008 data set, which quixotically introduced the Tiljander sediments, the modern portion of which was contaminated with bridge-building sediments".

The details of the statistical analysis in M&W (and the 13 commentaries) are quite complex and are only intelligible to specialists. This writer was not able to follow all the intricacies of the analysis, most of which were permeated with jargon. One possibility raised by M&W was:

[10] *climateaudit.org*.

"... it is possible that the proxies are ... too weakly connected to global annual temperature to offer a substantially predictive (as well as reconstructive) model over the majority of the instrumental period. This is not to suggest that proxies are unable to detect large variations in global annual temperatures (such as the differences that distinguish our current climate from an ice age). Rather, we suggest it is possible that natural proxies cannot reliably detect the small and largely unpredictable changes in annual temperature that have been observed over the majority of the instrumental period."

This appears to be the actual case. Proxies seem able to detect very large excursions in temperature, such as occur in transitions between Ice Ages and interglacials, but probably are not able to resolve small temperature changes within an interglacial. The conclusions reached by M&W include the following.

The problem of back casting historical temperatures from proxy measurements calibrated during a limited period of overlap between temperature measurements and proxy measurements is very complex and requires very sophisticated statistical analysis that might be beyond the capability of climate scientists. M&W attempted to bring to bear such a sophisticated statistical analysis on the problem:

"... we conclude unequivocally that the evidence for a 'long-handled' hockey stick (where the shaft of the hockey stick extends to the year 1000 AD) is lacking in the data. The fundamental problem is that there is a limited amount of proxy data which dates back to 1000 AD; what is available is weakly predictive of global annual temperature. Our back-casting methods, which track quite closely the methods applied most recently in Mann *et al.* (2008) to the same data, are unable to catch the sharp run up in temperatures recorded in the 1990s, even in-sample.... Consequently, the long flat handle of the hockey stick is best understood to be a feature of regression and less a reflection of our knowledge of the truth."

The main contribution of M&W was "to seriously grapple with the uncertainty involved in paleoclimatological reconstructions". According to them, the challenges include:

"(i) a short sequence of training data,
(ii) more predictors than observations,
(iii) a very weak signal, and
(iv) response and predictor variables which are both strongly auto-correlated.

The final point is particularly troublesome: ... the number of truly independent observations (i.e., the effective sample size) may be just too small for accurate reconstruction. Climate scientists have greatly underestimated the uncertainty of proxy-based reconstructions and hence have been overconfident in their models.... Proxy based models with approximately the same amount of reconstructive skill produce strikingly dissimilar historical back-casts: some of these look like hockey sticks but most do not. Natural climate variability is not well understood and is probably quite large. It is not clear that the proxies currently used to predict temperature are even predictive of it at the scale of several decades let alone over many centuries."

Thirteen independent groups or individuals wrote commentaries on the M&W paper. Evidently, there is very little objectivity in paleoclimatology, as evidenced by the facts that the establishment warmist climatologists vigorously defended Mann *et al.*, the statisticians made abstruse mathematical comments, and several climatologists exterior to the paleoclimatological *cabal* indicated support for M&W.

One aspect of this controversy is the use of PCA. One starts with a set of data from various proxies at various locations over various time periods. If one adds these up with equal weight, one obtains mainly mush—a smear of sparse data with neither direction nor structure. Then, PCA is applied. While one might naively treat all proxies equally, PCA assigns weights to the various proxies on the basis that those proxies with the least tendency toward trend are given low weight, and those proxies with the greatest tendency toward a trend are given greater weight. As M&M and Wegman showed, in the extreme case, MBH gave some proxies hundreds of times the weight of other proxies. The data set was very sparse to begin with, and PCA further reduces the dimensionality of the data set to put a microscopic focus on those few proxies that demonstrate a strong trend, some of which were suspect tree-ring proxies. How can a weak, sparse data set be improved by throwing out most of the data? The statisticians might respond by saying they have identified the proxies that generate the trend for the whole set, but, considering the uncertainty and unreliability of all proxies, this seems like a very biased, counter-productive approach. PCA gives climatologists and statisticians fodder to play with, but the whole process seems to add up to GIGO. In short, this writer thinks that the use of PCA as a method in this application is highly suspect.

Gavin A. Schmidt, Michael E. Mann, and Scott D. Rutherford, claimed that M&W used faulty data in their analysis but they provided little evidence that their own data set was adequate. In order to find any trends in the noisy data, they were forced to go to the 10th or 4th principal component—an extremely weak statistical inference. There is no way they can escape from the fact that the data are very sparse, noisy, and divergent. They seemed most intent on insisting the medieval warmth was minimal and that the decade 1997–2006 was the warmest in 1,000 years—the same old warmist mantra. It may well be true that 1997–2006 was the warmest decade since the height of the MWP, but the height of the MWP was about 1,200 years ago, and not in the past 1,000 years. Furthermore, the issue is not whether 1997–2006 was warm; we know that it was. The issue is whether its deviation from average was unprecedented, and the data are not adequate to answer this.

Amazingly, after all the discussion raised by M&M and Wegman on the problems inherent in centering and standardizing over the calibration period, rather than the full length of the data, Eugene R. Wahl and Caspar M. Ammann described centering and standardizing over the calibration period by MBH as a "reasonable judgment". They then went on to discuss technical aspects of PCA that are beyond the scope of this discussion, including taking PCA to the second or even the fourth component in a desperate effort to evoke a trend. Somehow, they came up with a silk purse starting with a sow's ear.

The responses by Alexey Kaplan; Peter Craigmile and Bala Rajaratnam; Doug Nychka and Bo Li; Jason E. Smerdon; Martin P. Tingley; Jonathan Rougier; Murali Haran and Nathan Urban; and Richard A. Davis and Jingchen Liu were abstruse and mainly of use only to specialists. Some of these were critical of methods used by M&W but they did not justify methods used by Mann *et al.* For example, Haran and Urban said that their criticisms of M&W equally applied to Mann *et al.*

L. Mark Berliner made some brief comments. He said:

> "I join the authors in expressing dissatisfaction with some paleoclimate analyses. I endorse their claim that there has been underestimation of uncertainty in paleoclimate studies. The implication that additional participation of the statistics community is needed is undeniable."

He questioned the assumption of linearity between proxies and temperature, even as used by M&W. He also described the reliance on principal components as "highly questionable".

Lasse Holstrom said: "The authors demonstrate convincingly that the data used in Mann *et al.* (2008) does not allow reliable temperature prediction using this approach and that purely random artificial proxy records in fact perform equally well or even better."

Jason E. Smerdon criticized some of the statistical methods used by M&W but he did not seem to offer any support for the methods of Mann *et al.*

Martin P. Tingley began with:

> "M&W find that under certain scenarios ... randomly generated series are as predictive of past climate as the commonly used proxies. They conclude that 'the proxies do not predict temperature significantly better than random series generated independently of temperature'.... If this assertion is correct, then M&W have undermined all efforts to reconstruct past climate, which are based on the fundamental assumption that natural proxies are predictive of past climate. I disagree with M&W's conclusion and provide an alternative explanation: [their procedure] is simply not an appropriate tool for reconstructing paleoclimate."

Note the pejorative tone in this opening statement as if M&W were akin to the Grinch who stole Christmas. But, as in the case of Smerdon, Tingley seems to have criticized some of the statistical methods used by M&W but he does not seem to offer any support for the methods of Mann *et al.*

M&W wrote a "rejoinder"—a sort of rebuttal—after 13 comments by other authors were sent in to the *Annals of Applied Statistics*. In this rejoinder, they said:

> "... Wahl and Ammann (WA) note 'there is an extensive literature contradicting McShane and Wyner (2011a)'s assertions about low or poor relationships between proxies and climate'. On the other hand, Tingley asserts 'each proxy is weakly correlated to the northern hemisphere mean (for two reasons: proxies generally have a weak correlation with local climate, which in

turn is weakly correlated with a hemispheric average)' and Davis and Liu (DL) state 'there is just not much signal present'."

How can one explain such a vast difference between these conclusions? M&W attempt to explain it:

> "This contrast can be explained at least in part by context. Our paper addresses the specific task of reconstructing annual temperatures over relatively short epochs during which temperatures vary comparatively little. Nevertheless, such contrasts are suggestive of the important frontiers for research and we hope our paper and this discussion will lead to advances on these fronts."

However, this explanation does not do justice to the reality that the warmists (e.g., Wahl and Ammann, 2007) are dedicated to the hockey stick and the hockey stick requires that proxies be representative of past temperatures, whereas objective scientists observe the weak connection between proxies and temperature.

M&W's rejoinder then goes on at length to rebut the various commentaries by 13 authors and the details are abstruse and only appropriate for specialists.

In their conclusion, M&W characterized the assumptions made by the hockey stickers (linearity, stationarity, data quality, etc.) as "questionable, perhaps even indefensible". They also said: "... we reiterate our conclusion that 'climate scientists have greatly underestimated the uncertainty of proxy-based reconstructions and hence have been overconfident in their models'." They closed with: "Finally, and perhaps most importantly, the NRC assumptions of linearity and stationarity outlined in our paper are likely untenable and we agree with Berliner in calling them into question."

Subsequent to the paper by M&W, Li *et al.* (2010) entered the fray with a lengthy and somewhat obscure discussion that involves combining a climate model with proxy data. This paper generated a number of published comments. Smith (2011) followed this with a paper of his own. These papers deal with complex aspects of wringing out statistical inferences from large noisy data sets. McIntyre[11] commented on these papers. He pointed out that none of this work dealt with the adequacy of the underlying data as to whether the tree-ring data (in particular) are truly representative of temperature. He said that "indices of tree growth ... in many cases, are more responsive to precipitation than temperature. Academics in this field are far too quick to assume that things are 'proxies' when this is something that has to be shown". McIntyre showed that the data used in these analyses, relying heavily on Graybill bristlecone chronologies, are highly suspect and provide the source of hockey stick form in the results. Thus, the most comprehensive, sophisticated statistical analyses are efforts in futility if the underlying proxies are not good measures of temperature. There are at least three issues here that go beyond complex statistics: (1) the relationship between proxy signals and temperature over the entire calibration period; (2i) the uncertainty introduced in applying the proxy-temperature relationship outside the calibration period; and (3) the lack of full regional and

[11] *http://climateaudit.org/2011/06/09/richard-smith-2011-and-the-graybill-bristlecones/*.

temporal coverage by the proxies. There has been, and remains, a notable lack of detail on the degree of conformity between proxies and temperature during the calibration periods. As we have pointed out previously, if one combines a set of noisy "proxies" that are highly random by any sophisticated statistical algorithm, one will end up with a relatively flat profile. Then, tacking on a rising measured temperature in the 20th century, one must obtain a hockey stick as the result.

2.4.5 Evidence for the MWP and the LIA

The term "*Little Ice Age*" (LIA) has been widely used to describe a period from roughly 1300 (or later) to about 1850 (or perhaps a bit later) that evidence indicates was, on average, colder than the preceding or following eras, particularly in the NH. Matthews and Briffa (2005) discussed the LIA in some detail. It is believed that there was no uninterrupted, centuries-long cold phase following a similar, uninterrupted, centuries-long *Medieval Warm Period* (MWP). While the LIA was marked by prevailing cold temperatures, there were fluctuations within the LIA both spatially and temporally, and this has provided ammunition for some purists to debate whether there actually was an LIA. Matthews and Briffa (2005) indicated that the term "*Little Ice Age*" (LIA) has been defined in various ways according to the context and there is no single widely accepted interpretation of the term. In particular, depending on whether the emphasis is on glacierization or climate, some differences will inevitably result. Some authors suggest that the term should be used cautiously, some say it should not be used at all, and some say it should be allowed to disappear from use or should be avoided because of its limited utility. However, the existence of the LIA is a barrier of sorts to those who subscribe to the *hockey stick* version of global temperature history, and those who disparage the LIA seem to have an alarmist view of global warming.

One serious problem with Matthews and Briffa (2005) is that temperatures for the LIA are compared with a standard based on the average for 1961–1990. Since LIA temperatures are clearly lower than 1961–1990 temperatures, this serves to unduly emphasize the temperature rise of the 20th century. But the thing that makes the LIA unique is that temperatures during the LIA were lower than the long period that preceded it as well as the 20th century that followed it, and that is why it is colloquially called an LIA. *From the perspective of global warming, the important point is how 20th century temperatures compare with the warm period prior to the LIA—not how they compare with temperatures during the LIA.*

As Matthews and Briffa (2005) pointed out:

> "The expanded state of glaciers, relative to today, during the last few hundred years is an incontrovertible fact ... glaciers on all continents, from the tropics to the polar regions, were characterized by glacier expansion and subsequent retreat. However, beyond the European Alps, and to a lesser degree in Scandinavia and North America, data on the precise timing of variations in glacier size during this broad time interval are still patchy. Consequently, several controversial issues remain, including: (1) the timing of the onset (and end) of

the LIA; (2) the amplitude and timing of glacier variations within the LIA; (3) the degree of synchroneity [*sic*] between glaciers from the different regions; and (4) the attribution of cause(s) in terms of large-scale climate forcing ... Greatest reliance must ... be placed on the geographically restricted evidence available from the European Alps, where the historical sources are sufficient in quality and quantity to answer not only the question of onset but also questions about when the '*Little Ice Age*' glaciers reached their maximum extent and what amplitude of glacier variations occurred within the LIA period."

Grove (1988) provided 1,000 pages of evidence for the LIA. Grove (2001) asserted that the greatest amount of information about the LIA can be gleaned from the Swiss Alps, where historical records are unusually rich and moraine dating is good. Many ice fronts extended below the current tree line and were in full view of settlements for hundreds of years, or even abutted onto farmland. Written records, paintings, and drawings made by both local observers and visitors are plentiful. Identification of the calendar dates at which many *in situ* trees were killed by advancing ice, together with their ages at death, has been made possible by multiple dendrochronological analyses.

The following paragraph is excerpted from Grove (2001):

"The most complete record of the Late Holocene fluctuation history of any glacier in the world comes from intensive investigations of the deposits of the Grosser Aletsch [Figure 2.28]. This reconstruction was based on a great variety of evidence. Some samples were taken from the outer rings of trees overrun by the advancing ice, which could be absolutely dated by reference to dendrochronological series. This chronology is accordingly more accurate than one based only on radiocarbon samples. Retraction during the MWP was ended by rapid advance, starting after 1250 and culminating around 1350. Although the MWP here was fragmented by a marked glacial advance between 1050 and 1150, this was not comparable in scale with those of later centuries and therefore

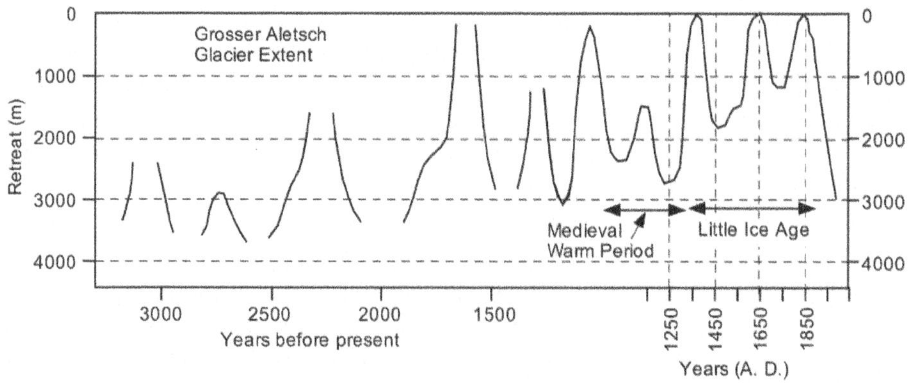

Figure 2.28. Retreat of the Grosser Aletsch Glacier, Swiss Alps, over the last 3,000 years based on documentary and proxy evidence (adapted from Matthews and Briffa, 2005).

has to be considered an interruption of medieval warmth rather than part of the LIA. The Grosser Aletsch advanced about 40 m per year after 1300, to reach its greatest extension around 1370–80. This 14th-century advance brought the front as far forward as did those in the seventeenth and nineteenth centuries. The major extension phases were separated by withdrawals that were insufficient to return the front to positions comparable with those of the warm period. It is evidently reasonable to view the whole period between the start of expansion in the thirteenth century and the great retreat of the late nineteenth and twentieth centuries as one complex LIA with each century scale fluctuation itself made up of smaller scale oscillations, such as those recorded during monitoring of the positions of glacier fronts."

Several aspects of Figure 2.28 are critical when using it to define the LIA in the Swiss Alps (Matthews and Briffa, 2005). First, the three glacier high stands of around 1350, 1650, and 1850 were remarkably similar in extent. Second, previous glacier maxima, including those in the 3rd, 7th, 9th, and 12th centuries, were less extensive. Third, the size of the glacier during the retreat phases between the LIA high stands remained much greater than in the earlier retreat phases. The data support the notion of a change toward a more glacierized region at the end of the 12th century, and so marking the onset of the central European LIA. This change has also been interpreted as marking the end of the "*Medieval Warm Period*", and a similar pattern and timing are supported on a centennial timescale by the somewhat less complete records from other Alpine glaciers. It must nevertheless be concluded that, even in the Swiss Alps, differences between the glacier variations during the LIA and those before the LIA were a matter of degree rather than of kind.

It can be seen from Figure 2.28 that, since the mid-19th century, mountain glaciers have been in a fairly steady retreat. Anon. (M) and Kotlyakov (1996) confirm that the retreat of the glaciers preceded large-scale buildup of CO_2.

Polissar *et al.* (2006) also noted three major glacial advances from Andes lake sediments from the period 1300–1750.

Grove (2001) pointed out that the Gorner Glacier has advanced and retreated in harmony with the Grosser Aletsch, showing only very minor differences. Grove also mentions that other large Alpine glaciers have been shown to have been similarly affected by early LIA advances. The Grindelwald was advancing around 1338. The Rhone Glacier advanced to a maximum between 1350 and 1400, which was slightly more extensive than any occurring later. Farther east, the Gurgl reached a maximum in about 1300. In the Austrian Alps, the Gepatschferner, Gurglerfemer, and Simonykees glaciers are also known to have advanced during this period.

Grove (2001) concluded that the LIA started in the second half of the 13th century and its three culminations, including the first in the 14th century, were very similar in scale. The LIA was not a unique event with a discrete beginning and a discrete end. Several phases of glacial expansion during the Holocene, comparable in scale with those of the LIA, have been traced in both the Swiss and Austrian Alps.

Grove (2001) also attempted to estimate the degree to which the LIA was applicable to regions other than the Swiss Alps. An extensive survey of available data for Canada, Greenland, Iceland, and Scandinavia provided fragmentary insights. It was concluded:

> "Though no reconstructions of the glacial history of the last millennium are as complete as the most detailed from Switzerland, evidence that the LIA had begun in the thirteenth century, with the first culmination in the fourteenth century, is widespread."

Joerin, Stocker, and Schluchter (2006) studied sub-fossil remains of wood and peat from six Swiss glaciers found in pro-glacial fluvial sediments. They discovered 12 phases of glacier recessions during the Holocene. Trees and mires grew where glaciers exist at present and, therefore, glaciers were smaller at that time than the present. The 12 major recession periods occurred at 9850–9600, 9300–8650, 8550–8050, 7700–7550, 7450–6550, 6150–5950, 5700–5500, 5200–4400, 4300–3400, 2800–2700, 2150–1850, and 1400–1200 years before present. They emphasized that "this natural variability of glacier extent, which occurs on a centennial timescale, is superimposed on a much longer term ... trend towards increased glacier extent culminating in the '*Little Ice Age*'".

A reconstruction covering the last 600 years based on selected tree-ring density series mainly from high-latitude land areas is shown in Figure 2.29. However, it is not clear how the tree-ring data were scaled. It appears to this writer that the data from tree rings might fit the measured temperatures better over its interval if the entire tree-ring curve is lifted about 0.1°C.

According to Matthews and Briffa (2005), Figure 2.29 demonstrates a distinct LIA climate from about 1570 to 1900 when NH summer temperatures (April to September) fell significantly below the 1961–1990 mean. Matthews and Briffa said:

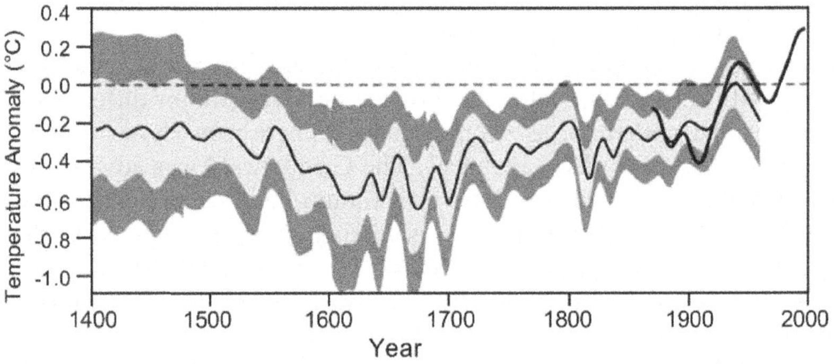

Figure 2.29. Tree-ring density reconstruction of NH (lands north of 20°N) summer temperatures (April to September) since AD 1400 (thin continuous line). Units are °C anomalies referenced to 1961–1990 mean (dashed line). Shaded areas show 68% and 95% confidence levels. Measured temperatures (thick line) are also shown (adapted from Matthews and Briffa, 2005).

"It would appear that there is a tenable statistical basis for belief in at least the main phase of the '*Little Ice Age*' as at least a hemispherical cold period ... and that, in terms of summer temperature, most of the seventeenth century was of the order of 0.5°C below the 1961–1990 mean. The question of whether the event was global remains more open."

In a later passage, Matthews and Briffa said:

"Indeed, we show here, for the first time in map form, that the majority of the Northern Hemisphere experienced a relatively low mean summer temperature for more than three centuries (1570 to 1900), and that the LIA was not merely or even mainly a European phenomenon."

But perhaps more importantly, Figure 2.29 is somewhat misleading because temperatures for the LIA are compared with a standard as the average for 1961–1990. As we previously pointed out, since LIA temperatures are lower than 1961–1990 temperatures, this serves to unduly emphasize the temperature rise of the 20th century. As long as one uses 1880 as a base year for comparison, 20th-century temperatures will show an increase. But is this a departure from the normalcy of the LIA, or was the LIA an aberration, with the 20th-century rise in temperature being a return to normalcy?

In Section 2.3.1, we provided a number of examples of proxy evidence for existence of the MWP and LIA.

While alarmists have denigrated the LIA as being minor and localized to Europe, there is some evidence that the LIA extended to tropical regions and the Southern Hemisphere. Lane *et al.* (2011) said: "Climate change during the LIA ... was once thought to be limited to the high northern latitudes, but increasing evidence reflects significant climate change in the tropics. ... Our results from Hispaniola further emphasize the global nature of LIA climate change. ..." Rabatel *et al.* (2008) reported on evidence of the LIA in Bolivia. Thompson *et al.* (2006) found evidence of the LIA in the Andes.

2.4.6 Asian climate records

Esper, Schweingruber, and Winiger (2002b) reported results on more than 200,000 ring-width measurements from 384 trees obtained from 20 individual sites that were analyzed to reconstruct regional climatic variation patterns in western central Asia since 618. A prolonged centennial trend toward better growing conditions was observed over the last 300 years (1700–2000) in the western central Asia tree ring records. This trend is of a lesser magnitude compared with conditions before 1100 and suggests the existence of a MWP. These early and recent benign periods were separated by a prolonged period of poor growth, which presumably reflects the LIA in western central Asia. These results support the hypothesis of the existence of the MWP and the LIA. The warmest decades since 618 appear to be between 800 and 1000, whereas the coldest periods were recorded in the first half of the 17th century.

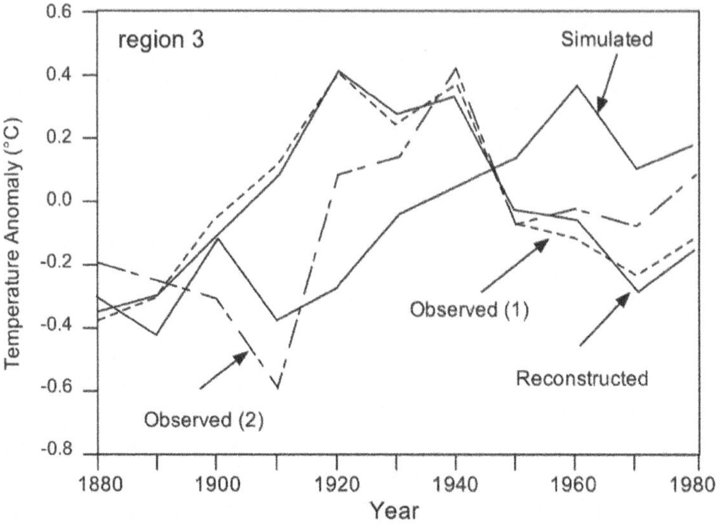

Figure 2.30. Basic data for the calibration period for Region 3 (South China) (adapted from Liu *et al.*, 2005).

Zhang *et al.* (2003) reported on a 2,326-year tree-ring chronology that is currently the longest annually resolved climate proxy record on the Qinghai–Tibetan Plateau and in China. Their results indicate that the climate on the plateau has undergone oscillations and, sometimes, very rapid swings during the last two millennia.

Liu *et al.* (2005) carried out a study of historical temperatures in China since 1550. China was divided into 10 districts that are relatively homogeneous. With proxy data from historical documents, tree rings, and ice cores, reconstructions of the temperature series were made for eight of these regions. The reconstructions of 10-year mean temperature anomalies were based on proxies via statistical techniques that rely on establishing empirical relationships between modern observations and environments during the calibration period (1880–1979). However, Liu *et al.* (2005) reported: "Unfortunately, this evidence is very uncertain." The basic data for one of the Chinese regions (Region 3) are shown in Figure 2.30. Note how the "simulated" data accentuate the rise during the 20th century. However, it is difficult for this writer to understand how the "simulated data" were derived.

Liu *et al.* (2005) provided "reconstructed" and "simulated" data from 1590 to 1980 for each of the eight regions. It is not clear to this writer how these were derived. Interestingly, when these data are compared with Figure 2.31 for Region 3, there doesn't seem to be much correlation.

Liu *et al.* then went through an empirical orthogonal analysis (EOF) that is a form of PCA[12] and ended up with Figure 2.31. The connection of this figure to Figure 2.30 appears to be "lost in translation".

[12] But the analysis was unintelligible to this author.

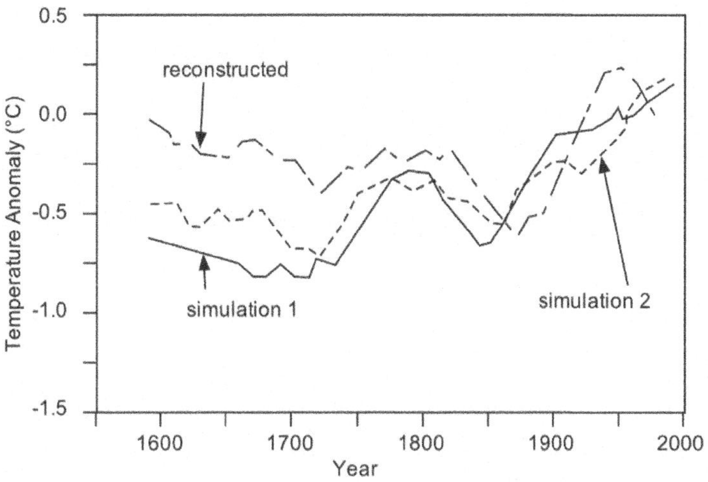

Figure 2.31. Longer-term data for Region 3 (adapted from Liu *et al.*, 2005).

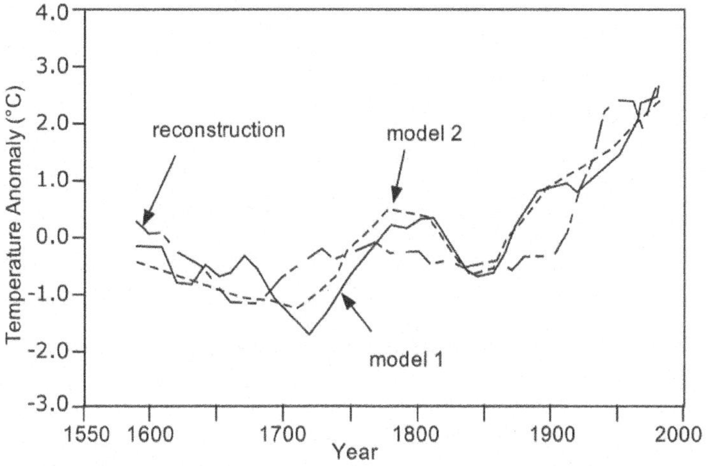

Figure 2.32. Estimated temperature profiles for all of China (adapted from Liu *et al.*, 2005).

The final result for the eight regions of China is given in Figure 2.32, showing a moderate *hockey stick* form. However, there is some evidence of a significant LIA, particularly between 1650 and 1750. Furthermore, the absolute values of the temperature anomalies seem very high in this figure. Without access to the original temperature data (1880–1980) from which the fundamental proxy relations were (presumably) derived, it is difficult to evaluate the veracity of this result. In particular, a concern is that, in statistical processing, they may have used the mean for the 1880–1980 period as the basis for calculating anomalies for the entire period 1590–1980, which could lead to exaggeration of the *hockey stick* syndrome. This writer is unable to penetrate the methodology enough to resolve this issue.

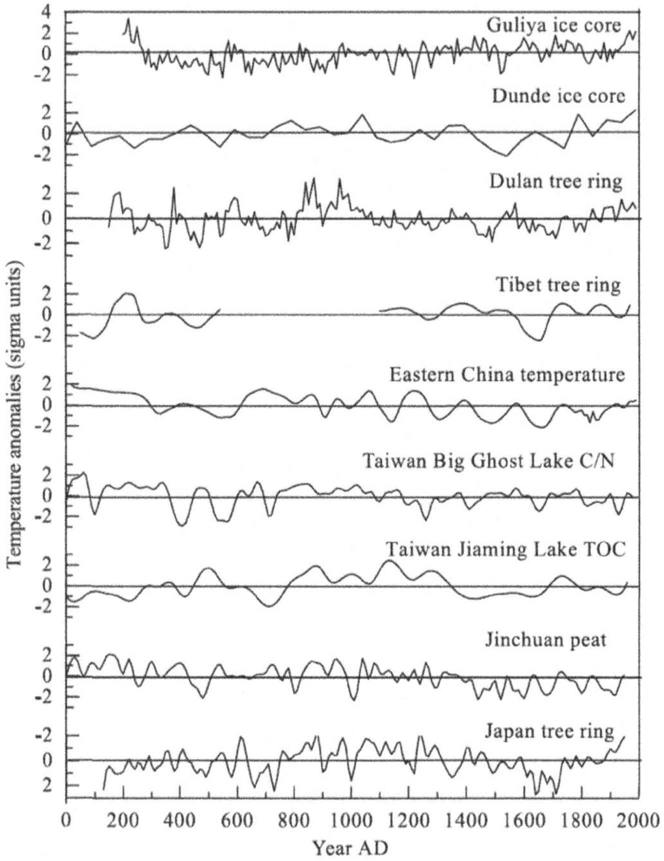

Figure 2.33. Two-thousand-year histories of average temperature for Chinese and vicinity areas (adapted from Yang *et al.*, 2002).

Yang *et al.* (2002) established three alternate China-wide temperature composites covering the last 2,000 years by combining multiple paleo-climatic proxy records obtained from ice cores, tree rings, lake sediments, and historical documents.

The basic proxy records are shown in Figure 2.33 for nine regions in and around China. As is usual with proxy data, there is considerable variation from region to region. Some of this could be real, while some could be the result of "noise" in the signal. There is a tendency for the time period around 1000 to be warm, but this is not uniform across all proxies. Similarly, there is a dip in temperature after 1600 in some of the proxies.

Yang *et al.* (2002) tried several methods to reconstruct regional paleo-temperature series from the nine data sets. One approach was to simply average all nine data sets with equal weighting. This was referred to as "Complete" China. A second approach ("Weighted" China) weighted the seven Chinese data sets in

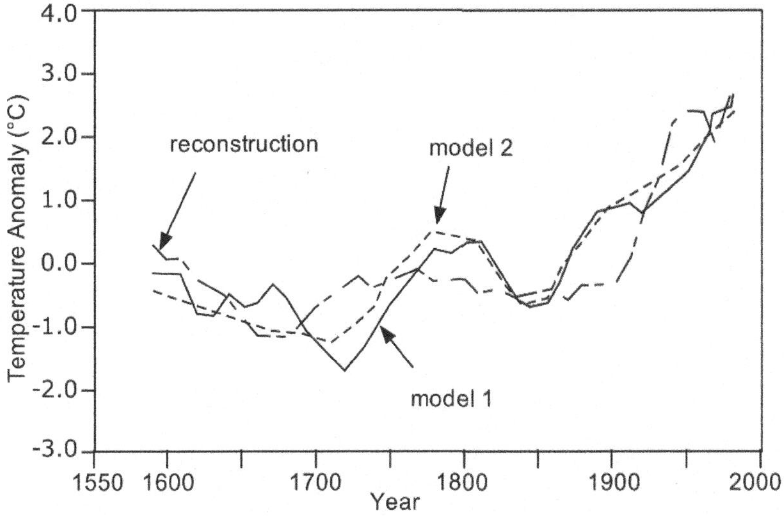

Figure 2.34. Comparison of several temperature reconstructions for China (adapted from Yang *et al.*, 2002).

proportion to the implied areas covered by the data. A third approach ("H-res" China) averaged only the higher-resolution Chinese data sets. These are shown in Figure 2.34.

There is good correlation between the three reconstructions. A cool period occurred between about 300 and 800. A MWP occurred between 800 and 1100. An LIA was pronounced from 1400 to 1700. A relatively flat neutral environment occurred from 1700 to 1900. The temperature rose again after 1900. Although the 20th-century temperature appeared to reach (or exceed) medieval levels around 2000, the number of temperature series dropped off sharply in the 20th century.

Yang *et al.* (2002) compared their results for China with several global temperature reconstructions, as shown in Figure 2.35. Yang *et al.* claimed good correlation with the global reconstructions, but the Chinese data show a much more pronounced LIA, a slightly more suggestive MWP, and a lesser *hockey stick* rise after 1900. This is exactly what one would expect, considering that the global studies used means for the 20th century (rather than for the whole data set) that skews the result and generates a *hockey stick* artifice (see Section 2.4.3.1).

Ge *et al.* (2010) reconstructed regional temperatures in China as far back as 500 to 2,000 years ago from proxies, principally tree rings, stalagmites, lake sediments, and historical documents. These results display considerable diversity from region to region. However, a few features are somewhat consistent across the regions. The LIA is shown roughly as a "double dip" in temperature from 1600 to 1700 and from 1800 to 1900. A warm period, at least comparable to today, occurred variably from about 1000 to 1200.

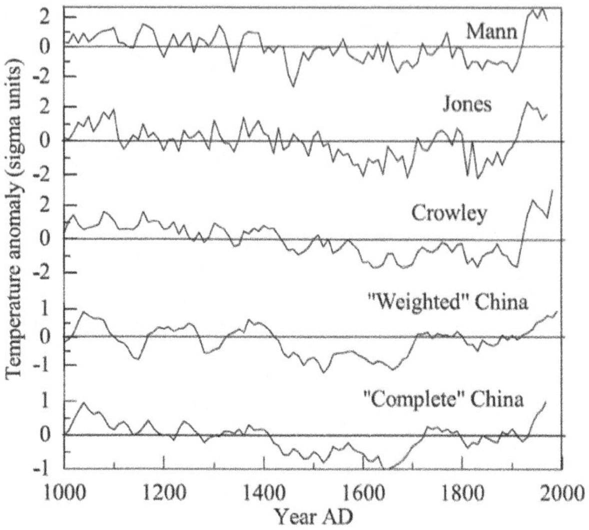

Figure 2.35. Comparison of several temperature reconstructions for China and the Northern Hemisphere (adapted from Yang *et al.*, 2002). Mann = Mann, Bradley, and Hughes (1999); Jones = Jones *et al.* (1998); Crowley = Crowley and Lowery (2000).

2.4.7 Regional approaches to the MWP and the LIA

One problem with attempting to deal with a single average global or hemispheric temperature is that most of the regional variations tend to cancel out, and one ends up with only small apparent net changes. One can place these changes under a magnifying glass and derive great concern from a global variation of a few tenths of a degree, but the utility of that concept seems to be limited. When the writer lived in Texas, if a cold front moved through, the temperature could drop by 25°C in one hour. Summer temperatures were over 40°C and winter temperatures dipped below freezing. There is no utility to a Texan in knowing that global temperatures changed by, say, 0.3°C. Understanding how the climate of the Earth has varied with time, and how the Earth responds to changes in the Sun as well as to anthropogenic effluents and activities, requires a more regional approach, as discussed in Soon and Baliunas (2003a, b). The changes in global average temperature don't even begin to describe the regional and local hardships and problems that derive from climate change.

The different sensitivities of proxies to climate variables and the potential time-dependence of the proxy-climate correlation require careful calibration and verification on a location-by-location basis; the emphasis on local results by Soon and Baliunas (2003a, b) avoided the difficulty of inter-comparing disparate proxies but did not generate a synoptic global view. Thus, Soon and Baliunas (2003a, b) gave up on quantitative synthesis of many proxies into single global or NH average temperatures, because, even for the same location, different proxies may yield different climate expressions simply because of their different sensitivities to local climatic variables. Soon and Baliunas (2003a, b) suggested that a compact mathematical representation of individual proxy variations (e.g., Mann*et al.*, 1998), without full understanding of proxy-climate calibration relations, may yield overconfident results. Furthermore, democratically combining large numbers of proxies of highly variable quality tends to dilute the reliable proxies, producing little more than noise.

Ljungqvist (2011) pointed out:

> "Considerable effort has been made during the last decade to reconstruct global or northern hemispheric temperatures for the past 1,000 to 2,000 years in order to place the observed 20th century warming in a long-term perspective. Less effort has been put into investigating the key question as to what extent earlier warm periods have been as homogeneous in timing and amplitude in different geographical regions as the present warming. It has been suggested (by some, particularly Mann *et al.*) that late-Holocene long-term temperature variations, such as the *Medieval Warm Period* (MWP) and the *Little Ice Age* (LIA), have been restricted to the circum-North Atlantic region (including Europe) and have not occurred synchronic in time with warm and cold periods respectively in other regions This view has, however, been increasingly challenged through the ever growing amount of evidence of a global (or at least northern hemispheric) extent of the MWP and the LIA that have become available. A main obstacle in large-scale temperature reconstructions continues to be the limited and unevenly distributed number of quantitative paleo-temperature records extending back a millennium or more. The limited number of records has rendered it impossible to be very selective in the choice of data. Paleo-temperature records used in a large-scale temperature reconstruction should preferably be accurately dated, have a high sample resolution and have a high correlation with the local instrumental temperature record in the calibration period. The number of long quantitative paleo-temperature records from across the globe, of which a majority are well suited for being used in large-scale temperature reconstructions, have been rapidly increasing in recent years. Thus, it has now become possible to make regional temperature reconstructions for many regions that can help us to assess the spatio-temporal pattern and the MWP and LIA. Only by a regional approach can we truly gain an understanding of the temperature variability in the past 1–2 millennia and assess the possible occurrence of globally coherent warm and cold periods."

Ljungqvist (2011) then went on to produce six regional reconstructions of temperatures over the past two millennia. These included:

(1) warm-season temperatures of Scandinavia north of 60°N;
(2) warm-season temperatures for northern Siberia;
(3) annual mean temperatures for Greenland;
(4) warm-season temperatures for the Alps region of Central Europe;

(5) annual mean temperatures for China;
(6) annual mean temperatures for the whole of the North American continent.

The specific proxies used by Ljungqvist are listed in his paper. "Only proxy records with reasonably high resolution (multi-decadal or better) were utilized and records with lower resolution were instead used for the purpose of verifying the reconstructions." Unlike many others who produced reconstructions, Ljungqvist did not use PCA methods that apply weights to the various proxies, thus further reducing an already sparse data set. Instead, he used a method that basically averages overlapping proxies ("composite-plus-scale"). According to von Storch *et al.* (2007), this method, though simpler, "clearly displays a better performance". His results for all six regions clearly show a significant MWP and a LIA. For Scandinavia north of 60°N, northern Siberia, and Greenland, the MWP was clearly warmer than current temperatures and the LIA bottomed out between 1600 and 1850. Current temperatures in Central Europe and China exceed those in the MWP. Current temperatures in North America are approaching those of the MWP.

Ljungqvist concluded:

> "Temperature changes, on centennial time-scales, occurred rather coherently in all the investigated regions—Scandinavia, Siberia, Greenland, Central Europe, China, and North America. ... Large-scale patterns as the MWP, the LIA and the 20th century warming occur quite coherently in all the regional reconstructions presented here but both their relative and absolute amplitude are not always the same. Exceptional warming in the 10th century is seen in all six regional reconstructions. Assumptions that, in particular, the MWP was restricted to the North Atlantic region can be rejected. Generally, temperature changes during the past 12 centuries in the high latitudes are larger than those in the lower latitudes and changes in annual temperatures also seem to be larger than those of warm-season temperatures. In order to truly assess the possible global or hemispheric significance of the observed pattern, we need much more data. The unevenly distributed paleo-temperature data coverage still seriously restricts our possibility to set the observed 20th century warming in a global long-term perspective and investigate the relative importance of natural and anthropogenic forcings behind the modern warming."

However, the rate of temperature increase in the 20th century appears to be greater than that of the MWP. Some of Ljungqvist's results for various regions are shown in Figure 2.36. A comparison of Ljungqvist's results for the NH with those of Moberg *et al.* (2005) and Mann *et al.* (2008) is shown in Figure 2.37.

Ljungqvist's results show a peak temperature around year 950 that is 0.12°C warmer than the temperature in year 2000. His estimated temperature during the depth of the LIA in year 1680 was 0.75°C lower than that in year 2000, and 0.87°C lower than that at the height of the MWP in year 950.

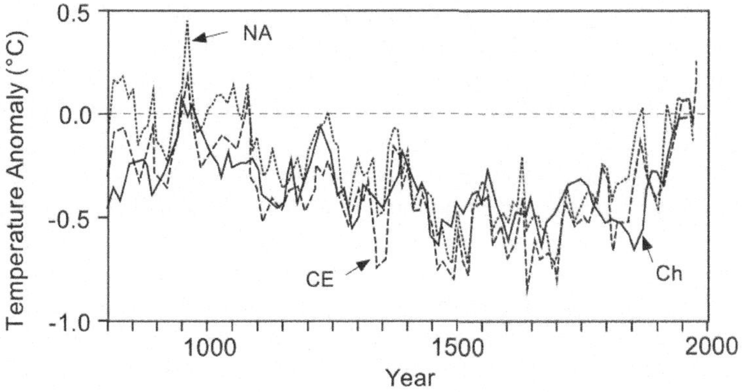

Figure 2.36. Annual mean temperature reconstructions for North America (dotted line); China (solid line); and Central Europe (dashed line) (adapted from Ljungqvist, 2011).

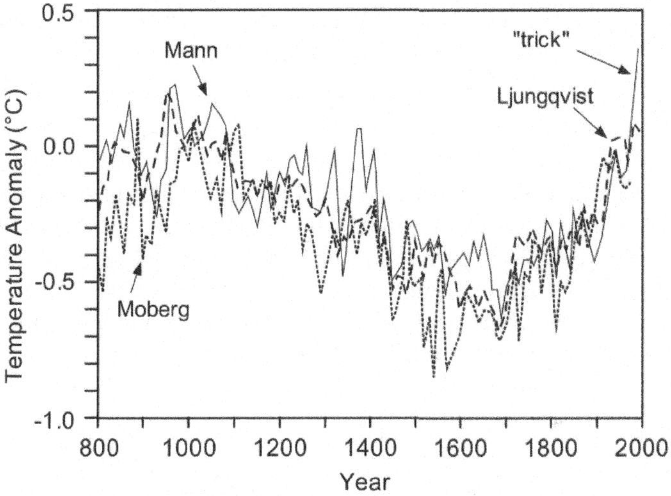

Figure 2.37. Temperature reconstructions for the NH as estimated by decadal means of Moberg *et al.* (2005); the "error-in-variables" (EIV) regression method variant of Mann *et al.* (2008); and and the extra-tropical Northern Hemisphere reconstruction by Ljungqvist (2011) (adapted from Ljungqvist, 2011). The "trick" was used by Jones and Mann to hide the discrepancy between measured temperatures (trick line) and proxies in the late 20th century.

2.4.8 Borehole measurements

Another approach for estimating past millennial temperatures is analysis of borehole temperatures. According to Pollack and Huang (2000), the fundamental concept behind subsurface temperatures as a climate proxy can be succinctly stated: if the Earth's atmosphere experiences a warming or cooling, the soil and rock in contact with the atmosphere will feel this change. Such temperature changes at the Earth's solid surface then propagate into the subsurface by heat conduction through the soil and rock. The process is analogous to the warming of a cold ceramic cup after hot tea is poured into it. The interior surface of the cup experiences an increase in temperature, which then propagates through the wall of the cup and can be sensed a short time later on the exterior surface. Similarly, variations of temperature at the Earth's surface associated with climate change can be thought of as a time-varying boundary condition on the upper boundary of the solid Earth. But, whereas heat conducts through a cup in just minutes, temperature fluctuations at the Earth's surface take several hundred years to penetrate the upper few hundred meters of the subsurface. The Earth filters out high-frequency energy fluctuations and retains only the long-term trends of surface energy imbalance, recording surface changes as perturbations of underground temperature as a function of depth. These changes in the energy balance at the Earth's surface are reflected in geothermal records whenever the underlying physical processes are sustained.

The present temperature distribution below the surface is the end product of variable surface temperature acting over past epochs of time to transfer heat by conduction down into the subsurface. The deeper that a temperature is measured the more it encompasses a longer duration of surface temperature effects:

> "Many thousands of boreholes around the world have been subjected to temperature logging in the course of determining the terrestrial heat flux. Thus, an abundance of observations exists, but because of the many investigators and different measurement practices and techniques, the data are heterogeneous. The heterogeneity arises from different borehole depths, different logging depth intervals, and variable information about thermo-physical properties, subsurface geological structure, and surface site characteristics. Even with such heterogeneity, however, quality data are sufficiently abundant and the analysis tools sufficiently flexible to allow credible climate reconstructions from these data at many sites around the world." (Huang *et al.*, 1997, 2000).

Kilty (1997) explained how borehole measurements are used to estimate the ground surface temperature (GST), say, over the past 1,000 years. The system involves circular reasoning because researchers begin with a model and use the observed borehole temperatures to find the parameters of the model:

> "However, a common problem with obtaining the history of surface temperature from borehole temperatures is that heat conduction destroys information regarding long past temperature quite completely, and, therefore, many different temperature histories explain the borehole data equally well. Quite a few of these histories oscillate in temperature wildly. By including a penalty for deviating from the initial model the program drives the final solution toward some unique result, and, if the initial model is smooth, the solution is also smooth." (Kilty, 1997)

One can allow either a loose or a tight constraint on variations from the original model. Kilty (1997) described the process of recovering past temperature from boreholes as an "ill-conditioned problem". Many doubts are raised about the veracity of borehole measurements in resolving past temperature variations. Only through circular reasoning by limiting variations from an assumed model can a unique result be obtained. Ogilvie and Jonsson (2001) also discussed the difficulties with borehole measurements, including the "uncertain relation between ground temperatures and atmosphere temperatures".

Jo Nova[13] provided a valuable review and summary. As Nova points out, there is "vastly more detail in ice cores, and we wouldn't bother with boreholes at all if there were glaciers conveniently located all over the world". ... "The pro side of boreholes is that there are thousands of measurements, and they are spread all over the land masses of the globe (all bar Antarctica). On the downside, it's hard to calibrate, and doesn't include the ocean". And they produce "highly smoothed past temperatures". Jo Nova reported that "it takes about 100 years for [a surface] perturbation to reach a depth of 150 m, and 1,000 years to reach 500 m depth. Boreholes are handy because they assess land areas that have few other proxies".

Nova reviewed a series of published papers by Huang and co-workers from 1997 to 2008. The latest paper is Huang *et al.* (2008). The borehole data show a broad peak temperature around year 1300 and a minimum around year 1770, corresponding roughly to the MWP and LIA. However, proxies suggest that the peak at the MWP was about 300 years earlier. Over the years, changes were reported in the relative height of the MWP and depth of the LIA borehole results. In addition, each study seemed to employ a different selection of borehole data. Noting that, as the years went by, the height of the MWP peak temperature declined, Nova raised the suspicion that the analyses were somehow tainted to make current temperatures higher than those during the MWP. The results of 1997–2000 indicated a much higher MWP peak temperature than the results of 2008. Data from Huang *et al.* (2008) are shown in Figures 2.38 and 2.39.

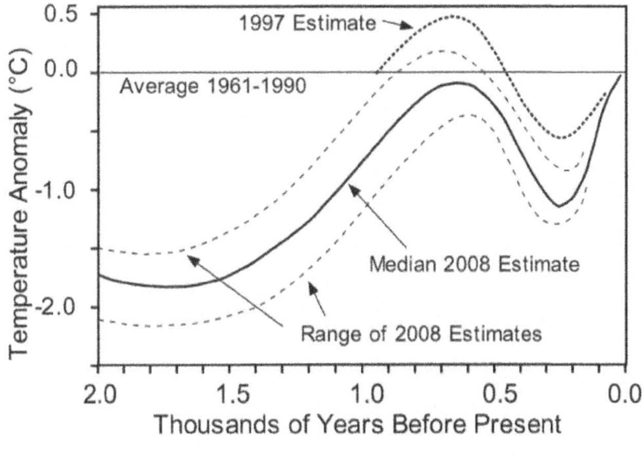

Figure 2.38. Borehole estimate of global average surface temperature over the past 2,000 years (Huang, 2008).

[13] *http://joannenova.com.au/2012/11/the-message-from-boreholes/#more-24964.*

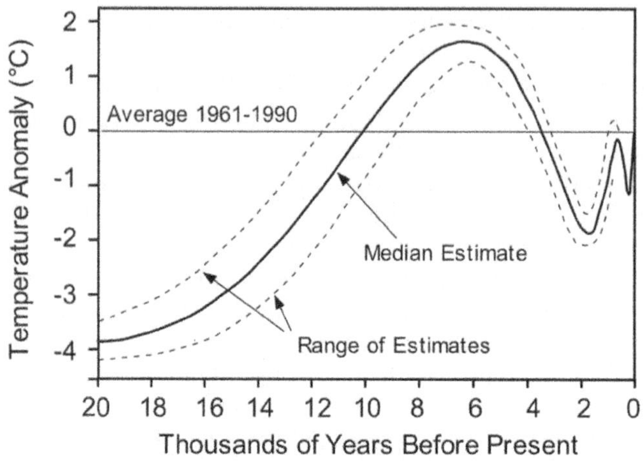

Figure 2.39. Borehole estimate of global average surface temperature over the past 20,000 years (Huang, 2008).

2.4.9 Arctic environment change

Overpeck *et al.* (1997) presented a compilation of paleoclimatic records from lake sediments, trees, glaciers, and marine sediments that provided a view of circum-Arctic environmental variability over the last 400 years. However, the instrumental record of Arctic climate change is brief and geographically sparse. In their article, the authors used the paleo-environmental record to assess the climate events of this century from the perspective of the last four centuries. They compiled a variety of complementary paleo-environmental indicators of climate from around the entire Arctic. This perspective permits the visualization of natural sub-decadal to century-scale climate variability in the circum-Arctic region. The results are shown in Figures 2.40a, b. It can be seen that, in general, the period from 1600 to 1925 was cold, although some intermittent warm periods were interspersed in some of these time series. Nevertheless, 22 out of 29 series were predominantly cold during this era, and the other 7 were variable. Therefore, there is clear evidence of an LIA.

The data after 1925 vary with the time series as follows:

(1) Strong warming: 12 series.
(2) Warming followed by moderate cooling (or variable): 13 series.
(3) Warming followed by strong cooling: 4 series.

In 17 out of 29 series, the predominant trend in the late 20th century was cooling. In 4 of the 17, the cooling was strong. We may therefore conclude that, in Arctic areas:

(1) the era 1600–1925 was relatively cold;
(2) compared with 1600–1925, the era that followed showed considerable warming;
(3) warming after 1925 was not consistent, and the majority of Arctic sites late in the 20th century were either in a cooling trend or were variable.

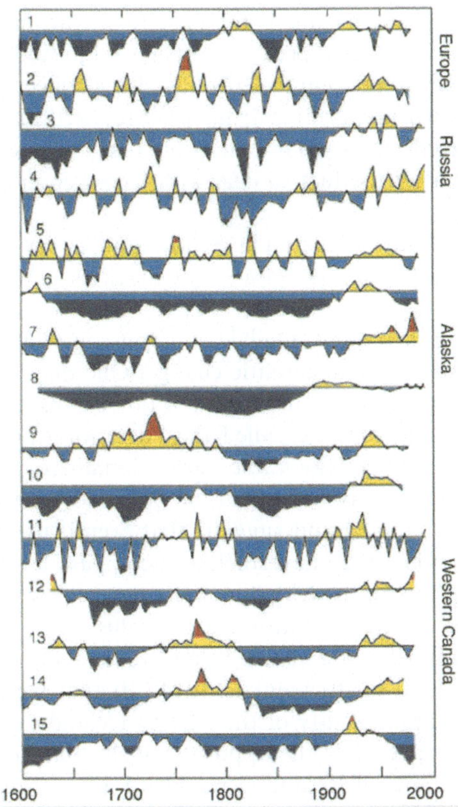

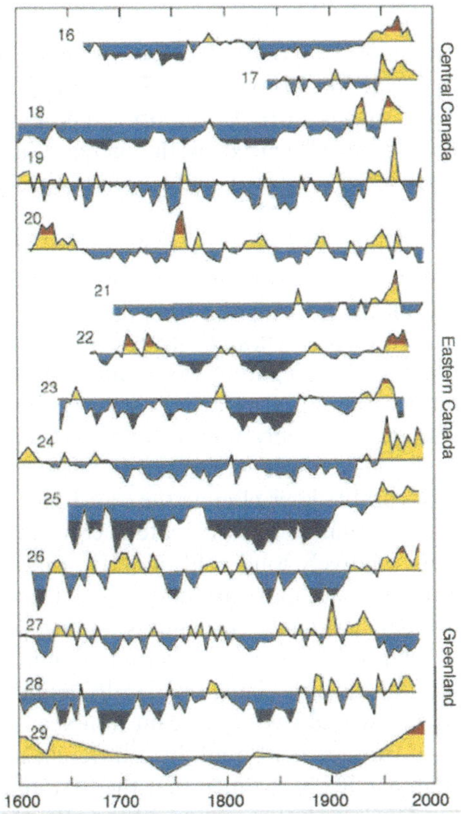

Figure 2.40a. (Left) Standardized 400-year proxy-climate records of surface air temperature. Red indicates temperatures greater than one standard deviation warmer than average for the reference period (1901–1960), whereas dark blue indicates at least one standard deviation colder than this average (based on Overpeck, *et al.*, 1997, with permission of *Science* J)

Figure 2.40b. (Right) Same as 2.43a but for sites in Canada east to Greenland. All series are presented as five-year averages except for sites 8 and 29, which are plotted at their original lower resolution. All time series represent surface air temperature except for site 29, which represents sea temperature (based on Overpeck, *et al.*, 1997, with permission of *Science* J)

Overpeck *et al.* (1997) seem to have been biased toward emphasizing global warming. They reported:

> "From 1840 to the mid-20th century, the Arctic warmed to the highest temperatures in four centuries. This warming ended the *Little Ice Age* in the Arctic and has caused retreats of glaciers, melting of permafrost and sea ice, and alteration of terrestrial and lake ecosystems."

However, they admitted that peak temperatures occurred around 1945, but did not discuss the cooling that occurred after that date. The data do not seem to support their conclusions.[14] Overpeck *et al.* said: "Half of the post-1840 warming (about 0.75°C) took place from 1840 to 1920. . . ." Overpeck *et al.* struggled to find a rationale for this warming. The cooling trend from 1950 to 1970 was also a concern. They suggested that "the observed slowdown in warming from 1950 to 1970 may have been influenced by the increase in Arctic tropospheric aerosols that occurred after 1950". Aerosols have been proposed after the fact to deal with recent cooling (see Section 3.4.5).

Miller *et al.* (2010b) wrote a review of "Temperature and precipitation history of the Arctic" in which they report large temperature fluctuations (as much as ±1°C during the past two millennia.

In their discussion of historical climate in Iceland, Ogilvie and Jonsson (2001) defined three separate viewpoints regarding Icelandic climatic change. One may be termed the uniformitarian view. As this name implies, it suggests a fairly constant climate, with only minor variations. A second may be called the "deterioration" view. This involves a transition from a relatively favorable early climate to an unfavorable later climate (i.e., an LIA). A third may be termed the "relative" view. This emphasizes that climate is constantly changing and suggests that even if there were relatively long cold or warm time periods, they nevertheless encompass large annual to decadal variability. Ogilvie and Jonsson provided an extensive review of journals written by Icelanders or Europeans from 1790 through the 1920s.

One 1914 book was definitely uniformitarian and claimed that "the climate of Iceland did not change markedly from Iceland's settlement to 1914". If that were true, it would cast significant doubt on the applicability of the LIA to Iceland.

Another 1914 book by a different author reached the opposite conclusion. For example, it was claimed that the first inhabitants of Iceland were able to grow grain but this later became impossible as the climate worsened. Several other writers (from Denmark and Norway) at a later date reached the uniformitarian conclusion. The sources of information become increasingly diffuse as one goes back in time. Ogilvie and Jonsson (2001) provide historical data on various aspects of Iceland. The sea ice index was plotted from 1600 to 1850, and from 1850 to 1990. Over the period 1600–1850, the sea ice index was a minimum during the 17th century and was highest from 1750 to 1850. Over the period 1850–1990, the sea ice index was highest before 1920, dropped to near zero from 1920 to 1967, and increased after 1967. The decrease starting around 1920 can hardly be due to greenhouse gases. The increases after 1967 are contrary to the expectation from greenhouse alarmists.

As in most studies of climate, Ogilvie and Jonsson (2001) provide many crosscurrents and conflicting indications that are highly stimulating but not fully revealing. Figure 2.41 shows the variation of Iceland temperatures since 1830.

[14] Some time ago, I was watching a boxing match on television. The announcer was enamored with the favorite and kept emphasizing how well he was doing, whereas my view indicated quite the opposite. Then there is the story of the boxer who returned to his corner between rounds and his trainer told him that his opponent never laid a glove on him. He complained that the referee must be hitting him because someone was administering a beating to him. Sometimes preconceived viewpoints affect what we perceive.

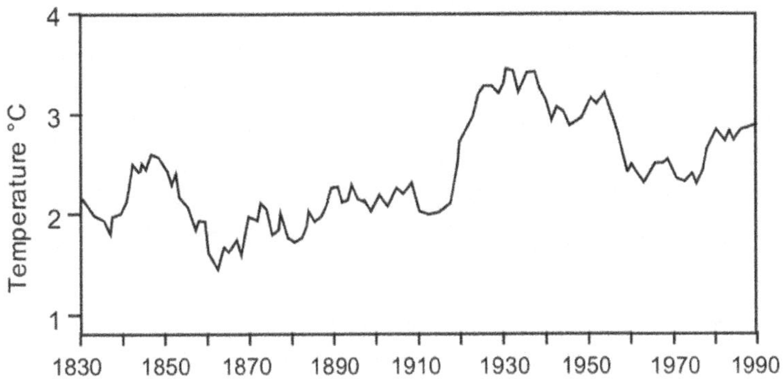

Figure 2.41. Ten-year running average Iceland temperatures (adapted from Ogilvie and Jonsson, 2001).

Table 2.2. Space–time matrix of temperature data.

Year⇓	Longitude = 0°					5°				360°				Proxies			
Lat⇒	0	5	10	...	70	0	5	...	70	0	5	...	70	1	2	...	N
850																	
851																	
852																	
...					Pseudo "real" data									Pseudo proxy			
1854																	
1855																	
1856						Real Data											
...																	
1980																	

Temperatures after 1855 were measured. Prior to that date, proxies were used. There was a rise around 1920, but, after 1950, temperatures fell. The rise around 1920 can hardly be due to greenhouse gases.

2.4.10 How reliable are proxy methods?

In order to test the robustness of methods for processing proxy data, Mann *et al.* (2005) set up a model for proxy-processing evaluation. Smerdon and Kaplan (2007) commented on this process. The description that follows is based on Smerdon and Kaplan (2007).

An artificial model is generated in which real temperature data are used for a calibration period, and a climate model is used to create real data prior to the calibration period. The climate model data are treated as if they were real for purposes of testing. A number of pseudo-proxies are created by selecting some of the real data for particular locations, and adding noise to them. The point here is that a proxy is treated as an absolutely accurate temperature series at one location, but with noise superimposed. Then, the challenge is to reconstruct the real data from the pseudo-proxies.

They considered a time period from 850 to 1980. For each year, a temperature is assigned to each spatial cell covering a 5° × 5° range of latitude and longitude. The latitude varies from 70°N to the equator and the longitude goes from 0° to 360° in 5° steps, as shown in Table 2.2. In actuality, we cannot know the actual temperatures over such a long time period. The true temperatures are known from 1855 to 1980 as shown by the rows 1855–1980 (for columns 0° to 360°) in Table 2.2. A global climate model is used to fill in the earlier data from 850 to 1854 (rows 850 to 1854 for columns 0° to 360°). Even though this data set is not real, it is treated as real for purposes of testing proxy processing. Then, a set of pseudo-proxies is constructed by selecting some of these time series, each pseudo-proxy representing one time series for a single latitude/longitude combination as shown in the far right columns. To emulate a real case, noise is added randomly to the various pseudo-proxies.

Then, the question is, given a set of pseudo-proxies as a starting point: how well does the data-processing scheme reproduce the original "true" data set?

Basically, in any statistical processing of proxy data, one begins by centering the data about its mean and normalizing it (scaling the deviations from the mean) in units of the standard deviation. This is called "standardization". There exists a calibration period during which there exist both proxies and actual measurements, and there is a longer period of time over which proxies are used to infer temperatures. One can standardize based on either the mean for the shorter calibration time period, or the mean over the entire data set. Smerdon and Kaplan (2007) found that, when standardization was carried out over the entire data set, the results were consistent and varied only slightly with the addition of noise to a data set. However, when the mean for only the calibration period was used for standardization, the results varied widely when noise was added. But Smerdon and Kaplan (2007) mentioned that, when standardization is carried out over the entire data set:

> "Such a decision may sound benign, [but] it amounts to knowing the mean and standard deviation of the target field prior to the calibration interval—a luxury that would obviate the need for a reconstruction in the first place."

Smerdon and Kaplan (2007) argued that the case where standardization is accomplished over the entire data set is artificial because one never has the full data set in reality. This would seem to suggest that the only practical way to process proxy data is based on standardization over the calibration interval, and this method is problematic because it produces variable results depending on the noise involved. As in the case of the M&M findings, the more noise that is added to the data, the more the result approaches a *hockey stick* form.

von Storch *et al.* (2007) also carried out a psuedo-proxy study. This was an outgrowth of their previous study (Zorita and von Storch, 2005). Their pseudo-reconstructions were performed in a climate simulation of the past millennium with the climate model ECHO-G. They said: "This is not the same simulation used by von Storch *et al.* (2004) in their pseudo-proxy analysis of the Mann *et al.* (1998) reconstruction method, but a simulation with the same model using different initial conditions." They admitted that "There still exists a large uncertainty in the amplitude of past TSI [total solar irradiance] at centennial timescales" but that doesn't matter because however inaccurate the climate model may be, it is still taken as truth in order to determine the ability of reconstruction models to replicate it. They tested three methods for reconstruction of the past climate from proxies. These included (1) the MBH method (inverse linear regression of proxies from principal components), (2) direct linear regression of principal components from proxies, and (3) the so-called "composite plus scaling" (CPS) that in some ways amounts to simple averaging of proxies. The results of the climate model were taken as factual. Noise was added to these representations of actual data to generate proxies, since proxies always contain a certain amount of noise. Given these proxies, the question was how well the reconstruction method reproduced the "factual" data. They concluded:

> "The results of the three reconstructions methods of the Northern Hemisphere temperature in the ECHO-G simulation for both pseudo-proxy networks from simple white-noise and red noise models ... are that all pseudo-reconstructions ... underestimate the past variations of the mean temperature, and the estimated temperature in the centuries previous to the instrumental period is too warm. The bias, however, depends on the noise model, on the calibration variant and on the size of the pseudo-proxy network.
>
> "However, the simple method Composite plus Scaling provides, in the conditions tested in this study, better results and is more robust against changes in the proxy network and noise characteristics."

The ultimate test for reliability of proxies is how well they track temperatures. Of all the many papers on proxies that I have reviewed, very few if any have provided such data in any detail. Briffa *et al.* (1998) is a notable exception. They compared tree ring proxies with temperatures at many sites in the NH from 1880 to 1980 as shown in Figures 2.42 and 2.43.

> "When averaged over large areas of northern America and Eurasia, tree-ring density series display a strong coherence with summer temperature measurements averaged over the same areas, demonstrating the ability of this proxy to portray mean temperature changes over sub-continents and even the whole Northern Hemisphere. During the second half of the twentieth century, the decadal-scale trends in wood density and summer temperatures have increasingly diverged as wood density has progressively fallen. The cause of this increasing insensitivity of wood density to temperature changes is not known ..."

Although Briffa *et al.* pointed out the discrepancy between tree ring data and temperature after 1950, their assessment that proxies tracked temperatures prior to 1950 might be somewhat optimistic. The time period was short and the variability of temperature was rather small.

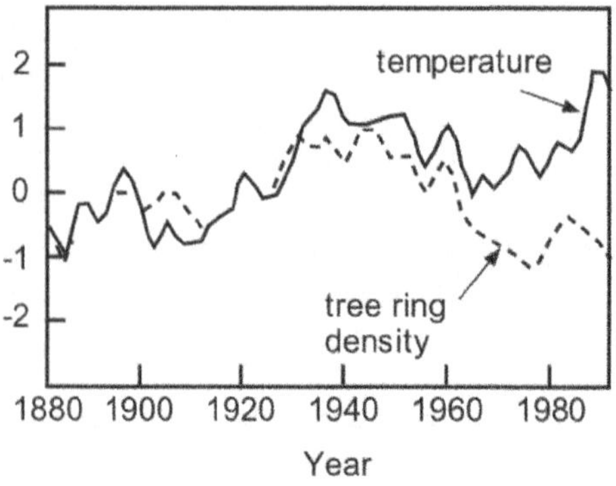

Figure 2.42. Comparison of tree-ring density with temperature for NH (Briffa *et al.*, 1998).

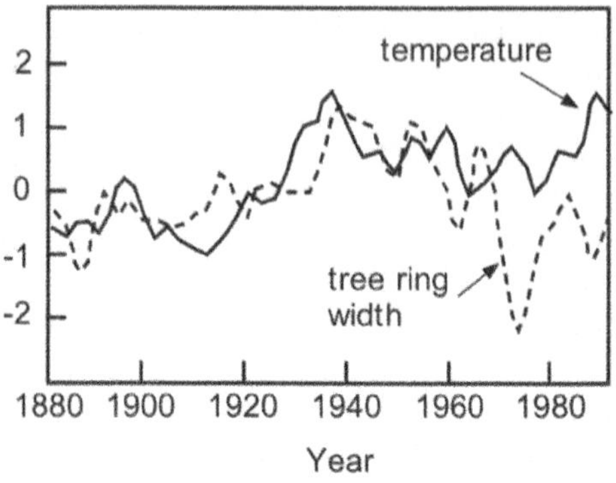

Figure 2.43. Comparison of tree-ring width with temperature for NH (Briffa *et al.*, 1998).

Since the publication of this data in 1998, a number of additional papers have appeared dealing in one way or another with tree ring proxies.

Jacoby *et al.* (2000) said:

> "Data from annual tree-ring widths are used to reconstruct May–September mean temperatures for the past four centuries. These warm-season temperatures correlate with annual temperatures and indicate unusual warming in the 20th century. However, there is a loss of thermal response in ring widths since about 1970".

Thus they admit to a divergence problem after 1970. However, when one examines their data prior to 1970, the correlation of tree ring data with temperature even prior to 1970 appears somewhat subjective to this writer.

D'Arrigo *et al.* (2006a) came to the defense of tree ring proxies. They began their paper with the usual orthodoxy that "recent warming in the Northern Hemisphere appears to have been unprecedented over the past millennium and that this warming is most likely a result of the anthropogenic release of greenhouse gases into the atmosphere"—which has not much to do with the reliability of tree ring proxies. They mention that D'Arrigo *et al.* (2006b) used simple averaging of tree-ring records (after accounting for differences in mean and variance over time), followed by linear regression". Simple averaging is a step in the right direction, but how good are the proxies as temperature indicators? As is usual in almost all papers on proxies, the data for the calibration period were not shown. They do allude to the divergence between tree ring proxies and temperature reported by Briffa *et al.* (1998) and others cited in their paper, where they said:

> "Theories for the cause (s) of this observed divergence, which may vary from site to site, include decreased temperature sensitivity due to warmer temperatures, drought stress, increased winter snowmelt and ozone effects. This divergence needs to be considered to avoid bias in dendroclimatic reconstructions; however it is not present everywhere. For example, tempera-

ture-sensitive elevational treeline sites in Mongolia and the European Alps exhibit dramatic growth increases in recent decades. Greater attention to site selection (e.g. avoidance of drought-prone sites) and careful comparison of adjacent sites with regards to their ecological characteristics can help circumvent this problem. [It has been] demonstrated that the divergence appears to be limited to the recent period (after ~1950) and to trees from some northern locations (at some sites within ~55-70°N), and that there is no evidence for a comparable divergence prior to this time (e.g. during the *Medieval Warm Period*). These observations suggest a unique, anthropogenic cause for the recent divergence and argue very strongly that tree-ring temperature reconstructions for the past millennium should not be called into question based on these recent observations".

One problem with site selection is that if one is attempting to estimate a global average temperature, one needs all the sites one can find. If only a few sites provide reliable data, how can one derive global or even hemispheric temperatures in the past? The claim that "there is no evidence for a comparable divergence prior to this time (e.g. during the *Medieval Warm Period*)" is unmitigated nonsense because there are no measured temperature data for that period and hence there is no way to ascertain whether such a divergence exists. D'Arrigo *et al.* (2006a) closed their paper with further homage to the orthodoxy of "unusual recent anthropogenic warming on a hemispheric to global scale" but their defense of tree ring proxies falls flat.

Wilmking and Singh (2008) discussed the "divergence effect" between measured temperatures and tree ring proxies in the 2nd half of the 20th century and pointed out that this "seriously questions the validity of tree-ring based climate reconstructions, since it seems to violate the assumption of a stable response of trees to changing climate over time". In their study they claimed to have

> "... eliminated the 'divergence effect' in northern Alaska by careful selection of individual trees with consistently significant positive relationships with climate (17% of sample) and successfully attempted a divergence-free climate reconstruction using this subset".

However, they did admit:

> "The majority of trees (83%) did not adhere to the uniformitarian principle as usually applied in dendroclimatology. Our results thus support the notion that factors acting on an individual tree basis are the primary causes for the 'divergence effect' (at least in northern Alaska)".

However, even the small subset of 17% of trees that are claimed to show good consistency with temperatures over the last century provide somewhat doubtful consistency. The diagram provided by Wilmking and Singh (2008) in their Figure 2 is a tiny little diagram that compresses the excursions between the temperature and tree ring curves. Nevertheless, accepting the claim that 17% of the trees show good correlation with temperature, the question arises as to whether it makes sense to select a subset of trees that happen to fit the temperature curve, and use these for estimating temperatures a thousand years or more ago. Apparently Wilmking and Singh suggest that there occurs a "mixture of trees with stable and non-stable climate growth relationships" and the ones with stable relationships provide a basis for estimating past climates. However, it may be equally likely that all the tree ring records are randomized by other variables than temperature, and by happenstance, about 17% of the records have correlation coefficients with temperature that satisfy the criterion adopted by Wilmking and Singh (which is not impressive to this writer). There is then no great reason to believe that even these 17% of trees would remain as accurate temperature indicators over much longer periods. In other words, select a theory, choose a small subset of data that agrees with the theory, ignore the majority of data that disagree with the theory, and claim the theory is verified!

2.4.11 The paleoclimatic *cabal*

Wegman, Scott, and Said (2006) have suggested that the field, temperature history of the Earth, is dominated by a cadre (*cabal*) that is vitally concerned about the potential impacts of global warming, and supports the *hockey stick* result, as well as the procedure used to derive it. Wegman, Scott, and Said (2006) said:

> "If there is a tight relationship among the authors, and there are not a large number of individuals engaged in a particular topic area, then one may suspect that the peer review process does not fully vet papers before they are published. Indeed, a common practice among associate editors for scholarly journals is to look in the list of references for a submitted paper to see who else is writing in a given area and thus who might legitimately be called on to provide knowledgeable peer review. Of course, if a given discipline area is small and the authors in the area are tightly coupled, then this process is likely to turn up very sympathetic referees. These referees may have co-authored other papers with a given author. They may believe they know that author's other writings well enough that errors can continue to propagate and indeed be reinforced."

It was concluded:

> "It is immediately clear that Mann, Rutherford, Jones, Osborn, Briffa, Bradley, and Hughes form a clique, each interacting with all of the others."

Other cliques were identified as well. It seems clear that this *cabal* has considerable control over which papers are approved for publication in journals on the topic of the history of temperature of the Earth.

As McLean (2007b) said:

> "The peer-review process was established for the benefit of editors who did not have good knowledge across all the fields that their journals addressed. It provided a 'sanity check' to avoid the risk of publishing papers which were so outlandish that the journal would be ridiculed and lose its reputation. In principle this notion seems entirely reasonable, but it neglects certain aspects of human nature, especially the tendency for reviewers to defend their own (earlier)

papers, and indirectly their reputations, against challengers. Peer review also ignores the strong tendency for papers that disagree with a popular hypothesis, one the reviewer understands and perhaps supports, to receive a closer and often hostile scrutiny. Reviewers are selected from practitioners in the field, but many scientific fields are so small that the reviewers will know the authors. The reviewers may even have worked with the authors in the past or wish to work with them in future, so the objectivity of any review is likely to be tainted by this association. Some journals now request that authors suggest appropriate reviewers, but this is a sure way to identify reviewers who will be favorable to certain propositions ... In 2002 the editor-in-chief of the journal *Science* announced that there was no longer any doubt that human activity was changing climate, so what are the realistic chances of this journal publishing a paper that suggests otherwise? The popular notion is that reviewers should be skilled in the relevant field, but a scientific field like climate change is so broad, and encompasses so many sub-disciplines, that it really requires the use of expert reviewers from many different fields. That this is seldom undertaken explains why so many initially influential climate papers were later found to be fundamentally flawed. In theory, reviewers should be able to understand and replicate the processing used by the author(s). In practice, climate science has numerous examples where authors of highly influential papers have refused to reveal their complete set of data or the processing methods that they used. Even worse, the journals in question not only allowed this to happen, but have subsequently defended the lack of disclosure when other researchers attempted to replicate the work."

In peer-reviewing manuscripts submitted to science journals, the situation is in some ways analogous to the criminal justice system. There are two extremes: (1) make sure that you catch all criminals even if, in the process, you convict some innocent people; or (2) make sure that you do not convict any innocent people even if some criminals escape. A sane justice system seeks some middle ground between these extremes. In a similar way, we could hypothesize two extremes in reviewing scientific manuscripts: (1) make sure that every publication is *bona fide* even if, in the process, some valid papers are rejected; or (2) make sure that every valid paper is published even if a few bad papers sneak through the review process. It seems evident that the *cabal* has adopted the philosophy that only *bona fide* papers should be published, and they have appointed themselves as sole arbiters of what is *bona fide*, based on whether the manuscript supports the alarmist position.

It is noteworthy that M&M submitted a letter to *Nature* about a flaw in the MBH procedure. After a long (eight-month) reviewing process, M&M were notified that *Nature* would not publish this letter. *Nature* concluded it could not be explained in the 500-word limit, and one of the referees said he found the material was quite technical and unlikely to be of interest to general readers. Instead, MBH were permitted to make a coy disclosure in their July Corrigendum. In an online Supplement (but not in the printed text itself), they revealed the non-standard method, and added the unsupportable claim that it did not affect the results. There is at least the appearance (and more likely the reality) that the *paleoclimatic cabal* was in complete control of the situation.

Mann *et al.* (2005) was written by the MBH team (1) to argue against von Storch *et al.* (2004), and (2) to claim that it makes little difference whether the proxy data are standardized by utilizing only the mean of the calibration period, or the mean of the entire historical data set. What would be humorous, if it were not sad, is the fact that the *paleoclimatic cabal* continues to relentlessly publish learned articles in defense of their methodologies, while pretending all the while that the valid criticisms of M&M do not exist.[15] Thus, for example, Mann *et al.* (2005) concluded:

> "Two widely used statistical approaches to reconstructing past climate histories from climate proxy data such as tree rings, corals, and ice cores are investigated using synthetic pseudo-proxy data derived from a simulation of forced climate changes over the past 1200 yr. These experiments suggest that both statistical approaches should yield reliable reconstructions of the true climate history within estimated uncertainties, given estimates of the signal and noise attributes of actual proxy data networks."

Mann *et al.* (2005) went on to say:

> "We find no evidence for the suggestion (e.g., von Storch *et al.*, 2004) that real-world proxy-based temperature reconstructions are likely to suffer from any systematic underestimate of low-frequency variability. Our findings suggest that both standard methods that have been used in proxy-based reconstruction are likely to provide a faithful estimate of actual long-term hemispheric temperature histories, within estimated uncertainties."

M&M were never mentioned. Mann *et al.* (2007) continued the relentless defense of flawed methods:

> "Our results reinforce previous conclusions that CFR methods, correctly implemented and applied to suitable networks of proxy data, should yield reliable reconstructions of past climate histories within estimated uncertainties."

Mann *et al.* (2008) continued the charade.

In late 2009 and early 2010, an extensive set of emails between principal figures[16] in the *paleoclimatological cabal* was made public (by unknown, but clearly illegal, means). These emails revealed an apparent deeply imbedded agreement amongst these climatologists to promulgate their orthodoxy that the Earth's climate has hardly wavered over the past 2,000 years, and that CO_2 was the principal cause of unprecedented global warming in the 20th century. As Mosher and Fuller (2010) pointed out, these climatologists:

[15] Why am I reminded of Frank Morgan as the Wizard of Oz admonishing Dorothy to "pay no attention to that man behind the screen ..."?

[16] Phil Jones and Michael Mann are the principal figures in the *cabal*, but several others were involved, such as Kevin Trenberth, Tom Wigley, Ray Bradley, Ben Santer, Gavin Schmidt, Jonathan Overpeck, and there are others.

"... ruthlessly suppressed dissent by ensuring that contrary papers were never published and that editors who didn't follow their party line were forced out of their position. When *Freedom of Information* requests threatened to reveal their misbehavior, the emails showed them actively conspiring to delete emails to frustrate legitimate requests for information. Worst of all, one scientist threatened to delete climate data rather than turn it over, and that data is still missing."

Some of the worst gaffes were committed by Phil Jones (Hadley Climate Research Unit), who said (amongst other things):

"And don't leave stuff lying around on anonymous download sites—you never know who is trawling them. McIntyre and McKitrick have been after the Climatic Research Unit ... data for years. If they ever hear there is a Freedom of Information Act now in the United Kingdom, I think I'll delete the file rather than send it to anyone.

"I've just completed Mike's Nature trick [Michel Mann's publication in *Nature* where he replaced tree-ring proxy data with actual data because the tree-ring data went in the 'wrong' direction[17]] of adding the real temperatures to each series for the last 20 years (*i.e.* from 1981 onwards) and from 1961 for Keith's to hide the decline." [See Figure 2.22.]

"We have 25 or so years invested in the work. Why should I make the data available to you, when your aim is to try and find something wrong with it?"

Mosher and Fuller (2010) provide great detail on this saga, colloquially known as "*climategate*". The Internet is full of commentary on this sorry situation in which scientists appear to have acted unprofessionally, unscientifically, and in some cases illegally. Kevin Trenberth, Senior Scientist at the National Center for Atmospheric Research, has emerged as a defender of the *cabal.*[18] He did admit to "lack of openness in sharing data and violations of the Freedom of Information Act" but he pointed out that five investigations failed to find any of the alleged misconduct. Unfortunately, these five investigations were conducted by friends of the *cabal.* He also asserted that "scientists would not make up stuff that could be disproven by others!" but the nature of paleoclimatic data is that they are not susceptible to proof, disproof, verification, or validation, and hence are a very safe field to work in. He cited an excerpt from a Phil Jones email:

"I can't see either of these papers being in the next IPCC report. Kevin and I will keep them out somehow—even if we have to redefine what the peer-review literature is!"

He implied that this was Jones's invention and he (Trenberth) had nothing to do with this. Whether this is true or not, this excerpt reveals the intellectual environment of the climate *cabal.* However, I have to agree with one slide in Trenberth's

[17] See: *http://climateaudit.org/2009/11/20/mike%E2%80%99s-nature-trick/.*
[18] *www.cgd.ucar.edu/cas/Trenberth/Presentations/ClimategateS.pdf.*

presentation that says the Internet is "An open sewer of untreated, unfiltered information and the American public is incapable of deciphering between facts, fiction and opinion".

The *cabal* refereed one another's papers submitted to journals, communicated improperly in a mutual back-scratch environment subverting the peer-review process, pressured journal editors not to publish papers contrary to the orthodoxy, conspired to write rebuttals to any papers that did slip through their barrier to publication of contrary views, and conspired to act in partnership to disparage and ridicule anyone with contrary findings. Several books present excerpts from the emails and provide interpretations of their implications, which are generally referred to as "climategate" (e.g. Mosher and Fuller, 2010).

One topic that annoys *cabal* members is the claim by some climatologists that persistent El Niños since 1976 were dominant in causing warming in the NH in the latter part of the 20th century. If this were true, it would suggest that the role of CO_2 in climate change may be far less than the orthodoxy believes. Thus, when the article by McLean *et al.* (2009) appeared in the literature suggesting an important role for El Niños as a dominant cause of warming in the NH in the latter part of the 20th century, it produced great animosity and consternation amongst the members of the *cabal*. This paper was reviewed and accepted by three independent referees. One referee commented in part: "I found the paper to be well-organized, well-written, and clear on the importance of the research ... The findings are likely to be of interest to a wide variety of readers." A second referee commented in part: "This very clear and well-written manuscript is an analysis of the relationship between MSU-derived and radiosonde-based tropospheric temperature variability and the Southern Oscillation, as modified by major tropical volcanic eruptions." After the paper was published, a flurry of emails was exchanged between *cabal* members, strategizing on how to carry out damage control for their orthodoxy by preparing a rebuttal. Soon afterwards, a group of *cabalists* (Grant Foster, James Annan, Phil Jones, Michael Mann, Jim Renwick, Jim Salinger, Gavin Schmidt, and Kevin Trenberth) decided to prepare a rebuttal, and, to ensure speedy publication, they pressured the editor of the *Journal of Geophysical Research* and suggested the following persons as possible reviewers for their submitted critique: Ben Santer, Dave Thompson, Dave Easterling, Tom Peterson, Neville Nicholls, and David Parker (with Tom Wigley, Tom Karl and Mike Wallace also mentioned.) All of these were professionally associated in some way to the Foster *et al.* group. Phil Jones commented: "All of them know the sorts of things to say—about our comment and the 'awful original', without any prompting." (They all subscribe to the same orthodoxy). McLean *et al.* describe the whole sordid story.[19] In their rush to rebut the original McLean article, the *cabal* posted their rebuttal on a website, in violation of JGR rules. Figures 3.33 and 3.35 to 3.39 provide clear evidence of the interaction of El Niños with climate.

[19] *scienceandpublicpolicy.org/originals/censorship_at_agu.html.*

2.4.12 The blogs

I note that, typically, whenever a climate scientist publishes a new paper or puts out a press release (however, in my day, no self-respecting scientist pushed his own wares via press releases), a flurry of websites appears reporting on, or commenting on, the original, but rarely providing a link to the original. These can be found via a Google search. Those websites that are configured to rank high in Google's prioritization scheme appear near the top, and finding a link to the original publication can range from difficult to impossible.

A similar thing happens when you are planning a trip. You seek a specific hotel on Google but you get many hotel compendium sites and it is difficult to reach the website of that specific hotel. In my opinion, there is something dreadfully wrong with Google's prioritization algorithm. Their principal basis for prioritizing a website seems to be based on how many other websites link to the website in question; they regard this as a vote of confidence by the public. What happens though, is that institutional sites and sites frequented by bloggers dominate, and sites run by individuals or specific commercial establishments, often containing the most content, get low priority. Thus, if you try to find any specific site that does not have extensive multi-links, it is likely to be buried in an ocean of heavily linked sites dealing with the general topic of the specific site you are trying to reach.

Blog sites are an amazing phenomenon. There are many websites on various topics (one important topic being climate change) that operate in the following way. Somebody (often anonymous, sometimes revealed) runs the site. That person posts a paragraph or an essay *du jour* on the blog, usually controversial, and, immediately, typically, hundreds of followers of that blog site send in cryptic remarks in response, sometimes one-liners, but sometimes considerably longer. Many climate blogs are dominated by followers of a single persuasion: alarmist or skeptic. Many of the responders to blogs hide behind the cloak of anonymity using a pseudonym. It is difficult to tell whether they are scientists or janitors. My experience with such sites is that the overwhelming majority of remarks seem to be pretty stupid or trivial, or both. Many get bogged down with trivia and tangential aspects. Some turn nasty and become personally abusive. On rare occasions, they may contain valid content. Some of these websites make false or misleading accusations against serious people, and their followers, hungry for any hint of impropriety or scandal, love it. In regard to climate change, many of the entries are nasty, insulting, insinuating, and unfriendly. We previously quoted Trenberth saying that the Internet is "An open sewer of untreated, unfiltered information and the American public is incapable of deciphering between facts, fiction and opinion".

Quite a number of blogs have evolved out of the controversies surrounding putative global warming. Some of these represent global-warming skeptics and some represent the *paleoclimatic cabal*. The great majority of these blogs are filled with short sound bites submitted by people who have not done the research needed in order to speak authoritatively on any subject, and, indeed, the bulk of the banter is nonsense. However, a few sites deserve special mention because the postings are often technically competent and worthy of review. (That does not necessarily imply that posting by followers in response to the original posting are competent.) One blog, which is anti-establishment, is *www.climateaudit.org*, run by Steve McIntyre (of M&M). This website is distinguished from many other climatological websites in that McIntyre has the skill and invests the time to actually investigate the minute details of published papers from their large data sets. Many published climatological papers depend on large data sets and it is impractical for most others to check the veracity of the manipulations of the data. However, McIntyre does this, and, in the process, he has discovered a considerable amount of improper analysis by highly recognized climatologists. In essence, McIntyre has become the *de facto* reviewer for complex statistical analyses of climate data, although most climatologists try to ignore his critiques—which appear to always be on target. Some of his attempts to obtain original data from climatologists have been thwarted by evasion, obfuscation, and illegal violations of the FOIA (typically with approval by institutional managers). At least one technical paper was withdrawn after McIntyre published a critique on his blog. However, McIntyre restricts his blog to a relatively narrow range of subjects utilizing large data sets.

McIntyre has penetrated into the data and details and presents authoritative analyses. He has evolved to become the established arbiter of published climate data. Most published climate papers are complex and typically utilize extensive data sets. Even if journal reviewers were objective (which they typically are not), they would not have time (and typically skill) to penetrate into the details of the data and the analysis. Thus, they content themselves with a more cursory overview of the paper and rubber-stamp it if it was written by seemingly competent people. Worse still, is the situation when the reviewers and the authors have a cozy relationship in which they approve one another's manuscripts for publication. Clearly, those who work in the field of climatology (typically professors at universities) are beset by the need to obtain funds, supervise student research, make presentations, and teach courses. They do not have time to provide the review needed to assure that climate papers are sound and credible. Only a retired person, dedicated to reviewing other people's work, has the time to do this. McIntyre acts to keep the field honest. However, the *cabal* nevertheless continues to manipulate the journals according to its agenda. The scientific method is only fulfilled in a report if the data and procedures are made available for checking and reproduction by others. McIntyre provides specific instances where important papers in climate change have not made the data available and his attempts to acquire the data were parried by the *paleoclimatic cabal*. McIntyre also presents a few examples where scientists responded fully and promptly to his inquiries for detailed data. In general, the business of processing very large extended data sets by means of sophisticated statistical procedures does not lend itself easily to review by others. But the efforts made by workers in the field to make their data available leave much to be desired.

McIntyre provides detailed accounts of his difficulties in acquiring original proxy data used in published papers by members of the *paleoclimatic cabal* (Osborn, Briffa, Jones, Crowley, Lowery, Esper, Moberg, Juckes, Mann, *et al.*). In general, obtaining proxy data from these authors required endless requests and cajoling, and even then was only sporadically successful. In one case, he could not get any satisfaction from the authors, and had to resort to 25 emails to the journal *Science*, and even then he was not able to recover the required data. He indicated that in several important cases, the authors utilized data from previous publications, but the data so acquired were not original, and had been processed (sometimes in ways that are suspect and difficult to trace). Seeking the original data required going back to multiple authors, leading inevitably to frustration.

Aside from the problem of acquiring the data, McIntyre found many inconsistencies and oddities in the handling of proxy data by the *cabal* members. He details these at length, although this writer found much of it difficult to absorb. One way or another, these manipulations of data all seemed to have a singular end result: amplification of the recent late-20th-century warming trend, and damping of the MWP, leading to the mantra that we are currently experiencing the warmest climate in over 1,000 years. It is particularly revealing to note some results of Briffa *et al.* (2001). Figures 2.20 and 2.21 show eight different reconstructions using various procedures with one preferred reconstruction. Note that all reconstructions decline in the second half of the 20th century while measured temperatures rise. The divergence is readily seen and Jones's "trick" of replacing the proxy data with measured data produces the hockey stick.

The major alarmist establishment blog is *http://realclimate.org*, run by Gavin Schmidt, which presents the viewpoints of the global-warming alarmists in a seemingly authoritative manner. Like its polar opposite (*climateaudit.org*), the *realclimate.org* blog is very extensive and detailed, and it is not possible to adequately describe all the material on this website. Only a few comments will be made.

Al Gore's film *An Inconvenient Truth* received a glowing review from *realclimate.org* including "admirable", and "for the most part he gets the science right". Al Gore received the Nobel Peace Prize, presumably partly based on this film. See Appendix I for a review of this film that provides the opposite view.

The *realclimate.org* blog provides a section entitled "Myth vs. Fact regarding the *hockey stick*". It is claimed that:

> "Numerous myths regarding the so-called *hockey stick* reconstruction of past temperatures, can be found on various non-peer reviewed websites, internet newsgroups and other non-scientific venues. The most widespread of these myths are debunked below."

Myth #0 *Evidence for modern human influence on climate rests entirely on the "hockey stick" reconstruction of Northern Hemisphere mean temperatures indicating anomalous late 20th-century warmth.*
The response to this putative myth on *realclimate.org* is vague and confused. A better response from the alarmist position would be this: Even if we ignore the *hockey stick* and accept that global temperatures varied significantly in the MWP and the LIA, there is a significant difference today from those periods: CO_2 and CH_4 concentrations are much higher. The major issue regarding modern human influence on climate is whether the climate models are credible that predict significant future temperature growth from increases in CO_2 and CH_4 concentrations.

Myth #1 *The hockey stick reconstruction is based solely on two publications by climate scientist Michael Mann and colleagues (Mann* et al., *1998, 1999).*
The *realclimate.org* response to this "myth" is that "this is patently false". To support their position, they mention: "nearly a dozen model-based and proxy-based reconstructions ... by different groups all suggest that late 20th century warmth is anomalous in a long-term (multi-century to millennial) context". However, the other publications typically utilized PCA with the mean chosen only for the calibration period, leading inevitably to some form of *hockey stick* if some of the proxies had an upward trend in the 20th century. It is not the number of papers that counts here. As Bob Foster emphasized, truth in science is not a matter of voting. The issue here is whether the reconstruction is correct, independently of whether the reconstruction was done in 2, 20, or 200 papers. This putative "myth" is irrelevant; there is no myth. Competent scientists do not doubt the hockey stick because it does not have enough publications to back it up. They doubt it because it has been shown to be based on incorrect math and inadequate data.

Myth #2: *Regional proxy evidence of warm or anomalous (wet or dry) conditions in past centuries contradicts the conclusion that late 20th-century hemispheric mean warmth is anomalous in a long-term (multi-century to millennial) context.*
This "myth" is presumably an allusion to the papers by Soon and Baliunas. The *realclimate.org* makes the point that Soon and Baliunas were rebutted by "a group of more than a dozen leading climate scientists" as though to say "our team is bigger than yours so it must be right". In addition, the "leading climate scientists" are members of the *paleoclimatic cabal.*

The rebuttal claimed that regional anomalies cannot characterize global anomalies. However, global anomalies are merely statistical averages of regional anomalies. Since temperature anomalies vary widely with location and time, a good deal of information is lost by averaging over all data because variations tend to get averaged out. As long as we understand that periods like the MWP and the LIA were not continuous and were not uniformly distributed, examination of regional anomalies one at a time can build up a much better and incisive picture of climate change than a concocted single average global temperature. And, from the regional studies, we learn that, while there were large spatial and temporal variations during the MWP and the LIA, the preponderance of the evidence suggests predominant warmth during the MWP and predominant cold during the LIA. Even in the 20th century, a century of predominant warming, one-third of all land measurement stations reported a decrease in temperature over that period. Climate is determined by a predominance of regional climates, not by unanimity of regional climates.

Myth #3: *The hockey stick studies claim that the 20th century on the whole is the warmest period of the past 1,000 years.*
The rebuttal claimed that "this is a mischaracterization of the actual scientific conclusions". It is claimed that "it is not the average 20th century warmth, but the magnitude of warming during the 20th century, and the level of warmth

observed during the past few decades, which appear to be anomalous in a long-term context". However, this response does not jibe with statements made by alarmists.

Mann, Bradley, and Hughes (1999) said: "... our results suggest that the latter 20th century is anomalous in the context of the last century. The 1990s was the warmest decade and 1998 the warmest year at moderately high levels of confidence."

The IPCC Report said: "The 1990s are likely to have been the warmest decade of the millennium in the Northern Hemisphere and 1998 is likely to have been the warmest year."

Singer and Avery (2007) quote a number of similar alarmist claims from various sources.

- "Nineteen ninety-nine was the most violent year in the modern history of weather. So was 1998. So was 1997. And 1996 ..."
- "A nine-hundred-year-long cooling trend has been suddenly and decisively reversed in the past fifty years. ... Scientists predicted that the Earth will shortly be warmer than it has been in millions of years."
- "A climatological nightmare is upon us. It is almost certainly the most dangerous thing that has ever happened in our history."
- "Climate extremes would trigger meteorological chaos-raging hurricanes such as we have never seen, capable of killing millions of people; uncommonly long, record-breaking heat waves; and profound drought that could drive Africa and the entire Indian subcontinent over the edge into mass starvation."

Myth #4: *Errors in the hockey stick undermine the conclusion that late 20th-century hemispheric warmth is anomalous.*
The *realclimate.org* response to this was: (1) the validity of the hockey stick is affirmed by the large number of researchers who agree with it; (2) the correction (see Mann *et al.*, 2004) was not an admission of the criticisms of M&M, but only admitted to very minor data issues; (3) spurious allegations made by M&M are of no value because (a) M&M are not paleo-climatic specialists, (b) their articles were not published in legitimate science journals, (c) as proof of their lack of veracity, their submitted article was rejected by *Nature*.

To this, I would reply that (1) the affirmations are by members of the *paleoclimatic cabal* who are in league with one another; (2) it is agreed that Mann, Bradley, and Hughes (2004) was not responsive to the criticisms of M&M, more to the discredit of the *paleoclimatic cabal*; (3a) the validity of the arguments by M&M does not depend on their field of endeavor—but rather their knowledge of PCA which appears to be better than that of the *paleoclimatic cabal*; (3b) the validity of the articles does not depend on where they were or were not published (furthermore, this is a self-serving response because the *paleoclimatic cabal* appear to have control over the publication of manuscripts that criticize their methods); (3c) was it *Nature* that rejected the article or was it the reviewers for *Nature* who are members of the *paleoclimatic cabal*?

There are a great many more articles on the *realclimate.org* website. It is not practical to review more of them in this book. Only one more will be cited here.

In their zeal to alert the world to their perceived dangers of global warming, the *paleoclimatic cabal* have promulgated the beliefs that the temperature variations in the MWP and LIA were minor, and it is claimed that even the Holocene Optimum was restricted to summers in the NH. As a result, unfounded claims have been made about the late 20th century being the warmest period in the past 1,000 years, and possibly for the entire Holocene (10,000 years).

The *realclimate.org* blog stated:

> "The [Holocene Optimum] is a somewhat outdated term used to refer to a subinterval of the Holocene period from 5000–7000 years ago during which it was once thought that the Earth was warmer than today. We now know that conditions at this time were probably warmer than today, but only in summer and only in the extra-tropics of the Northern Hemisphere."

NOAA also stated: "In summary, the mid-Holocene, roughly 6,000 years ago, was generally warmer than today, but only in summer and only in the northern hemisphere."

McIntyre rebutted these statements based in part on Stott *et al.* (2004). Vollweiler *et al.* (2006) determined that global temperatures about 7,700 YBP and 3,500 YBP were warmer than today's temperatures. Sundqvist *et al.* (2010) found "a large majority of the investigated temperature reconstructions indicate that temperatures were warmer at the mid-Holocene (6000 YBP $\pm$ 500 yrs) compared to the preindustrial period (1500AD $\pm$ 500 yrs), both in summer, winter and the annual mean. By taking simple arithmetic averages over the available data, the reconstructions indicate that the northern high latitudes were 0.9°C warmer in summer, 0.5°C in winter and 1.7°C warmer in the annual mean temperature at the mid-Holocene (6000 YBP) compared to the recent pre-industrial". Kaufman *et al.* (2004) measured the spatio-temporal pattern of peak Holocene warmth over 140 sites across the Western Hemisphere of the Arctic with "clear evidence for warmer-than-present conditions at 120 of these sites". At the 16 terrestrial sites where quantitative estimates were obtained, local summer temperatures were on average 1.6°C higher than the average of the 20th century. It is noteworthy that about 8,000 YBP, the CO_2 concentration was about 260 ppm, and, since that time, it rose essentially linearly to about 285 ppm in the pre-industrial era (Indermuhle, *et al.*, 1999). There is no correlation at all between temperature and CO_2 concentration during this period.

The blog run by Professor Judith Curry is unique in some ways because it attempts to steer a middle ground and provide a forum for diverse views. Her postings cover a very wide range of topics ranging from detailed technical issues to socio-economic and political aspects. Her postings are mostly intelligent and articulate. Her website provides a forum for discussion of all the major issues relevant to climate. However, like most blogs, it gets bogged down with hundreds of responses, many of which are *non sequiturs* (*http://judithcurry.com/*).

Other relevant climate websites include: *http://bobtisdale.blogspot.com/*; *http://*

pielkeclimatesci.wordpress.com/; *http://wattsupwiththat.com/*; *http://www.climate4you.com/*; and *http://www.drroyspencer.com*; *http://www.SEPP.com*.

At the other end of the scale is the worst climate blog of all: *http://deepclimate.org/*.

This blog is supported by rabid evangelical alarmists who post repeated idiotic messages, often making personal attacks on legitimate people, while hiding behind the cloak of anonymity. The blog is operated by a masked person called "DC" who is a scurrilous low-life.

2.5 CONCLUSIONS ON MILLENNIUM TEMPERATURE HISTORY

Scientists abhor a vacuum. They can't seem to shrug their shoulders and admit that we just don't know the answer to a vexing problem. They demand explanations, however speculative they may be. Thus, we have theories that have gelled into beliefs on how life started on the Earth, how life begins from inanimate matter, how the universe began, how much life exists in the universe, and how the climate of the Earth varied over past millennia.

There is evidence that the Earth has been primarily in a warming trend during much of the 20th century, although there was a definite hiatus in this rise from 1945 to 1978 and the warming has neither been continuous nor universal. The 20th century also saw a steady rise in CO_2 concentrations in the atmosphere, due presumably to the burning of fossil fuels, land clearing, and cement production. Many scientists (and others) have legitimately become concerned that the greenhouse effect due to this CO_2 increase may be responsible for some or most of this observed rise in temperature, and, if left unchecked, could possibly lead to disastrous consequences in the future. In principle, if a sufficiently good global climate model can be produced, the effect of rising CO_2 concentration on Earth temperature in the future can be calculated. Unfortunately, there are so many variables and unknowns in the Earth system that such estimates can only be made as very rough approximations. As in any detective story, if direct evidence is not available, one falls back on circumstantial evidence.

One central issue in this regard is a comparison of the observed temperature rise in the 20th century with estimated variations of temperature in the past millennium or so. If past temperature fluctuations were small compared with the temperature rise in the 20th century, it would suggest that the temperature rise in the 20th century might be unique, unprecedented, and likely to be due to factors unique to the 20th century (e.g., greenhouse gases). On the other hand, if past fluctuations prior to industrialization were as large as, or greater than, those observed in the 20th century, it might suggest that the temperature rise observed recently might (at least partly) just be another fluctuation such as has occurred in the past. While this argument is not ironclad in either direction, it does provide some valuable insights. Accordingly, the quest for better space–time resolution of historical temperatures over the past couple of millennia has become an important part of the effort to understand the causes of global warming.

A number of studies of historical temperatures were conducted in the past, either based on anecdotal records, models of solar variability and climate responses to variable solar intensity, or more likely, based on proxies for past temperature such as tree rings, ice cores, etc. Although numerous papers have pointed out the confounding factors inherent in proxies, "in the land of the blind, a one-eyed man is king."[20] Therefore, despite the problems inherent in the use of proxies, many studies of proxies have abounded in the literature. Some of these fragmentary glimpses of the past have evoked a picture of significant variations in the past climate, with a notable warm period during medieval times, and a relatively cold period called the LIA from about 1400 to about 1850, depending on the criteria used for selection.

The first major global, synoptic, encompassing study of historical global average temperatures from proxies was the "MBH" study reported in 1998 (Mann*et al.*, 1998). This was an audacious effort, encompassing over 1,000 proxies, which provided an unprecedented breadth to the study of historical temperatures. Nevertheless, the number of proxies diminished sharply going back in time, and the global coverage more than 400 years ago was minimal. To aid in processing all these data, a sophisticated statistical data-processing methodology (PCA)[21] was utilized. This was particularly remarkable because it was primarily the product of a Ph.D. dissertation by Michael Mann at the University of Massachusetts. This initial paper was followed by several more that extended the analysis further back into the past. The end result of these studies was a historical temperature profile that had the so-called *hockey stick* shape with a relatively flat profile for 1,000 years or more, followed by a sudden sharp rise in the 20th century. These papers were compact, full of jargon, and difficult to follow. Sufficient data for others to make independent checks were typically difficult to obtain. Nevertheless, they were impressive papers and outwardly seemed to be well done. As a result of this work, Michael Mann was rapidly catapulted from a newly graduated Ph.D. to a position of fame and renown and almost instantly became recognized as a world leader in paleoclimatology.

A number of climate scientists (and others), hungry for evidence of human-induced global warming, seized on the MBH results as a landmark. The *hockey stick* figure was reproduced and disseminated widely, being offered up as strong evidence of CO_2-induced global warming in the 20th century. The *hockey stick* was adopted by the Intergovernmental Panel on Climate Change of the U.N. (IPCC), Al Gore, and, in general, a majority of the paleoclimatology science community. The claim was made that the warming in the 20th century was unprecedented, that the 1990s was the hottest decade on record, and 1998 was the hottest year in at least the past millennium or two and maybe millions of years. Montford (2010) said:

> "Every home in Canada was sent a leaflet quoting the [IPCC hockey stick] and warning of the dangers of climate change. School books told children that the hockey stick meant that the world had to change. Politicians told voters that only they could save people from the threat. ... Insurers, newspapers,

[20] In modern terms, we may say: "It is the only game in town."

[21] Also known as "empirical orthogonal functions" (EOFs).

magazines, pamphlets, and websites were all in thrall to its message; the hockey stick swept all before it."

About five years later, M&M rained on the *hockey stick* picnic. M&M were experts in manipulating large noisy data sets, which is just the problem faced in reconstructing the Earth's climate from proxies. In a series of papers and informal reports, they clearly showed that:

(1) MBH made an innocent-looking mistake in the PCA by standardizing with a mean based only on the calibration period, instead of a mean based on the entire time period covered by the data. As it turns out, this unwittingly led to a chain of events that placed undue emphasis on a few highly suspect proxies that produced the *hockey stick* result while ignoring most of the proxy data in the study.
(2) Use of certain tree-ring data by MBH was unjustified because much of the observed growth in the 20th century was due to CO_2 fertilization and other factors, rather than a rise in temperature.
(3) When a proper recalculation of the MBH data is performed, the result shows that, although there was indeed a significant temperature increase in the 20th century, there were comparable high temperatures earlier in the past millennia.

Thus, the bases for the claim that the 20th century exhibited an unprecedented temperature rise, and that the 1990s and 1998 were the hottest in the past several thousand years, were undermined. In addition to this, there is another factor not usually discussed by the critics. The correlation of the proxies with measured temperature during the calibration period is usually poor, and the extrapolation backward in time for periods much longer than the calibration period is an unsupportable matter of faith. The use of all proxies democratically mixes in many poor ones with the few good ones, and produces mainly noise.

The responses to the findings of M&M are interesting. Instead of issuing a *mea culpa* and going on from there, Mann dug in his heels and protected turf from truth. He issued a response to the M&M charges that is a masterpiece of evasion and obfuscation, not even mentioning M&M or in any way dealing with their central issue. Most of the paleoclimatology community, which by and large adopted the *hockey stick* as its motif, cooperated by controlling which papers get published in the journals. In general, the U.N., Al Gore, and the climatological alarmists have simply ignored M&M and continued to vouchsafe the *hockey stick*, pretending that the criticism of M&M did not exist.[22] For those who are determined to raise the alarm to the world on the dangers of global warming, the *hockey stick* is too valuable as a public message to allow truth to interfere. In addition, if one standardizes against the calibration period, noisy data produce a *hockey stick* result if the data are rising during the calibration period. This raises questions about the reliability and utility of assembling proxy data into a global average temperature. Since the temperature was rising during the 20th century, use of 20th-century data for calibration or proxies can lead to misleading results. Use of PCA in the manner employed by MBH can exacerbate this problem.

[22] Think of *The Emperor's New Clothes.*

Anon. (M) presented a very detailed and generally objective review of surface temperature reconstructions for the past two millennia. Their conclusions are summarized below:

- "The instrumentally measured warming of about 0.6°C during the 20th century is also reflected in various proxy measurements."
- "Large-scale surface temperature reconstructions yield a generally consistent picture of temperature trends during the preceding millennium, including relatively warm conditions centered near 1000 (identified by some as the *Medieval Warm Period*) and a relatively cold period (or *Little Ice Age*) centered near 1700. The existence of a *Little Ice Age* from roughly 1500 to 1850 is supported by a wide variety of evidence including ice cores, tree rings, borehole temperatures, glacier length records, and historical documents. Evidence for regional warmth during medieval times can be found in a diverse but more limited set of records including ice cores, tree rings, marine sediments, and historical sources from Europe and Asia, but the exact timing and duration of warm periods may have varied from region to region, and the magnitude and geographic extent of the warmth are uncertain."
- "It can be said with a high level of confidence that global mean surface temperature was higher during the last few decades of the 20th century than during any comparable period during the preceding four centuries. This statement is justified by the consistency of the evidence from a wide variety of geographically diverse proxies."[23]
- "Less confidence can be placed in large-scale surface temperature reconstructions for the period from 900 to 1600. Presently available proxy evidence indicates that temperatures at many—but not all—individual locations were higher during the past 25 years than during any period of comparable length since 900. The uncertainties associated with reconstructing hemispheric mean or global mean temperatures from these data increase substantially backward in time through this period and are not yet fully quantified."[24]
- "Very little confidence can be assigned to statements concerning the hemispheric mean or global mean surface temperature prior to about 900 because of sparse data coverage and because the uncertainties associated with proxy data and the methods used to analyze and combine them are larger than during more recent time periods."[25]

[23] This statement reflects a generally prevailing implicit view that "the preceding four centuries" were normal, while the relatively higher temperatures at the end of the 20th century are comparatively abnormal. However, the preceding four centuries extend across the LIA, and therefore one might state the proposition differently: temperatures during the preceding four centuries were colder than they were at the end of the 20th century.

[24] This author cannot find any substantial evidence that temperatures were (as claimed) generally higher in the past 25 years than they were in 900.

[25] This author has very little confidence in estimates of temperature prior to 1600, let alone prior to 900.

3

Temperatures in the past century

3.1 NEAR-SURFACE LAND MEASUREMENTS

Since the Earth is ~70% covered by ocean, surface temperature measurements naturally are divided between land and ocean measurements.

3.1.1 Introduction

Surface temperatures on land are available dating back about 100 years at many sites, and as far back as ~200 years for a limited number of land sites. Meteorological scientists have methodically examined these data and attempted to derive the best overall data sets that the data permit. Studies of temperature change over land areas are routinely made by several groups based on measurements of the meteorological station networks. It is beneficial to estimate global temperature change from both the meteorological station data alone, and a combined analysis with ocean data, because the land and ocean data have their own measurement characteristics and uncertainties.

Hansen *et al.* (1999, 2001) limited their studies primarily to the period since 1880, because of the poor spatial coverage of stations prior to that time and the reduced possibility of checking records against those of nearby neighbors. Analyses for the earlier years need to be carried out on a station-by-station basis with an attempt to discern the method and reliability of measurements at each station. Data collected and recorded by thousands of individuals with equipment and procedures subject to change over time inevitably contain many errors and inconsistencies, some of which will be impossible to identify and correct. However, Hansen *et al.* (1999, 2001) concluded that the influence of errors was not dominant, because many of the errors in recording temperature were believed to be random in nature, and would therefore supposedly cancel out when a large sample is taken. Nevertheless, Hansen *et al.* (1999, 2001) examined the data quality to try to minimize local errors and obtain an indication of the nature and magnitude of any artificial sources of temperature change. The various records at any location were combined into a single record for that location. Regional and global temperatures were estimated by

D. Rapp, *Assessing Climate Change*, Springer Praxis Books,
DOI: 10.1007/978-3-319-00455-6_3

combining local temperature records. Homogeneity adjustments were made to local time series of temperature with the aim of removing non-climatic variations in the temperature record. The non-climatic factors include changes of the environment of the station, the instrument, or its location, observing practices, and the method used to calculate the mean temperature. The Climate Research Unit (CRU) at the University of East Anglia in the U.K. also maintained a detailed database of measured temperatures around the world[1] (other relevant references include Pielke *et al.*, 2007c; CCSP, 2005; and Brohan *et al.*, 2006).

The U.S. Historical Climatology Network (USHCN)[2] and the National Oceanic and Atmospheric Administration (NOAA) Climate Reference Network (USCRN)[3] are widely used temperature measurement networks (TMNs).

Highly touted and heralded in advance, the long-promised "Berkeley Earth Surface Temperature Analysis" (BEST) finally appeared (in preliminary PR form) in October 2011. Five relevant papers appeared on the Internet: Brillinger *et al.* (2011), Wickham *et al.* (2011), Muller *et al.* (2011a), Muller *et al.* (2011b), and Rohde *et al.* (2011). The stated objectives of the program were to:

(1) merge existing surface station temperature data sets into a new comprehensive raw data set with a common format that could be used for weather and climate research;
(2) review existing temperature processing algorithms for averaging, homogenization, and error analysis to understand both their advantages and their limitations;
(3) develop new approaches and alternative statistical methods that may be able to effectively remove some of the limitations present in existing algorithms;
(4) create and publish a new global surface temperature record and associated uncertainty analysis;
(5) provide an open platform for further analysis by publishing our complete data and software code as well as tools to aid both professional and amateur exploration of the data.

It should be noted at the outset that the work reported in October 2011 pertains only to land temperatures, and, since 70% of the globe is covered by water, this represents a minority of the Earth. As Judith Curry (a co-author) said:

> "In concluding, I will remind everyone that the REAL problem with the surface temperature data set lies with the ocean data. I hope that the Berkeley group will be able to extend their efforts to include ocean data." (*http://judithcurry.com/2011/10/20/berkeley-surface-temperatures-released/#more-5425*)

Following the lead of the Inter-governmental Panel on Climate Change (IPCC), the authors arbitrarily chose to emphasize the temperature change from 1956 to 2005. It is noteworthy that Earth temperatures were higher during the period from

[1] See *www.cru.uea.ac.uk/*.

[2] *cdiac.ornl.gov/epubs/ndp/ushcn/ushcn.html.*

[3] *www.ncdc.noaa.gov NESDIS NCDC.*

1940 to 1956 than they were around 1956, so choosing a baseline at 1956 served two purposes: (1) it maximized the reported rate of temperature rise of the mid-to-late 20th century, and (2) it made the temperature rise after 1956 appear somewhat more linear than it would have appeared had a baseline been chosen around 1940. It seems likely that the aim was to exaggerate the magnitude and consistency of the temperature rise of the mid to late 20th century. But this was unnecessary. By any yardstick, the Earth's temperature did go up substantially in the latter half of the 20th century. By choosing a yardstick that maximizes the apparent decadal rise, the authors lose some credibility for their supposed neutrality, thereby making the skeptical reader more alert to possible biases in the remainder of the work.

In late July 2012, the BEST Team released an update to their previous findings (Rohde, *et al.*, 2012).

3.1.2 Quality and reliability of land temperature measurement networks

There are a number of factors that must be taken into account in judging the reliability of global or hemispheric temperature averages derived from a network of ground measurement stations. (e.g., Pielke *et al.*, 2007c, c; Hoyt, 2006; Ball, 2007; Davey and Pielke, 2005). The issues of concern include the following:

(1) Are there sufficient data spatially and temporally distributed to provide a basis for a global average or hemispheric average temperature over any extended time period?
(2) How many stations are poorly sited to make good measurements?
(3) How many sites have changed over the years, with changes in the surroundings, changes in the instrumentation or recording plan, or actual movement of the site?
(4) How many sites were affected by the so-called "urban heat island" (UHI) effect due to heat stored in urban construction?
(5) If there is a preponderance of measurement sites at middle and upper-middle Northern Hemisphere (NH) latitudes, is proper account taken for the phenomenon of "arctic amplification"?
(6) Have the various teams that have developed estimates of historical Earth as far back as the 19th and even 18th centuries properly corrected for the above effects?

Pielke *et al.* (2007a) concluded: "The use of temperature data from poorly sited stations can lead to a false sense of confidence in the robustness of multi-decadal surface air temperature trend assessments." They suggested that there are problems in the existing temperature databases due to: (1) time-of-observation bias because, at many sites, the observing time has changed during the station's history, (2) changes in instrumentation at stations, (3) station moves or relocations, and (4) bias caused by station urbanization. In addition, three primary issues were identified relevant to

> "... land use/land clearing (LULC) and changes in LULC related to placement of climate stations. First, a station may be initially placed in what might be considered a poor LULC environment (e.g., near a highway or other

man-made environment that could influence the observed temperature based on day of week, holiday, etc.). Second, a station may have been initially located at what might be considered a good LULC environment only to have that environment change over time. And third, possibly due to one of the above situations, a station may be moved from one LULC environment to another."

Pielke *et al.* (2007c) discussed unresolved issues in using surface temperature trends as a metric for assessing global and regional climate change. The issues include warm bias in night-time minimum temperatures, poor siting of the instrumentation, effect of winds, effects of surface atmospheric water vapor content on temperature trends, uncertainties in the homogenization of surface temperature data, and influence of LULC change on surface temperature trends.

The observed average surface temperature at any site over land is computed by averaging observed daily maximum and minimum temperatures. While the daily maximum may be accurate, Pielke *et al.* (2007c) claimed that the nightly minimum temperature (typically about 1.5 m to 2 m above the land) is subject to variation, depending on surface characteristics (heat capacity) and wind speed. They pointed out:

> "As the boundary layer cools at night under light winds, the greatest decrease in temperature occurs near the surface. Unlike the daytime boundary layer where convective turbulence tends to reduce vertical gradients, in the nocturnal boundary layer the cooling suppresses turbulence and enhances vertical gradients. Thus, the vertical variation in temperature in light winds can be huge with temperature changes of 6°C or more often occurring within 25 vertical meters of the surface. This is why great care must be taken to avoid contaminating the climate record with measurements from sites that have changed even a meter or two in their height of observation."

Davey and Pielke (2005) conducted a study of temperature measurement stations in Eastern Colorado. They said:

> "It is important to know the site of stations relative to various structures and surfaces. Generally, near-surface air temperature observations should be representative of the free-air conditions over as much of the vicinity as possible, at a height approximately 1.5 m above the ground. The site should be level, without locally significant topographical variations or steep slopes or hollows, and should offer free exposure to both sunshine and wind (not too close to trees, buildings, or other obstructions). It thus becomes critical to conclusively determine how much of any potential regional change in observed air temperatures might be due to land-use changes at the site itself. These changes may include local-scale urban development around the site, changes in local vegetation characteristics, etc."

Local-scale exposure characteristics are therefore important in evaluating station data. Prior to the 1980s, a site sketch was available, illustrating the location of the weather station instrumentation and any nearby obstructions. Currently, however, only vague documentation regarding site exposure characteristics is typically available. This is particularly true for sites' terrain and surface features. Davey and Pielke (2005) visited 57 sites in Eastern Colorado, with emphasis on the 10 sites in the USHCN. Typical findings were (1) sensors close to buildings, (2) sensors over patchworks of different land coverings, (3) vegetation around sensor locations, or (4) urbanized sensor locations. Evidently, the TMN needs refurbishment.

According to McLean (2007ba):

> "In the early days thermometers could only show the temperature at the moment of reading and so the data recorded from that time was for just one reading each day. Later the thermometers were able to record the minimum and maximum temperatures, and so the daily readings were those extremes in the 24 hour period. Only in the last 20 or 30 years have instruments been available that record the temperature at regular intervals throughout the 24 hours, thus allowing a true time-based daily average to be calculated. The so-called 'average' temperatures both published and frequently plotted through time are initially based on only a single daily value, then later on the mathematical average of the minimum and maximum temperatures. Although time-based averages are now available for some regions, they are not generally used because the better instrumentation is not uniformly installed throughout the world and the historical data is at best a mathematical average of two values. The problem is that these averages are easily distorted by brief periods of high or low temperatures relative to the rest of the day, such as a brief period with less cloud cover or a short period of cold wind or rain."

In a 1999 report entitled *Adequacy of Climate Observing Systems*, the National Research Council said:

> "Climate researchers have used existing, operational networks because they have been the best, and sometimes only, source of data available. They have succeeded in establishing basic trends of several aspects of climate on regional and global scales. Deficiencies in the accuracy, quality, and continuity of the records, however, still place serious limitations on the confidence that can be placed in the research results."

Figure 3.1 is a plot of the mean temperature stations around the world. The upper plot shows all stations and the lower plot shows stations with 100 years (or more) of data.

Meteorological sites in the U.S. and Western Europe play a dominant role. In most of the world, we don't have adequate station coverage for temperature and even less for precipitation. Seventy percent of the world is ocean, for which the database is uncertain. The Arctic and the Antarctic have very skimpy data. African and South American stations are sparse. Figure 3.2 shows the same information for stations that measure maximum and minimum temperature. Figure 3.3 shows the longevity of existing stations, the rise in the number of stations in the 20th century, and the decrease in the number of stations after 1970.

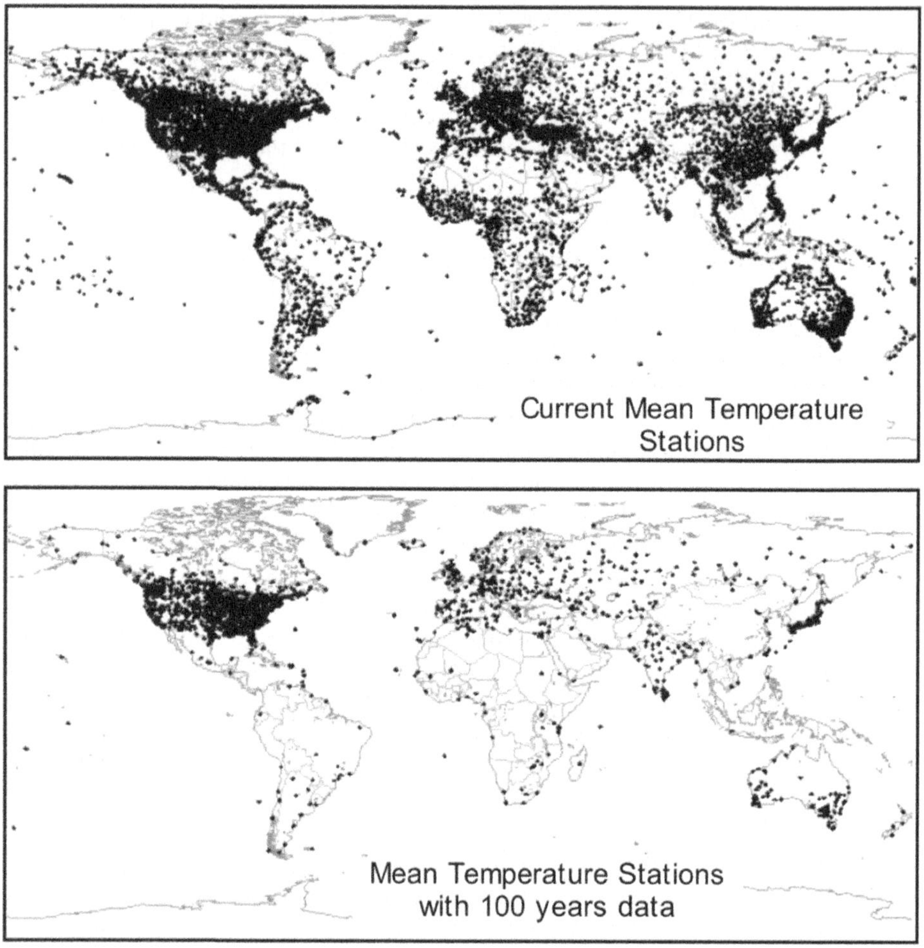

Figure 3.1. Locations of Global Historical Climatology Network (GHCN) mean temperature station locations. Upper: All stations. Lower: Stations with at least 100 years of data (adapted from Peterson and Vose, 1997).

Kevin Trenberth, a leading climatologist, was quoted on the Internet as saying:

> "It's very clear we do not have a climate observing system. ... This may come as a shock to many people who assume that we do know adequately what's going on with the climate but we don't."

McKitrick and Michaels (2007) summarized the situation:

> "The number of reliable monitoring sites around the world has fallen dramatically since the mid-1970s. The Global Historical Climatology Network reached a peak of 6,000 unique contributing sites in the late 1960s, but the number fell to fewer than 3,000 as of the late 1990s, with the most dramatic drop in the early 1990s when the number of stations fell by nearly half in four years."

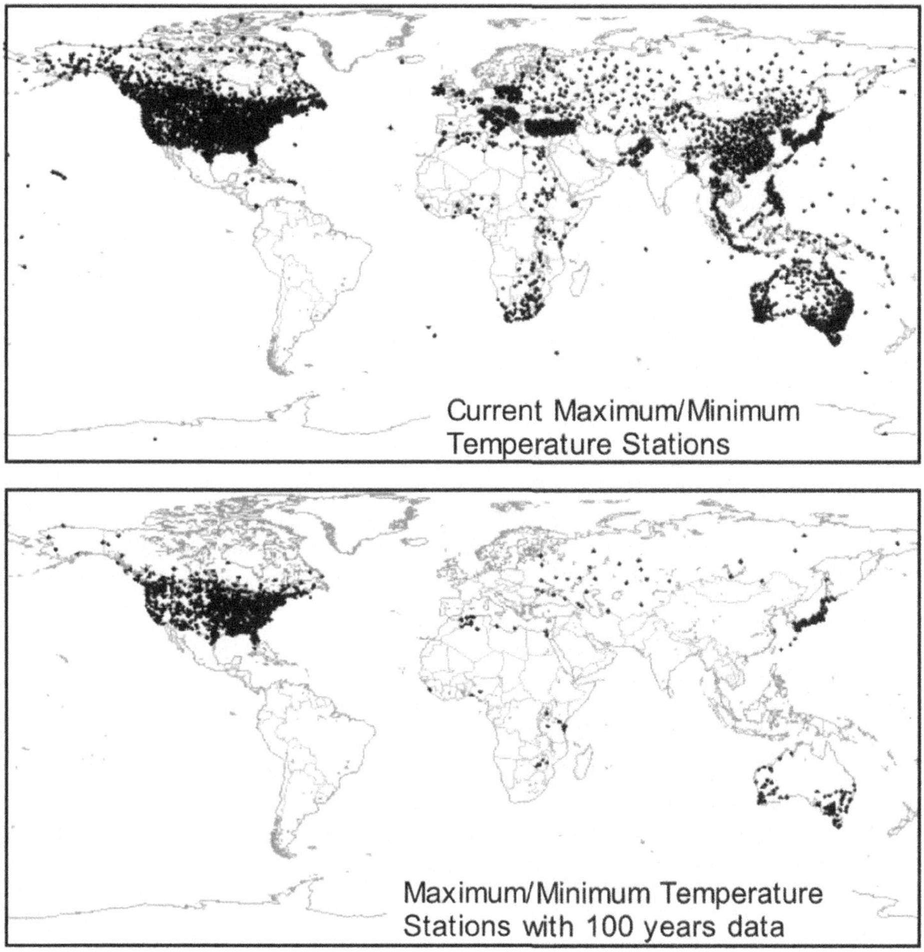

Figure 3.2. Locations of GHCN maximum/minimum temperature station locations. Upper: All stations. Lower: Stations with at least 100 years of data (adapted from Peterson and Vose, 1997).

Fall *et al.* (2011) surveyed 82.5% of the U.S. Historical Climatology Network stations and provided a classification based on exposure conditions of each surveyed station, using a rating system employed by the NOAA to develop the U.S. Climate Reference Network. This study:

> "... examined temperature differences among different levels of siting quality without controlling for other factors such as instrument type. Temperature trend estimates vary according to site classification, with poor siting leading to an overestimate of minimum temperature trends and an underestimate of maximum temperature trends, resulting in particular in a substantial difference in estimates of the diurnal temperature range trends. The opposite-signed

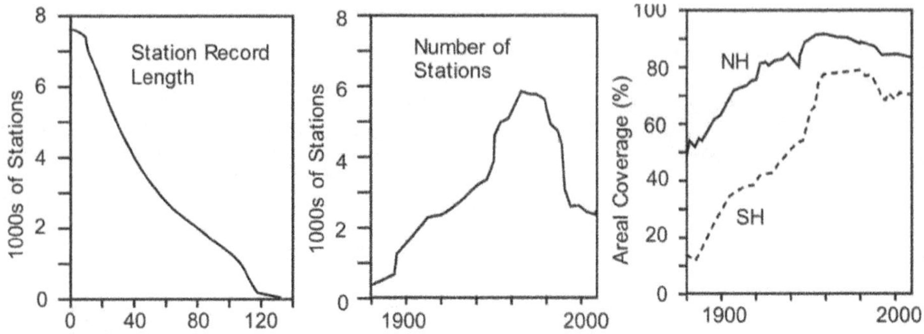

Figure 3.3. World temperature measurement stations. Left: Number of stations vs. record length. Middle: Number of stations vs. year. Right: Areal coverage by hemisphere (adapted from Ball, 2007).

> differences of maximum and minimum temperature trends are similar in magnitude, so that the overall mean temperature trends are nearly identical across site classifications. ... the most poorly-sited stations are warmer ... than are other stations, and a major portion of this bias is associated with the siting classification rather than the geographical distribution of stations. According to the best-sited stations, the diurnal temperature range in the lower [48] states has no century-scale trend."
>
> The classification scheme and percent of sites that fit each class are given below:

- CRN 1: (1.2%) A clear flat surface with sensors located at least 100 m from artificial heating and vegetation ground cover <10 cm high (error $<1^{\circ}$C).
- CRN 2: (6.7%) Same as CRN 1 with surrounding vegetation within 25 cm and artificial heating sources within 30 m (error $<1^{\circ}$C).
- CRN 3: (21.5%) same as CRN 2, except no artificial heating sources within 10 m (error $\geqslant 1^{\circ}$C).
- CRN 4: (64.4%) artificial heating sources within 10 m (error $\geqslant 2^{\circ}$C).
- CRN 5: (6.2%) sensor located next to/above an artificial heating source (error $\geqslant 5^{\circ}$C).

Based on the classification scheme of Fall *et al.* (2011), Muller *et al.* (2011b) concluded that the state of the U.S. station network is very poor, with 70% of the stations having projected errors $>2^{\circ}$C and 90% having projected errors $>1^{\circ}$C. Muller *et al.* (2011) presented a map of the sites, color-coded by quality (see Figure 3.4).

But the point that was made by Bell (2007) (brushing aside issues regarding whether it makes sense to even talk about a single global temperature or how one deals with the oceans), is how can we know what the global temperature rise was "since the late 19th century" when (1) there were hardly any stations in the late 19th century, (2) station sparseness was meager until about 1950 and still inadequate after that date, (3) the number of stations with record lengths of 100 years is minimal, (4) areal coverage was low until about 1950? The short answer is: we can't know.

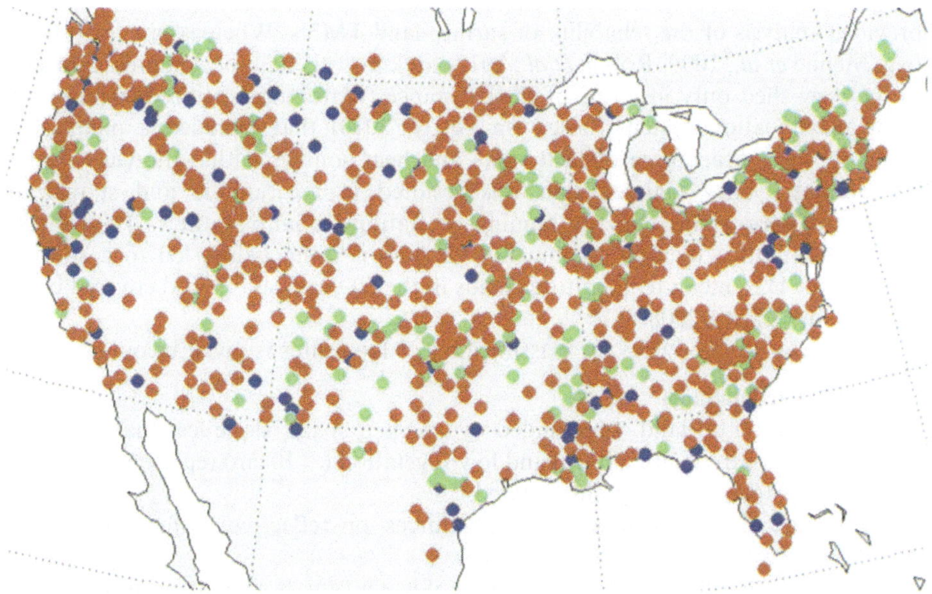

Figure 3.4. Distribution of temperature stations in the U.S. Stations are ranked 1 and 2 (blue)(errors < 1°C), ranked 3 (green) (errors > 1°C), and ranked 4 and 5 (red) (errors > 2°C or 5°C).

The decrease in the number of stations in the late 20th century is alarming. If this is due to "weeding-out" poorly equipped and situated stations, that might improve the quality of the residual database but it exacerbates the sparseness. Furthermore, it raises questions regarding the quality of the thousands of stations that were phased out late in the 20th century.

A more upbeat view of the network for temperature measurement on the Earth was given by Brohan *et al.* (2006). A new upgrade to the TMN was described. The database was derived from a collection of homogenized, quality-controlled, monthly-average temperatures for 4,350 stations. "Station normals" (monthly averages over the normal period 1961–1990) were generated from station data for this period where possible. However, "normals" were derived from station data "wherever possible" and various approximations were used in many cases. Using these "normals", anomalies were calculated as deviations from the means of the "normals". The discrete set of station anomaly data were then converted to a temperature distribution across a global grid by interpolation, typically at 5° × 5° resolution.

The entire situation regarding the reliability of the ground temperature network was temporarily jolted by the appearance of a "pre-print draft discussion paper" at the end of July 2012 by Watts *et al.*[4] This paper provides many references relevant to

[4] Watts, A., Jones, E., McIntyre, S., and Christy, J.R. (2012) "An area and distance weighted analysis of the impacts of station exposure on the U.S. Historical Climatology Network temperatures and temperature trends", *http://wattsupwiththat.com/2012/07/29/press-release-2/*

previous analysis of the reliability of surface land TMNs. Whereas previous studies (e.g. Menne *et al.*, 2009, Rohde *et al.*, 2012) took into account heat sources and sinks in the view-shed only in terms of their distance, Watts *et al.* utilized an improved siting classification system (based on Leroy, 2010) that included a method for including the surface area of heat sinks and heat sources within the view-shed of thermometer. They claimed that this resulted in a dramatic and statistically significant improvement in the binning of stations' quality ratings. They further claimed that these factors, combined with station siting issues, led to a spurious doubling of U.S. mean temperature trends in the 30-year data period covered by the study from 1979 to 2008.

The new rating system for classification of local sites is given below:

Class 1 (compliant)
Flat, horizontal land, surrounded by an open space, slope less than 1/3 (19°); Ground covered with natural and low vegetation (< 10 cm) representative of the region; Measurement point situated:

- more than 100 m from heat sources or reflective surfaces (buildings, concrete surfaces, car parks, etc.);
- more than 100 m from an expanse of water (unless significant of the region);
- away from all projected shade when the Sun is higher than 5°.

Class 2 (compliant)
Flat, horizontal land, surrounded by an open space, slope inclination less than 1/3 (19°); Ground covered with natural and low vegetation (< 10 cm) representative of the region; Measurement point situated:

- more than 30 m from artificial heat sources or reflective surfaces (buildings, concrete surfaces, car parks, etc.);
- more than 30 m from an expanse of water (unless significant of the region);
- away from all projected shade when the Sun is higher than 7°.

Class 3 (non-compliant, additional estimated uncertainty added by siting up to 1°C)
Ground covered with natural and low vegetation (< 25 cm) representative of the region; Measurement point situated:

- more than 10 m from artificial heat sources and reflective surfaces (buildings, concrete surfaces, car parks, etc.);
- more than 10 m from an expanse of water (unless significant of the region);
- qway from all projected shade when the Sun is higher than 7°.

Class 4 (non-compliant, additional estimated uncertainty added by siting up to 2°C)
Close, artificial heat sources and reflective surfaces (buildings, concrete surfaces, car parks, etc.) or expanse of water (unless significant of the region), occupying:

- less than 50% of the surface within a circular area of 10 m around the screen;
- less than 30% of the surface within a circular area of 3 m around the screen;
- away from all projected shade when the Sun is higher than 20°.

Class 5 (non-compliant, additional estimated uncertainty added by siting up to 5°C)
Sites do not meet the requirements of Class 4—that is, more than 50% of the surface within a circular area of 10 m around the screen. More than 30% of the surface within a circular area of 3 m around the screen.

The number of sites across the U.S. in each class was as follows:

Class	Number	%
Class 1	48	6
Class 2	112	14
Class 3	247	32
Class 4	277	36
Class 5	95	12
All	779	100

The results of Watts *et al.* for the time period 1979–2008 indicated that there was a universal tendency for Class 3/4/5 sites to report higher temperatures than nearby Class 1/2 sites with 3/4/5 sites showing a temperature gain from 1979 to 2008 that was about 60% higher than that seen by Class 1/2 sites. They concluded: "Taken *in toto*, these factors identified in this study have led to a spurious doubling of U.S. T_{mean} trends from 1979–2008." However, in a more recent posting,[5] it was noted that some problems in this analysis were discovered and it was currently being reworked. Since a reworked version of their paper never appeared, it seems likely that their conclusions may not be correct. In a subsequent posting, it was pointed out that the BEST Team compared the results of the sum of Classes 1, 2, and 3 with the sum of Classes 4 and 5 and found very little difference. While Watts argued on his website that BEST should have compared 1 + 2 with 3 + 4 + 5, this seems to be a trivial point. Nevertheless, it is difficult to believe that Class 4 and 5 sites would yield results as close to those of class 1, 2, and 3 sites as was reported by BEST.

Aside from all the other problems involved in monitoring the Earth's near-surface temperature (as discussed, for example, by Watts *et al.*), there is a question of what height to place the sensor above the ground. The current standard for monitoring stations is 1.5 m above the surface. This is supposed to be representative of a boundary layer of air. Presumably, air currents mix the boundary layer during the day, but, at night, it is relatively stable and striated. Most monitoring stations measure maximum and minimum temperatures diurnally. Presumably, these occur during the day and night, respectively. It has been observed that, over the past century or so, minimum temperatures have warmed nearly three times more than maximum temperatures and there is no satisfactory explanation for this effect. Climate models are not consistent with this observation.

[5] *http://wattsupwiththat.com/*.

McNider *et al.* (2012) developed an explanation based their model that indicated that slight increases in incoming long-wave radiation from greenhouse gases at night (as greenhouse gas concentrations rise) might destabilize the boundary layer of air near the surface. This will introduce some turbulence that will cause mixing of air from above with the boundary layer. At night, the surface cools by radiation to space, which in turn cools the boundary layer to temperatures below that of the air above it. If the air above the boundary layer mixes with the boundary layer at night, it will warm the boundary layer. In their model, they subjected a nocturnal boundary layer to an added increment of downward radiation (4.8 W/m^2) and compared the boundary layer with and without this forcing. They found that the 1.5 m height air temperature warmed substantially due to destabilization caused by additional downward IR irradiance. Most of the warming at 1.5 m height was due to the warm air mixed from aloft.

McNider *et al.* (2012) concluded:

> "Thus, it may be better for current climate models, when they test replication of past climates and to project future global warming, to only use maximum temperatures rather than the current metric of using the mean daily temperature, which contains the [amplified] minimum temperature."

In a very extensive study, Montandon *et al.* (2011) examined the locations of stations in the Global Historical Climate Network version 2 (GHCNv.2) surface temperature data set that is widely used for reconstructions of global average surface temperature anomalies. They analyzed the spatial distribution and LULC representation of GHCNv.2 stations using nightlight imagery, two LULC data sets, and a population and cropland historical reconstruction. They estimated the present and historical worldwide occurrence of LULC types and the number of GHCNv.2 stations within each. LULC has an important impact on surface temperatures. As Montandon *et al.* (2011) pointed out:

> "... LULC changes can cause large local and regional temperature changes. This effect on temperature can be the result of changes in surface roughness, vegetation amount and type, and the alteration of surface heat and moisture fluxes. LULC differ over the world and have been unequally modified so that their distribution and the extent to which they have been altered [by human activity] vary."

However, LULC data are limited, and furthermore the classification of land types into a limited number of subsets is only approximate.

The results of the study by Montandon *et al.* (2011) indicated that GHCNv.2 station locations are biased toward urban and cropland with underrepresentation of open shrub land, bare, snow/ice, and evergreen broadleaf forests, as well as non-urban areas that have remained uncultivated in the past century. They found that the temperature trends over the past century were smaller for the overrepresented LULC and larger for the underrepresented LULC. Thus they concluded: "This opens the possibility that the temperature increases of Earth's land surface in the last century would be higher than what the GHCNv.2-based analyses report." However, some of their results seem strange, such as the finding that the temperature trend for urban areas was lower than for broadleaf forests. Perhaps this is because they associated urban areas with high cropland levels.

Data on diurnal temperature range (DTR) provide some insights into the nature of measured temperature rise in the 20th century. Vose, Easterling, and Gleason (2005) reported data on maximum temperatures, minimum temperatures, and DTR over the equivalent of 71% of the total global land area. Over the period from 1950 to 2004, the average global minimum temperature increased more rapidly than the average global maximum temperature (+0.20°C/decade vs. +0.14°C/decade) , resulting in a significant DTR decrease (–0.066°C/decade). However, over the most recent period from 1979 to 2004, the trends in global maximum and minimum temperatures were nearly equal at +0.29°C/decade resulting in almost no change in the DTR. The changes in temperature and DTR were considerably greater in the NH than in the Southern Hemisphere (SH).

Leathers *et al.* (1998) analyzed variations in the DTR across the U.S. They found that the annual cycle of the DTR across the U.S. varies both spatially and temporally and is associated in one manner or another with moisture content of the atmosphere. In many areas, the DTR undergoes large abrupt changes in magnitude at specific times within the annual cycle.

Urbanization tends to reduce DTR because heat is emitted during the night by urban structures and roads. Degaetano and Allen (2002) found significant effects of urbanization, particularly in regard to unusually warm nights.

Stone and Weaver (2003) pointed out that variations in cloud cover are strongly correlated with those in the DTR. The higher albedo of clouds decreases the downward solar radiation during the day, and thereby reduces T_{max}. They also indicated that clouds produce more downward long-wave radiation, so increasing night-time cloud cover would increase T_{min} and thereby decrease the DTR. Soil moisture is also expected to influence the DTR, through control of evaporative cooling, the ground albedo, and the ground heat capacity. In addition, aerosols in the atmosphere will also affect the DTR. Stone and Weaver (2003) employed a global climate model (GCM) to investigate the influence of anthropogenic-induced forcing on the magnitude of the DTR. However, the results do seem to be clear-cut.

Braganza, Karoly, and Arblaster (2004) used comprehensive European and global temperature databases to examine the DTR since 1950. They found that the

> "... observed DTR over land showed a large negative trend of 0.4°C over the last 50 years that is very unlikely to have occurred due to internal variability. This trend is due to larger increases in minimum temperatures (0.9°C) than maximum temperatures (0.6°C) over the same period."

Like Stone and Weaver (2003), they found that GCMs do not predict changes in DTR very well, probably "due to poor representation of cloud changes over land".

While it is widely accepted that the DTR decreased during the 20th century, one report indicates that the decreasing trend reversed in Europe during the 1970s and 1980s (Makowski *et al.*, 2008).

It is perhaps a coincidence that most recent "corrections" carried out by global-warming alarmists to past climate data lead to an increase in global warming in the late 20th century compared with a century earlier.

3.1.3 Urban Heat Islands (UHIs)

The heat generated by human activities has traditionally been small compared with the massive heat flows involved in the Earth's climate. However, with the expansion of world population and higher energy usage in the 20th century, human activity is no longer negligible. Hoyt (2006) indicated that the total primary energy consumption in the U.S. in 2002 was 97.7 quadrillion Btu. This is equivalent to 97.7×10^{15} Btu = 1.03×10^{20} joules (J).

On a yearly average, this amounts to a continuous power generation of 3.27×10^{12} watts. The area of the U.S. is 9,640,000 km^2. If we divide the total power by the total area, we obtain about 0.34 W/m^2. However, most of the power generated is localized into urban areas. The power density is probably considerably higher than 1 W/m^2 in urban regions, and much higher than that in cities. These power densities are of the same general magnitude as those indicated by GCMs for greenhouse gas forcing.

The effect of urbanization on temperature measurements is significant. Figure 3.5 shows a photo taken from space of the U.S. at night showing the concentration of light in urban regions.

Parker (2010) provided the following description of UHIs:

> "The urban heat island is the elevation of air temperature within cities, and to a smaller extent within towns and villages, relative to the surrounding countryside. It is caused mainly by the retention of solar heat in the fabric of buildings and ground surfaces, and the obstruction and re-absorption of nighttime outgoing long-wave radiation by buildings that obstruct the sky view. Paved ground surfaces transport more solar heat downwards than soil; this heat is then available for release overnight. Reduced ventilation can hinder the dispersal of urban heat islands. An important contribution to some urban heat islands is the emission of heat by human activities such as transport, and by heating and air conditioning of buildings. Accelerated run-off of rainwater, along with reduced vegetation cover can reduce moisture availability in cities. This can enhance the urban heat island by reducing the fraction of solar energy converted into latent heat and increasing the fraction becoming sensible heat. Conversely, dry-climate cities with irrigated vegetation can have an urban cool island. Local enhancement of atmospheric aerosol concentrations may also affect urban temperatures. Generally, the urban heat island is strongest at night in high-rise city centers. ... During daytime, some of the solar heat absorbed by the urban structures is transferred to the overlying air; turbulent convection then mixes the warm air with cooler air aloft, so that the urban heat island is weak and may even be completely annulled locally by shading of the ground by tall buildings. Because both solar and outgoing long-wave radiation are restricted by cloud, the urban heat island is weakened in cloudy weather.

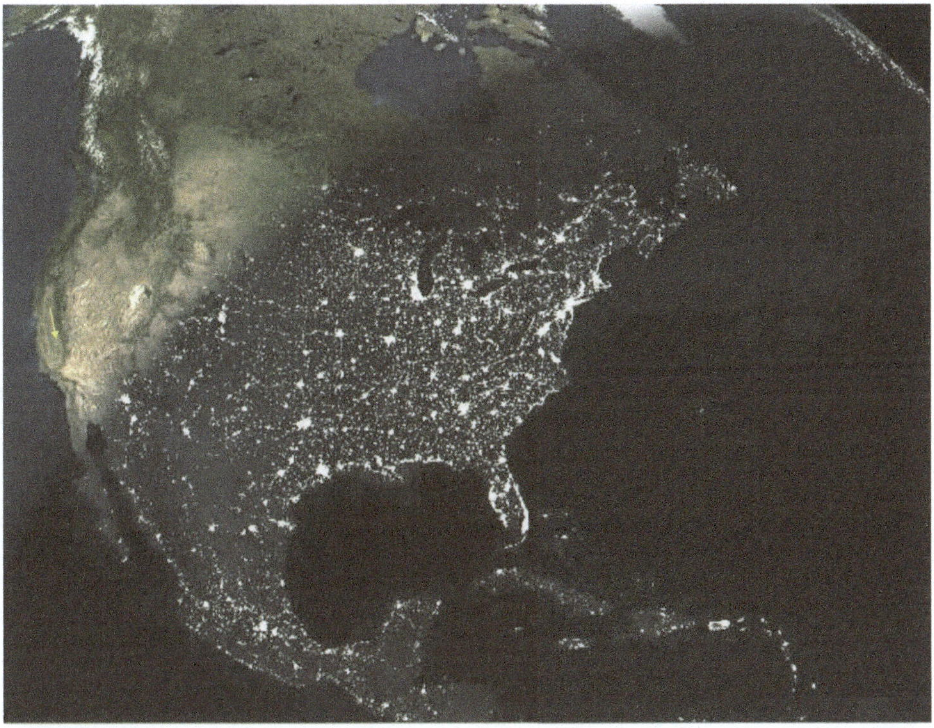

Figure 3.5. The U.S. at night, showing nightfall proceeding from right to left (*www.slideshare.net/rahul/photos-of-earth-by-sunita-williams*).

Because winds and the associated turbulence mix air both horizontally and vertically, the urban heat island is also weakened in windy weather."

Degaetano and Allen (2002) found significant effects of urbanization on U.S. temperatures in the 20th century, particularly in regard to unusually warm nights.

Parker (2004) presented a contrary view. He studied warming trends in the second half of the 20th century at 264 stations worldwide, and found no significant difference in warming on windy nights vs. warming on calm nights. Since it was expected that the effect of urban release of heat at night is diffused by wind, the lack of difference between windy and calm nights suggests that the observed warming trend from 1950 to 2000 may not be due to urban heating. Parker (2004) drew a firm conclusion:

"Here we show that, globally, temperatures over land have risen as much on windy nights as on calm nights, indicating that the observed overall warming is not a consequence of urban development."

But Parker (2004) went further:

"So, urbanization has not systematically exaggerated the observed global warming trends in T_{min}. The same can be said for poor instrumental exposure and microclimatic effects, which are also reduced when instruments are well ventilated."

While this analysis is certainly a point in favor of the argument that urban heating may not be a major factor in global warming, it is far from ironclad. While wind might be expected to reduce the urban heating effect in a small urban locality, it might not have much effect in a heavily urbanized region. Furthermore, the extent and quality of the station data appear to be limited. The emphatic conclusions reached by Parker (2004) appear to be part of the Hadley Center party line: "The reality and magnitude of global-scale warming ..."

Peterson (2003) came to a similar conclusion to Parker. His results were based on a comparison of urban and rural data for a very small time period (1989–1991). However, Peterson suggested that "industrial sections of towns may well be significantly warmer than rural sites," but urban meteorological observations may have been made in locally cool regions such as parks. However, these results are not in agreement with many observations (e.g., Figures 3.6 and 3.7).

Pielke, Sr.[6] performed a number of analyses of the global temperature measurements network and found it lacking, particularly in regard to the effect of UHIs and land-use/land-clearing change (LULCC) effects. In addition, Pielke, Sr. pointed out that, while some claim that independent analyses of global temperatures arrive at similar results, when he delved further into it, these analyses were not independent because they ultimately relied on almost the same data. In the jousting between the alarmists (who would like to believe the existing network is satisfactory) and the skeptics (who think that reported warming is amplified by urban heating effects and other defects of the measurement system), Pielke, Sr. and Parker have been key players. Parker (2009) wrote a comment on a Pielke, Sr. paper in which Parker claimed "we have provided an additional demonstration of the robustness of global and hemispheric land surface air temperature series". Pielke Sr. (2009b) responded with disagreement regarding LULCC effects and independence of separate analyses. Mahmood *et al.* (2010) joined the battle in regard to LULCC, emphasizing the importance of LULCC in the climate system. They concluded: "It has become clear from various ... that data used in existing long-term climate assessments, including the U.S. Historical Climatology Network, have undocumented biases that have not been corrected using data analysis and data adjustment techniques."

Parker (2010) extended his previous analysis of the effects of UHI on estimates of observed climate change. It is evident from the start that he was determined to justify the validity of the global temperature measurement system, and, in support of this, he pointed out that 70% of the Earth's surface is covered with water and these temperatures are unaffected by urban development. However, that is not true in the NH, where most of the global warming has been reported. Parker suggested: "...

[6] *http://pielkeclimatesci.wordpress.com/2009/01/23/reply-by-pielke-et-al-to-the-comment-by-parker-et-al-on-our-2007-jgr-paper-unresolved-issues-with-the-assessment-of-multi-decadal-global-land-surface-temperature-trends/.*

there are at least five ways to avoid, assess or compensate for possible effects of the urban heat island on estimates of global temperature trends. These are:

(1) Exclusion of sites showing urban warming (*However this would weaken the network in populated areas*).
(2) Adjustment of urban records to match nearby rural observations (*However, nearby rural observations may not be available or may suffer from their own defects*).
(3) Analysis of temperatures in windy, cloudy weather (*Here, the belief is that if the urban heat effect is found to be roughly independent of wind speed, that would indicate that the effect is small. However, this point is arguable.*)
(4) Analysis of, or comparison with trends in ocean surface temperatures (*It is not clear to this writer what the connection is from ocean temperatures to urban heat islands*).
(5) Use of atmospheric reanalyses. (*It is not clear to this writer how or why atmospheric reanalysis relates to urban heating*)."

He also said: "In future, high-resolution climate models are likely to provide further guidance on the large-scale influence of urban warming." However, it is not clear to this writer why or how climate models can tell us anything about meteorological data.

Parker (2010) pointed out that several previous analyses of temperature data attempted to correct for UHI effects and found these effects to be small. These included Hansen *et al.* (2001) as well as several technical reports by P.D. Jones *et al.* But these papers appear to have concluded that there were comparatively few urban sites, which does not jibe well with Figures 3.1 and 3.2, that suggest that a large number of land sites are located in Europe and the U.S., typically in populous areas. Parker also relied somewhat on the work by Peterson but that work was based on just a few years of observation. Parker's strongest point was his finding that urban heat effects were not strongly affected by wind, which led him to conclude that urban heat effects are small.

Mahmood *et al.* (2010) dealt more generally with LULCC, of which UHIa are only one aspect. They concluded:

> "It has also been established in the literature that biases, inaccuracies, and imprecision have been introduced to the climate monitoring systems because of meteorological station moves, instrument changes, improper exposure of instruments, and changes in observation practices. Hence, we also need strategies that will help us to detect and overcome these biases and thus lead to an improved understanding of the role of land use forcing within the climate system."

Hansen *et al.* (2010) provided a lengthy update of previous U.S. and global temperature measurements. It is noteworthy that this new revision increases the temperature rise of the 20th century by about 0.3°C. In doing this, Hansen *et al.* admitted:

> "A major concern about the accuracy of analyses of global temperature change has long been the fact that many of the stations are located in or near urban areas. Human made structures and energy sources can cause a substantial local warming that affects measurements in the urban environment. This local warming must be eliminated to obtain a valid measure of global climate change."

However, Hansen *et al.* relied heavily on Parker (2010) and claimed:

> "The urban influence on long-term global temperature change is generally found to be small. It is possible that the overall small urban effect is, in part, a consequence of partial cancellation of urban warming and urban cooling effects. A significant urban cooling can occur, for example, if a station is moved from central city to an airport and if the new station continues to be reported with the same station number and is not treated properly as a separate station in the global analysis."

In contrast to the claims of Parker and Hansen, Imhoff *et al.* (2010) and Zhang *et al.* (2010) provided results from Landsat and MODIS averaged over three annual cycles (2003–2005) "to assess the urban heat island (UHI) skin temperature amplitude and its relationship to development intensity, size, and ecological setting for 38 of the most populous cities in the continental United States". The UHI effect was found to be greatest on summer days. It was lower on summer nights and winter days, and negligible on winter nights. The UHI effect increased with city size, ranging from about 2°C for cities smaller than 10 km^2, to 8°C to 12°C for cities ranging from 20 to 1,000 km^2.

These studies found that the background ecology surrounding the city has a significant impact on the degree of the UHI effect. "Cities in forested areas have much larger heat islands than cities in arid areas. Cities in grassy or agricultural areas are somewhere in the middle" (Zhang *et al.*, 2010). Larger cities have greater UHI effects. For example, Philadelphia (population 1.6 million) had a measured UHI of 11.7°C whereas Lynchburg, VA (population 7,000), had a measured UHI of 5.5°C. "More compact, densely developed cities have warmer urban areas. Increased tree cover in surrounding rural areas makes these rural areas cooler [producing a greater relative UHI effect in the city center]" (Zhang *et al.*, 2010) Thus, Providence, RI, with rural tree cover of 84%, had a UHI of 12°C while Buffalo, NY, with rural tree cover of 15%, had a UHI of 7°C.

Yang *et al.* (2011) studied UHI effects in China. China provides an excellent opportunity to study these effects because of the rapid rate of industrialization and the growth of large cities over the past several decades. They divided weather measurements stations into groups, depending on their location: metropolis, large city, medium city, suburban, small city, and rural. Temperature measurements since 1980 were compared for these groups. Temperatures rose during the period 1980–2007 for all groups. The average measured temperature rise for each group was as listed below:

Metropolis	1.4°C
Large City	0.9°C

Medium City	0.8°C
Suburban	0.6°C
Small City	0.7°C
Rural	0.3°C

Yang *et al.* (2011) concluded that "Overall, UHI effects contribute 24.2% to regional average warming trends. The strongest effect of urbanization on annual mean surface air temperature trends occurs over the metropolis and large city stations, with corresponding contributions of about 44% and 35% to total warming, respectively". These conclusions follow from the fact that the majority of stations were rural, so the effect of higher rates of warming in built-up areas was diluted by the relative lack of stations. Nevertheless, it is clear that, for stations located in small cities and more populous areas, the majority of the observed warming was due to UHI effects, rather than climate change.

These results suggest that the UHI effect is much greater than that estimated by Parker and Hansen, and casts doubt on their results. Additional discussion and references on the UHI effect are given in NIPCC (2011).

We require surface temperature measurements for the purpose of defining the global climate. If a significant portion of measurement sites are located near local "hot spots" in the form of UHIs the data can be very misleading, particularly because urbanization has increased with time, and that would produce an artificial apparent increase in global temperature. In general, the effect of UHIs is to slightly decrease the daily maximum temperature (due to a variety of factors: reflection from buildings, shading, etc.) and to significantly increase the nightly minimum temperatures as stored solar energy (in concrete and structures) is released overnight. Thus, urbanization is expected to reduce the DTR as well as to increase the average temperature. For example, in the UHI of Houston, TX, between 1990 and 2000, it was found that the relative increase in air temperature had an average magnitude of 1.25°C at night but was largely absent during the day. The UHI had an area of about 1,000 km^2 (Streutker, 2003). Other studies report temperature increases in city centers of several degrees, with contours of decreasing temperature toward the suburbs.

Much (or even most) of this can be attributed to heat generated locally by human activity, rather than by global changes. This is evidenced by comparing the New York City data with data for a rural area in New York State, as shown in the same figure. Prior to 1900, the two curves are similar, but the New York City data increasingly diverge from the West Point, NY, data as the 20th century progresses. Anthropogenic effects can also cause a non-climatic cooling, as a result of land clearing, irrigation, and planting of vegetation in large farming areas, for example.

Corrections for urban heating are typically made by comparing urban areas with nearby rural areas and imposing trends from rural areas on the urban areas. Figure 3.6 shows the raw measured temperature data for Tokyo, Japan, in the upper panel. The middle panel shows the adjustment based on nearby rural areas. The lower panel compares the adjusted Tokyo temperature profile with that of nearby rural neighbors. Examples such as Tokyo illustrate that urban effects on temperature

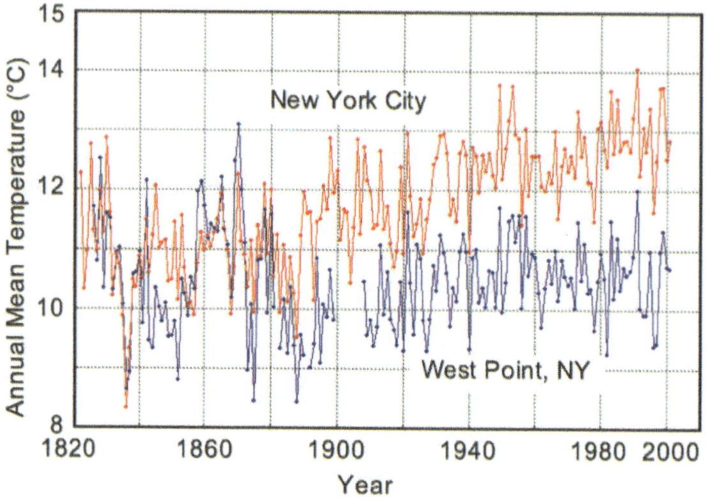

Figure 3.6. Comparison of annual average temperatures in rural New York State at West Point with temperatures in the center of New York City (in Central Park).

in specific cases can dominate over real climate trends. However, Hansen *et al.* (1999) claimed that there are far more rural stations than urban stations, so it is not essential to employ the urban data in analyses of global temperature change. They included adjusted urban station data in their standard analysis primarily for the sake of the last few years of the record, especially the final year. The fraction of reporting stations that are urban jumped from about one-quarter to one-third in the mid-1990s and to about one-half in the final year of the record. It is not clear to what extent this procedure amplified global-warming trends near the end of the 20th century. The problem of correcting for urban heating remains as a source of concern.

As we demonstrate in Figures 3.6 and 3.7, one approach for correcting for the urban heat effect is to compare the temperature trend over a period of years of an inner-city site with that in an outlying rural area, assume the rural measurement is correct, and adjust the city measurement to have the same trend as the rural measurement.

In doing this, one must have a direct correspondence between the city and its nearby rural area. Nevertheless, other factors may be important; for example, the city may be coastal while the rural area may be inland. Wickham *et al.* (2011) considered two sets of stations: "... a complete set [39,000 stations] and a set restricted to sites that are far from urban regions [16,000 stations]." However, they admitted that this classification scheme suffered from considerable subjectivity. They then fitted a straight line to the temperature record for each station; the slope of this line was called the "temperature trend" for that station. It does not seem to make sense to compare the entire set with the set of supposedly rural stations; it would make much more sense to compare urban stations with rural stations. But, more

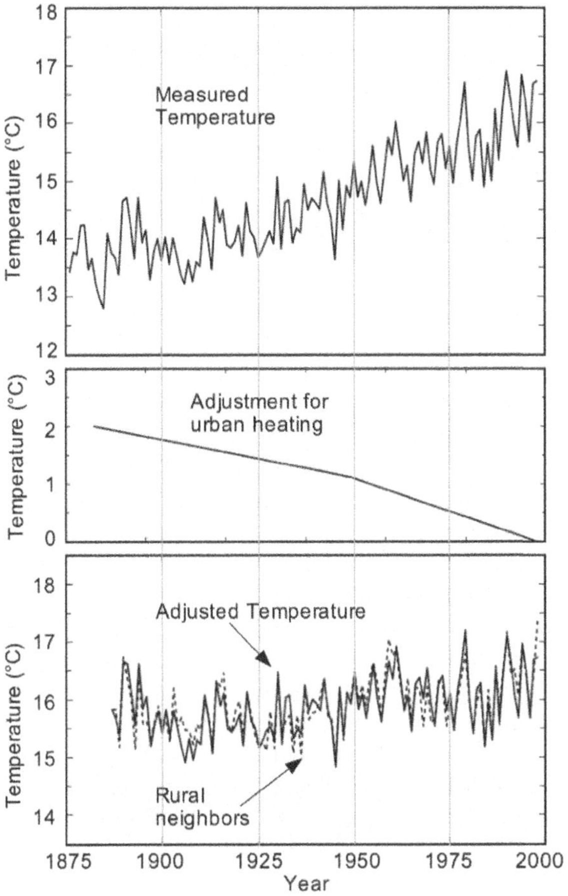

Figure 3.7. Adjustment of raw temperature data at Tokyo, Japan, for urban heating. The upper panel is raw temperature data. The middle panel is adjustment for urban heating obtained by comparing Tokyo data with data for rural surroundings. The lower panel is Tokyo data after adjustment (adapted from Hansen *et al.*, 1999).

importantly, this comparison should be made station by station as coupled pairs. By combining all the rural stations into a histogram and all the stations into another histogram, all they accomplished was to build up a broad histogram dominated by noise. It is therefore not too surprising that they found that the trend for rural stations was greater than that for all sites, which is in the opposite direction of the effect expected if urban heating was warming cities faster than rural areas. This bogus result casts doubt on the whole procedure.

Keenan[7] commented that the conclusion "that there has been some urban cooling ... contradicts over a century of research as well as common experience. It is

[7] *www.informath.org/apprise/a5700.htm.*

almost certainly incorrect". McIntyre[8] commented on corrections for UHI used by various investigators in the past involving "comparisons of supposedly 'urban' and supposedly 'rural' subpopulations in papers by Jones, Peterson, Parker and others purporting to prove that UHI doesn't 'matter' ". As he said: "Such papers set up two populations—one 'urban' and one 'rural', purport to show that the trends for each population are similar and claim that this 'shows' that UHI is a non-factor in trends". But McIntyre pointed out "... the two populations are very poorly stratified—with the 'rural' population all too often containing urban cities, sometimes even rather large cities". McIntyre also pointed out that, according to Oke's UHI concept, the UHI effect is proportional to log (population). He said: "If 'urbanization' is occurring in towns and villages as well as in large cities—which it is, then the contribution of UHI increase to temperature increase will depend on the percentage change in population (rather than absolute population). If proportional increases are the same, then the rate of temperature increase will be the same in towns and villages as in cities." In other words, UHI is artificially increasing the apparent rate of temperature rise for both "rural" and "urban" areas.

Hoyt (2006) discussed the differences between urban sites and rural sites. It was claimed that those who support the IPCC viewpoint consider a town with a population of less than 10,000 people to be rural and not to require any adjustment for urbanization. But, according to Hoyt (2006), quoting a 1973 paper by Oke, the temperature increase of an urban center compared with its surroundings is roughly approximated as 0.73 log (population), so that, according to this model, even a town of 1,000 people may have an urban temperature rise of 2.2°C.

McIntyre emphasized that the paper by Hansen *et al.* (2010) "purported to show that urban heat islands don't matter". This paper came out shortly before Zhang *et al.* (2010) reported that UHI effects can reach 7°C to 9°C in north-eastern U.S. cities. The effects of the *Krakatoa, El Chichon*, and *Pinatubo* volcanic eruptions are also evident in the data. Most recently, the effects of two major El Niños are also evident. In November 2012, Steven Mosher[9] wrote an analysis of the UHI effect based on papers by Imhoff *et al.* (2010) and Peng *et al.* (2012). He concluded that the apparent UHI effect (city–rural temperature) has a logarithmic dependence on city size:

$$\Delta T \sim 3.5 \log (\text{city area in km}^2) + 1.75\ [°C].$$

This implies that the UHI effect can be as large as 5°C for a city of 10 km^2, and 10°C for a city of 300 km^2.

Corrections for urbanization are especially difficult in China, where rapid growth has changed the demographics at a very rapid rate. From 1978 to 2000, China's Gross Domestic Product grew at an average annual rate of 9.5%, compared with 2.5% for developed countries and 5% for developing countries; the number of small towns soared from 2,176 to 20,312: the number of cities increased from 190 to 663; and the proportion of urban population rose from 18% to 39%.

[8] *http://climateaudit.org.*

[9] *http://stevemosher.wordpress.com/.*

Using rural–urban temperature differences to estimate the impacts of urbanization on climate in China may be inappropriate, for several reasons:

(1) Most Chinese stations are located in or near cities, with only a few in mountainous or remote regions or on small islands.
(2) Although China is comparable in size with the U.S., it has considerably fewer meteorological stations, and each city generally has only one station. For example, each of China's two biggest cities, Beijing and Shanghai, has only one station available in the Chinese network.
(3) It is impossible to find a corresponding rural station for most of the urban ones, especially in eastern and southern China.
(4) China's rapid urbanization in the past two decades could transfer a station from rural into urban in a very short period.
(5) Chinese cities have a much higher density of population and urban buildings than do cities in most developed countries (Zhou *et al.*, 2004).

Pielke, Sr.[10] reported on a new paper[11] that modeled summer climate effects of urban sprawl in the rapidly expanding sunbelt in Arizona. "Basically, the more concrete you have in terms of buildings and roads, the hotter it gets during the day. And, because concrete and asphalt absorbs and then releases heat, it cools off less at night." Thus, day temperatures rise and night temperatures rise even more as urban sprawl expands. Further expansion in Arizona could raise peak temperature several degrees.

The HadCRU group published a paper on quantifying uncertainties in global and regional temperature change (Morice *et al.*, 2012). They discussed uncertainties, and, in particular, they said:

> "... sufficient metadata are not available and studies over small regions are too few for uncertainties in land station homogenization, urbanization and exposure biases to be adequately described on an individual grid-box level. In a similar fashion, the characterization of spatial and temporal correlations in SSTs is limited by missing ship call-sign information prior to 1981."

As this book went to press, a preliminary version of a new study of UHI appeared on the *realclimate.org* website (Hausfather *et al.*, 2013). Their results indicate a rather modest effect of UHI on the overall database: "... urbanization accounts for 14% to 21% of the rise in unadjusted minimum temperatures since 1895 and 6% to 9% since 1960." The intent of this paper seems to be one of justifying the Goddard Institute for Space Studies (GISS) procedures for minimal correcting for UHI. Rather incredibly, this paper does not refer to commentaries by Pielke Sr., or the the papers by Georgescu *et al.*, Imhoff *et al.* (2010), Peng *et al.* (2012), and Zhang *et al.* (2010). It is difficult to be certain but this seems to be a rather one-sided view.

[10] *http://pielkeclimatesci.wordpress.com/*.
[11] Georgescu, M., Moustaoui, M., Mahalov, A., Dudhia, J., "Summer-time climate impacts of projected megapolitan expansion in Arizona", *Nature Climate Change*, *www.nature.com/nclimate/journal/vaop/ncurrent/full/nclimate1656.html*.

3.2 OCEAN TEMPERATURE MEASUREMENTS

About 70% of the Earth's surface is covered by ocean. Sea-surface temperatures play an important role in arriving at a global average temperature. In this connection, Soon and Baliunas (2003a, b) pointed out:

> "The sea surface temperature (SST) record also is complicated by a change in procedure. Prior to the 1940s, SST was determined by measuring the temperature of a dunked bucket of seawater with a thermometer. After the 1940s, SST was taken by measuring the temperature of the seawater at the intake to the engine cooling system. Large adjustments had to be made to the older data to make it compatible with the new data. These adjustments affected average SST by 0.10 to 0.45°C; the upper end of this range is three quarters of the observed change in global average surface temperature for the 20th century. Problems with defining a global-scale mean climatology for SST still exist ... as late as 1961–1990."

Houghton (2004) echoed this:

> "In the case of ships, the standard method of observation used to be to insert a thermometer into a bucket of water taken from the sea. Small changes of temperature have been shown to occur during this process; the size of the changes varies between day and night and is also dependent on several other factors including the material from which the bucket is made—over the years wooden, canvas and metal buckets have been variously employed. Nowadays, a large proportion of the observations are made by measuring the temperature of the water entering the engine cooling system."

Prior to the advent of the Argo network for monitoring subsurface sea temperatures about 2004, "the historical ocean subsurface dataset was quite sparse over much of the World Ocean" (Harrison and Carson, 2007). Harrison and Carson (2007) pointed out that various "global-scale studies typically involved removing a climatological ocean temperature field, followed by different amounts of interpolation, extrapolation, and averaging, or have otherwise depended on some analyzed field" and they implied that such manipulations might involve some artificiality. They thought it "worthwhile to look at subsurface thermal variability from a perspective that minimizes manipulations of the data themselves and also focuses on those depths and areas where statistically significant trends can be identified with minimum smoothing or averaging".

Harrison and Carson (2007) studied ocean temperatures down to 500 m depth over the 51-year period: 1950–2001. They found:

> "... highly structured patterns of 51-yr trends of alternating sign at 100-, 300-, and 500-m depth. ... Each of the ocean basins exhibits both warming and cooling trends over this 51-yr analysis period. ... The upper ocean evidently is replete with variability in space and time, and multidecadal variability is quite energetic. ... These results suggest that trends based on records of one or two

decades in length are unlikely to represent accurately longer-term trends. Further, the magnitude of the 20-yr trend variability is great enough to call into question how well even the statistically significant 51-yr trends identified here represent longer-term trends."

Over the period 2004–2007, the Argo system has been in operating to monitor ocean temperatures.

"Argo consists of a large collection of small, drifting oceanic robotic probes deployed worldwide. The probes float as deep as 2 km. Once every 10 days, the probes surface, measuring conductivity and temperature profiles to the surface. From these, salinity and density can be calculated. The data are transmitted to scientists on shore via satellite. The initial project goal was to deploy 3,000 probes, completed in November 2007. The Argo program was designed to operate on the same 10-day duty cycle to match the existing satellite measurements of the ocean's sea surface. These satellites, called Topex/Poseidon and Jason 1, measure changes in the surface topography of the ocean. With such measurements, information about temperature, mass redistribution, or surface currents can be inferred. The Argo floats measure subsurface changes in temperature and salinity, hence the float measurements are complementary to the altimetry."

The number of floats is continually changing as floats are lost or expire, while others are deployed. Nominally, some 750 floats are deployed each year to sustain the system. The floats have a nominal 300 km spacing, although the exact separations depend on the randomness of the float drift.

The Argo temperature and salinity measurements are yielding valuable information about the large-scale water properties and currents of the ocean, including the variability of these properties over time scales from seasonal to decadal. However, the short duration of this system does not allow any long-term trends to be confirmed.

Nevertheless, Brohan *et al.* (2006) concluded that uncertainties in SSTs are fewer than for land surface temperatures. Marine data used by Brohan *et al.* (2006) consist of a gridded data set made from *in situ* ship and buoy observations from the new International Comprehensive Ocean–Atmosphere data set. For each grid box, mean temperature anomalies, measurement and sampling error estimates, and bias error estimates are available.

Blending a SST data set with land air temperature data makes an implicit assumption that SST anomalies are a good surrogate for marine air temperature anomalies. It is claimed that this is the case, and that marine SST measurements provide more useful data and smaller sampling errors than marine air temperature measurements would. So it is claimed that "blending SST anomalies with land air temperature anomalies is a sensible choice". Over the period from 1850 to 1940, the predominant SST measurement process changed from taking samples in wooden buckets, to taking samples in canvas buckets, to using engine room-cooling water inlet temperatures. A bias correction was applied to remove the effect of these changes on the SSTs, but this adds some uncertainty. Brohan *et al.* (2006) said:

> "As with the land data, the uncertainty estimates cannot be definitive: where there are known sources of uncertainty, estimates of the size of those uncertainties have been made. There may be additional sources of uncertainty as yet unquantified."

Brohan *et al.* (2006) said:

> "If the gridded data had complete coverage of the globe or the region to be averaged, then making a time series would be a simple process of averaging the gridded data and making allowances for the relative sizes of the grid boxes and the known uncertainties in the data. However, global coverage is not complete even in the years with the most observations, and it is very incomplete early in the record. In general, global and regional area-averages will have an additional source of uncertainty caused by missing data. To estimate the uncertainty of a large-scale average owing to missing data the effect of sub-sampling on a known, complete dataset is used."

The monthly averages are dominated by short-term fluctuations in the anomalies; combining the data into annual averages produces a clearer picture, and smoothing the annual averages with a 21-term binomial filter highlights the low-frequency components and shows the importance of the bias uncertainties. The bias uncertainties are zero over the normal period by definition. The dominant bias uncertainties are those due to bucket correction and thermometer exposure changes, both of which are larger before the 1940s. The station, sampling and measurement, and coverage errors depend on the number and distribution of the observations, and these components of the error decrease steadily with time as the number of observations increases. These components also decrease with averaging to larger space and timescales, so they are smaller in the annual than in the monthly series, and smaller again in the smoothed annual series. The bias uncertainties, however, do not reduce with spatial or temporal averaging, and they are largest in the early 20th century; so the smoothed annual series, where the uncertainty is dominated by the bias uncertainties, also has its largest uncertainty in this period (Brohan *et al.*, 2006).

Brohan *et al.* (2006) discussed three types of errors in the database: (1) station error (the uncertainty of individual station anomalies due to measurement errors, changes in location, changes in time of measurement, or changes in instrumentation), (2) sampling error (the uncertainty in a grid box mean caused by estimating the mean from a small number of point values), and (3) bias error (the uncertainty in large-scale temperatures caused by systematic effects such as UHIs, and exposure changes). It is difficult to summarize the net result of all the contributing errors. Data presented in Brohan *et al.* (2006) (their Figures 5, 6, and 9) seem to suggest that potential errors in anomalies of 1°C to 2°C are commonplace, which exceeds the putative temperature rise of the 20th century. However, in their final summing-up of hemispheric and global temperatures, Brohan *et al.* (2006) indicated much smaller temperature uncertainties. But the really surprising aspect of this is that, according to Brohan *et al.* (2006), the uncertainty in SSTs is fairly flat at ± 0.1°C from 1850 to 2000. This estimate is difficult to believe. By contrast, the uncertainty in land

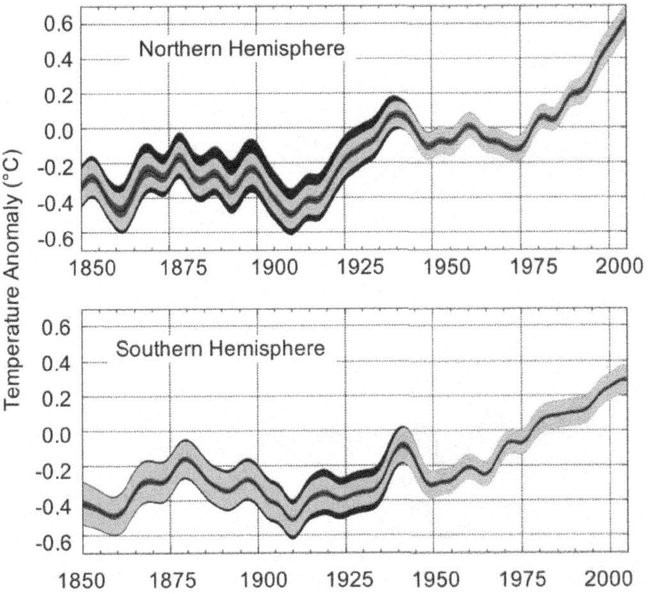

Figure 3.8. Smoothed monthly temperatures for the two hemispheres (adapted from Brohan *et al.*, 2006).

temperatures is claimed to range from about ± 0.4°C around 1850 to about ± 0.15°C in 2000. Since the oceans occupy about 70% of the globe's surface area, the overall uncertainty would be expected to depend more on the ocean data. According to Brohan *et al.* (2006):

> "A notable feature of the global time series is that the uncertainties are not always larger for earlier periods than later periods. The uncertainties are smaller in the 1850s than in the 1920s, at least for the smoothed series, despite the much larger number of observations in the 1920s."

This is also difficult to believe.

Figure 3.8 shows the resultant smoothed monthly temperatures for the two hemispheres and the globe as derived by Brohan *et al.* (2006). The widths of the swaths are measures of the uncertainties associated with each set of measurements. A comparison of the smoothed mean temperature anomalies for the NH and SH shows the difference in uncertainties between the two hemispheres. According to Brohan *et al.* (2006), the difference in the uncertainty ranges for the two series stems from the very different land/sea ratio of the two hemispheres. The NH has more land, and so has a larger station sampling and measurement error, but it has more observations and so a smaller coverage uncertainty. The bias uncertainties are also larger in the NH both because it has more land, and because the SST bias uncertainties are largest in the NH western boundary current regions. As before, I believe that uncertainties in the 19th century are underestimated.

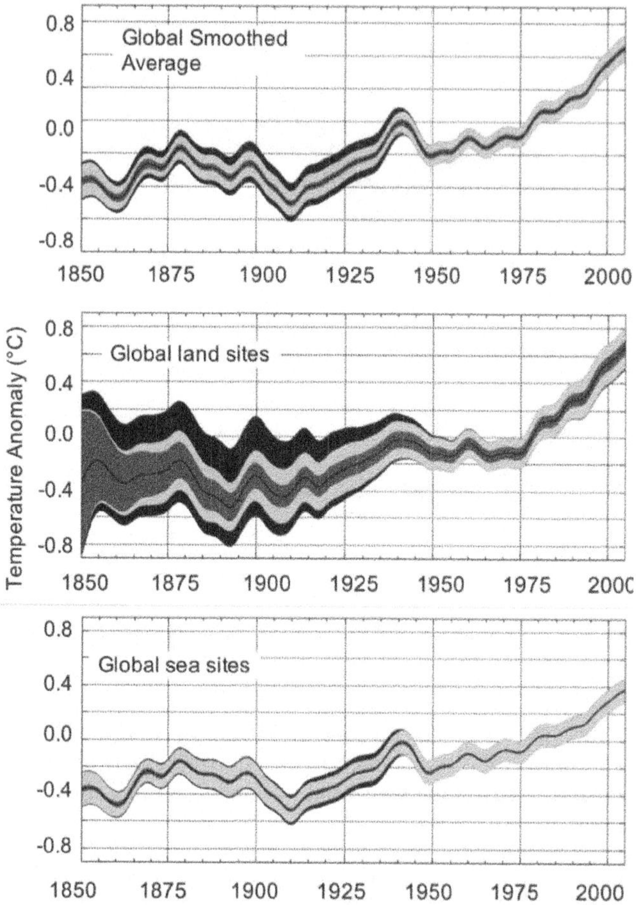

Figure 3.9. Global smoothed monthly sea-surface temperatures. Upper panel: All sites. Middle panel: Land sites. Lower panel: Sea sites (adapted from Brohan *et al.*, 2006).

Brohan *et al.* (2006) calculated the global temperature as the mean of the NH and SH series (to stop the better-sampled NH from dominating the average). Figure 3.9 shows the resultant smoothed monthly temperatures based on land measurements, sea measurements, and overall.

There are a number of aspects of the results of Brohan *et al.* (2006) that are sources of concern. First, consider the point that it is claimed that sea temperature measurements are more precise than land temperature measurements. On the one hand, this would seem to be intuitively correct because the sea measurements are not afflicted with site aberrations due to urbanization, shade or sun, exposure to winds, changes in land use, etc. However, doubts have been raised about the accuracy of sea temperature measurements due to the changes in measurement procedures over the years. Next, consider that, despite the huge preponderance of sites in the NH compared with the SH, and the fact that the number of SH sites probably falls off much more rapidly as one goes back in time, Brohan *et al.* (2006) claim that the measurements of temperature in the SH are more precise. This, of course, is based on the putative claimed accuracy of SSTs and the preponderance of ocean in the SH. Nevertheless, this seems counter-intuitive and worthy of further examination. But the strangest thing about Figures 3.8 and 3.9 is that they do not seem to link to one another very well. For example, the upper panel in Figure 3.9 is the global temperature anomaly, and it should be an arithmetic mean of the SH and NH curves in Figure 3.8. However, it is not. For example, at the far right of these graphs, the values are global = 0.62, NH = 0.62, and SH = 0.28. By averaging the NH and SH, one should get 0.45 for the global, not 0.62. Something is amiss here. Another problem is that, if we take 30% of the land curve and 70% of the sea curve in Figure 3.9, we do not end up with the global curve (upper panel in Figure 3.9). Once again, at the far right of the graphs, we find $0.7 \times 0.35 + 0.3 \times 0.67 = 0.45$ whereas the global curve indicates 0.62. It is also difficult to reconcile the width of the uncertainty in the upper panel.

The ultimate results on temperature profiles derived by Brohan *et al.* (2006) date back to 1850 as shown in Figures 3.8 and 3.9. Spatial coverage varied widely, with some areas (U.S. and Western Europe) having a higher density of stations, and the remainder of the world being quite sparse (see Figures 3.1 and 3.2). In converting station data to a gridded data set, all gridded cells are treated equally. Since the majority of the gridded cells pertain to regions with sparse station coverage, the high densities of stations in the U.S. and Western Europe are greatly diluted by the low density of stations elsewhere. Thus, the total number of stations (~4,350) becomes an irrelevant figure when the data set is converted to a grid. In actual fact, it is not the total number of stations that matter, but rather the distribution function for the number of stations per grid cell. This function will have strong peaks for cells in the U.S. and Western Europe, but will be very low for most of the world, leading to the inevitable conclusion that spatial coverage across the globe is very poor. Furthermore, very few of the ~4,350 stations dated back to 1850, or even 1900, as shown in Figure 3.3. In addition, it seems likely that the majority of stations that do date back to 1900 or 1850, are located in the U.S. and Western Europe. Hence, not only is spatial coverage poor throughout most of the world for all years, but spatial coverage prior to perhaps 1940 is even worse than it was after that.

Since Brohan *et al.* (2006) did not provide data on the distribution of the number of sites per gridded cell for all years, it is difficult to be certain just how flimsy the historical global temperature record is, but, clearly, the weaknesses of the TMN have been glossed over.

McKitrick and Michaels (2007) provided an extensive review of the literature on biases in surface measurements of temperature. They examined a 22-year gridded dataset that began in 1980 to estimate anthropological influences on measured temperatures. They said:

> "If measured surface temperatures are accurate, then the spatial pattern of grid cell temperature trends should be uncorrelated with variables like Gross

> Domestic Product, population density, average income, and other local, non-climatic factors. The presence of such correlations, on the other hand, would indicate that gridded surface climate data contain extraneous biases, thus measured climatic trends may be inaccurate and attempts to identify the climatic influences of greenhouse gases might misattribute the causes of apparent trends."

McKitrick and Michaels (2007) carried out a detailed statistical analysis to detect the dependences of measured temperatures on such socioeconomic variables as local population, per capita income, Gross Domestic Product, coal consumption, GDP density, average educational attainment, as well as the number of missing months in the observed temperature series over the interval. They pointed out that "the standard interpretation of global climate data is that extraneous effects, such as urbanization and other land surface effects, and data quality problems due to inhomogenieties in the temperature series, are removed by adjustment algorithms, and therefore do not bias the large-scale trends". However, their results indicated "that trends in gridded climate data are, in part, driven by the varying socioeconomic characteristics of the regions of origin, implying a residual contamination remains even after adjustment algorithms have been applied". These results suggest that there have been overstatements of the rate of global warming over land. They also suggest that some observed climate changes in recent decades may be more due to land surface modifications, rather than greenhouse gas emissions. However, the magnitude of such overstatements of temperature rise over land was estimated to be only about 0.1°C in most cases.

3.3 TROPOSPHERE TEMPERATURE MEASUREMENTS

Marsh (2002) pointed out:

> "The layer of the Earth's atmosphere from the ground up to an altitude of a few miles is called the troposphere, and the boundary between it and the rest of the atmosphere above is called the tropopause. The tropopause is about 11 miles high at the equator and only about five miles high at the poles. The troposphere is the part of the atmosphere that is responsible for the greenhouse effect, since it contains essentially all of the greenhouse gases. Because the Earth's troposphere, surface and boundary layer are closely coupled by air movement, they are considered to be a single thermodynamic system. For this reason, changes in radiative flux at the tropopause are used to express changes to the climate system."

According to the Wikipedia: "The troposphere ... contains approximately 80% of the atmosphere's mass and 99% of its water vapor and aerosols."

According to Gregory *et al.* (2004):

> "Heat is rapidly exchanged within the troposphere and at the surface; the troposphere, surface and upper ocean thus constitute a tightly coupled 'climate

system.' The stratosphere tends to equilibrate separately and on a timescale of only a few months."

Most recent GCM hindcasts and forecasts are consistent in depicting a tropical lower troposphere that warms at a rate about 1.3 times that of the surface. Thus, the models would predict that the trend for warming of the troposphere would be at a higher rate than the surface.

NOAA has flown about 10 microwave sounding unit (MSU) satellites since 1979. Each satellite has a life cycle of a few years but their timing is arranged so there is overlap between successive satellites. The MSU instruments measure thermal emission from atmospheric oxygen constituting the major component of the measured brightness temperature. Initial processing of the data for the past two decades indicated significant discrepancies between temperature trends for the troposphere vs. trends observed at ground level. Some work in the late 1990s revealed the presence of a spurious cooling trend introduced by neglect of the differential effects of satellite orbit decay on near-limb and near-nadir observations. Accounting for this led to an increase in the global trend of approximately 0.12°C/decade, bringing the lower troposphere observations in the direction toward GCM predictions and surface measurements. However, Christy, Spencer, and Braswell (2000) discovered additional diurnal and target temperature contributions that they claimed largely offset the orbit decay effect, leading to a tropospheric trend of +0.04°C/decade, thus continuing the surface/troposphere disconnect. Further processing of data uncovered a number of important sources of error, including inter-satellite offsets, diurnal warming with slow evolution in the satellite local equator crossing times, and the presence of a significant correlation between observed inter-satellite brightness temperature differences and satellite hot calibration load temperature. Impacts of these various contributions were gradually identified and corrected in successive versions of their merged data set, with resulting long-term trends indicating less warming relative to the surface, particularly in the tropics and subtropics. These results, combined with the even more rapid cooling in the modified data set MSU 2LT, sparked a lively debate regarding errors in the (TT) data, how to correct for them, and the relationship to inconsistency with general circulation model predictions (Mears *et al.*, 2003; Anon. (C), 2005).

Mears, Schabel, and Wentz (2003) carried out a complete reanalysis of systematic errors in the MSU data set because of its importance in climate change research. The details of this process are complex and intricate and are beyond the scope of the present discussion.

Several studies in the period 2003–2007 created greater confidence in the tropospheric measurements, and in particular, with the publication of Christy *et al.* (2007), TT measurements became widely accepted. There still remained some differences from ground measurements but these are believed to be real. Douglass and Christy (2009) presented evidence that past problems with calibration have been solved and they presented the latest TT measurements.

3.4 MEASURED EARTH, REGIONAL, AND LOCAL SURFACE TEMPERATURES

3.4.1 U.S. surface temperature measurements

NASA's GISS has analyzed U.S. and world temperature patterns since about 1880 in great detail. Global temperatures were estimated based on rural, small-town, and homogeneity-adjusted urban stations (Hansen *et al.*, 1999). Hansen *et al.* (2001) modified the previous work from Hansen *et al.* (1999) by improving on the corrections applied to urban stations using satellite photos of the Earth at night to determine areas of high power density from lighting observations, as well as a number of other corrections to the database. It is claimed that local human effects (urban warming) can be important even in suburban and small-town (SAT) records. They said: "We believe that this evidence is suggestive of a significant urban effect within the United States, but it requires further investigation." These data were updated by the GISS (Hansen *et al.*, 2010). Their estimate for the U.S. mean temperature history (after making adjustments) is shown in Figure 3.10. Steve McIntyre[12] commented briefly on this paper. He pointed out that, in this update, GISS made remarkably large changes from the August 2007 version, constituting "a rewriting of history that has increased the [temperature during] 2000–6 relative to the 1930s by about 0.3 deg C". Such large changes seem suspect, especially when they significantly increase the global warming reported by warmists. McIntyre also emphasized that Hansen *et al.* (2010) "purported to show that urban heat islands don't matter". This paper came out shortly before Zhang *et al.* (2010) reported that UHI effects can reach 7°C to 9°C in north-eastern US cities. Hansen's credibility must be called into question at this point. As of 2010, U.S. average temperatures are not much different than they were in the 1930s. Kunkel *et al.* (2008) provide an analysis of weather and climate extremes for North America dating back to the mid–19th century in some cases. While the tone of this report leans toward the alarmist persuasion, the data provide mixed results. A heat wave index was defined as the occurrence of warm spells of at least four days in duration with mean temperature exceeding the threshold for a 1-in-10-year event. The historical occurrence of heat waves is plotted in Figure 3.11. While this index has been increasing in the past few decades, it remains far below the peaks in the 1930s. In addition, Kunkel *et al.* (2008) showed that the area over which hot daily highs and lows were reached has been increasing over the past few decades, but the percentages of total area remain low at under 0.5%.

In a similar fashion, a cold wave index was plotted by Kunkel[13] as shown in Figure 3.12. This index has been decreasing over the past decade but there is no evidence of departure from historical norms.

However, there are data that suggest the warming of the past few decades may

[12] *climateaudit.org.*

[13] *www.worldclimatereport.com/index.php/2006/03/15/an-extreme-view-of-globalwarming/.*

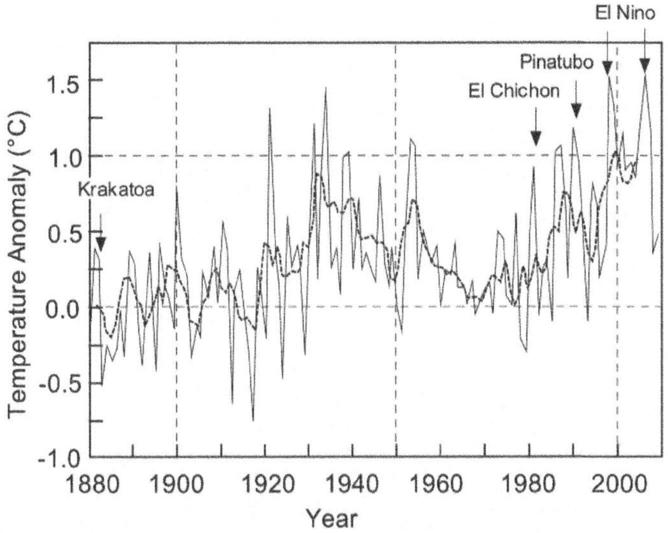

Figure 3.10. U.S. mean temperature anomalies (deviations from mean temperature). The dashed curve is a multi-year running average (adapted from Hansen *et al.*, 2010).

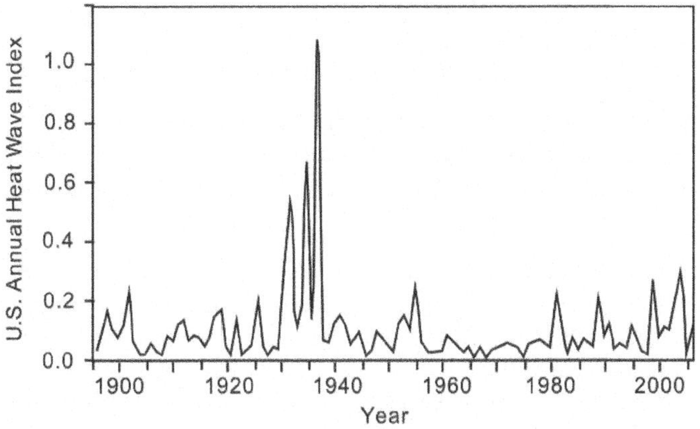

Figure 3.11. Historical variation of U.S. heat wave index (Kunkel *et al.*, 2008).

have a different character than that of the 1930s. For example, Kunkel *et al.* (2008) showed that the recent U.S. average frost-free season period was at least 10 days longer on average than the long-term average, and was significantly greater than that of the 1930s.

Hansen *et al.* (2012) said:

"The distribution of seasonal mean temperature anomalies has shifted toward higher temperatures and the range of anomalies has increased. An important change is the emergence of a category of summertime extremely hot

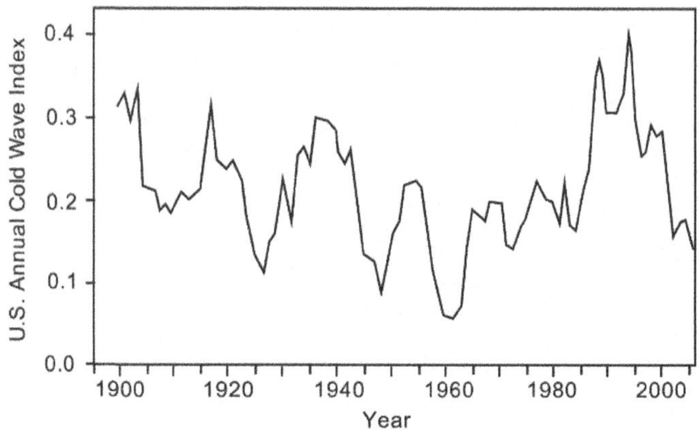

Figure 3.12. Historical variation of U.S. cold wave index (Kunkel *et al.*, 2008).

outliers, more than three standard deviations (3σ) warmer than the climatology of the 1951–1980 base period. This hot extreme, which covered much less than 1% of Earth's surface during the base period, now typically covers about 10% of the land area. It follows that we can state, with a high degree of confidence, that extreme anomalies such as those in Texas and Oklahoma in 2011 and Moscow in 2010 were a consequence of global warming because their likelihood in the absence of global warming was exceedingly small."

John Christy[14] rebutted this claim showing that, as usual, Hansen craftily chose to compare data and time periods that supported his claim, whereas a broader view of more data over extended time refutes his claim. One point is that Hansen chose to use data on $T_{mean} = (T_{max} + T_{min})/2$. As Christy pointed out:

"T_{max} represents the temperature of a well-mixed lower tropospheric layer, especially in summer. T_{min}, on the other hand, is mostly a measurement in a shallow layer that is easily subjected to deceptive warming as humans develop the surface around the stations.

"Since T_{max} represents a deeper layer of the troposphere, it serves as a better proxy (not perfect, but better) for measuring the accumulation of tropospheric heat, and thus the greenhouse effect. This is demonstrated theoretically and observationally in McNider *et al.* (2012). I think T_{max} is a much better way to depict the long-term temperature character of the climate."

The record highs recorded by Hansen were mostly due to highs in T_{min}, rather than T_{max}, and this can be attributed to a variety of causes other than greenhouse gas warming.

A second problem with Hansen's analysis pertains to his database and his choice of reference period. Hansen used a reference period from 1951 to 1980, and thus

[14] *www.drroyspencer.com/2012/08/fun-with-summer-statistics-part-i-*usa/.

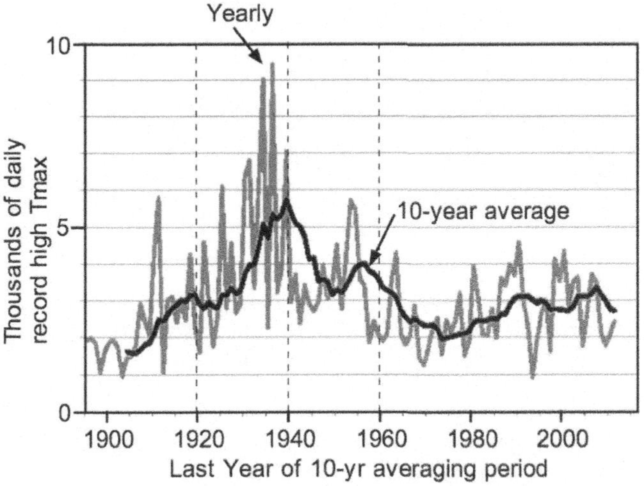

Figure 3.13. Number of daily high U.S. T_{max} records (1895–2011) based on 970 USHCN stations with 80 years of data (Christy, 2012).

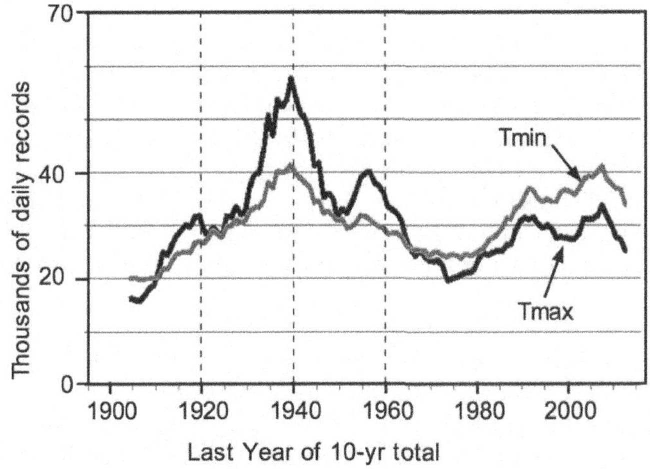

Figure 3.14. Number of daily U.S. T_{max} and T_{min} records (1895–2011). Ten year running totals based on 970 USHCN stations with 80 years of data (Christy, 2012).

missed the extremely warm period of the 1930s. This would lead him to conclude that the warmth of later years was more unusual than it was.

The issue at hand is to plot year by year, the relative number of station highs that set records relative to a reference period. Christy recalculated Hansen's model for the U.S. and for the globe. For the U.S., Christy plotted the number of daily record highs in T_{max} from 1895 to 2010 as shown in Figure 3.13. It can be seen that there were far more record highs in the 1930s than in recent years. In fact, record highs from 1910 to 1955 were generally higher than recent decades. Figure 3.14

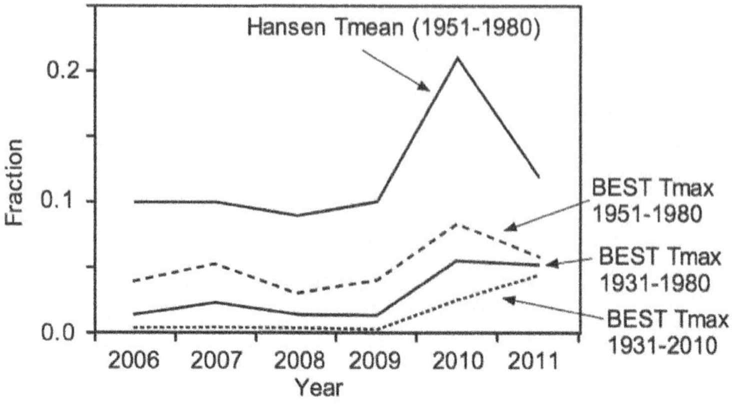

Figure 3.15. Fraction of monitored area that exceeds the 3σ threshold for reference periods and databases as indicated (Christy, 2012).

shows the number of daily record highs for both T_{max} and T_{min}. It can be seen that T_{min} has risen more than T_{max} since 1975.

Hansen estimated that, for the NH, typically 10% or more of the area experienced high temperatures above the 3σ threshold for the reference period (1951–1980). Of course, choosing this short reference period ignores the relatively hot period earlier in the century. Christy estimated the results shown in Figure 3.15. Replacing Hansen's data base with the BEST data base but retaining the short reference period, reduced the fraction of area experiencing high temperatures above the 3σ threshold by roughly a factor of two. It is not clear why this is so. But of even greater importance, when the reference period was extended, the fraction of area experiencing high temperatures above the 3σ threshold dropped even more.

Kunkel *et al.* (2008) also presented data on precipitation in the U.S. They plotted the percentage area of the U.S. in drought since 1895. Their Figure 2.6 shows that recent occurrences are in line with historical norms and are well below the drought periods of the 1930s and the 1950s. Their Figures 2.8 and 2.9 show that occurrence of intense precipitation has increased since 1895, but the data are skewed by a high peak around the El Niño events in the 1990s.

Winstanley and Wendland (2007) analyzed the climate record in the State of Illinois dating back to the mid-19th century. He found that regional temperatures peaked in the late 1930s and dropped sharply thereafter, with temperatures near 2000 comparable with those at the turn of the 20th century. Heavy precipitation events were more frequent in the 19th century than in the 20th century. The occurrence of extreme heat waves was greatest from 1850 to 1875, with a secondary high period in the 1930s. The occurrence of heat waves in the 1980s and 1990s was well below earlier peak periods.

3.4.2 Global and hemispheric surface temperatures

GISS (2007) estimated the global temperature change by dividing thc world into broad latitude zones, estimating temperature anomaly time series for each zone, and then weighting these zones by their area. The zones included were: northern latitudes (90°N–23.6°N), low latitudes (23.6°N–23.6°S), and southern latitudes (23.6°S–90°S), covering 30%, 40%, and 30% of the Earth's surface, respectively.

On a global scale, the temperature history is shown in Figure 3.16. Here, there is greater evidence of strong global warming since 1976. However, it is not clear how reliable the data are for the globe, and whether sufficient accounting has been taken of oceans (70% of Earth area), urban heating, and sparseness of stations. Houghton (2004) claimed that all of these problems were mitigated:

> "Careful analysis of the effects of these details on observations both on land and from ships has enabled appropriate corrections to be made to the record, and good agreement has been achieved between analyses carried out at different centers."

Hansen *et al.* (2010) assure us that the data are reliable and aberrations were small and corrected for. However, recent data on UHIs cast doubt on Hansen's view. Brohan *et al.* (2006) apparently believe that SST records are in good shape, but sea temperatures appear to this writer to remain a source of uncertainty. The 20th-century history of temperature rises in the northern and southern hemispheres, is shown in Figure 3.17. Temperature fluctuations in the north were greater than in the south. Three factors that could possibly affect these data are (1) much higher density of stations in the north, (2) greater concentration of urban warming in the north, and (3) there is a much greater preponderance of oceans in the SH.

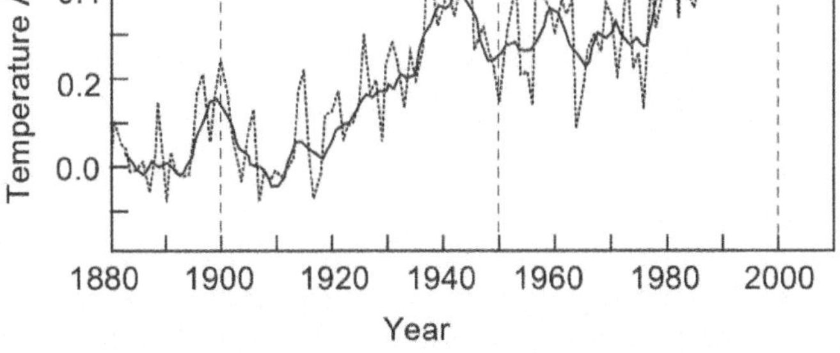

Figure 3.16. Global temperatures based on a combination of land and sea data shown with five-year moving average (solid line) (adapted from Hansen *et al.*, 2010).

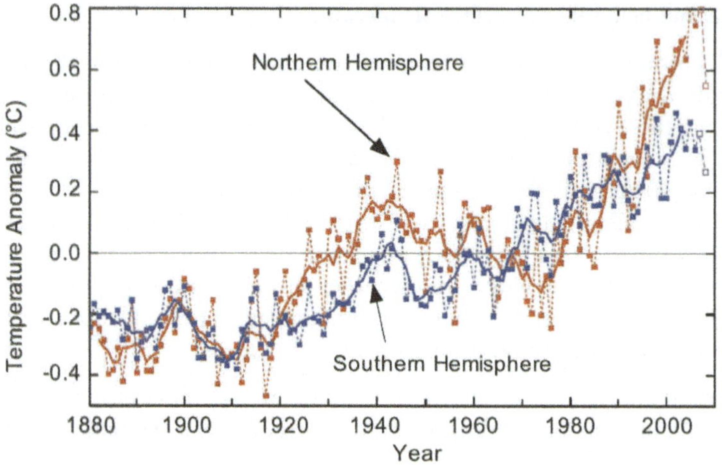

Figure 3.17. Global mean temperatures for the two hemispheres (adapted from GISS, 2007).

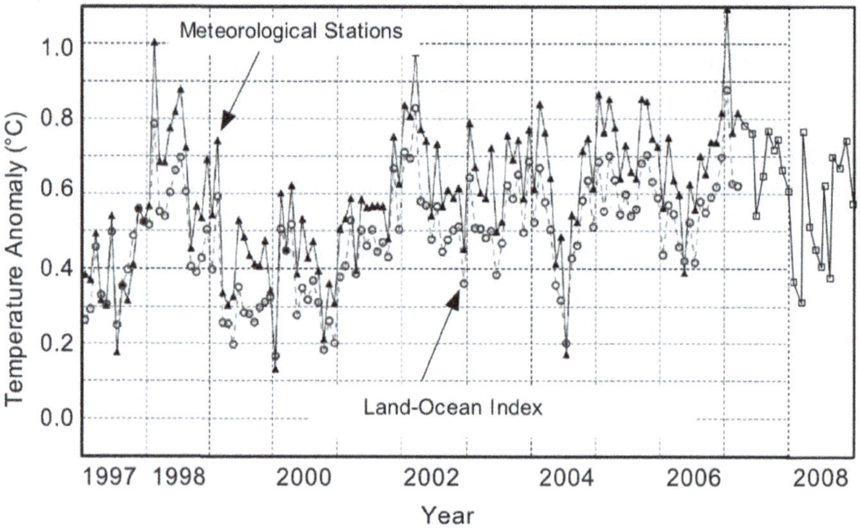

Figure 3.18. Global mean temperature anomalies after 1997 on a monthly basis (adapted from GISS, 2007, with updates).

The monthly global temperature record for the period 1997–2008 is shown in Figure 3.18. Temperatures during this period appear to have stabilized. There is no evidence of a continuation of the sharp run-up from 1980 to 1997.

The variation in 20th-century global temperatures partitioned by latitude ranges (with roughly equal surface areas) is shown in Figure 3.19. It can be seen that the greatest observed temperature rise in the 20th century was north of 23.6°N. This might be influenced by the density of stations in the north coupled with urban

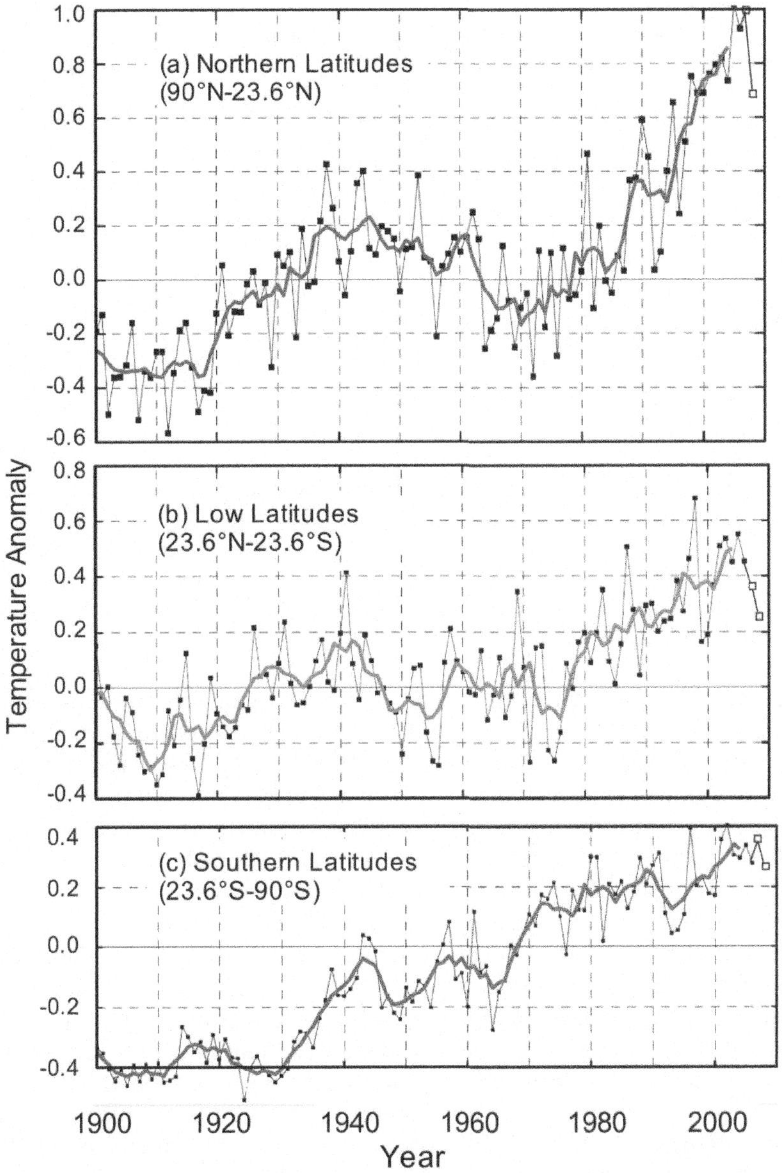

Figure 3.19. Global temperature anomalies for three latitude ranges with roughly equal surface areas (adapted from GISS, 2007, with updates).

warming and other local human urbanization effects as well as "Arctic amplification". In any event, a good deal of the observed recent global warming has been concentrated in the NH.

Figure 3.20 shows global temperature anomalies as a function of latitude and

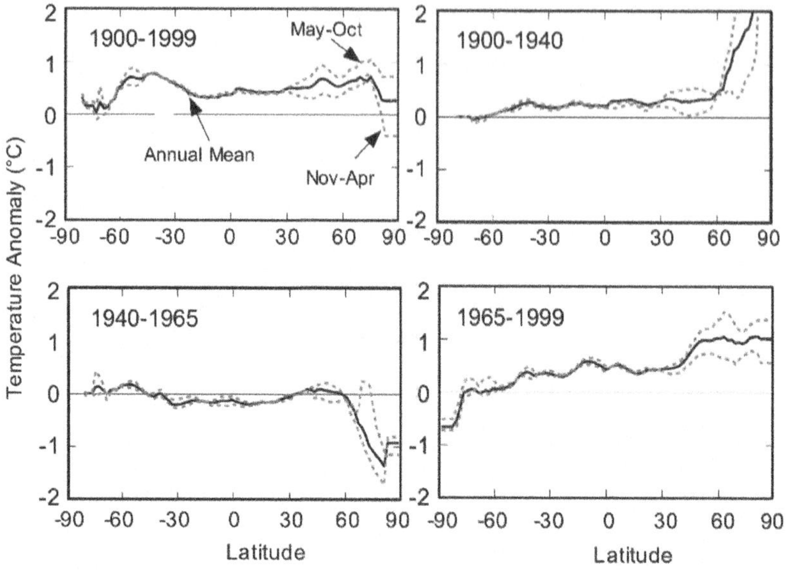

Figure 3.20. Mean temperature anomalies as a function of latitude and time period (Adapted from Hansen *et al.*, 2010).

time period. For the 20th century, the upper-left panel shows a moderate but consistent temperature rise. From 1900 to 1940 (upper-right panel), there was a rather drastic increase in temperature at high northern latitudes. This was followed by a decrease in temperature at high northern latitudes from 1940 to 1965 (lower-left panel). Temperatures across the globe elsewhere than north of 60°N were quite stable from 1900 to 1965. Temperatures increased globally after 1965, with the greatest increases at high northern latitudes (lower-right panel). More details on latitude dependence are provided by Jones *et al.* (1999).

It is not clear why there is such a significant difference between the histories of U.S. and global temperatures since 1880. Considering that the database for the U.S. appears to be more complete, presumably better corrected for urban heating, it seems likely that the U.S. history is more reliable. Oddly enough, since U.S. data are entirely based on land and are independent of ocean temperatures, one might have thought *a priori* that it would show greater variability than global data that include oceans.

One notable feature of all the global temperature measurements of the 20th century is the fact that the temperature rise of the 20th century was divided into two steps (from 1910 to 1940 and from 1978 to 2000) with an intermediate period (roughly 1940–1978) in which temperatures actually decreased slightly. Considering that CO_2 emissions from 1910 to 1940 were far lower than that needed to account for the observed temperature rise, and that the intermediate period (1940–1978) had decreasing temperatures while CO_2 emissions increased significantly, the correlation of the temperature rise of the 20th century with greenhouse gas emissions is less than convincing.

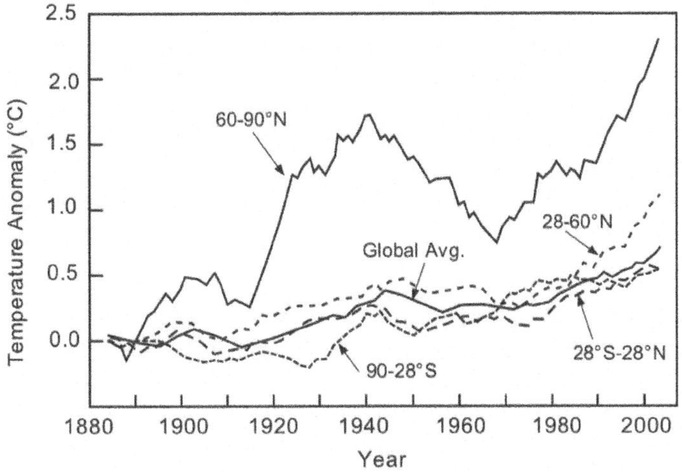

Figure 3.21. Area-weighted mean observed surface temperatures over the indicated latitude bands. The values are running nine-year means relative to the 1880–1890 mean (Shindell and Faluvegi, 2009).

These facts were seized on by skeptics as evidence of the inadequacy of GCMs that concluded that the 20th-century temperature rise was mainly due to rising CO_2 levels. However, several studies concluded that sulfate aerosols and particulates reflect incident sunlight, producing a cooling effect. The global-warming alarmists then claimed that the cooling observed from 1940 to 1978 was due to an increase in aerosol production that overwhelmed the greenhouse effect, but that clean-up of power plants starting around 1980 reduced aerosol production after that. This claim was challenged, and there remains a controversy regarding yearly production rates of aerosols and the impact of aerosols on global temperatures in the mid- to late 20th century. It is difficult to resolve this important issue at this time.

A breakdown of global temperature changes by latitude range is provided in Figure 3.21. This shows that by far, the greatest change in temperature was in the far north. The latitude range 28°N–60°N had a moderate temperature rise, and the tropics and SH had minimal increases in temperature.

It is evident that, as the Earth warms or cools due to various causes, the Arctic will experience greater than average warming or cooling, whereas tropical areas will undergo lesser changes. Warming in the SH is limited by the large heat uptake of the Southern Ocean, leaving the Arctic as the global location with the greatest warming. Increased warming in Arctic regions is usually referred to as "Arctic amplification". Figure 3.21 shows this very clearly.

Usually, Arctic amplification is regarded as the ratio of the Arctic-averaged SAT change to the global average temperature change during a period when the Earth's climate changes. In the past decade, a number of studies have been conducted on Arctic amplification during the warming that occurred in the 20th century, and projections have been made of continued warming in the 21st century.

Almost all of these begin by assuming that warming in the 20th and 21st centuries is predominantly due to increased concentrations of greenhouse gases, particularly CO_2, and then proceed to analyze the role that Arctic amplification will play in response to rising greenhouse gas concentrations in the 21st century. Several of these papers seem to have the intent of showing that anthropogenic emissions of CO_2 were the cause of the observed warming.

Miller *et al.* (2010a, 2010b) emphasized the sensitivity of Arctic temperatures to change via various feedbacks that can act to amplify or diminish incipient climate changes. These include:

- Ice-albedo feedback: "Changes in the seasonal and areal distribution of snow and ice exert strong influences on the planetary energy balance through their impact on Earth's albedo", primarily in summer. During the winter, Arctic areas receive almost no sunlight so albedo does not matter.
- Ice-insulation feedback—"Sea ice also causes a positive insulation feedback, primarily in winter. By insulating the cold polar atmosphere in winter from the relatively warm ocean, little of the ocean's energy can be transferred to the atmosphere."
- Vegetation feedbacks: "A warming climate can cause tundra to give way to lower albedo shrub vegetation or even dark-green boreal forest. The lower albedo of shrubs and boreal forest, especially in spring when high-albedo snow cover may still bury tundra, results in earlier warming and hence exerts a positive feedback on warming."
- Permafrost feedbacks: "... poorly understood feedbacks in the Arctic involve changes in the extent of permafrost, and how changes in cloud cover interact both with permafrost and with the release of carbon dioxide and methane from the land surface. Melting allows ancient plant debris to decompose, releasing greenhouse gases (CO_2 and/or CH_4) that mix globally, amplifying the initial warming by enhancing the planetary greenhouse effect."
- Feedbacks during glacial-interglacial cycles: "... slow positive feedbacks that operate on time scales of 10^4 years were important contributors to glacial-interglacial climate cycles." These included changes in albedo due to growth of ice sheets, changes in ocean volume changing land/sea areal balance, changes in vegetation, changes in water vapor, changes in dust levels, changes in greenhouse gas concentrations, etc.

Miller *et al.* (2010a) also discussed other feedbacks involving fresh water inputs to the Arctic seas and changes in thermohaline circulation.

According to NRC (2011):

> "A key mechanism driving increased warming in the polar regions is the albedo feedback effect caused by variations in sea-ice cover, snow cover, and in the Arctic (broadly defined herein to include northern tree line boreal vegetation), forest cover. In addition, changing atmospheric and oceanographic circulation patterns also lead to increased regional warming in the Arctic and Antarctic."

Serreze and Barry (2011) said:

> "Arctic amplification is now recognized as an inherent characteristic of the global climate system, with multiple intertwined causes operating on a spectrum of spatial and temporal scales. These include, but are not limited to, changes in sea ice extent that impact heat fluxes between the ocean and the atmosphere, atmospheric and oceanic heat transports, cloud cover and water vapor that alter the long wave radiation flux to the surface, and soot on snow and heightened black carbon aerosol concentrations."

They also pointed out that Arctic amplification is "stronger during the autumn and winter seasons, and is much weaker for spring and summer".

Holland and Bitz (2003) discussed attempts to characterize Arctic amplification by means of climate models. Climate model simulations have included ice-albedo feedbacks associated with variations in snow and sea-ice coverage, variations in the thickness of sea-ice, clouds, and changes in heat transported by the atmosphere and/or ocean. There is agreement among models that the Arctic warms more than sub-polar regions whereas high southern latitudes exhibit a minimum warming due to changes in ocean heat uptake. Figure 3.22 shows the range of estimates of Arctic amplification by various climate models.

As we have discussed, any change in global temperature will be amplified in the Arctic due to various feedbacks. In the last century or two, an additional factor tended to amplify warming in the Arctic, not by amplification of a global trend, but by direct increases of solar energy absorbed at higher latitudes. This is due to deposition of soot on high-latitude snow and ice from power plants and industry in the NH. This is discussed further in Section 7.3.

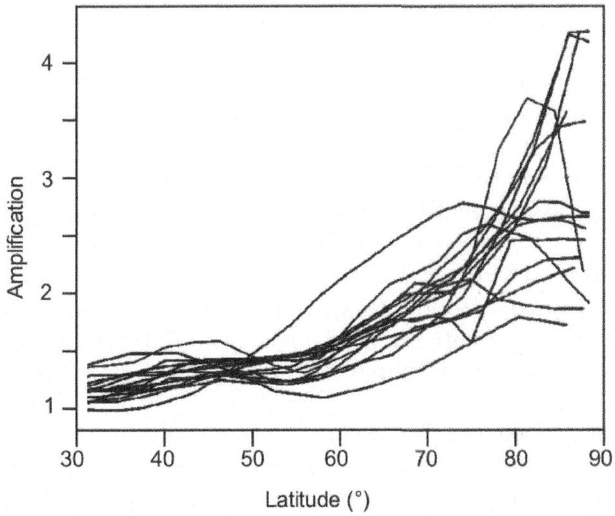

Figure 3.22. The relative amplification of air temperature 2 m above the surface as a function of latitude, as estimated by various climate models (Holland and Bitz, 2003).

Spielhagen *et al.* (2011) pointed out that "northward flowing Atlantic water is the major means of heat advection toward the Arctic and strongly affects the sea ice distribution". Continuous historical records reach back only ~150 years. They used marine sediments off Western Svalbard (79°N) to estimate a multi-decadal-scale record of ocean temperature variations during the past 2,000 years. They found that early-21st-century temperatures of Atlantic water entering the Arctic Ocean are unprecedented over the past 2,000 years. Whether this is a cause of, or a result of, Arctic amplification of global warming, remains unclear.

In late 2011, a team led by Richard Muller released a preliminary version of the BEST that estimated global land temperatures back to the year 1800. (Note that land covers only about 30% of the globe. Global temperatures will not change as rapidly as land temperatures because of thermal inertia of the oceans.) The worldwide distribution of stations was not provided, but Figure 3.4 shows the distribution for the U.S. The majority of U.S. stations were poor. However, they claimed that "the mean temperature trends are nearly identical across site classifications" for the U.S. They concluded that:

> "... poor station quality in the United States does not unduly bias estimates of land surface average monthly temperature trends. No similar study is possible for the rest of the world because we do not have indicators of good/bad station quality; however, the lack of a significant difference in U.S. stations suggests that such effects may be minimal."

Certainly, this is, at best, optimistic for the U.S.; for the world, it is pure speculation. Taken at raw face value, this suggests that the U.S. warmed at a much slower rate than the rest of the world. Since the U.S. presumably has the best station network in the world (or the least-worst network), this should be taken into account. But this does not seem to be very credible, considering the siting issues discussed in Section 3.1 and particularly the effect of UHIs discussed in Section 3.1.3.

It seems likely that station coverage for the period 1800–1880 was heavily weighted by U.S. and European locations. There are many unresolved problems with this analysis.

They noted: "One immediate observation is that for all categories, about 1/3 of the sites have negative temperature trends, i.e. cooling over the duration of their record."

As we pointed out in Figures 3.61 and 3.7, one approach for correcting for the urban heat effect is to compare the temperature trend over a period of years of an inner city with that in an outlying rural area, assume the rural measurement is correct, and adjust the city measurement to have the same trend as the rural measurement.

Keenan[15] said that the conclusion that "there has been some urban cooling ... contradicts over a century of research as well as common experience. It is almost certainly incorrect". McIntyre[16] commented on corrections for UHI used by various

[15] *www.informath.org/apprise/a5700.htm.*
[16] *http://climateaudit.org.*

investigators in the past involving "comparisons of supposedly 'urban' and supposedly 'rural' subpopulations in papers by Jones, Peterson, Parker and others purporting to prove that UHI doesn't 'matter' ". As he said: "Such papers set up two populations—one 'urban' and one 'rural', purport to show that the trends for each population are similar and claim that this 'shows' that UHI is a non-factor in trends." But McIntyre pointed out that "the two populations are very poorly stratified—with the 'rural' population all too often containing urban cities, sometimes even rather large cities". McIntyre also pointed out that, according to Oke's UHI concept, the UHI effect is proportional to log (population). He said: "If 'urbanization' is occurring in towns and villages as well as in large cities—which it is, then the contribution of UHI increase to temperature increase will depend on the percentage change in population (rather than absolute population). If proportional increases are the same, then the rate of temperature increase will be the same in towns and villages as in cities." In other words, UHI is artificially increasing the apparent rate of temperature rise for both "rural" and "urban" areas.

There are many challenges involved in reconstructing a best estimate of past global average land temperatures from station data. Data are available from a large number of sites with highly variable quality over highly variable time periods. Stations not only vary in quality (location, equipment, changes in population and industrialization, etc.), but also in elevation and other geographic features. Over the years, many changes may typically have occurred in the vicinity as populations grew and industrialization spread. These station data must be combined in some statistical model to perform an average. The methodology used was described by Rohde *et al.* (2011). The details are intricate and beyond the scope of this book. However, McIntyre[17] commented on the procedure introduced by Menne called "slicing" that "introduces thousands of breaks in noisy and somewhat contaminated data". This involves: "breaking time series into independent fragments at times when there is evidence of abrupt discontinuities, and adjusting the weights within the fitting equations to account for differences in reliability". McIntyre suggested that this procedure might "remove more negative steps than positive steps, [thus] increasing the trend of the final temperature series".

Two statisticians (D. J. Keenan and William M. Briggs) posted critiques of the methodology on the Internet.[18] Briggs argued that the "general point estimate is probably roughly in the ballpark, more or less, plus or minus, but the uncertainty bounds are far too narrow". He argued that even the stated uncertainties given by the BEST Team provide such a wide swath in the early 1800s that one cannot tell whether 20th-century temperatures were higher or lower. Briggs argued that the uncertainties "should be multiplied by at least 5 to 10". Nevertheless, it seems likely that, as Briggs put it, the "general point estimate is probably roughly in the ballpark".

In a more recent release by the BEST Team (July 2012), the data were extended

[17] *http://climateaudit.org/2011/10/31/best-menne-slices/*.

[18] *www.informath.org/apprise/a5700.htm* and *http://wmbriggs.com/blog/?p=4530*.

even further back in time. Rohde *et al.* (2012) reported an estimate of the Earth's average land surface temperature for the period 1753–2011. To address issues of potential station selection bias, they used a larger sampling of stations than prior studies. For the period post 1880, their estimate was similar to those previously reported by other groups, although they claimed smaller error uncertainties. They claimed that the land temperature rise from the 1950s decade to the 2000s decade was 0.87 $\pm$ 0.05° C (95% confidence). Both maximum and minimum daily temperatures increased during the last century. According to them, diurnal variations decreased from 1900 to 1987, and then increased; this increase is significant but not understood.

Figure 3.23 shows a 10-year moving average of global land temperatures derived by BEST. Figure 3.24 shows yearly data along with the timing of major volcanic eruptions with explosive indices shown on the graph.

The period from 1753 to 1850 was claimed to be marked by sudden drops in land surface temperature that were coincident with volcanism; they claimed that the response function was approximately 1.5 $\pm$ 0.5°C per 100 Tg of atmospheric sulfate. According to their interpretation, this volcanism, combined with a simple proxy for anthropogenic effects (logarithm of the CO_2 concentration), can account for much of the variation in the land surface temperature record and they claimed that the fit was not improved by the addition of a solar forcing term. Thus, for their very simple model, they concluded that solar forcing does not appear to contribute to the observed global warming of the past 250 years and they believe that the entire change can be accounted for by a sum of volcanism and anthropogenic proxies.

The large oscillations prior to 1900 were attributed to volcanoes. It is not clear why the apparent temperature effects of early volcanoes were so much greater than the effects of more recent volcanoes. Furthermore, the downward trends in the two

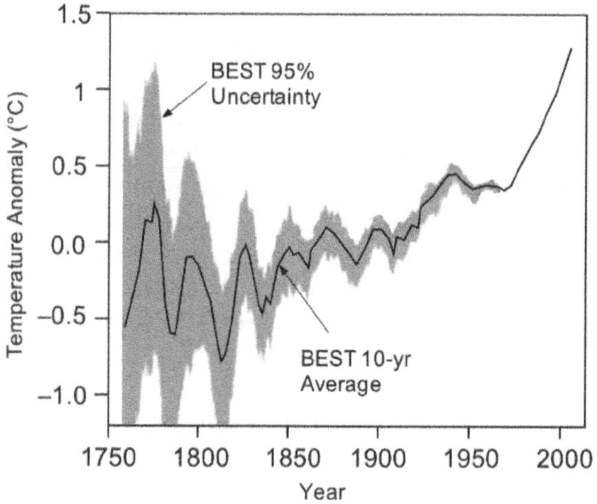

Figure 3.23. BEST 10-year moving average of global land temperatures.

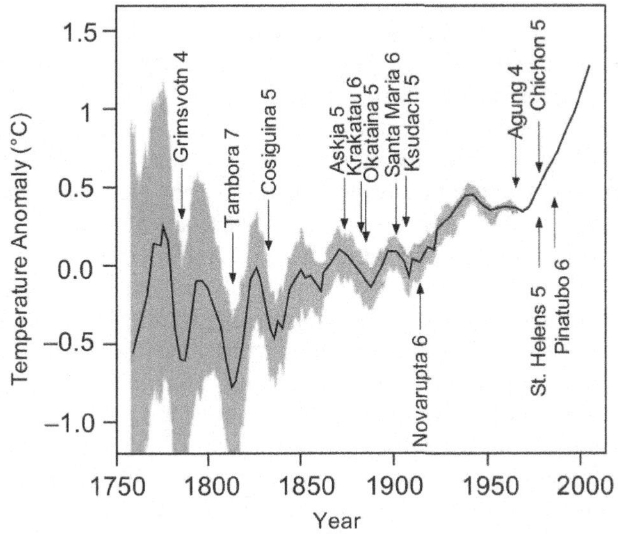

Figure 3.24. BEST 10-year average data showing dates of major volcanic eruptions with volcanic explosive indices.

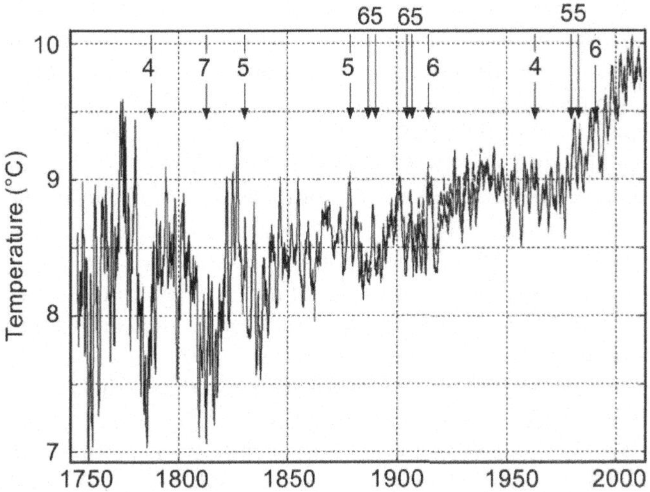

Figure 3.25. BEST yearly land temperature data compared to volcanic eruptions with indicated volcanic explosive indices.

earliest volcanoes seem to have begun prior to the eruptions (see Figure 3.25). The large oscillations prior to 1850 might be artifacts if the limited number of stations that were positioned in regions particularly susceptible to effects of volcanic eruptions. In particular, the eruption of 1783 (*Grimsvotn*) with an explosive index of only four seems to have had a far greater climatic effect than much more powerful eruptions that occurred later. It has been estimated that *Grimsvotn* released only 12.3 km^3 whereas *Tambora* (1815) released 160 km^3; yet they had similar climatic effects according to BEST. The volcanic explosive index is logarithmic so the volcanoes in the 19th and 20th centuries were 10–100 times greater than *Grimsvotn*; yet, according to BEST, *Grimsvotn* had a much greater climatic effect. This seems unlikely to be correct.

The folklore on the *Tambora* eruption is extensive. In New England, 1816 has been described as the "year without a summer". There are numerous references on the Internet to freezing and snow in June 1816, as well as crop failures. Yet, the sources of these myths do not seem to be very well substantiated. In a web posting,[19] W. Eschenbach cast doubt on these myths by showing that archival data in several European cities did not seem to display unusually cold temperatures in 1816. He also showed that food commodity prices were at a low in 1816, which is contrary to what one would expect from bad weather. There were 300 responses posted to Eschenbach's posting, the overwhelming majority of which had no value. One useful response pointed to a Hadley Centre Central England Temperature (HADCET) database that showed that the summers of 1816 and 1817 were indeed relatively cold compared with the extended period 1659–2012, with July 1816 being the coldest July in the whole set of 353 years. A website[20] provides considerable detail on unusually cold weather in the summer of 1816 in the eastern U.S. It is not clear where these data derives from and how valid it might be.

Another paper[21] used "recently recovered meteorological observations from 1816 onwards for three stations located in Portugal (Lisbon, Coimbra and Oporto) and also for a longer period for the Spanish stations of Madrid, Barcelona and San Fernando-Cadiz, [to produce] a better characterization of the anomalous climate for this peculiar period over southwestern Europe". They reported that "all available stations reveal a cold summer of 1816, mainly in July and August. In comparison to the 1871–1900 reference period, those two months were 2–3°C cooler, close to what has been reported for central Europe".

The BEST Team has offered no explanation for why the early volcanoes supposedly had a remarkable impact on climate, while later more powerful volcanoes had a minor effect. It seems likely that either the early data are wrong, or the explanation in terms of volcanoes is wrong.

Another aspect of the BEST analysis of their data is the implicit assumption that the climate would have been steady, had it not been for the influences of rising CO_2, solar variations, and volcanic eruptions. Thus, they neglect internal influences (the possibility that the climate will change without external forcing). Yet we know (for example) that an internal change occurred after 1976 in the El Niño–La Niña balance that produced a heating trend from 1976 to 1998. Furthermore, as we discuss in Chapter 5, there are no reliable estimates of solar forcing dating back 250

[19] *http://wattsupwiththat.com/2012/04/15/missing-the-missing-summer/.*

[20] *www.islandnet.com/~see/weather/history/1816.htm.*

[21] Trigo, R. M. *et al.*, (2009) "Iberia in 1816, the year without a summer", *Int. J. Climatol.* **29**, 99–115.

years. It does seem likely that solar variations cannot come close to accounting for the observed data; yet we still lack reliable solar data.

3.4.3 Troposphere temperatures

3.4.3.1 Measured troposphere temperatures

The current widely accepted TT measurements are shown in Figure 3.26.

The results shown in Figure 3.26 can be interpreted in more than one way. Alarmists assume that the prime driver for temperature change is rising CO_2 concentration, and therefore they draw an upward sloped line through the data in this figure as shown in Figure 3.27. An alternative interpretation is a step function with a fairly constant plateau prior to the great El Niño of 1998, with an upward step to a new higher plateau after 1998. A least-squares fit of the two fits to the data shows that both fit the data with roughly equal validity.

It is of particular interest to note that, since the great El Niño of 1998, global temperatures have meandered around a horizontal plateau and the average from 1998 through 2013 has essentially been constant, as shown in Figure 3.28.

As Carlin (2009) pointed out, these results are in conflict with the predictions of the IPCC that temperatures would escalate upward as CO_2 emissions increased during the past decade. Carlin said:

> "What's really rather remarkable, is that since 2000, the rates at which CO_2 emissions and concentrations are increasing have accelerated. According to Canadell *et al.* (2007), fossil fuel and cement emissions increased by 3.3 percent per year during 2000–2006, compared to 1.3 percent per year in the 1990s.

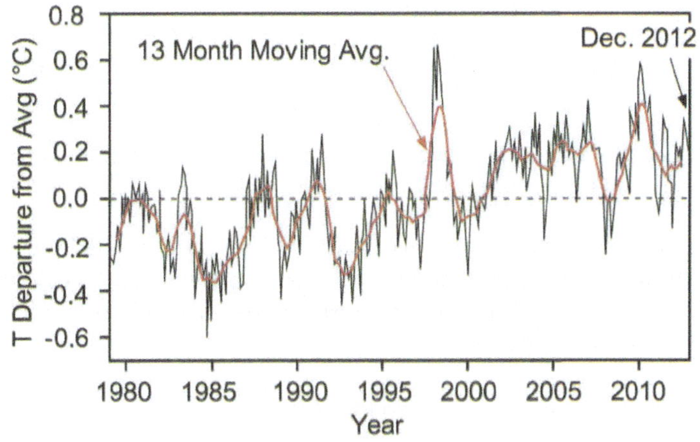

Figure 3.26. Monthly and 13-month running average of tropospheric temperature over three decades (adapted from Christy *et al.*, 2007, as updated by Spencer in 2012, *www.drroyspencer.com*).

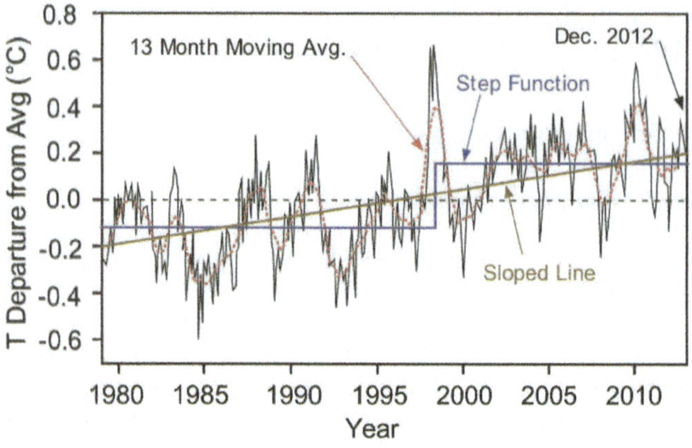

Figure 3.27. Globally averaged satellite-based temperature of the lower atmosphere (*www.drroyspencer.com*).

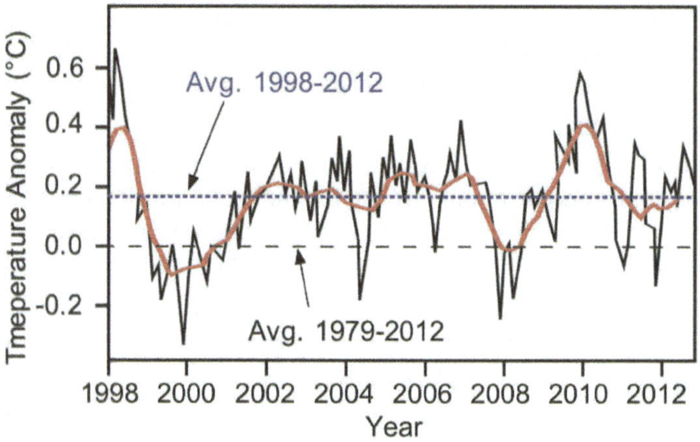

Figure 3.28. Globally averaged satellite-based temperature of the lower atmosphere from 1998 through 2012 (*www.drroyspencer.com*).

Similarly, atmospheric CO_2 concentrations increased by 1.93 parts per million per year during 2000–2006, compared to 1.58 ppm in the 1990s. And yet, despite accelerating emission rates and concentrations, there's been no net warming in the 21st century, and more accurately, a decline."

3.4.3.2 Temperature plateau: 1998 through 2012

Obviously, the lack of rising TT measurements during the 15-year period from 1998 through 2013 is contrary to predictions that rising CO_2 will produce constantly rising temperatures. Climate skeptics immediately claimed that this 15-year hiatus constituted a failure of climate models that predict ever-rising temperatures. On the other hand, it is possible that short-term weather fluctuations might hide the underlying rise in global average temperature, and, given a sufficiently long period of measurement, the 15-year hiatus might prove to be a temporary variation from the underlying trend. The question then is how statistically significant is a 15-year hiatus in rising temperatures?

In keeping with recent practice over the past few years in which alarmists promptly publish rebuttals to any papers that slip through their control of which manuscripts get accepted by climate journals, it was necessary for the alarmists to publish such a rebuttal to the claims of climate skeptics. Ben Santer took on this responsibility and the result was Santer *et al.* (2011). It is interesting, perhaps, that Santer included 16 co-authors in addition to himself; yet the nature of the work is such that it is difficult to imagine how 16 individuals could each contribute significant portions to the work. Pielke Sr.[22] commented: "This is an unusual number of co-authors for a technical paper, but I assume Ben Santer wants to show a broad agreement with his findings." In other words, many names were added to give the paper political endorsement; yet many co-authors made little or no technical contribution?

Santer *et al.* (2011) were concerned with a very basic problem in climatology: how to distinguish between long-term climate change and short-term variable weather in regard to TT measurements. They treated the problem in terms of signal and noise: the signal is assumed to be a long-term linear trend or rising temperatures due increasing greenhouse gas concentrations, that is obfuscated by short-term noise. However, the climate–weather problem is innately different from a classical signal/noise problem such as a radio signal affected by atmospheric activity. In that case, if the radio signal has a sufficiently narrow frequency band, and the noise has a wider frequency spectrum, the signal-to-noise ratio (S/N) can be improved with a narrow-band receiver tuned to the frequency of the radio signal. The radio signal and the noise are separate and distinct. By contrast, in the climate–weather problem, the instantaneous weather is the noise, and the signal is the long-term trend of the noise. The noise and signal are coupled in a unique way. Furthermore, there is no evidence that it is even meaningful to talk about a "trend", since there is no evidence that the variation of TT with time is linear. Remarkably, Santer *et al.* never referred to Christy *et al.* (2010) but based their analysis on older papers (e.g. Christy *et al.*, 2007). However, it should be noted that whereas Christy *et al.* (2010) used tropical lower troposphere (TLT) data (20°N–20°S), Santer *et al.* used global TT data (82.5°N–70°S). It is also important to note that Santer *et al.* performed their analysis based on a 10-year period of unchanging average temperature. Yet, including the latest data from Roy Spencer through December 2012, the period of unchanging average temperature is now 15 years.

Santer *et al.* used an average of several climate models to estimate the global TT over the period 1979–2011. A comparison of their estimate with the actual data is given in Figure 3.29. It is not clear to this writer what parameters were adjusted in

[22] *http://pielkeclimatesci.wordpress.com/category/climate-models/*.

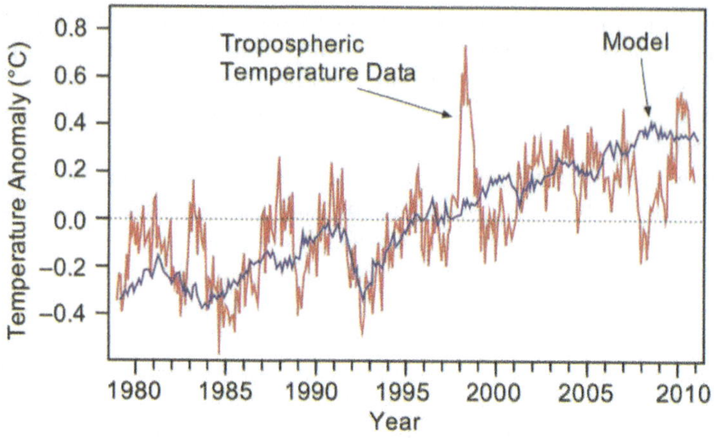

Figure 3.29. Comparison of global tropospheric temperature data with Santer *et al.* (2011) model.

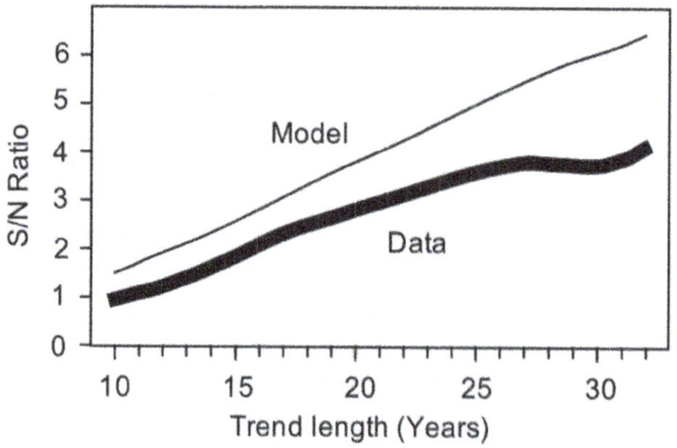

Figure 3.30. S/N estimates by Santer *et al.* (2011).

these models. It is evident that the models provide much smaller variations than the data.

Santer *et al.* (2011) were primarily concerned with estimating how many years of data are necessary to provide a good estimate of the putative underlying linear trend. They were also intent on showing that short periods with no apparent trend do not violate the possibility that over a longer term, the trend is always there. They derived S/N ratios for both the temperature data and the model average by assuming straight-line fits and estimating the standard deviation of the model and the data from the straight-line fits. These results are shown in Figure 3.30. Evidently, the modeled S/N is higher because the model has smaller excursions from its trend. (The model could not predict the wide excursions in the data due to El Niño and La Niña, although it did allow for effects of volcanic eruptions.) In essence, the higher S/N of the model merely indicates that the model agrees with a straight-line fit better than the data—which is exactly what skeptics have emphasized.

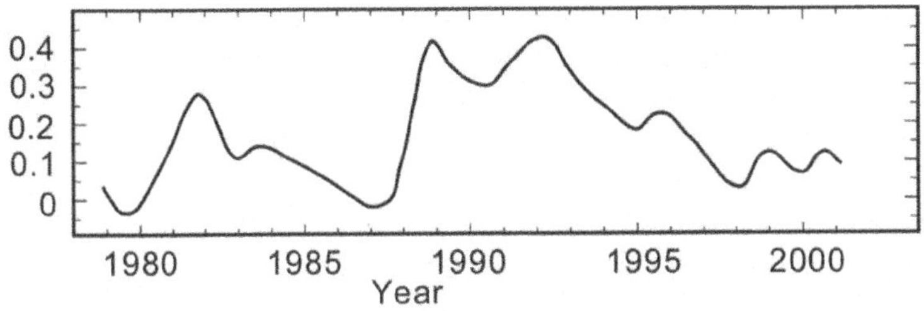

Figure 3.31. Trend (°C/decade) of TT vs. start year for 10-year durations. (Santer *et al.*, 2011).

As Santer *et al.* (2011) indicated, one can pick any starting date and any duration length and fit a straight line to that portion of the data curve of TT vs. time. They did this for various 10-year and 20-year durations. In each case, depending on the start date, they derived a best straight-line fit to the TT data for that time period. They found that the range of trends for 10-year periods was greater (–0.05°C to +0.44°C/decade) than for 20-year periods (+0.15°C to +0.25°C/decade) depending on which time period was chosen. The trends for various start dates for 10-year trends are shown in Figure 3.31. Clearly, the trend line was steepest for a start date around 1988 (ending in the giant El Niño year of 1998). Prior to 1988 and after 1998, the trends were minimal.

Santer *et al.* described use of longer durations as "noise reduction", which it is in a sense, provided that one assumes the overall signal is linear in time. It still was problematic that the trend was nil after 1998, which they rationalized by saying:

> "The relatively small values of overlapping 10-year TT trends during the period 1998 to 2010 are partly due to the fact that this period is bracketed (by chance) by a large El Niño (warm) event in 1997/98, and by several smaller La Niña (cool) events at the end of the ... record."

However, as Pielke, Sr. pointed out, the period from 1998 through 2010 was 13 years, not 10 as claimed by Santer *et al.*, and, furthermore, the period after 1998 had roughly equal periods of El Niño and La Niña and was not dominated by La Niñas as they claimed. What Santer *et al.* (2011) implied was that an unusual conflux of a large El Niño early on and multiple La Niñas later on caused the trend to minimize for that unique period as a statistical quirk.

In simplistic terms, the S/N ratio can be estimated as follows. For either 10-year or 20-year durations, the signal was the mean trend (i.e., slope of straight-line fit) derived by a straight-line fit to the TT data over that duration. The noise was the range of trends for different starting dates. For 10-year durations, the trend was 0.19 ± 0.25°C/decade. For 20-year durations, the trend was 0.20 ± 0.05°C/decade. The signal in each case is taken as the mean trend. The distribution of trends within these ranges was similar to a normal distribution. Thus we can roughly estimate the noise as ~0.7 times the full width of the range. Hence, the S/N ratio for 10-year durations can be crudely estimated to be S/N $\sim 0.19/(0.7 \times 0.5) = 0.5$ and for 20-year durations is S/N $\sim 0.2/(0.7 \times 0.1) = 2.9$. Santer *et al.* obtained S/N = 1 for 10-year durations and S/N = 2.9 for 20-year durations.

If it can be assumed that the signal varies linearly with time, one can then estimate what level of precision for the estimated trend can be obtained for any chosen duration. Santer *et al.* obviously believe that the underlying signal is linear with time for all time. In my discussion, I have relied entirely on the TT data and I have not included predictions of models. However, the paper by Santer *et al.* mixes up models with TT data and it is sometimes difficult to separate these. By some logic that escapes me, Santer *et al.* concluded:

> "Our results show that temperature records of at least 17 years in length are required for identifying human effects on global-mean tropospheric temperature."

This conclusion seems to be grossly exaggerated. A more proper statement might be as follows:

> *Assuming that the variability of TT is characterized by a long-term upward linear trend caused by human impact on the climate, and that variability about this trend is due to yearly variability of weather, volcanic eruptions, El Niños and La Niñas, and other climatological fluctuations, the recent data suggest that the trend can be estimated for any 17-year period with a S/N ratio of roughly 2.5.*

Finally, we get to the nub of the paper by Santer *et al.* that asserted: "Claims that minimal warming over a single decade undermine findings of a slowly-evolving externally-forced warming signal are simply incorrect." Here is where Santer *et al.* attempted to dispel the notion that minimal warming for a period contradicts the belief that, underneath it all, the long-term signal continues to rise at a constant rate. Pielke Sr. argued that this was an overstatement and should be replaced by:

> "If one accepts this statement by Santer *et al.* as correct, than what should have been written is that the observed lack of warming over a 10-year time period is still too short to definitely conclude that the model's are failing to skillfully predict this aspect of the climate system."[23]

However, I would go further than Pielke Sr. First of all, the period of minimal temperature rise was not 10 years, and is now 15 years. Second, there is no cliff at 17 years whereby trends derived from shorter periods are statistically invalid and trends derived from longer periods are valid. There is just a continuously improving S/N ratio as the duration increases. According to Santer *et al.*, a trend derived from a 15-year period is associated with a S/N $\sim$2 which, though not ideal, is good enough to cast some doubt on the validity of models.

In summary, Santer *et al.* (2011) assumed that the variation of TT with time over the past 32 years followed a long-term straight line (presumably due to forcing by greenhouse gases (the signal)) with superimposed yearly variations due to El Niños, volcanoes, and chaotic weather changes acting as noise. Within this time period, one can fit a straight line to the TT data for any duration and start date.

[23] *http://pielkeclimatesci.wordpress.com/category/climate-models/.*

They showed that the S/N ratio for such a segment of the timeline increases as the duration increases, and claimed that a segment of at least 17 years' duration is needed to obtain a "good" estimate of the long-term trend. However, it is not clear from the data that a straight line plus noise best represents reality. Another interpretation is that TT was relatively flat prior to the giant El Niño of 1998, jumped up after that El Niño, and then remained relatively flat afterward but at a higher level. The fact that TT was relatively flat for 15 years after 1998 suggests (but does not yet prove) that the model of a long-term straight line plus annual noise is probably not valid.

While climate models predict a steady rise in global average temperature of roughly 0.2°C/decade, the 15-year period starting in 1998 has been a period of meandering temperatures about a flat trend. This generated great glee in the skeptics camp, whereas the warmists have felt the need for damage control. Santer *et al.* (2011) was an attempt to explain away this period as a temporary fluctuation. Pielke Sr.[24] provided a set of commentaries by prominent climatologists regarding the lack of warming after 1998 (originally published by Paul Voosen). As he said:

> "For some scientists, chalking the hiatus up to the planet's natural variability was enough. Temperatures would soon rise again, driven up inexorably by the ever-thickening blanket thrown on the atmosphere by greenhouse gases. People would forget about it.
>
> But for others, this simple answer was a failure. If scientists were going to attribute the stall to natural variability, they faced a burden to explain, in a precise way, how this variation worked. Without evidence, their statements were no better than the unsubstantiated theories circulated by climate skeptics on the Internet."

One argument made by several warmists (including Santer *et al.*, 2011) is that the period after 1998 was unusual because it began with the warmth of a giant El Niño in 1998 and was followed by a La Niña in 2008. However, this is misleading because the period from 2002 to 2007 was one of continuing moderate El Niños. On balance, the climate of the past 30 years has followed the path of the Niño index (see Figure 3.32). Another argument was that this lack of temperature rise was due to a resurgence of aerosol emissions, particularly from China. There may be some partial validity to this claim but, at this point, it is vague and not substantiated in a credible manner.

3.4.3.3 Climate: El Niños and volcanic eruptions vs. CO_2

Douglass and Christy (2009) compared their satellite-measured temperatures with an El Niño index (*NINO3.4*). The *NINO3.4* index is one of several El Niño /Southern Oscillation (ENSO) indicators based on SSTs. This index is the average SST anomaly in the region bounded by 5°N–5°S, from 170°W to 120°W.

[24] *http://pielkeclimatesci.wordpress.com/2011/10/27/candid-comments-from-global-warming-climate-scientists/.*

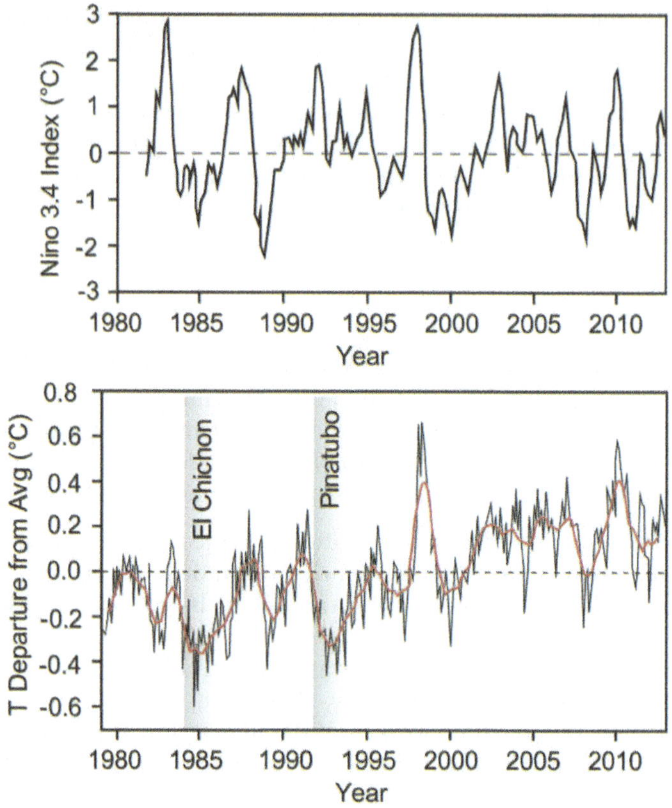

Figure 3.32. Measured tropospheric temperatures compared to Nino 3.4 Index with two volcanic eruptions also shown.

As Figures 3.32 and 3.33 show, satellite-measured temperatures have meandered during the past three decades and there is evident correlation of cooling periods with the aftermaths of volcanic eruptions, and heating trends with the occurrence of El Niños. In particular, the great El Niño of 1998 (one of the strongest ever recorded) produced an unusual spike in temperature that was widely heralded by alarmists as "proof" of their claim that increased CO_2 was producing runaway global warming. This figure suggests that it was the state of the surface waters of the Pacific Ocean (rather than CO_2 build-up) that controlled the Earth's climate since about 1980.

Christy *et al.* (2010) "updated tropical lower tropospheric temperature datasets covering the period 1979–2009" and assessed them for accuracy. As Christy *et al.* (2010) pointed out:

> "The temperature of the tropical lower troposphere (TLT, 20°S–20°N) figures prominently in discussions of climate variability and change because it (a) represents a major geographic portion of the global atmosphere (about one third) and (b) responds significantly to various forcings. For example, when the

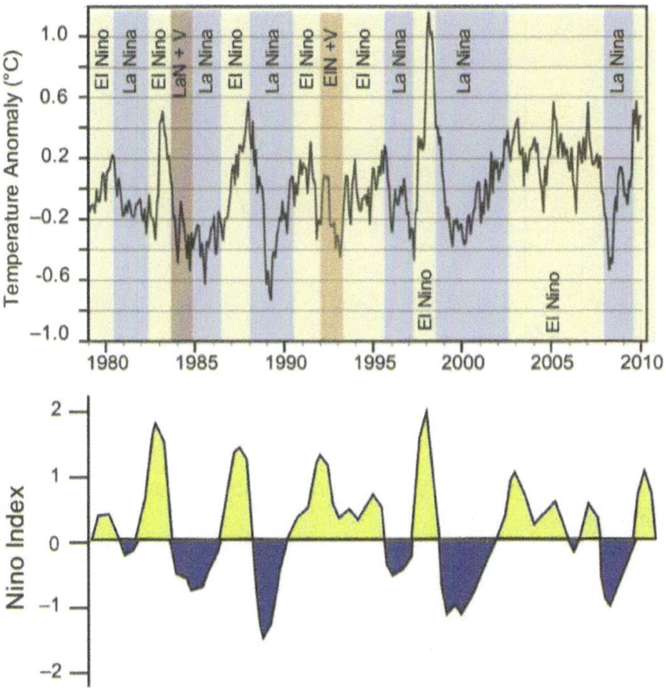

Figure 3.33. Global temperature anomaly compared to the Nino 3.4 index showing strong correlation of upticks in temperature with El Niños. Brown patches show where volcanic eruptions occurred.

> ENSO mode is active, TLT displays a highly coupled, though few-month delayed, response, with a general warming of the tropical troposphere experienced during El Niño events. The TLT also responds readily to the impact of solar scattering anomalies when substantial volcanic aerosols shade the Earth following major volcanic eruptions. ... In terms of climate change due to increasing greenhouse gases ..., climate models project a prominent warming of the TLT which in magnitude is on average twice as large ... as changes projected for the surface."

3.4.3.4 Tropospheric and surface temperatures

FORCING AT THE TROPOPAUSE VS. THE SURFACE DUE TO INCREASED CO_2

Quite a number of studies have estimated the forcing (additional downward back IR radiation) due to an increase in the CO_2 concentration from pre-industrial levels. Most of these estimated the forcing at the top of the atmosphere (TOA) for a doubling of CO_2. However, we are interested in the downward back IR radiation at the surface. The estimates given in this section do not include feedbacks.

Newell and Dopplick (1979) estimated that the IR flux change at the surface due

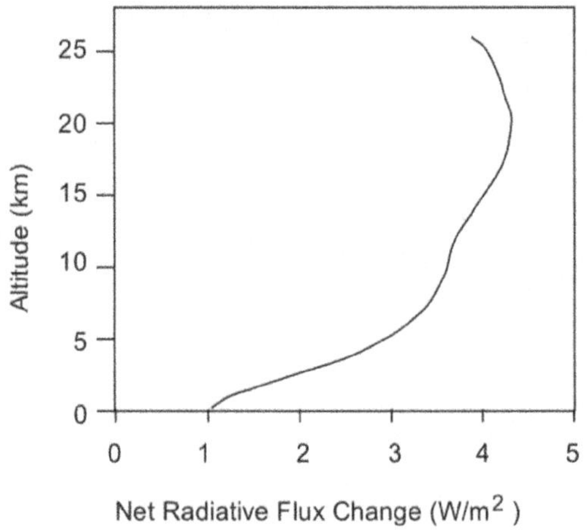

Figure 3.34. Forcing due to doubling CO_2 from pre-industrial value as a function of altitude (Ramanathan, 1981).

to increasing the CO_2 concentration from 330 ppm to 600 ppm would be latitude-dependent, and would vary from roughly 0.8 W/m^2 at low latitudes to roughly 1.5 W/m^2 at high latitudes. When the calculation was repeated for clear skies, this range increased to 1.1–2.6 W/m^2.

Ramanathan (1981) estimated the forcing as a function of altitude. He pointed out:

> "The troposphere as a whole is subjected to a net radiative heating of about 3.5 W/m^2 (and ~3 W/m^2 on a global average basis) for a doubling of CO_2, which is roughly a factor of 3 larger than the surface heating." (Note: He added an increase in the downward emission from the stratosphere by about 1.2 W/m^2 to obtain the total forcing at the tropopause; his profile for forcing from doubling CO_2 is shown in Figure 3.34.)

As Ramanathan (1981) put it:

> "It is commonly stated that CO_2 absorbs upwelling radiation and then re-emits it to the surface as back radiation. The CO_2 bands overlap with water vapor bands whose opacity is so large that most of the back radiation from CO_2 is absorbed by the intervening layer of H_2O. As a result, the CO_2 back radiation at the surface increases by only 1.2 W/m^2 as opposed to the 4.3 W/m^2 tropopause radiative forcing."

Lindzen (2007) analyzed several general circulation models and showed that "warming is strongly peaked in the tropical troposphere". He went on to conclude:

> "Roughly speaking, the warming [near the tropopause] in the tropics is ... twice to about three times larger than near the surface regardless of the sensitivity of the particular model. This is, in fact, the signature (or fingerprint)

of greenhouse warming. Stated somewhat differently, if we observe warming in the tropical upper troposphere, then the greenhouse contribution to warming at the surface should be between less than half and one third the warming seen in the upper troposphere. ... The modeling studies establish that the ratio of upper tropospheric tropical warming to surface warming is approximately 2.5:1 regardless of the model sensitivity."

Collins *et al.* (2006) asserted "The interaction of short wave and long wave radiation with an (idealized) atmosphere free of clouds and aerosols can be calculated to a very high degree of accuracy". They said "the introduction of clouds would greatly complicate the ... exercise and therefore clouds are omitted from [the calculations]". They said: "Flux is defined as flux for clear-sky and aerosol-free conditions and forcing" and was without stratospheric adjustment. They also said "the effects of adjustment on forcing are approximately –13% for CO_2". Collins *et al.* estimated clear sky forcing at (1) the top of the model, (2) at a pressure of 200 mb (surrogate for tropopause) and (3) at the surface. Their results are given in Table 3.1.

In a later paper, Iacono *et al.* (2008) improved the previous clear-sky estimates with better models that included more layers in the model atmosphere. Their results are shown in Table 3.2. In both cases (Collins *et al.* and Iacono *et al.*), the forcing at the surface is considerably less than at the troposphere or the top of the model.

Table 3.1. Clear-sky forcing at various altitudes for various changes in greenhouse gas concentrations (Collins *et al.*, 2006).

	Forcing (W/m^2)		
Change	*Top of model*	*200 mb*	*surface*
CO_2 goes from 287 ppm to 369 ppm	0.9	1.8	0.4
CO_2 goes from 287 ppm to 574 ppm	2.5	5.2	1.3
All greenhouse gases go from year 1860 to year 2000	2.2	3.0	1.3

Table 3.2. Clear-sky forcing at various altitudes for various changes in greenhouse gas concentrations (Iacono *et al.*, 2008).

	Forcing (W/m^2)		
Change	*Top of model*	*200 mb*	*surface*
CO_2 goes from 287 ppm to 369 ppm	1.1	2.0	0.6
CO_2 goes from 287 ppm to 574 ppm	3.0	5.7	1.7
All greenhouse gases go from year 1860 to year 2000	2.1	3.0	1.1

The effect of clouds can be surmised from the work of Schmitt and Randall (1991) who pointed out: "Clouds influence the surface energy budget through cloud shadows, by downward emission of infrared radiation from cloud base and by blocking downward infrared radiation emitted above the level of the cloud. Through these various effects, the clouds can modulate the CO_2 forcing." They included clouds in their climate model. They "evaluated the CO_2 forcing, [by running] the radiation code twice; once with a CO_2 mixing ratio of 330 ppm, and a second time with 660 ppm. The CO_2 forcing is then obtained as the difference in the long wave radiation fields between these two cases". Schmitt and Randall (1991) divided the atmosphere into 11 layers and assumed a distribution of clouds. They used a GCM to estimate the forcing due to a doubling of CO_2 from the 330 ppm level. Their test runs indicate that the forcing due to doubling CO_2 maximizes near the tropopause and decreases at higher and lower altitudes. Surface forcing is considerably lower than that at the TOA. Averaged over all latitudes, they found that forcing at the surface was of the order of ~1.5 W/m^2 for clear skies and 1.0 W/m^2 when clouds were included. At the tropopause, forcing was of the order of ~5.1 W/m^2 for clear skies and 4.4 W/m^2 when clouds were included. These results for clear skies are comparable to those of Iacono *et al.* (2008).

From this work, we may conclude that the forcing at the surface due to a doubling of CO_2 from the pre-industrial value of ~280 ppm would be about 1.0–1.2 W/m^2 when clouds are included. The forcing at the surface in going from the pre-industrial level of CO_2 to the present value (~395 ppm) is about 0.6 W/m^2. The average forcing over the past 55 years was roughly 0.4 W/m^2. The forcing at the tropopause is several times that at the surface.

Comparison of warming at tropopause with warming at the surface

Santer *et al.* (2005) emphasized that "a robust feature" of climate models is that increasing greenhouse gas concentrations will amplify warming in the middle and upper tropical troposphere (compared to the surface). It was then with some consternation that they noted that the data do not support this prediction; indeed, surface warming typically exceeds tropospheric warming. As Klotzbach *et al.* (2009) pointed out:

> "Santer *et al.* (2005) presented three possible explanations for this divergence: (1) an artifact resulting from the data quality of the surface, satellite and/or radiosonde observations, (2) a real difference because of natural internal variability and/or external forcings, or (3) a portion of the difference is due to the spatial coverage differences between the satellite and surface temperature data."

Evidently, the failure of data to support amplification of warming in the troposphere is a serious problem for the credibility of climate models and climate modelers would like to shift responsibility onto the data. Santer *et al.* focused on the second and third explanations, saying they were "more plausible" than "residual errors" that occurred in some data sets, and they suggested that the data that do show increased temperature in the troposphere are more reliable than those measured by the UAH group (c.f. Christy *et al.*, 2007). Klotzbach *et al.* (2009) presented considerable evidence that surface measurements over land often contain biases and effects due to their local surroundings. Indeed, one of the authors (Pielke, Sr.) has written extensively on this subject. The nature of most of these biases is to increase measured surface temperatures. Thus, Klotzbach *et al.* (2009) concluded that a significant factor in the discrepancy between climate models and measured temperature data may lie in the measured surface temperature data being too high.

Christy *et al.* (2010) asserted:

> "The magnitude of the trend in recent decades of TLT has become controversial because of differing views on ... whether the relationship between the observed temperature trend of TLT and the observed temperature trend of the surface (T_S) is faithfully reproduced by ... climate model simulations. These model simulations indicate that a clear fingerprint of greenhouse gas response in the climate system to date is that the trend of TLT should be [1.4 times] greater than [that of] T_S. There have been essentially two groups of publications on this contentious issue, one reporting that trends of TLT in observations and models are statistically not inconsistent with each other and the other reporting that model representations are significantly different than observations, thus pointing to the potential for fundamental problems with models."

There are two aspects of this result that are particularly important. One is simply that two long periods without a statistically significant increase in TLT would seem to contradict the view that continuously rising CO_2 is continuously driving up TLT. The second aspect deals with the scaling ratio of the trend of TLT to the trend of T_S in the tropics. Climate models consistently predict this ratio to be ~1.4; the TT is expected to rise faster than the surface temperature. However, as Christy *et al.* (2010) pointed out, the observed linear trend for TLT (0.9°C/decade) is only about 80% of the observed linear trend for T_S, so the observed scaling ratio is roughly 0.8, not the predicted value of 1.4. These results cast doubt on the veracity of climate models, and also suggest that a linear rate of temperature rise does not necessarily result from a linear increase in CO_2 concentration.

Thorne *et al.* (2011) said:

> "Over the past twenty years the vexatious issue of whether the troposphere is warming or not and, if it is, then whether it is warming at a rate consistent with climate model expectations, has spawned more than 200 research papers ... and has been a focus of reports ... of the IPCC. Over time, attention has shifted from the global mean to changes in the deep tropics, which are dominated by convective processes and where climate model behavior is strongly constrained [Santer *et al.*, 2005]. Here, any change in temperature at the surface is amplified aloft. The physical reasons for amplification are well understood. On month-to-month and year-to-year time scales, all climate models and observational estimates exhibit remarkable agreement with each other and with simple theoretical expectations."

Thorne *et al.* (2011) provided a detailed history of the evolution of measurements of TT, whether by radiosonde or from satellite instruments. Early satellite measurements indicated far less warming in the troposphere than was found at the surface by land-based thermometers. This caused challenges for climate modelers who predicted that TT would rise with surface temperatures, although the stratosphere would cool. As the years went by, adjustments and corrections of satellite measurement techniques reduced the gap between measured tropospheric and surface temperatures but a significant gap still remains. Radiosonde measurements seem to involve greater uncertainty and variability.

In interpreting the latest results, Thorne *et al.* (2011) seemed determined to minimize differences between tropospheric and surface temperatures, and emphasize warming in the recent part of the record. They said:

> "For the surface temperatures it shows (1) very good agreement between the three analyses [NOAA, NASA, and HadCRU]; and (2) the trend has remained quite stable over more than a decade."

Thorne *et al.* (2011) concluded:

> "In summary, the most recent versions of all data sets do not support the conclusion of a significant difference in trend between the surface and troposphere when considering (1) the structural uncertainty (as evidenced by the spread) in the tropospheric trend estimates, (2) the very likely remaining cold bias in the radiosonde trend estimates, and (3) the fact that the tropospheric trend has a small stratospheric cooling component."

In their conclusion, Thorne *et al.* (2011) said:

> "Overall, there is now no longer reasonable evidence of a fundamental disagreement between models and observations with regard to the vertical structure of temperature change from the surface through the troposphere. This is mainly due to a much better understanding of the real level of uncertainty in estimates of past changes and expectations from climate models. Ironically, elucidation of the true (large) degree of uncertainty in actual trends from observations and expected trends from models has led to greater confidence that they are not inconsistent."

Their Figure 10 shows:

(1) A constant upward trend in T_S of about 0.15°C/decade from about 1992 to 2009, whereas it is now widely accepted that there has been no trend from 1998 to 2012 (see Figures 3.27 and 3.28).
(2) There is a huge amount of scatter reported for tropospheric measurements. The most credible UAH measurements are claimed to show a negative trend prior to 2000, and a trend of about 0.05°C/decade in the first decade of the 21st century. However, the claimed negative trend prior to 2000 contradicts their claim of no major difference between surface and TT, and their claimed positive trend after 2000 is not in agreement with the data (see Figures 3.27 and 3.28).

3.4.4 Twentieth-century climate and El Niños

We have already seen a strong connection between climate and the El Niño index over the past 30 years in Figures 3.32 and 3.33.

Maasch (2009) provided a good description of the ENSO cycle:[25]

> "It has long been recognized that when sea-surface temperatures warm during El Niño, typically in December, rainfall also increases and the success of fisheries decreases along the north Peruvian coast. But not every summer sea-surface warming is the same. Some years are characterized by especially warm water that remains until May or even strong warm intervals. Local warming of waters off Peru coincides with positive SST (sea-surface temperature) anomalies over a much larger domain, namely the eastern half of the equatorial Pacific. La Niña is the opposite phenomenon, referring to abnormally cold SST in the eastern half of the equatorial Pacific. We now know that these changes are part of a much larger climate pattern and that the atmosphere and ocean are coupled in the equatorial Pacific."

In the "normal" state, the SST (and oceanic heat content) along the equator in the Pacific is warm in the west (near Asia) and cold in the east (near South America). The ocean surface waters are well mixed by wind stirring. Along the equator in the Pacific, this surface mixed layer is usually 150 m deep or deeper in the west, but it becomes shallower to the east until it essentially disappears near the South American coast. Sea level is also higher in the west. The trade winds, driving currents westward along the equator, feed and maintain the build-up of excess warm water on the western side (Cane, 1983).

Cane and Zebiak (1985) modeled El Niño events and concluded that the ENSO cycle is an oscillation of the coupled atmosphere–ocean system. Fedorov and Philander (2000) raised the question of whether global warming will affect the occurrence of El Niños, but found a diversity of predictions and concluded: "At this time, it is impossible to decide which, if any, are correct."

It has been observed that quasi-periodic variations occur about this normal state every few years, in which the Pacific climate changes from the so-called El Niño conditions to the so-called La Niña conditions. NOAA has described these episodes as follows. El Niño episodes reflect periods of exceptionally warm SSTs across the eastern tropical Pacific. La Niña episodes represent periods of below-average SSTs across the eastern tropical Pacific. These episodes typically last approximately 9–12 months. During a strong El Niño, ocean temperatures can average 2°C–3.5°C above normal between the Date Line and the west coast of South America. These areas of exceptionally warm waters coincide with the regions of above-average tropical rainfall. During La Niña, temperatures average 1°C–3°C below normal between the Date Line and the west coast of South America. This large region of below-average temperatures coincides with the area of well below-average tropical rainfall. For

[25] In 2012, Tisdale produced a remarkably extensive compendium of data, description, and analysis of El Niño phenomena in the form of a lengthy book: *http://bobtisdale.wordpress.com/*

both El Niños and La Niñas, the tropical rainfall, wind, and air pressure patterns over the equatorial Pacific Ocean are most strongly linked to the underlying SSTs, and vice versa, during December–April. During this period, the El Niño and La Niña conditions are typically strongest, and have the strongest impacts on U.S. weather patterns. El Niño and La Niña effect episodes typically last approximately 9–12 months. They often begin to form during June–August, reach peak strength during December–April, and then decay during May–July of the next year. However, some prolonged episodes have lasted two years and even as long as three to four years. While their periodicity can be quite irregular, El Niños and La Niñas typically occur every three to five years on average.

El Niños were originally recognized by fishermen off the coast of South America as the appearance of unusually warm water in the Pacific Ocean, occurring near Christmas and hence the name "El Niño", referring to the holy child. El Niños and La Niñas are the warm and cold phases of an oscillation referred to as El Niño/Southern Oscillation, or ENSO, which has typically had a period of about three to seven years. El Niño is thus one phase of a natural mode of oscillation that results from unstable interactions between the tropical Pacific Ocean and the atmosphere.

Although ENSO originates in the tropical Pacific Ocean–atmosphere system, it affects weather patterns over the entire world. According to a NOAA website, there is evidence that ENSO has been occurring for at least 125,000 years. El Niños vary in intensity and duration. The 1982 and 1997–1998 El Niños were particularly strong. In such a strong El Niño, the accumulation of excess heat in the eastern Pacific is about 10^{16} kWh—a very large amount of energy (total consumption of energy in the U.S. is about 10^{13} kWh/year). This affects the climate of a good part of the world, not merely the Pacific. For example, *realclimate.org* reports: "El Niños typically perturb the winter Northern Hemisphere jet stream in a way that favors anomalous warmth over much of the northern half of the U.S., the typical amplitude of the warming is about 1°C."

Cane (1983) described a severe El Niño that occurred in 1982:

> "In July 1982 conditions in the eastern equatorial Pacific were unremarkable; by October the sea-surface temperature (SST) was almost 5°C above normal and sea level at the Galapagos Islands had risen by 22 cm. The anomalies at depth were even greater: a huge influx of warm water had increased the heat content of the upper ocean at a rate that exceeded the climatological surface heat flux by a factor of more than 3, and the thickness of the warm layer was now greater than all previously observed values. Temperatures at the South American coast were near normal, but within a month they too would rise sharply."

Changnon (2000) described the remarkable El Niño that lasted from May 1997 to May 1998 as "the climate event of the century". From August 1997 to December 1997, SSTs rose by 1°C for the dateline to the South American coast, by 4°C east of 140°W, and by more than 5°C near the Galapagos Islands. These temperatures persisted until May 1998. As a result, during the winter of 1997–1998, temperatures across the northern states of the U.S. were 3°F to 10°F warmer than normal. Precipitation across the southeastern states was 40% to 50% above normal.

Temperature and precipitation records were broken in many areas. In the spring of 2008, precipitation in Los Angeles and San Francisco was triple the normal level, and similarly for Tampa, F. Spring 1998 temperatures in a corridor of 22 U.S. states ranging from the Midwest to the north-east were the warmest in 100 years. This created a field day for alarmists, who trumpeted the proclamation that 1998 was the hottest year in 100 years, which, though true, had little to do with carbon dioxide. However, Al Gore insisted it was due to greenhouse gases.

As Frauenfeld (2005) explained:

> "The Southern Oscillation (SO) is characterized by low atmospheric sea level pressure (SLP) in regions dominated by tropical convection, ascending air, and rainfall—such as the west Pacific–Australia–Indonesia region. Anomalously low pressure in these areas of convection corresponds to the concurrent occurrence of anomalously high pressure in the southeast Pacific region—characterized by dry conditions and descending air. This phase of the Southern Oscillation sets up an East–West pressure gradient across the equatorial Pacific, which creates enhanced trade winds as well as enhanced mass exchange between the East and West Pacific.
>
> "The Southern Oscillation Index (SOI) is quantified by the standardized SLP difference between Darwin, Australia—representing SLP over the western Pacific—and Tahiti, French Polynesia—representing the eastern Pacific.
>
> "SST in the eastern equatorial Pacific is normally lower than its equatorial location would suggest, mainly due to the influence of a cold oceanic current flowing equatorward along the coast of Chile, and due to upwelling of cold deep water off the coast of Peru. Upwelling occurs also along the equatorial Pacific as the easterly trade winds generate an easterly equatorial current, which is deflected northward in the Northern Hemisphere and southward in the Southern Hemisphere, due to the Coriolis force and Ekman transport. El Niño, the oceanic component of ENSO, is the occasional anomalous warming of the eastern equatorial Pacific ..."

The atmospheric signature, the SO, reflects the monthly or seasonal fluctuations in the air pressure difference between Tahiti and Darwin. The Southern Oscillation Index (SOI) is a measure of the large-scale fluctuations in air pressure occurring between the western and eastern tropical Pacific during El Niño and La Niña episodes. Traditionally, this index has been calculated based on the differences in air pressure anomaly between Tahiti and Darwin, Australia. In general, smoothed time series of the SOI correspond very well with changes in ocean temperatures across the eastern tropical Pacific. The negative phase of the SOI represents below-normal air pressure at Tahiti and above-normal air pressure at Darwin. Prolonged periods of negative SOI values coincide with abnormally warm ocean waters across the eastern tropical Pacific typical of El Niño episodes. Prolonged periods of positive SOI values coincide with abnormally cold ocean waters across the eastern tropical Pacific typical of La Niña episodes.

Estimates of the SOI have been made that date back to the mid-19th century (*www.cgd.ucar.edu/cas/catalog/climind/soi.html*). These results show roughly equal positive and negative fluctuations prior to about 1976 (see Figure 3.35). However, in

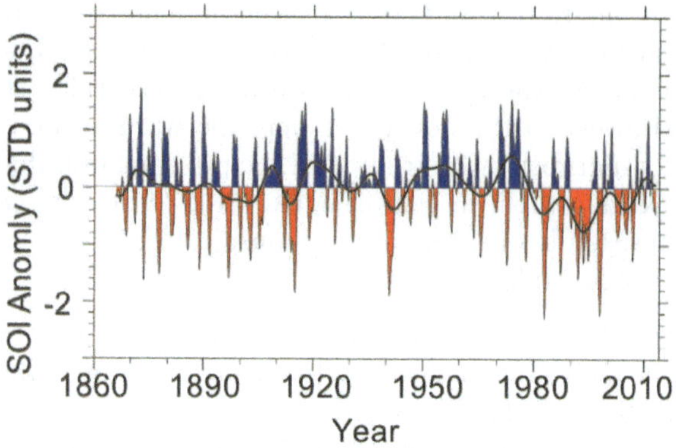

Figure 3.35. Long-term SOI index based on monthly values. Also shown are smoothed curves filtered to remove fluctuations of less than eight-months' duration. Values prior to 1935 should be used with caution (aapted from *www.cgd.ucar.edu/cas/catalog/climind/soi.html.*

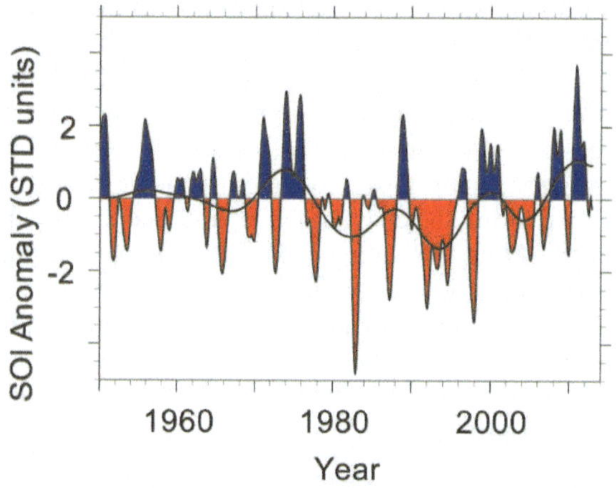

Figure 3.36. SOI Index since 1950 showing predominantly negative values after 1976 (adapted from *www.cgd.ucar.edu/cas/catalog/climind/soi.html.*

the last 25 years of the 20th century, the trend has been predominantly negative (McLean, 2007a, b). This is demonstrated in Figure 3.36.

This trend in the SOI has led to some controversy in connection with global warming. On the one hand, some climatologists have used computer simulations to infer that this trend in SOI is due to anthropogenic-induced global warming (e.g., Vecchi *et al.*, 2006). Power and Smith (2007) also suggest human activity as a cause of negative SOI but they admit that "there is currently no consensus amongst climate models concerning change in the behavior of ENSO in response to global warming".

On the other hand, two notable skeptics (John McLean and Bob Foster) have suggested that a rather sudden and decisive change in the circulation patterns and upwelling characteristics in the Pacific began around 1976 and continued until 1998. As can be seen from Figure 3.36, the Pacific was predominantly under El Niño

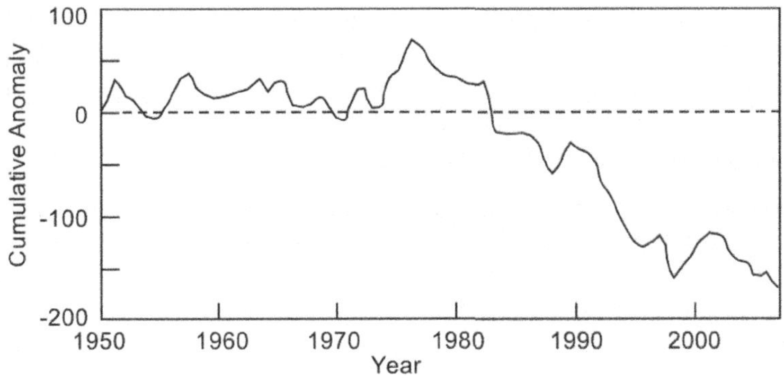

Figure 3.37. Integral of SOI anomaly since 1950 (adapted from McLean, 2007a, b).

conditions from about 1976 to 1998. According to McLean and Foster, this predominance of warm surface waters in the Pacific heated the Earth, particularly in the NH, and generated a rather abrupt upturn in global warming after 1976. According to McLean:

> "The abruptness of this change in upwelling appears likely to be related to some cataclysmic event in the region. Scientists would surely have noticed any shift in winds that was strong enough to cause a semi-permanent 25% reduction in the upwelling of eastern Pacific cold water so the answer is probably hidden in the ocean itself. The only cataclysmic event in the general region at that time was the Guatemala earthquake of February 1976 in which 250,000 people were killed, but any link is purely speculative at the moment." (*http://mclean.ch/climate/global_warming.htm*)

This is illustrated more dramatically in Figure 3.37. As is the case in most issues in climate change, there are diverse opinions. Did something changed in the Pacific Ocean that was heating the Earth, independently of greenhouse gases? Or did the effect of greenhouse gases produced a change in the Pacific Ocean?

Further corroboration of the abrupt change in the ENSO behavior around 1976/1977 was provided by other studies. For example, Meehl and Washington (1996) said:

> "Sea-surface temperatures in the tropical Pacific Ocean increased by several tenths of a degree during the 1980s and early 1990s, contributing to the observed global warming during this period."

Trenberth and Hoar (1997) said: "... the tendency for more El Niño and fewer La Niña events since the late 1970s is highly unusual and very likely to be accounted for solely by natural variability".

Frauenfeld *et al.* (2005) developed "a uniquely inter-decadal Pacific signal [IPS] embedded in the Pacific Ocean–Northern Hemisphere atmosphere system that, by its statistical construction, is representative of the interaction between the large scale atmospheric circulation of the Northern Hemisphere and SSTs of the Pacific

Ocean". They said "the time series of the IPS from 1949–2000 [was] dominated by the Pacific Climate Shift with negative anomalies prior to 1976/77 and almost exclusively positive anomalies since …". Their plots of the time series of the IPS show a sharp step function around 1977.

Guilderson and Schrag (1998) said:

> "Several studies have noted that the pattern of El Niño–Southern Oscillation (ENSO) variability changed in 1976, with warm (El Niño) events becoming more frequent and more intense. This '1976 Pacific climate shift' has been characterized as a warming in SSTs through much of the eastern tropical Pacific."

Desler, Alexander, and Timlin (1996) said:

> "A prominent decade-long perturbation in climate occurred during the time period [1970–1991] in which surface waters cooled by 1°C in the central and western North Pacific and warmed by about the same amount along the west coast of North America from late 1976 to 1988."

DiLorenzo *et al.* (2007) said:

> "Particularly dramatic physical and biological excursions occurred during the 1976–77 change in the Pacific Decadal Oscillation."

Hare and Mantua (2000) said:

> "It is now widely accepted that a climatic regime shift transpired in the North Pacific Ocean in the winter of 1976–77. This regime shift has had far reaching consequences for the large marine ecosystems of the North Pacific. Despite the strength and scope of the changes initiated by the shift, it was 10 to 15 years before it was fully recognized. Subsequent research has suggested that this event was not unique in the historical record but merely the latest in a succession of climatic regime shifts."

Wu, Lee, and Liu (2005) said:

> "The 1970s North Pacific climate regime shift is marked by a notable transition from the persistent warming (cooling) condition over the central (eastern) North Pacific since the late 1960s toward the opposite condition around the mid 1970s. … This large-scale decadal climatic regime shift has produced far-reaching impacts on both the physical and biological environment over the North Pacific and downstream over North America."

Kim and Miller (2007) studied "the 1976/1977 climate regime shift". They concluded that the thermocline warmed but did not deepen. Power and Smith (2007) emphasized that "the lowest 30-year average value of the June–December SOI just occurred in 1977–2006" along with "the highest tropical sea-surface temperatures on record [in] what appears to be a concurrent period of unprecedented El Niño dominance".

According to NOAA:

"El Niños are not caused by global warming. Clear evidence exists from a variety of sources (including archaeological studies) that El Niños have been present for hundreds, and some indicators suggest maybe millions, of years. However, it has been hypothesized that warmer global sea surface temperatures can enhance the El Niño phenomenon, and it is also true that El Niños have been more frequent and intense in recent decades. Recent climate model results that simulate the 21st century with increased greenhouse gases suggest that El Niño-like sea surface temperature patterns in the tropical Pacific are likely to be more persistent.

"A rather abrupt change in the El Niño–Southern Oscillation behavior occurred around 1976/77 and the new regime has persisted. ... However, it is unclear as to whether this apparent change in the ENSO cycle is caused by global warming."

McLean *et al.* (2009) concluded that the El Niño index:

"... is a dominant and consistent influence on mean global temperature. Shifts in temperature are consistent with shifts in the [El Niño index] that occur about 7 months earlier. The relationship weakens or breaks down at times of volcanic eruption in the tropics. ... Since the mid-1990s, little volcanic activity has been observed in the tropics and global average temperatures have risen and fallen in close accord with the [El Niño index] of 7 months earlier. Finally, this study has shown that natural climate forcing associated with ENSO is a major contributor to variability and perhaps recent trends in global temperature, a relationship that is not included in current global climate models."

According to their estimates, changes in the El Niño index could account for about 70% of the variance in the global TT over the past 50 years. The results of McLean *et al.* (2009) would seem to be a major stumbling block for alarmists who attribute most of the warming of the 20th century to greenhouse gases. It is therefore not surprising that the alarmists struck back with members of the *cabal* publishing Foster *et al.* (2010), who claimed that the results of McLean *et al.* (2009) "are seriously in error" and concluded "In fact, the general rise in temperatures over the 2nd half of the 20th century is very likely predominantly due to anthropogenic emissions of greenhouse gases". Foster *et al.* (2010) fell back on climate models that attribute only 15% to 30% of temperature variation in the 20th century to variability of the El Niño index. McLean *et al.*[26] attempted to rebut the criticism by Foster *et al.* (2010), but the *Journal of Geophysical Research* (JGR) refused to publish it. There are several important aspects of this episode that require further elaboration. These include (1) technical aspects, (2) attitudes and collusion amongst *cabal* alarmists, and (3) collusion of the JGR with *cabal* alarmists.

In regard to technical aspects, the issue revolves about methods used for filtering

[26] Their rebuttal, which will be referred to as "McLean2010", is available at either of these websites: *icecap.us/images/uploads/McLeanetalSPPIpaper2Z-March24.pdf*; *http://scienceandpublicpolicy.org/originals/censorship_at_agu.html*.

in statistical processing of data. Foster *et al.* (2010) appear to have made some valid criticisms of specific details, but these do not negate the strong correlation of the El Niño index with climate change. Perhaps the contribution of the El Niño index is less than the 70% claimed by McLean *et al.* (2009), but clearly the El Niño index is an important factor in climate change, as evidenced by Figures 3.32 and 3.33. It seems doubtful that climatology has sufficient data and analytical insight to pin down its quantitative share in influencing climate change. A number of authors, even members of the *alarmist cabal*, have admitted that climate models do not adequately account for El Niño effects.

McLean (2010) presented excerpts from the *climategate* emails that clearly show that the *alarmist cabal* regarded McLean *et al.* (2009) as a threat to their orthodoxy, and they colluded together to disparage McLean *et al.* (2009). The *cabal* regards itself as a police force to eradicate any contrary evidence or analysis that would refute their overemphasis on greenhouse gases.

The original paper by McLean *et al.* (2009) was approved by three independent reviewers for the JGR. One reviewer said:

> "I found the paper to be well-organized, well-written, and clear on the importance of the research. The abstract is informative, reference section is excellent, and the graphics are of high quality. The findings are likely to be of interest to a wide variety of readers."

The JGR published Foster *et al.* (2010), a rather vicious criticism of McLean *et al.* (2009), but refused to publish McLean's response. Evidently, the JGR is acting in collusion with the *alarmist cabal*, and probably regrets that McLean *et al.* (2009) "slipped through". McLean (2010) provides all the details.

Stammerjohn *et al.* (2008) showed that Antarctic annual sea-ice retreat and advance is related to variability of the ENSO and Southern Annular Mode.

Gergis and Fowler (2006, 2008) utilized a variety of proxies in attempting to reconstruct the historical occurrence of ENSO events dating back to 1525. They categorized such events according to a scale based on percentile range. Of particular interest were "exceptional" events (E = 90+ percentile) and "very strong" (VS = 70–90 percentile). They found that there was an unusual preponderance of VS and E events in the 20th century compared with the previous four centuries. For example, they found five E El Niños in the 20th century, but only a total of four E El Niños in the previous four centuries. They also noted: "30% of all extreme ENSO years occur post-1940 suggesting that recent ENSO variability may be anomalous in the context of the past five centuries." These authors seemed to assume that these changes in the 20th century were effects caused by human activity. ("These results suggest that ENSO may operate differently under natural (pre-industrial) and anthropogenic background states.") It is also possible that this change in ENSO activity contributed to climate change in the 20th century independently of human activity. There are a few anomalies in this work that are puzzling. For example, the well-known strong El Niño of 1998 was listed in their tables as a strong La Niña. In a presentation available on the Internet, Gergis said: "Multi-proxy reconstruction of ENSO is still in its infancy...

encouraging results but there are many opportunities for resolving more information on past ENSO behavior."

Douglass (2010) pointed out that commonly used El Niño indices contain an unwanted effect from the annual cycle that can be reduced by digital filtering. He then developed an improved El Niño index dating from 1856. He analyzed the occurrences of major El Niño and La Niña events in some detail, and pointed out the existence of a basic asymmetry favoring El Niños in recent times. While Douglass (2010) restricted his attention to peak values of the index, it is also instructive to integrate the El Niño index over time. The result is shown in Figure 3.38. This shows that, from ~1860 to ~1920, the El Niño and La Niña events were quite balanced. There was a predominance of El Niño events from about 1920 to ~1940, followed by a predominance of La Niña events from ~1940 to the late 1970s. Starting around 1980, El Niño events have been strongly dominant. The integral of the Niño index is more or less parallel to global temperature variability, suggesting that El Niño and La Niña events have been associated with most of the global temperature change in the past century and a half (see Figure 3.39). The index itself seems to correlate well with the TT.

Additional discussion and references on ENSO effects are given in NIPCC (2011) and particularly in Tisdale's book at *http://bobtisdale.wordpress.com/*.

Stevenson *et al.* (2012) plotted the temperature record at Nome, AK, from 1949 to 2009. They found that, over the period 1949–1976, the trend in mean annual temperature was slightly negative at –0.055°C/yr. In 1976, there was a sudden jump upward of over 4°C which they attributed to "the 1976 PDO shift". After 1976, Nome temperatures meandered quite a bit but the overall trend from 1976 to 2009 was negative at –0.030°C/yr. Yet, alarmists would say there was a constant upward trend from 1949 to 2009 of about +0.02°C/yr. The authors concluded:

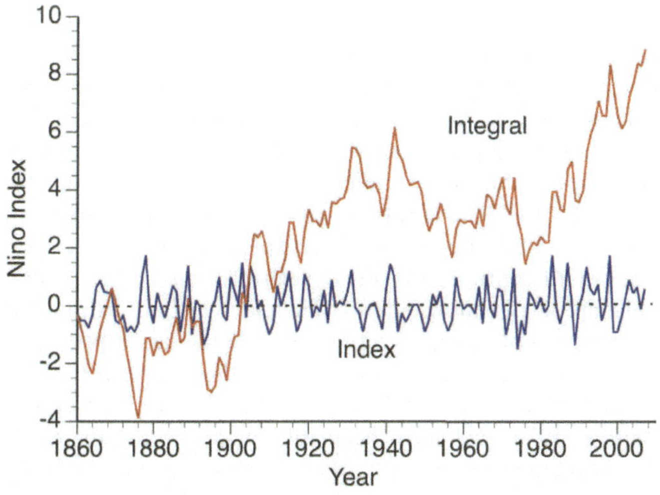

Figure 3.38. Integral of Douglass's modified El Niño index.

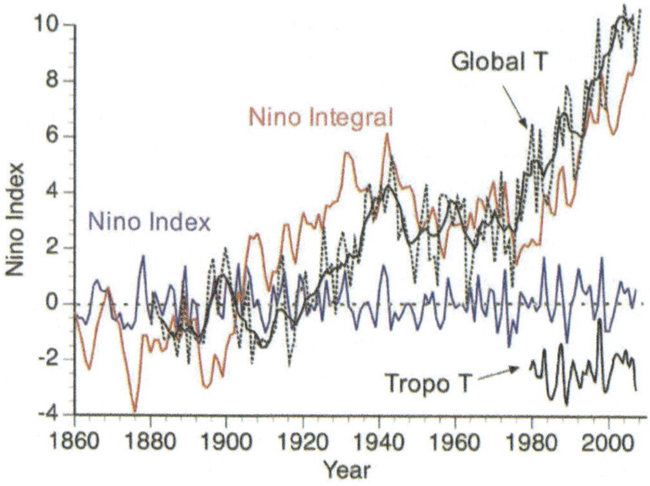

Figure 3.39. Comparison of Douglass's modified El Niño index and its integral with measured global surface temperature anomalies and tropospheric temperature anomalies.

> "Choice of time scale, reference dates, and statistical approach can severely impact the representation of climate change. In particular, special consideration must be given to major climatic events, such as the 1976 PDO shift, which has the potential to interact with nearby reference start dates and introduce erroneous information into reports. Having a time range that covered well before and after the 1976 PDO shift would help to minimize the effects of this event. ..."

3.4.5 The 1940–1978 temperature dip: effect of aerosols

Figures 3.14, 3.15, 3.17, and 3.19 show that global temperatures rose from about 1920 to 1940, dipped from 1940 to 1978, and then rose again after 1978. This effect was particularly accentuated in the NH. Since the period from 1940 to 1978 experienced increasing CO_2 emissions, these data would appear to conflict with the belief that the 20th-century temperature rise was due to the greenhouse effect. This fact was seized upon by skeptics as an indication that warming in the 20th century was not directly tied to CO_2 concentration. However, the alarmists responded by indicating that there was a build-up in sulfate aerosols as the number of power plants expanded in the mid-20th century, and these aerosols acted as a cooling force to counteract the build-up of greenhouse gases. The claim is that, as the 20th century wore on, greenhouse gases continued to build up while the developed nations began to control sulfur emissions, and, late in the century, the temperature began to rise again. For example, Overpeck *et al.* (1997) suggested that "the observed slowdown in warming from 1950 to 1970 may have been influenced by the increase in Arctic tropospheric aerosols that occurred after 1950". Chylek *et al.* (2007) argued that forcing due to aerosols is comparable with that produced by greenhouse gases (but of opposite sign).

Chin *et al.* (2009) provided a detailed review of the role of aerosols in affecting climate. Aerosols affect the Earth's energy budget by scattering and absorbing radiation and by modifying the radiative properties of clouds. Aerosols influence cloud properties through their role as cloud condensation nuclei (CCN) and/or ice nuclei. Aerosols can also affect clouds by absorbing solar energy and altering the environment in which the cloud develops, thus changing cloud properties. Such effects can change precipitation patterns as well as cloud extent and optical properties. The addition of aerosols to the atmosphere alters the intensity of sunlight scattered back to space, absorbed in the atmosphere, and arriving at the surface:

> "Atmospheric aerosols are suspensions of solid and/or liquid particles in air. Aerosols are ubiquitous in air and are often observable as dust, smoke, and haze. Both natural and human processes contribute to aerosol concentrations. On a global basis, aerosol mass derives predominantly from natural sources, mainly sea salt and dust. However, anthropogenic (manmade) aerosols, arising primarily from a variety of combustion sources, can dominate in and downwind of highly populated and industrialized regions, and in areas of intense agricultural burning. The term 'atmospheric aerosol' encompasses a wide range of particle ... compositions, sizes, shapes, and optical properties." (Chin *et al.*, 2009).

The effect of aerosols on climate is not known precisely. According to Chin *et al.* (2009), "Clearly there are still large gaps in assessing the aerosol impacts on climate through modeling". These authors provide a detailed analysis of what is and what is not understood about aerosols and climate.

Anderson *et al.* (2003) discussed two methods for incorporating aerosol forcing in climate models. In the "forward" approach, pertinent aerosol physics and chemistry are used to estimate the aerosol-induced forcing *a priori* and the climate model is based on this. In the "inverse" approach, the climate model is exercised using the aerosol forcing as an adjustable parameter, and the parameter is adjusted to maximize agreement with experimental temperatures, assuming that the other forcings are known better. However, as we discuss in Section 6.4.4, the forcings due to water vapor and clouds remain quite uncertain. Forward calculations suggest a higher forcing than is inferred from inverse models. Current climate studies typically use only the lower aerosol forcing values consistent with the inverse approach. If the larger range of aerosol forcings from the forward calculations were used, the total forcing from pre-industrial times to the present might be small or even negative. If this were correct, it would imply that climate sensitivity and/or natural variability (i.e., variability not forced by anthropogenic emissions) is much larger than climate models currently indicate. However, greenhouse gases accumulate in the atmosphere, whereas aerosols do not. Thus, as greenhouse gas emissions continue to increase in the 21st century, the forcing will turn positive provided aerosol emissions are controlled. Anderson *et al.* (2003) concluded: "... the possibility that most of the warming to date is due to natural variability ... must be kept open."

The translation of estimated sulfur emissions into aerosol concentration, and the subsequent conversion of this concentration to an effective forcing (W/m^2), has not been clearly elucidated in the literature. Mitchell and Johns (1997) estimated future temperatures in the 21st century based on assumed forcings from CO_2 and aerosols. They stated:

> "The scenarios are plausible, though there are large uncertainties in future greenhouse gas concentrations, and even greater uncertainties in those of sulfates and the associated direct radiative forcing. The effects of sulfate aerosols on clouds, and possible changes in other forcing including that due to industrial soot, tropospheric ozone, and biogenic aerosols, are ignored."

However, the actual figures used by Mitchell and Johns (1997) seem rather strange and even incredible. The CO_2 concentration in 2000 was assumed to be 522 ppm (and we know it was actually about 370 ppm), and the CO_2 concentration was assumed to rise to 1,414 ppm in 2100, which is 2.5–3 times what it is likely to be (see Figure 7.19). The aerosol forcing was assumed to rise from zero in the pre-industrial era to about 0.7 W/m^2 in 2000 to about 1.5 W/m^2 in 2100 but it is not clear how this was derived.

Lefohn, Husar, and Husar (1999) was the most authoritative source on aerosol production, but Stern (2005a, b), written about six years later, appears to encompass the earlier paper as well as adding more recent data. The estimate of world sulfur emissions by Stern (2005b) was compared with estimates made by others, particularly the data of Lefohn, Husar, and Husar (1999). The result is shown as Figure 3.40. Stern (2005b) suggested that some rebound in aerosol production would seem likely in the current business cycle (after 2000), especially as the downtrend in the 1990s was strongly affected by the collapse of the Soviet Bloc economies. Some recent data are available for the U.S. and China that suggest that increases in China

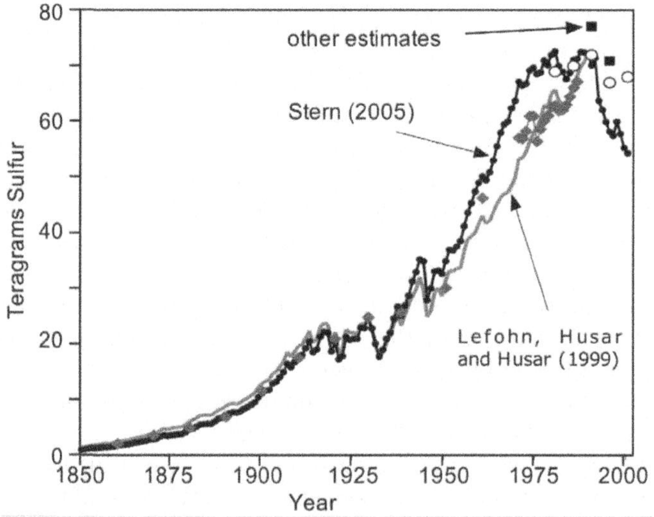

Figure 3.40. Estimate of sulfur emissions 1850–2000. Vertical scale is teragrams (10^{12} grams) of sulfur (adapted from Stern, 2005b, and Lefohn *et al.*, 1999).

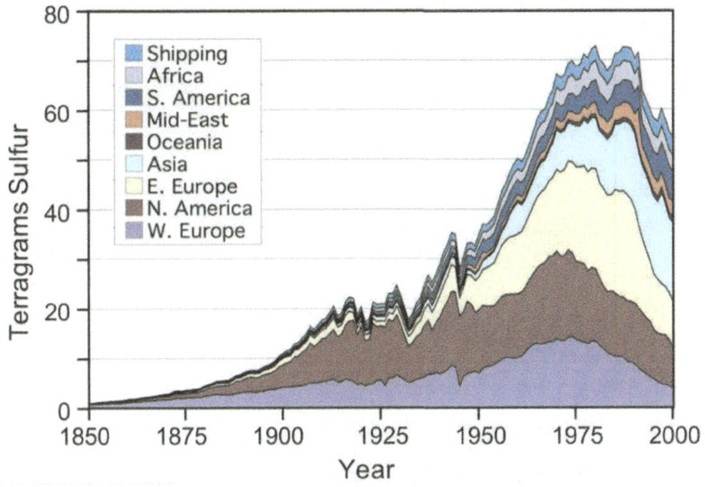

Figure 3.41. Estimate of sulfur emissions 1850–2000 (adapted from Stern, 2005a).

will outweigh decreases in the U.S. Stern (2005b) concluded that "global emissions likely declined through 2002 but in 2003 and following years emissions could be rebounding moderately".

While the focus of Stern (2005a, b) was on the decade 1991–2000, estimates were also made for the historical period 1850–2000. Data were not available for all locations over all time spans and some extrapolation was required. Figure 3.41 shows the estimate of sulfur emissions prepared by Stern (2005a). World emissions peaked around 1975–1985 and declined afterward. After 1989, a precipitous decline set in, only punctuated by the Kuwait oil fires, which contributed around 4.7 Tg of sulfur,[27] as the Soviet Union and Eastern European economies collapsed. Chinese emissions overtook U.S. emissions in 1987 to make China the largest single emitter. Chinese emissions were claimed to have peaked in 1996. However, it is known that, in 2007, China was adding one coal-fired power plant per week, and, since China seems to have little interest in pollution control of any form, this seems hardly believable.

Stern (2005b) estimated that the net radiative forcing in 1990 due to aerosols was about $-2.0 W/m^2$, and that the change in forcing from 1985 to 2000 was $\sim +0.35$ W/m^2. This would seem to imply that the peak aerosol forcing around 1987 was about -2.1 W/m^2 and dropped to about -1.7 W/m^2 by 2000. However, it is not clear how this was derived.

Smith *et al.* (2011) produced an extensive detailed update and review of global sulfur emissions by sector, and region as of 2005. They found:

> "Global emissions peaked in the early 1970s and decreased until 2000, with an increase in recent years due to increased emissions in China, international shipping, and developing countries in general. An uncertainty analysis was

[27] 1 teragram (Tg) = 10^{15} grams.

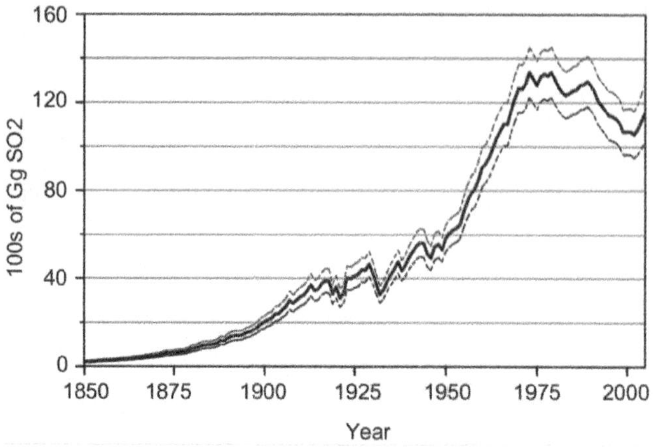

Figure 3.42. Global sulfur dioxide emissions. Dotted lines show range of uncertainty. (Smith *et al.*, 2011).

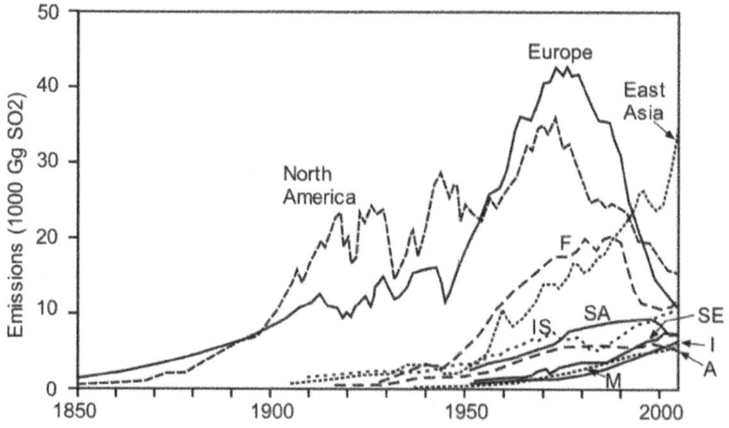

Figure 3.43. Sulfur dioxide emissions by region. (A = Africa; F = FSU; I = India; IS = international shipping; M = Middle East; SA = South & Central America; SE = Southeast Asia) (Smith, *et al.*, 2011).

conducted including both random and systemic uncertainties. The overall global uncertainty in sulfur dioxide emissions is relatively small, but regional uncertainties ranged up to 30%. The largest contributors to uncertainty at present are emissions from China and international shipping."

Their results for global SO_2 missions over the past 150 years are shown in Figures 3.42 and 3.43. The downtrend from about 1975 to 2000 seems to be reversing due primarily to the uptrend since 1975 by China and other Asian and African regions. China is now driving the world SO_2 mission trend.

Two related papers (Nagashima *et al.*, 2006; Nozawa *et al.*, 2005) carried out

GCMs that included the impact of aerosols on 20th-century climate change. However, these papers appear to raise more questions than they answer. Nozawa *et al.* (2005) did not actually isolate the effect of aerosols, but rather, carried out four calculations: (1) full (with all forcings), (2) natural (with only solar and volcanic forcing), (3) anthropogenic (with GHG and aerosols forcing), and (4) GHG (with only greenhouse gas forcing). However, by subtracting (4) from (3), the effect of the aerosols were claimed to be discerned. Nozawa *et al.* (2005) found that the natural model underestimated late-20th-century warming by a significant amount. The anthropogenic model underestimated mid- and late-century warming. The GHG model overestimated late-century warming. And the full model fit best. It can be noted that the difference between the anthropogenic and GHG curves in Nozawa *et al.* (2005) was 0.28°C in 1950, 0.38°C in 1975, and 0.60°C in 2000. These can be attributed as the temperature decrease due to aerosols alone. Nozawa *et al.* (2005) does not quite succeed in reproducing the observed temperature dip from 1940 to 1978 but the natural curve goes downward after 1940 including a sharp dip from 1960 to 1970, and this, rather than aerosol production fall-off, leads to a lull in temperature rise in that decade. However, the credibility of their natural curve is doubtful.

In a related activity, Nagashima *et al.* (2006) used a climate model to infer the effect of aerosols on climate. They examined the effect of sulfate aerosols and carbon aerosols separately. They found that, without aerosols being included, the climate model predicted too much warming in the 20th century. When both carbon and sulfate aerosols were included, the fit to experimental temperature data was better. However, it is not clear from this study what produced the temperature decline from 1940 to 1978, or how the model was contrived to fit the data.

The observed dip in global temperature from 1940 to 1978 does not seem to correlate well with timing of aerosol emissions. The aerosol emission peak occurred well after temperatures began rising again after 1978. The temperature dip from 1940 to 1978 remains only partly attributable to aerosols, and, in the absence of credible explanations remains as an unexplained issue.

Ghan and Schwartz (2007) and Chin *et al.* (2009) wrote comprehensive reviews of the status of modeling aerosols in various generations of GCMs, past, present, and future. Early approaches for incorporating the effects of aerosols simply adjusted the albedo as a crude approximation, while subsequent models used much more sophisticated representations of aerosols, but even these did not adequately account for interactions of aerosols with other climate elements:

> "It is now recognized that accurate representation of aerosol influences must take into account phenomena such as correlations of aerosol loading with meteorological variables and the influence of aerosol on clouds and precipitation, and hence that aerosol loading and those properties must be represented actively and interactively in climate models. It is this recognition that is driving much of the current effort to actively represent aerosol processes, properties, and effects in climate models." (Ghan and Schwartz, 2007).

The treatment of aerosols in future generations of climate models will rest on an improved understanding of the processes that control aerosol properties and the ways that they affect climate. Chin *et al.* (2009) provided a summary of unresolved issues in regard to aerosols and climate. Ghan and Schwartz (2007) provided a plan for systematically improving our understanding of aerosols in the future.

Lüdecke (2011) and Lüdecke *et al.* (2011) were primarily concerned with the question of how to determine how much of warming in the 20th century was due to anthropogenic influence, and how much was due to natural fluctuations in the climate. They "analyzed 2249 worldwide monthly temperature records from GISS (NASA) with a 100-year period covering 1906–2005 and the two 50-year periods from 1906 to 1955 and 1956 to 2005". All of their sites were on land, although they noted which ones were within 30 km from an ocean. They divided the data into groups based on the estimated population near the measurement site. Most of their sites were located in a rather narrow range of northern latitudes (30°N–50°N). The distribution of temperature changes from 1906 to 2005 was a bell-shaped curve centered at +0.58°C with a width at half height of about 1.9°C. About 25% of the sites were located in areas where the population was >100,000. He found that the mean of all stations indicated a temperature rise of 0.58°C from 1906 to 2005. About 85% of the stations were concentrated in the latitude range 40°N–50°N. Since climate models indicate that temperature changes are amplified at these latitudes by roughly 25% compared to the global average (e.g., Holland and Bitz, 2003), we might guess that, according to this result, the global average rose by perhaps 0.48°C. Restricting consideration to stations with a population of under 1,000 (and below 800 m in elevation) this figure drops to 0.41°C. However the scatter in the data vs. population and elevation was very large. They concluded that a strong UHI and elevation warming could be identified from this. About a quarter of all records indicated falling temperatures from 1906 to 2005. According to Lüdecke *et al.* (2011) this would be "an indication that the observed temperature series are predominantly natural fluctuations". He claimed that "the probability that the observed global warming was a natural fluctuation" lies between 80% and 90%, for the period of 1906–1955 and between 60% and 70% for 1956–2005.

These papers did not provide much detail on the temperature station data used by Lüdecke *et al.* (2011) or the proxies that were used by Lüdecke (2011). In fact, the treatment of the data acquisition was very cursory. They provide references to the sources of their data but they didn't really introduce any new discussion of the credibility, uncertainty, or completeness of the data. In fact, in both papers, the data are presented rather late in the papers.

These papers were mainly concerned with the question of whether one can discern a human influence in the station temperature data, as opposed to natural variability. There is quite an extensive literature attempting to deal with this question. From the start, both papers plunge into mathematics aimed at determining to what extent there is long-term correlation in the data. In other words, is there a detectable long-term trend? However, they are not very clear on the philosophy of their approach. Perhaps the closest they come is in the passage:

> "Temperature series are persistent—other notations are 'long-term correlated' or 'long-term memory'—which is a well-known phenomenon. A warm

day is more likely to be followed by another warm day than by a cold day, and vice versa. Short-term persistence of weather states on a time scale of about one week is caused by general weather situations. Longer-term persistence over several weeks is generally caused by blocking situations that arise when a high pressure system remains in place for many weeks. Persistence over many months, seasons, years, decades, and even longer periods is usually associated with anomaly patterns in sea surface temperatures, and even with the influence of long-term variations in the activity of the sun, but there is no universal explanation that can be used in all causes."

From what I can glean from these rather nebulous papers, the widely accepted underlying philosophy begins with the belief that natural variations are random and cannot generate long-term trends, whereas long-term trends are indicative of external impacts. Apparently, this belief was used by those of the alarmist persuasion (e.g., Barnett *et al.*, 2005a) to infer that, indeed, the effect of human influence is readily discernible in the data. I have the impression that, just as Barnett *et al.* (2005b) began with the belief that human generation of greenhouse gases has impacted the climate significantly, Lüdecke *et al.* (2011) began with just the opposite belief. To support this view, they show (for example) in their Figures 1 and 2 that, at least for some stations, there were downtrends from 1800 to 1900 followed by weaker uptrends after 1900. If the uptrends after 1900 were caused by an external influence (greenhouse gases), what caused the downtrends prior to 1900?

By some reasoning that I cannot follow, they concluded that "the impact of anthropogenic greenhouse gases is most probably a minor effect and—in view of the 19th century temperature fall of similar magnitude—not appropriate as an authoritative explanation for any temperature rise in the northern hemisphere during the 20th century". But then, by further reasoning that seems very far-fetched to me, they suggested: "The hypothesis expressed here suggests that the Sun could be predominantly responsible for the 100-year-long rises and falls in temperature over the last 2000 years."

These types of assertions, in this case by skeptics, and in other cases by both alarmists and skeptics, are not justified by the data and must be taken as mostly fluff.

3.5 MOUNTAIN GLACIERS AS CLIMATE INDICATORS

Holzhauser *et al.* (2005) discussed the advance and retreat of mountain glaciers as climate indicators. As they pointed out:

> "Glaciers in mountain areas are highly sensitive to climate changes and thus provide one of nature's clearest signals of warming or cooling and/or dry and wet climate periods. Thus, in a simplified manner, one can say that the [periodic] fluctuations of Alpine glaciers were driven by glacier-hostile (warm/dry) and glacier-friendly (cool/wet) periods."

However, they emphasized that "a detailed examination of historical data suggests a rather more complex situation". They provided examples where "specific and varying temperature and precipitation courses during winter and summer" were responsible for glacier advance and retreat. In particular, they claimed that "the LIA between 1300 and 1850 coincided with wetter summers and colder winters, i.e., two factors favorable to glacier extension and higher lake levels". They also suggested an influence of the North Atlantic Oscillation (NAO) index on glacier advance and retreat, and indicated this may have initiated "the impressive loss of the alpine glaciers since the last glacial maximum c. 1850/60".

A number of scientists studied the retreat of mountain glaciers in various localities since the 19th century, based partly on actual data, and partly on anecdotal evidence. In general, most of these studies emphasized the retreat of mountain glaciers as evidence of global warming. Most of these studies seem to miss a point: that the retreat of a number of mountain glaciers began in earnest as the LIA waned, long before CO_2 had built up. On the other hand, there is evidence that glacier retreat accelerated toward the end of the 20th century, when CO_2 concentrations built up.

One interesting case is the shrinking glaciers of Mount Kilimanjaro in Africa, used as a key example of the effects of global warming by Al Gore in *An Inconvenient Truth*. However, Mote and Kaser (2007) said:

> "... warming fails spectacularly to explain the behavior of the glaciers and plateau ice on Africa's Kilimanjaro massif, just 3 degrees south of the equator, and to a lesser extent other tropical glaciers. The disappearing ice cap ... which gets a starring role in *An Inconvenient Truth*, is not an appropriate poster child for global climate change. Rather, extensive field work on tropical glaciers ... reveals a more nuanced and interesting story."

Mote and Kaser went on to point out:

> "Another important observation is that the air temperatures measured at the altitude of the glaciers and ice cap on Kilimanjaro are almost always substantially below freezing (rarely above –3°C). Thus the air by itself cannot warm ice to melting by sensible-heat or infrared-heat flux. ..."

The mass balance of Kilimanjaro is primarily determined by precipitation vs. solar-induced sublimation. The reason that Kilimanjaro has been losing ice cover is that precipitation decreased and sunlight increased; it has nothing to do with global warming. It may be possible that changing wind patterns producing less precipitation and less cloudiness might be related to global warming, but that is only speculation. As Mote and Kaser conclude:

> "If the Kibo ice cap is vanishing or growing, reshaping itself into something different as you read this, glaciology tells us that it's unlikely to be the first or the last time."

Studies of two mountain glaciers are reviewed here (Zumbuhl *et al.*, 2008; Hall and Fagre, 2003; Pederson *et al.*, 2004).

Hall and Fagre (2003) studied the glaciers at Glacier National Park in Montana, U.S.A. As they pointed out:

"Since its establishment in 1910, Glacier Park has lost most of its glaciers. Over two-thirds of the estimated 150 glaciers existing in 1850 had disappeared by 1980. Furthermore, over that same time period, the surviving glaciers were greatly reduced in area."

In the course of their description, they pointed out that "nearly all the ice masses in Glacier National Park had been undergoing rapid recession since the turn of the century" and that "many of the glaciers in the western United States that formed during the *Little Ice Age* began retreating about 1850 to 1855". They indicated that mountain glaciers in the park retreated at a modest rate from 1917 to 1926, retreated at a rapid rate from 1926 to 1942, retreated at a slower rate after 1942, stopped retreating from 1950 to 1975, and began retreating again after 1975. The measured summer temperature increase in the Glacier National Park area from 1910 to 1980 was about 1.7°C, far greater than the average for the U.S. Hence, Glacier National Park has been particularly affected by global warming, and much of the glacial retreat occurred prior to the large-scale build-up of CO_2. Modeling suggested that increased summer temperatures were the main drivers for the retreat of these glaciers. However, it is not clear from their work what role variability in precipitation might have played. Pederson *et al.* found that reduced winter precipitation played an important role. Their projections for the 21st century are highly speculative. Their historical reconstructions for two major glaciers in the park are shown in Figure 3.44. The steep decline in Sperry occurred prior to 1925.

Pederson *et al.* (2004) used tree-ring reconstructions of North Pacific surface temperature anomalies and summer drought as proxies for winter glacial accumulation and summer ablation, respectively, over the past three centuries. They concluded that rapid early-20th-century retreat of GNP glaciers arose from a

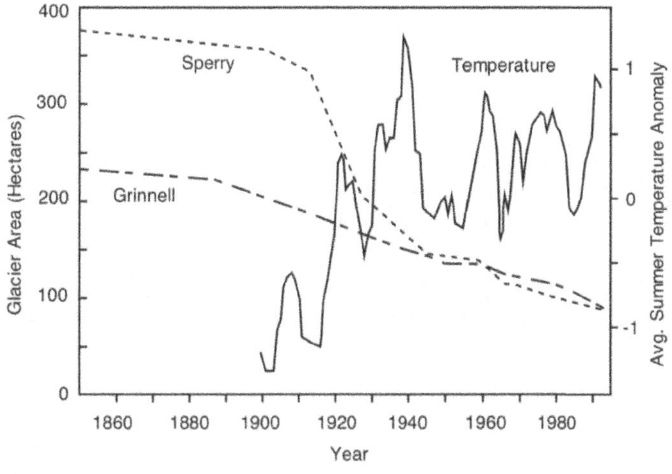

Figure 3.44. Retreat of two glaciers in Glacier National Park in the 19th and 20th centuries (Hall and Fagre, 2003).

combination of high summer temperatures and low winter accumulation. There was no evidence of an acceleration of retreat as CO_2 levels built up.

Zumbuhl *et al.* (2008) studied the evolution of Lower Grindelwald Glacier, Switzerland, and the Mer de Glace, France. These are examples of well-documented Alpine glaciers with a wealth of different historical sources (e.g., drawings, paintings, prints, photographs, maps) that allow reconstruction of glacier length variations for as long as 400–500 years. Their results for the two major European Alps glaciers are shown in Figure 3.45. There was a major retreat from 1860 to 1880, and a second retreat from 1925 to 1970, after which they appear to have stabilized.

Dyurgerov and Meier (1999) estimated the historical retreat of a number of mountain glaciers during the 20th century (see Figure 3.46). For most glaciers, the data were restricted to the period 1961–1997. However, eight glaciers were traced back to 1880. They concluded:

> "One can conclude that the present-day wastage of glacier volume is, on the average, part of a continuous process started in or before the 19th century, after the end of *Little Ice Age* maximum. Climate became warmer, and glaciers continued losing volume in response to this change. However, the rate of loss has been accelerating recently; this suggests that it is not just a simple adjustment to the end of an 'anomalous' *Little Ice Age*, as some have claimed."

Actually, the data suggest that glacier retreat for the most part is in fact "a simple adjustment to the end of an 'anomalous' *Little Ice Age*" and the claim to the contrary by Dyurgerov and Meier appears to be mainly pandering to the warmists.

There is a tendency for glacier retreat to reflect the trends in global temperature (sharp rise early in the 20th century, dip in mid-century, and rise again after the mid-1970s). This accounts for the acceleration observed in some glaciers after the mid-

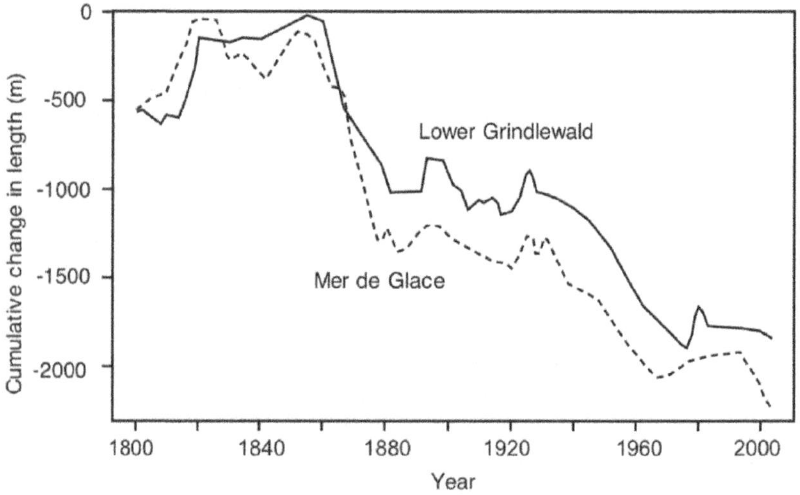

Figure 3.45. Retreat of two glaciers in the European Alps in the 19th and 20th centuries (Zumbuhl *et al.*, 2008).

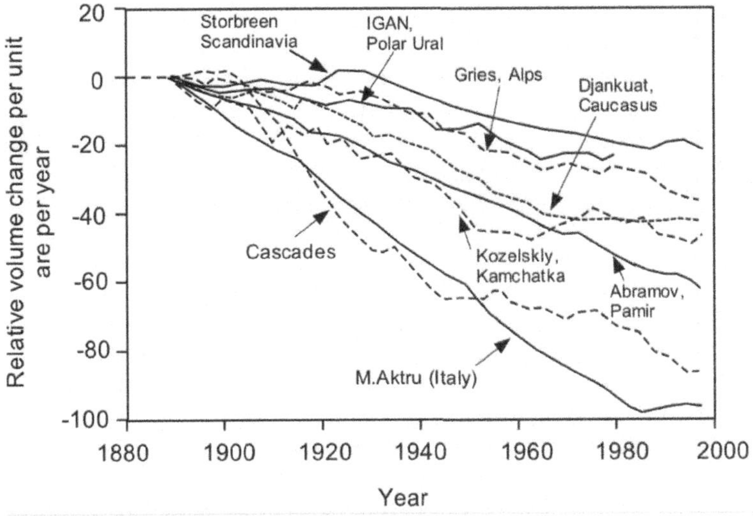

Figure 3.46. Changes in glacier volume (reported as changes in volume per glacier area per year) (Dyurgerov and Meier, 1999).

1970s. The data for the eight glaciers that were traced back to 1890 are shown in Figure 3.46. There is not much post-1970s' acceleration evident in these glaciers.

The glacier data from 1961 to 1997 were arranged by region. The results showed a good deal of variability. The glaciers in the Canadian Arctic, Mainland North America, and Alaska showed some acceleration after the mid-1970s as the NH warmed. For Scandinavia, the Alps, and other locations, about half of the glaciers actually expanded, while the other half retreated—with some acceleration in the post-1970s' time period.

Holzhauser *et al.* (2005) reported on glacier and lake-level variations over the last 3,500 years. Some of their results are replicated in Figure 3.47. These results clearly show a *Medieval Warm Period* and a *Little Ice Age* (LIA). However, the retreat of the glaciers in the most recent period appears to be sharper than previous retreats and reflects the global warming in the 20th century. This may be linked to rising CO_2. It is noteworthy that they claim a strong correlation of glacier advance and retreat with atmospheric residual ^{14}C, which, according to some theories, is a measure of activity of the Sun. Yet, this correlation appears to lie more in the eye of the beholder than in the data themselves.

Vuille *et al.* (2008) reviewed the retreat of South American Andes mountain glaciers during the late 20th century. They discussed the tropical glacier energy balance through its sensitivity to changes in temperature, as well as atmospheric humidity (which governs sublimation), precipitation (whose variability induces a positive feedback on albedo), cloudiness (which controls the incoming long-wave radiation), and atmospheric circulation. While the authors pointed out the impact of yearly ENSO variations, they did not seem to notice the prevalence of El Niño events after about 1976 as a source of continued glacier retreat at the end of the century.

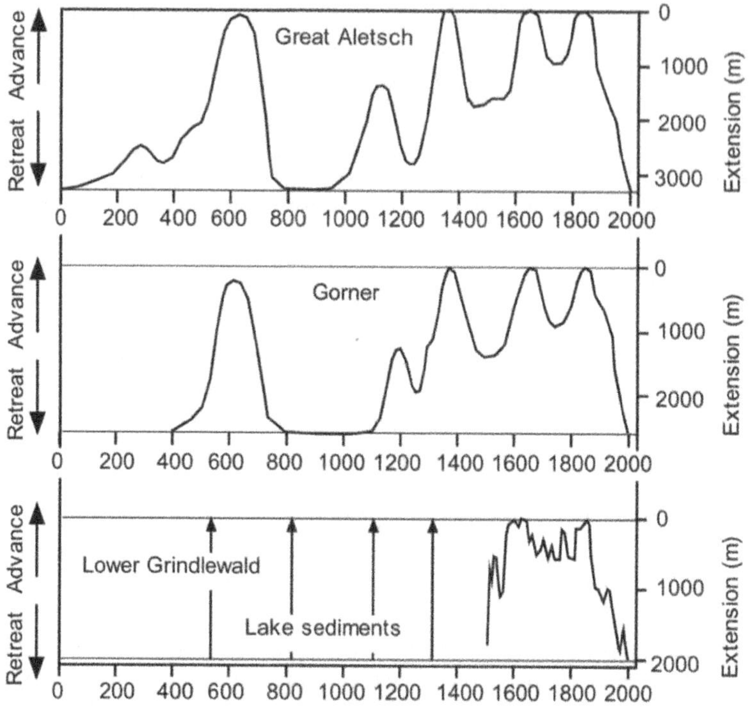

Figure 3.47. Reconstruction of "extension" of three Alpine glaciers in the Alps. Here, use of the word "extension" seems backwards. It should probably be "retraction" (Holzhauser *et al.*, 2005).

They suggested that "by the end of the 21st century, following the SRES A2 emission scenario, the tropical Andes may experience a massive warming on the order of 4.5–5°C" and predicted further demise of South American mountain glaciers. However, it would seem prudent to insert "or may not" after "may" in this sentence.

Zemp *et al.* (2009) reported on six decades of annual mass-balance data compiled by the World Glacier Monitoring Service:

> "In total, there have been 3,480 annual mass-balance measurements reported from 228 glaciers around the globe. However, the present data set is strongly biased towards the Northern Hemisphere and Europe and there are only 30 'reference' glaciers that have uninterrupted series going back to 1976. The available data from the six decades indicate a strong ice loss as early as the 1940s and 1950s followed by a moderate mass loss until the end of the 1970s and a subsequent acceleration that has lasted until now."

They also emphasized "the shortcomings of the available dataset" and "the relatively small set of current long-term observations with a strong bias towards the Northern Hemisphere and Europe".

Braithwaite (2009) echoed Zemp *et al.* in emphasizing the sparseness of the

Figure 3.48. The Aletsch glacier in 1856 (left) and 2001 (right) (Haberli and Hoelzle).

mountain glacier data set. He also pointed out that the 30 glaciers with 30 years of data "are biased towards wetter conditions than are typical for global glacier cover". They went on to say that "The mass-balance variations for these glaciers are therefore not representative of the global glacier cover" and "these 30 glaciers must be showing larger mass-balance changes than the global average". They concluded:

> "Current estimates of the glacial contribution to sea-level rise ... may therefore be too large. There is already a discrepancy between observed sea-level rise and estimated contributions from the different sources, and the discrepancy will increase if the contribution from glaciers is really overestimated."

Haberli and Hoelzle[28] reviewed data on several mountain glaciers. Of particular interest is the Aletsch glacier in Switzerland. These authors present pictures of this great glacier taken in 1856 and 2001, showing the great retreat of the glacier (Figure 3.48). The 20th-century history of the Aletsch glacier retreat is shown in Figure 3.49. In addition, the two-millennium history of the Aletsch glacier was reconstructed, as shown in Figure 3.50. The data in Figures 3.49 and 3.50 show that (1) the Aletsch glacier retreat began before build-up of CO_2 concentrations, and (2) there have been alternating advances and retreats for two millennia. On the other hand, retreat seems to have accelerated in the late 20th century as CO_2 built up.

[28] *www.zamg.ac.at/ALP-IMP/downloads/session_haeberli.pdf.*

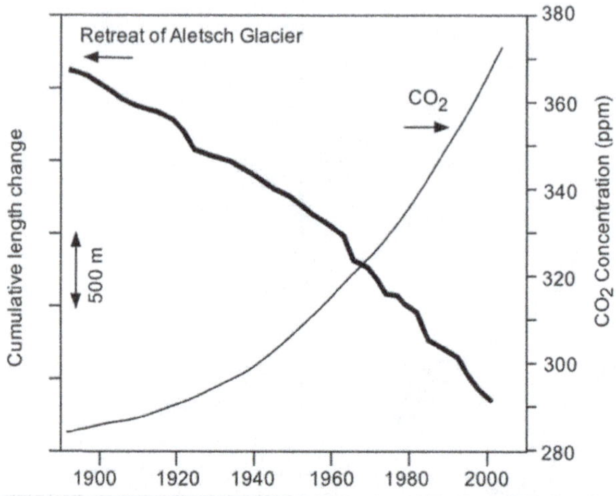

Figure 3.49. Retreat of Aletsch glacier compared to buildup of CO_2 (Haberli and Hoelzle).

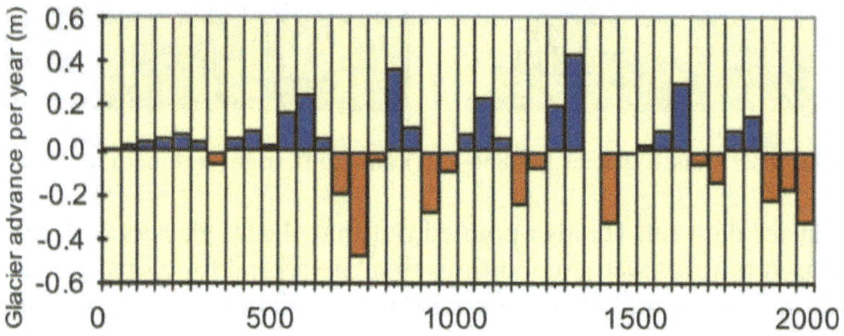

Figure 3.50. Advance and retreat of Aletsch glacier reconstructed over two millennia (Haberli and Hoelzle).

An interesting case occured in Glacier Bay National Park, Alaska. According to the *National Park Service* brochure:

> "Glacier Bay today is the product of the *Little Ice Age*, a geologically recent glacial advance in northern regions. The *Little Ice Age* reached its maximum advance about 1750. But when Captain George Vancouver sailed here about 45 years later, the glacier had melted back five miles into Glacier Bay, which it had gouged out. When conservationist John Muir traveled here in 1879 the glacier had retreated 40 more miles up the bay since Vancouver's visit. Today you must travel 65 miles up the bay to view tidewater glaciers—a far cry from the glacier's 1750 maximum."

Figure 3.51 shows the extent of the Glacier Bay glaciers at four time periods. Most of the glacier retreat occurred prior to build-up of CO_2 in the atmosphere.

Nature Magazine commented on glacier retreat in the June 10th, 2010 issue. This

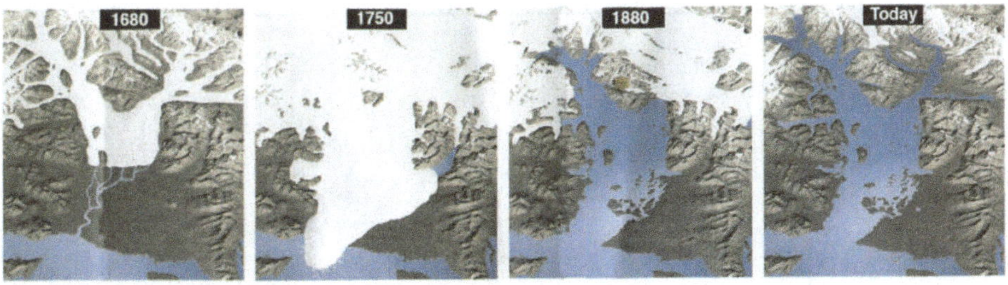

Figure 3.51. Extent of Glacier Bay Alaska glaciers at several time periods.

editorial was based on a recent Swiss study (Huss *et al.*, 2010)[29] that involved over 10,000 *in situ* observations of 30 Swiss mountain glaciers. The results suggest that Swiss glacier retreat was due to a combination of factors: (1) changes in precipitation patterns, (2) cyclic changes in North Atlantic SSTs (60-year cycle?), and (3) increased air temperatures. Since *Nature* is highly biased toward the extreme alarmist view, it is not surprising this editorial ended with some dire predictions for the remainder of the 21st century. Huss *et al.* concluded that "Strong glacier retreat is very likely in the 21st century!" Extrapolating forward from the 20th century, this seems to be a fairly safe prediction. The next several decades should provide a test of this prediction.

Roe (2010) provided a very insightful analysis of glacier movements and their relevance to climate change. He showed that large fluctuations in glacier extent occur from year to year even though the climate remains unchanged. Hence, glacier advance and retreat are classic examples of a low signal-to-noise parameter for detecting climate change. While glacier retreat during the past 100 years is consistent with global warming, it is not by itself proof of global warming by greenhouse gases.

3.6 ANTARCTIC AND ARCTIC TEMPERATURES

High-latitude regions of the Earth (the Arctic and Antarctic) have been considered as bellwethers in the detection of global climate change. According to IPCC (2001):

"Climate change in polar regions is expected to be among the largest and most rapid of any region on the Earth, and will cause major physical, ecological, sociological, and economic impacts, especially in the Arctic, Antarctic Peninsula, and Southern Ocean. Polar regions contain important drivers of climate change. Once triggered, they may continue for centuries, long after greenhouse gas concentrations are stabilized, and cause irreversible impacts on ice sheets, global ocean circulation, and sea-level rise."

As we discussed in Section 3.4.2, there are a number of feedback effects that amplify climate change in polar areas. Hence temperature changes in polar areas

[29] Huss, M., Jouvet, G., and Funk, M., *C2SM Climate Change Scenario Workshop*, ETH Zurich, March 2nd, 2010.

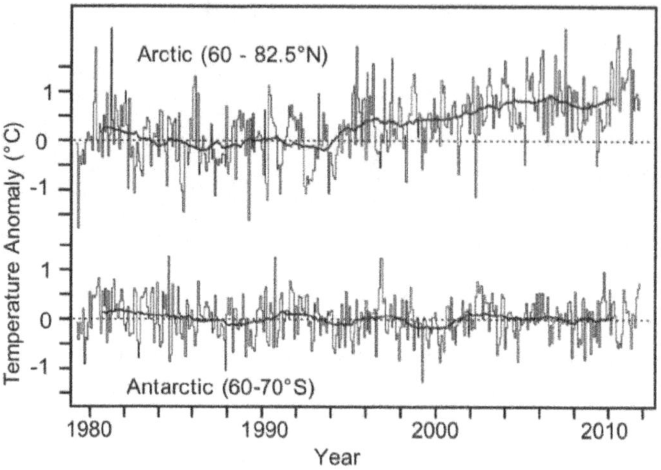

Figure 3.52. Arctic and Antarctic temperatures over the past 30 years (Humlum, 2011).

during decades or centuries of climate change tend to be much greater than in temperate zones.

Taylor (2006) said:

> "Global climate models (GCMs) suggest that polar regions should warm more quickly than temperate or tropical regions because the cold temperatures cause the air to be very dry. The low amounts of water vapor, the most important greenhouse gas, causes the relative effects of other gases, notably carbon dioxide (CO_2), to be greater. Thus climate change caused by an increase in the latter should be most evident in the polar-regions."

However, this argument does not seem to be valid. Whatever warming takes place due to rising CO_2 will tend to occur equally at all latitudes. The only exception is where some absorption bands are already populated by water vapor, as in the tropics, and therefore increased CO_2 will not have as great an effect. But, since CO_2 absorption bands are already saturated from CO_2, this argument does not (pardon the pun) hold water. The real reason that polar regions warm with greater amplitude as the Earth warms is due to the positive feedback of ice sheets, snow, and sea ice as discussed in Section 3.4.2.

Recent data on Arctic and Antarctic temperatures over the past three decades are shown in Figure 3.52. Antarctic temperatures have been relatively constant while Arctic temperatures increased after about 1995.

3.6.1 Antarctic temperatures

Antarctica is a very cold place (Anon. (A)). The coldest temperature recorded was (–89°C). Several factors combine to make Antarctica so cold:

- Unlike the Arctic region, Antarctica is a continent whose interior areas do not benefit from the moderating influence of water.
- With 98% of its area covered with snow and ice, the Antarctic continent reflects most of the Sun's input rather than absorbing it.
- The extreme dryness of the air causes any heat that is radiated back into the atmosphere to be lost instead of being absorbed by water vapor in the atmosphere.
- During the winter, the size of Antarctica doubles as the surrounding sea water freezes, effectively blocking heat transfer from the warmer surrounding ocean.
- Antarctica has a higher average elevation than any other continent on the Earth, which results in even colder temperatures.

In discussing the Antarctic continent, it is important to distinguish between the Antarctic Peninsula, which projects to lower latitudes, and has a low elevation, vs. the Antarctic continent that lies at higher latitudes and higher elevations (see Figure 3.53).

Most of the climatological measurement sites are located around the continent on the coast. Only two are inland.

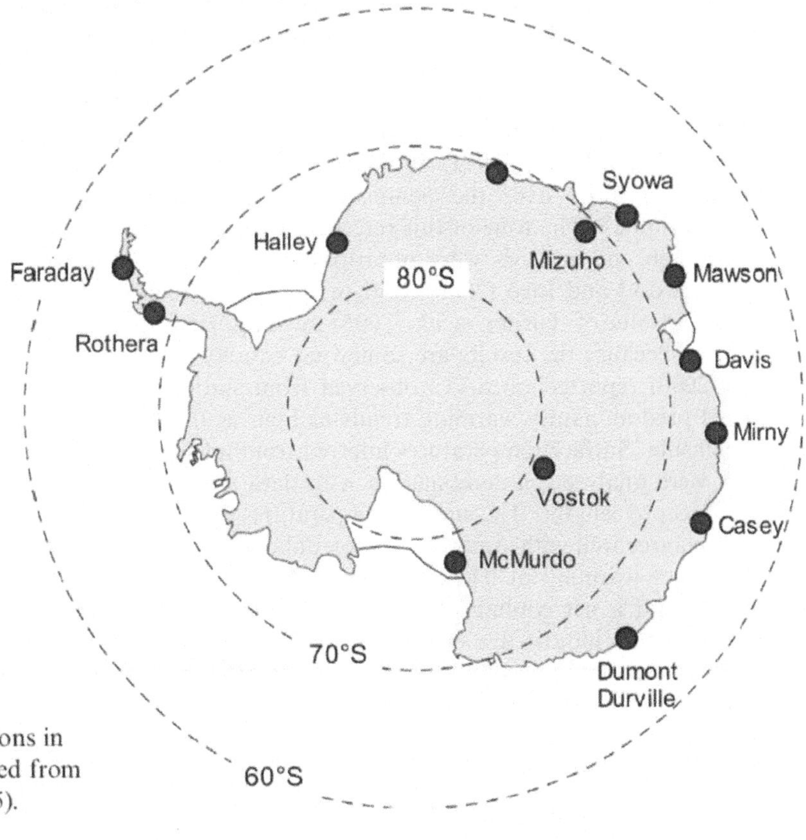

Figure 3.53. Stations in Antarctica (adapted from Turner *et al.*, 2005).

According to Doran *et al.* (2002), the average air temperature at the Earth's surface has increased by 0.06°C/decade during the 20th century, and by 0.19°C/decade from 1979 to 1998. Doran *et al.* (2002) said that models tend to indicate amplified warming in polar regions, as observed in Antarctica's peninsula region over the second half of the 20th century. Although previous reports suggested slight recent continental warming, Doran *et al.* (2002) found a net cooling on the Antarctic continent between 1966 and 2000, particularly during summer and autumn. They pointed out that although a previous study had found a warming trend from 1958 to 1978, this seems to have reversed itself after 1978. Doran *et al.* (2002) concluded: "Continental Antarctic cooling, especially the seasonality of cooling, poses challenges to models of climate and ecosystem change."

Turner *et al.* (2005) analyzed 19 long-term stations over the Antarctic continent. The peninsula experienced a major warming over the last 50 years. The warming at low elevations on the western coast of the Antarctic Peninsula was described as being "as large as any increase observed on Earth over the last 50 years, which at the Faraday (now Vernadsky) station amounted to about 2.5°C". However, Turner *et al.* (2005) said: "the region of marked warming is quite limited and is restricted to an arc from the southwestern part of the peninsula, through Faraday to a little beyond the tip of the peninsula". The 14 stations around the continent were equally divided between warming and cooling, leading to an uncertain result. The data also indicated that warming trends appeared to be greater for the 1961–1990 period compared to the 1971–2000 period, suggesting that warming tailed off after 1971. It is interesting that Turner *et al.* (2005) emphasized the warming over the peninsula and described the temperatures elsewhere as "variable". The tone of this reference suggests that it supports global warming, although the authors were appropriately cautious. However, Taylor (2006) and Sherwood and Idso (2006), both written by global-warming skeptics, referred to the results of Turner *et al.* (2005) as supportive of doubt regarding global warming, because the results are somewhat equivocal.

Comiso (2000) reported on SATs observed from stations in Antarctica. The results showed predominantly warming trends as high as 0.5°C/ decade along the Antarctic Peninsula. Surface temperatures inferred from infrared satellite data from 1979 to 1998 were analyzed in combination with data from 21 stations that have relatively long record lengths. The surface temperatures derived from infrared data were claimed to agree well with Antarctic station data. The 45-year record of station data showed a net warming of 0.012°C/yr (± 67%) while the 20-year record of station data showed a net cooling of 0.008°C (± 300%). The 20-year record of satellite measurements indicated a cooling of 0.042°C/yr (± 150%). The 20-year record length was claimed to be about the minimum length required for a meaningful trend analysis study. It was concluded:

> "The slight cooling of the entire ice sheet observed in both in situ and satellite records during the last 20 years is intriguing since during the same time period a general warming is being observed globally."

Thompson and Solomon (2002) indicated that recent trends in SH tropospheric circulation contributed to warming over the Antarctic Peninsula and cooling over the Antarctic plateau.

Scientists have long believed that a large part of Antarctica—the East Antarctic ice sheet—has been cooling despite the more prevalent global warming, although the Antarctic Peninsula has certainly warmed. Steig *et al.* (2009) estimated 50-year temperatures in the interior of Antarctica using a combination of surface data (mostly around the periphery of Antarctica), satellite data (IR radiated by the snowpack), and models to fill in missing data. It is difficult to appraise the credibility of the models used because the paper is mostly jargon and unreadable except to a very few specialists. They concluded that essentially all of Antarctica warmed in the last 50 years of the 20th century. However, several of the authors are noted alarmists, and one is the leading defender of the *hockey stick*. The details are intricate and beyond the scope of this book. Apparently, there is a statistically significant difference between satellite data and ground data, with satellite data showing more warming than ground data. As always, in the process of combining data from multiple sources over extended time periods, the results are noisy and unraveling the implicit signal to discover trends involves complex procedures.

The procedures used in this paper were discussed by O'Donnell *et al.* (2011). They concluded: "Though the general reconstruction concept has merit, it is susceptible to spurious results for both temperature trends and patterns." They investigated two alternate methods to improve the procedure. They concluded that warming over the period of 1957–2006 was concentrated in the peninsula (~0.35°C/decade). They also found that average trends for the continent, East Antarctica, and West Antarctica were half or less than that found by Steig *et al.* (2009). Subsequently, additional commentary appeared on Internet blogs. In particular, McIntyre[30] describes efforts by the Steig team to block publication of the O'Donnell team criticism. McIntyre said: "... whatever is new in Steig *et al.* (2009) is not only incorrect, but an artifact of flawed math and whatever is valid was already known." According to McIntyre: "Steig's methodology smeared warming from the Antarctic Peninsula into other parts of Antarctica." Yet the new edition of the IPCC Report refers generously to the Steig *et al.* paper and totally ignores the results of O'Donnell *et al.*

Frolov *et al.* (2009) presented Figure 3.54, which suggests that temperatures in Antarctica have been meandering for many years, although the peak reached around 1980 was somewhat higher than previous peaks.

Chapman and Walsh (2007) prepared a comprehensive analysis of SAT anomalies for Antarctica and the Southern Ocean at a grid resolution sufficient to capture regional variations of trends. Land surface station, automatic weather station, and gridded summarized SST data were utilized. Their results are shown in Figure 3.55, which shows the average Antarctic temperature change from any starting date to 2002.

Chapman and Walsh said: "The salient finding in the context of climate change is that the 45-year trends are small. ..." In fact, not only were the trends small, but

[30] /*www.climateaudit.org*.

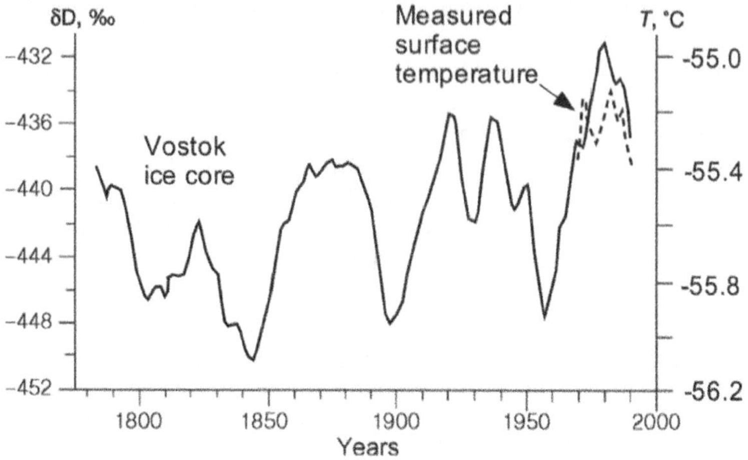

Figure 3.54. Change in air temperature in the Antarctic from isotopic composition of ice cores at Vostok Station. The dashed line indicates measured air temperature (Frolov *et al.*, 2009).

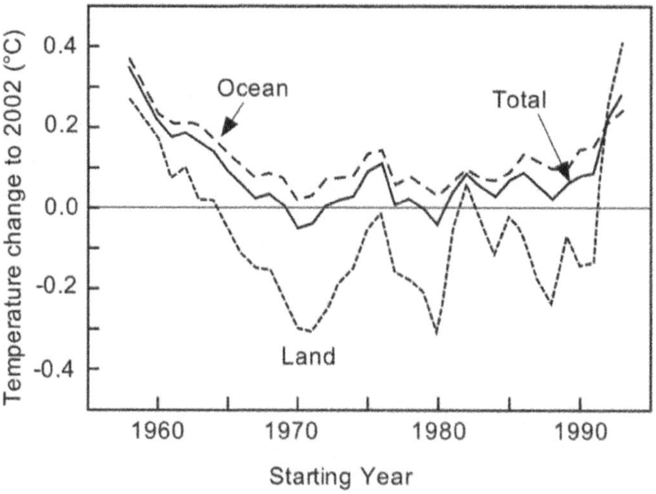

Figure 3.55. Change in Antarctic temperature from any starting date to 2002 (Chapman and Walsh, 2007).

the trends over land were negative for the 1970s and 1980s. Most of whatever warming occurred was over the Antarctic Peninsula. Despite this, Chapman and Walsh interpreted the data according to the alarmist agenda, projecting strong warming over Antarctica in the 21st century based on climate models.

Tedesco and Monaghan (2010) reviewed what has been learned about the melting of snow and ice over all of Antarctica since 1979, when routine measurement of the phenomenon via space-borne passive microwave radiometers

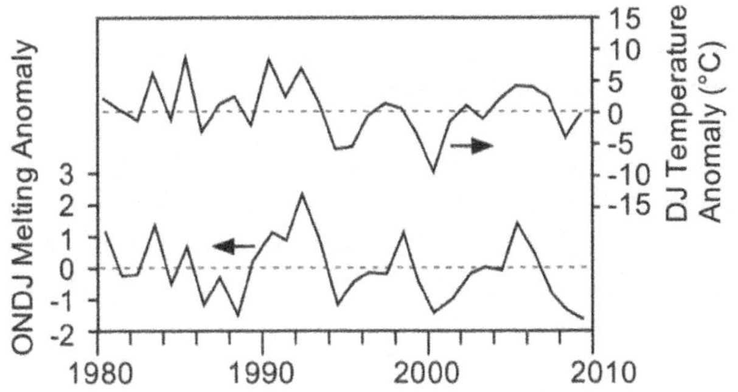

Figure 3.56. December–January temperature anomaly and October–November–December–January melting anomaly for Antarctica (Tedesco and Monaghan, 2010).

first began (NIPCC, 2011). Their results for the snow and ice melting trends averaged over 12 continent-wide sites over the past 30 years did not display any notable trends (see Figure 3.56). In other words, they found no evidence of significant melting and the climate in Antarctica has been stable. Nevertheless, they had to pay homage to the orthodoxy by concluding "Negative melting anomalies observed in recent year do not contradict recently published results on surface temperature trends over Antarctica". They also concluded that, in the future, "enhanced summer melting is likely to occur if the positive Southern Hemisphere Annular Mode (SAM) trends subside". This can be paraphrased as "enhanced summer melting is likely to occur if it occurs"—as in the Marx Brothers, "The first party is the party of the first part".

Zwally *et al.*[31] reported:

> "During 2003 to 2008, the mass gain of the Antarctic ice sheet from snow accumulation exceeded the mass loss from ice discharge by 49 Gt/yr (2.5% of input), as derived from ICESat laser measurements of elevation change. The net gain (86 Gt/yr) over the West Antarctic (WA) and East Antarctic ice sheets (WA and EA) is essentially unchanged from revised results for 1992 to 2001 from ERS radar altimetry. ... A slow increase in snowfall with climate warming, consistent with model predictions, may be offsetting increased dynamic losses."

3.6.2 Arctic temperatures

It is observed that, as the Earth warms or cools due to various causes, the Arctic will experience greater than average warming or cooling, whereas tropical areas will undergo lesser changes. Warming in the SH is limited by the large heat uptake of the Southern Ocean, leaving the Arctic as the global location with the greatest warming.

[31] "Mass gains of the Antarctic ice sheet exceed losses" *http://ntrs.nasa.gov/search.jsp?R=20120013495*.

Increased warming in Arctic regions is usually referred to as "Arctic amplification". Figure 3.22 shows this very clearly. There is evidence that Arctic amplification was significant as far back as the Phanerozoic Eon.

In the past decade, a number of studies were conducted on Arctic amplification during the warming that occurred in the 20th century, and projections have been made of continued warming in the 21st century. Almost all of these begin by assuming that warming in the 20th and 21st centuries is predominantly due to increased concentrations of greenhouse gases, particularly CO_2, and then proceed to analyze the role that Arctic amplification will play in response to rising greenhouse gas concentrations in the 21st century. Several of these papers seem to have the intent of showing that anthropogenic emissions of CO_2 were the cause of the observed warming.

According to Hanna and Cappelen (2003):

> "The Polar Regions are among the most interesting regarding the ongoing debate on global climate change because, due to several key climatic feedbacks, they are potentially extremely climatically sensitive. The best known is the ice-albedo feedback by which an initial perturbation (slight warming) melts some ice; this new more extensive darker melt-water area absorbs more incoming sunlight, which accelerates the warming and melting of surrounding ice. Therefore it is crucial to improve our understanding of current conditions and past history of the major ice sheets and sea ice, and to model how they are likely to behave in future (e.g. in response to man-made global warming). This requires not only glaciological observations but also meteorological ones. Unfortunately polar areas are noted for their dearth of observations. Greenland, the world's largest island,[32] hosts the world's second largest ice sheet (after Antarctica) and is important in both meteorological and glaciological terms. It is a vast heat sink, due to its predominantly ice-covered surface that reflects some 60–90% of incoming solar radiation. The GIS is a huge ice dome, 3 km thick in the middle, which locally rises to more than 3 km above sea level. Cold, dense air streams off the surface and fierce katabatic winds roar down ice fjords near the outer edges. If the ice sheet were to melt in its entirety, which would take some thousands of years at projected rates of global warming, it could contribute around 6–7 m to global sea level. Recent studies, especially airborne laser measurements of surface elevation of the ice, suggest that most of the interior of the ice sheet is relatively stable or growing slightly, whereas the margins have thinned substantially over the past few years. Most melting of the ice, and resulting loss by runoff of surface melt water, occurs within a relatively thin zone (about 100–200 km) around the edges, and depends strongly on summer temperatures, so the mass balance of the ice sheet is highly sensitive to climatic change. Likely future warming could also affect the way the ice flows outwards near the edges, and the rate of iceberg calving (formation), the latter another important mass loss process. Meanwhile, changes in precipitation and

[32] Other than Australia.

snowfall are paramount for snow accumulating on the ice sheet. ... Climatic change in Greenland has been important historically. The period 800–1100 was relatively warm and favored Viking expansion to Greenland. A climatic deterioration then set in from c. 1250 through 1350. The fourteenth century is considered the coldest in the last 1000 years in Greenland. The subsequent chill of the *Little Ice Age* over Europe may have been muted over western Greenland ... although there is evidence for a LIA cooling over central Greenland. A glacial advance that began in Greenland about 400 years ago continued until the eighteenth, nineteenth or even early twentieth centuries ... followed by a recession back towards the LIA level ... In contrast to general global warming, Greenland (also Iceland and the Faeroes) cooled significantly during the latter half of the twentieth century."

Hanna and Cappelen (2003) also said:

"Analysis of new data for eight stations in coastal southern Greenland, 1958–2001, shows a significant cooling (trend-line change −1.29°C for the 44 years), as do sea-surface temperatures in the adjacent part of the Labrador Sea, in contrast to global warming (+0.53°C over the same period). The land and sea temperature series follow similar patterns and are strongly correlated but with no obvious lead/lag either way."

Hanna and Cappelen (2003) went on to say:

"Different temperature and precipitation records, covering the period 1958–99, from six stations in Southern Greenland have been studied. ... Despite global warming over the past few decades, the SW marginal areas of southern Greenland seem to have actually cooled, especially daytime temperatures in winter. ... It will be intriguing to see if this trend continues, as it could substantially influence the mass balance (through changes in snow accumulation, surface melt runoff and iceberg calving) of the southern parts at least of the ice sheet. The overall cooling may also have caused a sharp increase in snowfall days at some of the coastal stations, where snow as a fraction of precipitation is critically dependent on temperature. These results demonstrate the regional vagaries of the global weather machine; climatic change is not a simple uniform process. Yet as one of these regions, Greenland greatly influences the surface heat budget, atmospheric circulation and (through the waxing and waning of its ice sheet) global sea level."

Przybylak (2002) provided a detailed analysis of intra-seasonal and inter-annual temperature variability for the whole Arctic for the period 1951–1990. Four temperature variables were used: average (T_{mean}), maximum (T_{max}), and minimum (T_{min}) temperatures, and the DTR. Ten stations were utilized, representing almost all climatic regions in the Arctic. Regional trends in intra-seasonal and inter-annual temperature variability were mixed and the majority of them were insignificant. Trends in intra-seasonal temperature variability were upward in the Norwegian Arctic and eastern Greenland and downward in the Canadian and Russian Arctic. The final conclusion was:

> "The absence of significant changes in intra-seasonal and inter-annual variability of T_{MEAN} T_{MAX}, T_{MIN} and DTR is additional evidence (besides the average temperature) that in the Arctic in the period 1951–1990 no tangible manifestations of the greenhouse effect can be identified."

According to Chylek, Box, and Lesins (2004):

> "A considerable and rapid warming over all of coastal Greenland occurred in the 1920s when the average annual surface air temperature rose between 2 and 4°C in less than ten years (at some stations the increase in winter temperature was as high as 6°C). This rapid warming, at a time when the change in anthropogenic production of greenhouse gases was well below the current level, suggests a high natural variability in the regional climate. ... Since 1940, however, the Greenland coastal stations data have [undergone] predominantly a cooling trend. At the summit of the Greenland ice sheet the summer average temperature has decreased at the rate of 2.2°C per decade since the beginning of the measurements in 1987. This suggests that the Greenland ice sheet and coastal regions are not following the current global warming trend."

Chylek, Dubey, and Lesins (2006) found that "current Greenland warming is not unprecedented in recent Greenland history". Warming experienced from 1995 to 2005 also occurred from 1920 to 1930, except that the rate of warming from 1920 to 1930 was about 50% higher than that in 1995–2005.

Zwally *et al.* (2005) extended the analysis of radar altimeter data from two European remote-sensing satellites to 90.0% of the GIS, 77.1% of the Antarctic ice sheet, and 81.8% of the Antarctic ice shelves. The estimated changes in ice mass from elevation changes derived from 10.5 years (Greenland) and 9 years (Antarctica) of satellite radar altimetry data indicated that the GIS is thinning at the margins (–42 Gt/yr below the equilibrium-line altitude (ELA)) and growing inland (+53 Gt/yr above the ELA) with a small overall mass gain (+11 Gt yr). The ice sheet in West Antarctica (WA) is losing mass (–47 Gt/yr) and the ice sheet in East Antarctica (EA) shows a small mass gain (+16 Gt/yr) for a combined net change of –31 Gt/yr. The contribution of the three ice sheets to sea level is $+0.05 \pm 0.03$ mm/yr. The Antarctic ice shelves show corresponding mass changes of –95 Gt/yr in WA and +142 Gt/yr in EA. Expected responses of the ice sheets to climate warming are growth in thickness of the inland ice areas, due to increasing precipitation, accompanied by thinning near the margins, due to increasing surface melting.

Soon and Baliunas (2003a, b) pointed out:

> "In considering the possible link of early 20th century warming to the rise in atmospheric CO_2 concentration, it should be noted that the Arctic-wide temperatures of Overpeck *et al.* (1997) began rising in the mid-19th century and peaked around 1940–1960, when the increase in the air's CO_2 content was less than 20–30% of the cumulative CO_2 increase to date."

Polyakov *et al.* (2003c) examined long-term Arctic SAT and pressure data for the period 1875–2000 poleward of 62°N. The Arctic air temperature and pressure displayed substantial variability on timescales of 50–80 years. They suggested that the origin of this variability might lie in the complex interactions between the Arctic and North Atlantic. The two periods of highest temperatures in the Arctic were: in the 1930s–1940s, and in recent decades. They speculated that global warming alone cannot explain the retreat of Arctic ice observed in the 1980s–1990s. They concluded:

> "The complicated nature of Arctic temperature and pressure variations makes understanding of possible causes of the variability, and evaluation of the anthropogenic warming effect most difficult."

Similarly, Polyakov *et al.* (2003a) found a high level of Atlantic temperature variability in and sea-surface salinity fluctuations on the Siberian Shelf and the Amundsen Basin. They suggested that there are "strong limitations on our ability to define amplitudes of anomalies by comparing recent synoptic measurements with [climate models], especially for [such] regions characterized by strong variability". This echoed the conclusion from an earlier paper (Polyakov and Johnson, 2000) that said:

> "Our results suggest that the decadal Arctic oscillation and multi-decadal low-frequency oscillation drive large amplitude natural variability in the Arctic, making detection of possible long-term trends induced by greenhouse gas warming most difficult."

Howat, Joughin, and Scambos (2007) found large year-to-year variations in ice loss from ice sheets. Shepherd and Wingham (2007) reviewed the results of more than a dozen satellite-based studies of ice loss from the Greenland and Antarctic Ice Sheets. They estimated that the East Antarctica Ice Sheet (EAIS) is gaining some 25 Gt/yr, the West Antarctica Ice Sheet (WAIS) is losing about 50 Gt/yr, and the GIS is losing about 100 Gt/yr. These trends provide a modest contribution to sea-level rise of about 0.35 mm/yr. However, these short-term results should not be extrapolated because of the cyclic behavior of ice sheet loss. Frolov *et al.* (2009) presented an extensive review of centennial ice cover observations in Eurasian Arctic Shelf seas. They emphasized:

> "The variability and state of the Arctic sea ice cover strongly depend on atmospheric conditions and on ocean dynamic and thermodynamic processes. A number of parameters influence the direction and intensity of these processes. The most significant are: the surface air temperature, wind, oceanic boundary layers and their stratification, and ocean circulation. In order to understand the causes of long-term changes in the ice cover, it is necessary to define the temporal and spatial relationships of the sea-ice cover with all [these factors]."

Frolov *et al.* (2009) utilized anomalies of mean annual SAT in the zone from 70°N to 85°N for the period from 1900 to 2003 to analyze climatic changes in the Arctic Seas throughout the last century. Their result for mean annual air temperature in the 70°N–85°N zone is shown in Figure 3.57. These authors emphasized the periodicity of cooling and warming events interpreting this figure in terms of a proposed 60-year cycle. Note that the dip in temperature from 1940 to

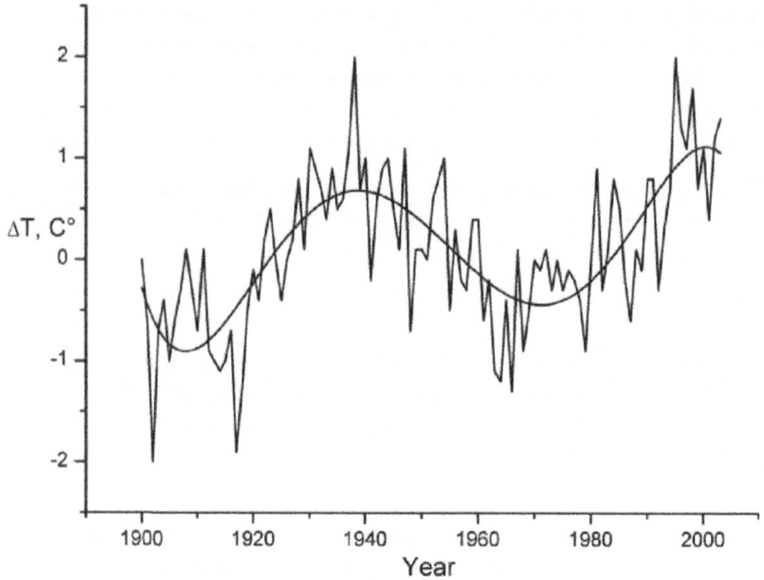

Figure 3.57. Changes in mean annual air temperature in the 70°N–85°N zone (Frolov *et al.*, 2009).

about 1975 coincides with a similar dip in temperature experienced at other latitudes. They found that the average location of the winter boundary of prevailing multi-year ice showed consistent southward displacement of the boundary from 1960 to 2000 even though there was a small negative trend in the region's sea-ice extent. The cause of this southward displacement was a noticeable weakening of the Arctic High toward the end of the 20th century, accompanied by a diverging ice cover and a decrease in ice export from the Arctic Basin to the Greenland Sea during the warm periods as compared with the cold periods. Frolov *et al.* (2009) drew quite a number of conclusions from their study of Eurasian Arctic Shelf seas. Only a brief summary of a few conclusions is given here.

There was a gradual quasi-linear decrease in Arctic sea-ice cover area from the beginning to the end of the 20th century, which can be expressed by a linear trend. However, the rate of decrease slowed in the second half of the 20th century for western seas. In addition to the linear trend, there were several cyclic trends superimposed, particularly one with a 50-year to 60-year period with an expansion of sea ice in the 1970s.

More detailed data for the Arctic and Antarctic regions since 1957 are provided in Figure 3.58. Antarctic temperatures have vacillated about a constant mean whereas Arctic temperatures began rising with the advent of strong El Niños in the 1980s.

The sharp temperature rise in the Arctic early in the 20th century remains as a problem for those who would blame all climate change on greenhouse gases. Wood and Overland (2010) revisited this issue. They said that "the recent widespread

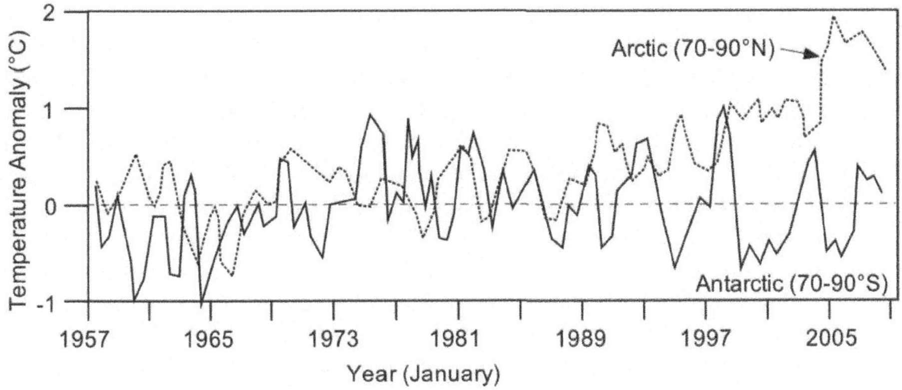

Figure 3.58. Temperature anomalies for Arctic and Antarctic regions since 1967 (three-year moving average) (Humlum, 2009).

warming of the Earth's climate is the second of two marked climatic fluctuations to attract the attention of scientists and the public since the turn of the 20th century," and the first of these, "the major early 20th century climatic fluctuation (~1920–1940)", has been "the subject of scientific enquiry from the time it was detected in the 1920s". Furthermore, they wrote that: "the early climatic fluctuation is particularly intriguing now because it shares some of the features of the present warming that has been felt so strongly in the Arctic".

These authors went back to early records to try to gain more information on the early warming episode. They said "there is evidence that the magnitude of the impacts on glaciers and tundra landscapes around the North Atlantic was larger during this period than at any other time in the historical period". They went on to say that "the ultimate cause of the early climatic fluctuation was not discovered by early authors and remains an open question", noting that "greenhouse gas forcing is not now considered to have played a major role". Thus, they suggested that: "the early climatic fluctuation was a singular event resulting from intrinsic variability in the large-scale atmosphere/ocean/land system and that it was likely initiated by atmospheric forcing". They concluded that "thus far, human influence does not stand out relative to other, natural causes of climate change".[33]

Greenhouse gases do not explain the sharp rise in Arctic temperatures prior to 1940 because carbon emissions were relatively low during that period. A more likely explanation centers on the role of black carbon (BC) emitted from power plants, vehicles, and factories in the U.S., Europe, and the former U.S.S.R. in the first half of the century, falling on Arctic snow and ice. By about 1940, BC emissions from mid-north latitudes diminished but sulfate aerosol emissions increased steeply after 1940. It seems possible that aerosols were dominant in producing a cooling effect from about 1940 to the mid-1970s. It has been theorized that BC played a secondary role during this period, being overwhelmed by the sharp rise in sulfate aerosol

[33] The above two paragraphs were paraphrased from *www.sepp.org*.

emissions. It is not so clear what was responsible for the renewed global warming that began in the 1970s. While BC emissions began to rise again during this period, the region of predominant emission moved from more northerly locations to Asian regions at lower latitudes. It is not clear how efficiently such emissions can be transported to higher latitudes. CO_2 emissions rose sharply after the 1970s. But another factor is the state change of the Pacific Ocean in the 1970s. There is a close correspondence between SATs and El Niño indices over the past 30 years.

Bengtsson *et al.* (2004) suggested that "four possible mechanisms, individually or in combination, could have contributed to the early twentieth century warming: anthropogenic effects, increased solar irradiation, reduced volcanic activity, and internal variability of the climate system". They concluded that "It seems unlikely that anthropogenic forcing on its own could have caused the warming, since the change in greenhouse gas forcing in the early decades of the twentieth century was only some 20% of the present". However, in considering anthropogenic effects, they dealt only with greenhouse gases, and did not consider the deposition of BC. They pointed out the uncertainties in reconstructing past solar irradiance, and they dismissed volcanic activity as the cause of this warming. Therefore, they sought an answer in terms of the atmospheric flow pattern that drives ocean circulation and results in the advection of warm water into the north-eastern North Atlantic. Johannessen *et al.* (2004) concluded that the warming of the 1920s and 1930s was due to "natural fluctuations internal to the climate system". Reductions in albedo due to decreasing sea ice induced by wind changes were attributed as the cause of this early warming. However, they claimed that more recent warming in the 1980s and 1990s was due to greenhouse gases. However, Polyakov *et al.* (2003b) concluded that the Arctic is subject to natural oscillatory variations, the principal driver for climate change, and that "[greenhouse] warming alone cannot explain the retreat of Arctic ice observed in the 1980s–90s". Their final conclusion was:

> "The complicated nature of Arctic temperature and pressure variations makes understanding of possible causes of the variability, and evaluation of the anthropogenic warming effect most difficult."

In Section 7.3, another possible factor in Arctic climate change is explored: deposition of BC on Arctic snow and ice.

The problem for those who attribute all climate change to greenhouse gases is that, if events occasionally occur independently of greenhouse gases that produce effects greater than or equal to predicted effects of greenhouse gases, how does one distinguish between the two?

While the case can be made that recent warming in the Arctic is not unprecedented, nevertheless there are indications that, in recent years, Greenland has been subject to significantly increased melting. Tedesco *et al.* (2010, 2011a, 2011b) used satellite data to infer surface melting over the GIS. A standardized melting index is defined as the number of melting days per year times the area subject to melting. Tedesco *et al.* presented their estimates for the melting index as shown in Figure 3.59. As Tedesco *et al.* pointed out, "Satellites data cannot produce estimates of runoff and liquid water content". However, these can be estimated with models.

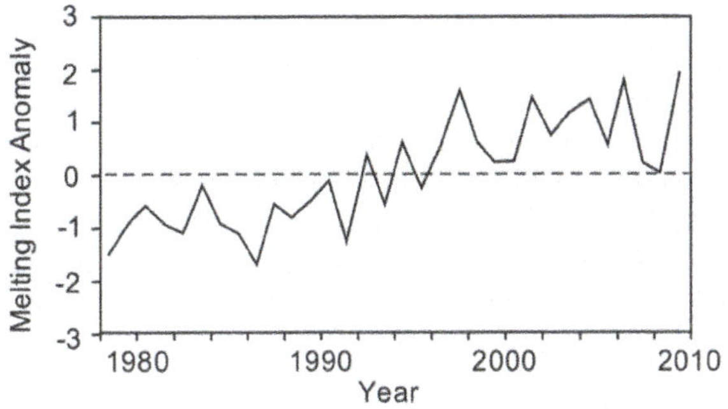

Figure 3.59. Standardized melting index for Greenland (Tedesco *et al.*, 2011a, 2011b).

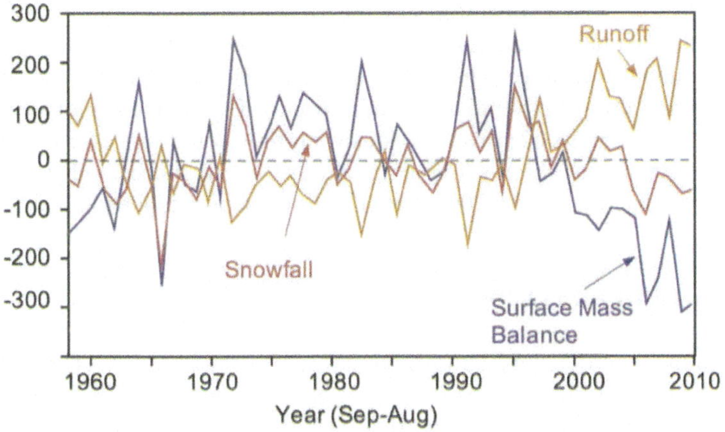

Figure 3.60. Snowfall and modeled run-off and surface mass balance for Greenland (Tedesco *et al.*, 2011a, 2011b).

The model used by Tedesco *et al.* led to the estimates shown in Figure 3.60. According to this, snowfall has been relatively constant over the past 50 years, although it took a dip from 2005 to 2010. Run-off increased substantially after 2000, leading to negative surface mass balances after 2000. Considering that the CO_2 concentration has been steadily increasing over this 50-year period, it is not clear why the run-off should suddenly increase after 2000. Over the next several years, it may become apparent whether the estimate of increased run-off is real and persistent, or whether it was a temporary fluctuation.

Using old temperature observations from early observers, Vinther *et al.* (2006) extended estimates of Greenland temperature records back into the late 18th century. They presented data for four three-month periods as well as annual data. While data were less complete as one goes back from about 1880 to about 1780, there were sufficient data for them to draw these conclusions: the period from 1780 to about 1910 was mostly cooler than the 20th century, although there was a warm period near 1850. We may think of this as the tail end of the LIA. The warmest year was 1941, while the 1930s and 1940s were the warmest decades. Greenland cooled somewhat after about 1950, but then began warming again in the 1990s. Two distinct cold periods, following an 1809 unidentified volcanic eruption and the eruption of *Tambora* in 1815, made the 1810s the coldest decade on record.

Chylek, Dubey, and Lesins (2006) analyzed temperature time series from various Greenland locations. They found that temperatures were relatively flat from 1900 to 1920. Starting around 1920, Greenland temperatures rose rapidly for 10 years. After that, temperatures meandered but drifted slowly downward, reaching 1920 levels again around 1990. Around 1992, temperatures began climbing rapidly again in a manner reminiscent of the 1920–1930 sequence. As of 2004, temperatures were comparable with those reached at the previous high in 1930. They did not find that 1941 was the hottest year but they did find that the 1930s and 1940s were warm decades. It was also noted that, generally, Greenland temperatures in the second half of the 20th century were colder than in the first half. These results do not support the belief that rising CO_2 is the main cause of global warming. While some alarmists have emphasized the run-up in Greenland temperatures after 1992 as an indication of CO_2-induced global warming, the fact that such a run-up occurred previously suggests some sort of cyclic behavior may have occurred rather than a long-term upward trend. Indeed, GISS data from January 2009 indicate that, since 2005, Greenland temperatures have fallen.

3.7 OCEAN TEMPERATURES

The Earth's atmosphere is very tenuous compared to the oceans. As Pielke (2003) discussed, the energy balance of the Earth can be estimated by either calculating the various forcings that act on the Earth at the TOA and thereby estimating the net flux (W/m^2) acting to heat the Earth, or by estimating the heat deposited into the Earth system (e.g., J/yr). In the former case, the argument is that the imbalance in energy flux at the TOA must be stored in the Earth system, and the only place it can go is the oceans; therefore, it must heat the oceans. Since the heat capacity of the Earth lies mainly in the oceans, one can estimate the heat gain of the Earth by measuring the heat gain of the oceans. In fact, measurement of thermal changes in the oceans is far more important than thermal changes in the air for understanding the Earth's energy balance. There are difficulties in both approaches. Of particular importance in regard to forcings are the unknowns relative to changes in humidity, aerosols, and clouds as the Earth warms and industrialization proceeds. The problem in estimating ocean heat content is that the oceans are so diverse, and the collective mass of the oceans is so huge that, even with a large heat flux imbalance, the temperature rise is small and difficult to measure.

It is common for climatologists to report heat gain by the oceans in units of W/m^2, but, in reading these papers, it is important to be certain which area they refer to. Some refer to the area of the Earth while others refer to the area of the oceans. A given heat gain based on the surface area of the Earth is equivalent to about 0.7 times the heat gain based on the surface area of the oceans.

Palmer and Haines (2009) estimated the global heat absorbed by the ocean to a depth of 220 m from 1970 to 2000 to be 0.25 W/m^2 (based on ocean area, not global area).

According to Levitus *et al.* (2012), over the 55-year period from 1955 to 2010, the heat content of the oceans down to a depth of 2,000 m increased by 2.4×10^{23} J. They claim that this corresponds to an average temperature rise for the 0–2,000 m layer of 0.09°C. Since the 0–2,000 m layer includes 48% of the volume of the oceans, its volume is about $0.48 \times 1.3 \times 10^9$ km^3 = 6.3×10^8 km^3. The average temperature rise over 55 years for this layer is thus estimated to be:

$$2.4 \times 10^{23}\ \text{J}/(6.3 \times 10^8\ \text{km}^3 \times 1.025 \times 10^{12}\ \text{kg/km}^3 \times 3993\ \text{J/kg/°C}) = 0.08\text{°C},$$

which is close to the value given by Levitus *et al.* (2012). The same calculation can be repeated for the 0–700 km layer. According to Levitus *et al.* (2012), over the 55-year period from 1955 to 2010, the heat content of the ocean down to depth 700 m increased by 1.67×10^{23} J. They claim that this corresponds to an average temperature rise for the 0–700 m layer of 0.18°C. Since the 0–700 m layer occupies 16% of the volume of the oceans, its volume is about $0.16 \times 1.3 \times 10^9$ km^3 = 2.1×10^8 km^3. The average temperature rise over 55 years for this layer is thus estimated to be:

$$1.67 \times 10^{23}\ \text{J}/(2.1 \times 10^8\ \text{km}^3 \times 1.025 \times 10^{12}\ \text{kg/km}^3 \times 3993\ \text{J/kg/°C}) = 0.19\text{°C}.$$

By subtraction, we can infer that, for the 700–2,000 m layer, the increase in heat content was 0.73×10^{23} J. The volume of this layer is roughly 4.2×10^8 km^3. The average temperature rise over 55 years for this layer is thus estimated to be:

$$0.73 \times 10^{23}\ \text{J}/(4.2 \times 10^8\ \text{km}^3 \times 1.025 \times 10^{12}\ \text{kg/km}^3 \times 3993\ \text{J/kg/°C}) = 0.04\text{°C}.$$

Thus, the average rates of temperature increase per year over the 55-year span were:

$$\text{0–700 m layer: } 0.19\text{°C}/55 = 0.0035\text{°C/yr}$$
$$\text{700–2,000 m layer: } 0.04/55 = 0.00073\text{°C/yr}.$$

On average, the heat input to the oceans per unit area can be estimated as follows:

> Over the 55-year period, the heat content of the ocean down to depth 2000 m increased by 2.4×10^{23} J. The volume of ocean below 2,000 m is 52%. The average temperature rise over the 55-year span below 2,000 m is not known but was probably around 0.01°C or less. Adopting as a guess the figure 0.01°C, the total heat gain of the entire ocean over the 55-year period is estimated to be 3.0×10^{23} J.

Since the surface area of the oceans is 3.61×10^8 km^2, the total heat gain per unit area of ocean was:

$$3.0 \times 10^{23}\ \text{J}/3.61 \times 10^8\ \text{km}^2 = 8.3 \times 10^{14}\ \text{J/km}^2.$$

Over the 55-year period, there were $55 \times 3.2 \times 10^7 = 1.76 \times 10^9$ seconds.

Therefore, on average, the rate of heat gain by the oceans according to Levitus *et al.* per unit area of ocean over the 55-year period was:

$$8.3 \times 10^{14}/1.76 \times 10^{9} = 4.7 \times 10^{5}\ \text{W/km}^2 = 0.47\ \text{W/m}^2.$$

It is noteworthy that Hamon *et al.* (2012) made corrections to the world ocean database and concluded that the results "show a fairly prominent trend in 0–700 m ocean heat content of 0.39×10^{22} J/yr between 1970 and 2008". By contrast, Levitus *et al.* found a trend of 1.67×10^{23} J/55 years = 0.30×10^{22} J/yr between 1955 and 2010 for the 0–700 m layer.

Loeb *et al.* (2012) reviewed more recent data on ocean warming. Their results are shown in Table 3.3.

Table 3.3. Rate of heat absorption for the 0–700 m ocean layer (W/m^2) per unit area of earth for two time periods. Note that the area used to calculate the heat flux is the area of the Earth, rather than the area of the oceans. The area of the Earth is 1.41 times the area of the oceans. Therefore, to compare these data with the previous given calculations based on surface area of the oceans, one should multiply the figures in the table by 1.41. (Loeb *et al.*, 2012).

	Measured W/m² in top 700 m of ocean (based on surface area of the Earth)		
Time period	PMEL/JPL/JIMAR	NODC	Hadley
1993–2003	0.66 ± 0.17	0.48 ± 0.23	0.40 ± 0.21
2004–2008	0.18 ± 0.60	0.10 ± 0.60	0.31 ± 0.57

The error bars for the 2004–2008 data are excessively large.

Church *et al.* (2011) estimated the heat gain by the oceans for the periods 1972–2008 and 1993–2008. By subtraction, we can also estimate the heat gain for 1972–1992. Their reported heat gains are shown in Table 3.4. From these data and the volumes of the layers, we can (as before) calculate the estimated temperature rise for each layer for each time period. This is given in Table 3.5. The rate of heat gain by the various layers for various time periods per unit area of ocean is given in Table 3.6.

Table 3.4. Heat gains in units of 10^{21} J by time period and ocean layer according to Church *et al.* (2011).

	1972–2008	*1993–2008*	*1972–1992*
0–700 m	112.6	45.9	66.7
700–3,000 m	49.7	20.7	29.0
3,000 m to bottom	30.7	12.8	17.9
Total	193.0	79.4	113.6
Years	27	18	11
Seconds	8.6×10^{8}	5.8×10^{8}	3.5×10^{8}

Table 3.5. Total temperature rise (°C) by time period and ocean layer based on data of Church *et al.* (2011).

	Volume (km³)	*1972–2008*	*1993–2008*	*1972–1992*
0–700 m	2.1×10^8	0.131	0.053	0.078
700–3,000 m	7.0×10^8	0.017	0.007	0.010
3,000 m to bottom	3.9×10^8	0.019	0.008	0.011
Total	1.3×10^9	0.036	0.015	0.021

Table 3.6. Rate of heat absorption by time period and layer (W/m²) per unit area of ocean based on data of Church *et al.* (2011).

	1972–2008	*1993–2008*	*1972–1992*
0–700 m	0.36	0.22	0.53
700–3,000 m	0.16	0.10	0.23
3,000 m to bottom	0.10	0.06	0.14
Total	0.62	0.38	0.90

Trenberth (2010) described ocean heat measurements prior to build-up of the Argo profiling float system in 2003–2005:

> "Before then, the bulk of the observations of the ocean were from expendable bathythermographs (XBTs) dropped from ships along their tracks as opportunities arose. As a result, coverage was spotty and irregular, and missing over many regions such as the Southern Ocean. The XBTs recorded temperatures, but the exact depth they were at was an estimate based on an assigned drop rate, which turned out to be sensitive to the exact design and character of the XBT probe. Recent careful comparisons with calibrated probes deployed from research vessels have shown the need for corrections. The severe under-sampling of the ocean until about five years ago, along with the variety of methods used to correct for problems and biases, has led to many estimates of how the temperatures in the ocean have changed over time. Of particular interest for climate is the vertically integrated ocean heat content. The reprocessing of XBT and Argo observations has resolved some issues ... but there remains a surprisingly large spread among different estimates of ocean heat content as discussed in the paper by Lyman *et al.* (2010). They delved into the origins of these differences, and compiled and reprocessed a common data set for the upper 700 m of the ocean."

Trenberth concluded that, for the 0–700 m ocean layer, the measured heat gain was 0.90 W/m² over the time interval 1994–2008 and 0.77 W/m² over the time interval from 2004 to 2008. His conclusions suggest a slowdown after 2003 when the data network was better. These results are per unit area of the oceans and are considerably higher than those obtained by Levitus *et al.* and Church *et al.* It is not clear why his values are so much higher than those who analyzed the data.

There seems to be considerable uncertainty in the measured rates of heat gain by the oceans over the past few decades, but the data appear to suggest a heat gain of roughly 0.5 W/m^2 (area of ocean).

Some climatologists have argued that, aside from the absolute rate of gain of heat of the oceans, the rate of heat gain has been accelerating in recent years. Some of this is tied to the rate of sea-level rise, but the recent discovery that ground water depletion is contributing significantly to sea-level rise adds a new wrinkle to this issue (see Section 8.3.4).

3.8 UTILITY OF A SINGLE GLOBAL TEMPERATURE

Almost all studies of global climate change deal primarily with a single global average temperature, or a single hemispheric average temperature. Global warming is expressed as a rise in such a putative temperature.

However, some have questioned whether the concept of a single average global or hemispheric temperature has much utility. At any single location, temperatures typically vary widely during a single day. Temperatures fluctuate from day to day and averages change significantly from season to season. Surface temperatures vary widely between the tropics and polar regions, and between lowlands and mountains. Additionally, the range of temperatures experienced at any location over the course of a day, a season, a year, or decades is typically very large. A lowland tropical region experiences far less difference in temperature over the course of a year than does a highland temperate region. Determining the year-to-year temperature change in one location does not indicate how temperature varies in a region with a variable geography.

From an operational point of view, the procedures by which one combines data from hundreds or thousands of meteorological sites, corrects for urban heating and other aberrations, takes into account the oceans, accommodates differing densities of stations with geography, and thereby arrives at a global or hemispheric average temperature, are complex (Hansen *et al.*, 2001; Pielke *et al.*, 2007c; CCSP, 2005; Brohan *et al.*, 2006).

One problem with attempting to deal with a single average global or hemispheric temperature is that most of the regional variations tend to cancel out, and one ends up with only small apparent net changes. One can place these changes under a magnifying glass and derive great concern from a global variation of a few tenths of a degree, but the utility of that concept seems to be limited. Understanding how the climate of the Earth has varied with time, and how the Earth responds to changes in the Sun as well as to man-made effluents and activities, requires a more regional approach, as discussed for example by Soon and Baliunas (2003a, b).

A single global temperature does not reveal the regional variations, the seasonal variations in different regions, or the day/night variability. If the putative global temperature rises, is that due to warmer days or nights, and how does it depend on season? Where did temperatures rise, and where did they fall? BEST found that one-third of all land sites reported a drop in temperature while the other two-thirds reported a gain.

As difficult as the problem is for defining a single global or hemispheric temperature based on current measurements, the problem is far more difficult in combining proxies to arrive at historical average temperatures. The problems in defining historical global average temperatures from proxies were discussed in Sections 2.2.3 and 2.4.

It appears that many proxies, by their nature, are so loaded with noise and uncertainty that the data merely produce a blurry result. Considering the noise inherent in such measurements, and the lack of spatial and temporal coverage, no historical global average temperature estimate is very precise (see Section 2.4).

Essex, McKitrick, and Andresen (2007) carried out a mathematical analysis of various attempts to derive a single global average temperature. It was concluded that, while it is possible to construct statistical averages for any given set of local temperature data, the calculated average can vary widely, depending on how the average is constructed.

Despite the fact that a single global average of measured temperatures is not a temperature in the thermodynamic sense, the concept could nevertheless have utility as an indicator of trends if the elements being averaged were not so widely disparate in magnitude. It would appear that there is much more utility in describing average temperatures over regions (e.g., Western Europe) than in trying to define a single global average temperature.

Even if one could convince oneself that the concept of a single global average temperature had utility, its magnitude would be dependent on averaging a worldwide network of temperature measurements. However, the adequacy of the worldwide TMN leaves much to be desired, as discussed at length in Section 3.1.2.

4

Global scares, subjective science, and climatologists

4.1 GLOBAL SCARES

Global-warming alarmism can be viewed from the broader perspective of global scares as one in a series of panics that rise and often fall; global warming is still in the rise period. Booker and North (2007) documented the details of the rise and fall of various global scares ranging from a wide variety of food scares, to mad cow disease, to dioxins, to the millennium bug, to lead and asbestos, to passive smoking, and finally to global warming. (They failed to cover excessive quarantine of returning astronauts or depletion of the ozone layer by sea-surface temperatures (SSTs). In the introduction to their book *Scared to Death*, Booker and North defined the unifying characteristics of global scares:

> "Each was based on what appeared at the time to be scientific evidence that was widely accepted. Each has inspired obsessive coverage by the media. Each has then provoked a massive response from politicians and officials, imposing new laws that inflicted enormous economic and social damage. But eventually the scientific reasoning on which the panic was based has been found to be fundamentally flawed. Either the scare originated in some genuine threat that had then become wildly exaggerated, or the danger was found never to have existed at all."

By now, however, the damage has been done. The costs have amounted in some cases to billions, even hundreds of billions, of pounds, imposing an enormous hidden drain on the economy. Yet almost all of this money has been spent, it turns out, to no purpose.

What does it say about the psychology of our time that such an extraordinary thing can happen, not just once, but again and again? When we examine the pattern behind these scares, we find further elements that each has in common:

- The source of the supposed danger must be something universal, to which almost anyone in the population might be exposed, such as eggs or beef, asbestos, or climate change.

D. Rapp, *Assessing Climate Change*, Springer Praxis Books,
DOI: 10.1007/978-3-319-00455-6_4

- The nature of the danger it poses must be novel—a threat that has never appeared in this form before.
- While the scientific basis for the scare must seem plausible, the threat must also contain a powerful element of uncertainty. It must in some way be ill-defined, maximizing the opportunity for alarmist speculation as to the damage it might cause.
- Society's response to the threat must be disproportionate. It is this more than anything which defines a true "scare"; that, even where the threat is not wholly imaginary, the response to it is eventually seen to have been out of all proportion to its reality.

In Ionescu's play *Rhinoceros*, written for the theater of the absurd, he explores the pressures on people to conform to trends and adopt expanding belief systems. As more and more people turn into rhinos, the pressure to conform by doing likewise becomes intense. Today, we witness just such a pressure on politicians, scientists, and the public to jump on the global-warming bandwagon. Even George W. Bush, who in his tenure as President of the United States had opposed every single effort to legislate even the most mild and moderate steps to improve or protect the environment, began to weaken on global warming in 2007. Former Vice-President Gore led a national campaign to raise consciousness about the dangers of global warming, based heavily on the *hockey stick* model. His efforts netted him the Nobel Peace Prize. The United Nations, through its Inter-governmental Panel on Climate Change (IPCC), has similarly taken an alarmist position, also dependent on the *hockey stick*. The Union of Concerned Scientists, and a number of U.S. governmental agencies have taken similar positions. In addition, quite a large number of scientists have also become very concerned regarding the potential impacts of global warming.

Matt Ridley wrote an essay on the history of apocalyptic predictions.[1] He said:

> "Best-selling economist Robert Heilbroner in 1974: 'The outlook for man, I believe, is painful, difficult, perhaps desperate, and the hope that can be held out for his future prospects seem to be very slim indeed.' Or best-selling ecologist Paul Ehrlich in 1968: 'The battle to feed all of humanity is over. In the 1970s ['and 1980s' was added in a later edition] the world will undergo famines-hundreds of millions of people are going to starve to death in spite of any crash programs embarked on now ... nothing can prevent a substantial increase in the world death rate.' Or Jimmy Carter in a televised speech in 1977: 'We could use up all of the proven reserves of oil in the entire world by the end of the next decade'.
>
> "Predictions of global famine and the end of oil in the 1970s proved just as wrong as end-of-the-world forecasts from millennialist priests. Yet there is no sign that experts are becoming more cautious about apocalyptic promises. If anything, the rhetoric has ramped up in recent years. Echoing the Mayan

[1] Ridley, Matt. (2012) "Apocalypse not", *www.rationaloptimist.com/blog/apocalypse-not.aspx*.

calendar folk, the Bulletin of the Atomic Scientists moved its Doomsday Clock one minute closer to midnight at the start of 2012, commenting: 'The global community may be near a point of no return in efforts to prevent catastrophe from changes in Earth's atmosphere'.

"Over the five decades since the success of Rachel Carson's Silent Spring in 1962 and the four decades since the success of the Club of Rome's The Limits to Growth in 1972, prophecies of doom on a colossal scale have become routine. ... The past half century has brought us warnings of population explosions, global famines, plagues, water wars, oil exhaustion, mineral shortages, falling sperm counts, thinning ozone, acidifying rain, nuclear winters, Y2K bugs, mad cow epidemics, killer bees, sex-change fish, cell-phone-induced brain-cancer epidemics, and climate catastrophes."

Ridley described exaggerated predictions of catastrophe due to synthetic pesticides, DDT in particular, air pollution, acid rain, loss of the ozone layer, 1976 swine flu panic, mad cow disease, Ebola, SARS, a virus from civet cats, 2005 bird flu, famine and overpopulation, resource depletion: oil and gas, metals, and species extinction.

Ridley concluded: "Over the past half century, none of our threatened eco-pocalypses have played out as predicted. Some came partly true; some were averted by action; some were wholly chimerical. This raises a question that many find discomforting: With a track record like this, why should people accept the cataclysmic claims now being made about climate change?" He quoted Rajendra Pachauri, head of the IPCC, who said in 2007 that "if there's no action before 2012, that's too late ... This is the defining moment".

He went on to say:

"So, should we worry or not about the warming climate? It is far too binary a question. The lesson of failed past predictions of ecological apocalypse is not that nothing was happening but that the middle-ground possibilities were too frequently excluded from consideration. In the climate debate, we hear a lot from those who think disaster is inexorable if not inevitable, and a lot from those who think it is all a hoax. We hardly ever allow the moderate 'lukewarmers' a voice: those who suspect that the net positive feedbacks from water vapor in the atmosphere are low, so that we face only 1 to 2 degrees Celsius of warming this century; that the Greenland ice sheet may melt but no faster than its current rate of less than 1 percent per century; that net increases in rainfall (and carbon dioxide concentration) may improve agricultural productivity; that ecosystems have survived sudden temperature lurches before; and that adaptation to gradual change may be both cheaper and less ecologically damaging than a rapid and brutal decision to give up fossil fuels cold turkey."

4.2 DEALING WITH SUBJECTIVE SCIENCE

4.2.1 Nature of subjective science—emergence of consensus

There are many phenomena in nature that, for one reason or another, are not susceptible to verification by independent testing. These typically include events that either occurred a long time ago or that occurred at distant sites not accessible to us, or both. Examples include the expansion of the universe after the Big Bang, the variations in climate of the Earth in the past or in the future, the origin of life on the Earth, putative existence of life elsewhere in the universe, the evolution of species on the Earth, continental drift, and other such topics. There is no way to go back into the past or travel great distances to directly verify hypotheses. Although the remnants of the past may be discernible to some extent in proxies that exist in the present, these tend to have significant limitations. For such phenomena that occurred long ago and/or in distant locations, scientists create hypotheses that would "explain" how these processes might have occurred in conformity with the known laws of science. If these hypotheses provide a reasonable explanation of phenomena and are in conformity with scientific laws, they acquire the elevated status of a theory. Such a theory is typically not unique and represents one viewpoint. It provides one possible explanation for events that cannot be directly verified. Scientists are put in the position of detectives trying to solve a crime by piling up circumstantial evidence. In the case of evolution and continental drift, the circumstantial evidence is very strong and any sensible person would accept that these theories are almost certainly valid.

Unfortunately, the foundations of almost every subjective aspect of climate change are weak. Conjecture for things improvable is a safe venture—no one can ever prove you wrong. It is far more dangerous to predict tomorrow's weather than it is to predict the climate 100 years from now—tomorrow's weather is subject to practical test. I call this kind of science "subjective science". It is not amenable to direct verification in the manner of say, the laws of motion.

Scientists do not seem to be able to shrug their shoulders and admit that we just don't know the answers to some questions. What happens is that one of the unprovable hypotheses in a subjective science gains popularity amongst scientists and is regarded by the majority as the most credible alternative. When a significant number of scientists agree, a consensus evolves. The consensus acts like a gigantic gravitational field, drawing in more and more scientists. Eventually, the consensus gels, and ultimately hardens into a belief system—an orthodoxy. The foundations are often weak, and not understood by the public. The emergence of the consensus as the essence of reality in science has replaced scientific skepticism, and, as Lindzen (2008) has noted: "... simulation and programs have replaced theory and observation, where Government largely determines the nature of scientific activity". As Lindzen (2008) emphasized, "the bulk of the educated public is unable to follow scientific arguments; 'knowing' that all scientists agree relieves them of any need to do so." Taking issue with the consensus "serves as a warning to scientists that the topic at issue is a bit of a minefield that they would do well to avoid". It should also be noted that many climatological publications are so full of jargon and so obscure that they are unreadable except to a very few narrow specialists. So, not only the general public, but even much of the science community is unable to digest these abstruse treatises.

The consensus acquires legitimacy in proportion to the number and prominence of the scientists and institutions that subscribe to it. As the consensus becomes firmly imbedded in culture, it acquires the respect usually accorded to fact. However, as Crichton (2003) said:

> "Let's be clear: the work of science has nothing whatever to do with consensus. Consensus is the business of politics. Science, on the contrary, requires only one investigator who happens to be right, which means that he or she has results that are verifiable by reference to the real world. In science consensus is irrelevant. What is relevant is reproducible results. The greatest scientists in history are great precisely because they broke with the consensus. There is no such thing as consensus science. If it's consensus, it isn't science. If it's science, it isn't consensus. Period."

Actually, he is not quite right here. There is nothing fundamentally wrong with forming a consensus of opinion on a subjective science topic. What is wrong is when the consensus hardens into an orthodoxy with inadequate scientific support.

Curry and Webster[2] wrote a review of the notion of consensus in science, with particular regard to climate change. The report by Curry and Webster follows two paths. One is the "consensus findings" of the IPCC regarding the role of greenhouse gases on warming over the past century, and the other is a review of philosophical views of consensus in science. They reviewed a number of philosophical questions regarding the role of consensus in science. They concluded:

> "Arguing from consensus to enforce conclusions does not work with the extended peer community. What is needed are serious attempts to engage the extended peer community with the modes of expert reasoning used to reach those conclusions."

There seems to be quite a bit of confusion in the world of climate science as to what the consensus is consenting to. For example, Curry and Webster focused on a IPCC conclusion that warming in the 20th century was primarily caused by anthropogenic generation of greenhouse gases. There might be considerable variability in the extent of widespread consensus on several beliefs as shown below (degree of consensus shown in brackets: 5 = greatest consensus; 1 = least consensus):

(1) The Earth's climate would have been steady and constant, were it not for the impact of anthropogenic activity on the climate system. [4]
(2) The climate of the mid-19th century was ideal. [3]
(3) The global average of Earth temperatures rose over the past ~120 years. [5]
(4) Rising concentrations of greenhouse gases warm the Earth's climate. [5]

[2] Curry, J. A. and Webster, P. J. (2012), "Climate change: No consensus on consensus", *http://judithcurry.com/2012/10/28/climate-change-no-consensus-on-consensus/*.

(5) Since the climate warmed over the past ~120 years, it must have been due to emissions of greenhouse gases by human activity. [4]
(6) Continued emission of greenhouse gases in the future will lead to further warming. [5]
(7) Global climate models, while not perfect, are good enough to predict future global temperature rise due to future increases in greenhouse gas concentrations. [2]
(8) The impacts of future temperature rise at any level are well understood. [3]
(9) The impacts of future temperature rise will be disastrous to the world unless we immediately drastically reduce the world rate of carbon emissions. [3]
(10) We can immediately drastically reduce the world rate of carbon emissions by a combination of introducing green energy, sequestration, and energy conservation at a more rapid rate, without severely impacting the world economy. [1]

There is no accurate way to estimate the degree of consensus on each of these relevant issues. Subjectively, my impression from reading papers, blogs, various press releases, and conversations with individuals is as follows. On a scale from 1 to 5, where 5 is the most widespread consensus of belief, and 1 is the least widespread consensus of belief, my impression is as given in brackets in the above list.

Let us consider these issues one by one.

(1) *The Earth's climate would have been steady and constant, were it not for the impact of anthropogenic activity on the climate system.* [4] There seems to be a fairly widespread belief in this postulate that the Earth's climate is steady unless acted upon by an outside influence. This is parallel to Newton's law of motion that a body remains at rest or in uniform motion unless acted upon by an outside force. This does not allow for changes in the Sun–Earth relationship that can produce major climate changes due to innate variability of the Sun, as well as variations in the Earth's orbit relative to the Sun. A problem with this belief is that the Earth is a very complex system with many feedback effects. Weather, ocean currents, and other aspects of the Earth system can vary widely, triggering feedbacks that produce longer-term variations (*i.e.* internally generated climate change). The historical record shows that, over hundreds of thousands of years, we have had Ice Ages and interglacials without human influence. More to the point, we have had small but significant climate fluctuations within our current interglacial period over the past couple of millennia. The alarmists have tried unsuccessfully to dispute this via the *hockey stick* picture of millennial climates.
(2) *The climate of the mid-19th century was ideal.* [3] I don't think that most people have really thought about this very much and the degree to which there might be consensus on this point is uncertain. But there seems to be widespread concern that we are presently warmer than they were the mid-19th century, which seems to imply that they wish we were back in the Little Ice Age. We can either argue that it is warmer now, or it was colder then. The majority seem to favor that we are warmer now.
(3) *The global average of Earth temperatures rose over the past ~120 years.* [5] This is very widely believed and it is a matter of fact.

(4) *Rising concentrations of greenhouse gases warm the Earth's climate.* [5] This is very widely believed and it is a matter of fact.
(5) *Since the climate warmed over the past ~120 years, it must have been due to emissions of greenhouse gases by human activity.* [4] This is a fairly widespread belief. But it is too digital. In retrospect, it seems likely that rising greenhouse gas concentrations contributed to warming over the past 120 years, but it also seems likely that natural fluctuations in climate (e.g., pulling out of the Little Ice Age) may have contributed as well. The degree to which each factor influenced the climate is unknown.
(6) *Continued emission of greenhouse gases in the future will lead to further warming.* [5] This is very widely believed and it is almost certainly correct.
(7) *Global climate models, while not perfect, are good enough to predict future global temperature rise due to future increases in greenhouse gas concentrations.* [2] I am under the impression that this is not widely believed. However, it is widely believed that there is a reasonable chance that the models provide us with a worst-case scenario.
(8) *The impacts of future temperature rise at any level are well understood.* [3] I am under the impression that this is believed to the extent of perhaps 50%. However, it is widely believed that the analyses at least provide us with a worst-case scenario.
(9) *The impacts of future temperature rise will be disastrous to the world unless we immediately drastically reduce the world rate of carbon emissions.* [3] I am under the impression that this is believed to the extent of perhaps 50%. However, it is widely believed that the analyses provide a worst-case scenario.
(10) *We can immediately drastically reduce the world rate of carbon emissions by a combination of introducing green energy, sequestration, and energy conservation at a more rapid rate, without severely impacting the world economy.* [1] I think that this is only believed by a limited number of rabid "greenies". I am under the impression that the overwhelming majority wish that this were so, but are very uncertain of its practicality.

While the role of greenhouse gases in warming over the past 100 years is somewhat relevant to the issue, the real issue we face is: For various future scenarios of world energy consumption by sector and fuel, what are the expected environmental consequences? Assuming that the more the future resembles business as usual, the environmental impacts are greater, what is the technical feasibility and cost of shifting from future scenarios with greater environmental impact to scenarios with lesser environmental impact? Is there a consensus that we know the answers to these questions? I think not.

Ward (2008) discussed the scientific method, for which he emphasized: "No hypothesis is considered proven until it has undergone rigorous scientific review and testing, and other scientists must be able to replicate the tests or experiments and achieve the same results." Unfortunately, his notion of scientific rigor seems to be that it is supported by a consensus. His report emphasizes the necessity of repeatability in the scientific method; for example, it says: "A scientist tells how he or she arrived at a conclusion in enough detail so that another investigator can follow the same trail, examine the same data, and get the same answer." Yes, repeatability is *necessary*, but not *sufficient*. If the method of processing the data is arbitrary or possibly improper, successive investigators can reproduce the results but they will be just as arbitrary or inaccurate as they were before they were replicated. Ward (2008) mixes up two requirements of the scientific method. There is a huge difference between, on the one hand, applying a hypothesis to a large array of independent phenomena vs. on the other hand, applying a hypothesis to a single case (with adjustable parameters) in such a manner that those who follow can duplicate the results by repeating the same steps. The problem with climate change is that very little in this field of endeavor is "proven" and most of the "accumulated advances in understanding of climate change" over the past 20 years range from arbitrary (climate model results) to improper (the so-called *hockey stick* result).

4.2.2 How consensus becomes orthodoxy

Orthodoxy is a belief system. Thus, in regard to global warming, we have believers (alarmists) and non-believers (skeptics). Otherwise intelligent people often discuss whether or not they "believe in global warming".

But, like any religion, scientific orthodoxies cannot tolerate disagreement with the orthodoxy. Therefore, the alarmists have politicized the science of climatology to enforce their views. Lindzen (2008) described the politicalization of climate science:

> "All such organizations, whether professional societies, research laboratories, advisory bodies (such as the national academies), government departments and agencies (including the National Aeronautics and Space Administration (NASA), the National Oceanic and Atmospheric Administration (NOAA), Environmental Protection Agency (EPA), NSF, etc.), and even universities are hierarchical structures where positions and policies are determined by small executive councils or even single individuals. This greatly facilitates any conscious effort to politicize science via influence in such bodies where a handful of individuals (often not even scientists) speak on behalf of organizations that include thousands of scientists, and even enforce specific scientific positions and agendas.
>
> "The temptation to politicize science is overwhelming and longstanding. Public trust in science has always been high, and political organizations have long sought to improve their own credibility by associating their goals with 'science'—even if this involves misrepresenting the science."

Occasionally, a counter-argument is published as, for example, the *New York Times* article[3] of Freeman Dyson's opposition to the orthodoxy on global warming, but this only serves to confuse the public a bit.

The world of science seems to have lost its foundation of skepticism. Instead of

[3] *www.nytimes.com/2009/03/29/magazine/29Dyson-t.html?_r=1&scp=1&sq=freeman%20dyson&st=cse.*

doubt and dialectic opposition, science has adopted orthodoxy and consensus. Scientists, like the public at large, seem unable to shrug their shoulders and simply admit that we just don't know the answers. The fierce competition for funding in an environment dominated by orthodoxy pressures scientists to bias their viewpoints. We note a significant rise in the number of news releases and papers by scientists with phrases such as "there might be …," or "it is possible that …". What science cannot seem to do these days is accept that:

> "Sometimes there is no alternative to uncertainty except to await the arrival of more and better data." (Wunsch, 1999).

It seems likely that scientific (or economic) progress in climatology will be impeded by the fact that data and models are routinely biased to adhere to a belief system. The IPCC has led the way with a plethora of conclusions and predictions regarding the role of CO_2 emissions on the Earth's climate and the potential impact on humanity. These represent mainly political, not scientific, conclusions. The majority of recognized climatologists have aligned like weather vanes to the prevailing wind, making it all but impossible to get contrary views published in journals. As a result, there has arisen a blogopolis in which contrary views are available on websites but not in the literature. While many of these blogs are populated by moronic entries, a few contain detailed analysis and data that you cannot find in the journals.

4.2.3 Counting adherents to the orthodoxy

To further the ends of the alarmists, some academicians have carried out counting studies where they sum up the number of scientists who subscribe to the alarmist persuasion. Oreskes (2004) built her argument in favor of anthropogenic global warming based on a list of who supports the hypothesis, rather than the scientific basis for it. She emphasized that the IPCC, the National Academy of Sciences, the American Meteorological Society, the American Geophysical Union, and the American Association for the Advancement of Science (AAAS) "all have issued statements in recent years concluding that the evidence for human modification of climate is compelling". In a study of 928 abstracts published in refereed scientific journals between 1993 and 2003, she did not "find one paper that disagreed with the consensus position." However, at the end, she provided an escape clause:

> "The scientific consensus might, of course, be wrong. If the history of science teaches anything, it is humility, and no one can be faulted for failing to act on what is not known … Many details about climate interactions are not well understood, and there are ample grounds for continued research to provide a better basis for understanding climate dynamics."

Three years later, she concluded that global warming due to greenhouse gases "is an established scientific fact" (Oreskes, 2007a). In 2007, she gave a presentation of 109 slides to the American Meteorological Society (Oreskes, 2007b). That a social scientist would have 109 slides' worth of information to convey about climate change speaks to the role of consensus as a force in science. Her first slide quoted no less an expert on climatology than Arnold Schwarzenegger. The *Terminator* said "the debate is over" (pun intended). Her second slide proclaimed: "There is a scientific consensus over the reality of anthropogenic global warming." Here she deals with belief #5 in my previously given list of 10 beliefs relevant to consensus. She then proceeded to quote various authorities to show that a consensus exists. Unfortunately for her, she ventured briefly into the science of climate change where she got in well over her head. Using the "*hockey stick*" result, she concluded that carbon dioxide as the cause of warming in the 20th century is "not just a correlation—it's a confirmation of a prediction—the scientific method". It is not clear which scientific method she refers to—certainly none that I am aware of. Of course, the real issue is not whether anthropogenic generated CO_2 contributed to warming in the past ~100 years; the real issue is what will be the consequence of further emissions in the 21st century? That is a totally different and more difficult question. I refer to Miyogi in the film *The Karate Kid*: "Answer only matter if ask right question." For the sake of argument, we might blandly accept beliefs #1 though #9 in my previously given list, but we are stuck with belief #10 with little evidence that it is true.

Oreskes, Conway, and Shindell (2008) wrote a 70-page treatise on CO_2 as the putative cause of global warming in which the entire argument is based on a comparison of the number and credentials of those who *believe in it* vs. the number and credentials of those who *disbelieve* it. It is worth noting that the issue here has degenerated down to what opposing camps *believe* rather than a question of what the data tell us. The discussion was highly biased. For example, the authors refer disparagingly to "challenges to climate science" as if those of the alarmist persuasion are climate scientists whereas the challengers are something other than climate scientists. These authors are social scientists rather than experts in the science of global warming. As it turns out, the majority of paleoclimatologists do accept the thesis widely promulgated by Al Gore, the U.N., NOAA, the National Academy of Sciences, and other predominant professional organizations, that CO_2 emissions were the cause of global warming in the 20th century, and this warming will increase in the 21st century in proportion to further emissions, causing great misery for humanity (beliefs #8 and #9). This is the orthodox institutional viewpoint that is taught to schoolchildren and widely promulgated by academia. As institutions and organizations continue to dominate over individuals in these matters, a consensus builds up on each topic.[4]

[4] An anecdote illustrates the point. At a recent NSF workshop, Reversing Global Warming: Chemical Recycling and Utilization of CO_2 I presented a talk showing why the *hockey stick* representation of past temperatures was incorrect. A representative of the NSF raised a question at the end of my talk. She asked: "Why should I believe you when the National Academy of Sciences says otherwise?" She was relying on the institution over the individual. Ignoring the data that I presented, she fell back on reliance on the consensus. The issue was no longer whether my data and analysis were accurate, but rather, whether more prestigious organizations took a contrary position. Ayn Rand must be turning over in her grave! While I was giving my talk, one attendee of the alarmist persuasion stomped out of the meeting hall, audibly cursing.

However, the degree of consensus has been exaggerated (Schulte, 2008). It is also noteworthy that a number of prominent European and Russian climatologists and scientists have aligned themselves with the skeptics.

The French journal *La Meteorologie* criticized a French scientist for taking "an opposing stance" to "*the prevailing ideology*". The French scientist said, "The term 'consensus', a term often employed by the IPCC, implies only that the 'good' keepers of the true faith are in the majority, meaning that mere weight of numbers (not necessarily synonymous with better quality) may control and dismiss discordant voices from publications and papers." Kondratyev *et al.* (2003) describe the IPCC Report as "less than scientific, speculative opinions" and describe climate models as being "forced to fit the observational data" through adjustment of parameters. They claim that reductions of greenhouse gas emissions based on such models "are senseless". Bischof (2000) said:

> "The reader should understand that in science, as in other sectors of public life, the outcome of a study is often guided, if not determined, by an *a priori* idea, a tenet. In the case of global warming, this belief was that, if enormous amounts of greenhouse gases are released into the atmosphere, a temperature rise must occur. This prior assumption has guided scientific thinking and triggered a true deluge of investigations all desperately trying to prove just that."

4.2.4 Climate science and determinism

Embedded in the concept of *cause and effect* is a belief in *determinism*: that events in a system are uniquely determined by the elements in the prior state of the system and its surroundings. We may say that some elements of the previous state "caused" the current state.

The following is excerpted from *Wikipedia*:

> "The philosophical analysis of causality extends over millennia. The deterministic world-view is one in which the universe is no more than a chain of events following one after another according to the law of cause and effect. We can distinguish between necessary and sufficient causes:
> *Necessary causes*:
> If x is a *necessary* cause of y, then the presence of y *necessarily* implies the presence of x. The presence of x, however, does not imply that y will occur.
> *Sufficient causes*:
> If x is a *sufficient* cause of y, then the presence of x *necessarily* implies the presence of y. However, another cause z may alternatively cause y. Thus the presence of y does not imply the presence of x."

In addition, *Wikipedia* mentions "*Contributory causes*: A cause may be classified as a *contributory* cause, if the presumed cause precedes the effect, and altering the cause alters the effect. A contributory cause may be neither necessary nor sufficient but it must be contributory".

It is common amongst non-scientific folk to attribute causes to many effects in their lives based on insufficient data. Most people I know are continually making assertions that "this caused that" in their lives in a totally unscientific way, with only one or two experiences to support the assertion. Most of the systems that govern our lives are complex, and it is impossible to separate out one factor as a cause (e.g., "Taking vitamin pills gave me good health").

Some complex systems are exceedingly sensitive to initial conditions, and it is technically impossible to pin down the initial conditions precisely enough to make deterministic predictions of the outcome in the future. The system appears to behave chaotically, as if it changes in ways that are unpredictable and erratic. Yet, random events may appear to have a pattern—for a limited time period. For example, if one tosses a coin repeatedly five times, there is a 1 out of 32 chance that you will get five heads in a row. If, while you are tossing the coin, you stand on one foot, you might think that standing on one foot *caused* the coin toss to be heads. The book *Fooled by Randomness* by Nicholas Taleb is:

> "... about luck disguised and perceived as non-luck (that is, skills) and, more generally, randomness disguised and perceived as non-randomness (that is, determinism). It manifests itself in the shape of the lucky fool, defined as a person who benefited from a disproportionate share of luck but attributes his success to some other, generally very precise, reason. Such confusion crops up in the most unexpected areas, *even science*, though not in such an accentuated and obvious manner as it does in the world of business. It is endemic in politics, as it can be encountered in the shape of a country's president discoursing on the jobs that 'he' created, 'his' recovery, and 'his predecessor's' inflation. We are genetically still very close to our ancestors who roamed the savannah. The formation of our beliefs is fraught with superstitions—even today (I might say, especially today). Just as one day some primitive tribesman scratched his nose, saw rain falling, and developed an elaborate method of scratching his nose to bring on the much-needed rain, we link economic prosperity to some rate cut by the Federal Reserve Board, or the success of a company with the appointment of the new president 'at the helm'." (Emphasis added)

We all fall prey to superstition at times. Taleb tells of a time when his futures trading was doing poorly. He arrived at work in a downpour of rain, and went quickly through the back door to avoid getting soaked walking around to the front door. He had the best day in months. After that, he never went in the front door again.

In general, most effects observed in climate appear to be highly chaotic and probably derive from a multitude of potential causes. The major challenge for climatologists is to ferret out which causes are most significant, and derive mathematical relationships between the putative causes and the observed effects. Implicit in this process is the belief that the system is deterministic and is not obscured by chaotic factors. However, there is typically no proof that the climate systems are deterministic. Even if they are deterministic, they may still be determined by so many conflicting contributing causes that attribution of the role of each putative cause is always very difficult, if not impossible. Most climatological analyses end up with a scatter plot in which the X–Y space is mostly filled with data points, and only a climatologist could believe that a valid trend could be extracted from this mess. Climatological analysis seems to almost always involve the challenge of resolving a putative signal from a very low signal-to-noise system.

The evangelists for climate alarmism continually use the phraseology that the skeptics are "attacking climate science", and all the evangelists want to do is "communicate the science" to the public. All of this presupposes that the alarmists have a lock on "climate science", and the skeptics are not scientists and their analyses do not constitute "climate science". However, there does not seem to be a succinct summary of just what "climate science" is. Maybe the latest IPCC Report might qualify in this regard but it is very diffuse and fragmented and does not present a cohesive summary of "climate science". Most climate scientists are specialists in one narrow area of climate science; that is required to succeed in academia. Not very many of them have a synoptic view of the whole field from one end to the other. It would be of great help to everyone if the evangelists for climate alarmism would put out a summary of exactly what "climate science" is. My understanding is that "climate science" constitutes mainly two things: (1) the hockey stick and (2) climate models, leading to the conclusion that most of the observed climate warming of the past ~120 years was due to increased greenhouse gas concentrations, and most of the greenhouse gas effect was due to rising CO_2.

The argument of the *hockey stick* goes something like this. It is claimed that the analysis of temperature proxies leads to evidence that the global average temperature of the Earth did not change much over the past 2,000 years. Yet, we observe that the temperature rose significantly in the past 120 years—a period when greenhouse gas concentrations were rising. Hence it is assumed that greenhouse gases made the difference between the past 120 years and the previous 2,000 years—and thus it is concluded that greenhouse gases caused the temperature rise of the past 120 years. As more than a few have put it: "What else could have caused this temperature increase?" There are several problems with this argument. The first is most important. Do the proxies work? Do they provide accurate renditions of the global average temperature over the past 2,000 years? There is considerable evidence that proxies are affected by many more variables than temperature, and, in the important case of tree rings, this is a particular problem. In general, proxies do not seem to be anywhere as good as they are claimed to be by their proponents. The prevailing implicit belief system in the work of Mann, Bradley, Jones, *et al.* is that proxies provide accurate renditions of temperature change at each locality. When one attempts to integrate large numbers of different proxies at various locations, the belief is that the measured temperature change at any locality is given by:

$$T_M = T_G + T_L$$

in which T_M is the measured temperature change at any locality, T_G is the average global temperature change, and T_L is the variation of the local temperature change from the global temperature change. T_G is presumed to be the same for all localities. The belief system is that T_L varies randomly from site to site, sometimes being >0

and sometimes being < 0. The belief system further assumes that, when one adds up T_M for many sites, the values of T_G will reinforce, but the sum of T_L will add up to zero due to cancellation of + and—values. As far as this writer knows, this belief system has never been demonstrated with data. If the measurements do not accurately reflect temperature changes, this procedure will not yield T_M. But if, regardless of how inaccurate the values of T_M are, they tend to be roughly equally distributed about zero, the end result of adding multiple proxies will be close to zero. Figures 2.9 and 2.16 show typical assortments of proxies. It is apparent that there is a wide diversity of patterns and there is no evidence of a consistent signal buried within the noise. Paleoclimatologists have used sophisticated statistical methods to extract a trend for T_G from such noisy data. However, there is no evidence for the assumption that these measurements of T_M include a consistent T_G plus a random distribution of T_L. What seems more likely, at least for temperature changes under 1°C over many centuries, is that all the data are representative of at least several factors, the signal due to temperature is inaccurate, and adding up these seemingly random patterns will always result in an almost flat profile. Figure 4.1 shows a cartoon of how adding random patterns produces a flat profile. As the number of patterns increases, the net result approaches null. It appears to this writer that a similar effect occurred in the work of Mann, Bradley, Jones, *et al.* Tacking on a rising measured T_G in the 20th century to a flat profile of noisy proxies for past centuries produces a hockey stick. We must ask ourselves why these various proxies are so widely divergent in form from one another. Are they (as paleoclimatologists would have you believe) true representations of local temperatures? If they are, it seems clear that local climates vary extremely widely, and there is no *a priori* reason to believe that they fit the $T_M = T_G + T_L$ paradigm. Local temperatures appear to be controlled by forces that are difficult to comprehend. It seems more likely, however, that these widely divergent results indicate that proxies are measuring something in addition to temperature. Either way, there is no reason to believe that a consistent signal for T_G can be extracted from such a set of proxies.

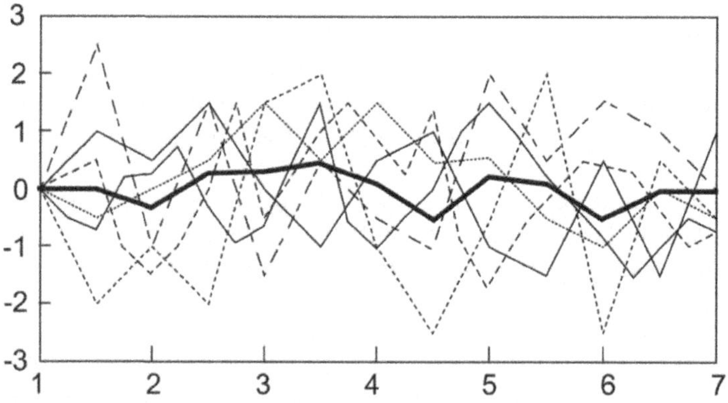

Figure 4.1. Cartoon showing several random patterns (thin lines) and their average (heavy line). The average of multiple random patterns is relatively flat.

Aside from the question of whether combining multiple proxies involves mainly adding up noise, there is considerable anecdotal evidence that there have been significant fluctuations in the Earth's climate over the past 1,000 years or so.

Unlike some skeptics, I do think that increasing CO_2 is worrisome, and we should try to understand its effects to better plan for the future. The problem is that the world economy is so deeply imbedded in fossil fuels that it will take many decades to bring about a significant reduction in CO_2 generation without a major world financial crisis. Meanwhile, CO_2 will continue to grow. If the alarmists like Hansen are correct that even 350 ppm is dangerous, there is no way to avoid the danger. Armour and Roe (2011) emphasized the lingering effects of equilibration of ocean warmth with the atmosphere, the longevity of CO_2 in the atmosphere, and the fact that reduction in production of aerosols will produce a short-term heating effect of unknown magnitude. We might as well just give up. Fossil fuels will gradually become depleted as the century wears on, the world population grows, and the developing nations industrialize. The world will do its best to incorporate solar and wind energy and, hopefully, nuclear as well. But there does not seem to be any way to get through the 21st century without continued use of fossil fuels and a consequent significant rise in CO_2. Even more worrisome is the 22nd century. What will 10 billion industrialized people use for energy?

4.2.5 Implications of uncertainty and extreme events

On Judith Curry's web log "Climate etc."[5] discussion threads were opened up on the case for global-warming alarmism based on the so-called "fat tail" of estimates of future increases in global average temperature predicted for the 21st century and beyond (e.g., see Figure 6.36) by alarmists. The argument has been phrased that the very uncertainty of future climate change dictates that we should act now to mitigate it. Whether a risk is worth insuring against depends on four things: (1) the probability of the event occurring (the uncertainty), (2) the cost if the event occurs, (3) the cost to insure against the event, and (4) the ability to pay the cost of insurance. The more uncertain a future catastrophe is, the smaller is the need to mitigate it. On the other hand, if the cost is extremely high for a low-probability event, it may still be worth insuring against. One thing is clear. If one insures, there will definitely be a cost—the cost for purchasing insurance. If one cannot afford that cost, one may be forced to forego purchasing insurance and hope for the best. A simplistic approach is to define a risk parameter as the probability times the cost of consequences if the event occurs.

The fat tail issue is mainly a matter of how rapidly the probability falls off in the outlying tail of major consequences of climate change. Nicholas Taleb discussed this in his books *The Black Swan* and *Fooled by Probability*. The question is whether the probability of an event falls off faster than the consequences rise in the tail of the

[5] *http://judithcurry.com/2011/10/12/the-case-for-climate-change-alarmism/#comments* and also http://judithcurry.com/2011/05/28/uncertainty-risk-and-inaction/.

probability distribution. A fat tail distribution falls off slowly enough that relatively improbable events still have enough residual probability that the risk parameter may be large for extremely costly outcomes. This simple argument can be elaborated with fancy mathematics[6] but the problem remains that we don't know the probabilities and we don't know the costs of occurrences. In fact, we don't even know the cost of insurance and, indeed, in regard to climate change, it seems likely that insurance might not be technically or economically feasible.

Where does the uncertainty derive from? The uncertainty derives from several sources. One source is uncertainty in estimating future use of fossil fuels in the 21st century. This depends on the world levels of economic activity, the development of higher-efficiency energy systems, and the gradual addition of renewable and nuclear energy to the energy mix. Nevertheless, reasonable estimates can be made of future CO_2 emissions for the remainder of the 21st century under several scenarios and, for scenarios in which significant fossil fuel usage persists in the 21st century, the uncertainty in estimation of emissions is not very great.

Another source of uncertainty is climatic effect of rising CO_2 levels. Here, the real uncertainty is not the spread of predictions by climate models. Rather, it is the fact that climate models don't have a clue as to how humidity, clouds, aerosols, and other effects result from rising greenhouse gas concentrations. Even if all of the models agreed on a single curve of temperature vs. CO_2, the uncertainty would more or less remain the same.

A third source of uncertainty is the cost to humanity (nation by nation) of putative climate changes induced by rising greenhouse gas concentrations.

A fourth source of uncertainty is whether it is technologically feasible and affordable to immediately ramp down CO_2 emission in a draconian fashion. What will be the cost, and what can we afford? The world is a poor place. Much of the world lives in poverty. The U.S. is a poor place. The census bureau reports that 46.2% of Americans live below the poverty line.[7] The U.S. federal debt is now about $50,000 per man, woman, and child in America. Our expenses outweigh our incomes and we can't stop spending. Even so, many of our government programs are underfunded and in trouble. If the world stops using fossil fuels, won't it enter a depression far worse than the 1930s? If we depend on renewable energy, won't we be beset by continual outages, and won't it take so long to establish renewable energy that CO_2 will build up in the interim? And who, pray tell, is going to get China, India, and the other Asian countries to stop using fossil fuels? It is not clear what the alarmist agenda is for dealing with their hypothesized future crisis, so it is not simple to estimate its cost. Yet, whatever it costs, it seems likely that, since we are a poor country in a poor world, we can hardly afford it.

[6] For example, Martin L. Weitzman (2009) on modeling and interpreting the economics of catastrophic climate change, "Fat-tailed uncertainty in the economics of catastrophic climate change", *The Review of Economics and Statistics*, **91**, 1–19; also *www.economics.harvard.edu/faculty/.../REEP2011%2Bfat-tail.pdf*.

[7] *www.nytimes.com/2011/09/14/us/14census.html?pagewanted=all.*

Global warming is the "goose that laid the golden egg" for agencies, social scientists, and economists. Even defense analysts seek a piece of the action:

> "Rather than justifying a lack of response to climate change, the emphasis on uncertainty enlarges the risk and reinforces the responsibility for pursuing successful long-term mitigation policy."[8]

4.3 HYPERBOLE ON IMPACTS OF GLOBAL WARMING

One of the favorite tactics used by alarmists is to focus on a short-term upward trend and predict doomsday from extrapolation of this trend into the future. Usually, the trend turns out be an upward lobe of an oscillation. The warming induced by the great El Niño of 1997–1998 provided grist for this mill and the alarmists were in their glory in the aftermath of that year, predicting wildly increasing temperatures and debilitating environmental consequences. As it turned out, the global average temperature remained essentially constant over the 15-year period starting in 1998 (see Figure 3.28).

Michaels and Balling (2009) (pp. xii and 127) provide good examples of selective presentation of short-term data to create the impression of a climatic trend.

Alarmists tend to see long-term danger in all short-term variations. In 1974, *Time Magazine* ran an article[9] entitled "Another Ice Age?" in which it was stated that:

> "... when meteorologists take an average of temperatures around the globe they find that the atmosphere has been growing gradually cooler for the past three decades. The trend shows no indication of reversing. Climatological Cassandras are becoming increasingly apprehensive, for the weather aberrations they are studying may be the harbinger of another ice age."

The article goes on to state that "the atmosphere has been gradually cooling for three decades". It cites expanding pack ice and snow cover, and changing polar winds. Ironically, it claims that humans may be involved in the cooling trend via dust released into the atmosphere from farming and fuel burning. Twenty years later, alarmists completely reversed their worries.

On April 26, 2007, James E. Hansen (perhaps the most well-known global-warming alarmist) gave testimony on the dangers of global warming to the Select Committee of Energy Independence and Global Warming of the U.S. House of Representatives. Hansen provided the case for alarmists in considerable detail. Only a few quotations are given here.

According to Hansen, the greatest near-term danger is sea-level rise. He said that "sea level is already rising at a rate of 3.5 cm per decade and the rate is

[8] Romig, A.D., Jr., Backus, G.A. and Baker, A.B., "A deeper look at climate change and national security", Sandia Report SAND2011-0039, March 2010.

[9] *www.time.com/time/magazine/article/0,9171,944914,00.html.*

accelerating" due primarily to "ice sheet disintegration". He said that "there is increasing realization that sea level rise this century may be measured in meters if we follow business-as-usual fossil fuel emissions," and that "adaptation to a continually rising sea level is not possible". Hansen concluded that "increasingly rapid changes on West Antarctica and Greenland ... are truly alarming".

One of the major slow feedback processes that Hansen identified is "the effect of warming on emissions of long-lived greenhouse gases", caused by the "melting of tundra in North America and Eurasia," which "is observed to be causing increased ebullition of methane from methane hydrates".

Hansen said that "continued business-as-usual greenhouse gas emissions threaten many ecosystems," and that "very little additional [climate] forcing is needed ... to cause the extermination of a large fraction of plant and animal species".

He also said that "Earth's history shows that climate is remarkably sensitive to global forcings" and that "positive feedbacks predominate", causing "the entire planet to be whipsawed between climatic states".

Summarizing, Hansen said that "The dangerous level of CO_2 is at most 450 ppm, and it is probably less ... Ignoring the climate problem at this time, for even another decade, would serve to lock in future catastrophic climatic change and impacts that will unfold during the remainder of this century and beyond". The Earth "is close to dangerous climate change, to tipping points of the system with the potential for irreversible deleterious effects ... The planet is on the verge of dramatic climate change". We "are forced to find a way to limit atmospheric CO_2 more stringently than has generally been assumed ... We cannot shrink from our moral responsibilities ... to preserve the planet for future generations".

Idso and Idso (2007) reviewed this testimony and provided a skeptical commentary and critique.

Alarmists have found it rewarding to engage in a contest of "Can you top this?" by issuing a constant barrage of press releases about what supposedly "may", "might", or "could" happen in the future as a result of putative global warming. If you go to Google and punch in "global warming", you get thousands of responses predicting disaster from global warming. These include claims such as:

1. the role of obese people in contributing to global warming by requiring extra resources;
2. "climate change may be century's greatest health threat";
3. "pets may be the latest victims";
4. "climate change may halve South Africa";
5. "increased incidence of tropical diseases, food shortages, natural disasters and heat waves threaten global ...";
6. "climate change may drive refugees to Australia";
7. "how climate change may be threatening national parks"; and
8. "climate change will overload humanitarian system".

—with many more like this.

In technical press releases, there is a strong bias to portray data in the worst possible light. Temperatures have meandered since the hot year of 1998 induced by a huge El Niño. As it turns out, 2008 was an unusually cold year in which much of the warming of the previous 20 years (or more) was mitigated. But, in early 2009, the U.S. National Weather Service (NWS) published a news release that said that "2008 was the 39th warmest year since 1895". Of course, 1895 was a very cold baseline year. Since 1930, 39 out of 79 years were warmer than 2008—hardly a basis for alarm. Yet the NWS made it seem as if 2008 was an exceptionally hot year! Similarly, the NOAA website reports the 2009 temperature as "the *n*th warmest year on record out of 130 years", where *n* is typically in the range 6 to 9. But the six to nine years that were warmer than 2009 all occurred since the hot year of 1998 induced by a huge El Niño; hence 2009 is stacking up as a relatively cold year compared with the previous decade!

In a situation where the global average temperature has obviously risen to a new level, continued mention that recent years are the sixth warmest, or the third warmest, or whatever, since records began adds absolutely nothing but alarm to the discussion. The fact is that we have been on a temperature plateau for the past 15 years and any and all of those years are among the warmest since the Little Ice Age. It is akin to the cartoon showing a mature person with height marks on the wall made while he was growing up, saying "This is one of the tallest years of my life".

Hansen *et al.* (2008) now claim that:

> "If humanity wishes to preserve a planet similar to that on which civilization developed and to which life on Earth is adapted, paleoclimate evidence and ongoing climate change suggest that CO_2 will need to be reduced from its current 385 ppm to at most 350 ppm."

In order to reduce the CO_2 content of the atmosphere to below 350 ppm in the 21st century, such draconian measures would be needed that it would imply the end of a modern technological world.

4.4 SPREADING THE GOSPEL

Ward (2008) reported on a series of workshops dealing with communication of science results to the public, with particular emphasis on climate change in which it was claimed that "the nation's top climate scientists and leading science and environmental journalists [met] together to discuss media coverage and communication of climate change science". The principal funder of the workshops project and its report was the Paleoclimate Program, Division of Atmospheric Sciences, National Science Foundation. Some financial support was also provided by grants from the U.S. Environmental Protection Agency's Office of Air Programs, and limited in-kind support was provided by the National Centers for Coastal and Ocean Science (NCCOS), the scientific research arm of the National Oceanic and Atmospheric Administration's National Ocean Service (NOAA/NOS) in the U.S. Department of Commerce, and by the NASA. David Verardo, Ph.D., of the National Science Foundation, was the primary person who enabled this science/

journalism project. While this report is portrayed as an effort to promote better communication on climate change, its real intent seems to be to promote the alarmist viewpoint. For example, it begins:

> "Climate scientists were frustrated by what they saw as a failure of the general public to understand and appreciate the seriousness of the climate change issue [i.e., the grave dangers according to the alarmist viewpoint]. Many scientists said they were frustrated that the accumulated advances in understanding of climate change over more than two decades of research had not led to a better-informed public [i.e., the use of climate models to predict large temperature increase in the 21st century]. ... The workshops focused in particular on what scientists call 'anthropogenic climate change'—that caused by human activities and not part of a natural cycle."

Like ("Fair and Balanced") Fox News on TV, this report suggests:

> "The preponderance of scientific evidence had since accumulated to a point where responsible reporters should give the scientific consensus on anthropogenic climate change much greater weight than dissenting claims challenging the mainstream scientific conclusions. The journalistic tenet of accuracy now demands that *the established science be given total or near total prevalence* in coverage of certain aspects of climate change science." (Emphasis added)

In other words, only the prevailing orthodoxy should be reported by the media. This attempt to muzzle the opposition is outrageously anti-scientific and anti-American. The report goes on to say:

> "Many participating reporters said they were having trouble convincing their editors of the virtues of reporting in an accurate and fair [i.e., alarmist], rather than quantitatively balanced fashion [i.e., roughly equal time to both sides]. Their reporting on new scientific findings often met with an editor's insistence that they also report the perspectives of climate science contrarians who lack comparable scientific expertise and standing, as if covering a political campaign or a public policy dispute [i.e., contrarians are thereby characterized as uneducated and occupying positions of less knowledge and importance compared with alarmists.]."

This viewpoint was echoed by Curry *et al.* (2006), who said:

> "Boykoff and Boykoff (2004) demonstrated that superficial balance in coverage of global warming by the U.S. "prestige press" (e.g., *New York Times, Washington Post, Los Angeles Times, Wall Street Journal*) can actually be a form of informational bias. Boykoff and Boykoff state that by giving equal time to opposing views, the major newspapers are significantly downplaying scientific understanding of the role humans play in global warming. Pitting what 'some scientists have found' against what 'skeptics contend' implies a roughly even division within the scientific community. In the media debate on global warming and hurricanes, greenhouse-warming deniers (which, in addition to scientists,

includes lawyers and others with at best minimal scientific credentials) are set side by side with scientists who have actually done the work and published papers on the subject."

Note that Curry *et al.* (2006) said that Boykoff and Boykoff "demonstrated" their claim. Curry *et al.* describe skeptics as "greenhouse-warming deniers" and are said to include some with "at best minimal scientific credentials". But intelligent skeptics do not deny the greenhouse effect. They merely doubt that we can quantitatively affirm feedback effects based on present knowledge. Furthermore, there are a great many strong supporters of climate alarmism who are clearly lacking in "scientific credentials". Naomi Oreskes, who has written articles and books and made presentations in favor of alarmism, is a prime example. In addition, the fact that a climate scientist may have done work and published papers means little in most cases. Most scientists are narrow specialists and do not have a synoptic view of the field. Many are immersed in their narrow slice of the pie. In fact, as Curry *et al.* (2006) emphasized in their paper, several of the logical fallacies that they attribute to deniers of the relation between SST and hurricanes, are the very same fallacies utilized over and over again by alarmists: *scientific credentials, appeal to authority, appeal to motive, begging the question, hasty generalization, and fallacy of the single cause.*

Now there is an entire website dedicated to improving communicating climate change and climate science to the public (i.e., the alarmist view of climate change).[10]

As climate change has evolved into a big business with plenty of funding, more and more peripheral (typically non-scientific) organizations have sought a piece of the action, introducing economic, social science, and communication aspects. A common theme of some studies (such as that of Ward, 2008) is to seek better ways of communicating *the* climate science (of the orthodoxy) to the public. For example, the Carsey Institute[11] posed the following three questions in a survey under the heading: "What do you personally *believe*?" Hence they framed the issue as a *belief* system:

"[On] the issue of global warming or climate change, how much do you feel you understand about this issue—would you say

- a great deal,
- a moderate amount,
- only a little, or
- nothing at all?

Which of the following two statements do you think is more accurate?

- Most scientists agree that climate change is happening now, caused mainly by human activities.

[10] *http://talkingclimate.org/guides/communicating-climate-science/*.

[11] Carsey Institute Issue Brief No. 26, "Climate change partisanship, understanding, and public opinion" *www.carseyinstitute.unh.edu*.

- There is little agreement among scientists whether climate change is happening now, caused mainly by human activities.

Which of the following three statements do you personally *believe*?

- Climate change is happening now, caused mainly by human activities.
- Climate change is happening now, but caused mainly by natural forces.
- Climate change is not happening now."

But these are all silly questions. Climate change has always occurred and is always occurring. Regardless of the *beliefs* of the public, human activity *is* contributing to climate change. The real questions are how additional emissions of greenhouse gases will affect the climate in the future quantitatively, what will be the impact on humanity, and what can we do about it? These questions are rarely posed and, indeed, there are no clear answers.

George Mason University and Yale University also conducted a similar study[12] asking: "Do you think that global warming is happening?" and similar questions. They categorized the public into six groups: alarmed, concerned, cautious, disengaged, doubtful, and dismissive in accordance with their "*beliefs*". They managed to fill up 57 pages with detailed breakdowns of *belief* systems.

Even the Catholic Church has gotten the message. They urge their brethren to: "Contact your members of Congress and urge greater U.S. leadership to address climate change, especially its disproportionate impact on poor and vulnerable people here and abroad".[13] The National Association for Advancement of Colored People (NAACP) has a paid-for listing on Google urging combating global warming, under the belief that global warming increase the prevalence of strong hurricanes that impact colored people.

An organization that calls itself "Transparency International—The Global Coalition Against Corruption" wrote a 400-page report entitled "Global corruption report: Climate change" with support from an investment bank and the German Ministry for Economic Development.[14] They began with:

> "We stand at the threshold of a global challenge: climate change. Governance lies at the heart of this challenge. Implemented with integrity and transparency, policies on climate change will make it possible for people around the world to understand, support and own the changes that will be required of them."

The message seems to be that we need more governance, more regulations, more policies, and, above all, more bureaucracies. The funny aspect of all this is that almost all of the corruption we have witnessed in regard to climate change was committed by noted alarmists.

[12] George Mason University, "Climate change in the American mind Americans' global warming beliefs and attitudes in May 2011", *environment.yale.edu/climate/files/ClimateBeliefsMay2011.pdf*, *environment.yale.edu/climate/files/SixAmericasMay2011.pdf*.

[13] United States Conference of Catholic Bishops, "Global climate change", February 2011.

[14] *www.transparency.org/whatwedo/publications/doc/gcr/*.

4.5 CLIMATOLOGISTS

The field of climatology, by its nature, deals with phenomena and data spread across the globe in vastly different environments. Much of these data are complex, and it is difficult to resolve cause–effect relationships because of so many confusing cross-factors. Furthermore, climate data needs to be long-term, typically 100 years or more, and most direct data are far shorter. Proxies are beset with a variety of problems. In this situation, it has come to pass, that climatologists tend to draw definitive conclusions from this checkerboard of inadequate, noisy data. A subtle compact has evolved whereby climatologists accept the results of one another, regardless of how flimsy the support base might be. Members of the club preserve this outward aura of scientific rigor, which in most cases is not justified. With the

Table 4.1. Personality traits (Weiler *et al.*, 2011).

Extraversion	*Intraversion*
Think out loud in discussions, talk more than listen	Process information internally, listen more than talk
Share ideas immediately	Share ideas after careful reflection
Sensing	*Intuition*
Focus on experience	Focus on theories
Build carefully and logically towards conclusions	Follow hunches to reach conclusions
Want details	Want big picture, become bored or impatient with details
Anchored in the present, relate to the past	Oriented towards the future
Prefer step-by-step information or instructions	Talk in general terms
Ask "what" and "how" questions	Ask "why" questions
Look for facts	Look for patterns and possibilities
Prefer practical, plain language to symbols, metaphors, theories, or abstractions	Use metaphors, analogies, and other symbolic language
Thinking	*Feeling*
Present information using cause-and-effect reasoning	Use personal situations, stories, and examples to communicate
Analytical	Empathetic
Need to know "why"	Connect with people
Judging	*Perceiving*
Prefer to make decisions quickly, come to closure and move on	Prefer to stay open to new information and last-minute options
Uncomfortable with free-flowing discussions	Feel confined by detailed plans and final decisions
Prefer focused discussion and options	Prefer open discussion to explore linkages between topics

Table 4.2. Comparison of personality traits of climate scientists with those of the general public (Weiler *et al.*, 2011).

Personality trait	*Climate scientists vs. public*
Extraversion/intraversion	Climate scientists similar to general public (roughly 50% extravert and 50% intravert)
Sensing/intuition	Climate scientists were far more likely to use intuition (82%) over sensing (18%) than the general public, who preferred sensing (73%) vs. intuition (27%)
Thinking/feeling	Climate scientists were somewhat more likely to use thinking (49%) over feeling (51%) than the general public, who preferred feeling (60%) vs. thinking (40%)
Judging/perceiving	Climate scientists were far more likely to use judging (73%) over perceiving (27%) than the general public but were more even with judging (54%) vs. intuition (46%)

advent of climate alarmism over the past couple of decades, climatologists seem to have taken on a more strident tone. Assertions today are presented as fact regardless of the technical underpinnings. This has created a great market for selling climate research, and climatologists have banded together to minimize and snuff out those who would raise the specter of "The Emperor has no Clothes".

Weiler *et al.* (2011) provided a very interesting insight into the personalities of climate scientists. Personality types of interdisciplinary, Ph.D. climate-change researchers were collected based on a Jungian type personality assessment (described below). Each person was characterized by four personality traits as shown in Table 4.1. Climate researchers were compared with the general public as shown in Table 4.2. One thing stands out. There is a huge statistical inversion between climate scientists vs. the public in that climate scientists greatly lean towards intuition whereas the public heavily leans towards sensing. This implies that the climate scientists "focus on theories" and "follow hunches to reach conclusions" whereas the public tends to "focus on experience" and "build carefully and logically towards conclusions". The strange thing is that one would expect that the very nature of the scientific method requires that scientists should focus on sensing, rather than intuition. In addition, there is also a much stronger tendency of climate scientists to prefer judging to perceiving, and there is a somewhat greater tendency of climate scientists to prefer thinking to feeling. Thus, climate scientists tend to "prefer to make decisions quickly, come to closure and move on". This is clearly evident in the many papers in climatology that utilize a penny's worth of data to draw a dollar's worth of conclusions. In fact, one might say in a Churchillian sense: Rarely have so many drawn so many conclusions from so little reliable data.

Martin (1979) wrote an interesting report in which he described the biases that inevitably creep into scientific research and reporting. According to Martin, scientists "do not disinterestedly look at the available evidence, do not make a balanced analysis, and do not present results in a neutral manner". Instead, he suggested "from the beginning [they] support or favor a particular conclusion, and in a number of ways organize their scientific work so as to selectively support this conclusion". He labeled this as "pushing the argument". Martin argued that pushing scientific arguments is inevitable, and therefore pushing should not reflect unfavorably upon the competence or integrity of the scientist. He said:

> "The important thing is not to eliminate pushing, which is impossible anyway, but to recognize that it exists. ... Neither does the existence of pushing automatically imply that a scientist's results or conclusions are unjustifiable or wrong. While a scientist's argument may be judged on the basis of current understanding to be pushed, it may eventually be vindicated. Or it may not."

He went on to say:

> "A scientist in developing an argument to support an hypothesis draws evidence from a number of sources. In presenting evidence one must always be selective—all the evidence and arguments cannot be presented. Often different authorities support different viewpoints, present different 'facts', and offer different interpretations of evidence. Depending on the field, a scientist may draw sound support for many points of view and find some support for nearly any view. Therefore it is easy for a scientist, knowingly or unknowingly, to push an argument by selective choice and use of available evidence."

Most of Martin's treatise was framed in terms of the debate during the early 1970s as to whether emissions from high-flying supersonic transports (SSTs) would destroy the ozone layer and thereby endanger the Earth's population by exposure to excessive radiation. For purposes of discussion, he presented detailed summaries and reviews of two prime scientific papers in the field with contrasting approaches. One paper was said to contain "the built-in assumption that the burden of proof lies with those who claim that SSTs are safe: that all that he must demonstrate is that there is at least some small possibility of danger". By contrast, the other paper used "the [implied] assumption that the burden of proof lies with those who claim that SSTs are dangerous to ozone: that all [they needed to] demonstrate was that the likelihood of significant danger was small".

Above all, Martin emphasized that scientists are human beings, motivated by various forces and factors in their lives. According to Martin:

> "People tend to selectively observe and interpret information in a way that supports their preconceived ideas. Because of this, the personal commitments of individual scientists can help to explain the link between the scientists' presuppositions and their pushing of arguments. ... In a scientist, this process might operate as follows. The scientist starts with an original idea or hypothesis, perhaps arrived at as a creative solution to a certain problem. In testing or validating the idea, the scientist will tend to notice and use supporting evidence and arguments. Data that seems mainly supportive will be studied, analyzed and applied so that every possible advantage can be drawn from it. Seemingly irrelevant or inconclusive items will be filtered from advantageous components,

> or interpreted in a way that promotes the argument. Evidence that seems mainly to contradict or challenge the argument at hand may be ignored completely or explained away or reinterpreted and twisted into support for the argument.
>
> "Some of the ways in which a person may deal with a challenging item of information are (1) flat denial of the item; (2) skepticism about the source of the item; (3) ascription of a motive to the source of the item; (4) isolation of the item from the context of one's attitude; (5) minimization of the importance of the item; (6) interpretation of the item to suit one's purpose; (7) misunderstanding of the item; and (8) thinking away or just forgetting the item."

According to Martin, one may often detect a deep-rooted personal commitment or bias of a scientist by examining a series of published papers and detecting a constancy of attitude that repeats itself from year to year. He claims:

> "The idea that scientists are often strongly committed or biased is quite compatible with the fact that scientists are human beings. ... That is, they are subject to motivations and failings similar to those of other people. They may strive for money, power and prestige; they may work for the satisfaction of a job well done or for revenge or to relieve boredom; they may make terrible blunders as well as have brilliant insights. It is sometimes said or suggested that scientists, at least when it comes to their work, live on a higher moral plane than other mortals. Don't believe it!"

In examining the literature on SST emissions and their impact on the ozone layer, Martin concluded:

> "From my point of view, the authors do not disinterestedly look at the available evidence, do not make a balanced analysis, and do not present results in a neutral manner. Rather, it appears to me that the authors from the beginning support or favor a particular conclusion, and in a number of ways organize their scientific work so as to selectively support this conclusion."

These claims made by Martin (1979) are backed up by lengthy and detailed discussions and analyses that seem quite credible to this writer.

Michaels and Balling (2009) devoted a chapter to "Pervasive bias and climate extremism". They began with an analogy. If one starts with a particular weather prediction for a locality, the next update of that prediction might forecast an increase or a decrease in the predicted temperature. There is an equal probability for either outcome. However, in regard to the climate–CO_2 connection and the predicted severity of future global warming, the overwhelming majority of new published papers find that the climate–CO_2 connection is strengthened and the predicted severity of future global warming is increased. Only rarely do papers get published with opposite conclusions. Alarmists might argue that the original hypothesis of CO_2-induced global warming becomes solidified as more data and better analyses accumulate. On the other hand, we have already demonstrated that *cabals* rule the review process for major journals, favoring those of the alarmist persuasion. We have also noted that the majority of published climatologists favor the alarmist position and this influences their perspectives. The alarmist position is also favorable to receiving funding for research. Michaels and Balling (2009) discussed the so-called "file-drawer problem" in which

> "Negative results are generally considered not noteworthy. ... Scientific journals are skewed by a prejudice for the publication of statistically significant, 'positive' results and prejudiced against findings of no relationship between hypothesized variables. ... For any given research area, one cannot tell how many studies have been conducted but never reported."

They also quote Stephen Jay Gould, who said publication bias results from "prejudices arising from hope, cultural expectation or ... a particular theory dictate that only certain kinds of data will be viewed as worthy of publication , or even documentation at all".

In the 30 years that have passed since Martin wrote his report, several major changes have taken place in the way that scientific information is distributed. With the advent of the Internet, the monopoly of scientific journals has been weakened. Other cultural changes have taken place. Of some relevance is the fact that scientists are now far more prone to issue press releases on their work prior to publication, and these tend to find their way onto many websites. Other scientists, disagreeing with the orthodoxy of the consensus, have difficulty getting published in the journals. A number of so-called web blogs dealing with climate change have emerged over the past several years, and these have become foci for discussions and commentary. Most blogs are rabidly one-sided and present forums for either alarmists or skeptics to agree with one another. Any moron can voice his or her opinion. Two blogs that stand out above the others are *climateaudit.org*, which has become a universal watchdog for reviewing statistical analysis of large data sets, and *judithcurry.com*, which provides an even-handed forum for both sides. The *judithcurry.com* blog has emerged in 2010–2012 as by far the best source of new ideas in climate science analysis, with many stimulating new posts by Judith Curry. Unfortunately, the responses on these blogs have become so numerous (typically many hundreds) that the wheat often gets lost in the chaff. It is particularly disappointing to observe that a limited number of adherents clog up the responses to Judith Curry's stimulating posts with mostly irrelevant, trivial, or nonsensical entries. Most of these responses are contributed with supercilious attitudes under psuedonyms. *pielkeclimatesci.wordpress.com/* is also a very informative website.

Ward (2008) extolled the peer-review system and

> "... warned about the growing number of unvetted publications being distributed through an expanding number of electronic and online outlets ... Publishing online ... has become an increasingly popular way to circumvent more rigorous peer review altogether ... the public and the media need to be attuned to these trends and distinguish them from highly respected professional peer-reviewed journals."

However, in many cases, peer review has become subject to political correctness, and assures that only one viewpoint will be heard. Michaels and Balling (2009) provided a number of examples of publication bias by the journals *Science* and *Nature*. Perhaps the most egregious example of politicization of science is the journal *Nature* that has become an alarmist propaganda medium. For example, the April 30, 2009, issue of *Nature* includes three articles that are essentially alarmist propaganda (Meinshausen *et al.*, 2009; Allen *et al.*, 2009; Schmidt and Archer, 2009). There is absolutely no doubt in these articles that CO_2 emissions were the prime cause of global warming in the 20th century. The only issue discussed is how rapidly CO_2 emissions must be reduced to save the world from disaster. The approach taken by these authors is statistical. The wide swath of modeled estimates of future CO_2 emissions and future temperatures are treated as votes, and the winning candidate is the result with the greatest preponderance of ballots. The conclusion is that, to save humanity, future CO_2 emissions must be draconically reduced—a recipe guaranteed to produce much more financial hardship in the world than global warming.[15]

Furthermore, most of the peer-reviewed articles in climatology are narrow, highly detailed, and represent new measurements or models. Most of the material that "circumvents peer review" is typically interpretive, synoptic, or in the nature of a review. Relatively little of it presents fundamental new measurements or calculations. One very important role for non-peer-reviewed reports is the activity of the *climateaudit.org* blog that checks out many of the publications passed by peer reviews, attempts to reproduce the results (usually they cannot), and reviews these papers to put them into perspective—a task ignored by the peer-review system. The *climateaudit.org* blog run by Steve McIntyre performs a valuable role in checking out the details of many published papers relevant to global warming—a task not done well (and usually not at all) by peer reviews. As it turns out, McIntyre has uncovered many errors and biases in these papers, most notably the Mann, Bradley, and Hughes "*hockey stick*" papers.

A significant characteristic of climatology is that, in general, data are very sparse and noisy, and of inadequate duration. Chaos seems to reign over the data and unless very good long-term data are available, the signal-to-noise ratio tends to approach zero. Nevertheless, a significant characteristic of climatologists is that they seem willing to draw incredibly firm conclusions from sparse noisy data.

A case in point is the vital issue of cloud feedback. When the Earth warms due to increases in greenhouse gas concentrations, does this warming produce deterministic changes in cloud cover that produce a feedback, and is the feedback positive (amplifying the greenhouse warming) or negative (opposing the greenhouse warming)? McIntyre discusses the debate on this issue at some length.[16] The effect of a change in greenhouse gas concentration is expressed as an equivalent forcing at the top of the atmosphere in W/m^2. Similarly, the feedback from cloud-cover changes is

[15] Perhaps unrelated to the politicalization of *Nature*, it is noteworthy that every journal except *Nature* has gladly and willingly given me permission to reproduce figures in my books, whereas *Nature* would have charged me as much as $700 for the right to reproduce a single figure.

[16] *http://climateaudit.org/2011/09/06/the-stone-in-trenberths-shoe.*

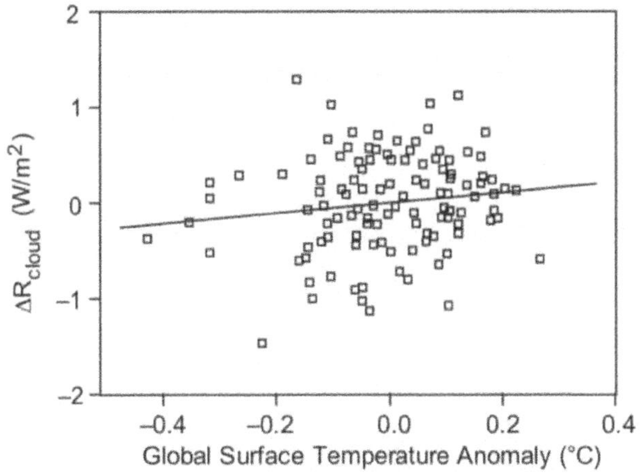

Figure 4.2. Measured cloud feedback vs. Earth surface temperature (Dessler, 2010).

also expressed as an equivalent forcing ΔR_{cloud} also in W/m^2. The debate between Spencer and Braswell (2010, 2011) and Dessler (2010) centers on whether ΔR_{cloud} is positive or negative. The raw data provided by Dessler (2010) are shown in Figure 4.2.

Unfortunately, as Dessler admits, this global cloud feedback data was taken "in response to short term climate fluctuations" where "the primary source of climate variations [was] the El Niño–Southern Oscillation (ENSO)". The data suffer from two lacks: (1) the data are short-term, and (2) the data are not based on greenhouse gas warming. While Dessler derived the straight-line fit shown in the figure, indicating a weak positive feedback, it seems evident that cloud cover was not driven by Earth surface temperature at all, and varied chaotically during this short period due to unknown factors. Had additional data been taken, it is possible that the slope of the linear fit might turn negative. More likely, the cloud cover varies widely and randomly, independently of temperature over short time intervals.

As McIntyre pointed out, an argument could be made that there is a time lag of several months between a change in temperature and a change in cloud cover, so he replotted the data with a four-month time lag and obtained Figure 4.3. While McIntyre now obtained a negative feedback, the scatter in the data is even greater than before. Only a climatologist or a statistician could believe there is any significant information contained in this plot.

Judith Curry posted[17] an interesting report on false positives in scientific research based on a paper[18] published in an on-line journal. Simmons *et al.* (2011) found that "flexibility in data collection, analysis, and reporting dramatically increases actual false-positive rates. In many cases, a researcher is more likely to

[17] *http://judithcurry.com/2012/01/12/false-positives.*

[18] Simmons, J.P., Nelson, L.D., and Simonsohn, U., "False-positive psychology: undisclosed flexibility in data collection and analysis allows presenting anything as significant", *dionysus.psych.wisc.edu/lit/articles/SimmonsJ2011a.pdf.*

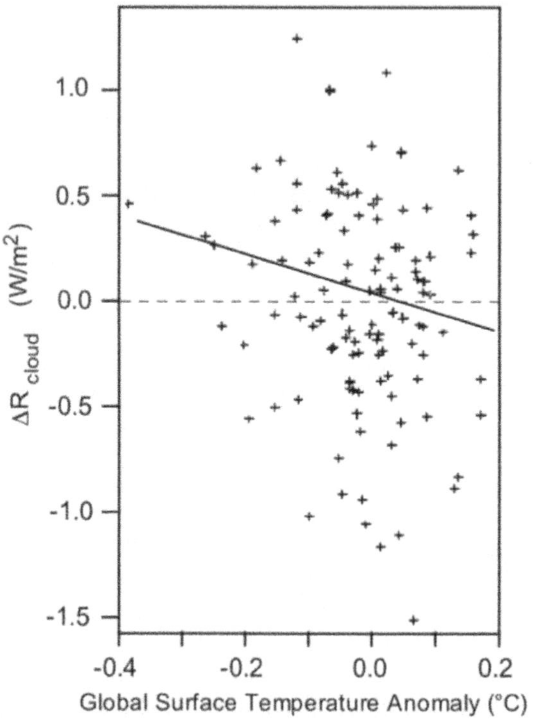

Figure 4.3. Cloud feedback plot with four-month time lag (*http://climateaudit.org/2011/09/06/the-stone-in-trenberths-shoe/*).

falsely find evidence that an effect exists than to correctly find evidence that it does not".

Over the past few years, the *www.judithcurry.com* website has emerged as the center point of discussion of a myriad of topics relevant to climate change and climatologists with quite a number of insightful articles attended by hundreds of blogger entries. (Unfortunately, the great majority of the blogger entries are not up to the high level of the original postings by Curry.) Nevertheless, there is much to learn from this website. An article appeared on August 3, 2012, written by Stephen Mosher on "post-normal science" (PNS). PNS deals with situations in which:

(1) facts are uncertain;
(2) values are in conflict;
(3) stakes are high;
(4) immediate action is claimed by one side to be required (this, according to Mosher, is the defining characteristic of PNS).

In the case of climate change, the stakes are high because of the extremity of the immediate action, the cost of that action, the impact of that action on our lives, the political difficulty in obtaining worldwide agreement to this action, and the technical difficulty in implementing it even if all the other problems could be overcome.

In normal science,

> "Because facts are uncertain, [both sides] listen to various conflicting theories. They try to put those theories to a test. They face a shared uncertainty and in good faith accept the questions and doubts of others interested in the same field. ... Because the field of personal values is never in play, personal attacks are minimized. Personal pride may be at stake, but values rarely are. In normal science, ... we can view the behavior of those doing science as puzzle solving. The details of a paradigm are filled out slowly and deliberately."

For example, in regard to continental drift, the facts are uncertain (yet far more confirmed than in climate change), and the values are not in conflict, the stakes are not high, and there is no need for immediate action. In regard to evolution of species, the facts are not absolutely certain (yet far more confirmed than in climate change), and the values are in conflict with certain religious yahoos. Yet the stakes are not high (they only pertain to textbook content in some redneck states), and there is no need for immediate action (at a large scale).

According to Mosher:

> "In all PNS situations it is almost always the case the one side sees the need for action, given the truth of their theory, while the doubters must of necessity see no need for immediate action. They must see no need for immediate action because their values are at risk and because the stakes are high. Another way to put this is as follows. When you are in a PNS situation, all sides must deny it. Those demanding immediate action, deny it by claiming more certainty than is present; those refusing immediate action, do so by increasing demands for certainty. This leads to a centralization and valorization of the topic of uncertainty. ... That is decidedly not normal science."

According to one blog entry in response to Mosher, anthropogenic global warming (AGW) proponents are justified in their alarm for the planet, but their view is that the public is always too selfish and parochial to appreciate the dire need to act. The IPCC and many climate scientists view the public as a bloc that needs to be manipulated into compliance. Thus, the IPCC manipulated the evidence to appear one-sided. It's this arrogant *noblesse oblige*, that serves as justification for their tribal mentality. Somewhere, the mission to convince the public ended up distorting the ability of the science to self-correct (paraphrased from blog entry).

4.6 THE GOLDEN RULE

One form of the Golden Rule may be stated: "He who has the gold rules." This is particularly true in climatology. The most prominent alarmists are typically college professors or researchers in National Laboratories, who derive most of their research funding from government agencies such as NSF, NOAA, NASA, DOE, etc. If global warming were not a major catastrophe facing mankind, why would these agencies want to invest heavily in climatology? At the same time, these agencies themselves are under scrutiny in these days of budget cuts and economic austerity. They need a *cause célèbre* to justify continued high-level funding of the agencies. Thus, we have a neat little mutual co-dependency between the *funders* and the *fundees* working to their mutual advantage. Global warming is the "goose that laid the golden egg" for the agencies and the climatologists.

At the other end of the scale, it is claimed that some skeptics were funded by special-interest groups. It is claimed on the Internet that Exxon funded the National Center for Policy Analysis and the Heritage Foundation to publicize anti-alarmist ideas. However, even if this were true, the amounts of funding involved ($75,000 and $50,000, respectively) were miniscule compared to the hundreds of millions doled out by government agencies to support the alarmist agenda.[19] *Nature* magazine[20] claimed:

> "The Heartland Institute plans to spend $1.8 million on its climate programme this year. Of that, $413,000 will go to supporting the Nongovernmental International Panel on Climate Change (NIPCC), a small group of skeptics who have set themselves up as a counterweight to the IPCC. Made up of ... a few dozen colleagues, the NIPCC mines the scientific literature for nuggets of contrary evidence and doubt—often the kind of uncertainties that scientists readily acknowledge in their publications. The NIPCC also ignores mountains of evidence about the adverse effects of global warming and instead strings together a confident story that makes rising carbon dioxide concentrations seem entirely beneficial."

The Heartland Institute is known as a right-wing conservative propaganda organization. However, the accusation made by *Nature* magazine is exactly transferrable to most of the college professors who espouse the alarmist viewpoint. They "mine the scientific literature for nuggets of *supporting* evidence—often *ignoring* the kind of uncertainties that scientists *should, but don't* acknowledge in their publications". (Emphasis added) The *alarmists* also ignore mountains of evidence and instead string together a confident story that makes rising carbon dioxide concentrations seem harmful well beyond what is understood.

Other claims exist on the Internet of funding of climate skeptics by oil and coal companies. As far as I can tell, some of this is probably true, but, again, such funding is trivial compared to government funding for alarmists. The stakes are much higher for alarmists. The U.S. governmental program in climate change[21] is $2.7 billion, allocated as follows:

[19] *www.guardian.co.uk/environment/2009/jul/01/exxon-mobil-climate-change-sceptics-funding.*

[20] *www.nature.com/news/2011/110727/full/475440a.html.*

[21] "Our changing planet: The U.S. Global Change Research Program for 2011—A supplement to the president's budget for FY 2011", *downloads.globalchange.gov/ocp/ocp2011/ocp2011.pdf.*

Focus Area	*Millions of $*	*Agencies & Departments*
Improving our knowledge of the Earth's past and present climate variability and change	1,429	USDA, DOC, DOE, DOI, NASA, NSF, SI
Improving our understanding of natural and human forces of climate change	549	USDA, DOC, DOE, DOI, DOT, NASA, NSF
Improving our capability to model and predict future conditions and impacts	281	USDA, DOC, DOE, HHS, DOI, USAID, NASA, NSF, SI
Assessing the nation's vulnerability to current and anticipated impacts of climate change	235	USDA, DOC, DOE, DOI, EPA, NSF, SI
Providing climate information and decision support tools	178	USDA, DOC, DOI, DOT, USAID, EPA, NASA, NSF, SI
Climate change communication and education	41	USDA, DOC, NASA, SI

4.7 THE LUNATIC FRINGE

Perhaps the most absurd aspect of the climate alarmist movement is the putative relationship between obesity and global warming. If you enter "obesity and global warming" into Google, you obtain 1,100,000 responses. Typical responses in the queue are: (1) Is obesity causing global warming? A new study has suggested that obesity is affecting the planet ... by raising carbon emissions ... ; (2) Do obese people aggravate global warming?—ABC News; (3) Scoop: Burning the Fat: Obesity and Global Warming, a study in the latest issue of the *International Journal of Epidemiology* by Phil Edwards and Ian Roberts plays out a grim scene: a world of overweight ... ; (4) thinner is better to curb global warming, study says—*CNN.com*; and there are thousands more like this. Some claim the effect is through excessive use of resources, while others blame it on increased flatulence.

In a February 2013 news announcement, it was revealed that a French cattle feed company (Valorex) developed a novel form of carbon credits aimed at providing incentive to farmers to stop cows from emitting climate-changing "farts". In France, cattle account for 5% of the country's carbon output. Valorex sells a trade mix that comprises corn, soy, lupin, and linseed, which it says means cows emit 64% less methane, deliver better-quality milk, and need fewer vet visits. A credit of 100 euros ($134) will be awarded for every tonne of CO_2-equivalent gas that is saved from entering the atmosphere. The new scheme, certified as bona fide by the French government and the U.N. Framework Convention on Climate Change (UNFCCC), has so far notched up 8,365 tonnes of averted carbon.

The website *www.numberwatch.co.uk/warmlist.htm* lists a huge number of ailments attributed by alarmists to global warming with links to the appropriate websites.

4.8 THE ROLE OF GOOGLE

Google has evolved to become the principal artery for finding information on almost any subject. As my granddaughter told me: "Why do I have to go to school to learn, when Google knows everything and has all the answers?"

Unfortunately, there are typically hundreds of thousands of responses to any query on Google, and, if a particular response appears, say, 7,000 down in the queue, it likely will never be found. In fact, if a response does not occur in the first two or three pages of a Google search, it is likely to be missed. Therefore, it is of great importance for any given website to have high priority in Google searches so it will appear in the first few pages of a search.

The exact algorithm used by Google to prioritize responses to a query is not public knowledge, but it is widely known that one very important element of the algorithm is the number of external web links to any particular website. Those websites with the greatest number of *links to them* by other websites will tend to rank higher in the queue of responses to a search. Thus, owners of websites are continually approached by other websites asking them for reciprocal links back and forth to enhance their places in search queues.

Unfortunately (or, depending on your viewpoint, perhaps fortunately), institutional websites have many more links to them than individual websites. Therefore, responses to a Google search strongly prioritize institutional websites high in the queue while leaving individual websites typically far down in the queue. As a result, institutional viewpoints are promulgated by Google, while individuals are often ignored. This brings up the question: Do institutions know more than individuals on any given subject? Or perhaps, more to the point, do institutional websites convey more and better information than individual websites?

For example, suppose you do a Google search for a specific hotel in a specific location. Google will return a queue that includes *tripadvisor.com*, *yelp.com*, *hotels.com*, *expedia.com*, *hotelscombined.com*, *travelweekly.com*, *orbitz.com*, *hotel guides.com*, *edreams.com*, etc. If you are lucky, you might find a website for your particular hotel on the third page.

In a Google search on "climate change", the queue was prioritized from the top as follows:

(1) Environmental Defense Fund: an alarmist propaganda site. (Note: this is evidently a paid-for site, since it appears at or near the top of all pages in the Google queue.)

(2) Climate Central: an alarmist propaganda site. (Note: this is evidently a paid-for site since it appears at or near the top of all pages in the Google queue.)
(3) National Association for the Advancement of Colored People: amazingly enough, this site came up third in the queue; its concern is the mistaken belief that climate change is producing more hurricanes detrimental to colored people. (Note: this is evidently a paid-for site, since it appears at or near the top of all pages in the Google queue.)
(4) Environmental Protection Agency: the EPA was created to clean up the environment, and actually did a pretty good job of it; just as its reason for being was beginning to wane, along came a great deal of publicity about climate change, and the EPA adopted climate change as its reason for continued existence.
(5) Wikipedia: this widely quoted source appears high in the queue of most searches.
(6) *The New York Times*: an article in the paper on climate change.
(7) *The Guardian*: an article in the paper on climate change.
(8) News for climate change: a Google routing page to various news articles in global newspapers.
(9) Intergovernmental Panel on Climate Change: an alarmist propaganda site.
(10) NASA: an alarmist propaganda site.
(11) New Scientist: an alarmist propaganda site.
(12) Real Climate: an alarmist propaganda blog.
(13) Nature: a journal that once in a long while publishes an alarmist paper on climate change.
(14) United Nations Environment Programme: an alarmist propaganda site.

I checked seven pages deep into the queue (about 70 responses) and did not find one response to a site that was not rabidly worried about extreme climate change. I have no idea how deep in the queue *judithcurry.com* might be, or if it is there at all.

5

Variability of the Sun

The Sun is the powerhouse that drives the Earth's climate. No understanding of climate change is possible without an understanding of the behavior of the Sun.

5.1 SOLAR IRRADIANCE

5.1.1 Introduction

We know from geological evidence that, over the past million years (and more), the Earth has gone though many thermal cycles of glaciation ("Ice Ages") and intervening warm periods. It is widely believed that variations in the Earth's orbit are associated with these climate changes, although the correlation of models with geological data is not perfect (Rapp, 2012). Other factors, including variation of solar irradiance, may also be involved. The last great Ice Age peaked some 20,000 years ago and we have been in a post-glacial warming period for the past 11,000 years.

We also have evidence that, over the past millennium or so, there have been smaller fluctuations in the Earth's climate. The *Medieval Warm Period* (MWP) and the *Little Ice Age* (LIA) represent fluctuations during this time period. Since these events occurred prior to the recent major industrialization, they were due to land clearing and naturally occurring phenomena, possibly variations in *total solar irradiance* (TSI), changes in ocean currents, and, for short periods, volcanic eruptions.

Today, we are in a period of global warming, and it is not immediately clear whether long-term variability of the TSI might have contributed significantly to this effect. The Earth is clearly warmer in the early 21st century than it was in any period over the past 500 years or so. How much (if any) of this could be due to an increase in TSI?

It has often been assumed in the past that the TSI is quite constant over long time periods, and, indeed, many papers and books refer to the "solar constant". However, there is no experimental basis for this assumption. Since 1978,

D. Rapp, *Assessing Climate Change*, Springer Praxis Books,
DOI: 10.1007/978-3-319-00455-6_5

measurements of TSI have commenced in space. While relative variations in TSI are measured with apparently high precision, absolute calibration has been more difficult (Scafetta, 2009). Nevertheless, we have seemingly good relative measurements of TSI over a period of about 30 years (see Figure 5.1). These measurements indicate that the TSI varied slightly (about 0.1%) over the past few 11-year solar cycles. However there are no data prior to 1978. As Krivova *et al.* (2009) emphasized: "Solid assessment of the solar forcing on the Earth's climate is still plagued, among other factors, by a shortage of reliable and sufficiently long irradiance records." If we simplistically extrapolate these data backward in time, such variations appear to be too small to explain the temperature variations in the MWP or the LIA. Woods (2008) reported recent results from a new satellite that monitors the Sun with significant new instrumentation (Total Irradiance Monitor (TIM) on the Solar Radiation and Climate Experiment (SORCE) satellite). Their absolute calibration indicated that the average TSI is about 1,361 W/m^2 as compared with a previously indicated value of 1,365 W/m^2. The new instruments measure spectral distribution in addition to TSI. A very interesting finding from the new instruments is that, during the course of an 11-year solar cycle, while the TSI only varies by about 0.1%, the irradiance in the UV part of the spectrum varies by about 10 to 30 times more than that (Krivova *et al.*, 2009). It is not immediately clear what effect this might have on the Earth's climate. Krivova *et al.* provided a good summary of what is known about spectral irradiance. Kopp[1] set up a calibration system for instruments that measure TSI. His estimates for the best approximation to TSI are shown in Figures 5.1 and 5.2. Kopp and Lean (2011) concluded:

> "The most accurate value of total solar irradiance during the 2008 solar minimum period is 1360.8 $\pm$ 0.5 W/m^2 according to measurements from the TIM ... and a series of new radiometric laboratory tests. This value is significantly lower than the canonical value of 1365.4 $\pm$ 1.3 W/m^2 established in the 1990s, which energy balance calculations and climate models currently use. Scattered light is a primary cause of the higher irradiance values measured by the earlier generation of solar radiometers in which the precision aperture defining the measured solar beam is located behind a larger, view-limiting aperture. In the TIM, the opposite order of these apertures precludes this spurious signal by limiting the light entering the instrument."

While this changes the absolute calibration of the TSI, the relative change across solar cycles remains as previously understood. A number of graphs in this chapter are tuned to the previous calibration of the TSI and, these should be scaled back by the factor 1361/1365. However, the absolute values of TSI are not very important. The changes in TSI over time are of the greatest interest.

[1] Global Change and the Solar-Terrestrial Environment, 12 June–17 June 2010, Aspen Global Change Inst., *www.agci.org/programs/past_scientist_workshops/about_the_workshop/sciSess_details.php?recordID=265; http://spot.colorado.edu/~koppg/TSI/.*

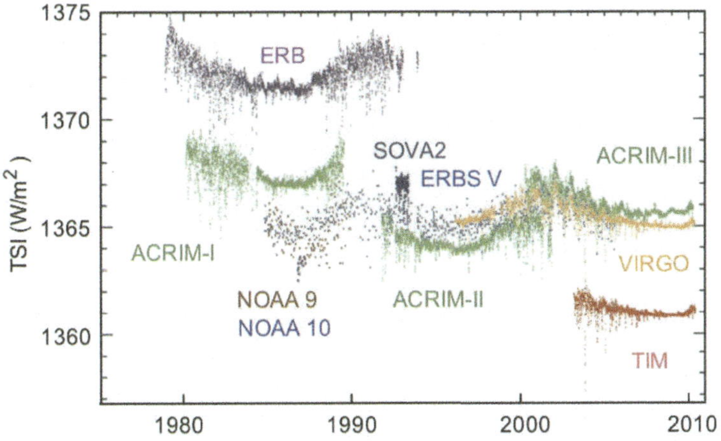

Figure 5.1. Result of the series of redundant, overlapping satellite TSI monitoring experiments that have provided a continuous record since late 1978 (adapted from Kopp, *http://spot.colorado.edu/~koppg/TSI/*).

There are two major reasons why the variability of TSI over the past millennium is important. One reason is to estimate whether such variability is sufficient to explain the putative climate variations of the MWP and the LIA. A second reason is to provide a comparison of the solar forcing of today's Earth's climate with solar forcing during previous periods in the past millennium. This would help in the critically important problem of distinguishing between natural (solar) variations as a cause of global warming vs. anthropogenic contributions (via greenhouse gas effects).

Fragmentary bits and pieces of evidence exist regarding past variations in solar activity in general, and TSI in particular. These include recorded variations in sunspot numbers and length of the solar cycle (see Sections 5.2 and 5.4). A number of attempts have been made to reconstruct variations in TSI over the past millennium using these data. In addition, there is evidence from cosmogenic nuclides, and comparison of our Sun with behavior of Sun-like stars (see Section 5.4.4). Depending on the assumptions made, a wide range of results is possible. These range from solar variations acting as the major warming factor in the 20th century to solar variations being a minor contributor to recent global warming. At this juncture, it does not seem possible to unequivocally resolve the past variations in TSI. The various methods and assumptions are discussed in considerable detail in Section 5.4.

For any given level of solar forcing of the Earth's climate, the impact of variations in TSI on global temperatures is an important consequence. This is discussed in Section 5.6.

There are many reconstructions of past TSI. However, even the best of these are quite speculative.

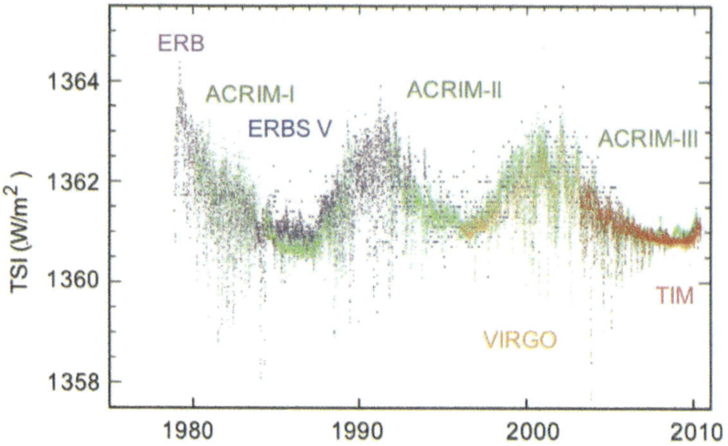

Figure 5.2. Composite of TSI measurements as developed by Kopp, *http://spot.colorado.edu/~koppg/TSI/*).

5.1.2 Measurements of TSI in space since 1978

Monitoring TSI from the ground is plagued by problems due to atmospheric absorption and scattering, weather, and other environmental factors. Variations in the extraterrestrial TSI were first measured by a new generation of electrically self-calibrating cavity sensors on extended spaceflight experiments that began in the late 1970s. The instruments typically utilize a pair of closely matched hollow cavities. One faces the Sun and is warmed by absorption of solar irradiance. The other faces away from the Sun and is electrically heated to bring it to the exact same temperature as the one facing the Sun. This provides an electrical power equivalent to the solar irradiance.

These instruments were capable of reducing the uncertainties of TSI monitoring by several orders of magnitude. The modern record began with observations by the Earth Radiation Budget (ERB) experiment on the National Oceanographic and Atmospheric Administration (NOAA) Nimbus-7 satellite (1978–1993). This was followed by the first experiment designed specifically for precision TSI monitoring, the Active Cavity Radiometer Irradiance Monitor (ACRIM1) on NASA's Solar Maximum Mission (SMM) (1980–1989).

The ACRIM1 experiment invoked a new mode of in-flight calibration that provided the first unambiguous detection of intrinsic solar variability on timescales from minutes to the sunspot cycle. The ACRIM1 experiment was followed by the Earth Radiation Budget Satellite Earth Radiation Budget Experiment (ERBS/ERBE) in 1984 and the Upper Atmosphere Research Satellite (UARS) ACRIM2 experiment in 1991. Three satellite TSI monitoring experiments were operational in early 2007: the Solar Heliospheric Observer (SOHO) Variability of solar Irradiance and Gravity Oscillations (VIRGO) launched in 1995, the ACRIMSAT/ACRIM3 launched in 1999, and the SORCE TIM launched in 2003. The ACRIMSAT/

ACRIM3 and SORCE/TIM experiments are operating at full capabilities but the SOHO satellite and VIRGO instrument have had some performance degradation occurrences and, after more than 10 years in operation, the SOHO/VIRGO experiment's remaining useful lifetime is uncertain. A summary of the current TSI monitoring database is shown as Figure 5.1. These data are corrected for the slightly elliptical orbit of the Earth and adjusted to 1 AU (astronomical unit) distance from the Sun.

As Krivova *et al.* (2009) pointed out, "Since none of the instruments survived over the whole period since 1978 and each of them suffers from its individual degradation, calibration or other problems, a construction of a composite TSI record is quite a challenge." The absolute accuracy of these instruments is far less precise than their relative accuracy. Kopp (*loc cit.*) attempted to correct for various deficiencies in the instruments and scaled them to form a continuous record, as shown in Figure 5.2.

Figure 5.2 shows that, over the 11-year solar cycle, the mean TSI varied by about 0.14% in going from solar minimum (S_{MIN}) to solar maximum (S_{MAX}). There are wider swings in TSI on a daily basis, as much as $\pm 0.3\%$ (Willson, 2006; Eddy *et al.*, 2004).

5.1.3 Short-term TSI models

The short-term variations in TSI described in Section 5.1.2 have been analyzed in terms of sunspots and faculae on the surface of the Sun, and it appears that such solar surface features account for most of the observed short-term variations in TSI. It is possible to extrapolate TSI to time periods prior to the 1970s for periods in which sunspots and faculae have been carefully measured, based on correlations between TSI and sunspot and faculae characteristics observed post 1978. However, this would not account for any slowly varying long-term secular components to the TSI that are not discernible from two or three cycles of data. We will have more to say about this in Section 5.4.

Solanki, Krivova, and Wenzler (2005) described models for predicting daily variations of TSI in terms of sunspots and faculae. The model includes a regression formula for the variation of TSI in terms of sunspot umbrae, penumbrae, and faculae measurements. An illustration of their results is shown in Figure 5.3. A more recent analysis of short-term solar variability was developed by Domingo *et al.* (2009).

As a consequence of the Sun's 27-day rotation, we cannot exclude the possibility that energy transport to the surface and its emission into space are anisotropic. Anisotropy is therefore an additional potential source of variability (Beer *et al.*, 2000).

5.1.4 Long-term TSI models

According to Beer, Mende, and Stellmacher (2000), the long-term energy flow from the Sun (irradiance) is determined by the rate of fusion of hydrogen to helium in the

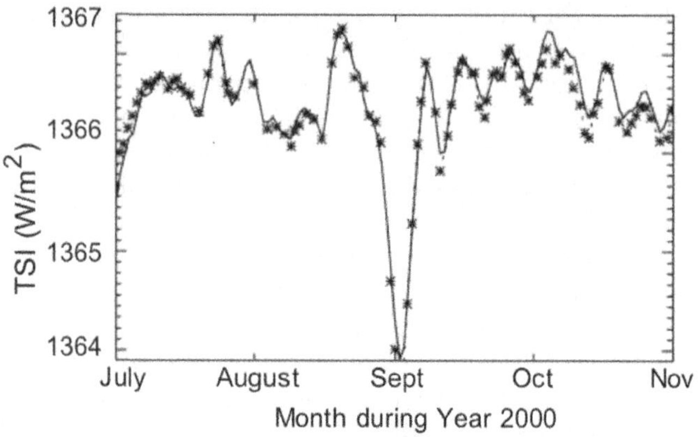

Figure 5.3. Comparison of calculated TSI (line) based on short-term correlations of TSI with sunspots and faculae with measured values (points) (adapted from Solanki *et al.*, 2005).

solar core on a billion-year timescale and follows a hyperbolic trend. However, the short-term deviations from this trend are extremely small. From physical models of the Sun's evolution, it was computed that the early solar radiation output four billion years ago was only about 75% of today's output. Since then, the energy output of the Sun steadily increased and will continue to do so for another four billion years.

The energy transport through the radiative zone (0.3 to 0.7 solar radii) is probably very stable on a million-year timescale and does not generate any measurable variability. However, the heat transport through the convective zone (0.7 to 1.0 solar radii) is related to convective and magnetic structures. It is probably the main source of variability of solar radiation on timescales of years to 100 kyr.

Irradiance variability is essentially due to the response of the outer solar layers to thermally or magnetically driven excitation near the bottom of the convection zone. This excitation also generates the conspicuous photospheric features such as sunspots and faculae. So, short-term irradiance and solar surface features have a common cause that explains their high correlation.

The rate of change of the fusion process is extremely small (10^4 W/m^2) per 1,000 years. Therefore, the variability potential of the fusion process is completely negligible in regard to current climate change. However, it is interesting to ask how has the Earth managed to remain a habitable planet with globally distributed liquid water during the past four billion years and why did it not turn into either an icy or a hot planet? This question is known as the "the early faint Sun paradox". It might well be that higher levels of greenhouse gases from active volcanism kept the Earth warm despite the much lower solar irradiance. However, this introduces other inconsistencies, and the effect of cosmic-ray flux has been proposed as an alternative explanation (Rapp, 2012).

5.2 ASPECTS OF SOLAR VARIABILITY

5.2.1 The solar cycle

As Beckman and Mahoney (1998) pointed out:

> "We are all familiar with the 11-year sunspot cycle. One familiar factor is the effect of solar activity on short-wave radio communications. During sunspot maximum high-energy protons and alpha particles from the Sun affect the ionosphere, reducing its effectiveness as a mirror from which short radio waves are reflected round the world, disrupting transmissions for days at a time. The association with the presence of large numbers of dark spots on the solar disc is widely known, and well understood, and it is also clear that such maxima repeat every 10 or 11 years, with minima between them. ... The key impulse here was the work of Eddy in the late 1970s, focused especially on the historical period between 1645 and 1715, for which there is evidence that sunspot activity was strongly suppressed or virtually absent. If this is accepted as true, it has strong implications for our ideas of how magnetic fields in stars are produced. Of more impact, it might go some of the way, even all of the way, to explaining the observed pattern of global warming of the Earth in the last decades of the 20th century."

5.2.2 Sunspots

Sunspots (Figure 5.4) are dark, planet-sized regions that appear on the "surface" of the Sun. Sunspots are "dark" because they are colder than the areas around them. A large sunspot might have a temperature of about 4,000 K. This is much lower than the 5,800 K temperature of the bright photosphere that surrounds the sunspots (Anon. (Q)).

Sunspots are only dark in contrast to the bright face of the Sun. If you could cut an average sunspot out of the Sun and place it in the night sky, it would be about as bright as a full moon. Sunspots have a lighter outer section called the penumbra, and a darker middle region named the umbra.

Sunspots are caused by the Sun's magnetic field welling up to the photosphere, the Sun's visible "surface". The powerful magnetic fields around sunspots produce active regions on the Sun, which often lead to solar flares and Coronal Mass Ejections (CMEs).

Sunspots form over periods lasting from days to weeks, and can last for weeks or even months. The average number of spots that can be seen on the face of the Sun is not always the same, but goes up and down in a cycle. Historical records of sunspot counts show that this sunspot cycle has an average period of roughly 11 years although it has varied widely over the past century.

Our Sun isn't the only star with spots. Just recently, astronomers have been able to detect "starspots" ("sunspots" on other stars). Some sunspots expand with time and evolve into "groups" (Figure 5.5).

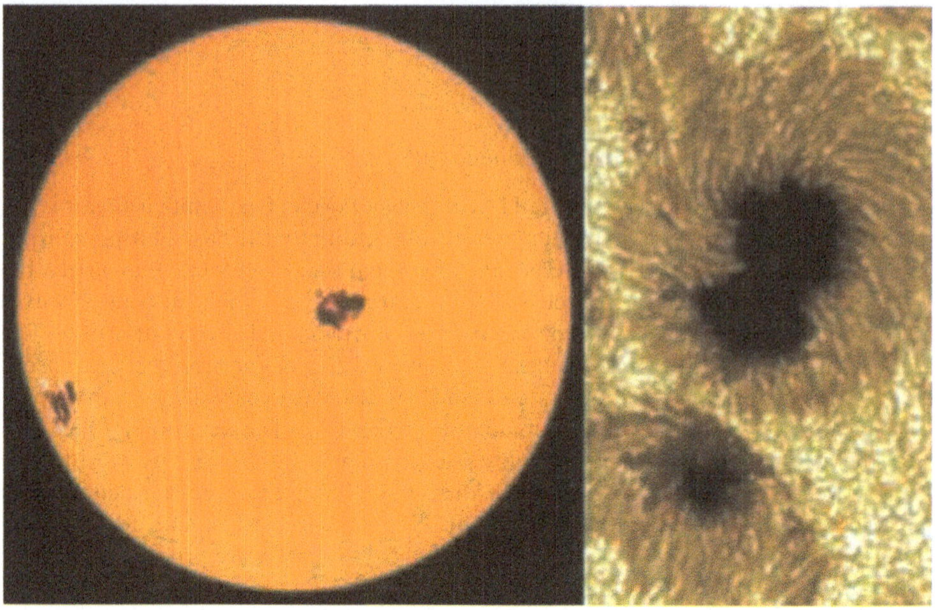

Figure 5.4. Sunspots on the Sun and close-up of sunspot (Anon. (Q)).

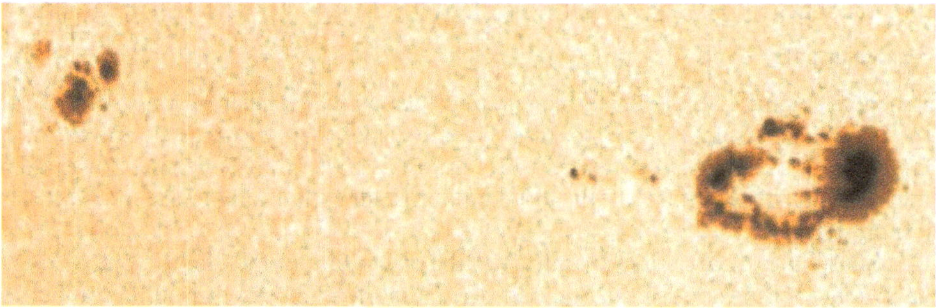

Figure 5.5. Sunspot groups.

5.2.3 Faculae

Faculae are hot structures in active regions of the Sun (Figure 5.6).

Faculae = plural of facula, Latin for "small torch". They were originally discovered near the limb in full-disk images of the Sun. Faculae are the small, bright, patterns around dark sunspots and in the "photospheric network":

> "Modeling of sunspots ... confirms that their darkness can be understood as thermal 'plugs' in which intense vertical magnetic fields divert heat flow from deeper layers. Magnetic brightenings act in the opposite sense, as local thermal leaks. They are bright because the effect of their smaller-scale magnetic field is to form small depressions in the photospheric surface, enabling radiation from

Figure 5.6. Sunspots (dark spots) and faculae (light-colored areas).

lower and hotter atmospheric layers to escape more easily ... Calculations with a thermal impedance model of solar luminosity variation show that the heat flux diverted by a spot is not merely shunted aside to reappear nearby. Instead, the blocked heat flux effectively remains stored in the Sun for hundreds of millennia instead of appearing elsewhere on the surface. ..." (Foukal *et al.*, 2006).

The Sun's remarkable ability to store energy, rather than quickly reradiating fluctuations in heat flow to the photosphere, is due to the fact that diffusion of heat within the Sun is rapid compared with the rate at which it radiates. The radiative equilibrium timescale is about 100,000 years.

5.2.4 Sunspot indices

Two major reconstructions of solar activity have been made on the basis of direct observations of the Sun. The International Sunspot Numbers (Wolf or Zurich sunspot numbers) have long served as the primary time series defining solar activity since the year 1700. This time series was derived by Rudolf Wolf in the 19th century and has been maintained by his successors. Wolf defined the sunspot number, R_Z, as

$$R_Z = k\,(10g + f),$$

where **g** is the number of sunspot groups, **f** is the number of individual sunspots, and **k** is a correction factor for each observer. For years prior to 1817, the number of missing days was so great that Wolf only tabulated monthly means. There are no observations for many months from 1749 to 1818, as well as for a few months after 1818. Wolf filled in these months by interpolation and using magnetic needle observations. Thus, the numbers $\mathbf{R_Z}$ are a mixture of direct sunspot observations and estimated values (Vaquero, 2007).

Hoyt and Schatten (1998) developed a reconstruction of solar activity from sunspot observations. This time series is known as the group sunspot number, $\mathbf{R_G}$, because it uses the observed number of sunspot groups. Hoyt and Schatten (1998) defined the group sunspot number as a sum of the number of sunspot groups observed, times a normalization factor, chosen to make the mean of $\mathbf{R_G}$ identical to the mean of $\mathbf{R_Z}$ for 1874 through 1976. Daily, monthly, and yearly means were derived from 1610 to the present. Hoyt and Schatten (1998) calculated daily values of solar activity on 111,358 days for 1610–1995, compared with 66,168 days for the international sunspot numbers, as well as tabulated estimates of their random and systematic errors. This series has complete or nearly complete coverage from about 1800 to 1995 and from 1645 to 1727 (Figures 5.7–5.9).

The group sunspot numbers are strongly recommended for analysis of sunspot activity before 1880. The Wolf and Group numbers are not fundamentally different alternative proxies of solar activity because the Group numbers are an upgrade of the Wolf numbers. Hathaway, Wilson, and Reichmann (2002) concluded that the $\mathbf{R_G}$ numbers are most useful for extending the sunspot cycle data further back in time and thereby adding more cycles and improving the statistics. However, the $\mathbf{R_Z}$ numbers are slightly more useful for characterizing the ongoing levels of solar activity.

The characteristics of the $\mathbf{R_Z}$ and $\mathbf{R_G}$ series are very similar for the 19th and 20th

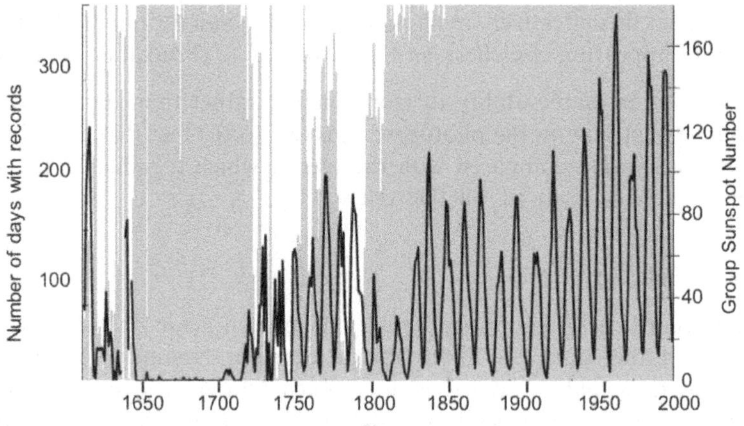

Figure 5.7. Estimated group sunspot numbers (right scale) since ~1600. Gray bars are days with records (adapted from Vaquero, 2007).

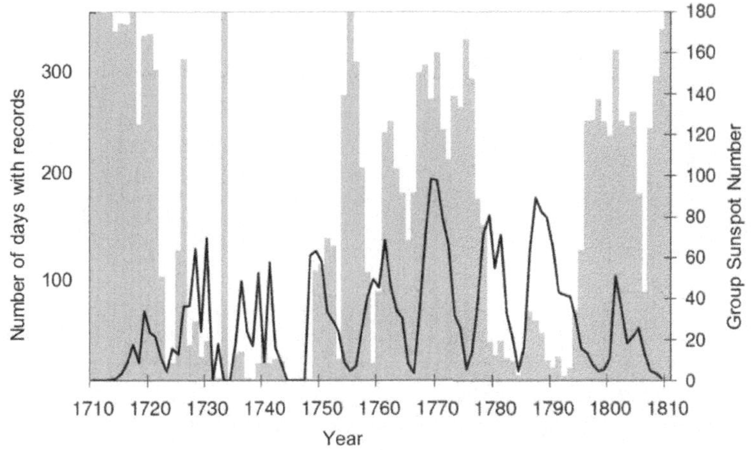

Figure 5.8. Estimated group sunspot numbers (right scale) from 1728 to 1799. There are many years with only sparse observations. For 1744–1745, and 1747, there exist no reports of sunspot observations (adapted from Vaquero, 2007).

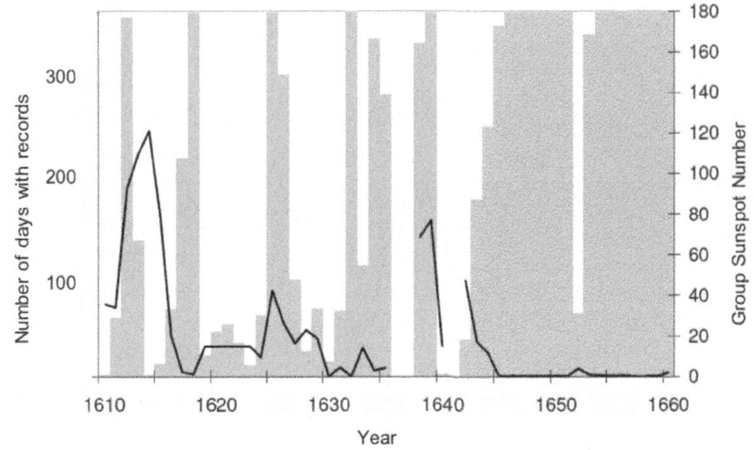

Figure 5.9. Estimated group sunspot numbers (right scale). From 1610 to 1644, there are many years with only sparse observations. For 1636–1637 and 1641, there exist no reports of sunspot observations. Lines indicate sunspot numbers observed (adapted from Vaquero, 2007).

centuries (see Figure 5.10). Twenty-five-year moving averages of the two indices are presented in Figure 5.11, showing the long-term differences between them.

The various solar cycles are numbered, with the cycle that reached S_{MAX} just after year 2000 designated as Cycle 23, and Cycle 1 being the cycle that reached S_{MAX} around 1760.

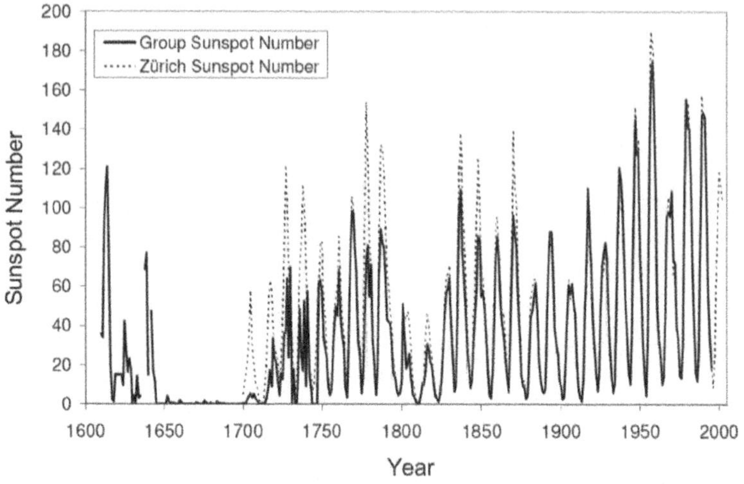

Figure 5.10. Comparison of R_G and R_Z series. Cycle 22 is the rightmost cycle in this plot (adapted from Vaquero, 2007).

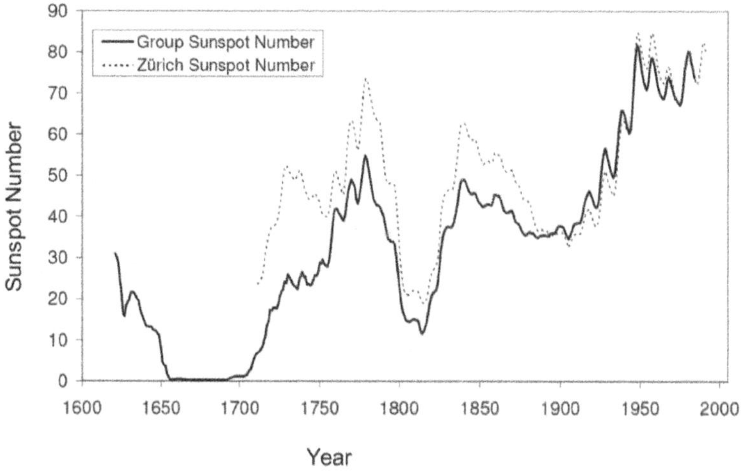

Figure 5.11. Comparison of 25-year moving averages of R_G and R_Z series (adapted from Vaquero, 2007).

5.2.5 Estimation of sunspot activity from proxies

Usoskin *et al.* (2003) carried out an analysis leading to estimates of sunspot number as far back as 850 based on the measured ^{10}Be concentrations in ice cores drilled at the Dye-3 site in Greenland (annual data for 1424–1985) and at the South Pole (roughly eight-year sampled data for 850–1900). The methodology is quite complex and depends on several sequential steps. The isotope ^{10}Be is formed in the Earth's atmosphere by the impact of cosmic rays on oxygen or nitrogen. The flux of cosmic

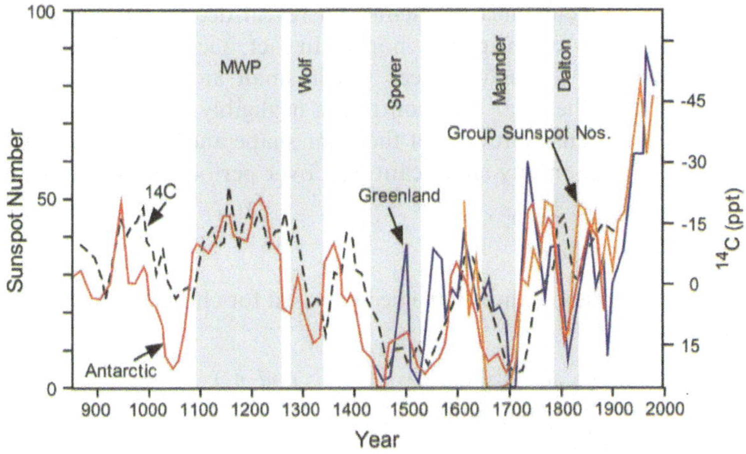

Figure 5.12. Sunspot number reconstructed from ^{10}Be concentrations in ice cores (adapted from Usoskin, 2003).

rays that impinges on the Earth's atmosphere is reduced during S_{MAX} and increases during S_{MIN} due to the effect of the higher magnetic fields between the Sun and the Earth during S_{MAX}. These magnetic fields provide a partial shield against energetic particles. The actual model is a three-step process for any year: (1) the ^{10}Be concentration is used to estimate the cosmic-ray flux, (2) the cosmic-ray flux is used to estimate the solar open magnetic flux, and (3) the sunspot number is estimated from the solar magnetic flux. The resultant reconstruction is shown in Figure 5.12. In this figure, data from Antarctica are shown in red and those from Greenland are blue. The orange curve shows the observed group sunspot number since 1610 and the dashed black curve shows the (scaled) ^{14}C concentration in tree rings, corrected for variation of the geomagnetic field. The gray areas indicate the times of minima and maxima.

The temporal lag of ^{14}C with respect to the sunspot number is due to the long attenuation time for ^{14}C. If these data are correct, it would imply that the sunspot numbers at S_{MAX} are higher today than they have been for 1,000 years.

5.2.6 Diameter of the Sun

According to Foukal *et al.* (2006), measurements of the solar diameter originally attracted attention because solar diameter variations seemed to offer a unique diagnostic of otherwise unobservable internal temperature variations. However, data from the Michelson Doppler Imager (MDI) instrument on the SOHO spacecraft show no evidence of secular trends in diameter, or variations attributable to the 11-year cycle at the level of a few milli-arcseconds (a few kilometers on the Sun). This is consistent with, but 10 times better than, upper limits from the best ground-based measurements of the Sun's diameter. However, conflicting detections nearly two orders of magnitude larger continue to be cited. Foukal *et al.* (2006) concluded that it is unlikely that diameter measurements can reveal deeper-lying sources of solar irradiance variation, as was originally hoped. In fact, such measurements put strict limits on the presence of such sources. Their small amplitude also implies that changes in the size of the solar disk contribute negligibly to the TSI. Rozelot *et al.* (2009) provided an extensive review of the solar shape and radius but it is not clear that this affects climate in any significant way over periods of, say, centuries.

5.2.7 Indices of solar activity

A number of geomagnetic indices have been devised for characterizing the activity of the Sun. These include:

(1) planetary magnetic activity indices (K_p, a_p, *and* A_p);
(2) planetary daily character figure (C_p, $C9$, C_i);
(3) Geomagnetic aa Index (*aa, Kpa*);
(4) Geomagnetic am Index (*am, an, as, Km, Kn, Ks, Am, An, and As*);
(5) Disturbance Storm-Time Index (D_{st});
(6) Bartels solar rotation number;
(7) International Sunspot number (R_z, R_i);
(8) Ottawa 10.7 cm solar radio flux adjusted to 1 AU ($F_{10:7}$);
(9) Auroral Electrojet Index (*AE*);
(10) Polar Cap Index (*PC*);
(11) Comprehensive Flare Index (*CFI*).

These are beyond the scope of the present discussion, but they are described at *www.spenvis.oma.be/spenvis/help/background/indices.html.*

Gray *et al.* (2010) provide data on several of these indices. The coronal source surface occurs where the solar magnetic field becomes approximately radial and lies at a heliocentric distance of about 2.5 solar radii. The total magnetic flux leaving the Sun, and thereby entering the heliosphere by threading through this surface, is the coronal source flux, $\mathbf{F_s}$. Lockwood and Stamper (1999) and Lockwood (2002) developed techniques for estimating $\mathbf{F_s}$ (see Section 5.4.7).

5.2.8 Effect of the Sun–Earth distance

The orbit of the Earth about the Sun is slightly elliptical. The Sun is located at one focus of the ellipse. The length of the semi-major axis of the ellipse is called the astronomical unit (AU), and is equal to 1.496×10^8 km. The eccentricity of the ellipse is presently 0.017, so the Earth–Sun distance varies from 0.983 AU around January 1 to 1.017 AU around July 1. Thus, the amplitude of the annual variation in the Earth–Sun distance is about 3.4%. Since the extraterrestrial TSI at the top of the Earth's atmosphere is proportional to the square of the distance to the Sun, the variation in TSI from January 1 to July 1 is about 6.8%. For a nominal TSI of $\sim$1,361 W/m^2, the annual variation (from high to low) due to the elliptical orbit is $0.068 \times 1{,}361 \sim 93$ W/m^2. However, when spread over the spherical surface of the Earth, this amounts to an effective forcing of 93/4 $\sim$23 W/m^2. That is a huge variation. Section 5.1 showed that the current variation from S_{MIN} to S_{MAX} is around 1.3 W/m^2. As Section 5.6 shows, a consistent long-term change in TSI of 93 W/m^2 would produce a large change in Earth temperature, leading to worldwide massive climate change. However, the time lag in establishing a new equilibrium when the TSI is changed reduces the impact of these annual fluctuations. Therefore, the Earth acts as a rectifier to damp out the annual variations (and even the 11-year variations) and, ideally, the Earth reacts to the average of TSI changes that occur annually. However, the 23 W/m^2 annual variation is so large, considering the chaotic nature of Earth feedback systems and climate response, that it could conceivably trigger global climate changes that could mask the effects of smaller long-term changes.

As Section 5.4 shows, a number of investigators have attempted to infer how the average value of TSI has varied in the past, before measurements were begun in 1978. The motivation for this was to try to explain inferred historical temperature variations prior to the industrial age. But the problem is made more difficult because one is trying to identify the effect of slowly varying (decades to centuries) changes of perhaps a few W/m^2 in TSI, while the solar cycle is producing changes of $\sim$1 W/m^2 every $\sim$11 years and the Earth's orbit is producing changes of 93 W/m^2 on an annual basis.

5.3 THE MAUNDER MINIMUM: JOHN EDDY'S STUDY

5.3.1 Historical telescope observations of sunspots

Eddy (1976) wrote an authoritative paper on the Maunder Minimum (MM) that has greatly influenced thinking about global climate variations ever since. In this paper, Eddy traced out the history of accumulation of data on sunspot observations and suggested that the reliability of the data may be graded into four epochs:

(1) reliable from 1848 to the present;
(2) good from 1818 through 1847;
(3) questionable from 1749 through 1817;
(4) poor from 1700 through 1748.

In addition, he traced out reports from various sources that indicated that the number of sunspots was minimal (or almost zero) for a 70-year period from 1645 to about 1715, and a more speculative indication that sunspot numbers may have been low for many years prior to 1645, which in turn might suggest that the era of higher sunspot numbers we have witnessed since 1715 may be an aberration from "normalcy". These sources were initially accumulated by E. W. Maunder around the turn of the 20th century, which is why the period of low sunspot numbers in the late 17th century is referred to as the "Maunder Minimum" (MM).

Eddy's paper is lengthy and detailed, and should be required reading for anyone interested in global climate change. Only a very brief summary is given here. Eddy provides the arguments that follow.

Observation of sunspots was well within the capability of observers with simple optical telescopes in the 17th century, and probably the latter half of the 16th century. Many reports of sighting an occasional sunspot as a rarity during the MM indicate their prevalent absence during that period even though specific counts may not have been available.

5.3.2 Historical records of aurorae

Eddy (1976) said:

> "Records of occurrence of auroras offer an independent check on past solar activity since there is a well-established correlation between sunspot number and the number of nights when aurorae are seen. Auroral displays are produced when charged particles from the Sun interact with the Earth's magnetic field, resulting in particle accelerations and collisions with air molecules in our upper atmosphere. Aurorae register, therefore, those particle-producing events on the Sun (such as flares and prominence eruptions) that happen to direct their streams toward the Earth. Since these events arise in active regions on the Sun, where there are also sunspots, we find a strong positive correlation between reported numbers of the two phenomena. Aurorae are especially valuable as historical indicators of solar activity since they are spectacular and easily seen, require no telescopic apparatus, and are visible for hours over wide geographic areas. ... An increase in the number of reported aurorae inevitably follows a major increase in solar activity, and a drop in their number can generally be associated with the persistence of low numbers of sunspots, with certain reservations. As with sunspots, aurorae will not be seen unless the sky is reasonably clear, and an absence of either on any date in historical records could be due simply to foul weather. ... The period between 1645 and 1715 was characterized by a marked absence of aurorae. Far fewer were recorded than in either the 70 years preceding or following."

Eddy (1976) went on to present quantitative data on aurora observations to back up these conclusions.

5.3.3 Historical visual observations of sunspots

Eddy provided information on historical observations of sunspots by eye prior to the advent of the telescope, including Asian reports dating back to AD 300. The data are fragmentary and sporadic, but, to the extent that they are credible (which is difficult to judge), they suggest that sunspot activity may have been relatively low for 2,000 years prior to the MM.

5.3.4 ^{14}Carbon in tree rings

Eddy also discussed evidence provided by the abundance of terrestrial ^{14}C in tree rings. This isotope of carbon is continuously formed through the action of galactic cosmic rays acting on ordinary ^{12}C in the CO_2 in the atmosphere. This process is modulated by solar activity. When the Sun is active, magnetic fields between the Sun and the Earth reduce the number of incoming galactic cosmic rays that reach the Earth. At these times, corresponding to maxima in the sunspot cycle, the amount of ^{14}C produced in the atmosphere is lower, and therefore less ^{14}C is found in tree rings formed at such times. When the Sun is quiet, terrestrial bombardment by galactic cosmic rays increases and the ^{14}C proportion in the atmosphere rises. Since tree rings are automatically dated as one year per ring, a yearly history of ^{14}C deposition into trees can be derived. However, there is a time lag of 10 to 50 years because of the finite time required for distribution of ^{14}C in the atmosphere and absorption by trees. Therefore, the ^{14}C abundance in any tree ring is not an instantaneous record of the solar activity in that year, but is somehow a representation of the previous few decades. During periods of several decades where solar activity is relatively constant, the tree rings should provide an indicator of relative solar activity. Eddy reported that the first major anomaly found in studies of ^{14}C history was a marked and prolonged increase that reached its maximum between about 1650 and 1700, in remarkable agreement in sense and date with the MM. The phenomenon, known in carbon dating as the "DeVries Fluctuation", peaked at about 1690. However, Eddy cautioned that one must take care in assigning any of the ^{14}C variations to a solar cause, for there are other important mechanisms that can affect the flux of incoming galactic cosmic rays. The trajectories of such charged particles are affected by the Earth's magnetic field, which has varied considerably over time. Unfortunately, the ^{14}C index is not useful in modern times because of fossil fuel combustion, which introduces CO_2 with different carbon isotopic abundance ratios—the so-called "Suess Effect".

5.3.5 The solar corona

According to Eddy (1976), historical accounts of the solar corona at total eclipse offer another possible check on anomalies in past solar behavior. He pointed out that the shape of the corona seen at eclipse varies with solar activity: when the Sun has many spots, the corona is made up of numerous long, tapered streamers that extend outward like the petals of a flower. As activity wanes, the corona dims and fewer and fewer streamers are seen. At a normal minimum in the solar cycle, the corona seen by the naked eye is highly compressed and blank. Coronal streamers are rooted in concentrated magnetic fields on the surface of the Sun, which are associated with solar activity and sunspots. As sunspots fade, so do concentrated surface fields and associated coronal structures. Eddy compared two cases:

(1) In a total absence of solar activity, he expected to observe a dim, uniform glow around the Moon at eclipse: the zodiacal light, or false corona, would remain,

since it is simply sunlight scattered from dust and other matter in the space between the Earth and the Sun.

(2) At times of normal solar activity, the corona seen at eclipse is a mixture of the true K-corona and the weaker glow of zodiacal light.

Eddy said that first-hand descriptions of total solar eclipses during the MM seem entirely consistent with an absence of the modern structured corona, but proof seems blurred by the customs of observing eclipses in the past and by the fact that scientists seldom describe what is missing or what is not thought to be important. Despite uncertainties in the historical record, Eddy concluded:

> "It thus seems to me more probable that, through much of the long period of the Maunder Minimum and the Sporer Minimum, extending between perhaps 1400 and 1700, the Sun was at such a minimum of activity that the K-corona was severely thinned or absent altogether. . . . In any case the corona as we know it may well be a modern feature of the Sun. It is an interesting question, and another important challenge for historians."

5.3.6 Beckman and Mahoney on Eddy's work

Beckman and Mahoney (1998) commented extensively on Eddy's work. The absence of sunspots for some 70 years in the 17th century was first pointed out by Spörer in 1887. A few years later, it was noted that this dearth of sunspots apparently coincided with an absence of terrestrial aurorae. Much later, Maunder (1922) found a note by Flamsteed, the first Astronomer Royal, describing a sunspot seen at Greenwich in 1684, in which Flamsteed said that it was the first he had seen since 1674. In spite of Spörer's, and especially Maunder's, advocacy, most specialists were fairly skeptical about the reliability of a prolonged sunspot absence. It was suspected that the main reason for few reports of sunspots from 1650 to 1715 was that people were not observing the Sun, or at any rate not systematically. Eddy showed that this was not the case.

Eddy (1976) succeeded in convincing many researchers that there was real evidence for the sunspot absence in the MM period. He also showed that the solar corona at eclipse during the period was strongly suppressed compared with its present exhibition of major streamers. In addition, he also looked at the tree-ring ^{14}C record, and confirmed a high incidence of ^{14}C during that period. While Eddy's arguments for a lull in solar magnetic activity during the MM seems to be widely accepted, some controversy has arisen out of his claim that the MM coincided in time with an era of colder weather and that, by implication, the absence of magnetic activity was accompanied by a net fall in TSI. An implicit corollary is that TSI in later years has been increasing, with a consequent warming of the Earth. However, the connection of variable TSI to climate change since the MM (if any) is difficult to resolve.

5.3.7 Eddy's conclusions

In addition to the ~70-year period of solar quiet during the MM (1645–1715), Eddy discerned a period of prolonged solar quiet between about 1460 and 1550 (which he called the Sporer Minimum) and a prolonged sunspot maximum between about 1100 and 1250 that is a century or two after the Medieval Warm Period (although some studies have placed the MWP a century or two earlier). He speculated that, if the prolonged maxima of the 12th and 13th centuries and the prolonged minima of the 16th and 17th centuries are extrema of a repetitive cycle of solar change, the cycle has a full period of roughly 800 to 1,000 years. If this change is periodic, he further speculated that the Sun might now be progressing toward a grand maximum that might be reached in the 21st to 23rd centuries. He pointed out that the overall envelope of solar activity has been steadily increasing since the end of the MM, giving some credence to this view. The coincidence of the MM with the coldest excursion of the LIA suggests possible relations between the Sun and the terrestrial climate. These coincidences suggest a possible relationship between the overall envelope of the curve of solar activity and terrestrial climate in which the 11-year solar cycle may be effectively filtered out or simply unrelated to the problem. The mechanism of this solar effect on climate may be the simple one of ponderous long-term changes in the total radiative output of the Sun, or "solar constant". These putative long-term drifts in solar radiation may modulate the envelope of the solar cycle through the solar dynamo to produce the observed long-term trends in solar activity. The continuity, or phase, of the 11-year cycle would be independent of this slow, radiative change, but the amplitude might be controlled by it. According to this interpretation, the cyclic coming and going of sunspots would have little effect on the output of solar radiation, or presumably on weather, but the long-term envelope of sunspot activity carries the indelible signature of slow changes in solar radiation which surely affect our climate. However, all of this is completely hypothetical and unproven.

Eddy made a strong case that, during the MM, solar magnetic activity was at a minimum and may have ceased altogether for periods of years. This might have caused a reduction in TSI during that period, but it is difficult to evaluate the magnitude of this putative decrease. (We shall have more to say about estimates of TSI during the MM in Section 5.4.) It seems possible that this caused colder weather, particularly at higher latitudes, but it is difficult to describe the climate in the late 17th century based on what we know today.

5.4 HISTORICAL SUNSPOT LEVELS AND RECENT SOLAR INACTIVITY

Figure 5.13 shows the sunspot record since 1750. As it turns out, 2008 was the "blankest year of the Space Age".

As of the end of September 2008, the Sun had been blank (i.e., no visible sunspots) on 200 days of the year, representing a 50-year low (Anon. (K)).

Livingston and Penn (2008) observed spectral changes in solar emission, as well

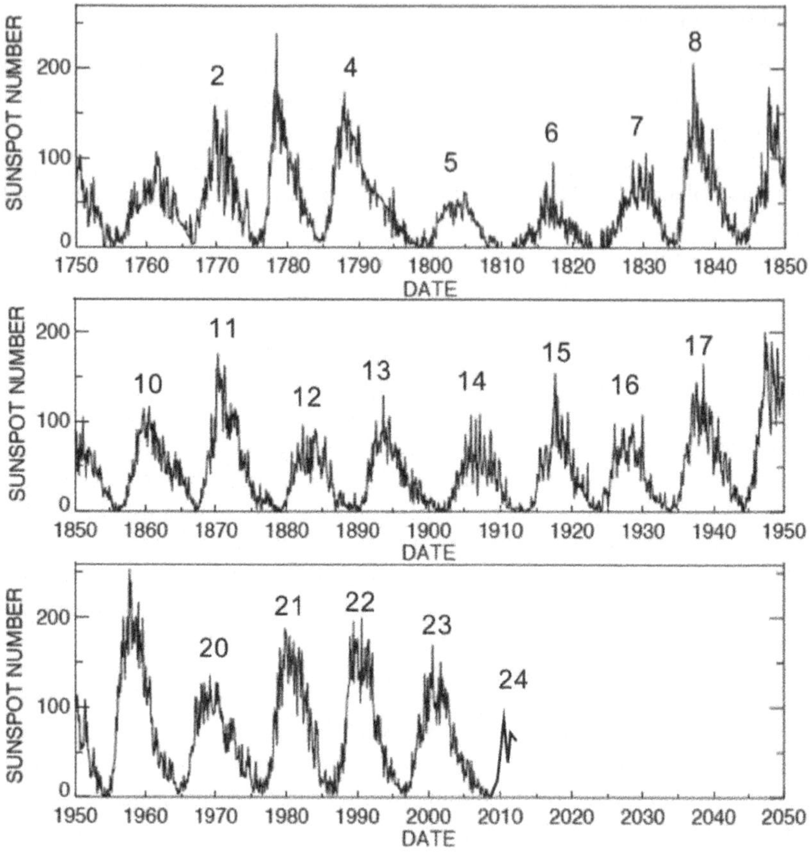

Figure 5.13. Historical record of sunspot activity (from a NASA site).

as in the continuum brightness of over 1,000 sunspot umbrae from 1990 to 2005. These measurements showed consistent trends in which the darkest parts of the sunspot umbra became warmer (45°C per year) and their magnetic field strengths have decreased (77 Gauss per year), independently of the normal 11-year sunspot cycle. They hypothesized that a linear extrapolation of these trends would suggest that relatively few sunspots will be visible after 2015.

Clilverd *et al.* (2006) used an analysis of cyclic variance of past sunspot activity to correctly predict low sunspot activity for Solar Cycle 24. Extrapolating further into the future, they predicted a period of quiet solar activity lasting until 2030. Their model also predicted a recovery during the middle of the 21st century to more typical solar activity cycles, with peak sunspot numbers around 120, followed by a return to low activity around 2100.

Archibald (2009) presented an interesting analysis of Solar Cycle 23 and its comparison with earlier sunspot cycles. Cycle 23 has been a very long cycle with low sunspot numbers at its tail end. As Figure 5.13 shows, Cycles 14–16 had relatively

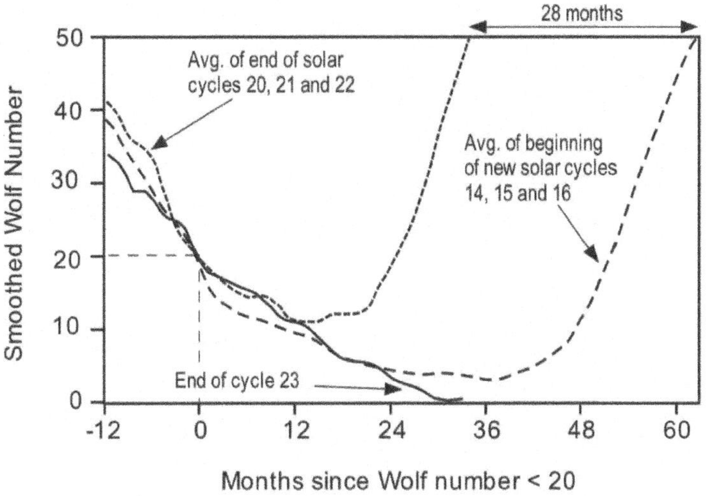

Figure 5.14. Comparison of end of Solar Cycle 23 with previous Cycles 20–22 and 14–16 (Archibald, 2009).

small peak sunspot numbers and long tails with low sunspot number at their endings, whereas Cycles 20–22 had high peaks and abbreviated tail ends. Figure 5.14 compares the endings of Cycles 20–22 with the beginnings of Cycles 14–16 (endings of Cycles 13–15). Cycles 20–22 did not bottom out at their endings, as did Cycles 13–15. What is interesting, though, is that Cycle 23 appears to have bottomed out with a lower, flatter shape than Cycles 13–15. Hence, Cycle 23 appears to be a progenitor of a period of reduced solar activity. The dearth of sunspots in 2008 was evident. Sunspot Cycle 24 has further proven to be relative low in sunspot count.

It is noteworthy that Cycles 5–7 represent the lowest sunspot activity since reliable records were kept. Figure 5.15 is based on a timescale measured from the previous cycle peak. On this basis, the tail end of Cycle 23 appears to be even weaker than Cycle 5, and at that time (2009) indicated that Cycle 24 could become a solar cycle with very low sunspot numbers. This is also illustrated in Figure 5.16, which suggests that cycles starting with Cycle 24 might resemble Cycles 5 and 6. This analysis by Archibald seems credible. The tail end of Cycle 23 does indeed suggest an imminent period of lower solar activity and so far, Cycle 24 is a relatively weak one.

It is worth mentioning that, back in 2004, Svalgaard *et al.* (2005) predicted that Solar Cycle 24 would be the "smallest in 100 years". Mackey (2007) reviewed a number of studies of solar cycles and predictions for Cycle 24. He suggested that, as a consequence of the low solar activity, there might be "thirty years of global cooling commencing in 2008 that would have adverse consequences for humanity". Schatten and Presnel (2007) predicted that Solar Cycle 24 would peak in 2013 at about 130 sunspots—about 65% of the maxima of Cycles 21 and 22. Choudhuri *et al.* (2007) predicted a peak for solar cycle less than half of that of Cycles 21 and 22, in the 2013 time frame. Bhatt *et al.* (2009) predicted a slightly weaker Solar Cycle 24 with a peak

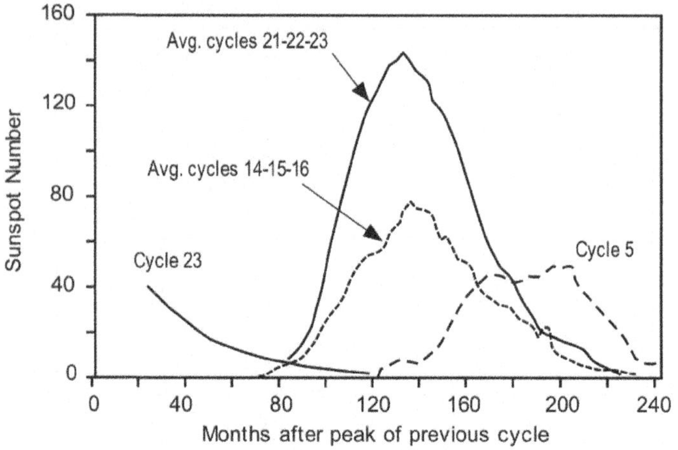

Figure 5.15. Comparison of end of Solar Cycle 23 with previous cycles (Archibald, 2009).

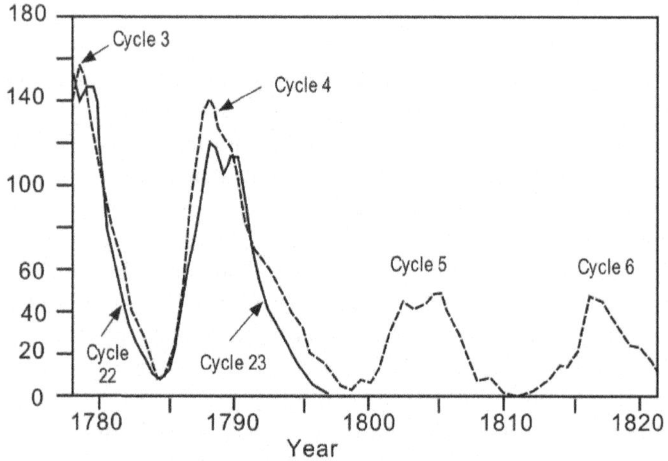

Figure 5.16. Comparison of end of Solar Cycles 22 and 23 with previous cycles (Archibald, 2009).

of 111 sunspots, compared with 120 for Solar Cycle 23. Others have predicted peaks as low as 50 sunspots (e.g., Badalyan *et al.*, 2001). Tsurutani *et al.* (2011) attributed the minimal sunspot activity to a combination of "exceptionally low solar (and thus low interplanetary) magnetic fields, ... the disappearance of equatorial and low latitude coronal holes and the appearance of mid-latitude coronal holes".

In an update released on May 8, 2009 a NOAA panel of experts predicted a peak of about 90 sunspots in May 2013 for Solar Cycle 24, which would make it the weakest cycle in a long time. If low solar activity is associated with enhanced cloud formation, this could suggest a cooling effect. However, NOAA cautioned: "Note,

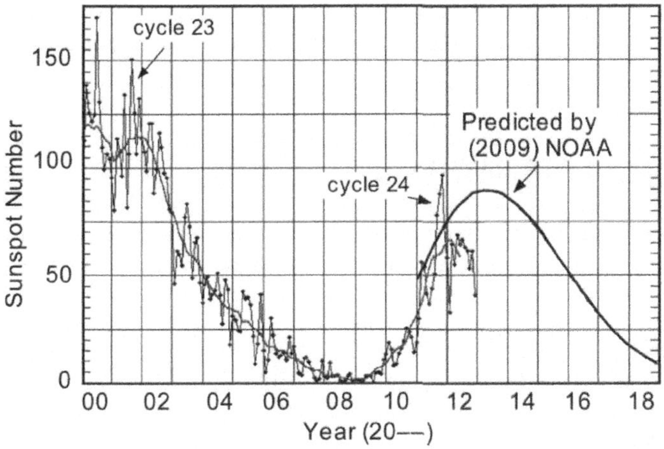

Figure 5.17. Measured sunspot numbers for Solar Cycle 23 and predicted sunspot numbers for Solar Cycle 24. Dashed line shows initial measurements of Cycle 24. Note the sharp upward jump at the beginning of the cycle.

this is a consensus opinion, not a unanimous decision. A super-majority of the panel did agree to this prediction." An update by NOAA as of January 2013 is shown in Figure 5.17.

By contrast, Dikpati *et al.* (2006) predicted: "that cycle 24 will have a 30–50% higher peak than cycle 23, in contrast to recent predictions by Svalgaard *et al.* and Schatten". It now appears that Dikpati *et al.* were way off the mark. Kane (2008) indicated that, depending on which method of correlation was used, the prediction for Solar Cycle 24 could vary over a wide range, but he seemed to predict that Solar Cycle 24 would be stronger than Solar Cycle 23, which is not occurring in reality. In December 2006, Hathaway and Wilson[2] were quoted with an untenable prediction as follows:

> "Evidence is mounting: the next solar cycle is going to be a big one. Solar cycle 24, due to peak in 2010 or 2011 'looks like its going to be one of the most intense cycles since record-keeping began almost 400 years ago' says solar physicist David Hathaway of the Marshall Space Flight Center. He and colleague Robert Wilson presented this conclusion last week at the American Geophysical Union meeting in San Francisco."

Typically, an alarmist website,[3] drew a graph showing the peak sunspot number rising to 170 in Solar Cycle 24 from 120 in Cycle 23.

Callebaut (2008) predicted a deep minimum for Solar Cycle 26 with a climate similar to that of the MM. His approach involves modeling of the progress of *global*

[2] *www.physorg.com/news86010302.html.*

[3] *http://astroblogger.blogspot.com/2007/03/sunspots-and-global-warming-oh-my.html.*

unipolar magnetic field regions (GMRs) since Solar Cycle 12, but is complex and beyond the scope of this book. He suggested that global temperatures during Solar Cycle 26 might drop by 1°C.

5.5 SOLAR CYCLE DURATION

Archibald asserted that there is a correlation between solar cycle length and surface air temperatures during the following cycle. This is attributed to the theory as stated by Archibald:

> "... weak solar activity causes a weak solar wind, which in turn increases the number of galactic cosmic rays penetrating the Earth's atmosphere. This increases low-level cloud formation and the Earth's albedo. The Earth cools as a consequence."

The evidence for this claim is based on data such as those shown in Figure 5.18. While this figure is suggestive, it is by no means convincing and the jury remains out in regard to judging this hypothesis. While Archibald (and other solar scientists) tend to attribute most climate variations to solar activity, it is difficult to separate out solar effects from climate variability due to other potential causes (oceans, winds, black carbon, greenhouse gases, etc.).

A number of other studies have been conducted on variability of solar cycle length. It is noteworthy that solar cycle length can be determined in a number of different ways, based on maxima or minima, and some subjectivity is always involved. Thejll and Lassen (2000) presented Figure 5.19, which appeared to suggest a correlation between solar cycle length and Northern Hemisphere (NH) temperature. Agee *et al.* (2010) provided Figure 5.20. They also showed that the

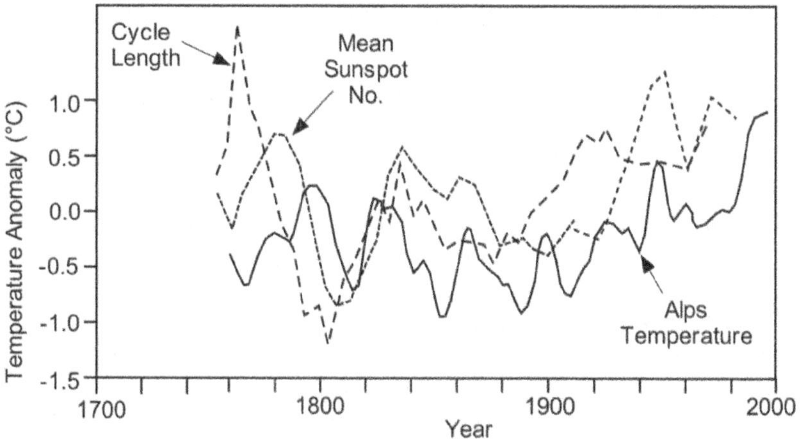

Figure 5.18. Comparison of Alps temperatures with sunspot number and solar cycle length (Brunetti, 2003).

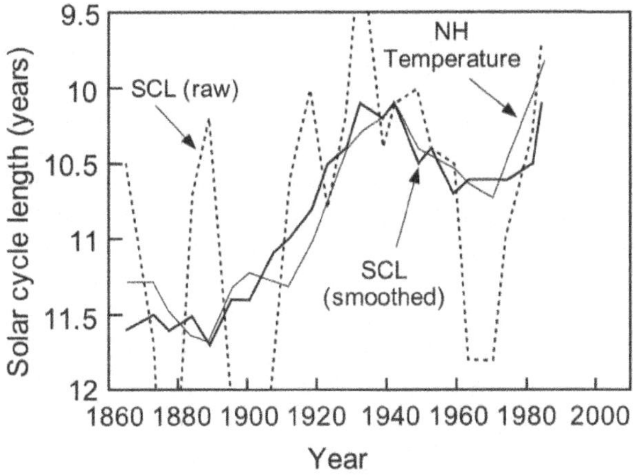

Figure 5.19. Variability of solar cycle length since 1860 based on minima (adapted from Thejll and Lassen, 2000).

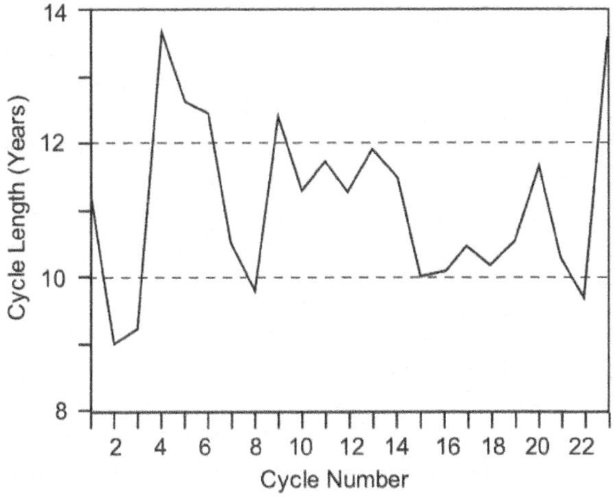

Figure 5.20. Length of "11-year" sunspot cycle from Cycle 1–23 (March 1755 to November 2009) (Agee *et al.*, 2010).

length of the solar cycle is closely related to the length of the "quiet period" defined as "the period from the first monthly sunspot number average of 10 to the last monthly sunspot number average of 10". However, Thejll (2009) provided Figure 5.21, which does not correlate at all with temperature. Rogers *et al.* (2006) claimed to have identified a 188-year cycle for solar changes in solar cycle length, with a minimum just prior to year 2000.

A number of studies attempted to estimate in advance the length of Solar Cycles 23, 24, and even as far out as Solar Cycle 29 based on various models of solar activity, as well as statistical correlations and extrapolations.

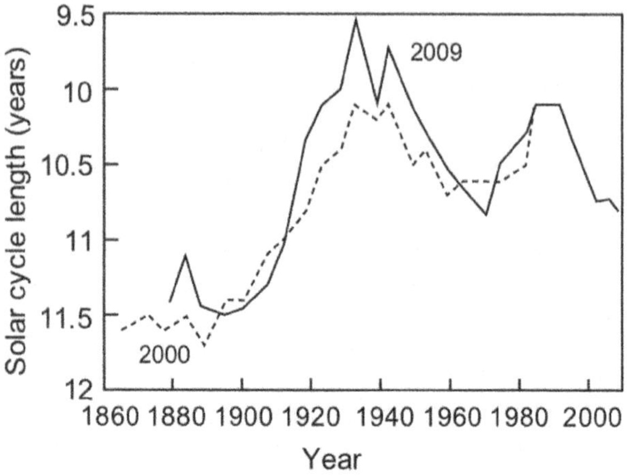

Figure 5.21. Comparison of estimates of smoothed solar cycle duration from Thejll and Lassen (2000) and Thejll (2009).

Feulner and Rahmstorf (2010) pointed out:

> "The current exceptionally long minimum of solar activity has led to the suggestion that the Sun might experience a new grand minimum in the next decades, a prolonged period of low activity similar to the Maunder minimum in the late 17th century."

They indicated that the "grand minimum of solar activity" that occurred in the MM might possibly occur again in the 21st century, and they investigated the climatic effects of such a hypothetical minimum using a climate model. However, a chain is only as strong as its weakest link, and the great unknown here is how the TSI might vary during an extended period of very low solar activity. Later in this chapter, we discuss the various reconstructions of historical TSI and we demonstrate that none of them is very trustworthy. The truth is that we simply do not know how TSI varies over centuries. Rahmstorf chose a reconstruction of TSI that leads to low solar forcing. Thus, it is no surprise that he ended up with the conclusion that heating from increased greenhouse gases will be much greater than cooling from reduced TSI.

Schrijver *et al.* (2011) asserted that there is a minimum state of solar magnetic activity that persists even when sunspots are absent. Hence, they argued that attempts to scale TSI to sunspot indices overestimate the reduction of TSI when sunspot indices are minimal. They pointed out that the period of low solar activity at the end of Solar Cycle 23 was not accompanied by a significant drop in TSI, and hence they argued that TSI is not much affected by variability of sunspot indices. They said: "If the 2008–2009 solar magnetic activity is indeed similar to the Maunder Minimum level as we argue here, then it would appear that drivers other than TSI dominate Earth's long-term climate change." This certainly was true for the brief period 2008–2009 but it is not clear how TSI will vary over a prolonged period of low solar activity, and we cannot be certain of the relationship between 2008–2009 and the MM.

5.6 RECONSTRUCTING TSI IN THE PAST

5.6.1 Reconstructions based on sunspots, solar cycles, and solar activity

5.6.1.1 Introduction

The observed TSI record was shown in Figure 5.2. The variation in TSI between S_{MAX} and S_{MIN} was roughly 0.14% over this ~30-year period. Over this time interval, the sunspot numbers ranged from well over 100 at sunspot maximum to fewer than 10 at sunspot minimum. Any model that relates TSI solely to sunspot number based on these data will inevitably conclude that the lowest value that the TSI can ever reach (or has ever reached in the past) is when the sunspot number goes to zero, in which case the TSI would be close to what it presently is at S_{MIN}, or about 0.14% less than it presently is at S_{MAX}. That being the case, it would be concluded that at no time in history has the TSI ever been less than it presently is at S_{MIN}, about 1,365.3 W/m^2.[4] These models are referred to by Hoyt and Schatten (1993) as "constant quiet Sun models" (CQSMs) because the TSI for a quiet Sun is taken as constant for all time. Several models have been developed on this basis. Some have been used to predict the daily variations shown in Figure 5.2 over a couple of decades. With enough empirical parameters, such predictions have been quite good. For example, Wenzler, Solanki, and Krivova (2005) modeled TSI for 2,055 days between November 1992 and September 2003 based on solar observations. Daily variations of TSI were well reproduced. The model used in these estimates was based on the measured position and flux density of magnetic features on the solar disk extracted from full-disk magnetograms. These magnetic features are classified as either sunspot umbrae, sunspot penumbrae, or faculae, based on their brightness in full-disk continuum images. The details on how this was done do not seem to have been revealed. Such models are interesting, but they do not seem to lead to estimates of TSI in the distant past prior to measurements of solar magnetograms. Other references reporting short-term modeling include Fligge *et al.* (2000) and Solanki and Krivova (2004).

As we have seen above, models based on current observations of TSI correlated with surface features (sunspot number, faculae, etc.) can reproduce the daily fluctuations in TSI observed by instruments such as ACRIM3. Unfortunately, as Reid (1997) amply demonstrated, such small variations in TSI are too small to account for historical variations in surface temperature. Any CQSM that is based

[4] Note that this was written prior to more recent analyses that suggest that TSI at S_{MIN} is closer to 1,361 W/m^2 as shown in Figure 5.2. Many graphs in the remainder of this chapter may need to be scaled to the lower estimate of TSI.

solely on the current solar cycle with an ultimate minimum TSI of 1,365.3 W/m^2 cannot possibly explain past climate variations.[5]

In addition, there is a very large variation in TSI that occurs yearly due to the Earth's elliptical orbit. This is usually neglected, based on the assumption that the Earth's thermal response is very slow compared with a yearly variation in TSI. However, the variation of 6.8% between January 1 and July 1 is a major perturbation and may not be entirely negligible.

5.6.1.2 Overview of Hoyt and Schatten

This section was excerpted from Hoyt and Schatten (1993).

TSI measurements in orbit have only existed since around 1978. No data are available prior to that date. Since the TSI measurements were made in Earth orbit, a number of investigators have developed relationships that allow one to use observations of sunspots and faculae to predict TSI.

Over the past two centuries, a number of scientists have hypothesized that the solar irradiance at the top of the Earth's atmosphere varies with time. Recent satellite measurements confirm that small variations exist at least on the timescale of the 11-year solar cycle. Most of the modeling of these secular variations in the solar "constant" was phenomenological, providing parameters that enable short-term TSI variations to be fit. Although the models do not answer questions concerning the basic cause of secular solar variations, they do allow us to examine the photospheric manifestations of these variations. To date, most of the solar constant secular variations observed (over a couple of decades) have been associated with photospheric blemishes (dark sunspots, bright faculae, and a bright network). At present, there seem to be only a few attempts to understand potential long-term secular trends. One view of active region physics suggests a positive correlation of the variations in the solar "constant" with solar activity if the active region process effectively transfers heat outward. In this view, the active regions are distinguished from the background photosphere by the influence of the magnetic field that produces down-flows at sunspots and up-flows at faculae. These effects were thought to be the origin of the positive correlation of solar activity with solar irradiance variations.

Even if we can understand the correlation of solar cycle with activity, these features may be merely photospheric blemishes and may not have a great influence on the longer-timescale "river of solar luminosity" flowing outward from the Sun's interior, but rather primarily serve only to divert the flow and/or temporarily store and release minor amounts of this vast energy flow. The perturbations from active surface regions may not necessarily extend to the deep interior to influence very-long-timescale solar luminosity and solar "constant" variations. To understand the long-term secular variations, the Sun might need to be viewed on a larger, more synoptic basis, with global observations (e.g., solar rotation, solar diameter, etc.).

On a timescale of decades to centuries, four classes of models were described by

[5] As stated previously, this value may be high and probably needs to be revised downward to about 1,361 W/m^2.

Hoyt and Schatten (1993) that postulate different variations of the Sun's output. These models can be called:

(1) The *constant quiet Sun model* (CQSM) postulates that the solar irradiance has only an 11-year cycle and all radiation changes can be explained by the presence or absence of active features. Since all solar minima are essentially the same in these models, it is called the constant quiet Sun model. In this model, the historical TSI when solar activity was at a minimum (as, for example, in the MM) would be set equal to the contemporary TSI at S_{MIN}. Examples of such models include Tapping *et al.* (2006), Foukal and Lean (1990), and Vaquero *et al.* (2006).
(2) The *solar diameter model* uses the solar diameter or its time rate of change as a proxy for solar irradiance variations. However, some controversy still exists about the history of solar diameter variations (Foukal *et al.*, 2006).
(3) The *activity envelope model* postulates that long-term solar irradiance variations follow the envelope of solar activity such as the so-called Gleissberg cycle (an 88-year overall modulation of the 11-year solar cycle), so that solar minima irradiances vary over time according to such a cycle.
(4) The *umbra/penumbra (U/P) variations model* is so-called because early models of this class used sunspot structure expressed as the ratio of umbral areas to penumbral areas as a proxy measure of solar irradiance. Subsequent studies have used solar equatorial rotation rate and sunspot cycle duration to derive similar models. The U/P variations model and the activity envelope model are similar except that they are out of phase with each other with variations occurring 20 years earlier in the U/P variations model. Hoyt and Schatten (1993) argued that the solar indices used in the U/P variations model are proxy indicators of long-term secular changes in convective energy transport. Additional proxies for the U/P variations model have been introduced, namely, the sunspot decay rate, the fraction of penumbral sunspots, the decay rate of the solar cycle, and the mean level of solar activity.

In addition to these models mentioned by Hoyt and Schatten in 1993, several models were developed in ensuing years. These include the following (named by this writer):

(1) The *MM temperature model* is based on (1) an estimate of the temperature lowering during the MM compared with today's temperatures, (2) an estimate of the reduction in TSI needed during the MM to produce that lowering of temperature, and (3) a linear scaling of the change in temperature from the MM to current times, with the change in sunspot number at S_{MAX} from the MM (zero) to today (over 100).
(2) The *stellar Ca HK index model* is based on (1) an estimate of the enhanced level of TSI vs. the solar Ca HK index model from recent measurements, (2) observation of the Ca HK index for non-cycling Sun-like stars, (3) the assumption that non-cycling Sun-like stars are representative of the Sun during the MM, and (4) linear extrapolation of the dependence of TSI from current

solar measurements to the expected Ca HK index for non-cycling Sun-like stars in order to estimate TSI during the MM.

(3) The *solar cycle duration model* utilizes the duration of the solar cycle from peak to peak rather than sunspot numbers as a measure of TSI.

(4) The *coronal source flux model* is based on correlations of coronal source flux with TSI. The coronal source flux, F_s, is the total magnetic flux leaving the Sun, and thereby entering the heliosphere.

5.6.2 Constant quiet Sun models

5.6.2.1 CQSM based on sunspot number

The period between 1700 and the present is important because there is a continuous record of sunspot number, which is a directly measured index of solar activity, of known pedigree with established relationships with other activity indices, and which antedates the rapid increase in anthropogenic greenhouse gases that began with the industrial revolution. A vital aspect of climatology studies is the issue of how the TSI varied over the past several centuries, particularly back to the MM (1645–1715). This will require some form of "bootstrap" model that associates recent measurements of TSI with parameters that have histories that go back in time. From past records of these parameters, the TSI in the past can be inferred. However, there is no way to verify the assumption that recent dependence of TSI on sunspot number held 400 years ago.

Group sunspot numbers of variable quality are available since 1610. Sunspot areas and white-light facular areas are available since 1874. Ca II plage areas are available since 1915. Various models use one or more of these proxies in order to reconstruct historical irradiance. Observations of sunspots and faculae go back to about 1954, and Lean and Foukal (1988) utilized these data to estimate TSI for the period 1954–1985. However, this procedure cannot estimate TSI over several centuries. There are a variety of proxies from which TSI might be inferred over long time periods (tree rings, ice cores, etc.) and these will be discussed in later sections. One parameter that has been used by a number of investigators to estimate TSI as far back as the 17th century is the group sunspot number. It is rather unfortunate that many of these papers have provided the end result: plots of TSI vs. year for hundreds of years, but they have not usually provided the detailed algorithms from which these data were derived. Since some of these are essentially one-parameter models, there must exist a functionality:

$$\text{TSI} = \text{function of } (N),$$

where N = group sunspot number. This functionality does not seem to appear in most of the relevant papers. In some cases, the models use long-term sunspot data and other proxies, in which case the functionality would take the form:

$$\text{TSI} = \text{function of } (N, a, b, \ldots),$$

where a, b, ... are parameters associated with proxies, but once again these functionalities are not typically stated very clearly:

"The history of solar irradiance variation is a critical component in understanding the solar–terrestrial climate connection and the relative role of the Sun in current climate change. However, direct measurements of solar irradiance currently cover only about three decades. Beyond that interval irradiance has to be estimated using available observations and activity indices. This entails three major difficulties: (a) the physical connection between the observed activity phenomena, such as sunspot number with irradiance is complex and difficult to quantify. Often the result is the need to use connections that are often largely empirical, (b) proxies might have to be used. In a sense this has some commonality with (a), except that here the physical connection is even less well understood, but a historical high correlation between the proxy and the desired quantity justifies its use, (c) having constructed a model which necessarily incorporates elements of (a) and (b), it has to be extrapolated substantially outside the parameter space that was used to set up that model." (Tapping *et al.*, 2006).

Tapping *et al.* (2006) developed a model for long-term TSI based on group sunspot number:

"Modeling irradiance is difficult. Firstly we do not fully understand the processes driving irradiance variations; the underlying physics is complex and multifaceted, and includes phenomena below and at the photosphere. Most of the relationships we have to work with are empirical, although the correlation coefficients between total irradiance and indices such as the 10.7 cm solar radio flux are high. For example, since total irradiance is highly correlated with sunspot number, it seems logical to plot irradiance against sunspot number and extrapolate back to zero sunspot number, and then conclude that the corresponding value of irradiance is the value that would be reached if solar activity remained low for an extended period. *This is almost certainly not the case*. Sunspots do not cause increases in irradiance; it is the accompanying active region structures, such as faculae and elements of the active network that do this. Although there might not be any sunspots, there are signs of activity during every observed minimum of the solar activity cycle. ... Sunspots might be a good indicator of magnetic activity when present, but they are not useful when activity is low. When examining solar activity during a sustained change in the solar activity cycle, or even a temporary cessation, one needs to examine two issues: firstly, does the nature of the process by which magnetic flux is processed change, and secondly, what is the solar activity machine below the photosphere doing? ... In the case of this investigation, the input to the model has to be sunspot number, which is the only direct index of solar activity available. In this paper we develop a model for the processing of solar magnetic flux and use it to model the historical record of total irradiance." (Tapping *et al.*, 2006).

Tapping *et al.* (2006) started out with a set of group sunspot numbers very nearly the same as that given in Figure 5.7. Then they carried through an involved

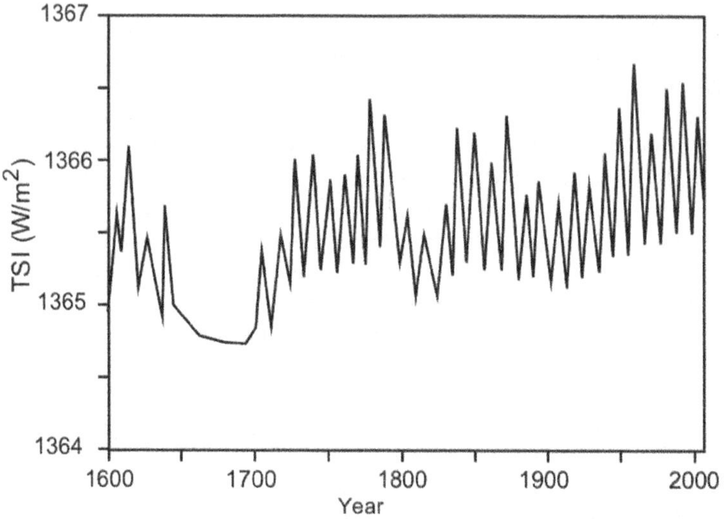

Figure 5.22. Modeled TSI through the MM up to the present (adapted from Tapping *et al.*, 2006).

procedure that is difficult to follow in detail, in which they estimated the TSI indirectly from the historical record of sunspot numbers. But, in the end, since their procedure involved only the single parameter of sunspot number as given in a chart similar to Figure 5.7, they must have implicitly developed a relationship of the form

$$\text{TSI} = \text{function of } (N),$$

although this functionality was not explicitly stated in their paper. However, we can surmise what that relationship must have been by comparing their estimate for TSI (as shown in Figure 5.22) with the sunspot record (as given in Figure 5.7). At each peak in the TSI plot, we note the peak TSI and the corresponding peak sunspot number. We then plot peak TSI vs. peak sunspot number as shown in Figure 5.23. We can also prepare a similar plot based on the sunspot minima as shown in Figure 5.24. Although Tapping *et al.* (2006) never actually revealed these functional relationships between TSI and sunspot number, they must have implicitly used relationships similar to those implied in Figures 5.23 and 5.24.

According to Figure 5.22, the best estimate for the TSI during the MM was about 1,364.7 W/m^2, compared with a current average of about 1,366 W/m^2, for a decrease of about 1.3 W/m^2 (about 0.1%) during the MM.[6]

[6] As stated previously, these values may be high and probably need to be scaled downward to about 1,361 W/m^2.

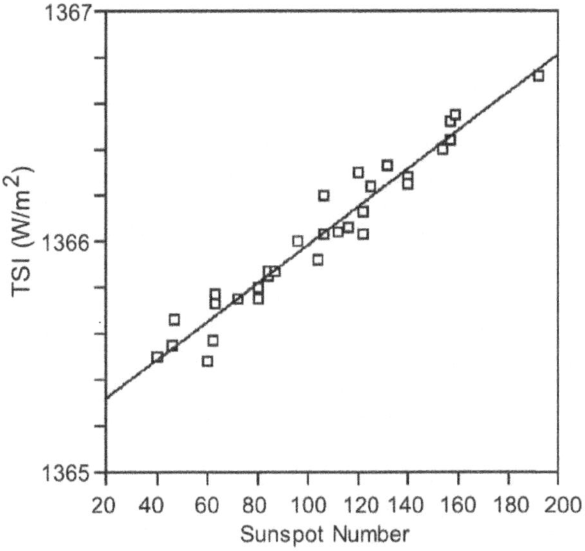

Figure 5.23. Relationship between TSI and sunspot number at the periodic peaks in the historical record of sunspots (S = 1,365.16 + 0.008 SSN).

The interesting thing about this model is that although Tapping *et al.* (2006) went to great pains to emphasize that:

> "... since total irradiance is highly correlated with sunspot number, it seems logical to plot irradiance against sunspot number and extrapolate back to zero sunspot number, and then conclude that the corresponding value of irradiance is the value that would be reached if solar activity remained low for an extended period. *This is almost certainly not the case.*"

Nevertheless, they went ahead and carried out a CQSM model anyway, leading inevitably to the result that TSI during the MM was a mere 0.1% lower than it is today—*which, as they said, " is almost certainly not the case".*

Foukal and Lean (1990) developed models for predicting TSI from data on sunspots and faculae. In the short term where good data on sunspots and faculae are available, the model appears to work very well. However, to estimate TSI at much earlier dates, they had to rely on sunspot numbers alone. It was noted that, while sunspots reduce the TSI, sunspot occurrence is always accompanied by the presence of bright faculae, and the net result is that TSI increases with increasing sunspot number.[7] For the few cycles examined by the ACRIM instrument, they were able to develop a relationship between TSI and sunspot number. However, this model relied on the assumption that the lowest TSI that is possible is the lowest TSI recorded by

[7] However, it is believed that, for very high sunspot numbers (150), the TSI might top out and eventually decrease with increasing sunspot number.

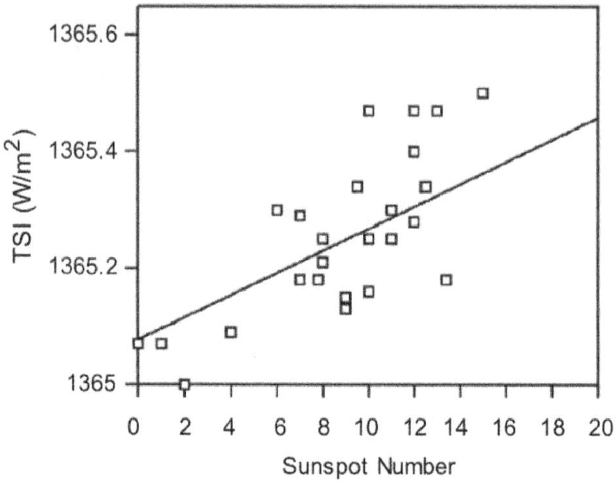

Figure 5.24. Relationship between TSI and sunspot number at the periodic minima in the historical record of sunspots (S = 1,365.08 + 0.019 SSN).

ACRIM for a quiet Sun at S_{MIN} (i.e., a CQSM). With an active Sun, the TSI could increase above this value in proportion to the sunspot number. At the time they wrote their paper, it was estimated that TSI for a quiet Sun was about 1,366.8 W/m^2, so they assumed that TSI never drops below this value. However, Foukal and Lean (1990) realized the futility of this approach:

> "... solar irradiance variations of the magnitude below 0.1% expected from photospheric activity are unlikely to have had significant influence on climate over the past century. ... Much larger variations in solar luminosity could be caused by global solar changes not associated with photospheric sunspots and faculae. A secular variation approaching even 1% over the past century cannot strictly be ruled out, given the low precision of pyrheliometry, until modern data became available from satellites beginning in 1978. The theory of the solar radiative interior and convection zone is also insufficiently accurate to rule out luminosity variations of such amplitude." (Foukal and Lean, 1990).

Krivova *et al.* (2009) reported on their SATIRE model—a CQSM that has been under development for a number of years. As they pointed out: "The assumption underlying SATIRE is that all changes in the solar irradiance on time scales longer than hours are solely due to changes in the solar surface magnetic flux as traced through surface features, such as spots and faculae." This has been shown to work quite well for the time period of the last few decades where TSI data are available. It has been estimated, for example, that in Solar Cycle 23, the TSI variation from S_{MIN} to S_{MAX} was estimated to be a darkening of ~0.8 W/m^2 due to sunspots and a brightening of ~1.7 W/m^2 due to faculae. Their model is pegged to sunspot data, and the underlying assumptions lead to a CQSM. Their results indicate that the TSI reached a minimum of 1,364.8 W/m^2 during the MM, and then rose over the following centuries to about 1,366 W/m^2.[8]

5.6.2.2 CQSM based on sunspot area and cycle duration

Vaquero *et al.* (2006) developed a correlation of TSI with yearly average sunspot area based on measurements of sunspot area made since 1832. However, the units of sunspot area were not specified. Vaquero *et al.* (2006) based their paper on previous work by Solanki and Fligge (1998, 1999).

The fundamental assumption made in Vaquero *et al.* (2006) and Solanki and Fligge (1998, 1999) is that the variations in TSI that we currently observe during the solar cycle are due to changes in sunspots and faculae in active regions of the Sun. Their combined effect on TSI is embodied in a term $\Delta(AS)$ representing the change in TSI due to such solar activity. Based on observations of the Sun since 1978, this term is typically quite small. The network (and possible changes in solar convection) provide the main contribution to TSI variations on timescales longer than the solar cycle, and secular changes in the network are denoted as variations in the quiet Sun $\Delta(QS)$. However, these long-term variations are presumably much larger than those due to solar activity, and therefore use of the term "quiet Sun" is misleading.

Thus, according to this concept, TSI at any epoch is the sum of three terms:

$$\mathrm{TSI} = \mathrm{TSI}(0) + \Delta(QS) + \Delta(AS),$$

where

$\mathrm{TSI}(0)$ = a constant to produce the correct absolute TSI.
$\Delta(QS)$ = additive term to account for long-term secular variations of the so-called "quiet Sun".
$\Delta(AS)$ = additive term to account for solar activity via the solar cycle.

Since $\Delta(QS)$ presumably varies slowly with time, measurements of TSI made in space since ~1980 over two decades can be assumed to include a current value of $\Delta(QS)$ corresponding to 1980–2000. The active Sun term $\Delta(AS)$ can be correlated with a parameter such as sunspot number or sunspot area by comparing two decades of TSI measurements with variations in such parameters. Solanki and Fligge (1999) made such a comparison with the daily sunspot number. Although the data were very noisy, they were able to arrive at a correlation by binning the data. The result was:

$$\mathrm{TSI}(0) = 1{,}365.4\ \mathrm{W/m^2}$$
$$\Delta(AS) = 0.0161\,(S_N) - 0.000055\,(S_N)^2$$

where S_N = Zurich sunspot number.

This function increases with S_N until about $S_N \sim 150$, and decreases for higher S_N. For low S_N, as S_N increases, the facular area increases faster than the sunspot area. However, for high S_N, the reverse occurs.

[8] As stated previously, these values may be high and probably need to be scaled downward to about 1,361 W/m^2.

Vaquero *et al.* (2006) performed a similar correlation based on sunspot area. They obtained:

$$\mathrm{TSI}(0) = 1{,}365.4\ \mathrm{W/m^2}$$
$$\Delta(AS) = 6.8 \times 10^{-4}\,(S_A) - 1.0 \times 10^{-7}\,(S_A)^2$$

where S_A = sunspot area (units not given).

In either case, whether based on sunspot number or area, $\Delta(AS)$ contributes a relatively small oscillatory term to TSI that varies with the solar cycle.

The key to evaluating TSI is estimation of the term $\Delta(QS)$. However, as Solanki and Fligge (1999) said:

> "We stress, however, that determining the quantitative long-term variations of the quiet Sun is highly speculative and subject to large uncertainties."

Solanki and Fligge (1998) generated two estimates for $\Delta(QS)$. Unfortunately, the descriptions of the procedures for doing this are rather murky to this writer. One procedure was based on an assumed linear relationship between chromospheric emission in the core of the Ca H and K lines and photospheric brightness, as will be discussed later in connection with Figure 5.32 (Lean *et al.*, 1998). The other is a linear correlation between the length of solar cycles and observed brightness of solar-type stars. Each of these correlations was apparently used independently to estimate past values of $\Delta(QS)$ from observations of either the Sun's cycle duration or HK measurements. Since such measurements are limited to post 1880, Solanki and Fligge (1998) only reconstructed TSI back to 1880. The results are shown in Figure 5.25.

Solanki and Fligge (1999) followed their 1998 paper with the intent to reconstruct TSI as far back in time as 1700. As before, Solanki and Fligge (1999) are equally obscure on the details of reconstructing $\Delta(QS)$. They mention Sun-like stars but it is not clear how stars relate to the procedure, which seems to involve either sunspot numbers or solar cycle duration. The procedure remains murky to this writer.

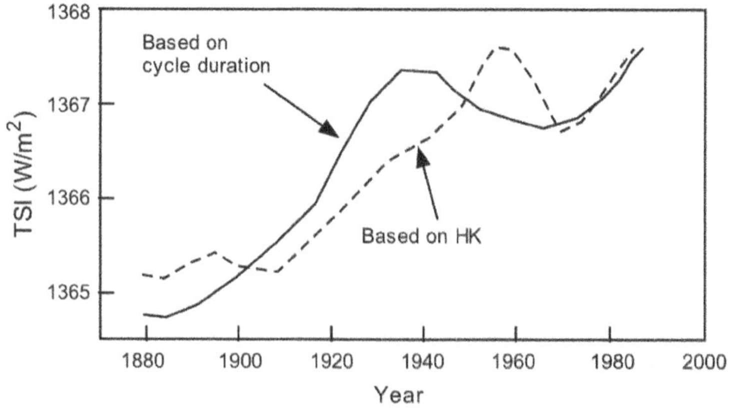

Figure 5.25. Reconstructed TSI since 1880 (adapted from Solanki and Fligge, 1998).

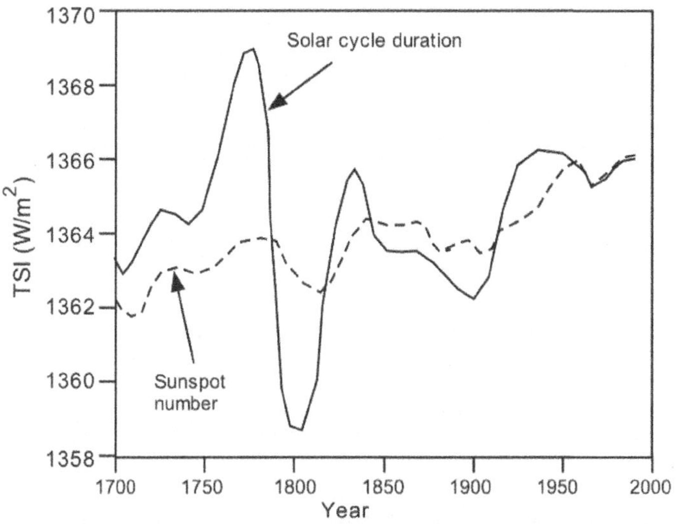

Figure 5.26. Reconstructed TSI based on sunspot number or cycle duration (adapted from Solanki and Fligge, 1999).

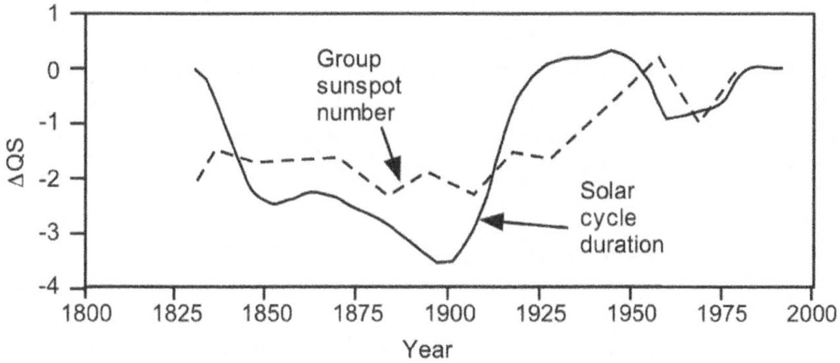

Figure 5.27. Estimates of Δ(*QS*) (W/m^2) based on sunspot number or cycle duration (adapted from Vaquero *et al.*, 2006).

The result is shown in Figure 5.26. The huge peak around 1770 and the dip around 1800 appear to be anomalous.

Vaquero *et al.* (2006) prepared a new estimate for Δ(*AS*) based on sunspot area, and combined this with the two estimates for Δ(*QS*) from Solanki and Fligge (1999). Vaquero *et al.* (2006) provided the Δ(*QS*) data for Figure 5.27, but these data were obtained from Solanki and Fligge (1999). The reconstructed TSI in Vaquero *et al.* (2006) is shown in Figure 5.28.

It is difficult to appraise the results presented in Figures 5.26–5.28 because the descriptions of how Δ(*QS*) was derived are so obscure. One obvious fact is that the correlations based on cycle duration lead to wider swings in TSI than correlations

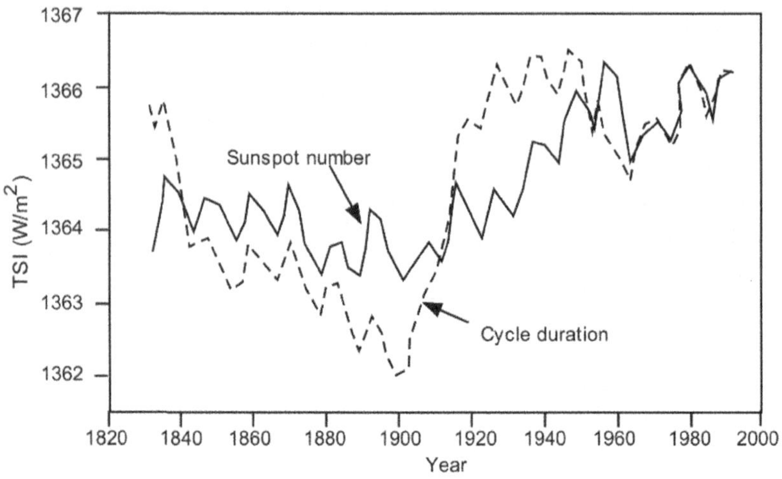

Figure 5.28. Reconstruction of TSI using two models for the "quiet Sun contribution" (aapted from Vaquero et al., 2006).

based on sunspot numbers. However, it seems likely that, although solar cycle duration varies significantly from cycle to cycle, this may not necessarily produce dramatic changes in TSI. The large variations in TSI obtained from the method based on cycle duration are likely to be artifacts. Nevertheless, as Archibald (2009) emphasized, the changes in solar activity associated with changes in cycle duration might affect cloud formation by modulating the penetration of galactic cosmic rays into the atmosphere, and that would affect climate.

5.6.3 The MM temperature model

Reid (1997) began by asserting that:

> "Any attempt to reconstruct the historical variation in TSI and to relate it to terrestrial climate variations is bound to be a highly speculative activity. Nevertheless, striking correlations between global-scale variations in temperature and solar-activity parameters suggest that such a relationship does exist."

Reid delineated two extreme scenarios that have been put forward in the past to reconstruct TSI variations. He referred to Foukal and Lean (1990) as an example of a CQSM model that assumed that the variation in irradiance is strictly linked to the Sun's surface activity, so that the 0.1% variation seen so far (during recent solar cycles) is close to the maximum that ever occurs, being representative of the change from sunspot maximum, when surface activity is high, to sunspot minimum, when it is virtually absent. The record of surface magnetic activity can then be used to reconstruct a history of irradiance variability that consists solely of 11-year cycles, with amplitude somewhat less than that observed during the recent period, since solar cycles have been generally growing in amplitude over recent decades. However, Reid argued that the inferred irradiance variations from such a CQSM model are too small to be climatically significant. At the other extreme, Reid quoted an earlier 1991 paper of his own, in which the 11-year cycles were smoothed out completely, and only the longer period variations in solar activity were retained. Reid's 1991 paper suggested that irradiance variations of several tenths of a percent may have occurred over the past several hundred years, and he estimated that the TSI during the MM period in the late 17th and early 18th centuries was about 0.65% below the present level. This is a considerably larger variance than other independent estimates of the MM TSI.

The approach taken by Reid (1997) was based on the assumption (inferred from historical records) that the prevailing temperature during the MM was about 1°C lower than it is currently, and this was assumed to be due to a weaker Sun. (It should be noted that climatologists favoring the *hockey stick* model would argue that this is an overestimate.) Reid (1997) used climate models to infer how much lower the TSI must have been during the MM to account for this putative 1°C drop. Then, given the known current levels of TSI and sunspot number during the MM, he was able to derive a relationship between long-term TSI and group sunspot number, assuming a linear relation. It is of the highest importance to distinguish between two very different models that might appear at first glance as if they were the same. One can analyze the current TSI data (as, for example, in Figure 5.2) in terms of a CQSM formula

$$\mathrm{TSI} = A + B(GSN),$$

where A and B are constants and GSN is the group sunspot number. This can be made to work for the period 1978–2007. The constant A is close to the minima in Figure 5.2 and the constant B is adjusted to make TSI approach the maxima in Figure 5.2 at S_{MAX}. This CQSM formula can never produce a variation in TSI greater than about 0.1%.

Reid (1997) discarded the solar cycle altogether and dealt with the average TSI vs. year over long time periods. He did this by filtering the sunspot data to acquire a 15-year moving average (see Figure 5.29). His climate model indicated that, in order to achieve a 1°C temperature drop during the MM compared with recent times, the TSI must have been about 0.65% lower in the MM than it was in 1980. He used an excessively high value of the 1980 TSI (1,372.6 W/m^2) based on early Nimbus-7 measurements, and based on the 0.65% reduction concluded that TSI during the MM was about 1,363.7 W/m^2 (i.e., about 8.9 W/m^2 lower than in 1980). Then he set

$$\mathrm{TSI} = 1{,}363.7 + 0.114(FSSN),$$

where $FSSN$ is the 15-year filtered group sunspot number. This formula may appear similar to the one above (in terms of A and B) but it is fundamentally different.

We adjusted his model because the Nimbus-7 measurements are too high as evidenced by Figure 5.1. Hence, we used a modified version of Reid's formula based on a current average TSI of 1,366.5 W/m^2:[9]

[9] As stated previously, this value probably now needs to be revised downward to about 1,361 W/m^2.

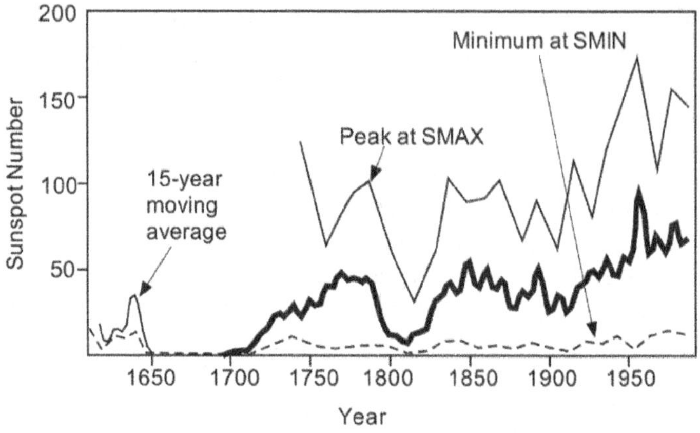

Figure 5.29. Reid's 15-year moving average of group sunspot variations (heavy line) (adapted from Reid, 1997).

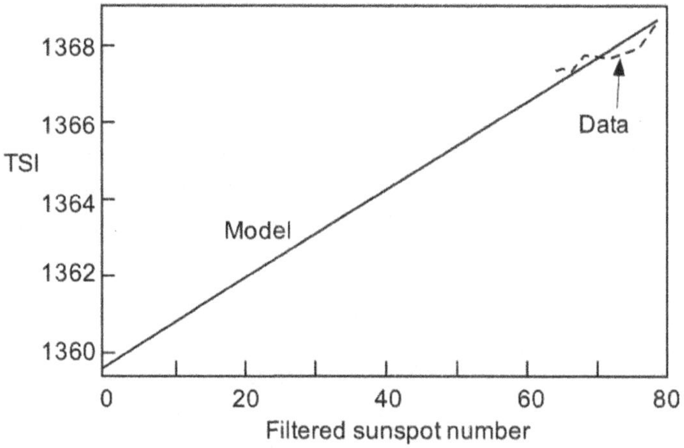

Figure 5.30. Dependence of long-term TSI on 15-year filtered sunspot number (adapted from Reid, 1997).

$$\text{TSI} = 1{,}357.6 + 0.114(FSSN),$$

This formula yields 1,366.5 W/m^2 when FSSN = 78 (for 1980). A plot of this formula is shown in Figure 5.30.

A modified version of Reid's estimated TSI vs. year is shown in Figure 5.31.

In his model, Reid included greenhouse warming for 1980, but not for the MM. This would imply that solar forcing accounted for only a part of the 1°C temperature rise from the MM to 1980. Reid estimated that greenhouse forcing in 1980 was about 1.5 W/m^2 due to CO_2, and another 1.2 W/m^2 due to other greenhouse gases for a total of 2.7 W/m^2. According to Figure 5.31, the increase in TSI since the MM was about 8 W/m^2, which would imply that solar forcing was 8/4 = 2 W/m^2 so that he

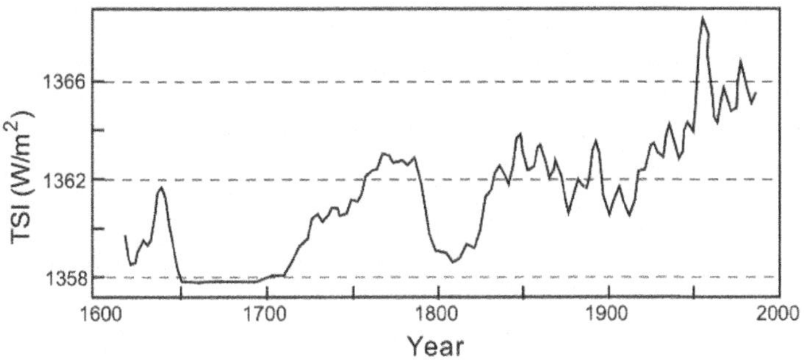

Figure 5.31. Modification of Reid's estimate of variation of TSI over four centuries using updated TSI (Reid, 1997).

estimated that, with solar forcing alone, the rise in temperature from the MM to 1980 would be $\sim$0.45°C.

Reid compared his results with the study by the 1995 Inter-governmental Panel on Climate Change (IPCC) Report. Reid pointed out that the 1995 IPCC Report assumed much smaller differences in TSI between the MM and 1980, and these small differences are much too small to explain the putative cold temperatures of the MM period. Reid said:

> "This ... essentially leaves the cold temperatures of the Maunder Minimum period to be explained by some unknown global forcing factor other than solar variability. While such a possibility cannot be ruled out, no alternative mechanism has yet been suggested."

However, the *hockey stick* advocates would simply deny that the MM was that cold. Reid concluded:

> "The irradiance reconstruction presented here suggests that solar forcing may have been a more important factor in recent climate change than the IPCC estimate would imply."

On the other hand, Reid's assumptions that the MM was 1°C colder than recent temperatures and this was caused by a weaker Sun cannot be verified and must be regarded as highly speculative.

5.6.4 Stellar Ca HK index models

Beckman and Mahoney (1998) reviewed the magnetic variability of nearby stars that are similar to our Sun. The hope was that we might be able to observe variations that could provide clues as to whether the solar cycles of other stars may vary with time. A good indicator of magnetic activity in the Sun is the emission fluxes of the H and K resonance lines of ionized calcium that maximizes at S_{MAX} for our Sun. Over the 11-year solar cycle, these vary with peak-to-peak amplitude of about 30%. Starting in the 1960s, a long-term observation program for the H and K resonance lines emitted by Sun-like stars has been underway at Mt. Wilson. There are now records of H + K intensity for over 100 stars covering 30 years. However, the interpretation of the data must take into account the fact that stars that rotate faster will have a stronger magnetic field, and also stronger H + K emission. Usually, the older a star, the slower it rotates. Thus, H + K emission has been used as an indicator of age:

> "If a star shows H + K variability, with amplitude similar to that of the Sun, say, a single measurement which catches the star at a maximum or minimum will give a misleading age estimate; what is required is a mean over a stellar cycle, or over a long enough period to cover short-term variations. In a cluster these effects can be cancelled even at a single epoch by averaging over its population, but for a single star this is not possible. The situation is worsened if stars have Maunder minima, because a measurement of H + K during a Maunder Minimum would give the false impression of very low activity, and very great age. Stars with low activity have indeed been found. A significant fraction, maybe 20% of the isolated solar-type stars of the Mt. Wilson survey have low, constant H + K levels. These might just be very old stars, but a similar situation is found in the stars of the open cluster M67, which is just a little younger than the Sun. This strongly suggests that solar-type stars do go through phases of low magnetic activity, and that these Maunder minima last some 20% of the time. Many doubts remain; only one star has possibly been 'caught' in transition from low to 'normal' activity, from a sample for which more than 5 such transitions could be expected during their period of observation. Possibly, the stars with low activity are always like this and are simply slow rotators. ... The whole question is open and is a subject of active investigation." (Beckman and Mahoney, 1998)

Starting in the 1960s at Mt. Wilson, a long-term study of magnetic cycles in cool stars was begun using the variable emission flux of the H and K resonance lines of ionized calcium, the appearance of which in emission is characteristic of activity in stellar chromospheres (Beckman and Mahoney, 1998). There is an excess of H and K emission in the faculae that surround sunspots, and epochs of sunspot maximum coincide with epochs of maximum H and K emission. If the Sun were a distant star, we could observe its 11-year cycle as a variation in integrated H + K flux with this period and with a peak-to-peak amplitude some 30% of the mean. Thus, while the Sun's TSI varies by perhaps 0.1% across the 11-year solar cycle, the H + K flux varies by perhaps 30% and is therefore far more easily detectable. As Beckman and Mahoney (1998) discussed, this project set out to determine whether stars of similar spectral type and luminosity class (i.e., similar surface temperature and mass) show comparable variations in cycling. This entailed a major project involving the monitoring of the fluxes of a group of stars for decades. The result is that there are now well-sampled records of H + K intensity for over 100 stars covering 30 years, plus samples of many hundreds more. One of the implications of these results is the possible detection of absence of solar cycling in some stars that might be analogous to MMs in Sun-like stars.

Beer, Mende, and Stellmacher (2000) reported on an investigation of 30 solar-type stars over the previous 30 years that showed that cyclic variability is typical for solar-type stars. The observed periodicities ranged from 7 to 20 years in analogy to the Sun's 11-year cycle. Most of these stars exhibit much larger brightness changes with cyclic peak-to-peak amplitudes of up to 1%, compared with the 0.1% irradiance amplitude observed for the Sun. This leads to the suggestion that the variability potential of the Sun might have been much greater during time periods other than the satellite-based observational period of the past few solar cycles. It is remarkable that the brightness of most monitored stars shows significant inter-cyclic variations; the minimum levels change from cycle to cycle by several percent. This raises the question of whether the Sun could develop similar variations over time in contrast to CQSM models:

> "... ~60% of Sun-like stars exhibit regular cycles akin to the Sun's (with the most well-defined cycles appearing typically in middle-aged stars as well as in K dwarfs with deep convective zones), ~25% vary irregularly (typically these are younger, faster rotators with presumably more vigorous dynamos), and ~15% are the so called 'flat activity' stars, which may be in periods analogous to the solar Maunder Minimum."[10]

In very broad terms, for two similar stars, the one that rotates faster will have a stronger magnetic field, and also stronger H + K emission. Again, in general terms, the older a star, the slower it rotates. Thus, H + K emission is an indicator of age, and has been calibrated against open stellar clusters, whose ages can be determined via collective photometry of their complete populations. One obvious problem here is that, if a star shows H + K variability, with amplitude similar to that of the Sun, a single measurement that catches the star at a maximum or minimum will give a misleading age estimate; what is required is a mean over a stellar cycle, or over a long enough period to cover short-term variations. In a cluster, these effects can be canceled even at a single epoch by averaging over its population, but, for a single star, this is not possible. The situation is worsened if stars have MMs, because a measurement of H + K during a temporary MMs would give the false impression of very low activity, and very great age. Stars with low activity have indeed been found. A significant fraction (perhaps 20%) of the isolated solar-type stars in the Mt. Wilson survey have low, constant H + K levels. These might just be very old stars, but a similar situation is found in the stars of the open cluster M67 that is just a little younger than the Sun. This suggests that solar-type stars may go through phases of low magnetic activity, and that these MMs perhaps last some 20% of the time. Many doubts remain; only one star has possibly been "caught" in transition from low to "normal" activity, from a sample for which more than five such transitions could be expected during their period of observation. Possibly, the stars with low activity are always like that and are simply slow rotators. Up to now, no observable correlation of activity with rotational period has been possible (the periods must be of order

[10] Hall, J.C. *et al.* (2011), "The stars as a sun: Secular variations of cycling and non-cycling stars", *Science White Paper for the Astro 2010 Decadal Survey.*

weeks or months, which requires extreme spectral resolution to measure). According to Beckman and Mahoney (1998), the whole question is open and is a subject of active investigation. In particular, direct measurements of indices of total stellar luminosity are being taken together with the H + K indices, to see whether, and to what degree, stellar (and thus solar) total power is correlated with magnetic field strength. The solar work provided the first clue, and the stellar work provides a framework to quantify, to predict, and eventually model the behavior of the Sun.

Lockwood, Skiff, and Radick (1997a) and Lockwood *et al.* (1997b) monitored the Ca II HK variations of 41 program stars and their 73 comparison stars (nearby, similar in color and brightness) from 1984 through 1995. The predominantly main sequence program stars spanned ranges of temperature and mean chromospheric activity centered on solar values. About 40% of all the stars showed measurable variability, typically at levels below 1%, on both night-to-night and year-to-year timescales.

Knaack *et al.* (2001) concluded that, compared with Sun-like stars, the TSI variations of the Sun over the solar cycle appear to be relatively small for its average activity level (variation of TSI of the Sun across a cycle is a factor of two to three smaller than variation of TSI for Sun-like stars). It has been suggested that part of the reason for this might be due to the special position of Earth-based observers in the ecliptic plane, who see the Sun almost equator-on, and earlier papers concluded that this could be the case. However, Knaack *et al.* (2001) found that, while measured TSI variations across a cycle for stars depend on inclination angle, the observed variation of TSI across a solar cycle will change by only ~40% as the viewing angle changes from 0° to 90° (highest at 90°). This can only account for part of the observed difference between the Sun and Sun-like stars. On the other hand, they found that the Ca II flux variation of Sun-like stars between activity minimum and maximum does depend significantly on the inclination of the rotation axis relative to an observer (change of about a factor of two in going from 0° to 90°). Lockwood *et al.* (2007) reviewed previous work on Sun-like stars. The variation of the mean of 32 primarily main sequence Sun-like stars was a factor of three greater than the variation of the Sun.

Lockwood, Skiff, and Radick (1997a) and Lockwood *et al.* (1997b) reported on the results of ongoing observations of Sun-like stars. It was pointed out that stellar photometric precision at the level of 0.05%–0.10% is needed to detect year-to-year variations as small as that of the Sun. This is a tough challenge, and it is admitted that the photometry:

> "... approaches but does not quite reach this precision, but provides useful comparisons with the solar example on a timescale comparable with the current solar TSI record."

However, the data appear quite noisy to this writer. In general, it is found that older stars tend either to vary in a smooth, cyclic fashion or have steady levels of HK emission, whereas young, active stars vary strongly but irregularly. Sun-like stars are characterized by their "chromospheric activity", which is the ratio of the Ca HK emission level to the innate luminosity of the star. Among Sun-like stars that have been studied, the Sun has somewhat lower than average chromospheric activity. However, chromospheric variability (i.e., periodic changes in HK emission) is slightly above average for its level of chromospheric activity. In general, the cyclic variability of TSI of stars increases with chromospheric activity. The Sun, as recorded by TSI measurements since 1978, has unusually low variability of TSI compared with Sun-like stars. Observations have not been conducted over enough time or with enough precision to determine how rapidly stars change in activity and variability.

Anon. (Q) studied the behavior of stars in the M67 cluster that are believed to contain Sun-like stars with similar age and metallicity as the Sun. This reference presented the results of a spectroscopic survey of the Ca II HK core strengths in a sample of 60 solar-type stars from the open cluster M67. They used the HK index, defined as the summed HK core strengths, as a measure of chromospheric activity. They compared the distribution of mean HK index values for the M67 solar-type stars with the variation of this index as measured for the Sun during the contemporary solar cycle. They found that the stellar distribution in the HK index is broader than that for the solar cycle. Approximately 17% of the M67 Sun-like stars exhibit average HK indices that are lower than that found in our Sun at S_{MIN}. About 7%–12% are characterized by relatively high activity in excess of S_{MAX} values, while 72%–80% of the solar analogs exhibit Ca II HK strengths within the range of the modern solar cycle. Thus, according to these findings, the Sun is quite an average star for its type in terms of chromospheric activity. However, it has above-average variability in HK emissions. What is not known is whether and how frequently Sun-like stars change cyclic behavior.

Based on a study of 30 Sun-like stars over 12–17 years, Radick *et al.* (2004) concluded: "The Sun's photometric variation still appears somewhat small in amplitude compared to the other stars in this sample with similar mean chromospheric activity."

Hall *et al.* (2007) reported on a "solar twin" star and found that it "exhibited luminosity variations remarkably similar to those of the Sun over the course of its activity cycle". Hall *et al.* (2009) reported on studies of 28 solar analog stars. They concluded that "The Sun does not appear to have unusually low photometric variability when compared with the most Sun-like inactive solar analogs", which is somewhat at odds with other studies that found the Sun to be typically less variable than other Sun-like stars.

In a series of related papers (Lean and Foukal, 1988; Lean, 2000; Lean*et al*, 1995; Lean *et al.*, 1997, 1998; Lean *et al.*, 1992; Fligge *et al.*, 1998; Lean and Rind, 1998; Rind *et al.*, 1999; Rind *et al.*, 2004), Judith Lean and co-workers estimated the variation of TSI over the past few hundred years based primarily on observations of Sun-like stars.

Lean and Foukal (1988) developed a simple model that included the reduction of total irradiance produced by the dark spots as well as the brightening caused by the faculae. Computation of the past behavior of the TSI with this model was limited by the available database of suitable indices of bright faculae. They were able to estimate both the sunspot darkening and the faculae brightening of the Sun from

1954 to 1984, and, by subtraction, the net variation of the TSI over that time period. This amounted to a CQSM for this recent period.

It was desired to find a way to estimate the TSI for previous centuries, particularly going back to the earliest sunspot observations around the time of the MM (1645–1715). Lean, Skumanich, and White (1992) carried out a procedure for doing this based on several assumptions that in retrospect appear to this writer to be quite speculative. They noted that, in the current solar cycles at S_{MIN}, although sunspots and faculae are at a minimum (or absent), nevertheless a "quiet" network of bright magnetic elements remains on the Sun. While the current solar cycle produces an oscillatory TSI with the rise and fall of the appearance of sunspots and faculae, the network remains. Thus, Lean, Skumanich, and White (1992) postulated that, in times past, this network might have disappeared, and if it did, that might have created a "grand minimum" TSI, below that currently experienced at S_{MIN}. In order to estimate what that "grand minimum" TSI might have been, they used measurements of Ca II H and K lines from the Sun as a measure of magnetic activity. At the time they wrote their paper, extraterrestrial measurements of TSI were only available for one solar cycle and absolute calibration was not as refined as it is today. Therefore, their absolute values of TSI require some adjustment, but their method can still be discussed. For the one solar cycle of TSI data available at the time, they carried out the following steps:

(1) They assigned a "quiet Sun" TSI (the TSI obtained at current S_{MIN}) to be S_Q = 1,366.9 W/m^2. This was based on the minimum of ACRIM1 shown in Figure 5.1 around 1985.
(2) They used a previous estimate of the sunspot blocking effect to estimate the reduction in TSI caused by sunspots to be ($P_S \times S_Q$), where P_S is a function of the number of sunspots, but P_S was not actually given in the paper. At S_{MIN}, $P_S \sim 0$.
(3) At any level of magnetic activity on the Sun, the TSI will be enhanced by an amount S_C that increases with magnetic activity. They allowed S_C to remain nonzero at S_{MIN}, thus providing for the possibility that a quiet network remains even at S_{MIN}.

Thus, the actual measured TSI at any time is:

$$\mathrm{TSI} = S_Q - (P_S \times S_Q) + S_C.$$

Therefore:

$$S_C = \mathrm{TSI} - S_Q + (P_S \times S_Q).$$

Lean and Foukal (1998) sought a relationship between S_C and the Ca II K index. However, inexplicably, they set as follows:

For the ACRIM1 data from 1978 to 1989, they plotted their version of S_C vs. a measure of the Ca IIi H and K lines as shown in Figure 5.32. They found a linear relationship between the brightening due to faculae and the Ca II K index.

A theoretical model suggested an estimate for the Ca II K index of our Sun in the absence of magnetic activity as K = 0.076, and Lean, Skumanich, and White

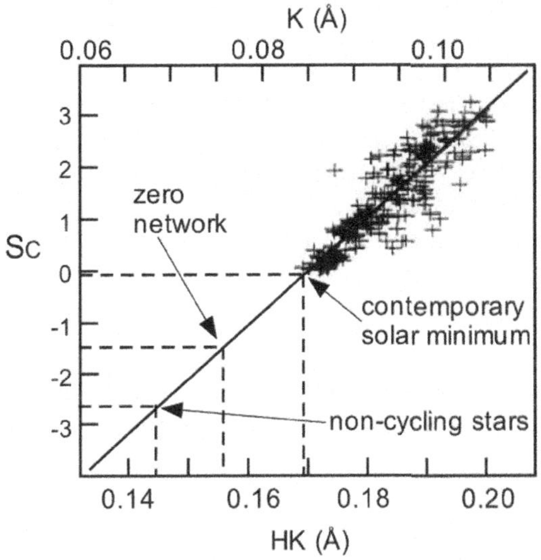

Figure 5.32. TSI data are plotted for ACRIM1 from 1978 to 1989 vs. Ca II H and K indices. (See text for note on the question of the sign of the term PS.) The "zero network" point is an estimate for the Sun in the absence of magnetic activity. The "non-cycling stars" point is based on observation of non-cycling stars. Lean, Skumanich, and White (1992) extrapolated the data to "non-cycling stars" to estimate TSI during the MM.

(1992) adopted this estimate for an estimate of our Sun when the entire surface is free of network brightening. In addition, they relied on studies of the Ca II K index from other stars. Some of these stars are non-cycling with reduced Ca II emission. Based on this work, an estimate was made of the mean Ca II emission for non-cycling stars, yielding K $\sim$0.069. The implication is that, in the absence of sunspots, faculae, and network, a non-cycling star will have further reduction (or elimination) of weak non-network magnetic fields. During the MM, there were essentially no sunspots, and magnetic activity on the Sun was at a minimum. Lean, Skumanich, and White (1992) assumed that the state of non-cycling stars would be representative of the MM. They then extrapolated the line relating SC to the Ca II K index to this lowest value of the K index. This corresponds to the point marked "non-cycling stars" in Figure 5.32.

This process appears to this writer to be rather fragile because there is no guarantee that the extension of the data in Figure 5.32 to a lower K index is linear, and the use of non-cycling stars to estimate the K index during the MM is a leap of faith.

This is not to diminish the imagination and innovation of Lean, Skumanich, and White (1992) who, in the absence of any firm ground to stand on, made some bold and intriguing assumptions to estimate TSI during the MM.

According to Figure 5.32, the TSI during the MM was about 2.8 W/m^2 lower than it is presently at S_{MIN} ($\sim$1,365.5 W/m^2). Thus, the TSI during the MM was estimated to be about 0.24% lower than the current TSI at S_{MIN} and about 0.33% lower than the current TSI at S_{MAX} (1,367.2 W/m^2).[11]

Lean, Beer, and Bradley (1995) was based on Lean, Skumanich, and White (1992). This required that the TSI during the MM be limited to a 0.24% decrease compared with the present TSI at S_{MIN}. (This is in contrast to Reid's estimate of a 0.65% decrease during the MM.) Using their estimate for the reduction of TSI in the MM, Lean, Beer, and Bradley (1995) concluded that annual variations in TSI account for 74% of the variance in the NH surface temperature anomalies from 1610 to 1800 and 56% of the variance from 1800 to the present. They further indicated that:

> "About half of the observed 0.55°C warming from 1860 to the present may reflect natural variability arising from solar radiative forcing, although since 1970, less than one-third of the 0.36°C surface warming is attributable to solar variability."

They concluded that:

> "Solar variability may have played a larger role in recent global temperature change than has hitherto been acknowledged."

Had Reid's model for TSI been used, for example, the solar contribution to temperature rise would have been much greater. However, Lean, Beer, and Bradley (1995) cautioned:

> "The longer term variability component [of TSI] is by necessity ... speculative."

Solanki and Fligge (2000) also utilized a variant of the approach used by Lean *et al.* (1998).

Scafetta and West (2006) utilized an estimate of the TSI from 1900 to 2000 in climatological models to estimate the contribution of solar variations to climate change during the 20th century. Faced with a variety of reconstructions of past TSI with variable amplitude from 1900 to 2000, they had to choose one, and decided that the conservative thing to do was to choose a reconstruction with intermediate amplitude. Therefore, they chose the reconstruction of Lean, Beer, and Bradley (1995) (see Figure 5.33).

Their estimate of global surface temperature in the 20th century is shown in Figure 5.34. Based on this analysis, they arrived at the results shown in Table 5.1. Had they used a reconstruction of TSI with greater amplitude, the percentage rise due to TSI variation would have been higher. Nevertheless, they concluded that the "impact of solar variation on climate seems significantly stronger than predicted by some energy balance models" (adapted from Lean, Beer, and Bradley, 1995).

[11] As stated previously, these values probably now need to be revised downward to about 1,361 W/m^2.

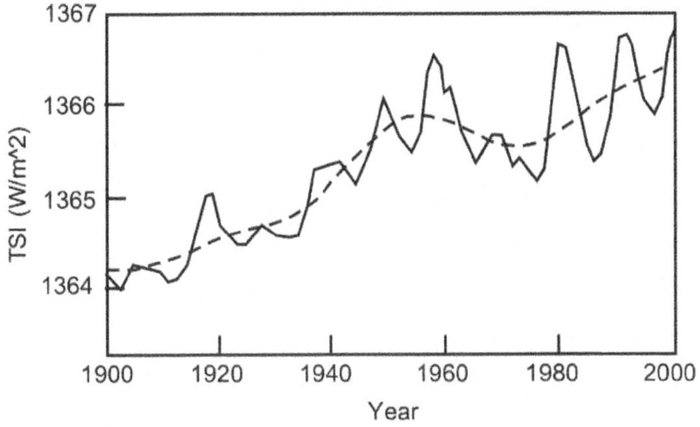

Figure 5.33. Estimated TSI for the 20th century (adapted from Scafetta and West, 2006).

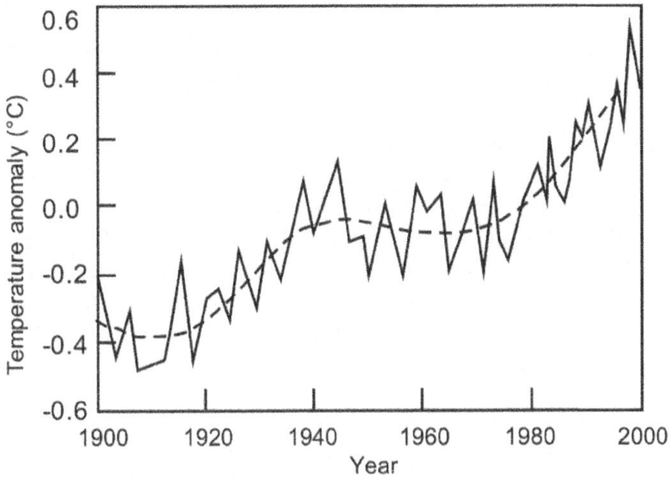

Figure 5.34. Global annual mean surface temperature anomalies (difference from mean) for the 20th century (adapted from Scafetta and West, 2006).

Table 5.1. Estimated TSI contribution to global warming (Scafetta and West, 2006).

Period	*Total T rise (°C)*	*T rise due to TSI (°C)*	*Percent due to TSI*
1900–2000	0.74	0.36	49
1900–1950	0.29	0.22	76
1950–2000	0.45	0.14	31
1980–2000	0.38	0.13	34

Foukal, North, and Wigley (2004) reviewed stellar observations of Sun-like stars that have low-activity phases during which the magnetic activity is even lower than during present minima in the solar sunspot cycle. Extrapolation of the Sun's radiometrically observed irradiance (as in Figure 5.32) to this low-activity level suggested that solar irradiance in the 17th century may have been 0.24% lower than today. Reconstructions of irradiance variations based on stellar evidence have been used in several climate studies. Additional evidence for large multi-decadal solar luminosity variations came from photometric studies of Sun-like stars, some of which exhibited cyclic variations three to five times those observed radiometrically in the Sun. This finding suggested that similarly large luminosity variations may have occurred on the Sun in the recent past. However, Foukal, North, and Wigley (2004) raised concerns as to whether the stars that were studied "were truly Sun-like—that is, with very similar mass, age, and chemical composition to the Sun—because the high-dispersion data required for such identification are relatively difficult to obtain". According to Foukal, North, and Wigley (2004), "only 18 Scorpii (HR6060) seems to be a sufficiently close solar analog for comparison with the Sun's present irradiance behavior".

In Figure 5.35, a CQSM model of TSI and temperature variations since 1600 is compared with a model that assumes that TSI was 0.24% lower during the MM (as in Figure 5.32). While Foukal, North, and Wigley (2004) cast doubt on the 0.24% lower model, they did not discuss the fact that this leads to all sorts of logical difficulties. Even the 0.24% lower model during the MM is insufficient to account for inferred temperature variations during the past millennium. If the CQSM curve in Figure 5.35 is adopted, that would wipe out all temperature variations in the past 400 years, including the LIA, and going back further probably the MWP as well, assuming that variations in temperature during that period were due to variations in TSI. There is no evidence that this assumption is correct.

Foukal, North, and Wigley (2004) raised a number of other objections to the process used by Lean and co-workers. They concluded that:

> "Data so far do not support the earlier conclusion that the Sun's irradiance variability of 0.08% is lower than that of similar stars. ... Any relationship on longer time scales must therefore remain speculative."

Foukal, North, and Wigley (2004) mentioned that Lean, Wang, and Sheeley (2002) apparently "accepted that the long-term irradiance variations used in [their] climate models in the past decade may be a factor of ~5 larger than can be justified".

In a more recent publication, Foukal *et al.* (2006) echoed the sentiments of their previous paper: "... this study [i.e., Lean, Beer, and Bradley, 1995)] has since been retracted ... [by Lean, Wang, and Sheeley (2002)]." Foukal *et al.* (2006) also said "Recent reconstructions from ^{14}C records disagree on whether the Sun might have experienced extended episodes of very high activity in the past 10,000 years, much above that encountered in the recent past."

The issues raised by Foukal, North, and Wigley (2004) appear to be significant. However, that does not imply that there have not been long-term secular changes in the TSI over the past millennium. If the MWP and the LIA involved significant

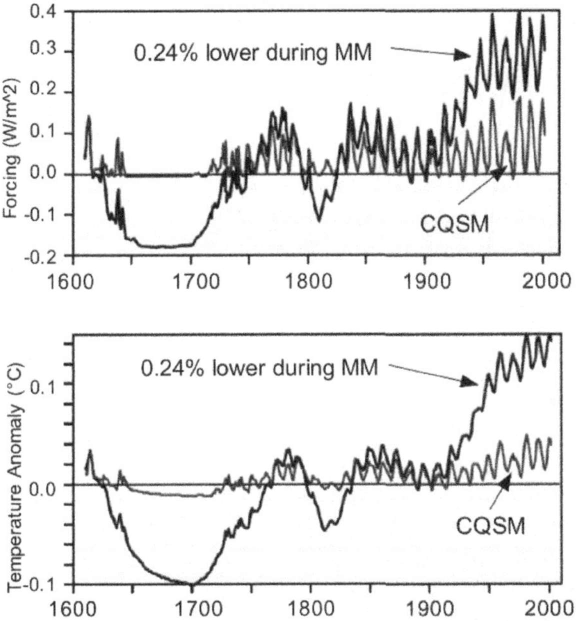

Figure 5.35. Illustration of the difference in modeled TSI and temperature variations since 1600 between a CQSM model and a model that assumes that TSI was 0.24% lower during the MM. Radiative forcing at the top of the atmosphere is one-quarter the change in TSI (adapted from Foukal *et al.*, 2004).

temperature excursions, and, if these temperature fluctuations in the past millennium were at least partly driven by changes in TSI, then there had to be significant fluctuations in TSI. The upper curves in Figure 5.35 might be an underestimate of the true variation; the lower curves are far worse. While previous studies of putative Sun-like stars may well be faulty, that does not lead to the conclusion that the Sun has been constant to within 0.1% for 1,000 years. The bottom line after all this speculation is that we have no reliable estimates of past variation of the TSI over hundreds or thousands of years and all of these models are based on assumptions that cannot be verified.

5.6.5 Solar cycle duration model

5.6.5.1 The "Sun Melody"

Beer, Mende, and Stellmacher (2000) summarized their view of other reconstructions of past TSI:

> "Most reconstructions of TSI for the last 250 yrs have been based on solar activity data such as the sunspot number. To estimate the irradiance, the amplitude, the solar cycle duration (or frequency) are often used. Several

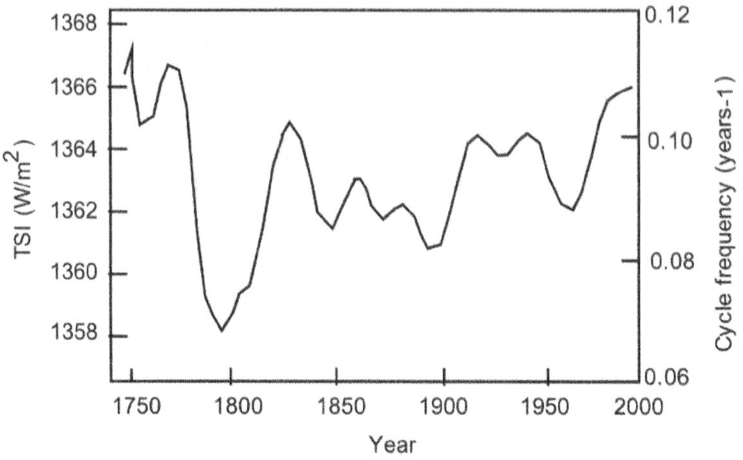

Figure 5.36. "Sun Melody" estimate of TSI based on variable duration of solar cycle (adapted from Beer *et al.*, 2000).

reconstructions have been proposed. They differ in details but agree in the overall shape, the generally increasing trend with local minima during periods of low solar activity (Maunder minimum: 1645–1715; Dalton minimum: 1800–1820; 1900 minimum: 1880–1900), and a slight decrease between 1940 and 1970. A sharp maximum around 1830 is connected with high solar activity. Superimposed on this long-term trend are short-term fluctuations caused by the 11-year Schwabe cycle." (Beer *et al., 2000)*.

However, these inferred past variations of TSI are quite speculative. Beer, Mende, and Stellmacher (2000) proposed an irradiance reconstruction that was based on the frequency of the Schwabe cycle because they claimed a better fit to temperature data is obtained than with sunspot count if they assumed a linear relationship between solar cycle frequency and irradiance. They estimated the frequency change in time by complex demodulation. This provided a continuous frequency modulation function. For comparison with temperature records, they used 14-year low-pass filtering in order to remove all periodic irradiance changes during one Schwabe cycle. This led to a relatively smooth frequency modulation function that they called the "Sun Melody".

Beer, Mende, and Stellmacher (2000) presented Figure 5.36 but no details are provided on how they obtained TSI from sunspot cycle duration (or its inverse: frequency).

It should be noted that the MM was not included in Figure 5.36 because solar cycle duration data were not available prior to about 1750. According to Figure 5.36, the low point in TSI just prior to 1800 was 0.6% lower than the current TSI. The MM was surely even lower. However, while the locations of ups and downs are fairly well correlated between sunspot numbers and inverse duration of the solar cycle, the amplitude of variation is much greater when based on solar cycle duration. The

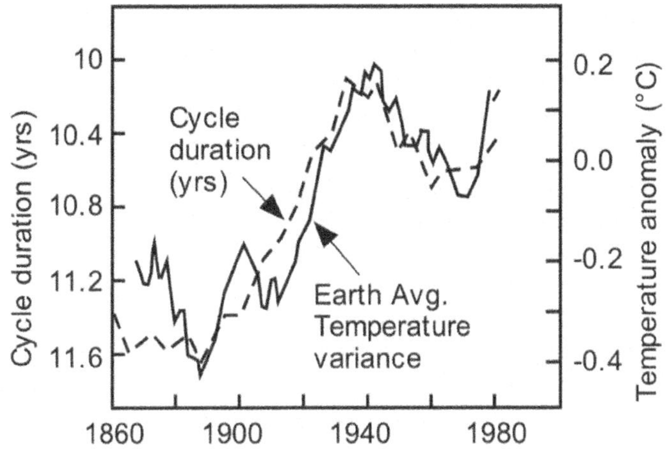

Figure 5.37. Comparison of solar cycle duration with temperature variance (measured from 1980) in the NH (adapted from Friis-Christensen and Lassen, 1991).

putative connection between solar cycle duration and TSI has not been adequately demonstrated and the ups and downs in Figure 5.36 are probably artifacts.

5.6.5.2 Danish Meteorological Institute studies

Friis-Christensen and Lassen (1991) presented Figure 5.37, taken from previous work. This figure compares the average temperature of the NH with the average solar activity defined by the interval between successive sunspot maxima.

The more active the Sun, the shorter the interval: the solar cycle runs more intensely. The dashed curve in Figure 5.37 illustrates the solar activity, which has generally increased over the past 100 years, since the cycle duration has decreased from around 11.5 years to almost 10 years. Within the same interval, the Earth's average temperature as indicated by the solid curve has increased by approximately 0.6°C.

Friis-Christensen and Lassen (1991) claimed that the varying length of the 11-year cycle is strongly correlated with long-term variations of the NH land surface air temperature since the beginning of systematic temperature variations from a global network (i.e., during the past 130 years). Although direct temperature observations before this interval are scarce, the correlation has been extended back to the 16th century using various series of proxy temperature data. Reliable sunspot data do not exist before 1750, but Friis-Christensen and Lassen (1991) derived epochs of minimum sunspot activity from auroral observations back to 1500 and combine them with direct observations to form a homogeneous series.

Based on Figure 5.37, Friis-Christensen and Lassen (1991) said:

> "Whereas the sunspot number returns to near zero at each 11-year minimum, the 11-year geomagnetic activity variations are superimposed on a long-term

variation of similar amplitude including a nearly monatomic increase from 1900 to 1950. This has been interpreted as a signature of an increase in the solar wind velocity through the century. The observed long-term variation in solar energy output by means of the solar wind suggests that similar long-term changes in other manifestations of solar energy output may have occurred."

To estimate this putative long-term variation in TSI, they utilized changes in the length of the approximately 11-year sunspot cycle.

The sunspot cycle duration record is subject to "noise", due to the fact that the start of a cycle cannot be easily defined in time because of the presence of short-term fluctuations in solar activity that provide obscuration. This is the case when the minimum activity in the "11-year" cycle is regarded as the start of the cycle, but it is even more difficult to define the start of a cycle by means of the time of maximum solar activity. Therefore, the cycle duration record must be filtered as shown in Figure 5.38.

A comparison of filtered data from Figure 5.38c with NH temperatures is given in Figure 5.39. The strong correlation between inverse sunspot cycle duration

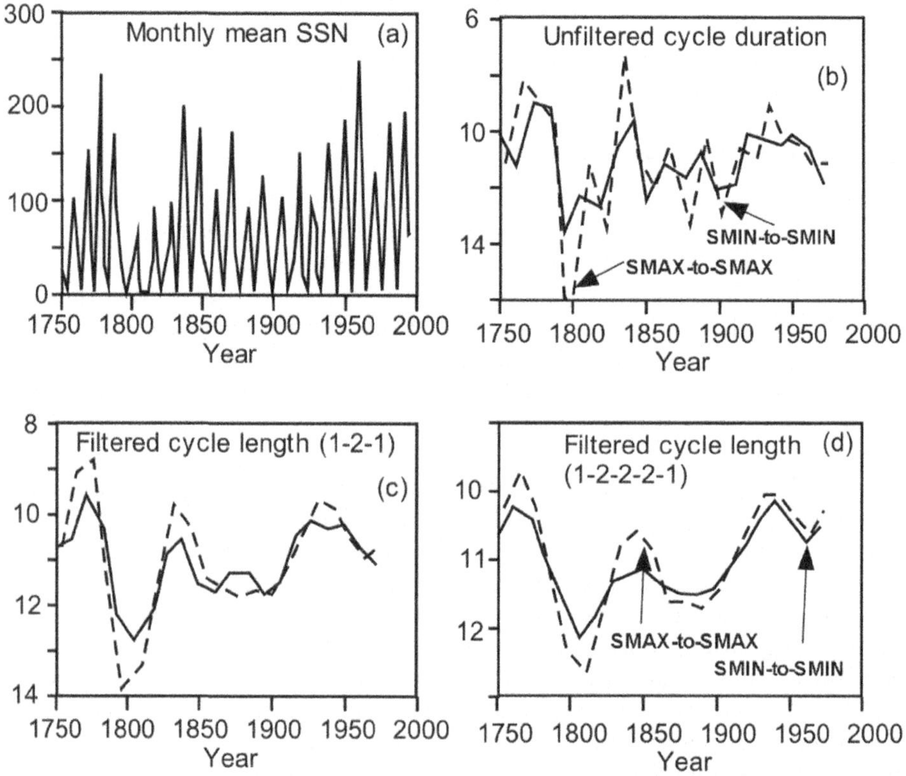

Figure 5.38. (a) Monthly mean of sunspot number, (b) unfiltered sunspot cycle duration (S_{MIN} to S_{MIN} and S_{MAX} to S_{MAX}), (c) 1-2-1 filtered cycle duration data, and (d) 1-2-2-2-1 filtered cycle duration data (adapted from Friis-Christensen and Lassen, 1991).

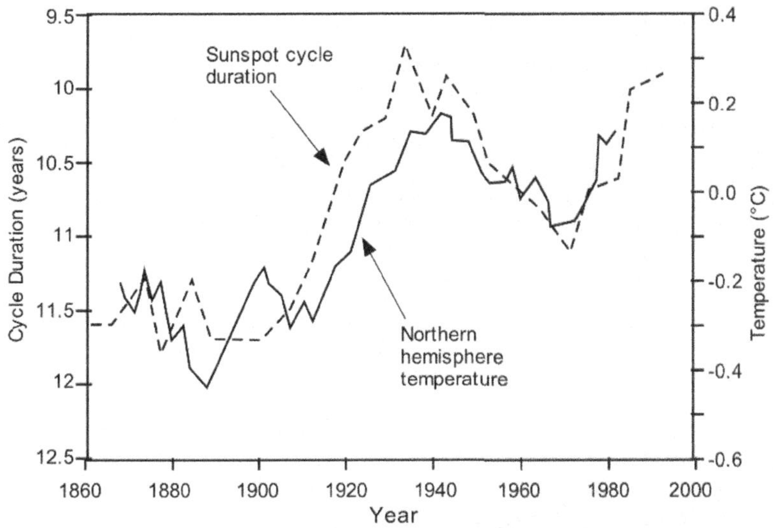

Figure 5.39. An 11-year running average of NH temperature compared to 1-2-1 filtered sunspot cycle duration (adapted from Friis-Christensen and Lassen, 1991).

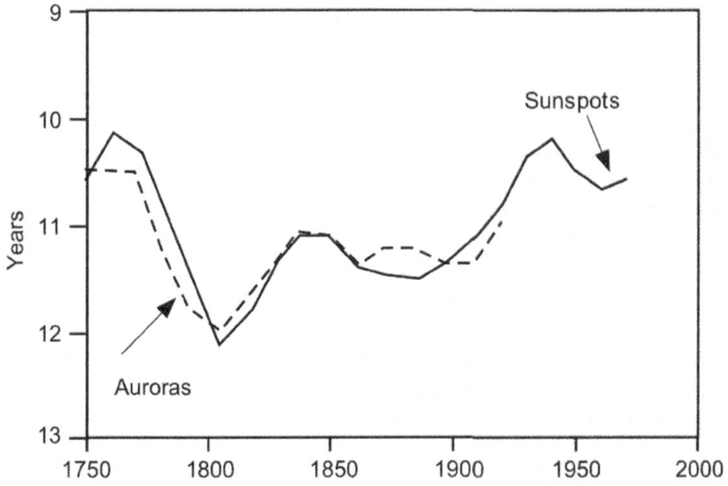

Figure 5.40. Comparison of sunspot cycle duration with frequency of aurora occurrence. The vertical scale for sunspots is the duration of the solar cycle and the vertical scale for auroras (scaled to fit) is the number of auroras seen per year (adapted from Friis-Christensen and Lassen, 1991).

variations and variations in NH temperature suggests that solar variations were the principal driver for temperature change since 1860. Friis-Christensen and Lassen (1991) also demonstrated a strong correlation between sunspot activity and frequency of aurora occurrence as shown in Figure 5.40.

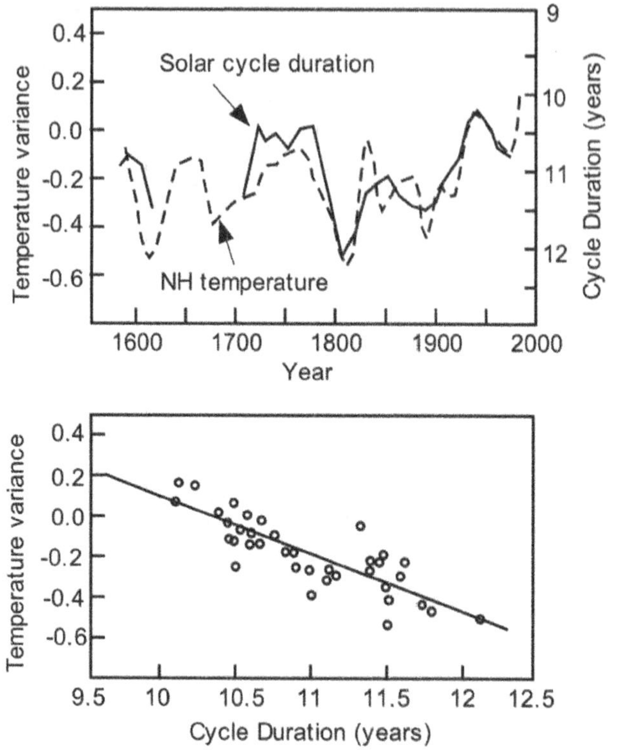

Figure 5.41. (a) Lower frame: dependence of NH temperature variance on solar cycle duration. (b) Upper frame: an 11-year running average of the annual mean values of reconstructed NH temperatures from 1579 to 1860 connected to corresponding measured values for 1851–1987, together with the smoothed values of the solar cycle duration from 1564 to 1989, with the exception of the interval 1641–1674, for which reliable data are missing (plotted inversely) (adapted from Friis-Christensen and Lassen, 1991).

Having demonstrated an inverse relationship between solar cycle duration and NH temperature, Friis-Christensen and Lassen (1991) went on to define the relationship quantitatively and they compared the resultant estimated temperature since about 1579 with an estimate of the temperature record, as shown in Figure 5.41.

Friis-Christensen and Lassen (1991) concluded that a comparison of the extended solar activity record with the temperature series confirms the high correlation between solar activity and NH land surface air temperature and shows that the relationship has existed through the whole 500-year interval for which data exist.

A corresponding influence of solar activity was also demonstrated for both the date of arrival of spring in the Yangtze River Valley as deduced from phenomenological data and the extent of the sea ice in the Atlantic sector of the Arctic Sea that correlate well with the length of the sunspot cycle during the last 450 years.

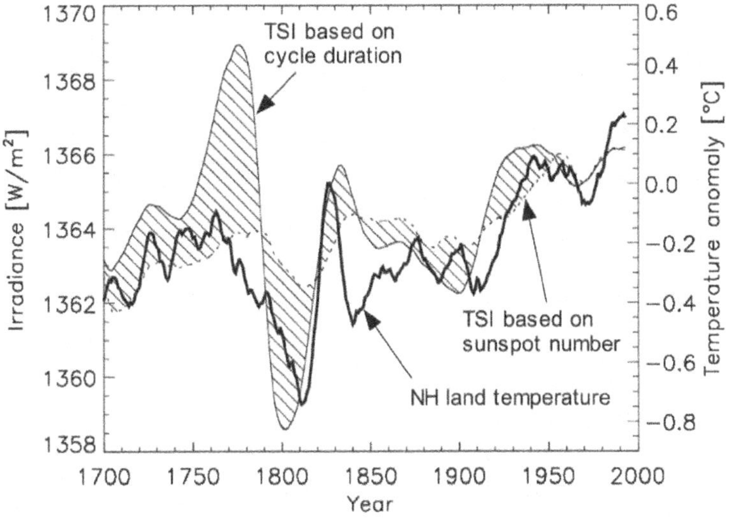

Figure 5.42. Reconstruction of TSI using cycle duration or sunspot indices. The shaded area shows the differential between reconstructions based on cycle duration and sunspot count (adapted from Solanki and Fligge, 2000).

Solanki and Fligge (2000) also presented a reconstruction of past TSI based on solar cycle duration correlations. However, few details were provided and it is not clear how they proceeded. Evidently, this work is derived from Solanki and Fligge (1999), as can be seen from Figure 5.26. However, that is a very compressed database. Their result is shown in Figure 5.42. The large peak around 1760 and dip around 1800 seem to be anomalous.

5.6.5.3 Hoyt and Schatten model

Hoyt and Schatten (1993) took the approach that the various solar indices, can be related to sunspot cycle decay rates, and, thus, the sunspot decay rate becomes the main determinant of TSI variations:

(1) fraction of spots that are penumbral;
(2) solar cycle duration;
(3) equatorial rotation rate;
(4) decay rate of the solar cycle; and
(5) mean level of solar activity, but prior to the MM temperature model or the Ca HK model, which are not included by Hoyt and Schatten (1993).

The results of their model are shown in Figure 5.43. For the time period 1700–1874, three indices were used: cycle duration, cycle decay rate, and level of solar activity. For 1875–1978, two additional indices were employed: solar rotation and fraction of penumbral sunspots. For the time period 1979–1992, they scaled their figures to Nimbus-7 measurements.

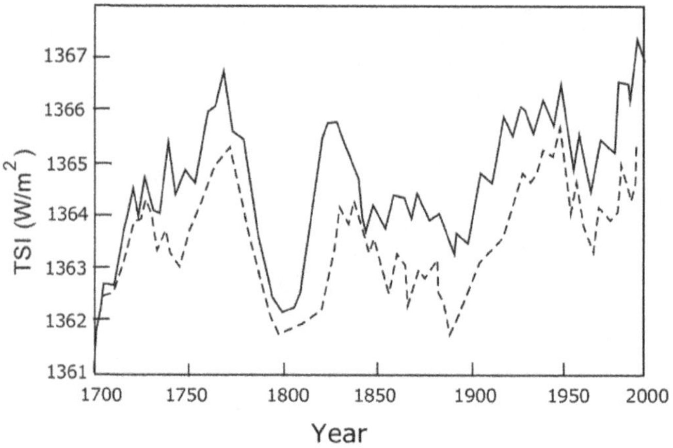

Figure 5.43. Modeled total solar irradiance based on sunspot decay rates. The solid line is an upper estimate and the dashed line is a lower estimate. The vertical scale was set to fit the latest ACRIM measurements (adapted from Hoyt and Schatten, 1993).

Although this correlation begins just around the end of the MM, extension of Figure 5.43 to earlier dates suggests that, according to this model, the TSI during the MM was more than 0.45% lower than it is today. If solar irradiance has varied with time as Hoyt and Schatten (1993) estimated, it can be expected to have had some significant effect on the temperature of the Earth. They examined two questions:

(1) Is the Earth responding in a manner consistent with an external forcing?
(2) Do the Earth's temperature variations and the model TSI variations correlate with each other?

A solar forcing will tend to cause the two hemispheres of the Earth to vary in parallel. The amplitudes of the responses will differ because the two hemispheres have different amounts and distributions of land and ocean. Hoyt and Schatten (1993) separated the Northern and Southern Hemisphere temperature records. Then, they compared the temperature history in the NH with the combined model for TSI history as shown in Figure 5.44; this figure suggests a close correlation of TSI with temperature. The modeled TSI and measured temperatures share many similarities. However, the temperature reconstruction used thermometer measurements predominantly from Europe, and it is not clear how reliable these estimates were. On the same axes, we now plot the results of the 2012 BEST temperature reconstruction as well. There is some resemblance between the Hoyt and Schatten TSI curve and the BEST temperature curve in the early years but not in the late 20th century.

Hoyt and Schatten (1993) claimed that, on a decadal timescale, the TSI model can explain 71% of the TSI variance during the past 100 years and 50% of the variance since 1700. They concluded that "There is plausible evidence for long-term changes in TSI".

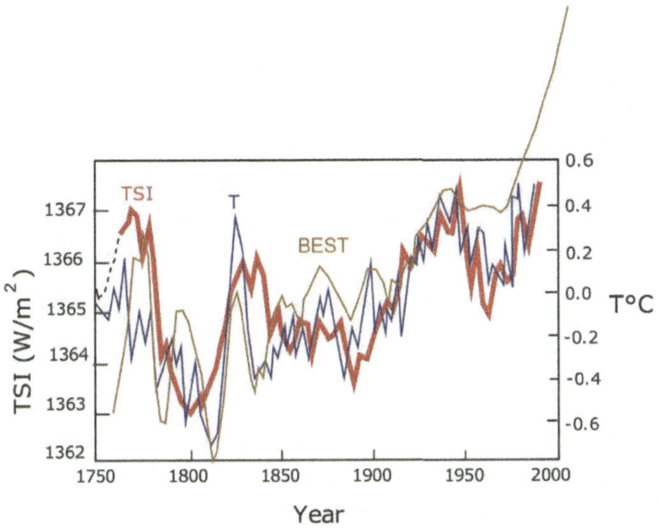

Figure 5.44. Comparison of 11-year moving mean of NH temperature with modeled TSI (adapted from Hoyt and Schatten, 1993).

Based on the energy stored in the Sun's convective zone ($\sim 10^{45}$ ergs), a change in solar emission of, say, 0.2% for a whole century amounts to 1 part in 40,000 of the stored energy. Thus, there is no *a priori* reason to rule out such variations based on energy. Hoyt and Schatten also argued that, from 1880 to 1940, the Earth warmed by about 0.5°C and climate models suggest that a change in TSI of 1% produces a change in temperature of about 1.67°C; therefore, a 0.5°C change should require a change in TSI of about 0.30%. They quoted some (CQSM) estimates that TSI changed by only 0.14% over this time interval and suggested that 0.14% is inadequate to account for the observed change in temperature.

5.6.6 Coronal source flux model

Lockwood and Stamper (1999) and Lockwood (2002) developed methods for estimating the coronal source flux, F_S. The coronal source surface occurs where the solar magnetic field becomes approximately radial and lies at a heliocentric distance of about 2.5 solar radii. The total magnetic flux leaving the Sun, and thereby entering the heliosphere by threading this surface, is the coronal source flux, F_S. One method depends on measurements of the geomagnetic *aa* index and another depends on measurements of the near-Earth interplanetary magnetic field (IMF). They studied the correlation of F_S with TSI. Lockwood (2002) said (regarding correlation of F_S with TSI):

> "Were it to reveal a real physical connection between TSI and F_S, this correlation would be very important—even though it is unlikely to be the result of a direct causal relationship."

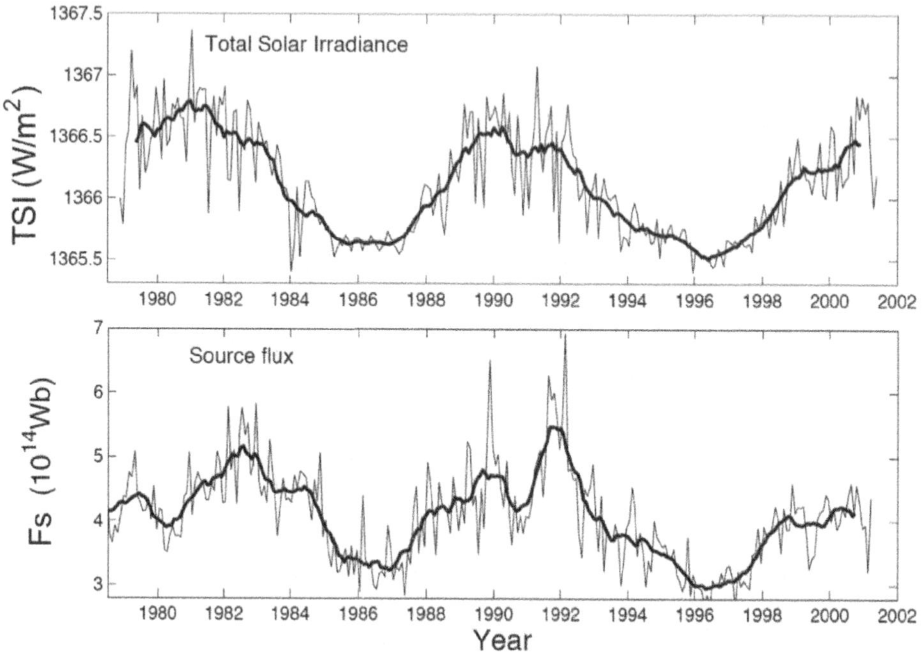

Figure 5.45. Variation of measured extraterrestrial TSI and calculated F_s since 1978. Thin lines are monthly means and thick lines are 12-month averages. Careful comparison of the peaks and valleys of the two monthly curves reveals that TSI tends to lag F_s by about a month or so (adapted from Lockwood, 2002).

A comparison of recent TSI data with F_s data is shown in Figure 5.45.

Lockwood (2002) developed a scatter plot of 12-month average F_s vs. 12-month average TSI and fitted a straight line to the scattered data resulting in

$$\text{TSI} = 1{,}363.5 + 0.645\ F_s\ (\text{W/m}^2),$$

although they found a best fit with the TSI data lagging by six months. Using this relationship, the predicted 12-month average TSI (from F_s) is compared with the actual 12-month average TSI in Figure 5.46. Note that, according to this algorithm, the TSI can never drop below 1,363.5 W/m^2 under any circumstances. Thus, this model is somewhat analogous to CQSM models in that it assumes that past variations in TSI are due only to factors discernible in the present solar cycle.

In an earlier paper, Lockwood and Stamper (1999) estimated F_s over a much longer time period, from about 1870 to the present. The lowest values of F_s occurred around 1880 and 1900 when F_s dipped to about 1.4×10^{14} Wb. According to the above algorithm, this would imply that the minimum TSI in those years was about $(1{,}363.5 + 0.645 \times 1.4) = 1{,}364.4$ W/m^2, which is about 0.15% lower than the present TSI at S_{MAX}. In essence, this is another CQSM except that the transducer for predicting TSI is F_s instead of sunspot number. This model does not account for any

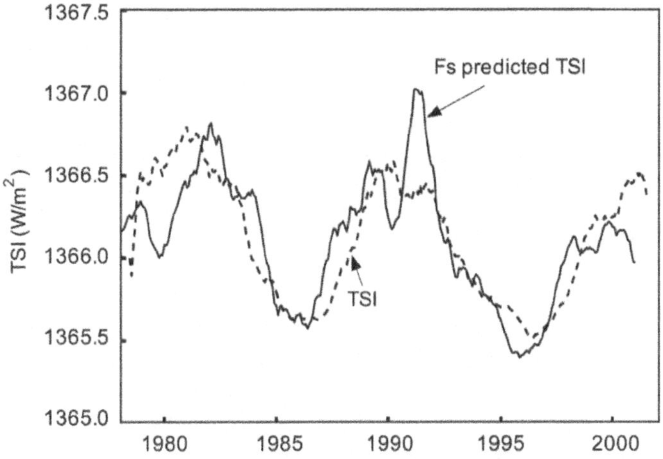

Figure 5.46. Comparison of the predicted 12-month average TSI (from F_s) with the actual 12-month average TSI (adapted from Lockwood, 2002).

long-term secular changes in solar output independently of solar magnetic activity within the framework of current solar behavior.

Wang, Lean, and Sheeley (2005) developed a variant of a CQSM that utilized the Sun's magnetic flux. Their paper is difficult to comprehend but, ultimately, it depended on utilizing the small variances between S_{MAX} and S_{MIN} recently measured, and extrapolating these back to year 1713. It is not surprising that they found that the cycle-averaged change in TSI since the MM was a mere ~ 1 W/m^2.

5.6.7 TSI reconstructions based on cosmogenic isotope proxies

5.6.7.1 Introduction

Galactic cosmic rays are continually impinging on the Earth's atmosphere, producing nuclear reactions that generate radioactive isotopes. These isotopes gradually settle through the atmosphere and may be incorporated into biota, ice deposits, and other media that preserve the "cosmogenic isotopes" for long periods of time. Of particular interest is the occurrence of ^{14}C in tree rings and ^{10}Be in high-latitude ice cores. The interpretation of such proxies has mainly been in reference to historical surface temperatures, but these proxies have also been used in a few instances to infer aspects of historical variation of TSI.

As Muscheler *et al.* (2005) pointed out:

> "The Sun influences the production rate of ^{14}C in the Earth's atmosphere by modulating the galactic cosmic-ray flux through its magnetic field. Increased magnetic field in the solar wind causes a stronger deflection of galactic cosmic rays and lower radionuclide production rates in the atmosphere, and vice versa."

Thus, the historical record of annual isotope production would have been influenced by the historical variation in the activity of the Sun. However, Muscheler *et al.* (2005) also pointed out that:

> "The atmospheric ^{14}C concentration also depends nonlinearly on the geomagnetic field intensity and the global carbon cycle. These factors and their uncertainties need to be carefully included in the reconstruction of solar activity."

Similar considerations apply to other proxies such as ^{10}Be in ice cores. Thus, the historical isotope record contains information about past solar activity, but unraveling that information is difficult due to a number of confusing factors.

The use of cosmogenic radioisotopes as proxies for TSI variation introduces uncertainties. The radioisotopes are not produced in the Earth's atmosphere directly by the Sun's flux of energetic particles. They are produced mainly by high-energy galactic cosmic rays, from outside the Solar System. Their modulation with the solar cycle is due to changes in the way these cosmic rays are shielded by the heliosphere (the solar wind). The efficiency of this shielding depends on solar plasma outflows from open magnetic fields in quiet regions and individual events such as flares and CMEs. Although this shielding increases roughly with the general level of solar activity, it is only very loosely proportional to the areas of the dark and bright magnetic structures that drive TSI. In view of this, it is unrealistic to expect a fixed relation between variations in ^{10}Be or ^{14}C production rate and TSI over the past millennium. The relation is complicated further by possible climate influences on the ^{10}Be and ^{14}C deposition rates, causing errors in the inferred ^{10}Be and ^{14}C formation rates (Foukal *et al.*, 2006).

Lean, Wang, and Sheeley (2002) dealt with the question of reconstructing TSI in the past based on models developed for solar magnetic fields. While the Sun's total magnetic flux (sunspots, faculae, network, etc.) has a significant effect on TSI, a relatively small fraction of the Sun's magnetic flux (whose footprint regions are observed as coronal holes) extends into the heliosphere. Open magnetic fields have a negligible effect on solar irradiance, but they are the source of the IMF. IMF variations modulate the terrestrial flux of cosmic rays that produce cosmogenic isotopes, and they also affect geomagnetic activity. When these fields are high, production of cosmogenic isotopes is reduced. In the model developed by Wang, Lean, and Sheeley (2002) and Lean, Wang, and Sheeley (2002), a hypothetical 110-year build-up of solar magnetic dipoles was simulated (with an 11-year solar cycle superimposed) and the consequences for the total magnetic flux and the open flux were estimated.

The main conclusion drawn by Lean, Wang, and Sheeley (2002) was that the open flux can build up with time due to an increase in dipoles, but the total surface flux may remain constant. Thus, they said:

> "Secular changes in terrestrial proxies of solar activity (such as the ^{14}C and ^{10}Be cosmogenic isotopes and the *aa* geomagnetic index) can occur in the absence of long-term (i.e., secular) solar irradiance changes. Increasing solar

cycle amplitudes produce a secular increase in open flux and interplanetary magnetic field (IMF), and can therefore explain variations like the cosmogenic isotope decrease from 1922 to 1965. Total magnetic flux, however, does not have an equivalent secular trend during minima. Since the primary sources of the total flux are features that modulate solar irradiance, this suggests that total solar irradiance may also lack significant secular trends."

In essence, Lean, Wang, and Sheeley (2002) argued in favor of a CQSM model for long-term solar variation. Foukal *et al.* (2006), building upon Lean, Wang, and Sheeley (2002), concluded that:

> "Overall, we can find no evidence for solar luminosity variations of sufficient amplitude to drive significant climate variations on centennial, millennial and even million-year timescales."

The reason that they found no evidence of sufficient solar luminosity variations is that their model does not provide for that possibility. In keeping with current trends in "political correctness" that have wiped out temperature variations for the past 1,000 years with *hockey stick* figures, Foukal *et al.* (2006) and Lean, Wang, and Sheeley (2002) have eliminated TSI variations as well. These conclusions do not seem to have very firm foundations.

5.6.7.2 Reconstruction of TSI from cosmo-nuclide production proxies

Bard *et al.* (2000) pointed out that previous reconstructions of historical TSI exhibited similar fluctuations in time, but they differed significantly in amplitude. The main differences are best illustrated by comparing the TSI decrease estimated or extrapolated for the deepest part of the MM ranging from 0.24% (Lean *et al.*, 1995), to 0.3% (Hoyt and Schatten, 1993), to 0.4% (Solanki and Fligge, 1999), to 0.5%–0.65% (Reid, 1997). The estimate of 0.24% by Lean, Beer, and Bradley (1995) is often used as a conservative view of TSI changes during the MM. Nevertheless, if there were no long-term trend, the TSI during the MM would obey a CQSM and be comparable with the current TSI at S_{MIN}.

Bard *et al.* (2000) argued that it is possible to use the high-frequency component of the solar magnetic variability as a proxy for production of cosmogenic nuclides such as ^{14}C and ^{10}Be. Magnetic fields of the solar wind deflect the primary flux of charged cosmic particles, which leads to a reduction of cosmogenic nuclide production in the Earth's atmosphere. In particular, it is claimed that the 11-year cycle modulates ^{10}Be production recorded in well-dated polar ice from Greenland and Antarctica. Very high cosmo-nuclide production (30%–50% above the modern value) has also been confirmed for the MM based on high ^{14}C content in tree rings and ^{10}Be in polar ice.

Bard *et al.* (2000) provided a profile of variation of ^{10}Be and ^{14}C during the past millennium. Their result is shown in Figure 5.47. The TSI is inversely related to cosmo-nuclide production, so the vertical scale is plotted inversely. The curve in Figure 5.47 is claimed to be representative of TSI, but the quantitative relationship of nuclide production to TSI is not obvious.

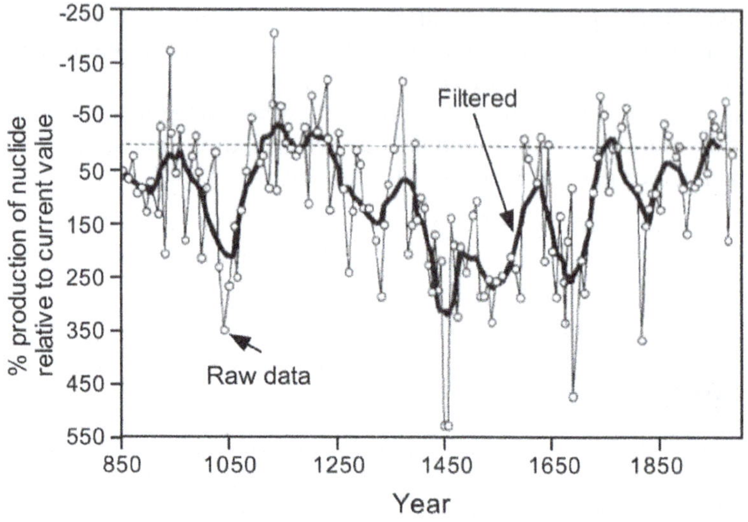

Figure 5.47. Cosmo-nuclide production as percent of present production based on ^{10}Be in polar ice and ^{14}C in tree rings (adapted from Bard *et al.*, 2000).

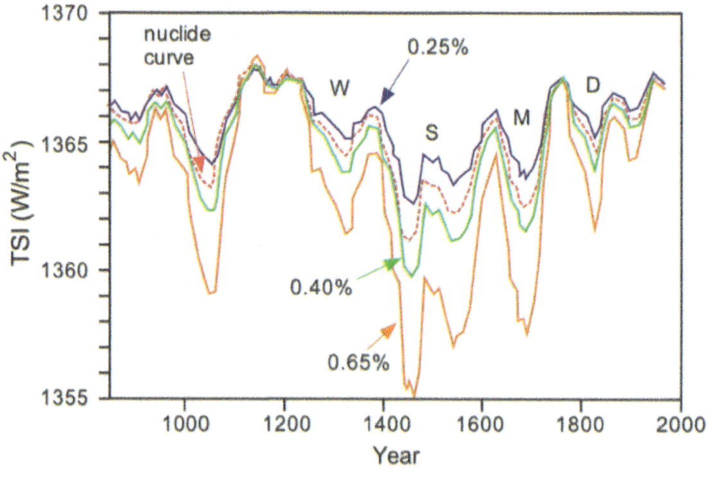

Figure 5.48. Modeled TSI during the period 850 to the present. The blue curve is the raw cosmo-nuclide data taken from Figure 5.47. The red, black, and orange curves are scaled to produce 0.25%, 0.40%, and 0.65% reductions in TSI during the MM as compared to today. Current TSI was set ~1367 W/m^2 (adapted from Bard *et al.*, 2000).

Bard *et al.* (2000) scaled the nuclide data by setting TSI equal to the known value (at that time, 1,367 W/m^2) in the 1990s, and then using a parameterized decrease in TSI during the MM to generate a family of curves for TSI over the past millennium for each assumed minimum during the MM (see Figure 5.48). However, in doing this, it appears that Bard *et al.* (2000) only scaled the nuclide production rate during periods of low TSI (high nuclide production) and assumed that, during periods of high TSI (low nuclide production), no scaling was needed. That is one possibility; it assumes that prior to (and after) the MM, during periods of relatively high TSI, the Sun acted as it does today. This model is the antithesis of the CQSM because in this model, the maximum TSI is fixed and the minimum TSI is varied and unbounded.

5.6.7.3 Projections for the Holocene

Solanki *et al.* (2004) carried out a reconstruction of the sunspot number over 11,000 years back in time, based primarily on archival concentration of cosmogenic isotope ^{14}C activity in the atmosphere obtained from high-precision ^{14}C analyses on decadal samples of mid-latitude tree-ring chronologies, but also including some comparison with ^{10}Be in ice cores from Antarctica and Greenland since 850.

The original data set used by Solanki *et al.* (2004) for sunspot number reconstruction is represented by the "wiggly" black line in Figure 5.49, based on ^{14}C data. The precision of the ^{14}C measurements is claimed to be generally 2%–3%, although, in the earlier part of the time series, it can reach up to 4%–5%. The long-term decline (indicated by the smooth curve) is caused by a reduction in ^{14}C production rate due mainly to an increase in the geomagnetic shielding of the cosmic-ray flux and does not necessarily indicate any changes in TSI. The short-term fluctuations (duration one to two centuries) are assumed to reflect changes of the production rate due to solar variability.

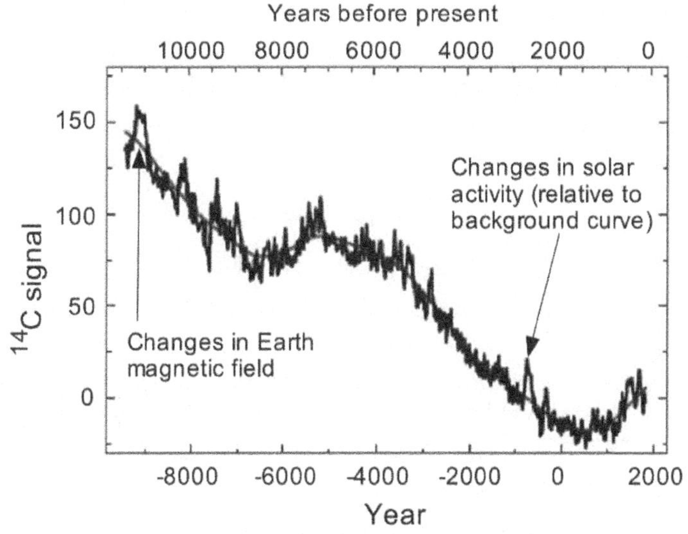

Figure 5.49. Atmospheric radiocarbon level ^{14}C (expressed as deviation, in , from the AD 1950 standard level) derived from mostly decadal samples of absolutely dated tree-ring chronologies (adapted from Solanki *et al.*, 2004).

According to Anon. (J), p. 36:

> "Whereas the long term trend in records of cosmogenic isotopes such as ^{14}C and ^{10}Be reflects, primarily, changes in the Earth's magnetic field that affect the interaction of cosmic rays with the Earth's atmosphere, the 'wiggles' superimposed on the smooth long term trend are believed to occur because of the modulation of the local cosmic-ray intensity by magnetic fields embedded in the solar wind, which varies in response to solar activity. Thus, enhanced solar activity corresponds to ^{14}C minima, and the mechanism proposed by Eddy (1976) for the apparent relationship between climate and the ^{14}C 'wiggles' involved changes in the total solar irradiance linked to the long term envelope of the 11 year sunspot cycle and reflected in the ^{14}C record."

This implies that most of the secular long-term variation in Figure 5.49 is due to changes in the Earth's magnetic field, and the smaller "wiggles" in Figure 5.49 represent the ups and downs due to changes in solar activity. On this basis, one could treat the smooth curve in Figure 5.49 as a reference and measure deviations in ^{14}C from this reference as the main signal. However, as Eddy (1976) suggested, there may also be a solar component to the secular long-term change in the ^{14}C signal, and it is not clear how Solanki *et al.* (2004) determined this component.

Thus, in order to estimate variability of TSI, one would have to flatten out the smooth curve in this figure to horizontal and deal only with variations about this horizontal line. Since the accuracy of the basic $\Delta^{14}C$ measurement is several percent, and the signal ("wiggles" about the trend line) amounts to only a fraction of the $\Delta^{14}C$ measurement, the accuracy of the signal will be limited.

As Solanki *et al.* (2004) pointed out:

> "The atmospheric ^{14}C level may also be affected by changes in the partition of carbon between the major reservoirs, that is, deep ocean, ocean mixed layer, biosphere and atmosphere. Variations in ocean circulation could influence ^{14}C via a variable uptake of CO_2 into the ocean or by the exchange of ^{14}C-depleted carbon from the deep ocean, but, owing to the rather small ^{14}C gradients among the reservoirs, strong changes in these processes need to be invoked."

For the Holocene, Solanki *et al.* (2004) claimed that:

> ". . . there is no evidence of considerable oceanic variability, so we can assume that the short-and mid-term fluctuations of ^{14}C predominantly reflect solar variability. This is supported by the strong similarity of the fluctuations of ^{10}Be in polar ice cores compared to ^{14}C, despite their completely different geochemical history."

However, Anon. (J) claims that:

> ". . . the extent to which cosmogenic isotope variations really indicate terrestrially relevant variations in solar energy outputs, either radiative or particle, and the scaling of the relationship over long times is poorly known; the paleo-climate record is similarly somewhat uncertain."

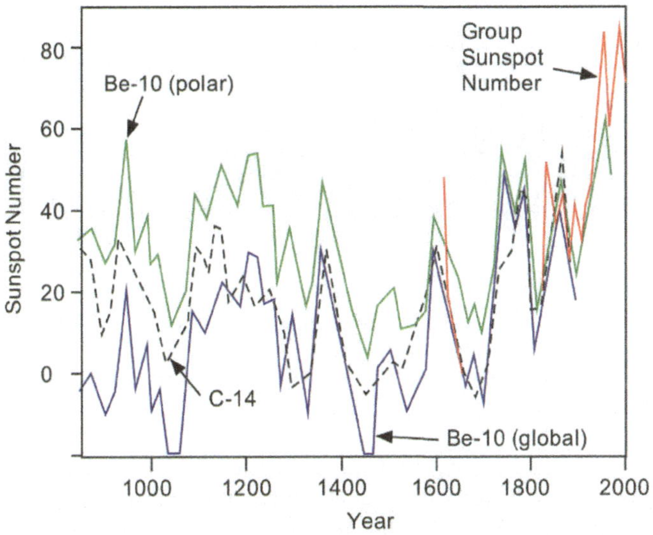

Figure 5.50. Comparison between directly measured group sunspot number (GSN) and SN reconstructed from different cosmogenic isotopes. The curves are: (a) SN reconstructed from ^{14}C, (b) the 10-year averaged group sunspot number since 1610, and (c) the SN reconstruction from ^{10}Be under the two extreme assumptions of polar and global production. Negative numbers should be treated as ~ 0 (adapted from Solanki *et al.*, 2004).

It is not clear to this writer how Solanki *et al.* (2004) processed the data in Figure 5.49. It appears possible that they simply flattened out the smooth curve to make it horizontal and took only variations about this line. However, later results do not necessarily agree with this interpretation. Solanki *et al.* (2004) used a rather involved procedure in multiple steps to estimate historical sunspot numbers from the ^{14}C production rate in Figure 5.49. This reconstruction method was previously applied to ^{10}Be data from Greenland and Antarctica. A comparison of their results with ^{10}Be results since 850 and observed group sunspot numbers since about 1600 is shown in Figure 5.50. Solanki *et al.* (2004) claimed that the similarity of the curves provides some measure of validation of the process. However, it is difficult to understand how they arrived at their results. These results suggest that the sunspot number is higher today than at any previous time in the past 1,000 years.

The results of Solanki *et al.* (2004) over a longer time period are shown in Figure 5.51. As before, it is difficult to understand how this figure follows from Figure 5.49. Presumably, each peak in Figure 5.51 should be associated with a down-wiggle in Figure 5.49 and the height of each peak in Figure 5.51 should be determined by the depth of the wiggle in Figure 5.49. It is difficult to make this connection. If the results presented in Figure 5.51 were correct, it would imply that current sunspot numbers are the highest they have been in the past 8,000 years.

An extension of the work of Solanki *et al.* (2004) was carried out by Muscheler *et al.* (2005). Solanki *et al.* (2004), like most ^{14}C studies, excluded the most recent 100

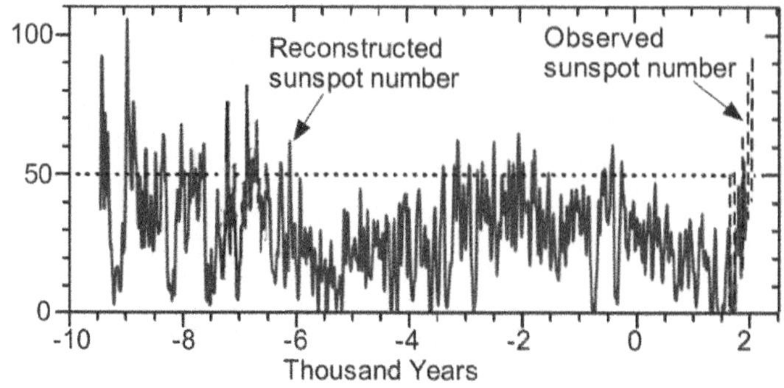

Figure 5.51. Reconstructed sunspot numbers and observed sunspot numbers (adapted from Solanki *et al.*, 2004).

years of the ^{14}C record, because they are influenced by ^{14}C-depleted fossil fuel emissions and atomic bomb tests conducted since 1950. However, Muscheler *et al.* (2005) claimed that they extended the analysis of the radiocarbon record from Solanki *et al.* (2004) to 1950, which allowed them to link the ^{14}C-based solar reconstruction to instrumental measurements of solar magnetic modulation that cover the past 68 years. Muscheler *et al.* (2005) said:

> "It is standard practice to model the observed dilution of the atmospheric isotope ratios caused by the addition of isotopically depleted carbon from fossil and land-use sources. Emissions from fossil sources are prescribed, emissions from land-use sources are inferred from the atmospheric carbon budget, and the two-way exchange fluxes between reservoirs are simulated. The dilution of ^{14}C is governed by the same processes that affect ^{13}C. The good agreement between modeled ^{13}C and ice-core data supports the reconstructed rate of ^{14}C production."

However, this description is too vague for this writer to understand how Muscheler *et al.* (2005) took into account the ^{14}C-depleted fossil fuel emissions and atomic bomb tests conducted since 1950.

According to Muscheler *et al.* (2005), the ^{14}C production record was transformed into a record of a solar-modulation parameter that describes the solar influence on galactic cosmic-ray deflection by normalizing it to neutron monitor and ionization chamber data covering the recent decades. Alternatively, balloon-borne estimates of galactic cosmic-ray deflection were used instead of the ionization chamber data. Like Solanki *et al.* (2004), Muscheler *et al.* (2005) assumed that natural variations in the carbon cycle were small during the past millennium, which is consistent with ice-core CO_2 and $^{13}CO_2$ data and models. However, the entire procedure for data reduction and modeling remains obscure to this writer.

The results of Muscheler *et al.* (2005) suggest that the ^{14}C-production rate tracks the group sunspot number fairly well, although the large peak in the late-18th-century ^{14}C production rate curve appears to be anomalous. Variations in ^{14}C

production rate appear to be well correlated with solar magnetic activity. Similar results for the solar-modulation parameter were obtained. However, Muscheler *et al.* (2005) cautioned:

> "The link between the visually based sunspot numbers and solar-modulation parameter is neither straightforward nor yet understood, and also that solar modulation must have reached or exceeded today's magnitudes three times during the past millennium."

Considering that the sunspot numbers are observed directly and the ^{14}C production rate depends on assumptions and intricate models, it appears possible that the sharp peaks in the ^{14}C production rate may be artifacts.

Muscheler *et al.* (2005) argued that the reconstruction by Solanki *et al.* (2004) implies generally less solar forcing during the past millennium than in the second part of the 20th century. However, Muscheler *et al.* (2005) then went on to provide the "politically correct" assurance:

> "In any case, as noted by Solanki *et al.* (2004) solar activity reconstructions tell us that only a minor fraction of the recent global warming can be explained by the variable Sun."

There is no known way to prove or disprove such an assertion.

Steinhilber *et al.* (2009) carried out a reconstruction of TSI over the past 9,300 years. They estimated the historical open solar magnetic field from the cosmogenic radionuclide ^{10}Be measured in ice cores. The conversion of solar magnetic field to TSI was accomplished by the method of Frohlich (2009) in which the recent TSI data over the past 30 years were compared with the solar magnetic field. However, this time period is too short to be certain that such a relationship existed over the past 9,300 years.

Vieira *et al.* (2011) carried out an estimate of TSI during the Holocene based on ^{10}Be measurements in ice cores. The method used was summarized by them as follows:

(1) First, the measured values of $\Delta^{14}C$ are converted, using a carbon cycle model, into a global ^{14}C production rate, $Q^{14}C$ using a standard multi-box carbon cycle model.
(2) Using geomagnetic field data and a model of ^{14}C production in the atmosphere, the production rate $Q^{14}C$ was converted into a cosmic-ray spectrum quantified via a modulation potential Φ.
(3) Using a standard theory of cosmic-ray transport in the heliosphere, the modulation potential Φ was related to the open magnetic flux.

By some kind of manipulation that is unintelligible to this writer, this led to estimates of TSI for the Holocene.

Shapiro *et al.* (2011) presented their approach for modeling historical values of TSI. They defined the TSI at any time *t* as follows:

$$TSI(t) = Q(t) + A(t)$$

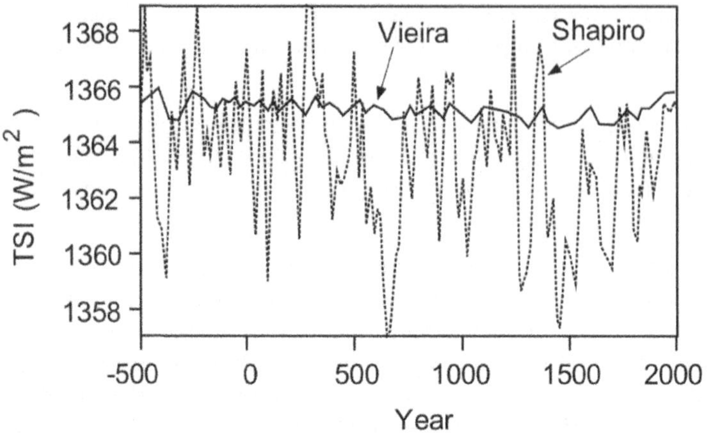

Figure 5.52. Reconstructions of TSI based on ^{10}Be proxy. The curve marked "Vieira" is the result of Vieira *et al.* (2011) while the curved marked Shapiro is due to Shapiro *et al.* (2011).

in which $Q(t)$ is the so-called "quiet Sun" contribution to TSI(t) attributed to that part of the solar surface that is unblemished, and $A(t)$ is the active contribution attributed to sunspots, plages, and network. The methods used to estimate $A(t)$ and $Q(t)$ are obscure to this writer,[12] but they depended on proxies for the ^{10}Be isotope. They ended up with the estimate of $TSI(t)$ shown in Figure 5.52. What is most fascinating about these reconstructions is the fact that different investigators obtain very different results using similar methods. If one compares Viera *et al.* (2011) with those of Shapiro *et al.* (2011), one finds vastly different results.

5.6.8 Temperature changes driven by the Sun

5.6.8.1 Global climate models

The impact of a change in TSI on the Earth is bound to be very complex. A number of groups have developed global climate models that analyze such effects (e.g., Meehl *et al.*, 2002; Doran *et al.*, 2002; Turner *et al.*, 2005). These studies typically examine the extent to which estimated solar, volcanic, and greenhouse forcing can account for past variations in temperature, as well as recent temperature rises in the late 20th century. Typically, such studies are mainly concerned with the potential impact of increased CO_2 emissions on global temperatures in the 21st century. Unfortunately, published estimates of historical TSI that are needed by these models suffer from a number of maladies as summarized in previous sections.

Ammann *et al.* (2007) employed a sophisticated global climate model to estimate the Earth's temperature history over the past 1,150 years. The time dependence of historical TSI was defined by an Antarctic ^{10}Be record. However, it was recognized

[12] Something may have been "lost in translation".

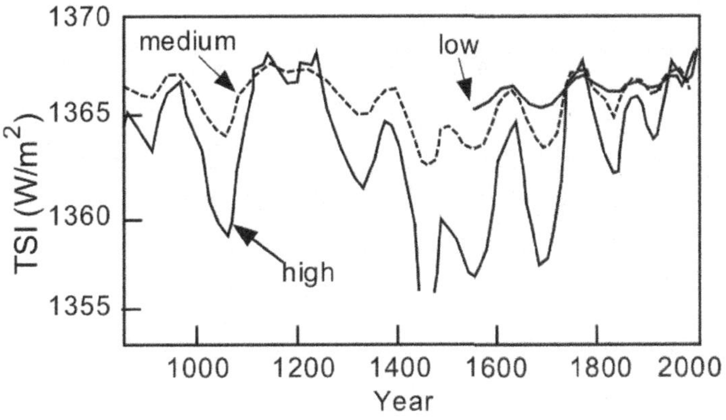

Figure 5.53. Estimated historical variation of TSI for three assumed differences between TSI at the MM vs. TSI today (high = 0.65%, medium = 0.25%, low = 0.1%) (adapted from Ammann *et al.*, 2007).

that there exist a range of estimates for past absolute variation of TSI and the essential unknown is how much lower was the TSI during the MM than it is at the current time? Ammann *et al.* (2007) treated this parametrically. Three cases were considered where TSI during the MM was lower by 0.65% (high), 0.25% (medium), and 0.1% (low) than at the present time (see Figure 5.47). Account was also taken of greenhouse gases and volcanism. Several sharp cooling episodes mark the response to very large volcanic forcing (e.g., 1258, 1453, 1815). The largest volcanic forcing was estimated for 1258 after what was probably the largest explosive eruption of the past few thousand years. Greenhouse gas forcing was based on a CO_2 equivalent of about 275 ppm in pre-industrial times, rising in the 20th century to about 425 ppm in 2000.

Figure 5.53 shows the three parameterized TSI functions. The procedure by which Ammann *et al.* (2007) scaled the historical TSI to obtain high, medium, and low profiles of TSI vs. time over the past 1,150 years (Figure 5.53) was based on Figure 5.47 that was generated by Bard *et al.* (2000).

Figure 5.54 shows the resultant temperature histories that result from the models.

As in the case of Figure 5.53, the relationship between curves in Figure 5.54 is confusing. How can the "high" curve ever exceed the "medium" curve? Common sense would dictate that the curves would be similar, with the "high" curve lying below the "medium" curve and the spacing between the two diminishing toward the 20th century. This is not the case in Figure 5.54.

One result that can be derived from Figures 5.53 and 5.54 is that, at the depth of the MM, the difference between the "high" and "low" curves for TSI is estimated to be about 8 W/m^2, while the difference in temperature anomaly is about 0.5°C. This would suggest a climate sensitivity parameter of $\lambda_S \sim 0.5/8 = 0.06$°C per W/m^2, or $\lambda \sim 0.06 \times 4 \sim 0.24$°C per W/m^2. However, if one divides the TSI by 4 to apply solar irradiance in space to the spherical Earth, λ approaches ~1°C per W/m^2 on this basis.

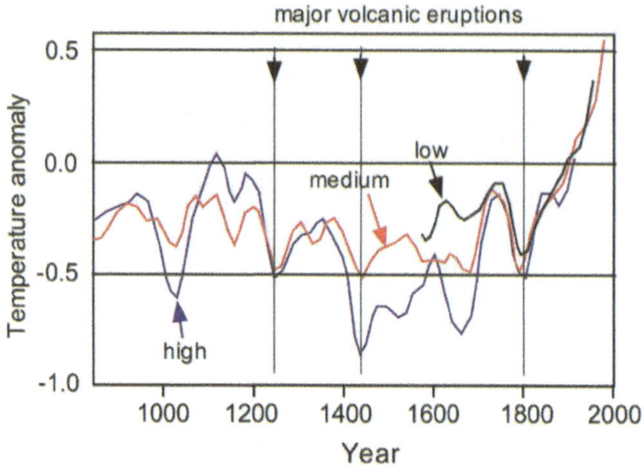

Figure 5.54. Estimated historical variation of temperature for three assumed differences between TSI at the MM vs. TSI today (high = 0.65%, medium = 0.25%, low = 0.1%). Note the temperature drops after each major volcanic eruption (adapted from Ammann *et al.*, 2007).

Goosse *et al.* (2005) used the TSI history estimates of Lean, Beer, and Bradley (1995) (basically, the "medium" case of Ammann *et al.* (2007) with 0.25% reduction in TSI during the MM) in a global coupled atmosphere–ocean–sea ice–land surface model of the Earth's climate. The model was forced over the period from 850 to 2000 using modeled time histories of solar irradiance, spatially estimated aerosol loading from explosive volcanism, the greenhouse gases (CO_2, CH_4, N_2O, CFC-11, and CFC-12), and anthropogenic sulfate aerosols with a recurring annual cycle of ozone and natural sulfate aerosol. The model's sensitivity to CO_2 doubling is a 1.8°C temperature rise, which is in the low range of coupled atmosphere–sea ice–ocean general circulation models (in fact, some models predict about double that temperature rise). Goosse *et al.* (2005) admitted (as they should) that "large uncertainties exist on forcing variations [i.e., variations in TSI] during the last millennium with consequences on simulated temperatures".

The result of running the model is shown in Figure 5.55. There is evidence of an MWP, an LIA, and a 20th-century warming beyond the warmth of the MWP. The estimated variations during the MWP and the LIA are small. This result follows from the forcings they input to their model which heavily emphasized toward greenhouse gases.

5.6.8.2 Climate sensitivity parameter

We are concerned here with approximations for the rise in global average temperature ΔT (°C) that is expected to result from an increase in TSI (W/m^2). It is common practice to define the resulting temperature change when an annually and globally averaged mean forcing function ΔF (W/m^2) is applied to the Earth at the top of the atmosphere

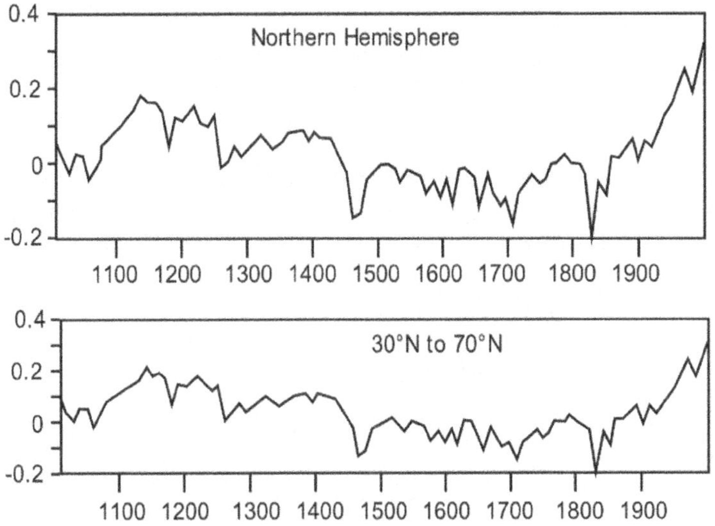

Figure 5.55. Estimated temperature anomaly for the last millennium (vertical scale = °C) (adapted from Goosse *et al.*, 2005).

$$\Delta T = \lambda\ \Delta F,$$

where λ is the so-called *climate sensitivity parameter*.

Most papers that deal with forcing are mainly concerned with forcing due to greenhouse gases. In this case, the forcing takes place within the Earth's atmosphere, and the area used to estimate ΔF is the surface area of the Earth $4\pi R^2$. Occasionally, some papers also deal with forcing of Earth temperatures caused by changes in TSI. However, there may be some ambiguity that needs to be resolved as to whether they refer to forcing at the TSI level ($\sim$1,366 W/m^2)[13] based on a plane in space facing the Sun, or at the Earth-input level (1,366/4 $\sim$342 W/m^2) based on the spherical surface area of the Earth. The factor of 4 arises because the area of the Earth as a disk facing the Sun is πR^2 whereas the spherical surface area of the Earth is $4\pi R^2$ so the TSI input of 1,366 W/m^2 is reduced to 342 W/m^2 when spread over the Earth's surface. In what follows, we will be concerned with ΔF representing a change in TSI at the in-space level. The difference between the two is that, for TSI:

$$\Delta T = \lambda_S\ \Delta\text{TSI}$$

and

$$\lambda_S = \lambda/4.$$

Beer, Mende, and Stellmacher (2000) used a simplistic radiation model to estimate that an increase in spherical forcing of 1 W/m^2 (at the top of the atmosphere) would cause an increase of the mean global temperature of 0.26°C. They showed that this

[13] As noted previously, this value now should be reduced to about 1,361 W/m^2.

forcing of the Earth energy balance by 1 W/m^2 (top of atmosphere) is equivalent to a 5.7 W/m^2 change in TSI. This was based on the area ratio (4) coupled with an assumed Earth albedo of 0.3. Hence, Beer, Mende, and Stellmacher (2000) suggest that:

$$\lambda_S = 0.26/5.7 \sim 0.046°C \text{ per } W/m^2 \text{ change in TSI.}$$

Reid (1997) estimated that a reduction of ~8 W/m^2 in TSI could have produced a global temperature change of ~0.45°C (i.e., $\lambda_S \sim 0.45/8 \sim 0.06$ °C per W/m^2).

Ramanathan *et al.* (1985) provided a summary of estimates of λ available in 1985. Eight estimates are listed that vary in a narrow range from 0.47°C to 0.53°C per W/m^2. For TSI, these values should be divided by 5.7 so the equivalent λ_S is ~0.08°C to –0.09°C per W/m^2.

Gerard and Hauglustaine (1991) indicated that: "A temperature response range of 1.1 to 2.3°C for a 1% solar total irradiance increase is predicted by climate models." This would suggest that:

$$\lambda_S \sim (1.1 \text{ to } 2.3)/13.7 \sim 0.08 \text{ to } 0.17°C \text{ per } W/m^2.$$

Senior and Mitchell (2000) estimated that a doubling of CO_2 would produce a downward forcing of ~3.5 W/m^2, leading to a global temperature increase of ~4°C suggesting a value of λ_S of ~(4/3.5)/5.7 = 0.20°C per W/m^2 for TSI forcing.

Gregory *et al.* (2004) appears to be an authoritative source for estimation of climate parameters. However, they were mainly concerned with CO_2 forcing and gave less emphasis to solar forcing. Nevertheless, their study is instructive. They used a relationship:

$$\Delta F = \Lambda \, \Delta T,$$

in which Λ, the *climate feedback parameter*, is actually the reciprocal of the *climate sensitivity parameter*, λ. Thus, $\Lambda = 1/\lambda$.

Gregory *et al.* (2004) assumed a positive forcing agent that is constant in time, generating an increased downward flux ΔF (W/m^2) at the tropopause. They assumed that the Earth environment responds to this perturbation with gradually increasing temperature and, thereby generating a gradually increasing upward flux ΔH(t) that is time-dependent. They assumed that H(t) is proportional to the temperature change ΔT(t) with $\Lambda = 1/\lambda$ being the proportionality constant. Thus, as time proceeds, the change in net flux downward is:

$$\Delta N(\mathrm{t}) = \Delta F—(1/\lambda)\,\Delta T(t).$$

At $t = 0$,

$$\Delta T(0) = 0 \text{ and } \Delta N(0) = \Delta F.$$

At some future time (t_F), a steady state is reached when $\Delta N(t_F) = 0$. At that point,

$$\Delta T(t_F) = \lambda \, \Delta F.$$

For our purposes here, we don't much care what t_F is; we are mainly interested in the ratio of the value of ΔT to ΔF (at the point where $\Delta N \to 0$). Thus, Gregory *et al.*

(2004) utilized a procedure in which they applied a constant forcing ΔF at the top of the troposphere, and used an atmosphere–ocean global climate model to calculate ΔN vs. ΔT for various times after the start. The slope of this line yields $(1/\lambda)$ and the y-intercept is ΔF.

For a forcing produced by doubling the CO_2 concentration in the atmosphere, Gregory *et al.* (2004) found that, depending on specifics of the model used, the values of ΔT were in the range 3.0°C to 3.8°C, and therefore the values of $\Lambda = (1/\lambda)$ were in the range 1.0 to 1.2 (W/m^2–°C).

For solar forcing, Gregory *et al.* (2004) found values of $(1/\lambda)$ in the range 1.3 to 2.0 (W/m^2-°C) based on forcing from the troposphere. The equivalent values of λ_S for TSI would then be in the range 0.09°C to 0.13°C per W/m^2.

Nozawa *et al.* (2007) provided a very brief discussion of climate feedback parameters. Using their climate model ("MIROC3.2"), they found that the appropriate values of $\Lambda = 1/\lambda$ for solar forcing (1.07 W/m^2-°C) and for CO_2 forcing (1.13 W/m^2-°C) were "consistent". However, the model used by Gregory *et al.* (2004) ("HadCM3") is claimed to lead to "inconsistent" results: $\Lambda = 2.0$ W/m^2-°C for solar and 1.26 W/m^2-°C for greenhouse gases. Using the value of $\Lambda = 1.07$ W/m^2-°C provided by Nozawa *et al.* (2007), the implied value of λ_S is then (1/1.07)/5.7 ~0.16°C per W/m^2.

Gregory (2004) provided a review of climate sensitivity and feedback parameters. They compared values of λ for various kinds of forcing. One model predicted that λ for CO_2 forcing is ~1.26 times λ for solar forcing. However, Gregory (2004) implied that this may be due to uncertainties in models and the two values of λ should be equal (note: as usual $\lambda_S = \lambda/5.7$). The value of λ_S was estimated to be about 0.125°C per W/m^2.

Cubasch *et al.* (2002) quoted a paper by Zorita *et al.* (2003) that provided four estimates of λ for TSI. The four estimates were made by comparing estimates of TSI with estimates of surface temperature over the period 1600–1900 AD. Unfortunately, neither the reconstructions of TSI nor those of temperature that underlie these estimates of λ appear to be trustworthy. However, for what they are worth, the four estimates of λ_S were:

(1) NCEP: 0.13°C per W/m^2;
(2) Jones *et al.* (1998) 0.13°C per W/m^2;
(3) ECHO-G: 0.11°C to 0.17°C per W/m^2;
(4) MBH99: 0.015°C to 0.08°C per W/m^2.

Ammann *et al.* (2007) mentioned that their "climate model results exhibited a response of 0.066°C in global temperature for each Watt per square-meter change in solar input" (i.e. $\lambda_S = 0.066$°C per W/m^2). The supplemental materials to this paper indicate that the calculation range was 0.062°C to 0.071°C per W/m^2.

Bony *et al.* (2006) summarized a number of estimates of λ based on water vapor in the atmosphere. The values of λ range from about 0.45°C to 0.8°C per W/m^2, with a mean value of about 0.6°C per W/m^2 which implies that λ_S ~0.11°C per W/m^2.

Hansen *et al.* (2005b) discussed the climate sensitivity parameter briefly. They pointed out that there is always a lag between sudden application of a steady climate forcing and the appearance of the full equilibrium Earth temperature rise. Indeed, as we pointed out previously, Gregory (2004) estimated λ by using global models to infer the (linear) rate of change of net forcing (constant forcing—reaction to forcing) and extrapolating to zero. Hansen *et al.* (2005b) claimed that this time lag depends on the rate of heat exchange between the ocean's surface mixed layer and the deeper ocean, and is a "sensitive function of the equilibrium climate sensitivity", (i.e., a sensitive function of λ) varying approximately as the square of λ. Hansen *et al.* (2005b) said:

> "The lag could be as short as a decade, if climate sensitivity is as small as 0.25°C per W/m^2 of forcing, but it is a century or longer if climate sensitivity is 1°C per W/m^2 or larger. Evidence from the Earth's history and climate models suggests that climate sensitivity is $\lambda = 0.75 \pm 0.25$°C per W/m^2, implying that 25 to 50 years are needed for the Earth's surface temperature to reach 60% of its equilibrium response."

If we can deduce an appropriate value of λ_S, we can convert any of the temperature reconstructions (prior to the 20th century) into equivalent variations in TSI, assuming the temperature changes prior to 1900 were caused mainly by variations in TSI (taking account of land clearing and volcanic eruptions). Similarly, we can reverse this process and convert reconstructions of past variations in TSI into equivalent changes in temperature. A number of estimates of λ_S have been made, as previously discussed. At the troposphere level, a rough guess for λ is 0.6°C per W/m^2, which would imply that, for TSI, the effective $\lambda_S \sim 0.105$°C per W/m^2.

Hansen (2004) said:

> "The composition of the ice age atmosphere is known precisely from air bubbles trapped as the Antarctic and Greenland ice sheets and numerous mountain glaciers built up from annual snowfall. Furthermore, the geographical distributions of the ice sheets, vegetation cover and coastlines during the ice age are well mapped. From these data we know that the change of climate forcing between the ice age and today was about 6.5 watts/m^2. This forcing maintains a global temperature change of 5°C, implying a climate sensitivity of 0.75 ± 0.25°C per watt per square meter. Climate models yield a similar climate sensitivity. The empirical result is more precise and reliable, however, because it includes all the processes operating in the real world, even those we have not yet been smart enough to include in the models."

However, his estimates of forcing and temperature change are uncertain, and others have deduced smaller sensitivities. While there is some significant divergence of opinion, it appears that a rough estimate for λ is 0.6°C per watt per square meter, and, on that basis, $\lambda_S \sim \lambda/5.7 \sim 0.10$°C per W/m^2. However, Schwartz (2007) deduced a lower value: $\lambda = 0.3$°C per W/m^2 from heat capacity arguments. But, if Schwartz's Earth time constant (five years) is doubled, his value of λ also doubles. However, this produced a flurry of comments (e.g., Foster *et al.* 2008) on this paper that led Schwartz to increase his estimate of λ to 0.5°C per W/m^2. As is the case with essentially all climate variables, there are many estimates, speculations, and assertions, but no credible answers.

5.6.8.3 Solar activity and climate change

Lean (2010) provided a review of solar activity and its relationship to climate change. Her viewpoint was upbeat:

> "With newly available modeled reconstructions of historical solar irradiance (*albeit with large uncertainties*), terrestrial studies are no longer relegated to using geophysically meaningless sunspot numbers as a proxy for solar irradiance, and direct comparisons of solar and other climate forcings are possible." (Emphasis added)

The last three solar cycles were observed from space and the small variations from S_{MIN} to S_{MAX} are well understood as being correlated with variability of sunspots and accompanying faculae, with the sunspots applying a darkening of ~ -1 W/m^2 and the faculae applying a brightening of ~ 2 W/m^2 in going from S_{MIN} to S_{MAX}. During this same period, global temperatures increased, but not uniformly, and certainly not in conformity with the steady rise in CO_2 concentration. As Lean (2010) pointed out: "Confounding expectations of a monotonically warming globe, the average warming rate from 2000 to 2008 subsided by almost an order of magnitude, and temperatures in 2008 were cooler than in 2002." Lean (2010) said:

> "There have been many attempts to reconstruct solar irradiance variations since the seventeenth-century Maunder Minimum, using the sunspot record, auxiliary evidence from Sun-like stars, direct associations with cosmogenic isotopes and geomagnetic activity, simulations with solar models of magnetic flux redistributed from sunspots to faculae and hypothetical solar interior structural changes. … In the reconstructions, the assumed longer-term irradiance changes manifest as slowly varying components underlying the 11-year activity cycle."

It is evident that Lean favors the CQSM models: "… solar irradiance cycles impart only modest global mean surface temperature changes (of $\sim 0.1°C$)."

de Jager (2008) provided an excellent description of the inner workings of the Sun. However, his discussion of "sun–climate relationships" suffers from a problem. The temperature profile he used for the period 1600–2000 was the *hockey stick*. The fact that the smoothed sunspot pattern for this period resembles the *hockey stick* in no way proves that there is a Sun–climate connection, particularly because the putative climate is not credible.

5.7 CONCLUSIONS ON TSI

Scientists abhor a vacuum. They can't seem to shrug their shoulders and admit that we just don't know the answers to important questions. They demand explanations, however speculative. Thus, we have theories on how life started on the Earth, how life begins from inanimate matter, how the universe began, what causes Ice Ages, and how the TSI varied over past millennia. There are many reconstructions of past TSI. Even the best of these are quite speculative. It could be argued that, in the absence of firm data on historical variations in TSI, even speculative models are better than nothing. As the saying goes: "In the land of the blind, a one-eyed man is king." And there is some merit to this argument. The one danger is we may unwittingly rely too heavily on speculations.[14] Thus, speculations tend to acquire a life of their own. The truth is that we just don't know how the Sun behaved in the past, although we have a few inklings.

One basic point is that, if significant temperature variations occurred on the Earth over the past millennium (MWP and LIA), it seems likely that such temperature changes were driven mainly by changes in TSI, land clearing, and volcanic eruptions, since these events occurred prior to industrialization. However, internally generated variability is also possible. The largest volcanic eruptions undoubtedly had a significant short-term impact but, overall, putative variations in TSI might have influenced the temperature changes experienced by the Earth prior to recent times. Thus, the histories of TSI variation and temperature changes are intimately entwined. On the other hand, some scientists have derived a *hockey stick* picture of past temperature variations which would minimize temperature variations during the MWP and the LIA. While there are good reasons for believing that the *hockey stick* result is fallacious, nevertheless controversy surrounds the uncertainty in historical temperatures as well as historical TSI. A necessary consequence of the *hockey stick* temperature profile is that past variations of TSI would have to be small. Thus, one can either believe the *hockey stick* temperature profile, in which case variations of TSI must have been very small during the past millennium, or one can believe that there were significant temperature excursions in the MWP and the LIA, driven at least partly by variability of TSI. The problem is that science is not supposed to rely on faith.

In attempting to reconstruct the TSI over the past several hundred years, or the past millennium, we only have bits and pieces of fragmentary evidence. Like a detective seeking to solve a crime from limited evidence, scientists have attempted to create models of past TSI based on fragmentary data. Unfortunately, although these models are often interesting and imaginative, they tend to be quite speculative and impossible to verify. However, there are a few things that we can be fairly sure of:

(1) We have had extraterrestrial measurements of TSI since about 1978. The Sun presently goes through periodic cycles of variable length, but approximately every 11 years. The TSI varies by about 0.1% during the 11-year solar cycle. At

[14] I am reminded of an event that occurred in the 1970s. At that time, there was some controversy regarding the altitude where the Earth's ionosphere transitioned from mainly O^+ to H^+. This was dependent on the rate of the reaction $O^+ + H \rightarrow O + H^+$. On the first day of a national meeting, a leading expert, Alec Dalgarno, made a presentation on this topic. At the end of his talk, someone asked him what the rate of the charge exchange reaction was. He said he didn't know. They pressed him to make a guess. So, he guessed. Three days later at the meeting wrap-up, Dalgarno was asked to present a summary. In doing this, he used the aforesaid reaction rate. Someone asked him where he obtained that figure. He replied: "I don't know. Someone provided it on the first day of the meeting!"

S_{MAX}, there are presently typically >100 sunspots and, at S_{MIN}, there are typically <10 sunspots.

(2) We have visual observations of sunspots that date back about 200 years. Sunspot data are characterized as (a) reliable from 1848 to the present, (b) good from 1818 through 1847, (c) questionable from 1749 through 1817, and (d) poor from 1700 through 1748. In addition, there is anecdotal evidence that the number of sunspots was minimal (or zero) for much of a 70-year period from 1645 to about 1715. Some properties of the Sun change significantly with time but it is not clear whether TSI varies significantly.

(3) The sunspot data clearly show that sunspot activity has been increasing since the start of the 20th century except for a brief dip around 1960 and, furthermore, sunspot activity is higher today than at any time in the past 400+ years, and possibly a lot longer. This shows that there have been significant changes in the behavior of the Sun, although it is not clear how this is reflected in changes in TSI.

(4) We have visual observations of the duration of the solar cycle that varied considerably since measurements were begun around 1750. Unfiltered cycle durations have swung widely from 8 to 14 years, and filtered cycle durations vary from roughly 10 to 12 years. The appearance of the Sun does change with time.

Another potential source of information is observations of putative Sun-like stars. Such studies indicate that the Sun's variability is slightly above average in regard to HK variations, but is well below average in regard to TSI variations. Attempts were made to estimate the TSI during the MM by comparison with the behavior of Sun-like stars. This required assumptions and extrapolation. A rather speculative approach for estimating the TSI of the Sun in the past was utilized by one group. In this approach, the HK index was correlated against recent measurements of TSI since 1978 and the regression line was extrapolated to the lowest HK index observed for non-cycling stars. With this assumption, the TSI during the MM was estimated to be 0.24% lower than the TSI at S_{MIN} in 1980. This work has had considerable influence and a number of climate studies adopted this value. Some stars do not cycle. It is not clear whether they are cyclic stars that have temporarily entered an MM-like period or whether they simply do not cycle. No definite conclusions can be drawn, although evidence exists for greater TSI variability in some stars. However, a recent paper has cast doubt on several aspects of these studies, including concerns that putative Sun-like stars may not resemble the Sun as closely as was thought.

A number of approaches have been used to reconstruct the past TSI over the last millennium. One approach that has been used by a number of investigators is called a "constant quiet Sun model" (CQSM) which is based on the current solar cycle. Reasoning that the TSI is presently $\sim 1{,}361.5\ W/m^2$ at S_{MAX} when there are >100 sunspots, and TSI is $\sim 1{,}360\ W/m^2$ at S_{MIN} when there are very few sunspots, one can assume a linear scale of TSI vs. number of sunspots, and use the sunspot observations of the past to estimate TSI in the past. In this approach, the past TSI can never drop below ~1,360 W/m^2 because that corresponds to ~0 sunspots. Thus, the lowermost possible TSI is defined, and TSI can increase slightly above this value in proportion to the sunspot count. This approach is implicitly based on the assumption that the Sun has behaved in the past as it does today, and no long-term secular changes have occurred. The only variations allowed to the Sun are those bounded by what we presently observe in the current solar cycle.

At the other extreme, Reid (1997) assumed that the Earth was about 1°C cooler during the MM, and attributed this as due to a lower TSI. His climate model indicated that, in order to achieve a 1°C temperature drop during the MM compared with recent times, the TSI must have been about 0.65% lower in the MM (about 1,352 W/m^2) than it was in 1980 (~1,361 W/m^2). He then assumed that, regardless of the small 0.1% fluctuations in TSI that occur during solar cycles, there was a secular change in TSI in proportion to a 15-year moving average of sunspot number. He was thus able to estimate TSI for all years from the MM to the present. His estimates of variability of TSI are much greater than those of other models. Reid found that the TSI bottomed out at around 1,352 W/m^2 during the MM, rose after 1700, dipped again just after 1800, and then rose back up and eventually increased during the 20th century.

It turns out that the duration of the solar cycle has varied considerably since measurements were begun around 1750. Unfiltered cycle durations have swung widely from 8 to 14 years, and filtered cycle durations vary from roughly 10 to 12 years. By developing a relationship between observed TSI and filtered cycle duration since 1978, some investigators have estimated TSI back to 1750 from the historical filtered cycle duration over that period. Solanki and Fligge (2000) did this but their procedure was not clearly defined, and the results seem to contain anomalies, although it does produce a dip shortly after 1800. The main problem appears to be that the database for TSI since 1978 is too short to arrive at a firm correlation with filtered cycle duration. This approach is limited by the same factors that limit CQSMs: there is no allowance for long-term secular changes in the Sun independently of changes in cycle duration.

Some estimates of TSI in the past have been made from measurements of cosmogenic isotopes. A good example is Bard *et al.* (2000). A curve for cosmogenic isotope production rate vs. year is derived for the past millennium or so. The question is how to convert isotope production rate to TSI. Bard *et al.* (2000) converted the isotope production rate to TSI on a parametric basis, by setting TSI equal to the current value for the most recent production rate, and using the reduction in TSI during the MM compared with the current TSI as an adjustable parameter to create a family of TSI vs. year curves for various assumed reductions in TSI during the MM (0.25%, 0.40%, 0.65%). This seems to be a rational procedure, but we still don't know which curve to choose.

Lockwood and Frohlich (2007) wrote a paper that shows that variations in TSI cannot possibly explain temperature changes over the past 20 years. This is clear from the outset and we did not need their 14 pages of analysis to prove this self-evident point. Nevertheless, there is no evidence that solar variability did not affect climate over longer periods of time.

In summary, it is clear that the appearance of the Sun has varied over the past millennium as evidenced by visual observations of sunspots and cycle duration, observations of Sun-like stars, and studies of cosmogenic isotopes. Unfortunately, it is not at all clear what these changes imply quantitatively for the range of variation of TSI.

5.8 SOLAR INTENSITY AT GROUND LEVEL

Wang *et al.* (2012) modeled solar intensity at ground level. As they pointed out, "The amount that is incident at the surface has been shown to undergo significant decadal variations by a surface network of radiometers". They made use of "Sunshine Duration" (SunDu), a measurement initiated 150 years ago, which records the time during a day that the direct solar beam irradiance exceeds 120 W/m^2. Most stations were located in a narrow band of latitudes in the U.S., Europe, and China. They found that Europe brightened perceptibly in the late 20th century, whereas China dimmed slightly. This is probably related to pollution control in Europe and lack thereof in China.

6

The Earth's heat balance and the greenhouse effect

6.1 THE GREENHOUSE EFFECT

6.1.1 Terrestrial examples

This section was abstracted from Gerlich and Tscheuschner (2007).

Radiation is the only means of heat transfer between bodies in the vacuum of space. The radiation emitted by a blackbody plane surface is a spectrum over many wavelengths. When integrated over all wavelengths, the emitted irradiance is:

$$I = \sigma T^4,$$

where $\sigma = 5.67 \times 10^{-8}$ W/(m^{2}K^4).

This integrated form of the radiation equation is only correct for an ideal blackbody. In the real world, radiation depends on geometry, and most bodies exhibit deviations from blackbody behavior, in that their emissivities and absorptivities deviate from unity in at least some parts of the spectrum. In many cases, the deviations are large.

In an atmosphere, heat can be transmitted between bodies via radiation, conduction, or convection. On the Earth, in the atmosphere at moderate temperatures, heat transfer between bodies and the atmosphere is typically dominated by convection, and radiation usually plays a secondary role.

The standard explanation for why a greenhouse (or equivalently, an automobile parked in sunlight) heats up is based solely on radiation. Gerlich and Tscheuschner (2007) provide many examples of these often used descriptions from books and articles. These descriptions go something like the following. Incident sunlight passes through the glass enclosing the greenhouse, and is absorbed by the interior, thus heating the interior. As the interior heats up, it radiates in the infrared (IR). However, the glass absorbs IR and thereby prevents the IR from passing through. Thus, the IR irradiance is trapped in the glass, which restricts the ability of the greenhouse to lose heat to the environment.

In actuality, the standard explanation for the heating of a greenhouse or an automobile in sunshine is fallacious. A greenhouse or an automobile gains heat from

D. Rapp, *Assessing Climate Change*, Springer Praxis Books,
DOI: 10.1007/978-3-319-00455-6_6

solar irradiance but loses heat to the surroundings mainly by convection to air surrounding the exterior glass. The reason that the greenhouse (or an automobile in the sunshine) heats up is because the thermal conductivity of the glass is sufficiently low that heat transmission from the hot air within the greenhouse through the windows to the exterior is slow, and the interior temperature must rise sufficiently so that the temperature difference across the window glass is high enough that conduction through the glass provides enough heat loss to balance the heat input from the Sun. Hence, the IR absorptivity of the glass has very little to do with why a greenhouse or an automobile in sunshine gets hot. In fact, the glass does not get hot, showing that it is not absorbing a great deal of IR.

Gerlich and Tscheuschner (2007) refer to examples where two insulated black boxes were placed in the Sun. One had a cover made of IR-absorbing glass and the other had a cover made of an IR-transmitting NaCl crystal. In order to make the two cases as compatible as possible, a glass was placed above both boxes, because otherwise the NaCl cover would transmit more of the incident solar IR into the box. The two boxes reached the same temperature (55°C). The IR absorptivity of the cover had nothing to do with the interior temperature. If this experiment could be repeated with both boxes surrounded by a vacuum instead of the Earth's atmosphere, the boxes could only lose heat to the surroundings by radiation. In this case, the box covered with NaCl would be cooler, since it can transmit IR through its transparent window and there would indeed be a greenhouse effect with the glass cover.

Gerlich and Tscheuschner (2007) provided the data shown in Table 6.1. Within a car parked in sunshine, surfaces exposed to the Sun can heat up to 71°C. This generates air currents within the car that warm the shaded areas within the car to as much as 39°C. By comparison, outside the car, the difference between being in the Sun or the shade is minimized by convective air currents. As Gerlich and Tscheuschner (2007) pointed out:

> "One can touch the car's windows and notice that the panes, which absorb the infrared light, are rather cool and do not heat the inside of the car in any way. If one holds his hand in the shade next to very hot part of the dashboard that lies in the Sun, one will practically feel no thermal radiation despite the high temperature of 70°C, whereas one clearly feels the hot air."

Table 6.1. Measured temperatures in and around a car parked in sunshine.

Location	*Temperature (°C)*
Inside the car, in direct sunshine	71
Inside the car, in the shade	39
Next to the car, in direct sunshine	31
Next to the car, in the shade	29

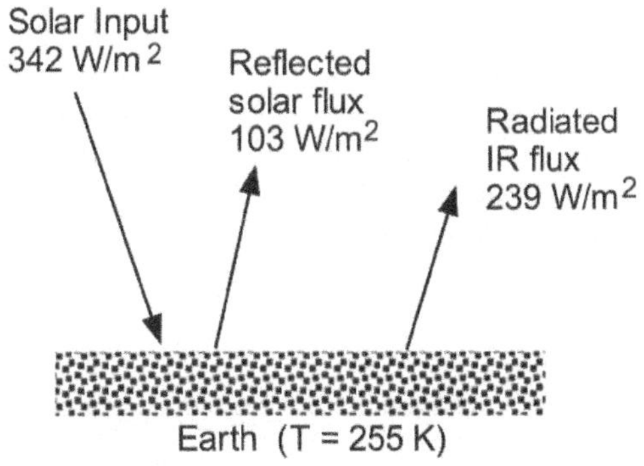

Figure 6.1. Radiant heat balance for a hypothetical Earth with no atmosphere.

6.1.2 Simplistic models of the Earth

The Earth can only lose heat to its surroundings by radiation. If the Earth had no atmosphere, its surface temperature would resemble that of the Moon. In the daytime, the mid-latitude temperature would rise to perhaps 100°C, and at night it would drop to perhaps −150°C. Temperatures at the poles would be decidedly lower than at the equator.

To illustrate the greenhouse effect, a number of books and websites present a simplistic average picture of the Earth without an atmosphere as a baseline, as shown in Figure 6.1.

The solar power approaching the Earth is about 1,367 W/m^2 passing through a plane above the Earth, facing the Sun. Since the area of this plane projected by the Earth is πR^2, where R is the radius of the Earth, and the surface area of the Earth is $4\pi R^2$, the solar power falling on the Earth would be 1,367/4 $\sim$342 W/m^2 per unit surface area of the Earth—if it impinged uniformly on all areas of the Earth. The Earth's albedo (reflection coefficient for solar irradiance) is estimated to be roughly 30%. If $\sim$30% of the solar irradiance on the Earth is reflected back into space, the net solar irradiance absorbed by the Earth (per unit surface area) is 0.7 × 342 = 239 W/m^2. This hypothetical Earth would warm up and radiate until the power radiated is equal to the power input from the Sun, whereupon equilibrium will be established. Thus the temperature of the surface of the Earth is calculated from:

$$239 \text{ W/m}^2 = \sigma\, T_E^4 = 5.67 \times 10^{-8} \text{ (W/m}^2\text{-K}^4)\, T_E^4$$
$$T_E = 255 \text{ K} = -18°\text{C}.$$

In this artificial model, the daily average solar input to an average element of the Earth's surface area is treated as a constant, whereas, in reality, it will vary widely from day to night. Nevertheless, the calculation is instructive. It suggests that the Earth would be (on average) a relatively colder place if it had no atmosphere. Actually, this model is very artificial because the atmosphere plays a major role as an intermediary between the Earth's surface and space.

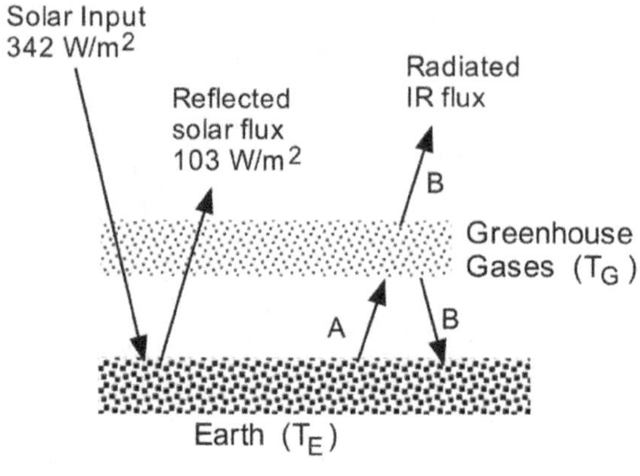

Figure 6.2. Simplistic model of effect of greenhouse gases on Earth temperature.

An artificial, simplistic averaged model is often used in books and web-sites to illustrate the "greenhouse effect" by introducing an overly simplistic atmosphere between the Earth and space that does nothing else but (1) transmit incident solar radiation, (2) absorb long-wavelength IR emitted by the Earth, and (3) reradiate long-wavelength IR both upward to space, and downward to the Earth. This model is illustrated in Figure 6.2.

The solar input to the Earth remains at 342 W/m^2 and is (according to this model) unimpeded by the fictional atmosphere. Of this, 103 W/m^2 is reflected back to space. The Earth radiates a flux A (W/m^2) that depends on its temperature (T_E), and the atmosphere absorbs all of this radiation reaching temperature T_G, and re-emits a flux B upward and a flux B downward, where $A = 2B$.

At equilibrium, the Earth and its atmosphere must emit as much radiation flux as it receives. Thus:

$$B = (342\text{–}103) = 239 \text{ W/m}^2.$$

Thus

$$T_G = 255 \text{ K}.$$

In this model, the surface that radiates to space has been moved from the Earth's surface to the top of the atmosphere. Since $A = 2B$, we find that $A = 478$ W/m^2 and therefore

$$T_E = (478/\sigma)^{1/4} = 303 \text{ K}.$$

This is considerably warmer than the previous baseline case with no greenhouse effect. According to this admittedly overly simplistic model, the presence of greenhouse gases (GHGs) provides significant heating to the Earth's surface, and, without the greenhouse effect, the Earth would be a very cold planet. This simple greenhouse effect provides a temperature increase of 48°C over the simple model without the greenhouse effect. This simple greenhouse effect is based on the assumption that there is an intermediate level between the Earth and space that absorbs radiation emitted by the Earth and reradiates upward and downward. However, the atmosphere is far more complex than this model allows for, and the transfer of heat between the surface and the atmosphere includes a major contribution from convection and latent heat of water vapor. In addition, the transmission of various wavelength bands through the atmosphere at different levels is quite variable and not amenable to simple description. It should also be noted that most of the greenhouse effect is produced by water vapor, not the so-called greenhouse gases.

6.1.3 More realistic description

In reality, the Earth's greenhouse effect does not actually operate in the manner described in Figure 6.2. A more realistic description was provided by Lindzen (2007). Lindzen and Choi (2011) provided a succinct description of the greenhouse effect:

> "This simply refers to the fact that the Earth balances the heat received from the Sun (mostly in the visible spectrum) by radiating in the infrared portion of the spectrum back to space. Gases that are relatively transparent to visible light but strongly absorbent in the infrared (greenhouse gases) will interfere with the cooling of the planet, thus forcing it to become warmer in order to emit sufficient infrared radiation to balance the net incoming sunlight. By the net incoming sunlight, we mean that portion of the Sun's radiation that is not reflected back to space by clouds and the Earth's surface. The issue then focuses on a particular greenhouse gas, carbon dioxide. Although carbon dioxide is a relatively minor greenhouse gas, it has increased significantly since the beginning of the industrial age from about 280 ppm to about 390 ppm, and it is widely accepted that this increase is primarily due to man's emissions."

A simple explanation for why increased CO_2 concentration leads to increased absorption of IR radiation emitted by the Earth is illustrated in Figure 6.3.

Carbon dioxide absorbs outgoing IR radiation emitted by the Earth primarily in the absorption band in the wavelength band between 13 mm and 17 mm. The absorption of any wavelength in the atmosphere is dependent on the integral of the absorptivity times the concentration on a vertical path through the atmosphere. This integral is called the absorption factor. Since the absolute amount of absorption depends on an exponential function of the absorption factor, an absorption factor of 3 corresponds to about 99% of complete absorption. As Figure 6.3 shows, at the pre-industrial level of 280 ppm of CO_2, the entire absorption band from 13 mm to 17 mm is fully saturated. Adding more CO_2 to the atmosphere does not increase the absorption significantly within this saturated region. Only at the "wings" of the absorption band is there any significant increase in absorption by the atmosphere when the CO_2 concentration is increased. Thus, at a CO_2 concentration of 280 ppm, absorption is saturated between vertical dashed lines A and C in Figure 6.3.

If, in the future, the CO_2 concentration is doubled (from the pre-industrial value) to 560 ppm, the absorption curve moves up by a factor of 2 on the vertical log scale, and the saturated region expands to the region between vertical lines *B* and *D*,

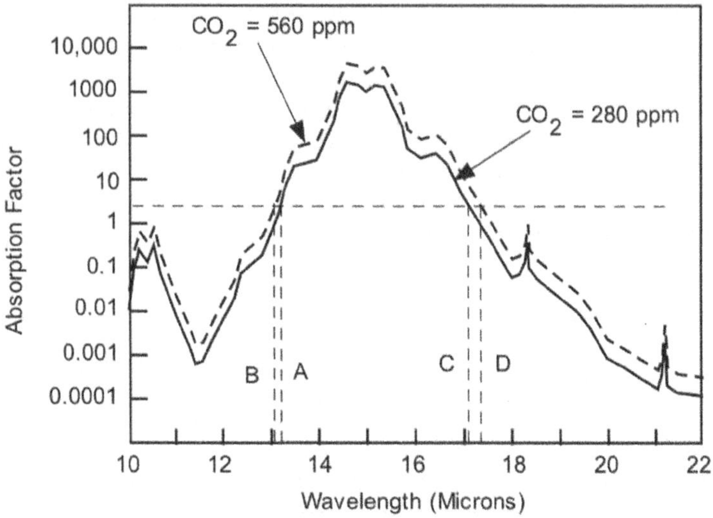

Figure 6.3. Absorption factor (absorptivity concentration integrated over vertical path through atmosphere) for CO_2 vs. wavelength. The horizontal dashed line corresponds to an absorption factor of 3 (essentially complete absorption along a vertical path through the atmosphere).

producing additional absorption of IR. However, the additional absorption (from absorption in regions between *A* and *B*, and between *C* and *D*) is much smaller than the original absorption in going from 0 ppm to 280 ppm (region between *A* and *C*). Thus, we see that, as more and more CO_2 is added to the atmosphere, the additional absorption decreases per unit amount of CO_2 added.

Lindzen (2007) described the effect of additional absorption by CO_2 and other greenhouse gases in the atmosphere. He pointed out that the simplistic description of the greenhouse effect is that greenhouse gases and clouds "inhibit cooling of the Earth by thermal radiation, and serves as a blanket which causes the Earth to be warmer than it otherwise would be". Lindzen objected to this description because, as he said:

> "… the surface of the Earth does not cool primarily by thermal radiation". … There is so much greenhouse opacity immediately above the ground that the surface cannot effectively cool by the emission of thermal radiation. Instead, heat is carried away from the surface by fluid motions ranging from the cumulonimbus towers of the tropics to the weather and planetary scale waves of the extratropics. These motions carry the heat upward and poleward to [altitudes] where it is possible for thermal radiation emitted from these levels to escape to space."

The lower atmosphere is relatively opaque to IR radiation but, as the altitude increases, the density decreases and transmission of IR improves. Lindzen defined an altitude sufficiently high that the optical depth for IR is about 1 so that transmission

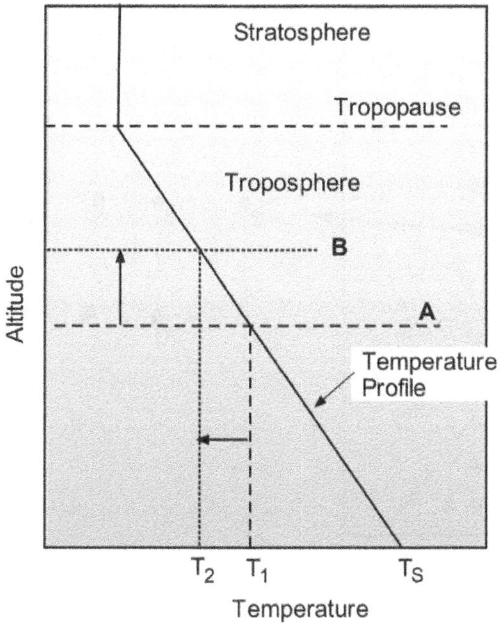

Figure 6.4. Lindzen's picture of how the greenhouse effect works.

of IR is attenuated roughly as e^{-1}. This is a region of the atmosphere that can radiate energy from the Earth to space. He provided a description of the effect of adding greenhouse gases to the atmosphere as shown in Figure 6.4.

The temperature of the atmosphere decreases with altitude to some level known as the tropopause. The height of the tropopause varies from about 16 km in the tropics to about 12 km at 30° latitude, and to about 8 km in polar regions. Before adding greenhouse gases to the atmosphere, the altitude at which the optical depth for IR was around 1 is denoted *A* in Figure 6.4. When greenhouse gases are added to the atmosphere, the level at which the optical depth is around 1 is raised in altitude due to the additional absorption by the greenhouse gas. As Lindzen pointed out, "because the temperature of the atmosphere decreases with altitude, the new characteristic altitude for emission is colder than the previous altitude". The new altitude for emission is *B*, and $T_1 > T_2$. Since IR emission is proportional to the fourth power of the temperature, emission from altitude *B* is reduced compared to what it was at altitude *A*. Thus the Earth cannot cool as effectively. The Earth will therefore warm until the temperature at altitude *B* approaches the original temperature. This is shown in Figure 6.5. The original temperature at the surface is T_S, leading to the vertical temperature profile shown as the slanted dotted line. Before the temperature rises, the temperature at the altitude for emission is T_2 at point *2*. After the surface temperature rises to $T_{S'}$, the new vertical temperature profile is the slanted solid line and the temperature at altitude *B* rises back to T_1. The Earth is now back in thermal balance.

Lindzen also pointed out that climate models indicate that, when the Earth warms due to the greenhouse effect, warming is strongly peaked in the tropical troposphere near the altitude where the optical depth $\tau \sim 1$ (see, e.g., Figure 3.34). He concluded:

> "Roughly speaking, the warming at $\tau = 1$ in the tropics is about two to three times larger than near the surface regardless of the sensitivity of the particular [climate] model. This is, in fact, the signature (or fingerprint) of greenhouse warming. Stated somewhat differently, if we observe warming in the tropical

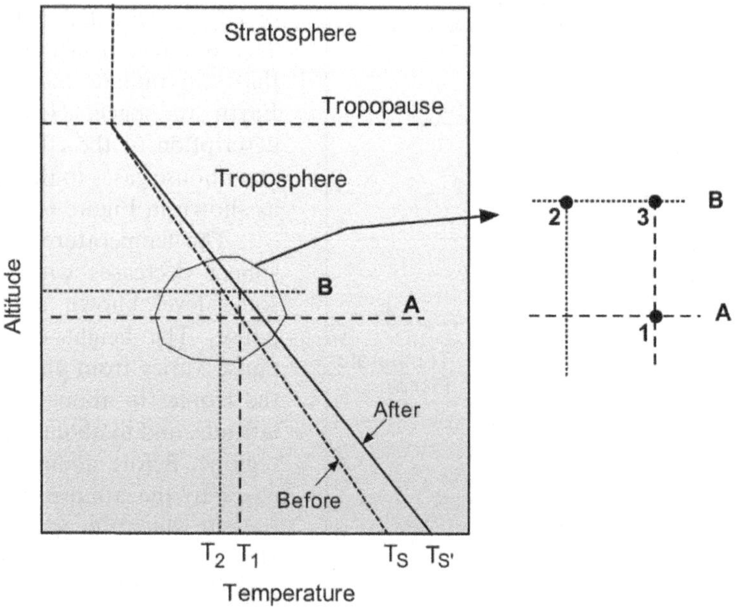

Figure 6.5. Lindzen's picture of how the greenhouse effect works, part two.

upper troposphere, then the greenhouse contribution to warming at the surface should be between less than half and one third the warming seen in the upper troposphere."

6.1.4 Absorption by greenhouse gases

A variety of climate models have been developed (see Section 6.4) with the primary purpose of predicting the effect of future greenhouse gas emissions on the climate in the 21st century. Most of this work has been carried out by climate alarmists, and it is not clear to what extent the findings influenced the scientists, or the extent to which scientists influenced the findings.[1]

The 2001 Intergovernmental Panel on Climate Change (IPCC) Report provides data on the various greenhouse forcings due to: CO_2, CH_4, N_2O, Halocarbons, Stratospheric O_3, Tropospheric O_3, aerosols, sulfate, carbon, biomass burning, mineral dust, aerosol indirect, contrails, land use, albedo, and solar variations. However, there does not seem to be much mention of water vapor or feedbacks, and it is not clear what to do about clouds. For each greenhouse gas, the IPCC has tabulated pre-industrial and current concentrations in the troposphere, lifetime of gases in the atmosphere, current rates of emission, and global-warming potential. The global-warming potential is a function of the absorptivity (per unit

[1] Just as earthquake specialists always seem to predict that "the big one" is coming, so one must wonder about the climatologists who predict a dire future for the planet.

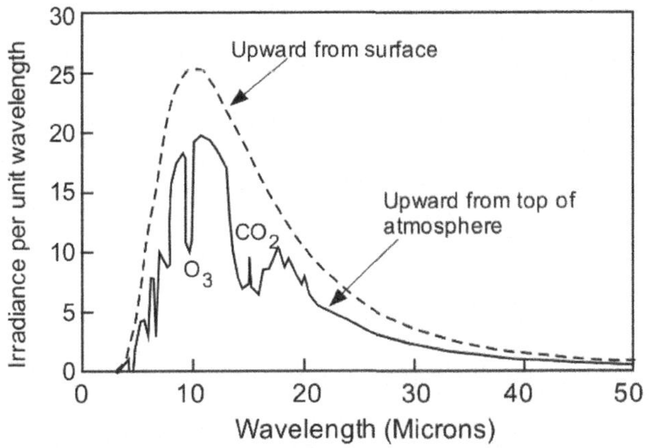

Figure 6.6. Comparison of spectral distribution of upward irradiance from the surface with that at the top of the atmosphere showing CO_2 and O_3 bands for assumed global cloudy conditions. Water absorption occurs throughout the spectrum in various amounts (adapted from Schimel *et al.*, 2001).

concentration of greenhouse gas) of IR radiation emitted by the Earth. The principal greenhouse gases are water vapor, carbon dioxide, and methane, although the combination of nitrous oxide plus a wide variety of halocarbons produces an effect comparable with that of methane. Figure 6.6 shows a comparison of spectral distribution of upward irradiance from the surface with that at the top of the atmosphere showing CO_2 and O_3 bands for assumed global cloudy conditions.

The term "radiative forcing" has been used in several contexts. According to Kiehl and Trenberth (1997):

> "The difference between the surface emission and the top-of-atmosphere emission defines the long-wave radiative forcing, which clearly illustrates that strong atmospheric absorption occurs at 15 microns by carbon dioxide and 9 microns by ozone, while the effects of water vapor are distributed throughout all wavelengths. Note that the radiative forcing centered at 15 microns extends from 12 to 18 microns, owing to numerous absorption bands of the vibration mode of the CO_2 molecule. Indeed, it is this series of bands centered at 15 microns that ensures that this band is not near saturation for present and future projected amounts of CO_2, although it further means that long-wave radiative forcing from increases in CO_2 is not linear but more closely approximates a logarithmic increase."

Figure 6.7 shows the downward radiant forcing (difference between curves in Figure 6.6) for assumed global cloudy conditions. Integrating over all wavelengths leads to a total long-wave radiative forcing of 155 W/m^2. As stated previously, the long-wave cloud forcing is 30 W/m^2. Thus the clear-sky radiative forcing is estimated in this case to be 125 W/m^2. The contribution of each absorber to the downward forcing was estimated by Schimel *et al.* (2001) as shown in Table 6.2.

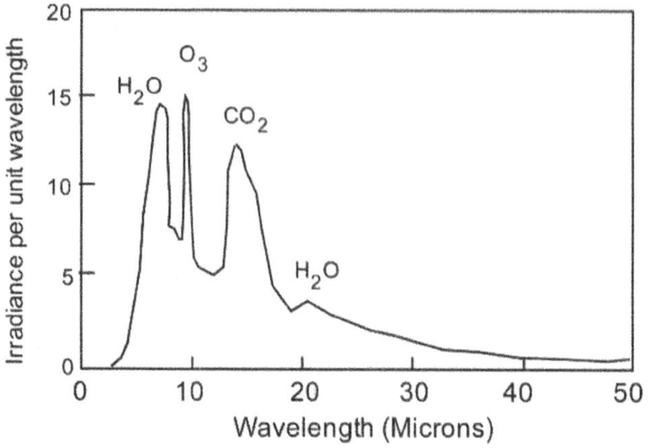

Figure 6.7. Downward radiant forcing (difference between curves in previous figure) for assumed global cloudy conditions (adapted from Kiehl and Trenberth, 1997).

Table 6.2. Percentage of downward forcing due to each absorber. (Schimel *et al.*, 2001).

Absorber	*Clear skies*	*Cloudy conditions*
Clouds	0	19
Water vapor	60	
CO_2	26	
Ozone	8	
Other gases	6	

However, in most papers in the literature, a different definition is used in which forcing refers to the additional downward radiant heat flux at the top of the atmosphere due to a change in a quantity that affects the Earth's heat balance, such as a change in greenhouse gas concentration, a change in solar irradiance, a change in aerosol concentration, etc.

More importantly, the claim that CO_2 is "not near saturation" is very misleading. The absorption in the 13–17-micron band of CO_2 is indeed saturated in the sense that essentially all of the upward radiation in this band is absorbed by the atmosphere. However, the excited CO_2 then reradiates in all directions. Thus, radiation in the 13–17-micron band traverses tortuous paths through the atmosphere with repeated absorption and re-emission, and about half reaches the top of the atmosphere. But increasing the concentration of CO_2 in the atmosphere will not change the fraction that reaches the top of the atmosphere very much because the atmosphere is already strongly absorbing in this wavelength range. Only at the wings of the absorption band where the absorption coefficient is much lower, can absorption be enhanced by higher CO_2 concentration. This is illustrated schematically in Figure 6.3.

6.1.5 Carbon dioxide as a greenhouse gas

A simplistic explanation for why increased CO_2 concentration can lead to increased temperature of the Earth was discussed in connection with Figure 6.3 in Section 6.1.3. In that section, we showed that, as more and more CO_2 is added to the atmosphere, the heating effect decreases per unit amount of CO_2 added.

Hansen *et al.* (2000) estimated the forcing on the climate due to changes in CO_2 concentration as shown in Figure 6.8. Estimating the quantitative shape of the curve in Figure 6.8 is of great importance for understanding the effect of future CO_2 emissions on future climate change. In Figure 6.8, vertical lines represent:

A = typical CO_2 at glacial maximum in an Ice Age;
B = typical CO_2 during an interglacial period between Ice Ages;
C = current CO_2 level due to human impact on environment; and
D = CO_2 level after it doubles compared to pre-industrial levels; the estimated forcings are shown as vertical double arrows:
F_1 = forcing in the transition from a glacial maximum to an interglacial period (~3 W/m^2);
F_2 = forcing due to CO_2 rise since before the industrial period to the present (~2 W/m^2); and
F_3 = forcing due to change from pre-industrial levels to doubled CO_2 (~4 W/m^2).

According to these calculations, the rise in CO_2 from pre-industrial times to the present has already produced about half the forcing that will result from doubling CO_2 from pre-industrial times.

However, as Chilingar *et al.* (2008) emphasized, transfer of heat from the Earth's surface to the troposphere occurs roughly 67% by convection, 25% by latent

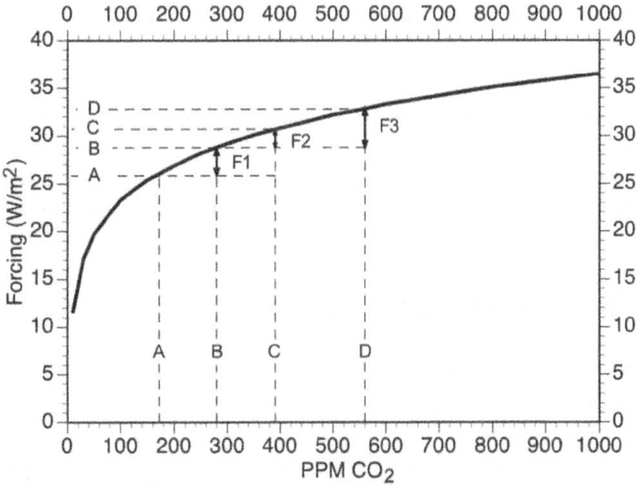

Figure 6.8. Estimated forcing of the climate due to changes in CO_2 concentration (Hansen *et al.*, 2000).

heat from condensation of water vapor, and only 8% by radiation. These authors therefore concluded that increasing CO_2 would only have a minor impact on this heat transfer, since they claimed that radiation plays a minor role in heat loss by the Earth's surface. However, the only way that the Earth loses heat to space is via radiation. Therefore, over some altitude range in the upper troposphere (UT), radiation must become the dominant form of heat transfer, and the absorption by CO_2 must act as an impediment to loss of heat by the Earth. The actual effect of additional absorption by CO_2 is a change in the vertical profile of temperature through the atmosphere, which results in a change in outward radiant emission from the atmosphere (Lindzen *et al.*, 1982; Lindzen, 1997, 2007).

6.1.6 Methane as a greenhouse gas

Ice-core records show that the concentration of methane in the atmosphere closely tracked the temperature changes that took place over the past several hundred thousand years, as the climate cycled between Ice Ages and interglacial periods. Typical methane concentrations during Ice Ages were roughly 400 ppb, and rose to 650–700 ppb during interglacials. These changes were presumably due to the effects of global temperatures on wetland emissions. During the 20th century, the methane concentration rose to about 1,700 ppb, presumably due to the combination of addition of anthropogenic sources.

Wuebbles and Hayhoe (2000) provided an extensive detailed review of the sources and sinks for methane in the atmosphere, as well as the role of methane in affecting the climate. The main natural source of methane is wetlands (~72%), with oceans, termites and other sources contributing the remainder. In 1995, the anthropogenic sources of methane were estimated to be extraction of fossil fuels such as natural gas, coal mining, and petroleum (~26%), waste disposal (~24%), rice cultivation (~16%), biomass burning (~11%), and ruminants (~23%). Reaction with atmospheric hydroxyl radicals (OH) is responsible for the removal of about approximately 90% of the methane that is destroyed. The remainder of the methane is removed through dry soil oxidation or transport to the stratosphere. OH is formed by photo-dissociation of tropospheric ozone and water vapor.

Methane has a lifetime in the atmosphere of about 10 years and is more effective than carbon dioxide (by about a factor of 40 per unit concentration) at absorbing IR emitted by the Earth. Fortunately, the concentration of methane is about 200 times lower than that of carbon dioxide. But both are increasing with time. Nevertheless, the rate of increase of atmospheric methane has slowed down considerably. From 1982 to 1992, the methane concentration increased by 1% per year, whereas, from 1992 to 2005, it increased by only ~0.3% per year (Robinson *et al.*, 2007; Khalil *et al.*, 2007; Idso and Idso, 2007). Khalil, Butenhoff, and Rasmussen (2007) showed that the methane concentration in the atmosphere reached a plateau in the 1990s and is no longer increasing. Figure 6.9 suggests that the trend in atmospheric methane concentration, "has been decreasing for the last two decades until the present when it has reached near zero," and that "it is questionable whether human activities can cause methane concentrations to increase greatly in the future" (Khalil *et al.*, 2007).

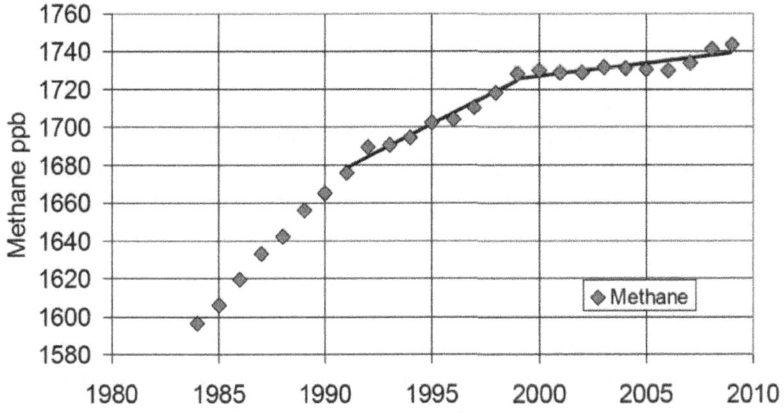

Figure 6.9. Atmospheric methane concentrations at Cape Grim (latitude 40°S). Values before 1991 are affected by fugitive natural gas from pipeline leakage (Quirk, 2012).

Simpson *et al.* (2012) reported that the rate of growth of methane concentration in the atmosphere had tailed off roughly linearly from about 15 ppb in 1980 to roughly zero in 2010, which they attributed to "sharply reduced skyward venting and flaring of methane from oil fields".

However, as Shakhova *et al.* (2010) reported, remobilization to the atmosphere of only a small fraction of the methane held in East Siberian Arctic Shelf (ESAS) sediments could release large quantities of methane, which they believe could "trigger abrupt climate warming". Nevertheless, they believe that "sub-sea permafrost acts as a lid to keep this shallow methane reservoir in place".

6.1.7 Water vapor as a greenhouse gas

Water vapor is by far the most important greenhouse gas, in the sense that it absorbs more irradiance from the Earth than all other greenhouse gases combined. As we showed in Table 6.2, Schimel *et al.* (2001) estimated that water vapor accounts for 60% of the total greenhouse effect. Leroux (2005) suggested the much higher value of "95.00%" but provided no basis for this figure and it is probably not credible.

A number of skeptic blogs have criticized the global climate modelers and alarmists (often the same people) who typically do not list water vapor as a greenhouse gas. For example, Leroux (2005) said that IPCC (2001) underemphasized the role of water vapor, and did not even include water vapor in its tables and figures depicting the effects of greenhouse gases. However, the reason for this is that climate modelers are not interested in the steady-state relatively unchanging role of water vapor as a greenhouse gas, but rather, they are interested in the changes induced by human activity.

Thus, according to the *realclimate.org* blog:[2]

[2] *www.realclimate.org/index.php?p=142.*

> "Whenever three or more contrarians are gathered together, one will inevitably claim that water vapor is being unjustly neglected by IPCC scientists. 'Why isn't water vapor acknowledged as a greenhouse gas?' 'Why does anyone even care about the other greenhouse gases since water vapor is 98% of the effect?' 'Why isn't water vapor included in climate models?' 'Why isn't water vapor included on the forcings bar charts?' etc. Any mainstream scientist present will trot out the standard response that water vapor is indeed an important greenhouse gas, it is included in all climate models, but it is a feedback and not a forcing."

The point here is that water-vapor concentrations are not being considered by modelers to increase directly by injection from anthropogenic sources, but rather, as a consequence of indirect processes due to feedback from other greenhouse gases. If climate models properly include such feedbacks,[3] then water vapor would be properly accounted for by climate models. However, widespread changes in land use with irrigation appear to be changing water-vapor inputs to the atmosphere over wide regions, so there may be some changes in water vapor aside from the feedback effect from greenhouse gas warming. Furthermore, water vapor is the predominant greenhouse gas and, whether it increases directly, or as a result of warming produced by other greenhouse gases, its action is still important.

Realclimate.org admitted that water vapor is the single most important absorber but suggested that it accounts for "between 36% and 66% of the greenhouse effect," and "together with clouds makes up between 66% and 85% of the temperature change induced [by greenhouse gas emissions]. CO_2 alone makes up between 9 and 26% ...". *Realclimate.org* went on to conclude that:

> "The maximum supportable number for the importance of water vapor alone is about 60–70% and for water plus clouds 80–90% of the present day greenhouse effect. (Of course, using the same approach, the maximum supportable number for CO_2 is 20–30%, and since that adds up to more than 100%, there is a slight problem with such estimates!)."

Realclimate.org could not find any support for the 98% figure, nor could this writer. Nevertheless, the degree of uncertainty regarding the relative importance of water vapor and carbon dioxide is a little unsettling. Estimates in the literature vary widely. According to Marsh (2002), estimates range from water vapor accounting for 64% of the greenhouse effect, carbon dioxide 21%, ozone 6%, and other trace gases 9%, to water vapor being responsible for 90% of the greenhouse effect, leaving 10% for carbon dioxide and the other greenhouse gases. Kiehl and Trenberth (1997) estimated 59% from water and 28% from CO_2. Van Dorland (1998) estimated that water vapor accounts for 70% of the greenhouse effect, clouds account for 10%, CO_2 accounts for 15%, and other gases account for 5%. Schmidt *et al.* (2010)[4] claim that water vapor accounts for 50%, clouds account for 25%, and CO_2 accounts for 20%.

[3] A consummation devoutly to be wished.

[4] Gavin Schmidt and *realclimate.org* are essentially one and the same.

Leroux (2005) claimed that the greenhouse effect cools the atmosphere. This reference also claims that the "so-called 'greenhouse effect' is fundamentally a 'water' effect," and chastises the IPCC "water vapor-free greenhouse effect". But ignoring these seemingly irrational meanderings, it is noteworthy that Leroux (2005) quoted some relevant figures from the IPCC 2001 Technical Summary of Group I,[5] which claimed that the following changes have occurred:

(1) lower stratosphere (above 18 km): 20% increase in water vapor since 1980;
(2) upper troposphere: no significant global trends since 1980; 15% increase in water vapor in tropics (10°N–10°S);
(3) troposphere: many regions with increased water vapor since 1960;
(4) near surface: widespread significant increases in water vapor in the Northern Hemisphere (NH) from 1975 to 1995.

These are significant changes. Leroux (2005) went on to criticize the IPCC by suggesting that it was these increases in water vapor (and not increases in CO_2) that may be producing putative global warming. But the viewpoint of the IPCC (and indeed all global-warming alarmists and modelers) is that these increases in atmospheric humidity are claimed to be a result of global warming induced by non-H_2O greenhouse gases, with the increases in H_2O acting as a feedback mechanism to amplify the effect of non-H_2O greenhouse gases.

If *realclimate.org* is correct in its claim that changes in humidity and clouds produce between 66% and 85% of the temperature change induced by greenhouse gas emissions, then pinning down changes in humidity and clouds as a consequence of CO_2 emissions is *the* crucial issue in predicting future climate change. The question of how credible climate models are in making such predictions then revolves about how well they account for changes in humidity and cloudiness.

According to Hall and Manabe (1999):

> "The latest IPCC assessment indicates that the likely equilibrium global-mean temperature response to a doubling of CO_2 ranges from 1.6°C to 4.6°C. The source of this uncertainty is our inability to quantify the role of feedback mechanisms in the climate system, including water vapor, cloud, lapse rate, and albedo feedback. Water vapor feedback has long been thought to be a positive feedback mechanism. This is due to the dependence of the saturation water vapor mixing ratio on temperature, as predicted by the Clausius–Clapeyron equation. Thus a CO_2-induced warming of the surface–troposphere system will lead to a water vapor increase in the atmosphere. Since water vapor is itself a greenhouse gas, this increase will ... make the warming larger than it would be otherwise. While there is a consensus that water vapor feedback is positive in the context of global warming, it remains unclear exactly how strong the effect is. Large uncertainties also exist regarding the magnitude (and sign, in the case of cloud feedback) of the other feedback mechanisms."

[5] *www.ipcc.ch/pub/wg1TARtechsum.pdf.*

Hall and Manabe (1999) carried out long-term (500-year) global climate model (GCM) analyses of a modeled Earth's climate in two ways: not allowing, or allowing water vapor to increase as greenhouse gases warm the Earth. In the former case, amplification by water vapor is eliminated and, in the latter case, it is included. They presented arguments in favor of a model that assumes that the relative humidity will not change as the Earth warms, resulting in a sizable increase in absolute humidity with temperature. The results of Hall and Manabe (1999) indicated that, without amplification by water vapor, the global temperature rise due to a doubling of CO_2 was 1.1°C but, with amplification by water vapor, it rose to 3.4°C. They also distinguished between two cases that might induce a warm anomaly in the surface–troposphere system: (1) greenhouse gas absorption, referred to as "external" forcing, and (2) "internally" generated forcing, such as a change in net cloud cover associated with atmospheric disturbances. They found that the amplification was greater for the greenhouse gas absorption case and explained why at some length; however, their explanation was not convincing to this writer.

Held and Soden (2000) discussed water-vapor feedback effects at some length. They derived an equation of the form:

$$dT/d\log[CO_2] = A/(1 - B - C)$$

in which the surface temperature (T) depends logarithmically on the CO_2 concentration in the troposphere with a coefficient A. To the extent that surface warming generates additional water vapor, and the water vapor contributes a feedback to more surface heating, the parameter B is inserted. In addition, surface heating could reduce the ice/snow albedo of the Earth, and the term C is included for this effect. Held and Soden (2000) pointed out that the combined effect of B and C could have much greater impact than either term alone. However, clouds will also form under these conditions, and the effects of clouds on radiation balance are not well understood. The feedback effects of water vapor in the tropics are complex. Held and Soden (2000) discussed aspects of this uncertainty but did not seem to reach any quantitative conclusions. Bauer *et al.* (2002) also studied water vapor feedback in this era.

Soden *et al.* (2002) studied the after-effects of the eruption of Mt. Pinatubo that produced a temporary cooling of the Earth. They also found that there was a reduction in the column water vapor during this period. Attributing this reduction in water vapor to the cooling event, they concluded that there is a positive water-vapor feedback. By analogy, during a period of warming, the water-vapor concentration should increase. Del Genio (2002) wrote a brief review of this work, saying:

> "A few cautionary notes are in order. Although the agreement shown by Soden *et al.* is good, volcanic eruptions are not perfect reverse proxies for greenhouse gas climate change. Volcanic aerosols affect incoming solar energy more than they do Earth's thermal radiation, whereas the reverse is true for greenhouse gases. Volcanic forcing decreases more from equator to pole than does greenhouse gas forcing. And both types of climate change reduce the rate at which temperature decreases with height from the surface to the upper

troposphere, even though one is a global warming and the other a global cooling".

Soden *et al.* (2005) reported "We use satellite measurements to highlight a distinct radiative signature of upper tropospheric moistening over the period 1982 to 2004". The reliability of these measurements is uncertain. But, more importantly, when one examines the data reported by this paper, one finds rather large yearly fluctuations in column integrated water vapor from year to year, presumably due in part to El Niño—La Niña cycles. As we have noted in regard to tropospheric temperatures (see Figure 3.34), the effect of the great El Niño of 1998 seems to have been long-lasting, producing a step-function change in humidity as well as temperature. Thus, Figure 1 of Soden *et al.* (2005) shows a pattern of humidity fluctuating about a mean prior to 1998, and fluctuations about a higher mean after 1998. Hence, the change in water vapor over the period 1982–2004 seems to have been driven by El Niño–La Niña cycles rather than greenhouse gases, and the entire thesis of El Niño–La Niña cycles relating humidity to greenhouse-driven temperatures seems to be fragile. Santer *et al.* (2007) continued this line of argument from Soden *et al.* They began by asserting that "Data from the satellite-based Special Sensor Microwave Imager (SSM/I) show that the total atmospheric moisture content over oceans has increased by 0.41 kg/m^2 per decade since 1988". This would seem to imply a steady rise in humidity over two decades. However, when their Figure 1 is examined, one finds essentially the same result as that of Soden *et al.* (2005); there are large oscillations in humidity with an apparent step-function in the mean after the 1998 El Niño. As in the case of Soden *et al.* (2005), Santer *et al.* (2007) utilized a very short period of data to infer great conclusions that remain very uncertain.

In most GCMs an initial warming caused by additional CO_2 and other greenhouse gases leads to enhanced evaporation at the surface and a general moistening of the atmosphere. Since water vapor is a strong infrared absorber, the added moisture causes further warming. The amplifying effect of water vapor can be quite large, with GCMs typically predicting an increase in the global average warming by 70%–90% compared with calculations that maintain fixed water vapor. Minschwaner and Dessler (2004) wrote:

> "Observational studies have attempted to verify the positive water vapor feedback by examining the response of atmospheric humidity to changes in surface temperature caused by inter-annual variability, the annual cycle, volcanic eruptions, and the El Niño–Southern Oscillation; however results have been inconclusive, with some studies yielding a positive feedback and others indicating a negative response. There are several plausible mechanisms for creating a negative water vapor feedback in the upper troposphere. One oft-cited mechanism invokes the drying effects of deep cumulus convection, arguing that the mean detrainment altitude of deep convection will be both higher and cooler in a warmer climate compared to the present. Because the water vapor content of air pumped into the upper troposphere by convection is governed by the saturation vapor pressure at the temperature of cloud detrainment, this

would imply a reduced supply of water at warmer surface temperatures, leading to drying and a negative feedback on climate."

Minschwaner and Dessler (2004) attempted to test the convective drying mechanism using a model specifically designed to examine the moisture content of the UT in the tropics. Implications for water-vapor feedback were also examined using measurements of relative and specific humidities in the tropical UT from microwave and infrared limb-viewing instruments in space. As the surface temperature rises, the saturation water-vapor pressure increases as required by the Clausius–Clapeyron equation, which implies that air can hold more water before it becomes saturated (i.e., the maximum possible absolute humidity increases). The relative humidity (ratio of absolute humidity to maximum possible absolute humidity) may vary in ways that are difficult to predict. In the absence of any better knowledge, some investigators assumed that, as the temperature changes, the relative humidity remains roughly constant, which then leads to an increase in absolute humidity because the maximum humidity increases with temperature. The effect of various assumptions regarding relative humidity were estimated by Pierrehumbert (2009) as shown in Figure 6.10. It can be seen that, depending on the assumption that is made regarding changes in humidity as the Earth warms, a very large feedback may ensue from changes in water vapor concentration.

According to Minschwaner, Dessler, and Sawaengphokhai (2006), this assumption (of constant relative humidity) leads to roughly a doubling of the effect of CO_2 (i.e., the change in global average temperature goes from 0.8°C with no increase in absolute humidity to 1.6°C with constant relative humidity). Minschwaner and Dessler (2004) found that the drying mechanism (mentioned previously) occurs weakly so that the relative humidity decreases slowly with increasing

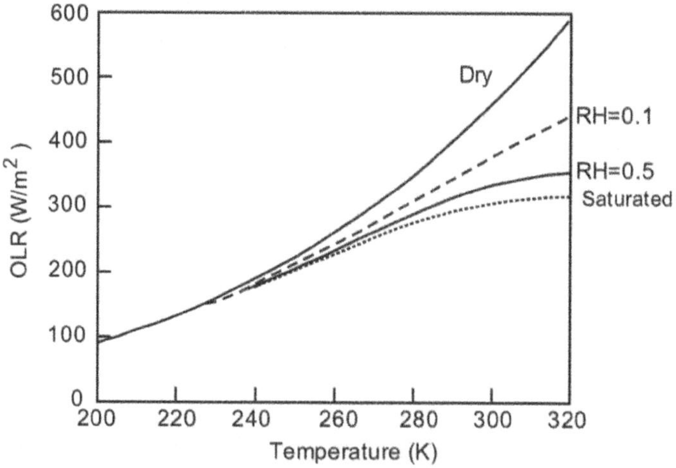

Figure 6.10. Calculated outgoing long-wave radiation (OLR) as a function of global average temperature, for various assumptions about relative humidity (Pierrehumbert, 2009).

temperature. Nevertheless, the absolute humidity does increase somewhat with increasing surface temperature. At the bottom line, Minschwaner and Dessler (2004) found that, with no water-vapor feedback, a doubling of CO_2 would produce a tropical surface warming of 0.8°C and, when water-vapor feedback is included, the increase in temperature is 1.2°C. Note the discrepancy with Hall and Manabe (1999), who concluded that, with constant relative humidity, the change in temperature would be 3.4°C.

Minschwaner, Dessler, and Sawaengphokhai (2006), in an update to Minschwaner and Dessler (2004), presented an analysis of the water-vapor feedback in the tropical UT as simulated by 17 coupled ocean–atmosphere climate models. The strength of the water-vapor feedback in the IPCC models was inferred using methods described by Minschwaner and Dessler (2004). The models indicated less drying than was found by Minschwaner and Dessler (2004). This would seem to imply a change in temperature due to doubling CO_2, including the effect of water vapor, between 1.2°C and 1.6°C.

According to a NASA press release (Anon. I):

> "A NASA-funded study found some climate models might be overestimating the amount of water vapor entering the atmosphere as the Earth warms. Since water vapor is the most important heat-trapping greenhouse gas in our atmosphere, some climate forecasts may be overestimating future temperature increases. In response to human emissions of greenhouse gases, like carbon dioxide, the Earth warms, more water evaporates from the ocean, and the amount of water vapor in the atmosphere increases. Since water vapor is also a greenhouse gas, this leads to a further increase in the surface temperature. This effect is known as 'positive water vapor feedback.' Its existence and size have been contentiously argued for several years.
>
> "The size of the positive water vapor feedback is a key debate within climate science circles. Some climate scientists have claimed atmospheric water vapor will not increase in response to global warming, and may even decrease. General circulation models, the primary tool scientists use to predict the future of our climate, forecast the atmosphere will experience a significant increase in water vapor.
>
> "Using Upper Atmosphere Research Satellite (UARS) data to actually quantify both specific humidity and relative humidity, the researchers found, while water vapor does increase with temperature in the upper troposphere, the feedback effect is not as strong as climate models have predicted. The increases in water vapor with warmer temperatures are not large enough to maintain a constant relative humidity."

An important analysis of the effect of CO_2 on climate was made by Lindzen (1997). Many aspects are covered in his paper; only a brief report on some parts is given here. Lindzen emphasized:

> "Water vapor, the atmosphere's main greenhouse gas, decreases in density rapidly with both height and latitude. Surface radiative cooling in the tropics,

which has the highest concentration of water vapor, is negligible. Heat from the tropical surface is carried upward by cumulus convection and poleward by the Hadley circulation and planetary-scale eddies to points where radiation can more efficiently transport the heat to space. Where radiation can more efficiently carry the heat depends on the radiative opacity and the motions themselves. In point of fact, without knowing the dynamical heat fluxes, it is clear that one cannot even calculate the mean temperature of the Earth. It is interesting, in this regard, to look at model intercomparisons of meridional heat flux, and their comparison with observationally based estimates. ... Such differences [are] roughly equivalent to differences in vertical fluxes of about 25 W/m^2—much larger than the 4 W/m^2 change that a doubling of CO_2 is expected to produce."

There are two points here: (1) the tropics lose a considerable amount of heat by processes other than radiation, and (2) meridional heat transfer is much greater than putative CO_2 forcing.

As we discussed previously, the prevailing view amongst climatologists is that global warming due to increased CO_2 is amplified by increased water-vapor content in the atmosphere. Lindzen provided a detailed discussion of several aspects of the regional distribution of water vapor in the atmosphere and its relationship to global warming induced by increased CO_2. Most climate models make the assumption that relative humidity does not change with global warming and, since warm air can hold more water vapor than cool air, a constant relative humidity implies an increase in absolute humidity as the Earth warms. The basis for the assumption that relative humidity does not change with global warming lies in some rather old radiosonde data that indicate that the average distribution of relative humidity (when plotted on altitude vs. latitude axes) does not change much from winter to summer. The argument then goes that, over the smaller temperature change characteristic of global warming, relative humidity would also not change. However, Lindzen raised serious questions about the accuracy of the radiosonde data. Clearly, the assumption of constant relative humidity rests on a weak foundation, and that assumption is critical to the alarmist position that doubling CO_2 produces unacceptable global warming due to increased absolute humidity.

But Lindzen went further than this. He emphasized that the degree of water-vapor feedback as a heating force in any region depends on absolute humidity. In desert regions with very low absolute humidity, an increase in humidity provides a significant heating force. However, in regions with high absolute humidity, an increase in humidity provides a very modest heating force (e.g., an increase in relative humidity from 10% to 20% produces a forcing of 1.5 W/m^2, whereas an increase in relative humidity from 50% to 60% produces a forcing of only 0.15 W/m^2). Tropical regions that already have high humidity do not gain much additional heating from an increase in humidity. And, as Lindzen pointed out:

> "Given the nonlinearity of the radiative effect of water vapor, the average radiative response to water vapor is not equal to the response of the average water vapor."

It has been estimated by climate modelers that a doubling of CO_2 implies a forcing at the tropopause of about 3.7 W/m^2. The question of climate sensitivity amounts to asking how much must the Earth's surface warm to compensate for this forcing. This requires estimation of the globally integrated total radiative flux at tropopause levels. A global change in the distribution of moist and dry regions can lead to a change in outgoing long-wavelength radiation (OLR) even in the absence of change in mean temperature. Changes in circulation and changes in temperature can both play a role in the moisture budget. Lindzen suggested:

> "... the interesting possibility that the primary feedback process might consist in the change in areal coverage of the very dry regions. Presumably, natural variations include a full range of such possibilities so that observed ratios of average temperature variations to variations in total OLR would show a significant scatter."

According to Lindzen, GCMs do not do a good job of estimating the coupling between tropical and extra-tropical regions and therefore do not allocate the global distribution of water vapor accurately; this has a profound effect on the putative heating effect of increased CO_2. It seems likely that global warming might decrease the humidity of air descending above desert areas of the Earth and, since these regions are by far the most sensitive to changes in humidity, they would counterbalance the smaller heating effect of increased humidity in regions where the humidity is higher. The regional effects of changes in humidity far outweigh the effects of changes in net global humidity. Drying of already dry regions is more important than net humidifying of the globe. Net moistening of the Earth could have a negative water-vapor feedback if most of that moistening occurs in already moist regions. Climate models that take an average humidity for the whole Earth are overly simplistic.

The effect of water-vapor feedback on amplifying global warming produced by increasing CO_2 concentration requires an understanding of the distribution of humidity changes resulting from warming; a global average of humidity change does not suffice. In addition, of course, even with a thorough understanding of the regional dependence of humidity change, one must still cope with the problem of changes in cloudiness. Lindzen and co-workers (2001, 2007) studied this and made four major points:

(1) The cloud and water-vapor feedbacks are intimately connected.
(2) Feedbacks are primarily associated with changing areas of moist and cloudy regions vs. regions that are dry and cloud-free (as opposed to mean humidity).
(3) Models must have spatial and temporal scales (5–10 km and hours) characteristic of clouds in order to evaluate feedbacks.
(4) The effect of cumulus activity must be included. A simplistic model that merely treats humidity as a global average that increases when surface temperatures rise, that ignores regional changes in humidity, and crudely treats clouds will always overestimate the temperature rise due to increased CO_2.

While most climate models deal with such elements as clear-sky humidity, average humidity, or differences between regions of high and low humidity, Lindzen and co-workers studied feedback involving changes in the relative areas of high and low humidity and cloudiness. Their results suggest that cloudy moist regions contract when the surface warms and expand when the surface cools. In each case, the change acts to oppose the surface change, and thus presents a strong negative feedback to climate change as a sort of Le Chatelier's Principle. They concluded that the relevant feedbacks are negative rather than positive, and very large in magnitude. Spencer *et al.* (2007) studied the effect of changes in clouds to changes in temperature in tropical regions and found a negative feedback of 6 W/m^2 per degree of temperature rise. This provides some support for Lindzen's hypothesis. Dessler *et al.* (2008b) attempted to derive water feedback sensitivity by comparing data on global temperature and humidity during the winter months of 2006–2007 and 2007– 2008. However, Christy demonstrated a strong correlation of global temperature with an El Niño index since 1978, and particularly for 2006–2008 (see Figures 3.32 and 3.33). Dessler *et al.* (2008b) also found a good correlation of global temperature with an ENSO index for 2006–2008. Hence, it seems clear that global temperature changes for 2006–2008 were driven primarily by changes in the oceans, and changes in humidity during that period were not a cause of global temperature change, but an effect. Masters (2012) showed clearly that none of these short-term analyses has any validity. As Masters said: "Overall, there is little correlation between the changes in the CRF [cloud radiative forcing] and surface temperatures on these timescales, suggesting that the net effect of clouds varies during this time period quite apart from global temperature changes."

The effect of changing CO_2 concentration and putative water-vapor greenhouse effect are buried in the noise of a much stronger signal due to El Niño variability during these years. Therefore, it is physically impossible to derive a water feedback sensitivity from data limited to these two winters. Yet, Dessler *et al.* claim that they have done so and quote a value in agreement with climate models. This seems impossible to this writer. They then reached the rather incredible conclusion that:

> "The existence of a strong and positive water-vapor feedback means that projected business-as-usual greenhouse gas emissions over the next century are virtually guaranteed to produce warming of several degrees Celsius."

This conclusion is utterly unsupportable from the analysis of a mere two winters' data controlled by El Niño activity.

Dessler *et al.* (2008a) analyzed a mere one month's data in 2005 to infer clear-sky top-of-atmosphere OLR and its relationship to humidity. It is not clear to this writer that this paper sheds any light on water feedback sensitivity.

Gettleman and Fu (2008) analyzed the changes in humidity produced by temperature changes from 2002 to 2007. As before, temperatures during this period appear to have been determined by El Niño variability, and changes in water vapor content appear to be effects of this temperature change. There is little or no connection to heating produced by CO_2 and water feedback sensitivity does not seem to be derivable from this work.

The data on oceanic water vapor are difficult to interpret (see Figure 6.11).

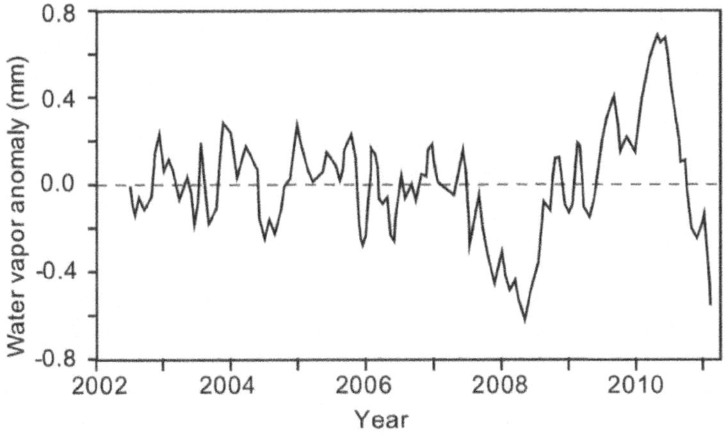

Figure 6.11. AMSR-E Global oceanic integrated water vapor variations (smoothed data) (*www.drroyspencer.com/*).

It is now becoming apparent that the warming experienced by the Earth from 1976 to 1998 were related to the prevalence of El Niño conditions, and with the balancing of El Niños with La Niñas after 1998, world temperatures flattened out from 1999 to 2013. While it is likely that growth in CO_2 contributed to global warming during the 20th century, it is clear that large chaotic fluctuations dictated by ocean conditions, aerosol emissions, humidity, cloudiness and unknown factors masked this putative CO_2 effect, making it very difficult to unravel the contribution of rising CO_2. The work by Soden *et al.*, Santer *et al.*, and Dessler *et al.* and others shows great ingenuity in ferreting out information from very limited amounts of data, some of which are of uncertain reliability. But, ultimately, the credibility of their results is limited by the scarcity of good long-term data. Parameters such as humidity and cloudiness vary widely from day to day and year to year even in the absence of any forcing. In attempting to determine how these parameters respond to a forcing, one must have data over very long periods to overcome the low signal-to-noise rations inherent in them. The same problem occurs in sea-level measurements. However, whereas climatologists studying sea level have emphasized the need for very long-term data (see Section 8.3.3), those who infer feedbacks from humidity and cloudiness seem to be content with very short-term data. Masters (2012) showed clearly that none of these short-term analyses has any validity. As Masters said: "Overall, there is little correlation between the changes in the CRF and surface temperatures on these timescales, suggesting that the net effect of clouds varies during this time period quite apart from global temperature changes."

Ban-Weiss *et al.* (2011) added another viewpoint on cloudiness change due to global warming. They pointed out that increased evaporation would produce a significant surface cooling effect that would be counterbalanced by an equal heating effect as moisture condensed in the atmosphere. Thus, there is no net heating effect.

However, the increased evaporation produces an increase in low-elevation cloudiness that increases the Earth's albedo, producing a net cooling effect. However, it is not clear how reliable their estimates of changes in cloudiness are.

6.2 THE EARTH'S HEAT BALANCE

6.2.1 Introduction

The net energy input to the Earth at the top of the atmosphere is determined by the incident irradiance from the Sun minus the reflected irradiance. However, note that, while many notable papers refer to the incident solar irradiance as "short-wave radiation", a significant portion of the Sun's spectrum extends into the near-IR. The net solar input flux at the top of the atmosphere is the difference between incident and reflected fluxes. When the Earth is in balance, this inflow of solar irradiance is equal to the flux of long-wave IR radiation emitted to space by the surface–atmosphere system. The outward-bound long-wave radiation that reaches the top of the atmosphere derives from the absorption and emission of long-wave radiation by gases throughout the atmosphere as well as convection and latent heat in the troposphere. Thus, very little of the long-wave energy that escapes to space represents emission directly from the surface.

6.2.2 Kiehl and Trenberth's analysis

The heat fluxes in the Earth–atmosphere system were estimated by Kiehl and Trenberth (1997) (to be denoted "KT97") and updated by Trenberth, Fasullo, and Kiehl (2009) (to be denoted "TFK09").

KT97 was based on use of the 1976 U.S. standard atmosphere and 1990 IPCC estimates for concentrations of greenhouse gases. KT97 adopted the best measurements and analyses available to produce a balanced heat flow chart. However, there were considerable variations in specific estimates of individual heat flows by various investigators. Hence, the results contained significant implicit uncertainty. KT97 mainly considered the average Earth with its cloud cover (estimated by KT97 to be about 62%). In some places, however, they attempted to determine the effect of clouds by considering a cloud-free Earth. KT97 relied heavily on observations from the NASA Earth Radiation Budget Experiment (ERBE) from 1985 to 1989. TFK09 utilized more recent data and models.

A principal interest is the degree to which the Earth is in power balance (the comparison of input power from the Sun with OLR). KT97 slightly adjusted the ERBE data to produce a balance. Some of the more recent data indicated a significant imbalance that TFK09 felt was not credible, and they adjusted the data to fit a GCM estimate by Hansen *et al.* (2005a) that the Earth is out of balance by 0.85 W/m^2 (the Earth is absorbing more power than it is emitting). It seems evident that, while TFK09 might have pinned down the basic power flows as well as the data permitted, significant uncertainties remain in some of the quantities.

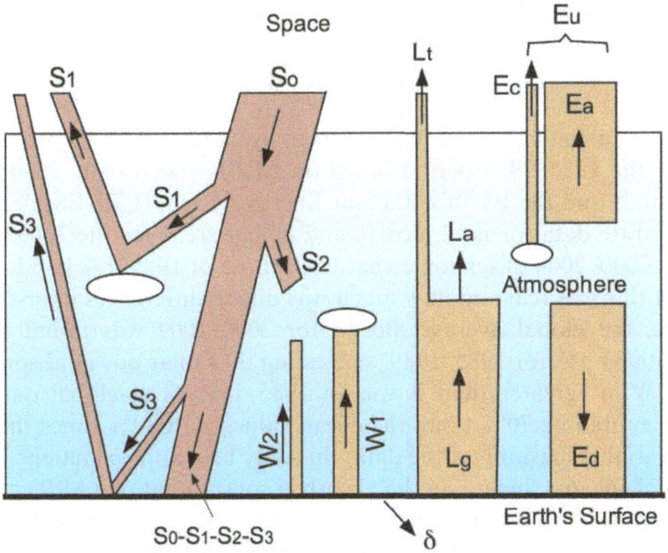

Figure 6.12. One-dimensional model of power fluxes between the Earth's surface, the atmosphere, and space. Short-wave fluxes are pink while long wave fluxes are orange. Clouds are shown in white.

Consider the one-dimensional model shown in Figure 6.12.

S_o represents the net incoming short-wave radiation from the Sun (342 W/m^2).[6] Of this, S_1 is reflected back into space from clouds and the atmosphere, S_2 is absorbed by the atmosphere, and S_3 is reflected back from the Earth's surface. The remainder, $S_o - S_1 — S_2 — S_3$, is absorbed at the Earth's surface.

The atmosphere radiates upward and downward in amounts E_u and E_d. E_u is divided into two parts: E_a emitted by the atmosphere and E_c emitted by clouds.

The Earth's surface transfers heat upward in two modes: long-wave radiation (L_g), and non-radiative transfer (W_1 and W_2: latent and sensible heat) to the atmosphere. A portion of L_g is transmitted through the atmosphere to space (L_t) and a portion (L_a) is absorbed by the atmosphere. If there is an imbalance (δ) in the Earth's power budget so that a net amount of power is delivered to the Earth, this would be determined from:h

$$\delta = E_d + (S_o — S_1 — S_2 — S_3) — W_1 — W_2 — L_g.$$

Short-wave fluxes: KT97 pointed out that the ERBE provided a wealth of data on heat flows. Based on ERBE, the average total long-wave outgoing emission from the top of the atmosphere ($E_u + L_t$) was estimated to be 235 W/m^2 for the Earth (with its partial coverage by clouds) over the time period of observations (1985–1989). In order to

[6] Since then, this figure has been revised downward to 340 W/m^2.

balance the incoming and outgoing radiant heat fluxes, it was necessary to choose the Earth's albedo as 31%; it then follows that the solar power input to the Earth is

$$(S_o—S_1—S_3) \sim 0.69 \times (342 \text{ W/m}^2) = (235 \text{ W/m}^2)$$

for the average Earth (with its partial coverage by clouds). TFK09 provided separate estimates for the 1985–1989 period based on ERBE and for the 2000–2004 period based on Clouds and the Earth's Radiant Energy System (CERES) data. They also provided separate data for land, ocean, and global areas for the 2000–2004 period. The data for 2000–2004 differ somewhat from those of 1985–1989 and it is not clear how much of this was real and how much was due to differences in instrumentation. For example, the global average albedo for 2000–2004 was found to be 29.8% compared with 31.3% for 1985–1989, suggesting that solar power absorbed in 2000–2004 was 5 W/m^2 greater than it was in 1985–1989. The global data should be obtained by combining 70% times the ocean values with 30% times the land values but, due to variable masking of the data, this only held approximately. These results are shown in Table 6.3, including the absorbed solar radiation (ASR) and the OLR. The net solar power input to the Earth is δ = ASR—OLR. This quantity could not be measured because it is determined by the difference between large numbers, each of which has much greater uncertainty than the difference between them (TFK09 estimated that uncertainties are 5–10 W/m^2 at the top of the atmosphere and 10–15 W/m^2 at the Earth's surface). Instead, TFK09 set δ equal to the value estimated by Hansen *et al.* (2005a) based on a climate model.

Table 6.4 provides data on heat flows at the Earth's surface. Note that the solar input to the Earth was higher in 2000–2004 (due to the reduced albedo) and so was the upwelling long-wave radiation.

Radiant transmission models for the Earth's atmosphere led to estimates of S, the flux absorbed by the atmosphere, and S_o—S, the flux at the Earth's surface

Table 6.3. Global power balance for 1985–1989 (based on ERBE) and 2000–2004 (based on CERES) according to TFK09 (W/m^2).

	Solar in	*Solar reflected*	*Albedo (%)*	*ASR*	*OLR*	*NET down*
	S_o	S_1+S_3		$S_o-(S_1+S_3)$	L_t+E_u	ASR–OLR
Global (FT08) 1985–1989	341.3	106.9	31.3	234.4	234.4	0.0
Global (TFK09) 2000–2004	341.3	101.9	29.8	239.4	238.5	0.9*
Global (Van Dorland, 1998)	343	103	30.0	240	240	0
Land (FT08) 1985–1989	330.1	118.0	35.8	212.1	228.7	–16.6
Land (TFK09) 2000–2004	330.2	113.4	34.4	216.8	232.4	–15.6
Ocean (FT08) 1985–1989	345.3	102.9	29.8	242.2	236.4	6.0
Ocean (TFK09) 2000–2004	345.4	97.8	28.3	247.7	240.8	6.9

* This quantity was not measured, but was based on a climate model by Hansen *et al.* (2005a).

Table 6.4. Surface components of the annual mean power budget for the globe, global land, and global ocean for 1985–1989 (based on ERBE) and 2000–2004 (based on CERES) according to TFK09 (W/m^2) (land and ocean data were not available for 1985–1989).

	S	$S_o - S_1 - S_2 - S_3$	S_3	W_1	W_2	L_g	E_d	$L_g - E_d$	δ
Global (FT08) 1985–1989	67	168	24	78	24	390	324	66	0
Global (TFK09) 2000–2004	78.2	161.2	23.1	80.0	17	396	333	63	0.9*
Global (Van Dorland, 1998)	80	160	20	Sum = 95		395	330	65	0
Land (FT08) 1985–1989	Not available								
Land (TFK09) 2000–2004	78.0	145.1	39.6	38.5	27	383.2	303.6	79.6	0.0
Ocean (FT08) 1985–1989	Not available								
Ocean (TFK09) 2000–2004	78.2	167.8	16.6	97.1	12	400.7	343.3	57.4	1.3*

* This quantity was not measured, but was based on a climate model by Hansen *et al.* (2005a).

(with the Earth's partial coverage by clouds). In the absence of clouds, KT97 estimated that the clear-sky value of S_o would increase from 235 W/m^2 to 285 W/m^2 corresponding to an albedo of 0.17 and a cloud forcing of –50 W/m^2. Atmospheric absorption would decrease from S = 67 W/m^2 to 60 W/m^2.

Emission by the Earth's surface: KT97 estimated L_g to be 390 W/m^2 based on radiation from a black body at 15°C. Radiant transmission models were used to estimate how this was partitioned between a window to space (L_t) and that which is absorbed by the atmosphere (L_a). The discussion by KT97 is somewhat obscure. They apparently claimed that the flux from the surface within the atmospheric window is 99 W/m^2 for a clear sky, and 80 W/m^2 for the cloudy case, "showing that there is considerable absorption and re-emission at wavelengths in the so-called window by clouds". But this is difficult to interpret because the amount of radiation emitted by the surface in the window depends only on its temperature. KT97 then assigned a value to L_t for the cloudy case as (100%—62%) = 38% of the flux "for the clear sky case, corresponding to the observed cloudiness of about 62%". But 38% of 99 is 37.6 and, unaccountably, KT97 assigned a value of 40 W/m^2 to L_t for the cloudy Earth. Thus, L_a was estimated to be 350 W/m^2. More importantly, if S_o rises to 285 W/m^2 in the cloudless case, the Earth would presumably warm above 15°C and thus all of the estimates by KT97 for long-wave fluxes in the clear-sky case would have to be revised.

Emission by the atmosphere: In the absence of clouds, ERBE data indicated a clear-sky value of $(E_u + L_t) \sim 265$ W/m^2 and a value of 235 W/m^2 for average cloudy skies for 1985–1989. KT97 then introduced a rather arbitrarily chosen cloud model

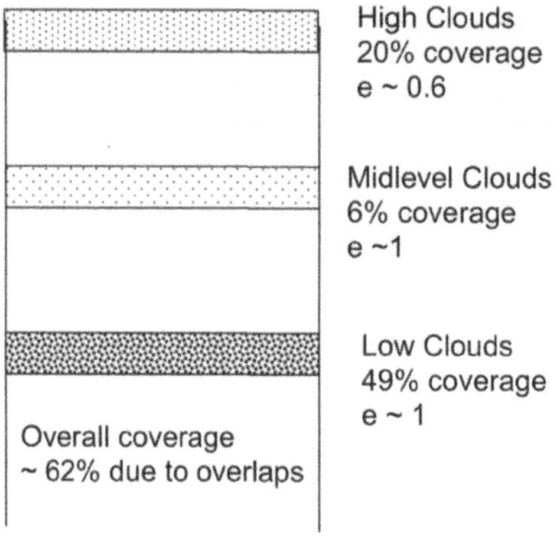

Figure 6.13. Distribution of clouds chosen by KT97 showing coverage and emissivity of each layer.

to account for this 30 W/m^2 difference between clear-sky and average cloudiness conditions (see Figure 6.13). Nevertheless, they found by test that, if they kept total cloud cover at 62%, varying the distribution between layers had only a small effect. With the assigned value of $L_t = 40$ W/m^2, the estimated value for E_u then became 195 W/m^2 in the cloudy case. KT97 carried out radiant transfer models to estimate radiant fluxes at the Earth's surface. The downward emission from the atmosphere (E_d) was estimated to be 324 and 278 W/m^2, with and without clouds present, respectively.

Greenhouse gas forcing: KT97 defined "greenhouse forcing" as the difference between emission from the surface and the OLR at the top of the atmosphere $(L_g—OLR) = (L_g—E_u—L_t)$. They estimated this to be 125 W/m^2 in the absence of clouds, and 155 W/m^2 with clouds. The clear-sky "greenhouse forcing" was 60% due to water vapor, 26% due to CO_2, and the remaining 14% due to other greenhouse gases. By contrast, Van Dorland (1998) estimated 70% due to water vapor, 15% due to CO_2, 10% due to clouds, and 5% due to other greenhouse gases.

The results of KT97 for the cloudy case for 1985–1989 are shown in Figure 6.14. It is difficult to arrive at a corresponding figure for the clear-sky case. As previously indicated, the solar input to the Earth would increase due to a reduced albedo, and S_o would increase from 235 W/m^2 for the cloudy case to 285 W/m^2 for the clear-sky case. S would decrease from 67 W/m^2 for the cloudy case to 60 W/m^2 for the clear-sky case. Presumably, the Earth's surface and atmosphere would warm in the clear-sky case, resulting in increased radiant fluxes from the surface and atmosphere in order that the Earth would radiate 285 W/m^2 to space. As we pointed out previously, if S_o rises to 285 W/m^2, presumably the Earth would warm above 15°C and thus all of the estimates by KT97 for long-wave fluxes in the clear-sky case would be questionable.

The comparable figure for 2000–2004 is shown in Figure 6.15.

It seems evident from the foregoing discussion that clouds have a major impact on the Earth's energy balance. According to the analysis of TFK09 and KT97, the global average albedo decreased from 31.3% for 1985–1989 to 29.8% in 2000–2004, suggesting that solar power absorbed in 2000–2004 was 5 W/m^2 greater in 2000–2004 than it was in 1985–1989. This is a much greater forcing than that attributed to greenhouse gases. Does this represent a diminution of cloudiness due to global warming (as for example Dessler insists); is it a fluctuation independent of greenhouse gases; or is it just noisy data?

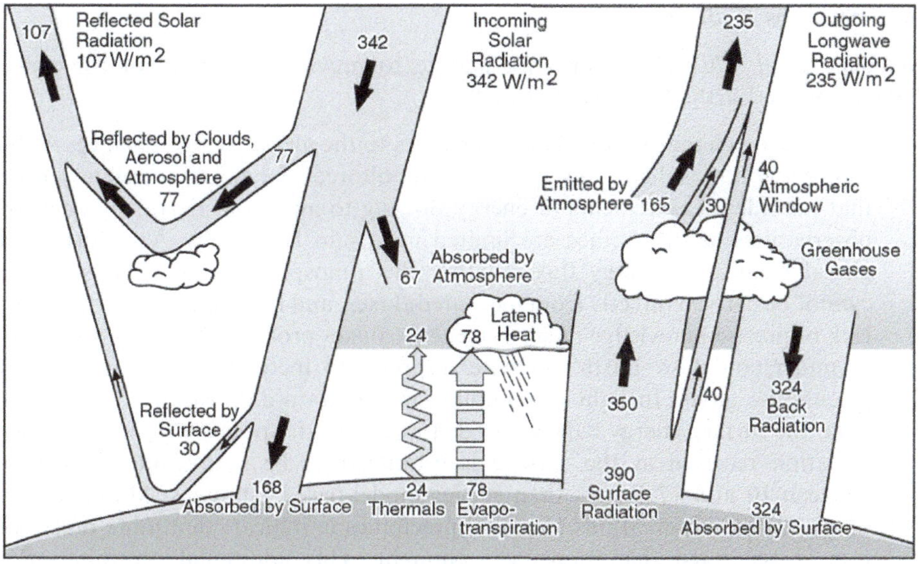

Figure 6.14. Heat flows in the Earth-atmosphere system during average cloudiness during 1985–1989 (based on Kiehl and Trenberth (1997) by permission of the *Bulletin of the American Meteorological Society*).

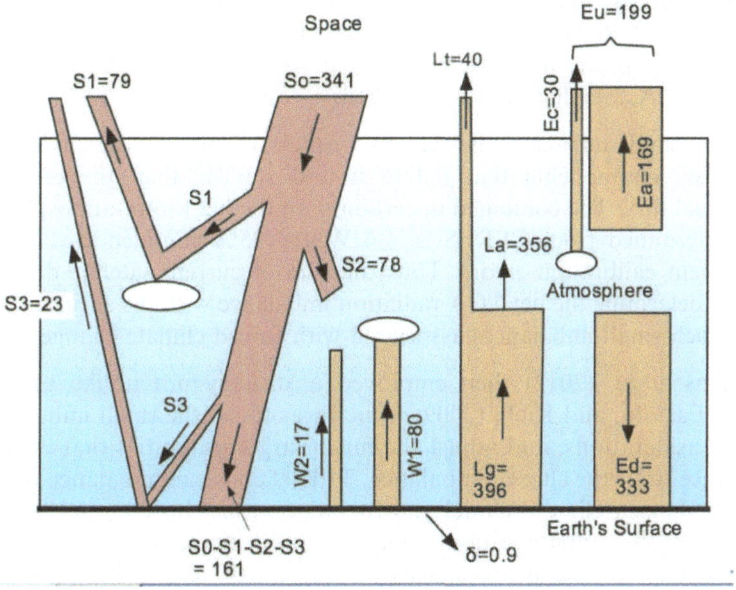

Figure 6.15. Heat flows in the Earth's atmosphere during 2000–2004 (W/m^2) (TFK09).

6.2.3 Stephens *et al.*'s model

Stephens *et al.* (2012) published an update to previous estimates of the power balance of the Earth. They claimed that:

> "Climate change is governed by changes to the global energy balance. At the top of the atmosphere, this balance is monitored globally by satellite sensors that provide measurements of energy flowing to and from Earth. By contrast, observations at the surface are limited mostly to land areas. As a result, the global balance of energy fluxes within the atmosphere or at Earth's surface cannot be derived directly from measured fluxes, and is therefore uncertain. This lack of precise knowledge of surface energy fluxes profoundly affects our ability to understand how Earth's climate responds to increasing concentrations of greenhouse gases. In light of compilations of up-to-date surface and satellite data, the surface energy balance needs to be revised. Specifically, the long wave radiation received at the surface is estimated to be significantly larger, by between 10 and 17 W/m^2, than earlier model-based estimates. Moreover, the latest satellite observations of global precipitation indicate that more precipitation is generated than previously thought. This additional precipitation is sustained by more energy leaving the surface by evaporation—that is, in the form of latent heat flux—and thereby offsets much of the increase in long wave flux to the surface."

In other words, one of the principal goals of carrying out an input–output power analysis of the Earth is to ascertain any small imbalance that may exist between the solar power input to the Earth and the radiant power output from the Earth to space. As Stephens *et al.* remarked:

> "This small imbalance is over two orders of magnitude smaller than the individual components that define it and smaller than the error of each individual flux. The combined uncertainty on the net top-of-atmosphere (TOA) flux determined from CERES is ± 4 W/m^2 (95% confidence) due largely to instrument calibration errors. Thus the sum of current satellite-derived fluxes cannot determine the net TOA radiation imbalance with the accuracy needed to track such small imbalances associated with forced climate change."

Stephens *et al.* (2012) then employed a strategy not unlike that used by Trenberth, Fasullo, and Kiehl (2009), which was to set the small imbalance based on other considerations, and adjust the much larger quantities that contribute to the imbalance to fit the chosen imbalance. TFK09 chose an imbalance of 0.9 W/m^2 based on a Hansen climate model and Stephens *et al.* chose an imbalance of 0.6 W/m^2 based on their interpretation of recent data on changes in ocean heat content. (Since almost all of the heat capacity of the Earth resides in the oceans, any imbalance in the Earth's power budget must end up in the oceans.) However, it is not clear that the choice 0.6 W/m^2 is appropriate. A slightly lower value might be more credible.

Unfortunately, the paper by Stephens *et al.* is written in a very terse manner and they do not explain the details of their results shown in their Figure B1. In considering the fate of the 340.2 W/m^2 of solar input to the Earth, they assert that the total of reflected solar flux is 100.0 ± 2 W/m^2, leaving 240.2 W/m^2 absorbed by the overall Earth system. Yet, it is difficult to piece together the individual components of reflection to get them to add up to 100 W/m^2. Their figures are $S_3 = 23$ W/m^2 and $S_1 = 47.5 + 27.2 = 74.7$ W/m^2, where the 47.5 figure is attributed to reflection by clouds and the 27.2 figure is "clear sky reflection". Adding these figures yields 97.7 W/m^2. Additionally, there is (in their diagram) a picture of a cloud with numbers 5 ± 5 W/m^2 inside of it, and no explanation is given of what this means. Perhaps, by means not clear to this writer, Stephens *et al.* may have interpreted this to imply an additional reflection of 2.5 W/m^2, bringing the total reflection to 97.7 + 2.5 = 100.2 W/m^2, although this additional term makes neither physical nor mathematical sense to this writer. A comparison of some of their figures with those of TFK09 is given in Tables 6.5 and 6.6.

Whether the Earth is warming or cooling depends on an energy flux balance at the top of the atmosphere (TOA). Hansen and co-workers have claimed that the Earth is out of balance by about 1 W/m^2; that is the outgoing long-wave radiant flux is about 1 W/m^2 less than the incoming solar flux. Lyman *et al.* (2010) and Stephens *et al.* (2012) assumed that the net warming imbalance is about 0.6 W/m^2 based on estimates of ocean warming. Knox and Douglass (2010) argued that one can obtain very different estimates of the imbalance ranging down to zero, depending on the time period chosen for analysis, the theory being that any imbalance in the Earth energy budget at the TOA must end up as ocean warming. The Earth's imbalance cannot be calculated from the measured energy fluxes. The uncertainties in the various energy fluxes are large compared to the claimed imbalances. The imbalance would be calculated as the difference between large numbers. Thus, the estimate of the imbalance is like weighing the captain by weighing the ship plus the captain and subtracting the weight of the ship. In actuality, the Earth's heat balance depends heavily on cloud cover. The heat balance is constantly shifting as the Earth's albedo meanders. The Earth's heat balance can only be assessed over a period of several decades to average out these fluctuations. Yet, the data reported in Tables 6.5 and 6.6 are only for the decade 2000–2010.

Table 6.5. Solar power balance by Stephens *et al.* compared to that of TFK09 (W/m^2).

	Solar in	*Solar reflected*	*Albedo (%)*	*ASR*	*OLR*	*NET down*
	S_o	$S_1 + S_3$		$S_o - (S_1 + S_3)$	$L_t + E_u$	ASR–OLR
Global (TFK09) 2000–2004	341.3	101.9	29.8	239.4	238.5	0.9
Stephens *et al.* (2012) 2000–2010	340.2	100.2	29.5	240.2	239.7	0.5

Table 6.6. Surface power balance by Stephens *et al.* compared to that of TFK09 (W/m^2).

	S	$S_o - S_1 - S_2 - S_3$	S_3	W_1	W_2	L_g	E_d	$L_g - E_d$	δ
Global (TFK09)									
2000–2004	78.2	161.2	23.1	80.0	17	396	333	63	0.9
Stephens *et al.*									
(2012) 2000–2010	75	165	23	88	24.7	398	345.6	52.4	0.5

6.2.4 Miskolczi's model

Miskolczi (2007, 2010) produced an alternate model of radiant fluxes through the atmosphere. Because Miscolczi's papers are obscure and extremely difficult to follow,[7] Van Andel (2008) wrote an explanatory paper for the theory (although his paper is also difficult to follow). Miskolczi began with a figure somewhat like Figure 6.13. We will use the same nomenclature as in Figure 6.13. Van Andel described the "standard theory" prior to Miskolczi (actually the results of Van Dorland, 1998), but provided slightly different results than those of KT97 (see Tables 6.3 and 6.4). Actually, KT97 alluded to a potential need to make some of these changes, such as increasing S from 67 W/m^2 to around 80 W/m^2.

Miskolczi (2007, 2010) analyzed radiosonde data over a wide range of locations and seasons and thereby obtained data covering a wide range of atmospheric conditions. He found that empirically, $L_a = E_d$ over a wide range of these variables. In his earlier paper, he attempted to attribute this to a law of nature (Kirchoff's Rule) but that is neither necessary nor correct, and, in his 2010 paper, he provided a simpler explanation. In an atmosphere that is in thermal equilibrium with the Earth's surface, each layer of atmosphere is in equilibrium with its neighboring layers, so the downward emission from the atmosphere to the surface must be equal to the upward emission from the surface to the atmosphere. ("The $E_d = L_a$ relationship holds because the contribution of a layer to the downward emittance is equal to the absorbed surface upward radiation in the same layer".) He claimed that the radiosonde data support this concept at all altitudes. However, in his Table 2, Miskolczi (2010) contradicted himself by reporting average values to be $L_a = 321$ W/m^2 and $E_d = 309$ W/m^2 with a 12 W/m^2 difference between these quantities. He did not explain why there is such a difference, considering that he insisted repeatedly in his papers that L_a must equal E_d. Note that KT97 found average values to be $L_a = 350$ W/m^2 and $E_d = 324$ W/m^2 with a 26 W/m^2 difference between these quantities. However, it is apparent that these quantities are fairly close. His results are shown in Table 6.7.

[7] Perhaps something was lost in translation from Hungarian?

Table 6.7. Comparison of Miskolczi's data with other models.

	$S_o - S_1 - S_3$	S_2	$W_1 + W_2$	L_g	L_a	L_t	E_u	E_d
KT97	235	67	102	390	350	40	195	324
TFK09	239	78	97	396	356	40	199	333
Van Dorland	240	80	95	395	355	40	200	330
Stephens *et al.*	240	75	112	398				345.6
Miskolczki	250			382	321	61	189	309

According to Van Andel, Miskolczi also found another empirical result: $L_g = 2E_u$. This result is in accord with all the other models. As Miskolczi noted,

$$L_t = L_g \, exp\,(-\tau_a)$$

where τ_a is the greenhouse gas optical thickness. Using KT97 data for L_t and L_g, we find that $\tau_a = 2.28$, whereas Miskolczi's data for L_t and L_g lead to $\tau_a = 1.83$ (although he claimed it is 1.87). This discrepancy between KT97 and Miskolczi is mainly due to their different estimates of the transparency of the atmosphere to long-wave radiation (KT97 had $L_t = 40$ W/m^2 and Miskolczi had $L_t = 61$ W/m^2).

After passing through seemingly endless rambling paragraphs, it is difficult to resolve just what Miskolczi's theory is. Van Andel (2008) claimed that the "theoretical and measured" optical depth is 1.86 (note that Miskolczi claimed 1.87 but his data seem to support 1.83). Van Andel then went on to claim that "removal of all CO_2 [from the atmosphere] brings us back to 1.73 or a perturbation of –7%, [while] a 100-fold CO_2 concentration [increase] causes [an increase to] 2.29, a perturbation of merely +23%". This writer has read both of Miskolczi's papers and Van Andel's paper twice and cannot find any support for these claims in the papers. Nevertheless, if these claims were correct, which seems very unlikely, it would suggest that variable CO_2 concentration has little effect on the Earth's climate, which seems to be impossible.

6.2.5 Heat balance of the Earth from 1950 to 2005

Murphy *et al.* (2009) estimated the annual heat balance of the Earth from 1950 to 2005. They estimated the forcing due to greenhouse gases over this period. It is not clear from their paper how this was done. They found that about 60% of total forcing was due to CO_2, about 20% was due to CH_4, and the remainder was divided amongst halocarbons, ozone, and N_2O. Their calculated forcing due to greenhouse gases, year by year, is shown in Figure 6.16. According to them, the forcing reached about 2.5 W/m^2 in 2005 when the concentration of CO_2 reached about 380 ppm (compared to 300 ppm in 1950). This is unusually high considering that the consensus of climate models is a forcing of about 3.7 W/m^2 results from a CO_2 concentration rise from 280 ppm to 560 ppm. If Murphy *et al.* were correct, they would obtain a forcing of about

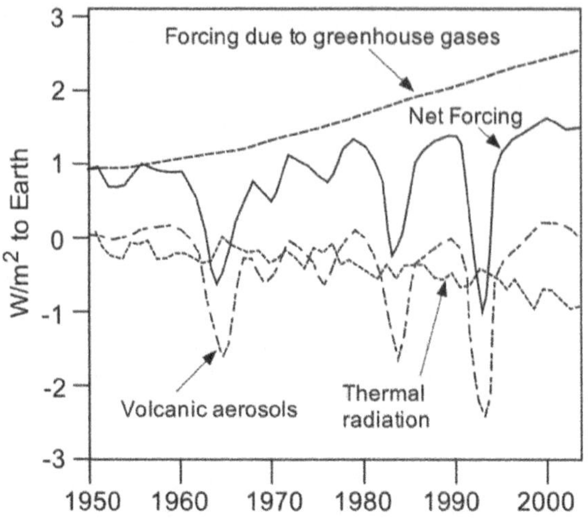

Figure 6.16. Factors contributing to heat balance of the Earth. The net forcing is the arithmetic sum of the forcings due to greenhouse gases, volcanic aerosols, and thermal radiation increase (adapted from Murphy *et al.*, 2009).

$$[(560–280)/(380–300)]\ 2.5\ \mathrm{W/m^2} = 8.75\ \mathrm{W/m^2}$$

for a doubling of CO_2 from pre-industrial values, which is more than double the widely accepted value. The estimate by Murphy *et al.* (2009) implies that radiation emitted by the Earth would be reduced by about 2.5 W/m^2 in 2005 compared to 1950, were it not for other factors. One factor is that the Earth was about 0.55°C warmer in 2005 than in 1950. This would increase radiation emitted by the Earth by about 1 W/m^2 in 2005 compared to 1950. The forcing due to increased radiation level is shown year by year in Figure 6.16. Another factor is the periodic occurrence of volcanic eruptions that inject aerosols into the atmosphere that reflect incoming sunlight, producing a cooling effect. Estimates of this forcing are also shown in Figure 6.16. The net forcing due to greenhouse gases is the sum of the gross forcing, the radiation forcing, and the volcanic forcing (the latter two being negative). This is also shown in Figure 6.16.

There are additional forcings that affect the Earth's heat balance. These revolve about water vapor in some form (aerosols, water vapor, clouds, etc.). Furthermore, Murphy *et al.* did not seem to consider variable humidity or cloudiness over the time period 1950–2005, but they did allow for anthropogenic aerosols. Murphy *et al.* did not attempt to estimate this *a priori*. Instead, they separately estimated the yearly change in ocean heat content of the Earth (the oceans hold most of the heat content of the Earth—land and atmosphere heat content are minor). This result was presented as an eight-year moving average (see Figure 6.17). Then they took an eight-year moving average of the "net forcing" in Figure 6.16 (also shown in Figure 6.17) and noted that it had a similar shape to the ocean heat content curve except it was considerably higher than the ocean heat content curve. If the Earth is being

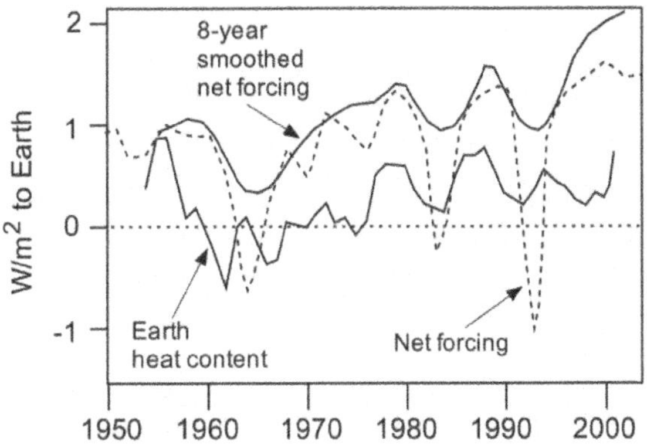

Figure 6.17. Comparison of net forcing due to greenhouse gases with increase in ocean heat content (adapted from Murphy *et al.*, 2009).

heated by greenhouse gases and the calculated net forcing does not appear in the oceans, there is a considerable amount of heat that is not accounted for. Murphy *et al.* (2009) assumed that anthropogenic aerosols reflected enough incident sunlight that the difference between the eight-year smoothed net forcing and the ocean heat content in Figure 6.17 was due to this effect. According to them, only 10% of the greenhouse gas forcing resulted in heating of the Earth. The other 90% of heating induced by greenhouse gases was counterbalanced by the following factors:

- increased radiation to space from a warming Earth: 20%;
- rejection of incident sunlight by volcanic aerosol emissions: 20%;
- rejection of incident sunlight by anthropogenic tropospheric aerosols: 50%.

According to Murphy *et al.* (2009), had there been no volcanic activity or anthropogenic aerosols over the period 1950–2005, the ultimate net forcing would have been considerably higher. This would have warmed the Earth a good deal more than the actual temperature rise of 0.55°C, resulting in further increases in radiation to space from a warming Earth. If the factors of rejection of incident sunlight by volcanic and tropospheric aerosols are ignored, the greenhouse warming that heats the Earth would rise. At the same time, the radiation to space would also increase. A rough guess is that removing the aerosol terms would change the increased radiation to space from a warming Earth from 20% to perhaps 40%, so that the percentage of the greenhouse gas forcing resulting in heating of the Earth would rise from 10% to 60%, a six-fold increase. Hence, according to Murphy *et al.*, were it not for volcanic and tropospheric aerosols, the Earth would have warmed from 1950 to 2005 not by 0.55°C, but by 3.3°C. Even the most extreme alarmists do not claim that the rise in CO_2 concentrations of the 20th century would produce a 3.3°C temperature rise.

None of the results of Murphy *et al.* (2009) seems very credible. First of all, their estimate of greenhouse gas forcing appears to be more than double the widely accepted value of 3.7 W/m^2 for a doubling of CO_2 from 280 ppm to 560 ppm. Secondly, they imply an extreme sensitivity of the Earth to CO_2 concentration that would have produced an enormous heating had it not been for fortuitous volcanic eruptions and anthropogenic aerosol generation.

It is not clear from the work of Murphy *et al.* (2009) what the roles of changing humidity and cloud cover might be. Solomon *et al.* (2010) reported:

> "Stratospheric water vapor concentrations decreased by about 10% after the year 2000. Here we show that this acted to slow the rate of increase in global surface temperature over 2000–2009 by about 25% compared to that which would have occurred due only to carbon dioxide and other greenhouse gases. More limited data suggest that stratospheric water vapor probably increased between 1980 and 2000, which would have enhanced the decadal rate of surface warming during the 1990s by about 30% as compared to estimates neglecting this change. These findings show that stratospheric water vapor is an important driver of decadal global surface climate change."

As we point out in Section 6.2.3, TFK09 found that the global average Earth albedo decreased from 31.3% in 1985–1989 to 29.8% in 2000–2004. Evidently, variability of cloudiness and stratospheric water vapor are important factors, and not well understood.

Wielicki *et al.* (2002) estimated the long-wave and short-wave radiation budgets for the tropical zone from 20°S to 20°N during the 1980s and the 1990s. They found these budgets to be quite stable from 1985 to 1990. Long-wave emission from the top of the atmosphere rose after 1990. The Mount Pinatubo eruption in 1991 produced a sharp spike in reflected short-wave radiation that faded out over the next two years, and reflected short-wave radiation continued to decrease in the 1990s. They pointed out that climate models do not account for this variability. They concluded:

> "We caution against interpreting the decadal variability as evidence of greenhouse gas warming. Whether the changes seen in the radiative balance in the last two decades are the result of natural variability or are a response to global change remains to be determined."

Chen *et al.* (2002) found similar results:

> "Satellite observations suggest that the thermal radiation emitted by Earth to space increased by more than 5 W/m^2, while reflected sunlight decreased by less than 2 W/m^2, in the tropics over the period 1985 to 2000, with most of the increase occurring after 1990."

There are conflicts between estimates of global warming after 1980. While Hansen *et al.* (2010) show a relentlessly rising surface temperature, the satellite measurements of tropospheric temperature show a very different pattern can be interpreted (see Figure 3.27). It is difficult to understand how temperatures can continue to rise if the results of Chen *et al.* and Wielicki *et al.* are correct. While decreased albedo would produce a warming effect, increased long-wave emission more than compensates. In any event, there is no evidence of greenhouse gases driving climate change since 1998.

6.2.6 Albedo and emissivity of the Earth

6.2.6.1 Albedo of the Earth

The effective albedo (reflection coefficient) of the Earth for incident solar irradiance is an important factor in the global heat balance. With a solar irradiance of 342 W/m^2 spread uniformly over the area of the Earth, a 1% difference in albedo would produce a sizable forcing of 3.4 W/m^2. For lower latitudes where the solar irradiance is higher than the average, the effect of albedo is even greater. Ice and snow have high albedos (0.5 to 0.9) while land (0.3) and oceans (0.1) have lower albedos. Clouds have an important effect on albedo and the presence of clouds tends to reduce differences in net albedo above land and oceans. Human-induced changes in aerosols and clouds can cause an enhanced albedo and hence cooling ("negative forcing"). Changes in land use/land clearing (LULC) also have a significant effect on albedo:

> "Many methods have been used to estimate albedo, which cannot be measured directly. These methods differ in their scattering geometries, calibration accuracy, and in spectral, space, and time coverage. The different modes of observation include measurements of earthshine reflected from the Moon, broadband radiometer data from low orbits around Earth, geostationary cloud-cover observations, deep space radiometry, and surface radiometry. All of these methods require a theoretical model for relating the measured parameters to albedo, and they all rely on different assumptions. ... To date, the results from different measurement and modeling approaches are inconsistent among themselves and with each other. The magnitudes of the inconsistencies exhibited by both measurements and models of albedo changes and effects are as large as, or larger than, the entire enhanced greenhouse gas effect when compared in terms of the albedo change equivalent of climate forcing." (Charlson *et al.*, 2005)

Figure 6.18 shows a comparison of several estimates of Earth albedo. The differences between various estimates amount to over 3%, which translates into a difference in radiative forcing of over 10 W/m^2. Hence, the uncertain overall average albedo of the Earth is a major factor in our lack of understanding of heat flows in the Earth's climate.

As we showed in Table 6.3, measurements of the Earth's heat budget by satellite suggest an average albedo of 34% to 36% over land, 28% to 30% over water, and 30% to 31% globally. Presumably, the presence of clouds reduces the innate difference between land and oceans.

A vital parameter for global climate analysis is the albedo at the TOA which determines the net energy input to the Earth. High latitudes are typically very cloudy, especially in maritime areas, and it is expected that cloud cover would increase the TOA albedo over the open ocean, while making little difference over sea ice or snow. A distinction is therefore drawn between surface albedo and TOA albedo. The radiative effectiveness (RE) of snow or ice is defined as the change in TOA albedo for an ice/snow concentration change between 0% and

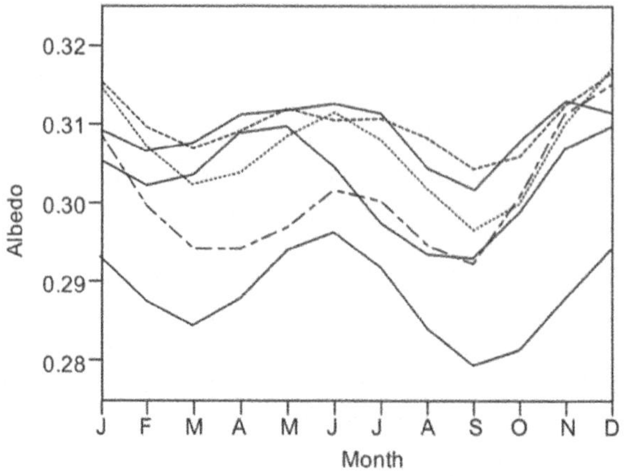

Figure 6.18. Comparison of several estimates of Earth albedo.

100%. It is diminished by the presence of clouds and altered by ice/snow properties. Based on data, Gorodetskaya *et al.* (2006) found RE values of about 0.2 for sea ice or land snow cover. This implies, for example, that, if a 400 W/m^2 solar flux impinges on a polar area above the atmosphere, the flux reflected back to space at the top of the atmosphere from that area would increase by roughly 80 W/m^2 as the Earth's surface in the area changed from 0% to 100% ice/snow cover. Thus, even at RE ~20%, snow/ice cover provides a powerful feedback mechanism for climate change.

Large-scale changes in land use have undoubtedly had an effect on the Earth's climate through changes in the average albedo of the Earth. The effects of conversion of forest to farmland and land clearing on albedo are discussed in Section 6.2.10.5.

The effect of land clearing and land use for agriculture on a large scale has contributed a net cooling to the areas most affected (mainly in the NH). However, in areas that were irrigated, the increased moisture likely produced more cloudiness, which would have increased temperatures at night. The bulk of the deforestation took place in two steps: (1) from 1000 to 1700, and (2) from 1700 to 1940. Land use appears to have made a moderate contribution to the cooling in the *Little Ice Age* (LIA) (from 1400 to 1850). Although the cooling effect of deforestation continued through the 20th century, the strong increase in urbanization and industrialization led to urban heat islands that spread out from urban centers. Thus, it is possible that, as we progressed through the 20th century, urban heating overtook the effects of land clearing and human intervention eventually produced a net heating at ground level.

Table 6.3 shows that, apparently, the Earth's albedo decreased from 1985–1989 (31.3%) to 2000–2004 (29.8%). This may be real or it may represent experimental uncertainty. If it were real, it provides a net forcing of 1.5% of 342 $W/m^2 = 5.1\ W/m^2$, which would have provided much more warming than that observed during that period.

6.2.6.2 *Ocean emissivity*

Volz (2006) discussed the role of the emissivity of seawater in the Earth's energy balance. It was mentioned that the word "emissivity" does not appear once in IPCC (2001). Yet, emissivity is a multiplicative factor in the equation for radiant flux emitted by a body. The average emissivity of the Earth is believed to be in the range 0.92 to 0.965. The thermal radiant flux emitted by the oceans (70% of the Earth's surface) depends on its emissivity. Volz (2006) estimated that the emissivity of seawater is about 0.93 with no wind and about 0.96 with a 15 m/s wind. Thus, wind increases the capability of the oceans to radiate thermal energy. The difference in radiant flux emitted at the top of the atmosphere was estimated to be about 7 W/m^2 across the extremes of no wind vs. a 15 m/s wind. Any incipient climate change that produces a change in wind velocities can trigger changes in ocean emissivity, which can then act to add to or subtract from the ongoing trend. A change in ocean emissivity of 1% would produce a significant forcing of the climate.

6.2.7 Simple models

Several climatologists have developed relatively simple models for the Earth's climate primarily based on energy flows within the Earth climate system. These have value in providing insights into some of the important processes without getting lost in a hopeless sea of computer complexity.

6.2.7.1 *Lindzen's adaptive infrared iris—cloud feedback*

As we discuss in several places in this book, two of crucial unknowns in modeling climate change are the questions of how global humidity and cloud cover change as the Earth warms—presumably due to rising greenhouse gas concentrations.

Lindzen *et al.* (2001) pointed out that, whereas orthodox climate models provide simplistic models for how global average atmospheric humidity changes within the moist and dry regions, Lindzen *et al.* provided evidence of feedback differences between moist and dry air responses to changes in surface temperature. They found a strong positive feedback between cloud coverage and the mean surface temperature of cloudy regions. As the temperature increases, so does cloud cover, which acts to oppose further increases in surface temperature. They argued that a plausible interpretation for this effect is a temperature dependence for the cirrus detrainment from cumulus towers. They concluded that:

> "This dependence appears to act as an iris (by analogy with the eye's iris) that opens and closes dry regions so as to inhibit changes in surface temperature (in contrast to the eye's iris, which does the same in order to counter changes in light intensity)."

Their point is that it is not appropriate to simply assume that, as the Earth warms, the absolute humidity goes up everywhere, but rather, one must take into account that increasing humidity in wet areas has a small effect on climate, while expanding dry areas have a large effect—and that effect is to oppose the increase in surface temperature.

As Chou *et al.* (2002) pointed out:

"Lin *et al.* (2004) reassessed this iris phenomenon by analyzing the radiation and clouds inferred from the Tropical Rainfall Measuring Mission (TRMM) Clouds and the Earth's Radiant Energy System (CERES) measurements. They found a weak positive feedback between high-level clouds and the surface temperature. [Their] feedback factor ranged between 0.05 and 0.10 instead of between –0.55 and –1.10 as found by Lindzen *et al.* (2001). The difference in the feedback factor is due to a larger contrast in albedos and a smaller contrast in the outgoing long-wave radiation (OLR) between the high-level cloudy region and the surrounding regions as derived by Lin *et al.* when compared with that specified in Lindzen *et al.* (2001)."

Chou *et al.* (2002) went on to say:

"It appears that the approach taken by Lin *et al.* (2004) to estimate the albedo and OLR is not appropriate and that the inferred climate sensitivity is unreliable."

Thus, a controversy emerged in which Lindzen and co-workers argued for a strong negative feedback from clouds, while Lin and co-workers argued for a small positive feedback. In both cases, the data pertained to the equatorial region (20°S to 20°N) that was subdivided into three regions: dry; clear and moist; and cloudy and moist.

Lin *et al.* (2004) extended their previous analysis. They used a much lower figure for the coverage of the cloudy/moist region and a much higher albedo for this region, as shown in Table 6.8. Dessler also used short-term data in an attempt to estimate cloud feedback but the database is too short to support any credible conclusions. Fu *et al.* (2002) obtained a negative feedback, but of smaller magnitude than that of Lindzen *et al.* Rapp *et al.* (2005) also contributed to the debate with their analysis based on a mere 20 months of data. They implied their position in the first paragraph of their paper when they quoted the IPCC projection that Earth temperatures could rise by as much as 5.8°C in the next 100 years. To no one's surprise, they sided with Lin *et al.*

The problem for all these analyses is that the Earth does not cooperate and merely respond with cloud changes in sole response to a simple long-term surface

Table 6.8. Comparison of alternative parameters for cloud feedback. The short-wave and long-wave terms are outgoing radiant heat fluxes in W/m^2.

	Lin *et al.*			Lindzen *et al.*		
	Dry	*Clear/moist*	*Cloudy/moist*	*Dry*	*Clear/moist*	*Cloudy/moist*
Coverage %	50	40	10	50	28	22
Albedo	0.154	0.258	0.510	0.211	0.211	0.349
Short wave	338.7	297.1	196.2	315.9	315.9	260.6
Long wave	287.7	253.9	154.8	303.1	263.1	137.7

warming. The duration of the data is too short and other significant changes take place during such data intervals, such as strong El Niños.

Since July 1983, global cloud cover has been monitored by the International Satellite Cloud Climatology Project (ISCCP). A website[8] provides some of their data. One interesting result is that the total column water-vapor concentration in the atmosphere remained flat from 1983 to 1998, and then decreased after 1998. This is in contrast to the prediction of climate models that the water-vapor concentration would increase as the Earth warms. Figure 6.19 shows that relative humidity at 3 km and 9 km altitude decreased as the Earth warmed in the late 20th century, whereas climate models have assumed that the relative humidity would increase. The near-surface humidity meandered but has not increased since 1950.

ISCCP measurements indicate that total global cloud cover was in the range 67%–69% from 1983 to 1990, decreased to about 65% from 1990 to 2000, and varied from 65% to 67% from 2000 to 2010. Hence, cloud cover did not appear to vary in any consistent way. Figure 6.20 indicates that, on balance, warmer temperatures were associated with lower cloud cover. However, there is very wide scatter in the data and, furthermore, most of the data at high cloud cover were prior to 1990, whereas most of the data at the high end were after 1990. Factors other than temperature may well have caused the changes in cloud cover. Here, we have a chicken and egg problem. Were the increased temperature driven by greenhouse gases and reduced cloud cover a consequence of greenhouse-induced warming? Or did the cloud cover change for other reasons, resulting in warmer temperatures when there was less cloudiness?

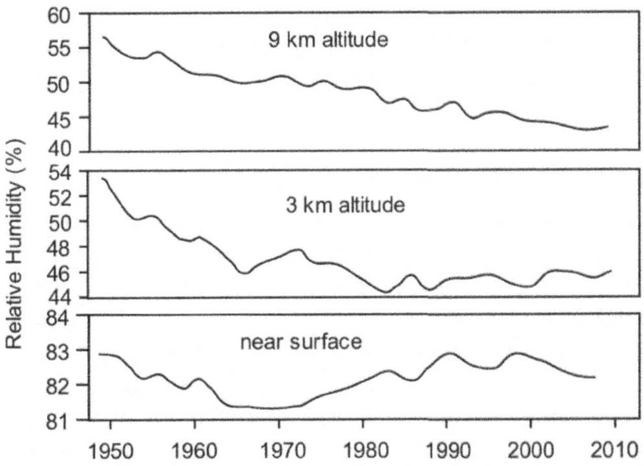

Figure 6.19. Relative atmospheric humidity (%) at three different altitudes in the lower part of the atmosphere (three-year running average) (data source: Earth System Research Laboratory (NOAA); adapted from *www.climate4you.com*).

[8] *www.climate4you.com.*

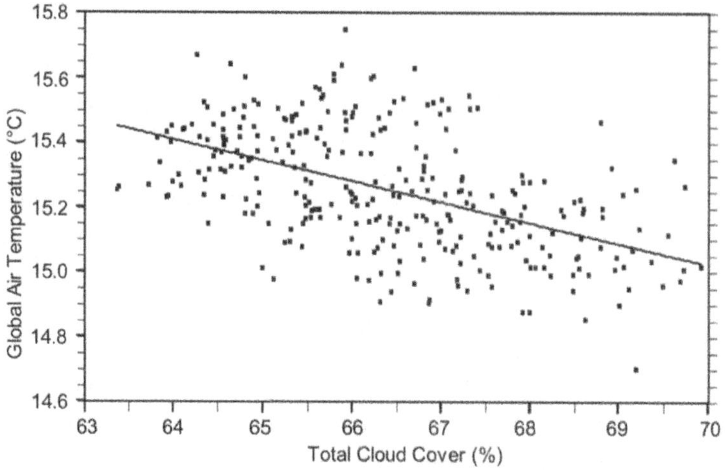

Figure 6.20. Cloud cover vs. temperature since 1983 (data sources: the International Satellite Cloud Climatology Project and University of East Anglia's Climatic Research Unit; adapted from *www.climate4you.com*).

The feedback from clouds is important but the Earth is too complicated to derive this feedback from short-term data while other factors (than surface temperature) contribute to cloud change. One thing is notable. Inevitably, in such matters where the data can be interpreted in opposite ways, the warmists always seem to come up with conclusions that support enhanced warming from clouds while skeptics always seem to come up with results that clouds exert a negative feedback. Is there any credibility left in climate science? Do climatologists of either persuasion have no shame?

6.2.7.2 Willis Eschenbach's thermostat model

Willis Eschenbach[9] emphasized the long-term stability of the Earth's climate over billions of years. Four billion years ago, the early Sun was believed to emit only about three-quarters as much radiant energy as it does today, but there is no evidence that the Earth was incredibly cold during that period. It is widely believed that a much higher concentration of greenhouse gases during that period produced adequate warmth despite the weak Sun, but Eschenbach suggested that correspondence of greenhouse gases with solar intensity over billions of years is too great a coincidence to be likely. He suggested that "a much more likely candidate is some natural mechanism which has regulated the Earth's temperature over geological time".

He pointed out that previous studies, in particular those of Bejan and Reis (2005) and Ou (2001), had modeled the Earth's climate as a heat engine, with the

[9] *wattsupwiththat.com/2009/06/14/the-thermostat-hypothesis.*

ocean and the atmosphere being the working fluids. The tropics are the hot end of the heat engine. Some of that tropical heat is radiated back into space. Work is performed by the working fluids in the course of transporting the rest of that tropical heat to the poles. There, at the cold end of the heat engine, the remaining heat is radiated into space. It was postulated (via some rather vague logic) that a quasi-steady state is reached when the rate of production of entropy is maximized. Eschenbach built his model on the shoulders of these earlier studies. He presented a model that outlined a typical day in the tropics. As the Sun comes up, evaporation produces cumulus clouds that reflect some sunlight, somewhat mitigating solar heating. But solar heating continues until cumulonimbus clouds form. As Eschenbach described it:

> "The columnar body of the thunderstorm acts as a huge vertical heat pipe. The thunderstorm sucks up warm, moist air at the surface and shoots it skyward. At altitude the water condenses, transforming the latent heat into sensible heat. ... At the top, the air is released from the cloud up high, way above most of the CO_2. In that rarified atmosphere, the air is much freer to radiate to space. By moving inside the thunderstorm heat pipe, the air bypasses most of the greenhouse gases and comes out near the top of the troposphere. During the transport aloft, there is no radiative or turbulent interaction between the rising air and the lower and middle troposphere. Inside the thunderstorm, the rising air is tunneled through most of the troposphere to emerge at the top."

Eschenbach also said that "In addition to reflecting sunlight from their top surface, and transporting heat to the upper troposphere where it radiates easily to space, thunderstorms cool the surface in a variety of other ways, particularly over the ocean". As Eschenbach pointed out, these cooling mechanisms include:

(1) wind-driven evaporative cooling: winds around the base increase evaporation;
(2) wind driven albedo increase due to white spray, foam, etc.;
(3) cooling by falling rain and entrained cold wind;
(4) increased reflective area of thunderstorm reflects sunlight along its entire length, particularly in late afternoon;
(5) modification of UT ice crystal clouds;
(6) enhanced night radiation: Cumulus and cumulonimbus generally die out and vanish as the night cools, leading to typically clear skies at dawn, which allows greatly increased night-time surface radiative cooling to space;
(7) delivery of dry air to the surface: the air being sucked from the surface and lifted to altitude is counterbalanced by a descending flow of replacement air emitted from the top of the thunderstorm; this descending air has had the majority of the water vapor stripped out of it inside the thunderstorm, so it is relatively dry; the dryer the air, the more moisture it can pick up for the next trip to the sky; this increases the evaporative cooling of the surface.

Unfortunately, this entire presentation is mainly qualitative and does not seem to lend itself to quantitative detail. Clearly, most of the heat deposited on the Earth from the Sun is delivered in the tropics, and the tropics have a built-in cooling mechanism that increases as the temperature rises. This is certainly a factor in maintaining stability of the Earth's temperature, but it is not clear how this relates to the effects of doubling CO_2.

6.2.7.3 Spencer and Braswell's model

There seems to be a fundamental unstated postulate underpinning the alarmist view of climate that is: the climate of the Earth is deterministic. Like Newton's first law of motion, the Earth's climate will persist in its present state unless acted upon by external forces. Any significant change in climate must be attributable to external forces.

This is why it is so important for the alarmists to minimize the magnitude and extent of the Medieval Warm Period (MWP) and the Little Ice Age (ala Mann, Jones, *et al.*). With a relatively constant climate persisting for thousands of years, a sudden change to a persistent warming would be a strong indicator of human influence, and what else could that be but rising greenhouse gases?

Spencer and Braswell (2011) put it very succinctly:

> "The sensitivity of the climate system to an imposed radiative imbalance remains the largest source of uncertainty in projections of future anthropogenic climate change. Here we present further evidence that this uncertainty from an observational perspective is largely due to the masking of the radiative feedback signal by internal radiative forcing, probably due to natural cloud variations. … It is concluded that atmospheric feedback diagnosis of the climate system remains an unsolved problem, due primarily to the inability to distinguish between radiative forcing and radiative feedback in satellite radiative budget observations."

Spencer and Braswell (2007, 2010, 2011) pointed out that, in addition to external forces, "internal factors" operating near the Earth's surface may affect global temperatures and heat flows. These might include random variations in cloud cover in the Earth's climate system, "brought about through circulation-induced changes in tropospheric wind shear, frontal system behavior, precipitation efficiency, trade wind inversion strength, or any other of the myriad processes that can potentially affect cloud formation other than feedback upon temperature" and non-radiative forcing of temperature change such as tropical intra-seasonal oscillations in the rate of heat transfer from the ocean to the atmosphere.

What this means, is that there is too much scatter to determine whether changes in cloudiness are directly due to changes in surface temperature vs. innate changes in cloudiness in the internal climate system that produce changes in temperature. In other words, as the Earth goes through it chaotic variability, it is possible that it can generate its own long-term trends due to feedback effects. For example, if there is a random change in cloudiness or wind pattern or whatever that persists for quite a few years, this could set into motion feedback effects that could move the climate persistently in one direction, and the whole thing got started internally. Hence, the detection of a long-term trend is not necessarily proof of an external influence as the driver. All of this remains unproven.

Spencer and Braswell (2007, 2010) developed an analysis of feedback effects in response to a forcing of the Earth's climate. The first paper is rather terse but the second paper provides a clearer discussion. In general, they were concerned with the Earth being subjected to a forcing F (W/m^2) at the top of the atmosphere causing temperature departures from equilibrium. Specifically, they were concerned with a positive forcing due to rising greenhouse gas concentrations. The deviation from normal global average surface temperature (T) rises with time in response to this forcing. If no feedback changes occurred, the equation representing the changing T would be:

$$C\ dT/dt = F,$$

where C represents the effective heat capacity of the Earth (essentially the oceans), and the temperature would continue to rise under the forcing.

However, as the temperature rises, other changes occur. The Earth emits more radiation as it warms, and changes occur in the hydrological cycle (evaporation, cloud formation, water–vapor distribution, etc.). Some of these factors act in opposition to the forcing to reduce the temperature rise resulting from the forcing (e.g., increased long-wave emission by the Earth), and some may either resist further temperature rise or enhance it (increased water-vapor concentration and some types of clouds would provide an enhanced greenhouse effect by absorbing radiant emission from the Earth, while some types of clouds would reflect incident sunlight, thus providing a negative feedback). These feedback factors will become stronger as T increases; hence, as a first approximation, it is assumed that they are proportional to the temperature. Thus, a feedback factor (β) is defined:

$$\beta = \beta_{SW} + \beta_{LW},$$

where SW = short-wave and LW = long-wave, and the equation for temperature change is now:

$$C\ dT/dt = F - \beta\ T.$$

The minus sign in front of β in this equation is arbitrary, since β can be positive or negative. The increase in long-wave emission by the Earth as it warms provides a positive contribution to β_{LW} that we can designate as β_{SB} (where SB represents the Stefan–Boltzmann emission increase). If there were no other feedbacks except for increased radiant emission by the Earth, β would equal β_{SB} and T would rise until $F - \beta\ T = 0$, at which point $dT/dt = 0$ and a new equilibrium would be established at a higher temperature. At this point,

$$T = F/\beta_{SB} = \lambda_{SB}\ F,$$

where $\lambda = (1/\beta)$ is the climate sensitivity parameter (change in temperature per unit forcing).

Most climate models estimate that a doubling of CO_2 concentration would produce a forcing or roughly 4 W/m^2. However, there is a wide range of estimates of the climate sensitivity parameter. If $\lambda_{SB} \sim 0.3$°C per W/m^2), corresponding to $\beta_{SB} \sim 3.3$ (W/m^2)/°C, it would be estimated that $T \sim 1.2$°C for a doubling of CO_2 in the absence of other feedback effects. However, estimates for λ_{SB} vary.

Some clouds have a principal effect of reflecting incident sunlight, thus increasing β_{SW}. Other clouds have a principal effect of making β_{LW} negative by absorption of long-wave radiation. In general, increased water-vapor concentrations produce greater negative values of β_{LW}. Most GCMs estimate that, due to increased water-vapor concentration and increased clouds, the net value of β is reduced to a less positive value than β_{SB} and therefore λ is increased to a value greater than λ_{SB}. Typical estimated values of λ by climate models tend to be around 0.8°C per W/m^2), thus leading to a value of T for doubling CO_2 of perhaps 3°C.

Spencer and Braswell (2007, 2010, 2011) pointed out that in addition to the aforementioned feedback factors (Stefan–Boltzmann emission, increased clouds and water vapor due to rising temperature), "internal factors" operating near the Earth's surface may affect global temperatures and heat flows. These might include random variations in cloud cover in the Earth's climate system, "brought about through circulation-induced changes in tropospheric wind shear, frontal system behavior, precipitation efficiency, trade wind inversion strength, or any other of the myriad processes that can potentially affect cloud formation other than feedback upon temperature" and non-radiative forcing of temperature change such as tropical intra-seasonal oscillations in the rate of heat transfer from the ocean to the atmosphere.

According to Spencer and Braswell, if these internal factors are relatively small, then measurements of the Earth's heat budget at the top of the atmosphere should provide radiative fluxes proportional to surface temperature. Spencer and Braswell (2007, 2010, 2011) analyzed heat budget data and found little correlation of heat flows with surface temperature, suggesting that internal factors were obfuscating the other heat flows, making it impossible to experimentally determine feedback factors.

The gospel of the orthodoxy, for example as preached by Dessler (2011), is that CO_2 is the driver of climate change. In this view, the climate would remain quite constant as long as the CO_2 concentration remains constant. The degree of cloudiness would remain within narrow confines dictated by the CO_2 concentration. If the CO_2 concentration were to rise, cloudiness would change as a response to the increase in CO_2 (via a change in global average temperature). Thus, according to this viewpoint, one should be able to plot a measure of cloudiness vs. a measure of global average temperature, and find a direct correlation within limited scatter. However, Dessler did not examine how cloudiness varies with CO_2 concentration or how cloudiness varies with increased surface temperature due to rising CO_2. Instead, he relied on relatively short-term data from volcanic eruptions and El Niño events to examine how cloudiness varied with temperature during these events. He found a very large amount of scatter, which is not surprising. However, he persevered by finding the best straight-line fit to the widely scattered data, and concluded that cloudiness decreased slightly as the temperature rose. It seems more likely that, over this short time period, temperature and cloudiness are uncorrelated, and the data are the result of stochastic variations, rather than a cause–effect relationship. Spencer and Braswell (2011) put it very succinctly:

> "The sensitivity of the climate system to an imposed radiative imbalance remains the largest source of uncertainty in projections of future anthropogenic climate change. Here we present further evidence that this uncertainty from an observational perspective is largely due to the masking of the radiative feedback signal by internal radiative forcing, probably due to natural cloud variations. … It is concluded that atmospheric feedback diagnosis of the climate system remains an unsolved problem, due primarily to the inability to distinguish between radiative forcing and radiative feedback in satellite radiative budget observations."

What this means, is that there is too much scatter in short-term data to determine whether changes in cloudiness are directly due to changes in surface temperature vs. innate changes in cloudiness in the internal climate system.

Aside from the science and pseudoscience involved in these analyses, there are social issues as well. The alarmists refer to their interpretation of climate science as simply "climate science". It is not one interpretation of climate science. It is **THE** CLIMATE SCIENCE—in their view. We see evidence of their arrogance in many publications and press releases. In particular, in regard to the effect of clouds, Dessler said: "In recent papers, Lindzen and Choi (2011), and Spencer and Braswell (2011) have argued that … clouds are the cause of, and not a feedback on, changes in surface temperature. If this claim is correct, then significant revisions to *climate science* may be required". In other words, he regards "climate science" as that which the orthodoxy subscribes to. It is not his interpretation of climate science—IT **IS** CLIMATE SCIENCE!

Another bizarre aspect of Dessler's publication was discussed by Pielke, Sr.[10] He said: "Dessler's paper was received 11 August 2011 and accepted 29 August 2011. This is some type of record … and indicates that the paper was fast-tracked. This is certainly unusual"—to say the least. Pielke, Sr. said:

> "It is not clear whether the Editor of *GRL* included Roy Spencer as one of the referees, [and if they did not] they were derelict in their responsibilities". Dessler's paper should have been submitted to *Remote Sensing* as a Comment [on Spencer's paper]. Then Roy Spencer would submit a Reply."

We are now witnessing a phenomenon in climatology publications that is occurring repeatedly. The climatology orthodoxy has united into an informal cabal dedicated to (1) prevent contrary analyses and interpretations from being published, and (2) to quickly respond to those few contrarian publications that slip through their net with vitriolic attacks on the paper on orthodoxy blogs, and in the literature via rapid rebuttal publications such as that of Dessler (2011). It is evident that many Editors are in cahoots with the orthodoxy; certainly the Editor of *GRL* is, and the Editor of *Remote Sensing* who let Spencer and Braswell's paper through the net, suddenly resigned for unclear reasons.

[10] *http://pielkeclimatesci.wordpress.com/2011/09/06/comments-on-the-dessler-2011-grl-paper-cloud-variations-and-the-earths-energy-budget/.*

6.2.7.4 Trends in middle- and upper-level tropospheric humidity

An important factor in estimating the rise in temperature produced by a doubling of CO_2 concentration is the upper atmosphere humidity. If this humidity increases as CO_2 increases, it will add a positive feedback to drive temperatures up even more than that due to the CO_2 increase alone. Radiosonde-derived humidity data are available for the period 1973 to 2007, during which the CO_2 concentration rose from about 330 ppm to about 375 ppm (Paltridge *et al.*, 2009). However, as Paltridge *et al.* emphasized, "radiosonde-derived humidity data must be treated with great caution" because they are "notoriously unreliable". However, Paltridge *et al.* claimed that, while the data are somewhat dubious, nevertheless, they may not be as bad as their reputation. In any event, these data are very important and, for what they are worth, Paltridge *et al.* analyzed the data and found that, in many cases, upper-level humidities actually decreased during the 1973–2007 period, as opposed to the increase predicted by climate models.

Dessler[11] took issue with this and claimed that there are abundant data that water vapor feedback is "strong and positive". However, the evidence he cited (very short-term data from volcanic eruptions and El Niño events) is irrelevant to the argument. Paltridge *et al.* did not argue that increased water vapor does not increase the greenhouse effect. They argued that the water-vapor concentration did not rise as CO_2 increased. So Dessler's argument is irrelevant to the question of the effect of rising greenhouse gases on humidity. This topic is discussed further on a blog.[12] It seems that Paltridge is neutral and seeks to find answers from sparse data, whereas Dessler is biased and knows the answer in advance, and seeks out data to support his position.

6.2.8 Heat capacity, time constant, and sensitivity of the Earth's climate system

The heat capacity of the Earth is almost entirely due to the oceans. The following estimate of the Earth's heat capacity is excerpted from Schwartz (2007).

The Earth's climate system involves a radiative balance between absorbed short-wave (solar) radiation Q and long-wave (thermal infrared) radiation emitted at the top of the atmosphere E:.

$$Q \approx E \tag{6.1}$$

The global annual mean absorbed short-wave irradiance is:

$$Q = \gamma\ J, \tag{6.2}$$

where γ is the mean planetary co-albedo (one minus albedo) and J is the mean solar irradiance at the top of the atmosphere $\approx$ 343 W/m^2.

Satellite measurements indicate Q $\approx$ 237 W/m^2, corresponding to $\gamma \approx 0.69$ (albedo $\approx$ 0.31). The global annual mean emitted long-wave irradiance is:

[11] *http://pielkeclimatesci.wordpress.com/2010/01/06/guest-post-by-andrew-dessler-on-the-water-vapor-feedback/*; also see Dessler and Davis (2010).

[12] *http://judithcurry.com/2011/09/25/trends-in-tropospheric-humidity/#more-5048*.

$$E = \varepsilon\, \sigma\, T^4, \qquad (6.3)$$

where ε is the effective planetary long-wave emissivity, σ is the Stefan–Boltzmann constant, and T_s is the global mean surface temperature.

An energy imbalance *Q–E* arising from a secular perturbation in *Q* or *E* results in a rate of change of the global heat content given by

$$dH/dt = Q - E, \qquad (6.4)$$

where *dH*/*dt* is the change in heat content of the climate system (the oceans). But the definition of the heat capacity of the Earth is given by the equation:

$$dH/dt = C\, dT/dt,$$

where *C* is an effective heat capacity that reflects only that portion of the global heat capacity that is coupled to the perturbation on the timescale of the perturbation. In the present context of global climate change induced by changes in atmospheric composition on the decade to century timescale, the pertinent heat capacity is that which is subject to a change in heat content on such time scales. Measurements of ocean heat content over the past 50 years indicate that this heat capacity is dominated by the heat capacity of the upper layers of the world's oceans. The lower layers require more time to readjust to changes at the surface.

Combining Equations (6.2)–(6.4), we have:

$$C\, dT/dt = \gamma\, J - \varepsilon\, \sigma\, T^4. \qquad (6.5)$$

Equation (6.5) can be solved for a few simple special cases. For the case of a step-function forcing in which *F* = (*Q–E*) makes a step-change from one constant level to another constant level, the forcing *F* is defined as the difference between these two levels. In this case, the change in temperature (from the original temperature prior to the step-change) at a time (*t*) after the step-change is

$$\Delta T(t) = (\tau/C)\, F\, [1 - exp(-t/\tau)], \qquad (6.6)$$

where τ is a constant of integration that characterizes the e-folding time over which the system readjusts to a new temperature.

For large *t*, the exponential becomes negligible, and

$$\Delta T(t) \rightarrow (\tau/C)\, F \text{ as } t \rightarrow \infty. \qquad (6.7)$$

But this is the climate sensitivity relation. Therefore, we can identify the climate sensitivity parameter as:

$$\lambda = (\tau/C) \qquad (6.8)$$

$$\Delta T(\infty) = (\lambda)\, F \qquad (6.9)$$

and λ represents the ultimate temperature rise produced by a step-function forcing *F*.

It turns out that:

$$\tau = C\, T_o/(4\, J\, \gamma) \qquad (6.10)$$

so that

$$\lambda = T_o/(4\ J/\gamma), \tag{6.11}$$

where T_o is the initial value of T.

As Schwartz (2007) showed, with $T_o = 288$ K, $J = 343$ W/m^2, and $\gamma = 0.69$, the estimated value for $\lambda = 0.3$ °C per W/m^2. Feedback factors can increase this value, which was calculated in the absence of feedback.

Schwartz (2007) also discussed the case where the forcing is not a step function, but a continuous ramp-up, as, in $F = \beta$ t, the solution of Equation (6.5) becomes

$$\Delta T(t) = (\beta\ \lambda)\ [(t - \tau) + \tau\ exp(-t/\tau)] \tag{6.12}$$

For t >> τ, the exponential is negligible, and we obtain

$$\Delta T(t) = (\beta\ \lambda)\ t, \tag{6.13}$$

showing that the temperature increases continuously in proportion to the forcing.

Equation (6.8)—or (6.10)—defines a relation between C, λ, and τ. If any two are known, we can calculate the third. Schwartz (2007) estimated τ by two methods. One was based on rates of decay of impacts of volcanic eruptions, and the other was an abstract technique based on the range of observed fluctuations of temperature over the past century, which is limited by the rate at which the system equilibrates to a perturbation. His resultant estimate was $\tau \approx 5$ years. Both the values of τ and λ are lower than values widely used by warmists. Since λ is proportional to τ, if τ was actually 10 years (instead of 5), the value of λ would double to 0.6°C per W/m^2.

Based on $\tau \approx 5$ years and $\lambda \approx 0.3$°C per W/m^2, Schwartz (2007) estimated the heat capacity of the Earth to be $\approx$ 16 W-years/°C per m^2.

The rather short time constant of the climate system determined by this analysis implies that the climate system is in near equilibrium with applied forcings. Hence, the total forcing of the climate system (F) over a given time period can be estimated empirically from knowledge of the change in global mean surface temperature (ΔT) over that period as (λ F) with little error resulting from lag of the temperature response to forcing.

Climate models have estimated that the forcing due to a doubling of CO_2 concentration is about 3.7 W/m^2. If Schwartz's estimated value of $\lambda \approx 0.3$°C per W/m^2 is correct, the estimated temperature rise due to the greenhouse effect would be 1.1°C. However, this does not include feedback effects.

As it turned out, Schwartz's paper produced a flurry of comments by warmists (e.g., Foster *et al.*, 2008) that induced Schwartz to increase his estimate of λ from 0.3°C to 0.5°C per (W/m^2) which implies $\tau \sim 8$ years and the estimated temperature rise due to the greenhouse effect (without feedback) would be then be 1.8°C.

Lea (2004) argued that:

> "The tropical oceans, because they are removed from the direct climate impact of continental ice sheet buildup during glaciation, provide an analog for equilibrium climate response to varying CO_2 and the hypothesis that CO_2 variations are the dominant source of radiative forcing in tropical ocean regions."

However, he did mention that "not all of the studies explicitly separate out the effect of CO_2 from other factors". A number of studies have attempted to estimate temperature changes in the tropical oceans during glaciation with variable results, typically with small values. Climate: Long range Investigation, Mapping, and Prediction (CLIMAP) was supposed to put this matter to rest but its results later proved to be somewhat embarrassingly and unacceptably low. More recently, "the development of two sensitive geochemical temperature proxies—alkenone unsaturation ratios and Mg/Ca in planktonic foraminifera" has greatly improved our understanding of cooling of the tropical oceans during the Last Glacial Maximum (LGM) which has been estimated at around 3°C. Lea (2004) used "a 360-kyr-long SST record from the Cocos Ridge, in the northeastern tropical Pacific" in an attempt to "1) to test the hypothesis that atmospheric CO_2 variations are the dominant control on tropical SST on orbital time scale, and 2) to compute a (tropical) climate sensitivity based on the tropical SST response to known greenhouse forcing". In regard to the first goal, he demonstrated that tropical sea-surface temperatures (SSTs) followed a very similar path as measured CO_2 concentrations over the past 400,000 years, encompassing four interglacial periods and three glacial maxima. Lea therefore concluded that changing CO_2 concentrations were the cause of variability of tropical SSTs over this time interval. However, Lea provided no explanation for why the CO_2 concentration varied as it did and, indeed, there is presently no fully satisfactory explanation for this effect. Hence, it is not clear that some other factor(s) caused CO_2 variations and tropical SST variations. There is no cause–effect relationship established between CO_2 variations and tropical SST variations, although CO_2 variations provide a prime suspect for at least part of the effect. In regard to his second goal, Lea plotted the tropical SST anomaly vs. CO_2 concentration for his data points over the past 400,000 years. Assuming that CO_2 alone caused the warming of tropical seas, a linear fit to the data scatter plot provides an estimate for the average temperature change induced by a CO_2 concentration change. He obtained:

$$\Delta T = 0.0226\ [\Delta CO_2] + (\text{constant}).$$

The relationship between forcing (W/m^2) and [CO_2] is logarithmic, but linearizing this relationship over the range of [CO_2] from 180 ppm to 280 ppm, I find:

$$\Delta F / \Delta CO_2 = 0.0238.$$

Hence,

$$\Delta T = 0.95\ [\Delta F] + (\text{constant}).$$

Note that Lea somehow derived a value of 1.4 instead of 0.95. Either value is higher than many other estimates. Lea admitted that "A potential weakness of the approach just described is that it does not take into account other factors that could influence tropical SST, such as ice volume or atmospheric dust" and, in addition, ocean transport of cold water from high latitudes is another important factor. For example, Ganapolski *et al.* (1998) found that "ocean circulation in the Atlantic" is an important contributor to cooling the tropics. To the extent that these factors contributed to the measured temperature changes of tropical oceans, the temperature change attributable to CO_2 alone would be reduced. It seems likely that Lea's estimate of climate sensitivity is too high by at least a factor of two.

6.2.9 Heat content of the oceans

While many climate studies focus on temperature changes in the atmosphere at the surface or in the troposphere, the climate of the Earth is ultimately determined by the temperatures of the oceans. The oceans have a heat capacity about 1,000 times greater than the atmosphere and land surface. Although air temperatures may change much more rapidly than ocean temperatures, it is the ocean temperature distribution that will ultimately determine the climate of the Earth.

The "holy grail" of climatology that many climatologists are searching for is the expected rise in global average temperature if the CO_2 concentration is doubled from the pre-industrial value of roughly 280 ppm (*aka.* climate sensitivity). If we could predict the increase in global average ocean temperature resulting from a doubling of the CO_2 concentration from the pre-industrial value, we could infer the climate sensitivity of the Earth. Unfortunately, like almost every issue in climatology, it is difficult to reduce uncertainties in estimates of future trends.

There is considerable evidence that SSTs have warmed significantly over the course of the 20th century (e.g., Tisdale, 2009). There is also evidence that on average, the bulk oceans have warmed over this time period but the temperature gains were smaller (e.g., Levitus *et al.*, 2012). Nevertheless, the oceans are so vast that this represents a very large amount of heat. While the data on ocean warming leave much to be desired, there seems to be little doubt that the oceans have acquired a significant amount of heat over the last five decades (see later part of this section). Aside from the still unsettled issue of specifying the ocean warming to higher accuracy, the question arises as to what factors caused the ocean warming. Had there been long-term systematic increases in solar intensity and/or decreases in cloud cover, that would certainly have contributed significantly to ocean warming. As we discuss in Chapter 5, there are several models that attempt to estimate how the solar intensity has varied during the 20th century, but they all make assumptions that cannot be validated and therefore they remain highly speculative. Estimates of global variation of cloud cover have been made by a number of investigators. A slightly upbeat review was provided by Dai *et al.* (2006), who found that cloud cover decreased over the past several decades. But even these authors admit "large inadequacies in monitoring long-term changes in global cloudiness with surface and satellite observations" and the sub-title of their paper is "A tale of monitoring inadequacies". Eastman *et al.* (2011) concluded that long-term secular changes in cloud cover over the oceans may have occurred during the period 1954–2008. Norris (2005) reported decreases in cloud cover over the oceans from 1952 to 1997. However, Evan *et al.* (2007) pointed out problems with cloud data and suggested that the ISCCP data set of cloud amounts "may not be appropriate". Norris and Slingo (2009) discussed the inadequacies of cloud observation systems. It should be noted that cloud cover over the oceans exceeds that over land by a considerable amount (~68% vs. ~54%).

Spencer and Braswell (2010, 2011) pointed out that "internal factors" operating near the Earth's surface may affect global heat flows. These might include random variations in cloud cover in the Earth's climate system, "brought about through circulation, induced changes in tropospheric wind shear, frontal system behavior, precipitation efficiency, trade wind inversion strength, or any other of the myriad processes that can potentially affect cloud formation other than feedback upon temperature and non-radiative forcing of temperature change such as tropical intra-seasonal oscillations in the rate of heat transfer from the ocean to the atmosphere".

While we remain uncertain about past variability of solar intensity and cloud cover, and variability due to internal factors remains speculative, one thing we do know is that, over the past ~120 years, the CO_2 concentration in the atmosphere has risen fairly steadily. Climate models show that there is a resulting decrease in upward flux of long-wave radiation at the TOA that produces an imbalance in the Earth's energy balance. The Earth then warms to regain equilibrium. Concurrently, the net downward back radiation from increased levels of greenhouse gases in the troposphere increases, due to the increase in CO_2 concentration as well as the increase in tropospheric temperature. It is widely theorized that much of this heat flux ends up in the oceans as ocean warming. Some estimates of the Earth's imbalance indicate that it exceeds the gains in heat content of the oceans, thus giving rise to discussions in the blogosphere about "the missing heat". At the other end of the scale, Tisdale (2009) claims that, since the IR radiation is absorbed in the top few microns of the ocean, it cannot heat the ocean and merely leads to greater evaporation. While we cannot claim to provide a full explanation of the warming of the oceans due to all factors, we think it is possible to at least partly clarify the contribution of the greenhouse effect to ocean warming.

Three aspects of the warming of the oceans by IR include:

(1) the excess flux of down-welling back radiation IR that impinges on the ocean surface for several levels of CO_2 concentration compared to the pre-industrial value (i.e., the *forcing* at the surface);
(2) the heating effect in the oceans produced by increased down-welling back radiation IR flux from the sky;
(3) a comparison of the estimated rate of warming of the oceans from (1) and (2) with the measured change in ocean heat content over the past few decades.

The effects of back radiation from the sky on oceans were studied as early as the 1920s and continued in the 1960s to the 1980s by a number of investigators. Since then, the ocean heating process has been incorporated into massive climate models and the physics of the process has been buried in these models. Several climatologists, notably James Hansen, have emphasized their estimates of an energy flux imbalance at the TOA. It has then been argued that this net energy flux into the Earth must end up somewhere, and the only place that could be is the oceans. There has been some consternation because measured increases in ocean heat content appear to be lower than predictions based on the estimated energy flux imbalance at the TOA. Meanwhile, the process by which energy accumulates in the ocean may require further elucidation.

A number of websites currently claim that back radiation from greenhouse gases cannot heat the oceans. For example, one website[13] asserts that "since the LWIR [long-wave infrared] re-radiation from increasing 'greenhouse gases' is only capable of penetrating a minuscule few microns (millionths of a meter) past the surface and no further, it could therefore only cause evaporation (and thus cooling) of the surface 'skin' of the oceans". Another web page[14] says that "It is impossible for a 1.7 W/m^2 increase [predicted by the IPCC due to man-made greenhouse gases] in downward 'clear sky' atmospheric LWIR flux to heat the oceans". Another website[15] says that "Infrared radiation from 'greenhouse gases' causes evaporative cooling of the oceans rather than heating". Yet another website[16] ridicules Livermore scientists for claiming that rising greenhouse gases warm the oceans. Finally, another website says that "Since the ocean is on average warmer than the atmosphere, the energy flux across the ocean/atmosphere interface is on average carrying heat from the ocean to the air. ... So given the general direction of the motion of the energy, how can infrared energy be pushed into the ocean, when it can't penetrate the surface further than its own wavelength?".[17]

If these claims were correct, then any warming of the oceans would have to be attributed to increases in solar intensity or decreases in cloud cover. In Appendix II, we attempt to clarify (to the extent possible) the role of forcing due to increased CO_2 in the atmosphere on ocean warming. There seems to be little doubt that increases in greenhouse gas concentrations will in fact heat the oceans.

Before proceeding to discuss ocean warming, it is useful to present some basic quantities:

- area of oceans ~ 361 million km^2;
- volume of oceans ~ 1.3 billion km^3;
- average depth of oceans $\sim 1.3 \times 10^9 / 3.61 \times 10^8 = 3.6$ km $= 3{,}600$ m;
- heat capacity of ocean water: 3,993 J/kg/°C;
- density of ocean water $\sim 1.025 \times 10^{12}$ kg/km^3;
- heat capacity of oceans: $1.025 \times 10^{12} \times 3{,}993 \times 1.3 \times 10^9 = 5.3 \times 10^{24}$ J/°C.

If 1.0 W/m^2 is added to the oceans, the total heat input would be $1.0 \times 361 \times 10^6 \times 10^6 = 3.6 \times 10^{14}$ W (equivalent to J/s) (note: we used the area of the oceans, not the area of the Earth).

In the course of a year, there are $3{,}600 \times 24 \times 365 = 3.2 \times 10^7$ s.

[13] *http://hockeyschtick.blogspot.com/2010/08/why-greenhouse-gases-wont-heat-oceans.html.*

[14] *http://hockeyschtick.blogspot.com/2010/08/energy-environment-full-special-issue.html.*

[15] *http://theantislave.wordpress.com/2012/09/04/man-made-co2-is-not-the-driver-of-global-warming/.*

[16] *http://stevengoddard.wordpress.com/2012/06/11/livermore-moves-reality-out-of-the-physical-world/.*

[17] *http://tallbloke.wordpress.com/2011/03/03/tallbloke-back-radiation-oceans-and-energy-exchange/.*

In a year, the heat input to the oceans would be:

$$3.6 \times 10^{14}\ \text{J/s} \times 3.2 \times 10^{7}\ \text{s} = 1.2 \times 10^{22}\ \text{J}.$$

The temperature rise of the oceans in one year would then be:

$$1.2 \times 10^{22}/5.3 \times 10^{24} = 2.3 \times 10^{-3} = 0.0023^{\circ}\text{C}.$$

Of course, the ocean is heated in layers and more heat would appear in the upper layers than the lower layers. The distribution of ocean volume vs. depth is shown in Figure 6.23.

Estimates of the heat content of the oceans (0 m to 700 m depth) have been made by several groups. Levitus, Antonov, and Boyer (2005) presented estimates of the variability of ocean heat content based on hundreds of thousands of temperature profiles. They also provided estimates of the components of the Earth's heat balance as shown in Table 6.9. However, errors have been found in some of the ocean instrumentation, and several authors have attempted to correct for these. Domingues *et al.* (2008), Wijffels *et al.* (2008), and Ishii and Kimoto (2009) presented updated estimates with their corrections to the instrumental data. Levitus *et al.* (2009) reviewed this work and made their own revised corrections. These results were updated by Levitus in 2010.[18] Levitus *et al.* (2009) emphasized that their work "represents an attempt to correct for observed [instrument] biases and that more work remains to solve this problem". They also pointed out that complications make this challenging. A comparison of various estimates is given in Figure 6.21.

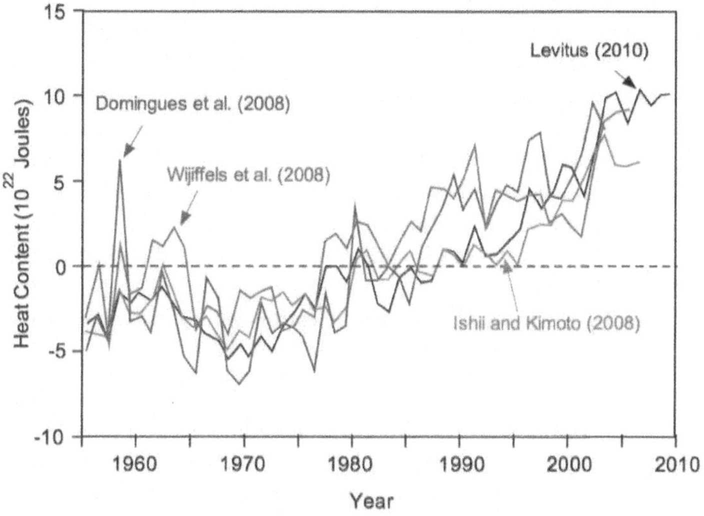

Figure 6.21. Changes in global ocean heat content to 700 m depth (Levitus *et al.*, 2009 and Levitus, 2010).

[18] *ftp://ftp.nodc.noaa.gov/...HEAT_CONTENT/.../heat_content_differences.pdf.*

Table 6.9. Estimated components of the Earth's heat balance (data from Levitus *et al.*, 2005).

Heat flow element	*Heat flow for period 1955–1998 (10^{22} J)*
Heat absorbed by oceans	14.5
Heat absorbed by continents	0.9
Heat to melt continental glaciers	0.8
Heat absorbed by atmosphere	0.7
Heat to reduce Antarctic sea ice	0.3
Heat to melt mountain glaciers	0.1
Heat to melt NH sea ice	0.005
Heat to melt Arctic sea ice	0.002

Although the overall average trend was predominantly upward, the heat content of the oceans was not uniform and various regions experienced very different changes in heat content.

A number of improvements have been made in the past decade in measurements to assess the changes in the global average ocean heat content anomaly (OHCA). The Argo global array of 3,000 free-drifting profiling floats measures the temperature and salinity of the upper 2,000 m of the ocean. This allows, for the first time, continuous monitoring of the temperature, salinity, and velocity of the upper ocean, with all data being relayed and made publicly available within hours after collection. Argo deployments began in 2000 and, by the end of 2006, the array was over 90% complete. However, this network requires continual servicing and replenishment. Expendable bathythermographs (XBTs) have been used by oceanographers for many years to obtain information on the temperature structure of the ocean to depths of up to 2,000 m. The use of XBTs to measure the ocean's subsurface has significantly increased over the past decade. NOAA's XBT program (SEAS) currently supports about 80 voluntary observing ships. More than 14,000 XBT observations are made each year. However, the XBT data have been found to be error-prone, and the various trends shown in Figure 6.22 show the range of efforts to deal with this problem.

Lyman *et al.* (2010) reviewed the various estimates of changing heat content in the oceans to 700 m depth from 1993 to 2009. They said:

> "A host of choices must be made when computing ocean heat content anomaly (OHCA) curves. These choices include how to quality-control the data, which mapping technique to use, which baseline mean climatology to use, which annual cycle to remove, how to treat unsampled or under-sampled areas, and how to correct biases in data from XBTs and other instruments. The several teams working on the problem around the world make their own choices and have produced apparently different OHCA curves."

As Lyman *et al.* (2010) pointed out, there are not only differences in magnitude, but differences in sign between various results. Particularly strange is the fact that

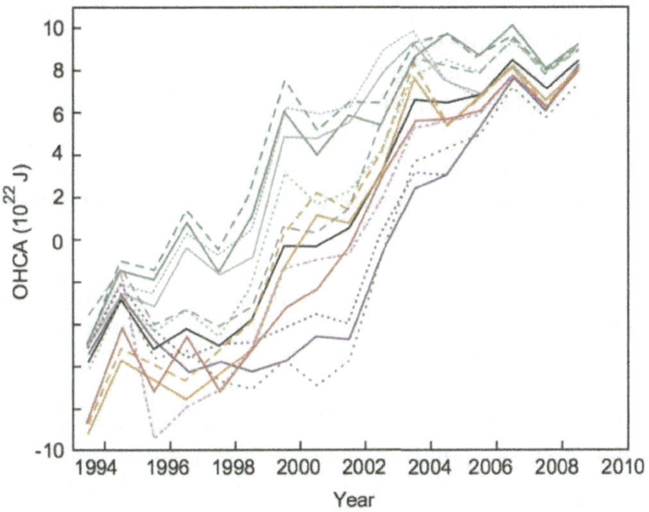

Figure 6.22. Ocean heat content anomaly (OHCA) curves by different methods (Lyman *et al.*, 2010).

some find warming and some find cooling during the very large El Niño of 1997–1998. Lyman *et al.* (2010) emphasized that "the individual OHCA curves all flatten out after around 2003", and that "sea surface temperatures have been roughly constant since 2000". Yet, sea level continued to rise during this period. "The amount of water added to the ocean by melting continental ice in recent years may account for most of this rise even with very little change in ocean heat content. However, this leaves the global energy budget with a large residual for this time period, because it takes less energy to melt ice than to warm the ocean for the equivalent sea-level rise."

Lyman *et al.* (2010) pointed out that the flattening of OHCA curves also occurred around the time (2004) that the Argo array of autonomous profiling floats first achieved near-global coverage and became the primary source of OHCA data. They suggested "the fact that this transition occurred at the same time as the flattening could be coincidental, but also raises the possibility of a yet-undiscovered bias in the observing system".

The Earth's atmosphere is very tenuous compared to the oceans. As Pielke (2003) discussed, the energy balance of the Earth can be estimated by either calculating the various forcings that act on the Earth at the TOA and thereby estimating the net flux (W/m^2) acting to heat the Earth, or by estimating the heat deposited into the Earth system (e.g., J/yr). In the former case, the argument is that the imbalance in energy flux at the TOA must be stored in the Earth system, and the only place it can go is the oceans; therefore, it must heat the oceans. Since the heat capacity of the Earth lies mainly in the oceans, one can estimate the heat gain of the Earth by measuring the heat gain of the oceans. There are difficulties in both approaches. Of particular importance in regard to forcings are the unknowns relative to changes in humidity, aerosols, and clouds as the Earth warms and industrialization proceeds. The problem in estimating ocean heat content is that the oceans are so diverse, and the collective mass of the oceans is so huge that, even with a large heat flux imbalance, the temperature rise is small and difficult to measure. Nevertheless, a number of studies of ocean warming have focused on the heat gain per unit surface area, assuming that this is equal to the imbalance in energy flux at the TOA. In reviewing the data, one must be careful to distinguish between per-unit-area data based on the surface area of the Earth ($\sim 5.16 \times 10^8$ km^2) vs. data based on the area of the oceans ($\sim 3.61 \times 10^8$ km^2).

Palmer and Haines (2009) estimated the global heat absorbed by the ocean to a depth of 220 m from 1970 to 2000 to be 0.25 W/m^2 (based on ocean area, not global area).

According to Levitus *et al.* (2012), over the 55-year period from 1955 to 2010, the heat content of the oceans down to a depth of 2,000 m increased by $2.4 = 10^{23}$ J. They claimed that this corresponds to an average temperature rise for the 0–2,000 m layer of 0.09°C. Since the 0–2,000 m layer occupies 48% of the volume of the oceans (see Figure 6.23), its volume is about $0.48 = 1.3 = 10^9$ km^3 $= 6.3 = 10^8$ km^3.

The average temperature rise over 55 years for this layer is thus estimated to be:

$$2.4 = 10^{23}\ \text{J} / (6.3 = 10^8\ \text{km}^3 = 1.025 = 10^{12}\ \text{kg/km}^3 = 3993\ \text{J/kg/°C}) = 0.08°\text{C},$$

which is close to the value given by Levitus *et al.* (2012). The same calculation can be repeated for the 0–700 km layer. According to Levitus *et al.* (2012), over the 55-year period from 1955 to 2010, the heat content of the ocean down to depth 700 m increased by $1.67 = 10^{23}$ J. They claim that this corresponds to an average temperature rise for the 0–700 m layer of 0.18°C. Since the 0–700 m layer occupies 16% of the volume of the oceans, its volume is about $0.16 = 1.3 = 10^9$ km^3 $= 2.1 = 10^8$ km^3. The average temperature rise over 55 years for this layer is thus estimated to be:

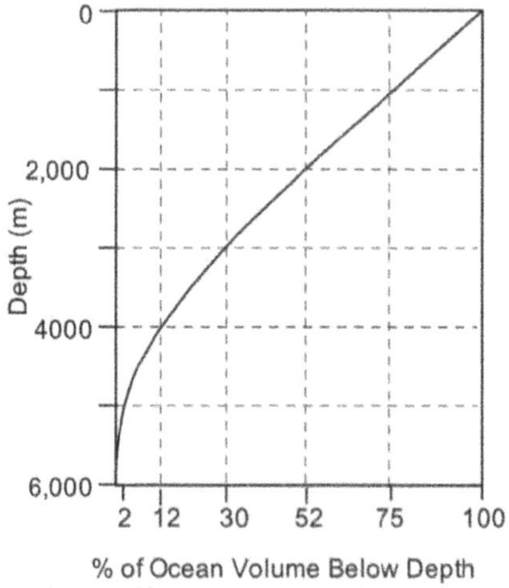

Figure 6.23. Percent of world ocean volume below any depth.

$$1.67 = 10^{23}\ \text{J}\ /(2.1 = 10^{8}\ \text{km}^{3} = 1.025 = 10^{12}\ \text{kg/km}^{3} = 3993\ \text{J/kg/°C}) = 0.19\text{°C}.$$

By subtraction, we can infer that, for the 700–2000 m layer, the increase in heat content was $0.73 = 10^{23}$ J. The volume of this layer is roughly $4.2 = 10^{8}$ km^3. The average temperature rise over 55 years for this layer is thus estimated to be:

$$0.73 = 10^{23}\ \text{J}\ /(4.2 = 10^{8}\ \text{km}^{3} = 1.025 = 10^{12}\ \text{kg/km}^{3} = 3993\ \text{J/kg/°C}) = 0.04\text{°C}.$$

Thus, according to Levitus *et al.*, the average rates of temperature increase per year over the 55-year span were:

0–700 m layer: 0.19°C/55 = 0.0035°C/yr
700–2000 m layer: 0.04/55 = 0.00073°C/yr.

On average, the heat input to the oceans per unit area can be estimated as follows.

Over the 55-year period, the heat content of the ocean down to depth 2,000 m increased by $2.4 = 10^{23}$ J. The volume of ocean below 2,000 m is 52%. The average temperature rise over the 55-year span below 2,000 m is not known but was probably around 0.01°C or less. Adopting as a guess the 0.01°C figure, the total heat gain of the entire ocean over the 55-year period is estimated to be $3.0 = 10^{23}$ J.

Since the surface area of the oceans is $3.61 = 10^{8}$ km^2, the total heat gain per unit area of ocean was:

$$3.0 = 10^{23}\ \text{J}/3.61 = 10^{8}\ \text{km}^{2} = 8.3 = 10^{14}\ \text{J/km}^{2}.$$

Over the 55-year period, there were:

$$55 = 3.2 = 10^{7} = 1.76 = 10^{9}\ \text{seconds}.$$

Therefore, on average, the rate of heat gain by the oceans per unit area of ocean was

$$8.3 = 10^{14}/1.76 = 10^{9} = 4.7 = 10^{5}\ \text{W/km}^{2} = 0.47\ \text{W/m}^{2}.$$

It is noteworthy that Hamon *et al.* (2012) made corrections to the world ocean database and concluded that the results "show a fairly prominent trend in 0–700 m ocean heat content of 0.39×10^{22} J/yr between 1970 and 2008". Levitus *et al.* found a trend of $1.67 = 10^{23}$ J/55 years $= 0.30 \times 10^{22}$ J/yr between 1955 and 2010 for the 0–700 m layer.

Loeb *et al.* (2012) reviewed more recent data on ocean warming. Their results are shown in Table 6.10.

The error bars for the 2004–2008 data are excessively large.

Church *et al.* (2011) estimated the heat gain by the oceans for the periods 1972–2008 and 1993–2008. By subtraction, we can also estimate the heat gain for 1972–1992. Their reported heat gains are shown in Table 6.11. From these data and the volumes of the layers, we can (as before) calculate the estimated temperature rise for each layer for each time period. This is given in Table 6.12. The rate of heat gain by the various layers for various time periods per unit area of ocean is given in Table 6.13.

Table 6.10. Rate of heat absorption for the 0–700 m ocean layer (W/m^2) per unit area of the Earth for two time periods. Note that the area used to calculate the heat flux is the area of the Earth, rather than the area of the oceans. The area of the Earth is 1.41 times the area of the oceans. Therefore, to compare these data with the previous given calculations, one should multiply the figures in the table by 1.41. (Loeb *et al.*, 2012).

	Measured W/m^2 in top 700 m of ocean		
Time period	*PMEL/JPL/JIMAR*	*NODC*	*Hadley*
1993–2003	0.66 ± 0.17	0.48 ± 0.23	0.40 ± 0.21
2004–2008	0.18 ± 0.60	0.10 ± 0.60	0.31 ± 0.57

Table 6.11. Heat gains in units of 10^{21} J by time period and ocean layer according to Church *et al.* (2011).

	1972–2008	*1993–2008*	*1972–1992*
0–700 m	112.6	45.9	66.7
700–3,000 m	49.7	20.7	29.0
3,000 m to bottom	30.7	12.8	17.9
Total	193.0	79.4	113.6
Years	27	18	11
Seconds	8.6×10^8	5.8×10^8	3.5×10^8

Table 6.12. Total temperature rise (°C) by time period and ocean layer based on data of Church *et al.* (2011).

	Volume (km^3)	*1972–2008*	*1993–2008*	*1972–1992*
0–700 m	2.1×10^8	0.131	0.053	0.078
700–3,000 m	7.0×10^8	0.017	0.007	0.010
3,000 m to bottom	3.9×10^8	0.019	0.008	0.011
Total	1.3×10^9	0.036	0.015	0.021

Table 6.13. Rate of heat absorption by time period and layer (W/m^2) per unit area of ocean based on data of Church *et al.* (2011).

	1972–2008	*1993–2008*	*1972–1992*
0–700 m	0.36	0.22	0.53
700–3,000 m	0.16	0.10	0.23
3,000 m to bottom	0.10	0.06	0.14
Total	0.62	0.38	0.90

Trenberth (2010) described ocean heat measurements prior to buildup of the Argo profiling float system in 2003–2005:

> "Before then, the bulk of the observations of the ocean were from expendable bathythermographs (XBTs) dropped from ships along their tracks as opportunities arose. As a result, coverage was spotty and irregular, and missing over many regions such as the Southern Ocean. The XBTs recorded temperatures, but the exact depth they were at was an estimate based on an assigned drop rate, which turned out to be sensitive to the exact design and character of the XBT probe. Recent careful comparisons with calibrated probes deployed from research vessels have shown the need for corrections. The severe under-sampling of the ocean until about five years ago, along with the variety of methods used to correct for problems and biases, has led to many estimates of how the temperatures in the ocean have changed over time. Of particular interest for climate is the vertically integrated ocean heat content. The reprocessing of XBT and Argo observations has resolved some issues ... but there remains a surprisingly large spread among different estimates of ocean heat content as discussed in the paper by Lyman *et al.* (2010). They delved into the origins of these differences, and compiled and reprocessed a common data set for the upper 700 m of the ocean."

Trenberth (2010) concluded that, for the 0–700 m ocean layer, the measured heat gain was 0.90 W/m^2 over the time interval 1994 to 2008 and 0.77 W/m^2 over the time interval from 2004 to 2008. His conclusions suggest a slowdown after 2003 when the data network was better. These results are per unit area of the oceans and are considerably higher than those obtained by Levitus *et al.* and Church *et al.* It is not clear why his values are so much higher than those who analyzed the data. Furthermore, according to a graph at *http://oceans.pmel.noaa.gov/*, the heat content of the 0–700 m layer of ocean was essentially constant from 2003 through 2012. Their graph shows the 0–700 m heat content was flat from 1993 through 1999, rose from 2000 to 2003, and then flattened out again after that.

There seems to be considerable uncertainty in the measured rates of heat gain by the oceans over the past few decades. Overall, the data appear to suggest a heat gain of about 0.5 W/m^2 (area of ocean).

Loeb *et al.* (2012) estimated the imbalance to be 0.50 ± 0.43 W/m^2, which is more in line with the findings in this book than previous estimates that were roughly double this amoung.

The Levitus *et al.* paper presents integrated heat content over the past 55 years, although the monitoring system for the oceans to depth 700 m was not fully in place until the mid-1990s. If one differentiates the integral curve of Levitus *et al.*, an approximate curve for the annual increase in heat content is obtained (units of 10^{22} J/yr) as shown in Figure 6.24.

There is no evidence of acceleration or a "hockey stick" in these data. As Eschenbach pointed out, the temperature changes corresponding to these changes in heat content amount to roughly 0.0012°C/yr. He questioned whether such measurements can be made accurately. Since the integral curve used five-year

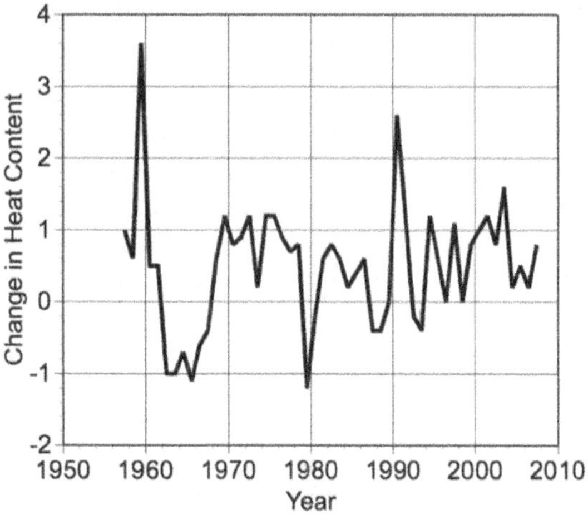

Figure 6.24. Yearly changes in heat content of the oceans 10^{22} J/yr (obtained by differentiating integral curve of Levitus *et al.*, 2012).

smoothing, these annual changes are not actually year-to-year changes in the change in heat content, but rather year-to-year changes in the five-year average. Therefore, one presumes that the annual changes in heat content were considerably greater than that shown in the figure. This would make sense considering how small the temperature changes actually were. As Eschenbach said: "Why not show the actual annual data? What are the averages hiding?"

6.2.10 Influence of human activities

6.2.10.1 Differences between surface temperatures and tropospheric temperatures

It is difficult to separate the anthropogenic influences on climate due to forcing by greenhouse gases vs. changes in land use, such as urbanization, industrialization, and agriculture. Surface temperature measurements have been modified to correct for urbanization by comparing temperature measurements in cities with those in surrounding rural areas but the results differ significantly depending on whether population data or satellite measurements of night light are used to classify urban and rural areas (Kalnay and Cai, 2003).

Kalnay and Cai (2003) compared the trends in observed surface temperatures at 1,982 sites in the continental U.S. over 50 years with the corresponding trends in satellite observations of tropospheric temperatures over a 2.5° square grid, to estimate the impact of land-use changes on surface warming. First, they carried out an in-depth study of Baltimore, MD. The two temperatures (surface and troposphere) track one another quite closely in shape over the 50-year period, with almost all of the "ups and downs" in both data sets showing nearly the same time variance.

However, it was found that, whereas the two sets of temperatures were nearly identical in the 1950s, with the passage of time, the surface temperatures increasingly diverged above the tropospheric temperatures. By the late 1980s and into the 1990s, the surface temperatures exceeded the tropospheric temperatures by a significant amount. They then extended this analysis to the continental U.S. Here, the data are much noisier, but, nevertheless, the surface temperature trend again exceeded the tropospheric temperature trend at the rate of 0.03°C per decade, or 0.3°C per century. Kalnay and Cai (2003) attributed this difference to land-use changes, mainly industrialization:

> "Although it is not possible definitively to attribute the differences between the observation and the [tropospheric] temperature trends solely to land use, including urbanization, agriculture and irrigation, our results are compatible with such an interpretation. The well-known 'urban heat island' effect actually takes place at night, when buildings and streets release the solar heating absorbed during the day. At the time of the maximum temperature the urban effect is one of slight cooling, owing to shading, aerosols, and to thermal inertia differences between city and country that are not currently well understood. The effect of agricultural development, increasing evaporation during the day, would also tend to decrease the maximum temperature: irrigation would increase the heat capacity of the soil, thus increasing the minimum temperature. Therefore, both urbanization and agriculture effects could be consistent with the general increase in the minimum temperature and slight decrease in the maximum temperature, and contribute to the reduction in the diurnal temperature range shown in our estimates east of the Rockies. This implies that the comparison of urban and rural stations without including agricultural effects would underestimate the total impact of land-use changes."

Trenberth (2004) raised several objections to the procedure used by Kalnay and Cai (2003), arguing that changes in cloudiness and surface moisture are likely the main sources of the discrepancies along with deficiencies in the tropospheric measurements. Vose *et al.* (2004) raised the objection that systematic effect due to changes in observing time at many sites, changes in instrumentation at stations, and station moves or relocations, could "overwhelm" the bias caused by station urbanization. In addition, Vose *et al.* (2004) claimed that the trend difference between tropospheric and surface temperatures has been decreasing (rather than increasing) with the passage of time. They believed that the tropospheric measurements "are not accurate". Kalnay and Cai (2003) responded to the two criticisms in the same issue of *Nature*, saying (among other things) that they deliberately used raw (unadjusted) surface data because the multiple non-climatic adjustments are uniformly positive, so their estimate of the difference should be considered to be a lower limit.

It is not clear how recent updates of tropospheric temperature data (*www.drroyspencer.com*) would affect the results of Kalnay and Cai (2003).

6.2.10.2 Correlation of surface temperatures with CO_2 sources

There appear to be significant differences between temperature measurements near the surface and in the troposphere. de Laat and Maurellis (2004) analyzed temperature measurements at the surface and in the troposphere, and correlated them with surface CO_2 emissions. Tropospheric temperatures have not displayed temperature increases as high as those indicated by surface sources. Therefore, de Laat and Maurellis (2004) raised the question of whether the higher surface heating could be, at least partially, due to surface heat production, rather than global greenhouse gas forcing:

> "The differences between troposphere and surface temperatures could arise if the heating were to occur only in the atmospheric boundary layer (ABL), i.e., the observed surface temperature changes were to arise primarily from surface- or ABL-related processes. In addition, if industrial processes were to drive surface temperature changes, then the most important expected observable should be that these changes occur primarily in regions with the largest industrial surface CO_2 emissions."

Thus, de Laat and Maurellis (2004) sought a connection between enhanced surface temperatures and CO_2 production—not from the greenhouse perspective—but rather using CO_2 production as an indicator of localized or regional industrial activity. Global temperature data (actually temperature anomalies from the mean) at various sites were divided into two groups according to whether the CO_2 emissions were higher or lower than a selected threshold value at each site. The average temperature anomaly in each group was then plotted as a function of the CO_2 threshold value. In each case (at the surface or in the troposphere), the temperature anomalies were significantly greater (more positive) for the group of sites where surface CO_2 emissions exceeded the CO_2 threshold. In addition, the temperature anomalies near the surface were much greater than those in the troposphere (de Laat and Maurellis, 2004).

The majority of the industrial CO_2 emissions occur at northern mid-latitudes. Therefore, if these industries are responsible for regional heating, the effect should be greater in the NH than the Southern Hemisphere (SH). So the hemispheres should show different temperature trends. In fact, the rate of surface temperature rise from 1976 to 2000 was about 2.4 times greater in the NH than the SH. The differences over the sea were smaller. This concurs with the rapid increase in industrialization that has occurred in the NH during the latter part of the 20th century. Regardless of the exact warming mechanism, one does not necessarily expect that the same degree of warming will occur simultaneously in the troposphere and at the surface, since the ocean's heat capacity will introduce significant delay in the SST response. However, the higher temperature trends at the surface than in the troposphere (as well as a lack of agreement with climate models) suggest a hitherto-overlooked driver of regional surface temperature increases, linked to the degree of industrialization. This lends strong support to other indications that surface processes (possibly changes in land use or the urban heat effect) are crucial players in observed surface temperature changes. Although the exact mechanisms have yet to be determined, the findings of de Laat and Maurellis (2004) suggest that a significant part of the observed surface warming may be related to processes other than enhanced greenhouse warming (Kalnay and Cai, 2003).

de Laat and Maurellis (2006) was a follow-on to de Laat and Maurellis (2004) in which additional analysis was used to confirm the previously discovered correlation between temperature change and industrialization. Near-surface temperature trends are found to be higher on average for regions with higher industrial CO_2 emissions. But GCMs show no such increase in trends for regions with higher emissions. It was not claimed that a physical causality exists between the CO_2 emissions themselves and the temperature trends, but rather that some underlying process (urban heating?), for which CO_2 emissions functioned as a proxy, was at work.

However, IPCC (2007) disputed these findings:

> "The locations of greatest socioeconomic development are also those that have been most warmed by atmospheric circulation changes, which exhibit large-scale coherence. Hence, the correlation of warming with industrial and socioeconomic development ceases to be statistically significant."

Thus, IPCC (2007) dismissed the notion that warming trends over land are strongly correlated with geographical patterns of industrial and socioeconomic development. However, McKitrick (2008) pointed out that IPCC (2007) "cited no supporting evidence for this assertion". He performed a detailed statistical analysis of gridded temperature trends to seek correlation with industrialization. However, the details are difficult to follow. McKitrick (2008) refuted the IPCC's assertion, claiming that the correlations with industrial activity are "quite robust to the inclusion of atmospheric circulation". He concluded that:

> "A substantial fraction of the post-1980 trends in gridded climate data over land are likely not 'climatically real' but result from data quality problems and local environmental modifications."

6.2.10.3 Urban heat islands

Section 3.1.3 provides a review of studies of urban heat islands where it is shown that the urban heat island (UHI) effect can be as large as 5°C for a city of 10 km^2, and 10°C for a city of 300 km^2.

6.2.10.4 Heat generation by urbanization

The heat generated by human activities has traditionally been small compared with the massive heat flows involved in the Earth's climate. However, with the expansion of world population and higher energy usage in the 20th century, human activity is no longer negligible. Hoyt (2006) indicated that the total primary energy consumption in the U.S. in 2002 was 97.7 quadrillion Btu. This is equivalent to:

$$97.7 \times 10^{15}\ \text{Btu} = 1.03 \times 10^{20}\ \text{joules.}$$

On a yearly average, this amounts to a continuous power generation of 3.27×10^{12} watts. The area of the U.S. is 9,640,000 km^2. If we divide the total power by the total area, we obtain about 0.34 W/m^2. However, most of the power generated is localized into urban areas. The power density is probably higher than 1 W/m^2 in urban regions, and much higher than that in cities. These power densities are approaching those indicated by GCMs for greenhouse gas forcing in the 20th century.

Hoyt (2006), quoting a 1973 paper by Oke, suggested that the temperature increase of an urban center compared with its surroundings is approximated as 0.73 log (population):

> "In 1900, the world population was 1 billion and in 2000, it was 6 billion for an increase of a factor of six. If the surface measuring stations are randomly distributed and respond to this population increase, it would equal 2.2 log (6) or 1.7°C, a number already greater than the observed warming of 0.6°C. However urban heat islands occur only on land or 29% of the Earth's surface, so that the net global warming would be $0.29 \times 1.7 = 0.5$°C which is close to the observed warming. It is not out of the realm of possibility that most of the twentieth century warming was [due to] urban heat islands.
>
> "Since satellite measurements began in 1979, the world's population has approximately doubled leading to an UHI signal of 0.67°C over land and 0.19°C globally. The observed surface warming is about 0.36°C over the same time period, so a substantial portion may be just uncorrected UHI effects. Other effects include land-use changes, increased darkness of vegetation, direct heat from fossil fuel burning, a brighter sun, changes in cosmic ray intensity, soot on snow, more soot in the atmosphere, and greenhouse gases (and this list is not exhaustive). There are many competing theories for the recent warming and some of them do a better job at explaining the observations than do greenhouse gases.
>
> "The land surface stations were designed to provide local climatology. They were not designed to detect climate change. Quality control of the surface network is inadequate." (Hoyt, 2006)

6.2.10.5 Effects of land-use/land-clearing changes

Annual world land clearing amounts to over 100,000 km^2. A comparison of global maps showing land use (crop land, grazing land, evergreen forest, savannah, grassland, steppe, open shrub land, deciduous forest, and hot desert) reveals major changes worldwide from 1700 to 1900, and further changes from 1900 to 1990 (Pielke *et al.*, 2007b). Global cropland area increased from 2.6 million km^2 in 1700 to 15 million km^2 in 1980. Over the same period, the global extent of forest and woodland decreased by 17%, from 61 million km^2 to 51 million km^2 (see Section 2.1.2.1). A significant portion of the natural vegetation of the world has been cleared to grow crops. In much of India, eastern China, the forests and shrub lands of Europe, the steppes of Asia, and the Great Plains of North America, over 75% of the land is now cropland. Bonan (2002) wrote:

> "Changes in land cover alter net radiation at the surface, the partitioning of this available energy between sensible and latent heat, and the partitioning of precipitation into runoff and evapo-transpiration. This arises from differences among vegetation in albedo, roughness, leaf area index, root distribution, and stomatal conductance and changes in soil texture. For example, vegetation generally has a lower albedo than bare soil, and forests have a lower albedo than pastures or croplands. As a result, overgrazing grasslands and clearing forests for croplands increases surface albedo. ... Clearing forests for cropland or pastureland also reduces surface roughness. This reduces mechanical turbulence and sensible and latent heat fluxes. The partitioning of net radiation into sensible and latent heat is influenced by leaf area, stomatal conductance, root distribution, and soil texture. It is also affected by the overall hydrologic cycle. Evapo-transpiration is an important regulator of global climate.
>
> "The impact of historical land cover changes on climate is noticeable on a global scale, is comparable in magnitude to other climate forcings, and is important to reconstructing historical climate change. In particular, land cover changes in Europe and North America have likely cooled the climate of the Northern Hemisphere. If so, this land use change would have dampened the warming from increasing CO_2 and other greenhouse gases. Cooling from deforestation of Europe and North America may have contributed to the anomalously cold temperatures during the *Little Ice Age* from 1550 to 1850." (Bonan, 2002)

A number of studies have been conducted in which a land surface model is used to define the properties of the surface for various forms and distributions of vegetation on the surface. The properties relevant to climate modeling include albedo, solar absorptance, roughness, sensible heat/latent heat ratio, and heat capacity of the soil. Typically, these properties are derived from sophisticated vegetation models. The results of these surface models can be used as boundary conditions for GCMs, and the temperature at the surface can thereby be estimated for varying types of vegetation. By comparing the estimated temperature for the period prior to human intervention (land clearing, deforestation, establishment of croplands, etc.) with that after such activities, one can estimate the change in climate for any region that was induced by such human intervention.

Bounoua *et al.* (2000) used a coupled biosphere–GCM model to examine the sensitivity of global and regional climates to variations in the "normalized difference vegetation index" (NDVI) over a nine-year period. Their results indicated a significant net cooling effect from increased vegetation.

Bonan (1997) combined a land model with a GCM to examine the effects of land use on the climate of the U.S. through three five-year model integrations using natural vegetation, modern vegetation, and maximum agricultural (i.e., maximum deforestation) scenarios. Land-use practices have replaced much of the natural needle-leaf evergreen, broadleaf deciduous, and mixed forests of the eastern U.S. with crops. The climate of the U.S. with modern vegetation is different from that with natural vegetation. The results of Bonan (1997) provide three important climate signals caused by modern vegetation:

(1) a 1°C cooling over the eastern U.S. and a 1°C warming over the western U.S. in spring;
(2) summer cooling of up to 2°C over a wide region of the central U.S.; and
(3) moistening of the near-surface atmosphere by 0.5 g/kg to 1.5 g/kg over much of the U.S. in spring and summer.

These changes in surface temperature and low-level moisture extend well into the atmosphere, and affect other aspects of the simulation. There are large changes in zonal wind. Regions with higher atmospheric moisture in spring and summer also have increased cloudiness, decreased boundary layer height, decreased net radiation, decreased sensible heat flux, increased latent heat flux, increased rates of infiltration, and increased soil water. Maximum agriculture reinforces these signals. Other studies have also found that the climate of the U.S. is sensitive to the specification of vegetation types.

Hale *et al.* (2006) used several GCMs in conjunction with land models over the period 1983–1992. They ran the models for the NH and the whole globe, considering:

(1) land effects only;
(2) CO_2 greenhouse effects only; and
(3) combined land effects and CO_2 greenhouse; their results are summarized in Table 6.14.

Table 6.14. Calculated temperature change (°C) from 1983 to 1992 due to deforestation and CO_2 greenhouse effect according to various GCMs (from Hale *et al.*, 2006).

Model	*Deforestation only*		*CO_2 only*		*Deforestation and CO_2*	
	Global	*NH*	*Global*	*NH*	*Global*	*NH*
1	–0.24	–0.31	0.52	0.58	0.29	0.25
2	–0.13	–0.23	0.27	0.36	0.17	0.14
3	–0.14	–0.19	0.42	0.51	0.37	0.35
4	–0.25	–0.36	0.34	0.40	0.10	0.05
5	–0.17	–0.23	0.62	0.68	0.46	0.47

According to these models, deforestation offset a significant fraction of putative CO_2-induced greenhouse warming, although there is some diversity in the results. As always, the effect of clouds remains an issue.

Hale *et al.* (2006) studied temperature trends over the past 30 years at 183 sites in the U.S. prior to, during, and after significant periods of LULC activity in these localities.

Station minimum temperatures: Prior to LULC activity, 22 stations showed a significant warming trend and 27 showed a significant cooling trend. After LULC, 78 had significant warming trends and only 2 had significant cooling trends. The most common type of dominant LULC conversion was from forest to urban. The post-change trend was 1.4°C/decade.

Station maximum temperatures: Prior to LULC activity, only seven stations showed a significant warming trend and five stations showed a significant cooling trend. After LULC, 76 had significant warming trends and only 4 had significant cooling trends. The post-change trend was 2.1°C/decade.

Although these observations do not necessarily establish a cause–effect relationship between LULC and temperature, they are suggestive.

Lim *et al.* (2005) investigated the sensitivity of surface climate change to land types in the NH by subtracting tropospheric temperatures from the observed surface temperatures under the belief that tropospheric temperatures represent large-scale climate changes due to greenhouse gases and atmospheric circulation, and are less sensitive to regional surface processes associated with land types. These temperature differences then provide suggestive evidence of local or regional surface warming from non-greenhouse processes. The results suggest that urban heating contributes to observed warming trends.

6.3 CLIMATE VARIABILITY

6.3.1 El Niños and climate

Sections 3.4.3.3 and 3.4.4 provide an extensive discussion of the relationship between El Niños and climate.

6.3.2 Interruption of warming

The data in Figure 3.28 indicate that global temperatures have meandered about no change over the past 15 years. Obviously, these temperature measurements are not supportive of climate models and also are contrary to the alarmist viewpoint that decrees that temperatures should rise continuously as CO_2 increases. Hence, in keeping with recent practice over the past few years, whereby alarmists promptly publish rebuttals to any papers that slip through their control of which manuscripts get accepted by climate journals, it was necessary for the alarmists to publish such a rebuttal. Ben Santer took on this responsibility and the result was Santer *et al.* (2011). It is interesting, perhaps, that Santer included 16 co-authors in addition to himself; yet the nature of the work is such that it is difficult to imagine how 16 individuals could each contribute significant portions to this work, which, on the face of it, does not appear to require more than a few man-days of effort. Pielke Sr. commented: "This is an unusual number of co-authors for a technical paper, but I assume Ben Santer wants to show a broad agreement with his findings."[20] In other words, many names were added to give the paper political endorsement?

[20] *http://pielkeclimatesci.wordpress.com/category/climate-models/.*

Santer *et al.* (2011) were concerned with a very basic problem in climatology: how to distinguish between long-term climate change and short-term variable weather in regard to temperature measurements. They treated the problem in terms of signal and noise: the signal is assumed to be a long-term linear trend of rising temperatures due increasing greenhouse gas concentrations, that is obfuscated by short-term noise. However, the climate–weather problem is innately different from a classical signal/noise problem such as a radio signal affected by atmospheric activity. In that case, if the radio signal has a sufficiently narrow frequency band, and the noise has a wider frequency spectrum, the signal-to-noise ratio (S/N) can be improved with a narrow-band receiver tuned to the frequency of the radio signal. The radio signal and the noise are separate and distinct. By contrast, in the climate–weather problem, the instantaneous weather is the noise, and the signal is the long-term trend of the noise. The noise and signal are coupled in a unique way. Furthermore, there is no evidence that it is even meaningful to talk about a "trend", since there is no evidence that the variation of temperature with time is linear. Remarkably, Santer *et al.* never referred to Christy *et al.* (2010) but based their analysis on older papers (e.g. Christy *et al.*, 2007).

Santer *et al.* (2011) were primarily concerned with estimating how many years of data are necessary to provide a good estimate of the putative underlying linear trend. They were also intent on showing that short periods with no apparent trend do not violate the possibility that, over a longer term, the trend is more discernable. They derived signal-to-noise (S/N) ratios for both the temperature data and the model average by means that are not clear to this writer.

As Santer *et al.* (2011) showed, one can pick any starting date and any duration length and fit a straight line to that portion of the curve of temperature vs. time. They did this for various 10-year and 20-year durations. In each case, depending on the start date, they derived a best straight-line fit to the temperature data for that time period. They found that the range of trends for 10-year periods was greater (–0.05 to +0.44°C/decade) than the range for 20-year periods (+0.15 to +0.25°C/decade). The trends for various start dates for 10-year trends are shown in Figure 6.25. Clearly, the trend line was steepest for a start date around 1988 (ending in the giant El Niño year of 1998). Prior to 1988 and after 1998, the trends were minimal.

Santer *et al.* described use of longer durations as "noise reduction", which it is in a sense, provided that one assumes the overall signal is linear in time. It still was problematic that the trend was nil after 1998, which they rationalized by saying:

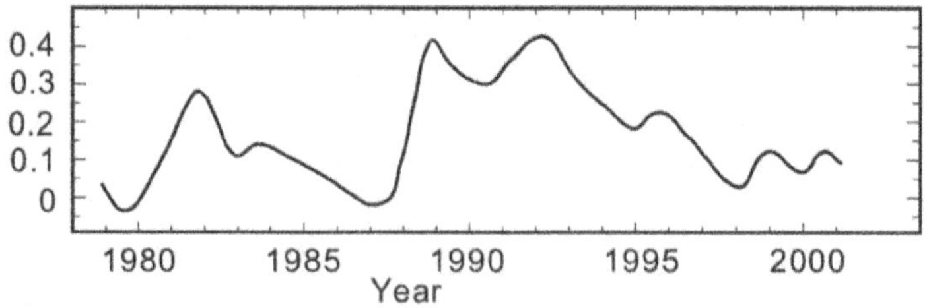

Figure 6.25. Trend (°C/decade) of TT vs. start year for 10-year durations. (Santer *et al.*, 2011).

"The relatively small values of overlapping 10-year temperature trends during the period 1998 to 2010 are partly due to the fact that this period is bracketed (by chance) by a large El Niño (warm) event in 1997/98, and by several smaller La Niña (cool) events at the end of the ... record."

However, as Pielke pointed out, the period after 1998 at that time (it is now 16 years) was 13 years, not 10, and, furthermore, the period after 1998 had roughly equal periods of El Niño and La Niña and was not dominated by La Niñas as Santer *et al.* claimed. What Santer *et al.* implied was that an unusual conflux of a large El Niño early on and multiple La Niñas later on caused the trend to minimize for that unique period as a statistical quirk.

In simplistic terms, the so-called signal-to-noise ratio can be estimated as follows. For either 10-year or 20-year durations, the signal was the mean trend derived by a straight-line fit to the temperature data over that duration. The noise was the range of trends for different starting dates. For 10-year durations, the trend was 0.19 ± 0.25°C/decade. For 20-year durations, the trend was 0.20 ± 0.05°C/decade. The signal in each case is taken as the mean trend. The distribution of trends within these ranges was similar to a normal distribution. Thus we can roughly estimate the noise as ~ 0.7 times the full width of the range. Hence, the S/N ratio for 10-year durations can be crudely estimated to be S/N $\sim 0.19/(0.7 \times 0.5) = 0.5$ and for 20-year durations is S/N $\sim 0.2/(0.7 \times 0.1) = 2.9$. Santer *et al.* obtained S/N = 1 for 10-year durations and S/N = 2.9 for 20-year durations. If it can be assumed that the signal varies linearly with time, one can then estimate what level of precision for the estimated trend can be obtained for any chosen duration. Santer *et al.* obviously believe that the signal is linear with time for all time. By some logic that escapes me, Santer *et al.* concluded that:

"Our results show that temperature records of at least 17 years in length are required for identifying human effects on global-mean tropospheric temperature."

This conclusion seems to be grossly exaggerated. A more proper statement might be as follows:

Assuming that the variability of TT is characterized by a long-term upward linear trend caused by human impact on the climate, and that variability about this trend is due to yearly variability of weather, El Niños and La Niñas, and other climatological fluctuations, the recent data suggest that the trend can be estimated for any 17-year period with a S/N ratio of roughly 2.5.

Finally, we get to the nub of the paper by Santer *et al.* that asserted: "Claims that minimal warming over a single decade undermine findings of a slowly-evolving externally-forced warming signal are simply incorrect." Here is where Santer *et al.* attempted to dispel the notion that minimal warming for a period contradicts the belief that underneath it all, the long-term signal continues to rise at a constant rate. Pielke Sr. argued that this was an overstatement and he concluded:

"If one accepts this statement by Santer *et al.* as correct, than what should have been written is that the observed lack of warming over a 10-year time period is still too short to definitely conclude that the models are failing to skillfully predict this aspect of the climate system." (*http://pielkeclimatesci.-wordpress.com/category/climate-models/*).

However, I would go further than Pielke Sr. First of all, the period of minimal temperature rise is now approaching 16 years. Second, there is no cliff at 17 years whereby trends derived from shorter periods are statistically invalid and trends derived from longer periods are valid. According to Santer *et al.*, a trend derived from a 15-year period is associated with a S/N ~2.0 which, though not ideal, is good enough to cast some doubt on the validity of models.

If one carries out a least squares comparison of linear and step-functions to the temperature, the best fits are:

Linear: temperature = −0.2 + 0.00156 (Y—1979) [RMS error = 0.026]
Step: temperature = −0.14 (Y < 1998); temperature = 0.18 (Y = 1998) [RMS error = 0.023].

Hence, the step function fits the data slightly better than the linear function. A function based on a Niño index would undoubtedly fit much better than either of these fits.

In October 2012, the Climate Research Unit (CRU) of the University of East Anglia, one of the important organizations that monitor Earth temperatures from networks, reported on an update of their estimates of global average temperatures (including land and sea). Their results showed that, during the period 1997 through 2012, the average temperature meandered up and down, but on average was unchanged over this period of 16 years. Skeptics pounced on this announcement with glee. Presumably, Ben Santer, with 20 or 30 alarmists as co-authors, will soon write a paper explaining this to be a temporary fluctuation.

The continued almost religious belief by alarmists that the temperature always rises linearly and continuously is evidently debatable. If the alarmists would only reduce their hyperbole and argue that rising greenhouse gas concentrations produce a warming force that is one of several factors controlling the Earth's climate, and there are periods during which the other factors overwhelm the greenhouse forces, perhaps we would have a rational description. Instead, the alarmists continue to find linear trends over various time periods, in some cases when they are not there.

6.3.3 North Atlantic climate variability and ocean oscillations

Marshall *et al.* (2001) wrote an extensive review of variability in Atlantic Ocean circulation, and the effect of these on the North Atlantic regional climate. The subject is lengthy and complex, and only the briefest of reports is given here. It was concluded that three major interconnected phenomena contribute to climate change in the North Atlantic region:

(1) Tropical Atlantic Variability (TAV): a fluctuation of tropical Atlantic SST and trade winds;

(2) North Atlantic Oscillation (NAO): a fluctuation in sea-level pressure difference between the Icelandic Low and the Azores High;
(3) Atlantic Meridional Overturning Circulation (MOC): fluctuations in the Atlantic's thermohaline circulation that may possibly play a role in abrupt climate change.

The NAO is the periodic change in the difference in atmospheric pressure between a low-pressure center around Iceland and a high-pressure center around the Azores Islands, a positive NAO (greater pressure difference) resulting in a stronger jet stream. During the winter, the jet stream "blocks" Arctic air masses from entering the lower latitudes. Winters in the U.S. are much harsher when this "blocking" mechanism is weaker (i.e., the NAO is negative).

Marshall *et al.* (2001) concluded:

> "The NAO exerts a dominant influence on the winter time temperatures of the Northern Hemisphere. ... Surface air temperature and sea surface temperatures in wide regions across the North Atlantic basin, in eastern North America, the Arctic, Eurasia and the Mediterranean, are significantly correlated with NAO variability."

They also concluded that changes in circulation patterns of the North Atlantic are directly tied to changes in regional wintertime precipitation leading to increased precipitation in some regions, and drought in others. The NAO is also linked to changes in "storm tracks over the Northern Hemisphere from North America to Eurasia and the Mediterranean".

The higher heating rate drives up tropical atmospheric pressures compared with high-latitude atmospheric pressures. This pressure difference has a significant impact on climate. The difference in surface pressure between the Azores High and the Iceland Low is used as a quantitative indicator of this differential for the North Atlantic. The magnitude of this difference—called the North Atlantic Oscillation (NAO)—controls strength and direction of westerly winds and storm tracks across the North Atlantic. However, unlike the El Niño phenomenon in the Pacific Ocean, the NAO is largely an atmospheric phenomenon. The NAO index is observed to vary in an oscillatory way with the passage of years. It is primarily measured for the December–March period when it is a maximum. Annual average measurements of the NAO are typically plotted as deviations from the long-term average (see Figure 6.26).

Prior to about 1980, the NAO index did not display any consistent trends. However, after 1980, it has been mainly and consistently positive. According to Hurrell and Dickson (2004):

> "The NAO exerts a dominant influence on wintertime temperatures (and related changes in rainfall and storminess) across much of the NH. Surface air temperature and sea-surface temperature (SST) across wide regions of the North Atlantic Ocean, North America, the Arctic, Eurasia, and the Mediterranean are significantly correlated with NAO variability. When the NAO index is positive, enhanced westerly flow across the North Atlantic during winter moves relatively

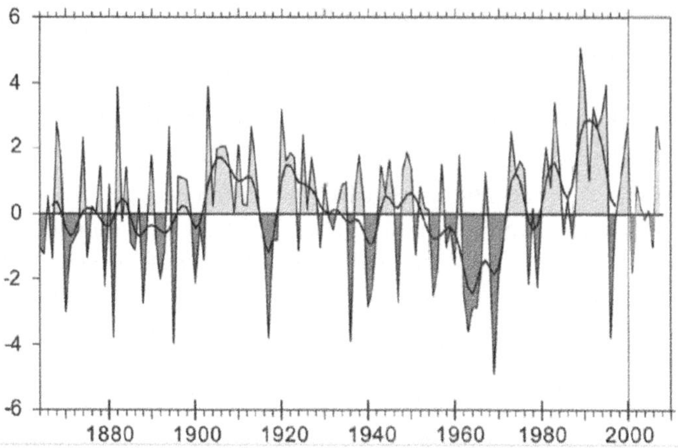

Figure 6.26. Winter (December–March) index of the NAO based on the difference of normalized sea level pressure between Portugal and Iceland from 1864 through 2000 (Hurrell and Dickson, 2004). Data subsequent to 2000 were added based on *www.cgd.ucar.edu/cas/jhurrell/indices.html.*

warm (and moist) maritime air over much of Europe and far downstream across Asia, while stronger northerlies over Greenland and northeastern Canada carry cold air southward and decrease land temperatures and SST over the northwest Atlantic. Temperature variations over North Africa and the Middle East (cooling), as well as North America (warming), associated with the stronger clockwise flow around the subtropical Atlantic high-pressure center are also notable. The pattern of temperature change associated with the NAO is important. Because the heat storage capacity of the ocean is much greater than that of land, changes in continental surface temperatures are much larger than those over the oceans, so they tend to dominate average NH (and global) temperature variability. Given the especially large and coherent NAO signal across the Eurasian continent from the Atlantic to the Pacific, it is not surprising that NAO variability explains about one-third of the NH inter-annual surface temperature variance during winter, and that *the trend towards the positive NAO phase in recent decades has contributed significantly to observed global warming.*" (Emphasis added.)

The Atlantic Multidecadal Oscillation (AMO) is a periodic shift in the temperature of the North Atlantic that affects the North American climate. Data from 1860 to the present suggest a 65-year period for the oscillation. Warm periods were 1860–1900, 1930–1965, and post 1995. Cool periods were 1900–1930 and 1965–1995. Knight *et al.* (2005) suggest that we are currently just past the peak in the current warm trend, and cooling can be expected over the next 20–30 years.

The fact that both the El Niño phenomenon in the Pacific Ocean and the NAO in the Atlantic have both been in a strong warming trend since about 1980 is very notable. Most commentators have assumed that these phenomena are by-products of global warming induced by increased greenhouse gas concentrations (e.g., Hester and Harrison, 2002; IPCC, 2007). However, Hurrell and Dickson would seem to suggest quite the opposite: observed global warming may be due to positive El Niño and NAO indices since 1980—at least to a considerable extent. Seager *et al.* (2010) showed that the climate of the NH is significantly affected by NAO and El Niño variability.

6.3.4 Volcanic eruptions

Volcanic eruptions have had a significant—but not sustained—effect on climate (Robock, 2004; Prohom *et al.*, 2003). Volcanoes can be classified as to size by various indices. In regard to climate change induced by volcanic eruptions, the most important factor is the radiative forcing that is produced by injection of sulfurous material into the stratosphere. This is not always directly related to the explosiveness of the eruption. For example, the Mt. St. Helens eruption (1980) emitted from the side of the mountain and relatively little material was injected into the stratosphere, so it had only a minor effect on global climate. The volcano explosivity index (VEI) is measured from 1 to 8, with 8 being the most powerful. Each increase of 1 unit in the index results in an increase of about a factor of 10 in magma volume emitted, as shown in Table 6.15. The largest volcanoes in the past 250 years are listed in Table 6.16.

Table 6.15. The *volcano explosivity index* (VEI) (de Silva, 2003).

VEI	*Description*	*Plume height*	*Magma volume*	*Example*
0	Non-explosive	< 100 m	1,000 m^3	Kilauea, Hawaii
1	Gentle	100–1,000 m	10,000 m^3	
2	Explosive	1–5 km	0.001 km^3	Galeras, Colombia, 1992
3	Severe	3–15 km	0.01 km^3	Ruiz, Colombia, 1985
4	Cataclysmic	10–25 km	0.1 km^3	Galunggung, 1982; Chile, 1993
5	Paroxysmal	> 25 km	1 km^3	Mt St Helens, USA, 1980
6	Colossal	> 25 km	10 km^3	Krakatau, Indonesia, 1883 Millenia Huaynaputina, Peru, 1600 Pinatubo, Philippines, 1991
7	Super-colossal	> 40 km	100 km^3	Tambora, Indonesia, 1815
8	Mega-colossal	> 40 km	1000 km^3	Toba, Indonesia, 71,000 YBP

Volcanic eruptions inject finely comminuted magma as ash and gases into the atmosphere. Ash consists of frozen magma shards and fragments and pulverized rock fragments from the conduit of the volcano, and crystals. Volcanic gases are dominated by water vapor (~80%), and CO_2 (~10%), and the rest is made up of N_2, SO_2, H_2S, CO, H_2, HCl, and HBr. Volcanic aerosols, formed by sulfur species injected into the stratosphere—and not the ash—produces most of the resultant climate effects. These stratospheric aerosols affect global radiation by absorbing and

Table 6.16. The largest volcanoes of the past 250 years (from Robock, 2000).

Volcano	*Year of eruption*	*VEI*
Grimsvotn (Lakagigar), Iceland	1783	4
Tambora, Sumbawa, Indonesia	1815	7
Cosiguina, Nicaragua	1835	5
Askja, Iceland	1875	5
Krakatau, Indonesia	1883	6
Okataina (Tarawera), North Island, New Zealand	1886	5
Santa Maria, Guatemala	1902	6
Ksudach, Kamchatka, Russia	1907	5
Novarupta (Katmai), Alaska, U.S.	1912	6
Agung, Bali, Indonesia	1963	4
Mount St. Helens, Washington, U.S.	1980	5
El Chichon, Chiapas, Mexico	1982	5
Mount Pinatubo, Luzon, Philippines	1991	6

back-scattering incoming solar radiation; the former leads to stratospheric heating, while the latter leads to surface cooling. The volcanic aerosols also catalyze a reduction of stratospheric ozone through heterogeneous chemical reactions. As ash, water vapor, SO_2, HCl, and other gases are injected into the atmosphere, the ash falls out of the atmosphere very rapidly, in a matter of minutes to weeks. Ash therefore has little global climate impact beyond local cooling and reduction of the diurnal cycle. Moreover, water vapor and soluble gaseous components like HCl condense as temperatures drop in the UT and these components often rain out before they enter the stratosphere. Thus, tropospheric particles are removed from the atmosphere in several weeks by frequent precipitation (de Silva, 2003).

Volcanic eruptions can inject into the stratosphere tens of teragrams of chemically and micro-physically active gases and solid aerosol particles. Large volcanic eruptions inject sulfur gases into the stratosphere, which convert to sulfate aerosols. These gradually diminish, dropping by a factor of about one-third each year. Large ash particles fall out much more quickly. The radiative and chemical effects of this aerosol cloud affect the climate system. By scattering some solar radiation back to space, the aerosols cool the surface but, by absorbing both solar and terrestrial radiation, the aerosol layer heats the stratosphere. Because the sulfate aerosol particles have an effective radius of about the wavelength of visible light, they interact more strongly with the short-wave incident solar radiation than the long-wave radiation emitted by the surface and atmosphere. Some incident sunlight that is back-scattered is reflected to space, cooling the planet. Some is forward-scattered, depleting the direct beam downward radiation, but increasing the downward diffuse radiation. Figure 6.27 shows the impact of two recent volcanic eruptions on direct and diffuse irradiance at a location (Mauna Loa, Hawaii) a considerable distance from the volcanoes. The major impact on solar irradiance occurs in the first year, extending somewhat through the second year. The increase in diffuse irradiance is roughly 70% of the decrease in direct irradiance, producing a

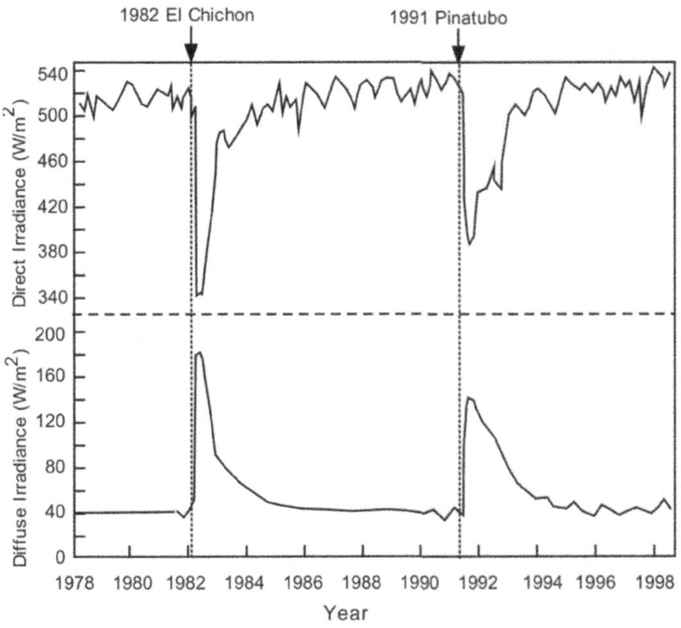

Figure 6.27. Change in direct and indirect insolation following volcanoes.

maximum deficit in total solar irradiance (TSI) of about 40 W/m^2 for both El Chichon and Pinatubo. This produces a global cooling. Although the Pinatubo eruption produced a larger stratospheric input, the center of the El Chichon cloud passed directly over Hawaii, while only the side of the Pinatubo cloud was observed at Hawaii (Robock, 2000).

Robock *et al.* (2009) revisited the question of whether the Toba eruption might have contributed to the budding Ice Age 74,000 years ago. They carried out investigations with a climate model of how the Toba eruption may have affected the Earth's climate. Robock *et al.* (2009) were not certain how much SO_2 was injected into the stratosphere so they parameterized this quantity as a multiple of the known amount emitted by the much smaller volcano, Pinatubo. Mt. Pinatubo released approximately 0.02 Gt of SO_2. Robock *et al.* (2009) considered the possibility that Toba emitted 33, 100, 300, and 900 times this amount without interactive chemistry, and 300 times with interactive chemistry (see Figure 6.28).

Using a model that did not include full interactive atmospheric chemistry, they found that a Toba-sized eruption produces a severe impact on the global climate with huge amounts of global cooling (up to $\sim 15°C$ for 300X Pinatubo) and reduced precipitation (up to 45% reduction). However, these effects only lasted for a limited time and diminshed after about a decade as shown in Figure 6.29. No matter what the amount of SO_2 was assumed, there was no evidence for Ice Age initiation. Although snow persisted for several summers in the mid-latitudes of the NH, it melted as the aerosols left the atmosphere and full insolation returned. When they

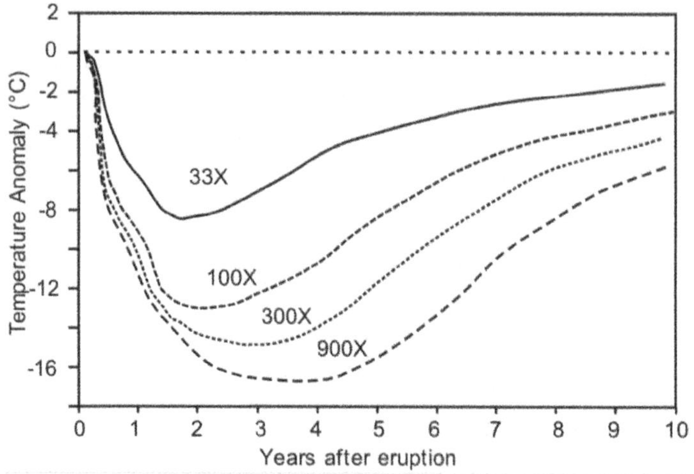

Figure 6.28. Global temperature change after Toba eruption (smoothed) for various magnitudes times Pinatubo (adapted from Robock *et al.*, 2009).

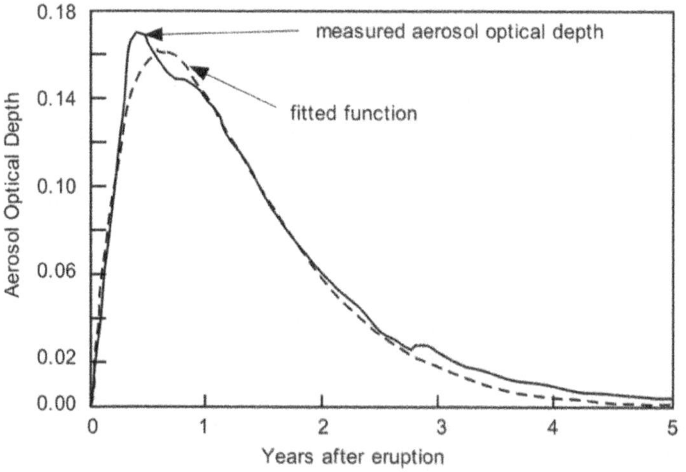

Figure 6.29. Aerosol optical depth following the 1991 Pinatubo eruption. The fitted function is AOD = 0.697 t exp (-t/0.63) where t is the time after eruption in years (adapted from Robock *et al.*, 2009).

used a model with full atmospheric chemistry for 300 Pinatubo, they found a larger ($\sim$18°C temperature anomaly) and longer-lasting response (more than 20 years), but still no evidence of Ice Age initiation. They then answered the question of whether a Toba-like eruption could produce an Ice Age today, and the answer was given as "no". Nevertheless, they pointed out that a "volcanic winter following a super-volcano eruption of the size of *Toba* today would have devastating consequences for humanity and global ecosystems".

What Robock *et al.* did not answer was the question of whether the Earth, if it was already in a state leaning toward glaciation, might have been driven toward further glaciation by an eruption of such a large volcano. As they put it:

> "Clearly, a volcanic eruption is not required to produce a glaciation, so it is obvious that if the climate system was poised to cool dramatically anyway, a slight nudge could have sped it along. With lower CO_2 concentrations, different solar activity, or even different vegetation patterns producing a different planetary albedo, the sensitivity of the climate system to massive radiative forcing might have been higher and maybe more prone to switch."

They indicated that further studies would be aimed in this direction.

Douglass and Knox (2005) (to be denoted "DK05") used the following equation to characterize Earth temperatures following a volcanic eruption:

$$\lambda\ C_p\ (d\Delta T/dt) + \Delta T = \lambda\ F(t). \tag{6.14}$$

As DK05 pointed out, ($\lambda\ C_p$) has the dimensions of time and acts as a time delay ($\tau = \lambda\ C_p$) whereby the change in temperature ΔT lags the forcing $\lambda\ F(t)$, where as usual λ is the climate sensitivity parameter and C_p is the specific heat of the ocean mixed layer. $F(t)$ is the forcing function due to the after-effects of the volcanic eruption. If $F(t)$ varies for some time but eventually steadies down to zero, $d\Delta T/dt$ will go to zero, ΔT approaches zero for large t after the eruption.

Because of the difficulties in estimating λ for steady-state climate models, it has been conjectured that the eruption of a large volcano might provide a forcing $F(t)$ that could be used to integrate Equation (6.14), and hopefully thereby derive a value for λ. In the case of events following a volcanic eruption, $F(t)$ is negative, and is primarily due to generation of aerosols and particulates in the atmosphere that reduce transmission of incident solar radiation to the Earth, as evidenced by a transient increase in the aerosol optical depth (as shown in Figure 6.29). DK05 fitted the simple function to the aerosol optical depth after the Pinatubo eruption:

$$\mathrm{AOD} = 0.697\ t\ \exp\ (-t/0.63) \tag{6.15}$$

with t in years, and used the estimate made previously by Hansen that the forcing $F(t)$ is -21 W/m^2 per unit increase in optical depth. DK05 then integrated Equation (6.14) and obtained an expression for ΔT as a function of time. This expression contains two parameters: τ and λ. They fitted this expression to observed temperatures in the lower troposphere and concluded that the best fit was $\tau \sim 5.8$ months and $\lambda \sim 0.18$ °C per W/m^2. This estimate of λ is about a factor of 4 smaller than estimates provided by climate models.

This result caused great consternation in the alarmist community because, if the estimated value of λ were correct, it might seem to imply that a doubling of CO_2 would produce a temperature change of merely $3.7 \times 0.18 = 0.67$°C, whereas, if λ is, say, ~ 0.8, then the temperature change due to doubling CO_2 would be $3.7 \times 0.8 \sim 3$°C. The prevailing view amongst climate modelers is that the temperature rise due to doubling of CO_2 is roughly 3°C. Robock (2005) wrote a criticism of DK05 in which he correctly pointed out poorly worded statements in DK05 (which were corrected in a revised version). Nevertheless, Robock's criticism merely amounts to saying that, if λ is as small as 0.18°C per W/m^2), the temperature rise due to doubling of CO_2 would be much smaller than the alarmists believe it would be. However, the value of λ appropriate for the first couple of years after a major volcanic eruption may not be typical of ordinary conditions. Wigley *et al.* (2005) also emphasized that climate models predict higher values of λ.

Spencer[21] utilized the ERBE measurements of radiative flux, which represent $[F(t) - \Delta T/\lambda]$. He then used measurements of tropospheric ΔT and the previously given estimate of $F(t)$ (from AOD measurements) to estimate $\lambda \sim 0.27$°C per W/m^2). This corresponds to a ΔT of ~ 1°C for a doubling of CO_2.

Bender, Eckman, and Rodhe (2010a) applied 10 general circulation models to the aftermath of the Mt. Pinatubo eruption. However, the details of how these models employ AOD, temperature, and ERBE data are obscure to this writer, and it is difficult to ferret out how these data were manipulated by the models. It appears that the models differ widely, with estimated values of λ ranging from 0.23°C to 0.5°C per W/m^2). We are nevertheless assured by the authors that the "equilibrium temperature increase due to a doubling of the CO_2 concentration, is between 1.7°C and 4.1°C", even though this would seem to imply higher values of λ. The main concern here seems not to model the Pinatubo eruption, but to somehow justify the alarmist position that doubling CO_2 produces large values of ΔT. Unfortunately, the paper is so obscure that I cannot assess it easily.

Van Dorland (1998) obtained a value of $\lambda = 0.3$°C per W/m^2) without water-vapor feedback, and $\lambda = 0.54$°C per W/m^2) with water-vapor feedback.

The reductions of TSI produced by volcanic eruptions lead to sharp drops in global temperature for a year or two after each eruption, as shown in Figure 6.30.

The reduction in net TSI (and surface temperature) produced by a major eruption is huge and produces a very strong reduction in solar forcing. If the reduction in TSI (and surface temperature) observed during the first year after an eruption persisted for many years, the Earth would almost certainly enter an Ice Age. Fortunately, the heat capacity of the Earth is great enough that the reduction in global temperature produced by a Pinatubo eruption (VEI = 6) is only about 0.3°C. However, regional temperature reductions can be greater. The Tambora eruption of 1815 led to the "year without a summer" in 1816 in the NH. Tambora's eruption activity peaked April 10–11, 1815:

> "*Tambora's* eruptive force and atmospheric impact exceeded anything like it on Earth in the past 10,000 years. The catastrophic eruption was so powerful that it sheared *Tambora* nearly in half, from 4,300 to 2,850 meters. *Tambora's* dust funnel pumped 200 megatons of dust, rock, and aerosols into the stratosphere. In the immediate aftermath, violent winds blew throughout the area and pumice chunks 20 centimeters long rained down on the surrounding region. Floating ash islands formed in the sea and a tsunami ravaged nearby shorelines. ... All of this destruction occurred far from European and American eyes." (Soon and Yaskell, 2007)

[21] *www.drroyspencer.com/2010/06/revisiting-the-pinatubo-eruption-as-a-test-of-climate-sensitivity/.*

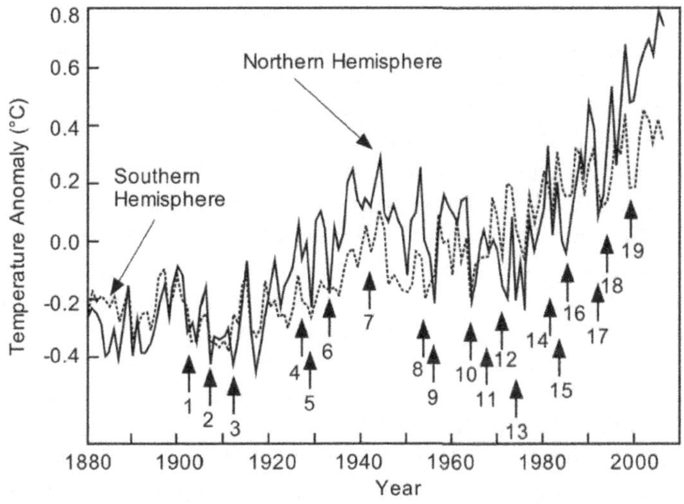

Figure 6.30. The relationship between major volcanic eruptions and hemispheric temperature variations. Volcanoes are identified by number in Table 6.17. This figure was inspired by a similar figure in Leroux (2005).

Table 6.17. List of volcanoes used in Figure 6.30.

Number	*VEI*	*Volcano(es)*
1	6	1902: Santa Maria (Guatemala), Mount Pelee (Martinique), Soufriere (Guadeloupe)
2	5	1907: Ksudach (Kamchatka)
3	6	1912: Katmai (Alaska)
4	3	1928: Paluweh (Indonesia)
5	3	1929: Reventador (Ecuador)
6	5+	1932: Cerro Azul (Argentina)
7	4	1942–1947: Hekla (Iceland)
8	4	1954: Mount Spurr (Alaska)
9	4+	1956: Bezymianny (Kamchatka)
10	4	1963: Gunung Agung (Indonesia)
11	4	1966: Awu (Indonesia)
12	4	1971: Hekla (Iceland)
13	4	1974: Fuego (Guatemala)
14	5	1980: Mount St. Helens (U.S.)
15	5	1982: El Chichon (Mexico)
16	3	1985: Nevada del Ruiz (Colombia)
17	6	1991: Pinatubo (Philippines)
18	4	1992: Mt. Spurr (Alaska)
19	3	2000: Mt. Usu

But the cold resulting from this eruption hit America with extreme effect in the next year and a half with destructive crop yields and winter weather in summer. Trigo *et al.* (2009) estimated that the summer of 1816 was 2°C to 3°C cooler than the average for the 1871–1900 reference period. However, since 1816 was deeper into the LIA, it is not clear that all of this was due to Tambora.

The greatest known eruption of the past 100,000 years was the Toba eruption of about 71,000 years ago. While the principal effects of Pinatubo lasted about two years, it is probable that the effects of Toba lasted considerably longer. Individual large eruptions certainly produce global or hemispheric cooling for a few years. As Figure 6.29 shows, the Earth typically returns to normal behavior about three years after a fairly major volcanic eruption. However, Robock (2000) raised the question of whether longer-term climate changes could be induced or enhanced by either (1) the impact of an extremely large volcano (e.g., Toba (VEI $\sim$ 8), or (2) the cumulative effect of a series of large volcanoes (VEI $\sim$5 to 6) over some time period.

As Rapp (2012) showed, 71,000 YBP to 75,000 YBP was a pivotal turning point where the climate turned from a "mild phase" to "a full glacial world". Could the eruption of Toba be implicated in this transition? It seems possible. Robock *et al.* (2009) used a climate model to show this might have been possible.

6.3.5 Cosmic rays as a source of climate change

Benestad (2005) provided an extended discussion of a theory that cosmic rays, controlled by the Sun's magnetic field, produce changes in cloud formation that affect the Earth's climate. He provided many references. Only a brief report is given here.

The theory here is that, as variations in solar activity take place, the solar magnetic field and the solar wind change, and they control the flux of galactic cosmic rays from deep space that enter our solar system and penetrate the Earth's atmosphere. The solar wind thus acts like a control grid on an old-fashioned triode vacuum tube where the cosmic rays provide the "current to the anode". The theory then claims that cosmic rays enhance cloud formation by producing charged atmospheric aerosols that act as nuclei for cloud formation. Thus, according to this model, an increased flux of cosmic rays due to lower solar activity produces a cooling effect on the Earth. So, it is claimed that a putative correlation of solar activity with climate is an indicator of solar wind effects that in turn affect cosmic ray penetration, which affects cloud formation, which in turn produces cooling. Several versions of this concept have been proposed.

Patterson (2007) asserted all of this as if it were self-evident and a proven fact.

Svensmark and Friis-Christensen (1997) compared the variation in low- to mid-latitude total cloudiness between 1984 and 1990 with the cosmic ray flux (which is inversely dependent on solar activity). During the period of minimum solar activity in 1986, total cloudiness was 3%–4% higher than near solar maximum in 1990. From this they suggested that cosmic rays might enhance cloudiness possibly through a mechanism involving an increase in atmospheric ionization and formation of cloud condensation nuclei. Such an increase in cloudiness would produce a cooling effect. Over a sunspot cycle, the cosmic rays varied by 15%–20%, and this correlated strongly with a 3% (absolute) variation in cloud cover over that same period. Since total cloud cover is about 63%, this is about a 5% relative change in cloud cover (about 3% absolute).

Kernthaler, Toumi, and Haigh (1999) disputed the results of Svensmark and Friis-Christensen (1997) on the grounds that, if higher-latitude data are included, the correlation between cosmic rays and cloudiness is weakened. But a greater concern is that the short period involved in the study is statistically inadequate to draw firm conclusions.

Svensmark (2000) extended previous work. He showed that the production of radiocarbon-14 in the Earth's atmosphere was inversely related to the pattern of Earth temperature over the past 1,000 years, with low production of ^{14}C during the MWP and high production during the LIA. The production of ^{14}C decreased sharply in the 20th century along with global warming. Svensmark said:

> "In 1900 the cosmic rays were generally more intense than now and most of the warming during the 20th Century can be explained by a reduction in low cloud cover. Going back to 1700 and the even higher intensities of cosmic rays, the world must have seemed quite gloomy as well as chilly, with all the extra low-level clouds."

Lockwood and Fröhlich (2007) published a rebuttal to Svensmark's theory. They admitted that, over the 20th century, the trend in ^{10}Be has been downward as the temperature trend was upward, which supports the Svensmark theory. However, they claimed that:

> "Over the past 20 years, all the trends in the Sun that could have had an influence on the Earth's climate have been in the opposite direction to that required to explain the observed rise in global mean temperatures."

Svensmark and Friis-Christensen (2007) responded to Lockwood and Fröhlich (2007). In this rebuttal, they pointed out that the use of running means of global temperature data over about 10 years obfuscated the fact that temperatures stopped rising after 1998. In addition, discrepancies between tropospheric temperature trends and surface temperature trends lead to different conclusions on temperature variations over the past few decades. Using tropospheric temperatures without averaging, and allowing for effects of El Niños and volcanic eruptions, Svensmark and Friis-Christensen (2007) found good anti-correlation between cosmic-ray levels and global temperatures over the past few decades. It is also noteworthy that the bias of observers toward (or against) the alarmist position on global warming produced by CO_2 may have crept into the arguments. Lockwood and Fröhlich (2007) said:

> "Our results show that the observed rapid rise in global mean temperatures seen after 1985 cannot be ascribed to solar variability, whichever of the mechanisms is invoked and no matter how much the solar variation is amplified."

Svensmark and Friis-Christensen (2007) took the opposite position:

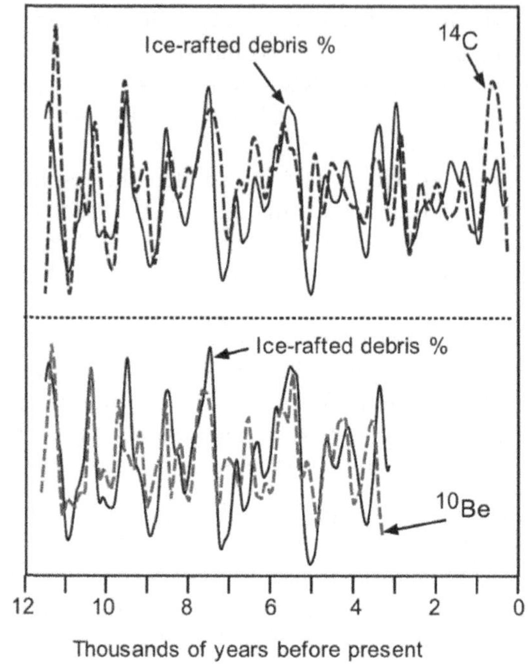

Figure 6.31. Comparison of radionuclide fluxes with relative amount of ice-rafted debris over the past 12,000 years (Bond *et al.*, 2001).

> "The continuing rapid increase in carbon dioxide concentrations during the past 10–15 years has apparently been unable to overrule the flattening of the temperature trend as a result of the Sun settling at a high, but no longer increasing, level of magnetic activity. Contrary to the argument of Lockwood and Fröhlich, the Sun still appears to be the main forcing agent in global climate change."

Kniveton and Todd (2001) found a close correspondence between the cosmic ray flux and global precipitation efficiency.

Bond *et al.* (2001) found close correlations of the extent of ice-rafted debris in the North Atlantic with fluxes of nuclides produced by galactic cosmic rays over the past 12,000 years. Figure 6.31 illustrates their results. Higher levels of ice-rafted debris are expected to reflect warmer temperatures. According to Bond *et al.* (2001), high levels of ^{10}Be reflect lower solar activity and therefore lower TSI. However, this author is unable to find evidence supporting this assumption. It does seem likely that high levels of ^{10}Be may be indicative of greater cloud formation. In either case, higher levels of ^{10}Be would be associated with lower temperatures, not higher temperatures. Therefore, the increase in ice-rafted debris seems to be associated with lower temperatures, not higher temperatures, contrary to the assertions of Bond *et al.* (2001).

Unfortunately, there does not seem to be a great deal of analysis over time spans of several hundred thousand years. However, Kirkby *et al.* (2004) analyzed the level of ^{10}Be over the past 220,000 years using ocean sediments with chronology set by tuning. They found that, during the past 220,000 years, the rate of ^{10}Be production was predominantly higher than today, although there were four periods when the ^{10}Be rate was as low as the current rate. These periods included: 230–190 thousand years before present (KYBP), 135–110 KYBP, 85–75 KYBP, and 50–43 KYBP. The period 230–190 KYBP was associated with a moderate interglacial period, 135–110 KYBP was associated with a major deglaciation, 85–75 KYBP was associated with a short-term spike in temperature, and the period 50–43 KYBP was associated with climatic instability. Thus, there are at least some indications that the level of cosmic ray flux might be affecting climate over the time span of hundreds of thousands of years. Kirkby *et al.* (2004) actually proposed "a new model for the glacial cycles in which the forcing mechanism is due to galactic cosmic rays, probably through their effect on clouds". They based this on "the accumulated experimental evidence of the last few years as well as new results presented here on a 220,000-year record of GCR [galactic cosmic ray] flux obtained in deep-ocean sediments". They concluded that the evidence was "sufficient to propose the GCR model for the glacial cycles, [but] clearly insufficient to establish it".

Kirkby (2008) reviewed the status of the cosmic ray theory over several time periods from the past few thousand years to hundreds of millions of years. While he concluded that "numerous paleoclimatic observations, covering a wide range of time scales, suggest that galactic cosmic ray variability is associated with climate change", he also admitted that there is considerable uncertainty in the mechanisms and the significance of the effect.

This topic should alert us to the possibility that complex processes may be at work in the Earth's climate that depend on factors seemingly unrelated to our climate. Most recently, Agee *et al.* (2011) reviewed the proposed hypothesis that "galactic cosmic rays (GCRs) are positively correlated with lower troposphere global cloudiness". They emphasized that Marsh and Svensmark (2000) and Svensmark (2007) utilized "lower troposphere cloud cover" rather than total cloud cover, and

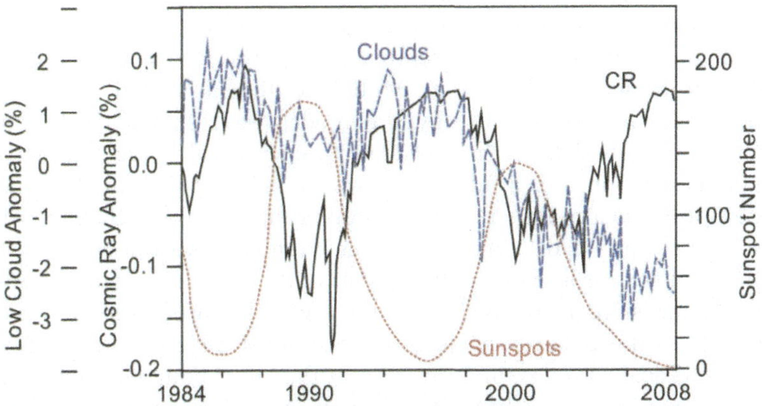

Figure 6.32. Comparison of lower troposphere cloud-cover anomaly with cosmic ray anomaly over the past two sunspot cycles (adapted from Agee *et al.*, 2011).

this may be more appropriate to the theory. However, several published papers have questioned the validity of the cloud data. Agee *et al.* (2011) examined the recent period between solar Cycles 23 and 24 during which solar activity was very low, leading to "record high levels of GCRs" by correlating data on GCR levels with measurements of lower troposphere cloud cover. Figure 6.32 shows very poor correlation between cosmic rays and clouds, which seems to cast doubt on the GCR theory.

6.4 GLOBAL CLIMATE MODELS

6.4.1 Description of GCMs

Global climate models divide the atmosphere into many small three-dimensional cells with typically 10 to 20 vertical layers and 50,000 to 100,000 horizontal cells. Land areas are divided into surface cells, and oceans are also divided into multiple vertical layers combined with many horizontal cells. Within each atmospheric cell, the various parameters such as temperature, humidity, barometric pressure, and wind velocity are uniform, but are updated frequently, typically every 30 minutes. Each cell interacts with its neighbors according to physical laws that are expressed as mathematical equations. Solar energy impinges from above, and radiation is emitted outward. The rotation of the Earth is also taken into account.

The following four numbered paragraphs are excerpted from CCSP (2007):

(1) Modern climate models are typically composed of four primary components: the atmosphere, the land surface, the ocean, and sea ice components.
(2) Atmospheric models calculate the state variables of the atmosphere, such as temperature, pressure, humidity, kinetic energy, etc., as a function of space and time. The set of model equations is formulated by using geophysical fluid dynamics theory and physical laws governing the exchanges of the mass and energy. The atmosphere is divided into discrete vertical layers, which are then overlaid with a two-dimensional horizontal grid, producing a three-dimensional mesh of grid elements. A set of primitive equations is then solved as a function of space and time on this mesh. Models differ in spatial resolutions and configuration of model grids. Typical models have spatial resolution of 200 km in the horizontal direction, and 20 vertical levels below the altitude of 15 km.
(3) The ocean component is a fully four-dimensional primitive equation model and is coupled to the atmosphere and ice models through the exchange of fluxes of heat, temperature, and momentum at the boundary between components. Like the atmosphere, the horizontal dimensions of the ocean are much larger than the vertical dimension, again resulting in separating the processes that occur in the vertical from those that occur in the horizontal. Unlike the atmosphere, which only has to deal with terrain differences at the lower boundary, the ocean has a much more complex, three-dimensional boundary, with continents and submarine basins and ridges. Further, the fluid behavior of seawater is very different from that of air.

(4) Clouds reflect solar radiation to space, cooling the Earth–atmosphere system. Clouds also trap infrared radiation, keeping the Earth warm. The net effect depends on the height, location, microphysical and radiative properties of clouds, and their appearance in time with respect to the seasonal and diurnal cycles of incoming solar radiation. Cloud feedback refers to the changes in cloud amounts and properties that can either amplify or moderate a climate change. Uncertainties of cloud feedbacks in climate models have repeatedly been identified as the leading source of uncertainty in model-derived estimates of climate sensitivity. The fidelity of cloud feedbacks in climate models is therefore important to the reliability of their prediction of future climate change.

Global climate models have been used to extrapolate global temperatures backward in time as much as 1,000 years (e.g., Foukal *et al.*, 2006; Reid, 1997; Goosse *et al.*, 2005, 2006; Bauer *et al.*, 2003; Brovkin *et al.*, 2006; Jones *et al.*, 1998; Rind *et al.*, 1999; Rind *et al.*, 2004). Global climate models have also been used extensively to try to explain the underlying physical reasons for the global climate changes that we have experienced over the past century or so (Lean and Rind, 2008). Global climate models have been used to analyze the climatic effects of volcanoes (Douglass and Knox, 2005; Robock, 2005) and Ice Ages (Charbit *et al.*, 2007; Dyke *et al.*, 2002). However, the main purpose of GCMs is typically to predict the future climate of the Earth and how it depends on future scenarios for greenhouse gas emissions. IPCC (2001, 2007) provides many references to various models.

Global climate models use physical laws to represent interactions between adjacent cells in the model, but they also require boundary conditions that specify how the TSI varies over the time duration of the model, the rate and spatial distribution of greenhouse gas and aerosol emissions, and the occurrence of special events such as major volcanic eruptions. Generally, the TSI is taken from some model. The model for historical TSI in Lean, Skumanich, and White (1992) has been used rather widely. However, as discussed in Chapter 5, this model is quite speculative. Bard *et al.* (2000) treated the TSI parametrically, dependent on how much lower the TSI was assumed to be during the MM. But as discussed in Chapter 5, we can only speculate about TSI in the past, and make wild guesses for the future. There exist reasonable models for future greenhouse gas emissions and aerosol production. However, when papers are published that utilize GCMs, it is not always clear which models are used for TSI, greenhouse gas emissions, and aerosol emissions, and these are the crucial factors that determine the outcome of the model.

6.4.2 The IPCC view of climate models

Chapter 8 ("Climate models and their evaluation") of the 2001 IPCC Report presented a positive view of GCMs:

> "Climate models are based on well-established physical principles and have been demonstrated to reproduce observed features of recent climate and past climate changes. There is considerable confidence that Atmosphere–Ocean General Circulation Models (AOGCMs) provide credible quantitative estimates

of future climate change, particularly at continental and larger scales. Confidence in these estimates is higher for some climate variables (e.g., temperature) than for others (e.g., precipitation)." (IPCC, 2001).

The IPCC Report tended to emphasize the improvements that have been made in models, and provided a mainly upbeat evaluation:

- Expanded evaluation efforts makes it less likely that significant model errors are being overlooked.
- Climate models are being subjected to more comprehensive tests. This more diverse set of tests increases confidence in the fidelity with which models represent processes that affect climate projections.
- Substantial progress has been made in understanding the inter-model differences in equilibrium climate sensitivity. Cloud feedbacks have been confirmed as a primary source of these differences, with low clouds making the largest contribution. ... The magnitude of cryospheric feedbacks remains uncertain, contributing to the range of model climate responses at mid to high latitudes.
- There have been ongoing improvements to resolution, computational methods, and additional processes (e.g., interactive aerosols) that have been included in more of the climate models.
- Most GCMs no longer use flux adjustments, which were previously required to maintain a stable climate. The uncertainty associated with the use of flux adjustments has therefore decreased, although biases and long-term trends remain in GCM control simulations.
- Progress in the simulation of important modes of climate variability has increased the overall confidence in the models' representation of important climate processes.
- The ability of GCMs to simulate extreme events, especially hot and cold spells, has improved. However, the frequency and amount of precipitation falling in intense events are underestimated.
- Simulation of extra-tropical cyclones has improved.
- Systematic biases have been found in most models' simulation of the Southern Ocean. Since the Southern Ocean is important for ocean heat uptake, this results in some uncertainty in transient climate response.
- The possibility that metrics based on observations might be used to constrain model projections of climate change has been explored for the first time. Nevertheless, a proven set of model metrics that might be used to narrow the range of plausible climate projections has yet to be developed.
- To explore the potential importance of carbon cycle feedbacks in the climate system, explicit treatment of the carbon cycle has been introduced in a few climate AOGCMs.
- Earth System Models of Intermediate Complexity have been evaluated in greater depth than previously. Coordinated inter-comparisons have demonstrated that these models are useful in addressing questions involving long timescales or requiring a large number of ensemble simulations or sensitivity experiments.

The 2001 IPCC Report asked the question:

> "How reliable are the models used to make projections of future climate change?"

Their response to this question is summarized below:

> "There is considerable confidence that climate models provide credible quantitative estimates of future climate change, particularly at continental scales and above. This confidence comes from the foundation of the models in accepted physical principles and from their ability to reproduce observed features of current climate and past climate changes. Confidence in model estimates is higher for some climate variables (e.g., temperature) than for others (e.g., precipitation). Over several decades of development, models have consistently provided a robust and unambiguous picture of significant climate warming in response to increasing greenhouse gases." (IPCC, 2001).

This is rather a strange statement, for several reasons. One is that it is difficult to see how one can conclude that models "provide credible quantitative estimates of future climate change" because the future is yet to occur and, at best, one would have to extrapolate from past success. But have the models been successful in the past? The IPCC seems to think so. The IPCC Report claims:

> "Models are routinely and extensively assessed by comparing their simulations with observations of the atmosphere, ocean, cryosphere and land surface."

The IPCC Report claims that models are able:

> "... to reproduce features of past climates and climate changes, such as the warm mid-Holocene of 6,000 years ago or the last glacial maximum of 21,000 years ago. They can reproduce many features (allowing for uncertainties in reconstructing past climates) such as the magnitude and broad-scale pattern of oceanic cooling during the last ice age."

However, we must distinguish between use of models with adjustable parameters that can be tweaked to approximate trends discerned from proxies, as opposed to *a priori* predictions of trends without one eye on a known result. Thus, models, tend to fall into the syndrome that they can explain everything, predict nothing. The attempt to explain how Ice Ages occur using climate models is still in an early emergent stage (Abe-Ouchi *et al.*, 2007). Broecker's famous "Angry Beast" article (Broecker, 1995) likened the Earth's climate to an angry beast. He said:

> "No one understands what is required to cool Greenland by 16°C and the tropics by 4 ± 1°C, to lower the mountain snowlines by 900 m, to create an ice sheet covering much of North America, to reduce ... CO_2 by 30%, or to raise the dust rain ... by an order of magnitude. If these changes were not documented in the climate record, they would never have entered the minds of the climate dynamics community. Models that purportedly simulate glacial climates do so only because key boundary conditions are prescribed (the size

and elevation of ice sheets, sea ice extent, sea surface temperature, CO_2 content, etc.). The current climate models do not explain and cannot reproduce the severe and abrupt climate changes in the proxy climatic record."

The IPCC Report also claimed:

> "Models can simulate many observed aspects of climate change over the instrumental record. One example is that the global temperature trend over the past century can be modeled with high skill when both human and natural factors that influence climate are included."

This is a rather optimistic view of the situation. As in the case of modeling Ice Ages or the Holocene, modelers can adjust parameters to seek fits to known trends, although even that doesn't workout perfectly. Two related papers (Nagashima *et al.*, 2006; Nozawa *et al.*, 2005) carried out GCMs that included the impact of aerosols on 20th-century climate change. However, these papers appear to raise more questions than answers.

The fact that the models are based on "accepted physical principles" is a *necessary* requirement but not necessarily *sufficient* to assure that models deal adequately with the complexities of the Earth's climate.

Finally, the fact that models indicate significant warming due to greenhouse gases has nothing whatever to do with their reliability.

On the positive side, GCMs provide a basic framework for modeling the Earth's climate, and, although these models are still rather primitive in many respects (clouds, aerosols, rain, inadequate spatial resolution, land use, lack of consideration of regional variations in humidity, etc.), these are all issues that are amenable to improvement in the future, and such improvements can be incorporated into the frameworks of models that have been developed.

6.4.3 Uncertainties and limitations of GCMs

> "The climate models developed in the U.S. and around the world show many consistent features in their simulations and projections for the future. However, they have not fully converged, since different groups approach uncertain aspects of the models in distinctive ways. This absence of convergence is one useful measure of the state of the science of climate simulation; convergence is to be expected once all climate-relevant processes are simulated in a convincing physically-based manner." (CCSP, 2007).

However, even after the various models converge, will they converge to the right answer?

The majority of studies with climate models have addressed the question of how much the future global temperature will rise as a result of a putative future doubling of carbon dioxide concentration in the atmosphere (doubling compared with the pre-industrial level of $\sim$280 ppm). The equilibrium response, the response expected if one waits long enough (several hundred years) for the system to re-equilibrate, is the most commonly quoted measure. The range of equilibrium climate sensitivity to doubling of CO_2 predicted by various models is typically a temperature rise of 1.5°C to 4.5°C, although some models predict higher values. The difficulty in simulating the Earth's clouds and their response to climate change are given as the fundamental reason it has proven difficult to reduce the range of uncertainty in model-generated climate sensitivity. Uncertainty still remains considerable and is not decreasing rapidly, due in part to the difficulty of cloud simulation, but also to uncertainty in the rate of heat uptake by the oceans and a variety of other factors (CCSP, 2007).

One major problem with GCMs is that it is difficult to test them against actual data. CCSP (2007) claimed that models are able to simulate the 20th-century global mean temperature record in a plausible way. Sterl *et al.* (2009) used a large ensemble of climate models to examine the signal-to-noise problem for climate models and concluded that "in large parts of the world the observed warming over the last 60 years is statistically indistinguishable from the estimated warming caused by increased greenhouse gas concentrations". However, variations in TSI are speculative, and uncertainties about the effects of issues. As Lindzen (2008) has emphasized, the treatment of humidity as a global average can produce serious errors. Explanations for the temperature dip from 1940 to 1978 are not fully satisfactory, as we discussed in Section 3.4.5. While some models claim to reproduce measured temperatures in the 20th century, these seem to be contrived after the fact, with forcings chosen to obtain agreement. Such methods explain everything and predict nothing. Hoyt (2006) pointed out that the global temperature rise of ~0.6°C during the 20th century was lower than would be predicted by GCMs based on the known increase in CO_2 concentration. However, other factors (aerosols, land use, etc.) may have affected 20th-century temperatures significantly:

> "Uncertainties in the climatic effects of man-made aerosols (liquid and solid particles suspended in the atmosphere) are a major stumbling block in quantitative attribution studies and in attempts to use the observational record to constrain climate sensitivity. We do not know how much warming due to greenhouse gases has been cancelled by cooling due to aerosols. Uncertainties related to clouds increase the difficulty in simulating the climatic effects of aerosols, since these aerosols are known to interact with clouds and potentially change cloud radiative properties and cloud cover." (CCSP, 2007).

Haerter *et al.* (2009) analyzed the uncertainty in climate models due to aerosols and found that the uncertainties were large..

Atmospheric carbon dioxide has increased ~30% over the industrial period. Radiative forcing of climate change by increased concentrations of CO_2 and other long-lived greenhouse gases has been estimated to be 2.4 ± 0.2 W/m^2 relative to the pre-industrial era. However, some models predict that a doubling of CO_2 would result in a forcing of about 4 W/m^2. The most recent IPCC Report estimated that this would produce a temperature rise in the range 1.5°C to 4.5°C, based on a climate sensitivity factor in the range 0.38°C to 1.13°C per W/m^2. The factor-of-three uncertainty in present estimates is certainly unacceptably large for planning for mitigation or adaptation (Schwartz, 2003).

The increase in global mean temperature to date, relative to the pre-industrial era has been estimated to be 0.6 ± 0.2°C, suggesting a sensitivity of 0.25 ± 0.09 °C/(W/m^2), well below the low end of the IPCC range. But such an empirical estimate assumes that climate response is near equilibrium and that forcing is due entirely to long-lived greenhouse gases. The equilibrium assumption is claimed to be valid. However, forcings other than by greenhouse gases, particularly the cooling influence of anthropogenic aerosols due to their scattering of solar radiation, are thought to offset much of the greenhouse gas forcing on a global basis, resulting in a much lower total forcing and consequently much greater sensitivity. Present uncertainty in total forcing is so great as to preclude a meaningful empirical estimate of climate sensitivity from the temperature record and forcing over the industrial period (Schwartz, 2003).

Schwartz (2004) concluded:

(1) A consistent finding of such studies is that the change in global mean temperature per forcing (i.e., climate sensitivity), is, to a good approximation, independent of the nature of the forcing (e.g., forcing because of changes in CO_2 mixing ratios, mixing ratios of other GHGs, aerosol direct forcing, or the solar constant), and independent as well of the geographical distribution of the forcing.
(2) An immediate consequence of the forcing–response paradigm is that forcings are additive. This hypothesis provides a path forward to calculating radiative forcing over the industrial period by adding the forcings.

However, feedback effects can significantly alter the additivity. In discussing climate models, Schwartz (2004) said:

> "In practice, these models embody numerous assumptions, parameterizations, and approximations of the variables and phenomena being represented—water vapor, clouds, precipitation, snow and ice, radiation, transport of heat and water on all scales—the list goes on. The resolution of models is limited, typically, at present to 300 km. . . ."

Schwartz (2004) went on to say:

> "Considerations such as the foregoing would seem to call into question confidence that can be placed in statements such as the following from the IPCC 2001 assessment of climate change:
>
> - Simulations that include estimates of natural and anthropogenic forcing reproduce the observed large scale changes in surface temperature over the 20th century.
> - Most model estimates that take into account both greenhouse gases and sulfate aerosols are consistent with observations over this period.
> - The large-scale consistency between models and observations can be used to provide an independent check on projected warming rates over the next few decades under a given emissions scenario.
> - Detection and attribution studies comparing model simulated changes with

the observed record can now take into account uncertainty in the magnitude of modeled response to external forcing, in particular that are due to uncertainty in climate sensitivity."

Schwartz (2004) concluded:

"The sensitivity of global mean temperature change to an increase in atmospheric carbon dioxide (CO_2) is not well established. The complexity of the climate system precludes calculation of the response of Earth's climate to a change in a radiative flux component (forcing) from well-established physical laws. Consequently, determination of global climate sensitivity is a subject of intense research. This work is reviewed from time to time by pertinent national and international bodies. One such landmark review was that of a 1979 National Research Council panel, which concluded: 'We estimate the most probable global warming for a doubling of CO_2 to be near 3°C, with a probable error of 1.5°C.' More recently, the IPCC concluded that 'Climate sensitivity [to CO_2 doubling] is likely to be in the range 1.5–4.5°C.' These estimates must be considered somewhat subjective. They are based mainly on calculations with climate models constrained, especially for the IPCC estimate, by observation of the extent of warming over the industrial period and concurrence of modeled and observed warming. Neither the Charney panel nor the IPCC quantitatively specified the meaning of their uncertainty bounds, but in the case of the Charney estimate, a National Research Council panel three years later expressed its understanding that 'the Charney group meant to imply a 50% probability that the true value would lie within the stated range.' Remarkably, despite some two decades of intervening work, neither the central value nor the uncertainty range has changed. The large uncertainty range, a factor of 3, in present estimates of climate sensitivity renders such estimates not particularly useful from the perspective of developing policy regarding either reduction of greenhouse gas (GHG) emissions or adaptation to a new, increasingly warm climate."

Leroux (2005) discussed the limitations and problems associated with GCMs in a 21-page diatribe. The "imperialism" of models is described. It was concluded that the virtues of "models are greatly overestimated".

Schwartz, Charlson, and Rodhe (2007) reviewed the IPCC (2007) assessments of climate change based on results from "an ensemble of 58 runs with 14 climate models" and concluded:

"[IPCC] estimates [of] total anthropogenic forcing [were] 0.6 to 2.4 W/m^2 (595% confidence range). This factor of four range greatly limits the ability to evaluate the 'skill' of climate models in reproducing past temperature changes and to infer climate sensitivity from observed change because a given temperature increase might result from a large forcing and low climate sensitivity or alternatively from a small forcing and high climate sensitivity."

By comparison, the range of predicted temperature rise (0.4°C to 0.8°C) from 1910 to 2000 was only a factor of 2. Schwartz, Charlson, and Rodhe (2007) raised questions whether the temperature estimates spanned the full range of potential forcing, and concluded:

> "The narrow range of modeled temperatures gives a false sense of the certainty that has been achieved."

Schwartz, Charlson, and Rodhe (2007) never raised the question as to whether the purveyors of GCMs "fudged" their models with one eye on the known (or at least the believed) 0.6°C temperature rise from 1910 to 2000. However, this seems to be a likely possibility.

Ghan and Schwartz (2007) pointed out:

> "The practice of climate modeling has become tied to the production schedules for periodic international assessments of the science of climate change by the IPCC."

Each successive generation of IPCC Reports at roughly six-year intervals relies on simulations that are about three years old, which, in turn, are based on climate models from the prior year (four years before the IPCC Report). Inevitably, there is a lag from gaining understanding of processes to representing that understanding in climate models. Thus, there is a lag from understanding processes, to representing this understanding in models; and there is a further lag in representing these models in GCMs that are used in IPCC assessments, which can be as long as a full IPCC cycle or more (six years). There is a further, similar lag of a full IPCC cycle between the representation of various processes in GCMs and the use of the results of that generation of models in scenario assessments (paraphrased from Ghan and Schwartz (2007)).

Meehl *et al.* (2007) provided an upbeat report on climate models, claiming that projections of the future warming pattern are "robust". However, Stainforth *et al.* (2007) discussed the credibility of atmosphere/ocean global circulation models. They discussd two aspects of model credibility: uncertainty and inadequacy. Model uncertainty deals with uncertainty as to which parameterizations to use and which values of the parameters are best. "Model inadequacy captures the fact that we know *a priori*, there is no combination of parameterizations, parameter values and initial conditions which would accurately mimic all relevant aspects of the climate system." Finally, they concluded:

> "Complex climate models, as predictive tools for many variables and scales, cannot be meaningfully calibrated because they are simulating a never before experienced state of the system; the problem is one of extrapolation. It is therefore inappropriate to apply any of the currently available generic techniques which utilize observations to calibrate or weight models to produce forecast probabilities for the real world. To do so is misleading to the users of climate science in wider society."

McWilliams (2007) concluded that: "climate models are structurally unstable in various ways that are not yet well explored, and this implies a level of irreducible imprecision in their answers that is not yet well estimated".

Rind (2008) reviewed the uncertainties in climate models with particular emphasis on the ability of models to predict relative climate sensitivities in the topics and at high latitudes. They concluded that little progress has been made in resolving latitudinal variations of climate sensitivity, even after 30 years of model development.

It has been claimed that a major problem for climate models is the disparity between the temperature trends observed at the Earth's surface and the much smaller trends observed in the lower troposphere that is just the opposite of what GCM models predict. Douglass *et al.* (2007) compared tropical temperature trends with climate model predictions for temperatures in the so-called "characteristic emission layer" (CEL) (2–6 km altitude) where the role of water vapor is most important. Over the period from 1979 through 2004, the models predicted a rising temperature trend of roughly 0.2°C to 0.3°C per decade, whereas satellite temperature measurements indicate essentially no increase below 10 km altitude, and a negative trend above 10 km. This was cited as evidence of the inadequacy of current climate models.

It has come to pass that a few determined skeptics (Douglass, Lindzen, McLean, Spencer, McIntyre, *et al.*) continue to publish contrarian papers (in those rare cases where the *cabal* does not succeed in censoring publication) and, immediately thereafter, a flurry of emails is exchanged between cabal members (Mann, Jones, Schmidt, Trenberth, *et al.*) castigating the skeptics, and strategizing to achieve damage control to protect their orthodoxy that rising CO_2 is essentially the sole cause of global warming. The most pugnacious and aggressive of these is Michael Mann. It is ironic that his own research, responsible for the *hockey stick*, is far less believable than the work of those he would criticize. We have already described one example of this in regard to McLean *et al.* (2009).

After publication of Douglass *et al.* (2007), the cabal came forth with Santer *et al.* (2008) as a rebuttal. This paper begins with the sentence: "There is now compelling scientific evidence that human activities have influenced global climate over the past century" which aside from the fact that the statement is not true, reveals the belief system to which the authors subscribe religiously. The details of the statistical processing of large data sets are complex. The issue is whether tropical tropospheric temperatures have risen more than surface temperatures as climate models would predict for the effect of greenhouse gases on climate. Douglass *et al.* (2007) concluded that models and data disagreed to "a statistically significant extent". Santer et al. (2008) claimed to achieve a "partial resolution of the long-standing 'differential warming' problem" although they also said:

> "We may never completely reconcile the divergent observational estimates of temperature changes in the tropical troposphere. We lack the unimpeachable observational records necessary for this task. The large structural uncertainties in observations hamper our ability to determine how well models simulate the tropospheric temperature changes that actually occurred over the satellite era. A truly definitive answer to this question may be difficult to obtain."

Yet, this did not prevent Santer *et al.* from producing a so-called "Fact Sheet"[22] that

[22] *https://publicaffairs.llnl.gov/news/news.../NR-08-10-05-factsheet.pdf.*

said "We've gone a long way towards such a reconciliation" [between climate models and tropical tropospheric temperatures].

In 2009, McIntyre[23] pointed out that when the data used by Santer *et al.* (2008) that ended in 1999 is extended through 2008, the discrepancy reported by Douglass remains, and "the claim by Santer *et al.* (2008) to have achieved a 'partial resolution' of the discrepancy between observations and the model ensemble mean trend is unwarranted". McIntyre also noted the difficulty in obtaining data from Santer *et al.*, and indicated that the *International Journal of Climatology* was stalling in responding to him. It appears that this article will never pass through the cabal's lock on the IJC, and McIntyre had to be content with merely archiving his article.[24] Yet, alarmists continue to refer to Santer *et al.* (2008) as evidence that climate models have been adequately tested.

The Goddard Institute for Space Studies (GISS) provides a website that describes their climate change simulations.[25] Their process is also described by Hansen, *et al.* (2007). Their model depends on forcings produced by various phenomena, some natural, and some induced by human activities, per year over the time period 1880–2003. The model has also been used to predict climate into the future out to 2050. Each forcing is an equivalent heat flow (positive or negative) at the top of the atmosphere measured in W/m^2. These forcings consist of:[26]

- *Greenhouse gases*: Hansen, *et al.* (2007) estimated the forcing due to greenhouse gases, with particular emphasis on CO_2. The forcings due to changes in CO_2 concentration were shown in Figure 6.8.
- *Ozone*: ozone is a greenhouse gas and its depletion in the atmosphere produces a cooling effect.
- *Stratospheric water*: in Section 6.2.5, we mentioned that Solomon *et al.* (2010) found that stratospheric water vapor acts as a potent greenhouse gas, and changes in this concentration may have contributed 25%–30% to the total greenhouse effect from 1980 to 2010.
- *Solar*: this is an estimate of how the total solar irradiance varied, year by year from 1880 to 2003. The GISS model adopted the model of Lean (2000) that is a modified CQSM. It resulted in a 0.24% increase in TSI from the Maunder Minimum to 2000, and about 0.15% increase from 1880 to 2000. CQSM are discussed in Chapter 5 of this book, where it is shown that such models are highly speculative. Lean (2000) emphasized this in her paper, saying: "Estimating the magnitude of long-term solar irradiance variability is speculative, at best, since it is presently not known whether the long-term irradiance variations actually do occur." She evidently thought the model may have overestimated solar variability but, of course, it equally well might have

[23] *http://climateaudit.org/2009/01/27/submited-article-on-tropical-troposphere-trends/*; *http://climateaudit.org/2009/04/14/tropical-troposphere-march-2009/*.
[24] *http://arxiv.org/abs/0908.2196*.
[25] *http://data.giss.nasa.gov/*.
[26] *http://data.giss.nasa.gov/modelforce/*.

underestimated solar variability. Hence the entire GISS model, like a chain, is not stronger than its weakest links, and solar variability is a weak link. To compound the felony, the GISS model assumed that the future TSI (out to 2050) will be the same as it is today—an assumption that has little or no basis. Realizing that the model for variable TSI is highly speculative, the GISS studies also treated variable TSI as an adjustable parameter to estimate how TSI variability would affect climate.

- *Land use*: the effects of human modification of the surface of the Earth through land clearing, deforestation, croplands, cities, irrigation, etc. are complex. A major factor is the increase in albedo from land clearing resulting in a cooling effect. Section 6.2.10.5 of this book provides a short review of some of the literature in this field. The GISS model used estimates at the small end of the range of previous estimates.
- *Snow albedo*: this is the effect of deposition of soot on ice and snow at high latitudes with a consequent decrease in albedo, resulting in higher absorption of incident solar energy. The GISS model was based on data from a 1985 reference, and did not take into account more recent data. In Section 7.3 of this book, more recent data and analyses are summarized. These publications suggest that the snow albedo effect might be considerably greater than that assumed in the GISS model. They also suggest that the snow albedo effect may have peaked from 1900 to 1940, whereas the GISS model has it peaking at the end of the 20th century. It is also interesting that Shindell and Faluvegi (2009), who work at GISS, estimated that a sizable fraction of the temperature rise in the 60°–90° latitude region of the NH in the 20th century was due to the snow albedo effect.
- *Stratospheric aerosols from volcanic eruptions*: these provide relatively short-term intense cooling forcings due to reflection of incident solar irradiance.
- *Tropospheric aerosols*: tropospheric aerosols in the GISS model include sulfate, nitrate, black carbon, and organic carbon. While these aerosols absorb some solar irradiance, their principal effect is to reflect incoming solar irradiance, resulting in a significant cooling effect.
- *Indirect effects of aerosols*L this forcing is mainly the effect of aerosols in affecting cloud formation, and is very difficult to estimate precisely. Nevertheless, the GISS model assigns a significant cooling effect.
- *Clouds*: the GISS team admitted that it cannot estimate the variability of clouds and therefore they treat cloud variability as an adjustable parameter. But cloud variability is likely to be a major factor in climate change, so it seems that almost any result can be obtained, depending on the assumptions that are made.

There are good things and bad that can be said about this model. On the positive side, the GISS team has made an attempt to integrate a number of phenomena that act to affect the Earth's climate. On the negative side, it appears to this writer that the uncertainties remain excessively large, and the results are dubious. Hansen *et al.* (2007) more or less admit this, implying that this is work in progress and they "aim to provide a benchmark against which the effect of improvements in the model, climate forcings, and observations can be tested". They

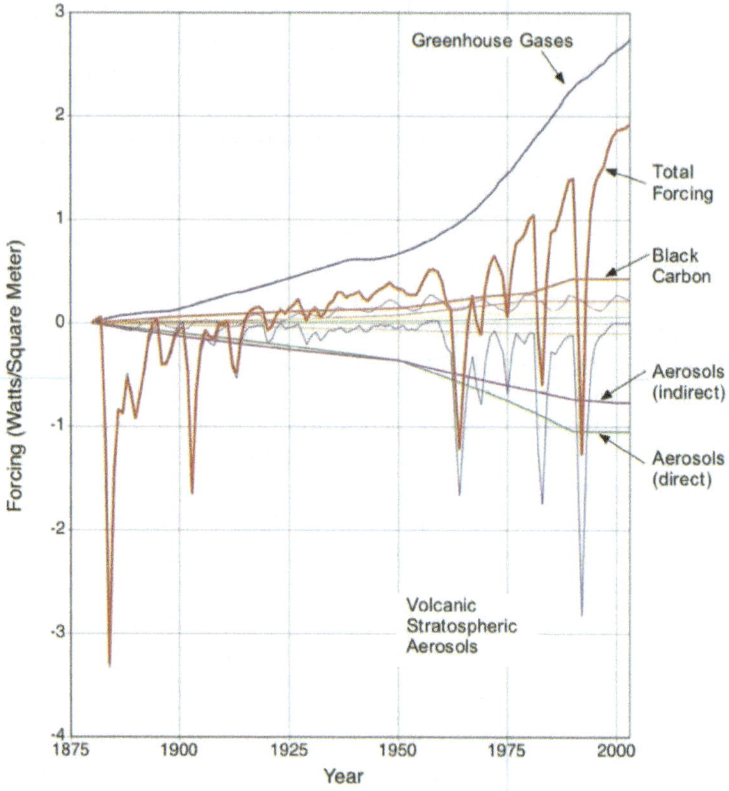

Figure 6.33. GISS estimates of forcings for the period 1880–2003.

go on to say: "Principal model deficiencies include unrealistically weak tropical El Niño-like variability and a poor distribution of sea ice, with too much sea ice in the Northern Hemisphere and too little in the Southern Hemisphere. Greatest uncertainties in the forcings are the temporal and spatial variations of anthropogenic aerosols and their indirect effects on clouds."

The GISS estimates for the period 1880–2003 are shown in Figure 6.33. According to this model, the four largest forcings were: greenhouse gases, volcanic stratospheric aerosols, tropospheric aerosols (direct), and tropospheric aerosols (indirect). The GISS model suggests that, had it not been for tropospheric aerosols, total forcing would have been about 3.6 W/m^2 in 2003. Hence, according to this model, without the tropospheric aerosols, the temperature rise from 1875 to 2003 would have been double what it actually was. Thus, the GISS model utilizes an Earth that is highly sensitive to forcing, particularly greenhouse gas forcing, but the model mitigates the calculated temperature rise during the 20th century based on several negative forcings, particularly those based on tropospheric aerosols.

One of the admitted weaknesses of the GISS model was "unrealistically weak tropical El Niño-like variability". Spencer and Braswell (2011) estimated the

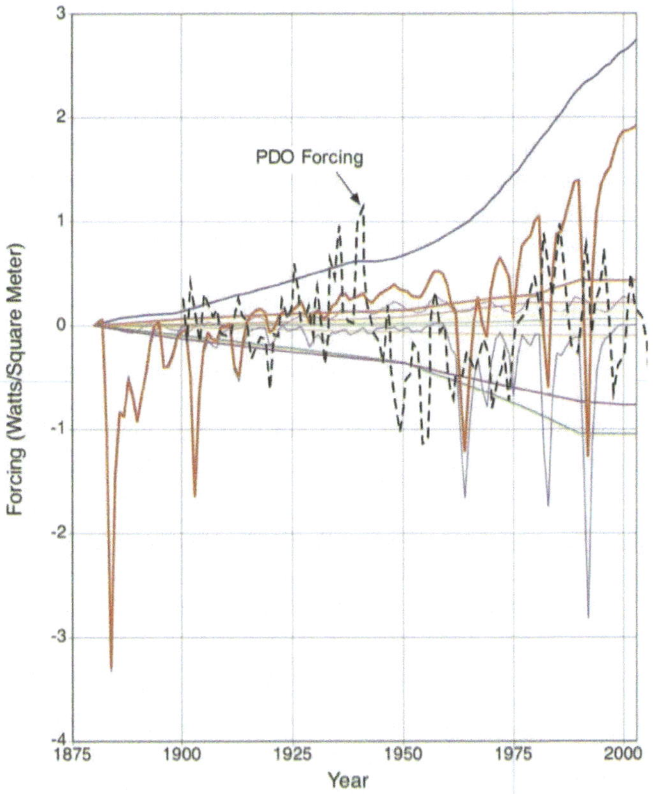

Figure 6.34. Overlay of Spencer's estimate for Pacific Decadal Oscillation (PDO) forcing on the GISS model.

radiative forcing of the Earth due to variability of the El Niño index from 1900 to 2003. He made use of data during the period 2000–2009 from the Terra/CERES satellite that monitors OLR and reflected solar radiation from the Earth. By a means unintelligible to this writer, Spencer estimated the long-wave radiative forcing during the period 2000–2009 and compared it to an El Niño index during that period. It is not clear to this writer why or how he could attribute this forcing due solely to variability of the El Niño index. From this, he inferred a relationship between radiative forcing due to variability of an El Niño index, and assumed this held true for the entire 20th century. He used this to convert measured values of the El Niño index from 1900 to 2003 to estimates of radiative forcing generated by variability of the El Niño index from 1900 through 2003. Spencer attributed this forcing to the effect of the El Niño index on global cloud cover but did not seem to elucidate the details of how this works. If Spencer's estimate is overlaid on Figure 6.33, we obtain Figure 6.34. This forcing was positive early in the 20th century, negative from about 1940 to 1975, and positive after 1975; this is exactly what we observe in NH temperature variability. If Spencer's estimate is reasonably accurate, variability of the El Niño index would add to the forcing after 1975, requiring that the Earth must have been less sensitive to greenhouse forcing than the GISS model would suggest. Spencer also indicated that the GISS model requires large negative forcing from tropospheric aerosols in the period 1940–1975 to reduce the forcing during this period when global temperatures did not rise, but the estimates of tropospheric aerosol forcing are quite speculative. Spencer showed that, if forcing attributed to tropospheric aerosols by GISS is omitted, and replaced by Spencer's estimate of PDO forcing, a better fit to 20th-century temperatures is obtained. However, Spencer's estimate of PDO forcing appears to this writer to be at least as speculative as the GISS estimates of tropospheric aerosol forcing.

Knutti (2008) wrote a review entitled: "Should we believe model predictions of future climate change?". As he put it: "To what extent should we trust the numbers that come out of our models?" He attempted to "explain why it is so difficult to quantify model performance" and pointed out that "the limiting factor is probably our understanding rather than the computational capacity". Quoting Stainforth *et al.* (2007), he went on to say:

> "Uncertainty in model projections arises from boundary and initial condition uncertainty, our incomplete theoretical understanding of the system, parameter uncertainty and the fact that our models are imperfect."

Knutti (2008) emphasized the uncertainties in how to treat clouds. He emphasized the difficulty in verifying models, saying that models cannot "be validated in the sense of being shown to accurately represent—both at present and for all future times—the processes responsible for the observed behavior of the real system". He then said:

> "So the best we can hope for is to demonstrate that the model does not violate our theoretical understanding of the system and that it is consistent with the available data within the observational uncertainty. For climate projections the situation is more difficult. Model calibration is strictly impossible in this case, as projections of future climate change relate to a state never observed before. Making a projection for the climate in the year 2100 and waiting a century for the data to evaluate the projection is unfeasible; also, a single realization of the climate may not tell us much anyway. The problem is that the life cycle of a model is much shorter than the time scale over which a prediction can be evaluated by observations."

The difficulty in evaluating models is that the data sets used for evaluation are essentially the same as those used for calibration, resulting in circular reasoning. Knutti (2008) also pointed out that:

> "Agreement between model and data does not imply that the modeling assumptions accurately describe the processes producing the observed climate system behavior; it merely indicates that the model is one (of maybe several) that is plausible, meaning that it is empirically adequate."

Knutti pointed out:

> "The downside is that the model spread into the future is often not decreasing, e.g. the spread of climate sensitivity or transient climate response in [the latest models] is almost the same as in the previous model generation. In addition, many problems seem to be similar across families of models, because models make similar assumptions."

The point is that, even if all the modelers could agree on parameterization and get the same results, it would not assure that the result was correct. An Internet site plotted global temperature vs. acts of piracy in the late 20th century and found good correlation. Does that prove that acts of piracy cause global warming?

The ultimate problem for climate models may be summarized by a quotation attributed to Ken Cruickshank:

> "If a committee is allowed to discuss a bad idea long enough, it will inevitably vote to implement the idea simply because so much work has already been done on it."

In their passages where they defend climate models, Stainforth *et al.*, McWilliams and Knutti all seem to fall back on the work that was put in, rather than the results that come out. They all seem to end up with the question: How can we go forward? The answer always seems to be to do more of the same.

Norris and Slingo (2009) pointed out how little we know about variability of cloudiness and how poorly GCMs account for cloudiness.

Huybers (2010) provided a very insightful review of climate models. He pointed out that Schwartz *et al.* (2007) identified "an important interdependence between the radiative forcing and climate sensitivity across the CMIP3 models" in that, "while twentieth-century changes in radiative forcing differs by a factor of 4 across the models, the resulting temperature spread differs by only a factor of 2". He also pointed out that Kiehl presented evidence that "this narrow temperature range results from an anti-correlation between radiative forcing and climate sensitivity". He then suggested that:

> "Inter-model compensation between climate sensitivity and radiative forcing underscores that the models are not based purely on theory but are also conditional upon observations and, possibly, expectations."

Huybers (2010) showed that the treatment of clouds was the "principal source of uncertainty in models". Indeed, his Table I shows that, whereas the response of the climate system to clouds by various models varied from 0.04 to 0.37 (a wide spread), the variation of net feedback from clouds varied only from 0.49 to 0.73 (a much narrower relative range). He then examined several possible sources of compensation between climate sensitivity and radiative forcing. He concluded that:

> "Model conditioning need not be restricted to calibration of parameters against observations, but could also include more nebulous adjustment of parameters, for example, to fit expectations, maintain accepted conventions, or increase accord with other model results. These more nebulous adjustments are referred to as 'tuning'."

He suggested that one example of possible tuning is that "reported values of climate sensitivity are anchored near the $3 \pm 1.5°C$ range initially suggested by the *ad hoc* study group on carbon dioxide and climate (1979) and that these were not changed because of a lack of compelling reason to do so".

He went on to say:

> "More recently reported values of climate sensitivity have not deviated substantially. The implication is that the reported values of climate sensitivity are, in a sense, tuned to maintain accepted convention."

This seems to be an example of the "Goldilocks Principle". Judith Curry discusses this on her website.[27]

> "The Goldilocks principle states that something must fall within certain margins, as opposed to reaching extremes. The Goldilocks principle is derived from a children's story The Three Bears in which a little girl named Goldilocks finds a house owned by three bears. Each bear has their own preference of food, beds, etc. After testing each of the three items, Goldilocks determines that one of them is always too much in one extreme (too hot, too large, etc.), one is too much in the opposite extreme (too cold, too small, etc.), and one is 'just right'."

Additional quotations from Huybers's paper are given below:

> "Although substantial changes to GCM cloud parameterizations have been implemented since 1990, it is not clear that a general increase in their accuracy is the sole explanation for the present trend toward convergence. It may be that current models are producing similar errors, while the earlier models produced different errors."

> "Tuning climate sensitivity to lie within the observed spread across the CMIP3 models is a sufficient explanation for the origins of the compensation between [clouds] and the other feedbacks."

> "The covariance between the CMIP3 model feedbacks may be symptomatic of the uneven treatment of outlying model results."

> "The specter of tuning leading to a curtailment of the inter-model spread in climate sensitivity is difficult to dismiss."

> "Convergence between model results, if not truly driven by a decrease in model uncertainty or clearly understood as a result of calibration, could have the unfortunate consequence of lulling us into too great a confidence in model predictions or inferences of too narrow a range of future climates. To the extent that it occurs, tuning the models based on expectation or convention renders the modeling process a partially subjective exercise from which it is very complicated to derive a statistical interpretation."

[27] *http://judithcurry.com/2012/12/22/the-goldilocks-principle/*.

> "Focusing on maximally inconsistent possibilities seems more likely to lead to scientific discoveries and to uncover climate surprises. A maximally inconsistent ensemble of state-of-the-art model realizations would also have the advantage of suggesting outer bounds upon the range of climate sensitivity and, therefore, be complimentary to existing estimates".

Translated into simple terms, the implication is that climate modelers have been heavily influenced by the early (1979) estimate that doubling of CO_2 from pre-industrial levels would raise global temperatures 3 ± 1.5°C. Modelers have chosen to compensate their widely varying estimates of climate sensitivity by adopting cloud feedback values countering the effect of climate sensitivity, thus keeping the final estimate of temperature rise due to doubling within limits preset in their minds. Had they not done this, the spread in estimates of temperature rise would be much greater. Thus, they have imposed their preconceived notions of the expected temperature rise on the models to make them come out "right".

A fundamental characteristic of GCMs is that they predict a positive feedback due to water vapor and clouds that adds to the predicted temperature rise due to increasing CO_2 concentration. Most of this positive feedback is due to trapping of OLR, and this increases the climate sensitivity to rising CO_2 above that due solely to CO_2. Pinning down a good estimate of the feedback, and hence the climate sensitivity is one of the most important things needed in climate science.

Lindzen and Choi (2009) examined data on the outgoing radiation budget from the Earth Radiation Budget Experiment (ERBE) in the tropics in an attempt to determine whether observations of the Earth's radiation imbalance can be used to infer feedbacks and climate sensitivity. As Lindzen and Choi (2009) pointed out, "such an approach has, as we will see, some difficulties ... ," but they attempted to overcome these problems. Their results indicated that the feedback is negative, rather than positive as predicted by climate models. Therefore, they believed that the climate sensitivity is considerably smaller than the values predicted by climate models. Their estimate of climate sensitivity depends on the ratio of change in outgoing long wave flux at the top of the atmosphere to a change in average sea surface temperature. They found that if one plots the flux change vs. the equilibrium climate sensitivity, the shape of the curve is asymptotic so that one can infer a climate sensitivity from Δ(flux)/Δ(SST) for sensitivities less than about 2°C (for doubling of CO_2) but for higher sensitivities the curve is flat, and one cannot determine the sensitivity if it exceeds 2°C. Later, Lindzen and Choi admitted: "This work was subject to significant criticism by Trenberth *et al.* (2009), much of which was appropriate". As a result, they wrote another paper (Lindzen and Choi, 2011) that was "an expansion of the earlier paper in which the various criticisms are addressed and corrected. In this paper we supplement the ERBE data for 1985–1999 with data from CERES for 2000–2008". As might be expected, Lindzen and Choi (2011) found that feedbacks were primarily negative, resulting in relatively low climate sensitivity. This is contrary to the alarmist position that feedbacks are positive leading to higher climate sensitivity (and therefore a greater increase in global temperature as greenhouse gas concentrations increase). The manuscript by Lindzen and Choi was rejected by the Proceedings of the National Academy of Sciences (PNAS), and the revelation of the reviewers and their comments led to a very extensive series of blog entries at climateaudit.org.[28] In the course of these blog entries, we find (along with the usual trivia) several nuggets of information worth mentioning. Lindzen is a member of the NAS and it is very rare that a paper submitted by a member would be rejected (96% are accepted). In a highly unusual move, the PNAS rejected Lindzen's suggestion for reviewers, and instead chose reviewers who were obviously antagonistic to Lindzen's viewpoint. The reviews of this paper were incredibly detailed and penetrating. It seems likely that papers expressing the warmist agenda glide through the review process with little friction and no depth of review. One blog contributor was a reviewer for the paper by Wahl and Ammann (2007). His review was discarded by the Journal of Climate because it was not in conformity with the warmista agenda. It appears that most of the papers in climatology are based on inadequate data: lacking in spatial and temporal coverage. The sophisticated data processing used to cover this up, whether filtering, smoothing, use of principal components, or otherwise, hides the fact that the foundations are typically very weak. Had other landmark papers in climatology that are repeatedly referred to in biblical tones been given the same kind of penetrating review as Lindzen's manuscript, they would also have been rejected. Indeed, most of the literature in climatology would have to be cleared out. Finally, the Lindzen and Choi paper was published in the *Asia-Pacific Journal of Atmospheric Science*.

Roe and Baker (2007)[29] pointed out that, if one plots a histogram of number of estimates vs. estimated equilibrium climate sensitivity (to doubling of CO_2), one obtains a function that rises sharply above 2°C, peaks around 3°C, and has a long tail that extends out beyond 10°C. They then posed the question: Why is uncertainty not diminishing with time? To answer this, they provided the following model. A reference climate system is perturbed by a forcing, ΔR. After equilibrium is achieved, the change in system temperature is ΔT (upper part of Figure 6.35). In the case of doubling CO_2, many models would agree (within a moderate range) that:

$$\Delta T_o = K\,\Delta R,$$

where R = the forcing due to doubling of CO_2 ($R \sim 4$ W/m^2) and K is the climate sensitivity to CO_2 doubling without feedback $\{K \sim 0.3$ [°C per W/m^2)]$\}$. Thus, they would calculate that $\Delta T_o \sim 1.2$°C without feedback.

Any number of feedbacks may operate in the system to impose additional forcings (positive or negative) that are proportional to ΔT (lower part of Figure 6.35). These feedbacks enter the equation as follows:

$$\Delta T = K\,(\Delta R + c_1\,\Delta T + c_2\,\Delta T + \ldots),$$

where ΔT is different from ΔT_o due to the feedback. This can be rearranged as:

[28] *http://climateaudit.org/2011/06/10/lindzens-pnas-reviews#comments.*

[29] Also see: *science.larc.nasa.gov/ceres/STM/2007-11/ce0711151330Roe.pdf*; *yosemite.epa.gov/ee/epa/eerm.nsf/vwAN/EE-0564-17.../EE-0564-17.pdf.*

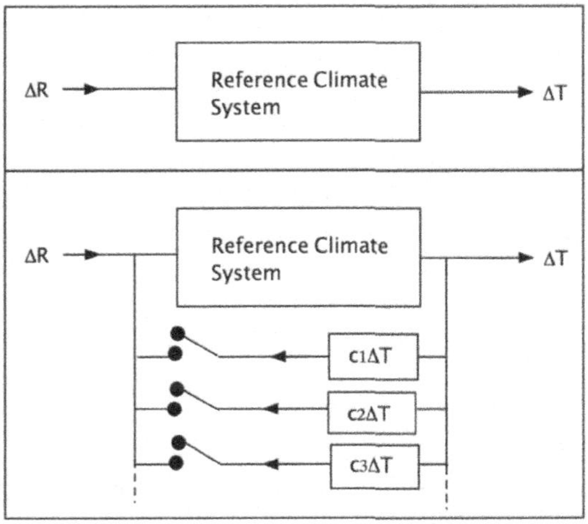

Figure 6.35. Model used by Roe and Baker (adapted from Roe and Baker, 2007).

$$\Delta T = K\,(\Delta R)/(1 - c_1\,\Delta T - c_2\,\Delta T - \ldots)$$

and, if we define the feedback factor as:

$$f = c_1\,\Delta T + c_2\,\Delta T + \ldots$$

we obtain:

$$\Delta T = \Delta T_o/(1-f) = G\,\Delta T_o,$$

where the "gain" is $G = 1/(1-f)$. The literature contains a number of estimates of f, and therefore G. As Roe and Baker (2007) pointed out, if one plots ΔT vs. f, the curve rises steeply when f exceeds about 0.7 ($\Delta T \sim 3°C$) (see Figure 6.36). There are a number of estimates of f, each presumably with some uncertainty function that describes the unknowns in the model, as in function A in Figure 6.36 (drawn for a most probable estimate of $f \sim 0.65$). This corresponds to an uncertainty distribution in ΔT as shown as function B in Figure 6.36. There is a long tail extending to high temperatures. The nature of feedbacks was further elaborated by Roe (2009).

Roe and Baker (2007) thus pointed out that, if feedback is small, one can estimate the ΔT that results from the feedback quite precisely but, if the feedback is great, the uncertainty in ΔT rises sharply. Many climate models predict feedback factors f in the range 0.6 to 0.7, whereas Lindzen and Choi (2009) found a much lower value.

Roe and Baker (2007) then went on to statistically analyze the probability that f lies within various bounds, based on the various climate model estimates that have been made. In doing this, there is an implicit assumption (not stated) that all the climate models are basically correct, but they approximate uncertain parameters in different ways. One can then arrive at probabilities that ΔT lies within certain bounds. However, if (as seems likely to this writer) none of the climate models is

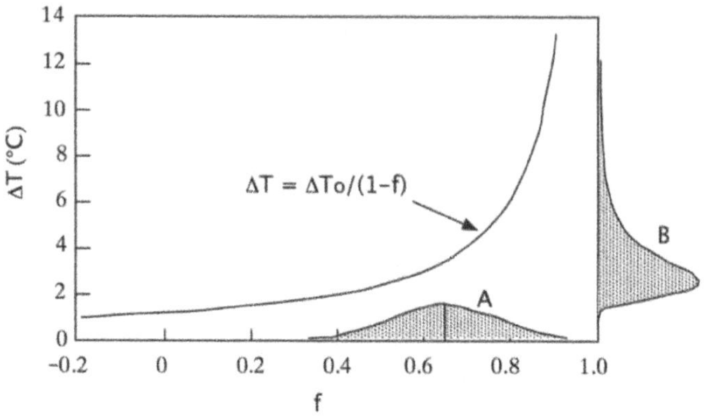

Figure 6.36. Dependence of ΔT on f. For any estimate of f with its probability distribution function A, there is a corresponding distribution of ΔT shown as function B (adapted from Roe and Baker, 2007).

credible, then statistical correlation of their results is not meaningful. It is a simple case of "GIGO".[30]

The 2007 paper was followed by Baker and Roe (2009) in which the authors defined *climate sensitivity* as "the equilibrium response of the global ... surface air temperature to a doubling of carbon dioxide over preindustrial values" and distinguished this from the "transient climate change" that is delayed and damped by ocean heat uptake. Figure 6.36 is implicitly based on an assumption that the CO_2 concentration is instantly doubled in a step-function, and the change in temperature is the long-term equilibrium response to this change. The 2009 paper explored the transient temperature change along the way toward the new equilibrium. They concluded that the transient response is more predictable than the ultimate equilibrium response. However, they did not treat the realistic case of a gradually evolving rise in CO_2.

Reacting to doubt expressed by alarmists, Roe and Armour (2011) defended the asymmetrical nature of the dependence of ΔT on f.[31]

Roe[32] distinguished between *climate sensitivity, transient climate response* (see above), and *climate commitment*, which is "is a measure of the climate change we already face because of emissions that have already occurred" (the implicit climate change yet to come from past emissions as the transient response gradually approaches a new equilibrium). Armour and Roe (2011) discussed climate commitment. These authors place great faith in the IPCC climate models. According to their viewpoint, the effect of greenhouse gases over the past century or so would

[30] "Garbage In—Garbage Out.

[31] Also note: "Comment on 'Another look at climate sensitivity'" by Zaliapin and Ghil, (2010), *Nonlin. Processes Geophys.*, **17**, 1–3.

[32] *yosemite.epa.gov/ee/epa/eerm.nsf/vwAN/EE-0564.../EE-0564-117.pdf.*

have been much greater, had it not been for the aerosols generated along with greenhouse gas emissions, with the aerosols acting to mitigate the forcing due to greenhouse gases. Thus, an immediate and total cessation of emissions would produce "an immediate and significant warming following the cessation of emissions as aerosols are quickly washed from the atmosphere, and the large uncertainty in current aerosol radiative forcing implies a large uncertainty in the climate commitment". The longevity of CO_2 in the atmosphere is estimated to be hundreds of years, so its forcing will slowly diminish over many centuries. Hence, Armour and Roe (2011) concluded that, after cessation of emissions, temperatures will rapidly rise significantly and remain high for millennia. The amount of additional temperature rise built into the system beyond that already experienced over the past century was estimated using several assumptions. It was estimated that this temperature rise is most probably 0.6°C with a very wide uncertainty range of 0.3°C to 6.3°C. While exploring the wonders of climate models, it is useful to keep one's eye on historical reality. As we show in Section 6.5, a CO_2 concentration of 4,000 ppm to 5,000 ppm might possibly be construed to be associated with a global temperature rise of as much as 7°C to 8°C. Thus, it would take a much greater increase in CO_2 concentration than merely increasing from 280 ppm to 390 ppm to raise the global average temperature by 6.3°C.

Reifen and Toumi (2009) said:

> "With the ever increasing number of models, the question arises of how to make a best estimate prediction of future temperature change. The Intergovernmental Panel on Climate Change (IPCC) Fourth Assessment Report (AR4) combines the results of the available models to form an ensemble average, giving all models equal weight. Other studies argue in favor of treating some models as more reliable than others."

Some models simulate global mean, Siberian, and European 20th-century surface temperature with a lower error than the total ensemble for some periods. These authors tested these models to see how good their predictions were for subsequent periods. It was found that they were no better than the ensemble. Therefore, justifying a climate model on the basis of its agreement with data over a limited period might not tell much about its validity in general.

As in almost every aspect of analysis of analysis of climate change, any numbers can play. Alarmists believe that the Earth is sensitive to forcing from greenhouse gases and require the presence of negative forcings to avoid models predicting too much heating in the 20th century. Skeptics believe that the Earth is not so sensitive to forcing by greenhouse gases and point to variability of the El Niño index as another positive forcing not included in most climate models.

According to Steve McIntyre,[33] the MET Office in the U.K. is a major player in climate modeling. They periodically update predictions for near-term future climate change. Their predictions from 2010, 2011, and 2012 are shown in Figure 6.36a.

[33] *http://climateaudit.org/2013/07/15/nature-hides-the-decline/#more-18160.*

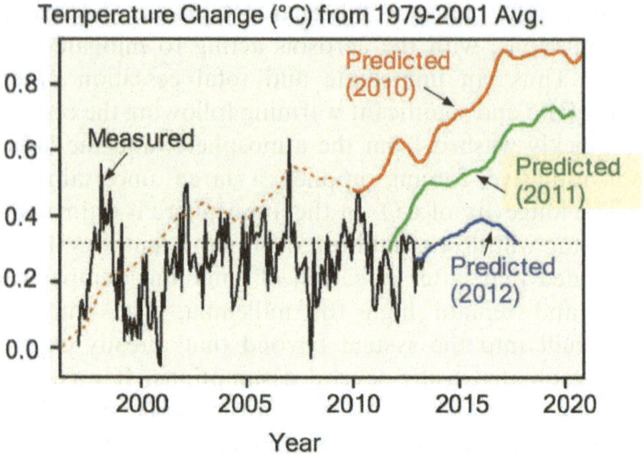

Figure 6.36a. Predictions by the MET Office. Red dotted line is hindcast to supposedly show validity of 2010 forecast. The solid red curve appears in the latest IPCC Report.

As Steve McIntyre pointed out, even though the *Nature* article was written in 2013, the curve presented in it "is not the most recent Met Office decadal forecast, but a variation of the older 2011 forecast".

Conclusion: *Nature* did not want to show a lower prediction.

6.4.4 Effect of clouds

A NASA website points out that:

> "As we all know from days at the beach, clouds block much of the solar energy and reflect it back to space before it can be absorbed by the Earth, the atmosphere, or the sunbather! The more plentiful and thicker the clouds are, the cooler the Earth. At the same time, clouds also act like greenhouse gases—they block the emission of heat to space and inhibit the ability of the planet to release its absorbed solar energy [to space]. [We all know that clear, cloudless winter nights produce the deepest freezes.] To complicate matters further, the altitude of clouds changes the amount of thermal infrared blocking. This effect is the result of the decrease in temperature with altitude—high clouds are colder and more effective at absorbing the surface-emitted heat in the atmosphere, while they emit very little to space because of their cold temperatures! So it turns out that clouds can either act to cool or warm the planet depending on how much of the Earth they cover, how thick they are, and how high they are. Low clouds made of spherical water droplets reflect much of the sunlight that falls on them, but have little effect on the emitted energy. Thus, low clouds act to cool the current climate. High clouds made up of ice crystals reflect less energy, but trap more of the energy emitted by the surface, and thus act to warm the current climate."

Ramanathan (1988) said that, as surface warming occurs, the increased moisture from the warmer oceans should alter cloud distributions and characteristics but the nature of these cloud changes and the ways in which they affect radiative heating are unclear. He said:

> "As a result, cloud feedback is one of the largest sources of uncertainty in the theory of climate change."

Ramanathan (1988) claimed that the albedo of the Earth would be about 10% if there were no clouds, but clouds bring it up to about 30%. An increase in the planetary albedo of 1% would zero out the predicted temperature rise for a doubling of CO_2 based on GCMs.

Leathers *et al.* (1998) emphasized that clouds will decrease daytime temperatures and increase night-time temperatures, thus decreasing the diurnal temperature range.

Clouds play an important role in the heat balance of the Earth. Climate models have included the effects of clouds in various ways. However, most climate modelers agree that uncertainties in regard to the effects of clouds is a principal cause of disagreement between models. According to Bony *et al.* (2006):

> "Clouds strongly modulate the earth's radiation budget, and a change in their radiative effect in response to a global temperature change may produce a substantial feedback on the earth's temperature. But the sign and the magnitude of the global mean cloud feedback depends on so many factors that it remains very uncertain. Cloud feedbacks have long been identified as the largest internal source of uncertainty in climate change predictions, even without considering the interaction between clouds and aerosols."

Bony *et al.* (2006) said that climate models exhibit "systematic biases" in simulating clouds that "restrict their ability to predict the magnitude of cloud feedbacks". They therefore concluded that:

> "Defining strategies for evaluation of cloud feedback processes in climate models is thus of primary importance to better understand the range of model sensitivity estimates and to make climate predictions from models more reliable."

Although they adopted the upbeat view that "progress has been made during the last few years in our understanding of processes involved in these feedbacks", there still remain many uncertainties in modeling clouds.

Held and Soden (2000) said that considerable controversy exists regarding the radiative treatment of clouds in climate models, and that the "treatment of clouds in climate models presents greater obstacles to quantitative analysis of climate sensitivity than does the treatment of water vapor".

Kiehl and Trenberth (1997) assumed that clouds exist in three layers and these layers are assumed to randomly overlap. In developing their model for the Earth's heat balance, they concluded that the average cloudy-sky TOA outgoing radiative flux was 235 W/m^2, as shown in Figure 6.14. They adjusted the parameterization of clouds in a climate model to achieve this flux. Three cloud levels were introduced into their model: (1) a low cloud layer between 1 km and 2 km with fractional area of 49%, (2) a mid-level cloud cover between 5 km and 6 km of fractional amount of 6%, and (3) a high cloud cover between 10 km and 11 km of 20%. They assumed random overlap leading to their previous estimate of 62% total cloud cover. The emissivity of the low- and mid-level clouds was assumed to be 1, while the emissivity of high-level clouds was set to 0.6. This led to the aforementioned outgoing flux of 235 W/m^2. In their model, the total long-wave radiative forcing for clear conditions was 125 W/m^2 and, when clouds were included, this added 30 W/m^2, bringing the total to 155 W/m^2. Clouds not only absorb outgoing thermal radiation, but also contribute to cooling by reflecting more incoming sunlight than they absorb. It was estimated that short-wave forcing by clouds amounts to –50 W/m^2—a cooling effect. Thus, the estimated net effect of clouds is 30—50 = –20 W/m^2—a net cooling effect.

Marsh (2002) provided slightly different figures (clear-sky forcing = 146 W/m^2 and cloud effect = 33 W/m^2). Marsh (2002) quoted the IPCC Report as estimating the net effect of clouds as 31 W/m^2 for long-wave and –44 W/m^2 for short-wave, with a net of –14 W/m^2. This is considerably smaller than the estimate by Kiehl and Trenberth (1997). Other modelers have estimated a wide range of values.

Gorodetskaya *et al.* (2006) discussed the effects of clouds in polar areas and oceans.

Rising global air traffic and its associated contrails have the potential for affecting climate via radiative forcing. Minnis *et al.* (2004) estimated increases in cirrus clouds due to jet contrails over heavily traveled air routes and used a GCM to infer the resultant temperature change of air near the surface. It was concluded that the increasing cirrus over the past few decades led to a U.S. temperature increase of about 0.3°C per decade from 1975 to 1995. However, a note to the *Journal of Climate* claimed that this estimate was high by an order of magnitude because it extended local effects to a wide region. The lead author of Minnis *et al.* (2004) answered by suggesting that the estimate in Minnis *et al.* (2004) was probably an upper limit, while the smaller estimate provides a lower limit (Shine, 2005). Yang *et al.* (2010) reviewed subject of the effect of jet contrails on climate. They showed that fractional cirrus cloud cover over several areas of the U.S. increased from about 1950 to 1990, suggesting that jet contrails might be contributing to this. They also discussed optical properties of ice crystals in contrails. However, the bottom line is that the magnitude of the effect of aircraft contrails remains very uncertain.

Norris and Slingo (2009) reviewed measurements and models for variability of the Earth's cloudiness and it effect on the Earth's radiation budget (ERB) and the Earth's climate. Their overall assessment was that our knowledge of past variability of cloudiness is very poor, our present capabilities for monitoring global cloudiness are weak, and the prospects for future measurements are worse. Global cloudiness is a very important factor in determining the Earth's climate. As Norris and Slingo noted, it is difficult to resolve any putative effects of greenhouse gases on climate when we don't understand how cloudiness varies, because variability of cloudiness has a much greater forcing effect than changes in greenhouse gas concentration:

"Small cloud changes are important because they can exert more leverage over ERB than equivalent changes in greenhouse gases."

While the effect of greenhouse gases is expected to be spatially coherent, changes in cloudiness will vary widely from region to region and it is necessary to make measurements at many sites globally to obtain a global average. In addition, Norris and Slingo said:

"Since various cloud types have strikingly different radiative effects, it is not a simple matter to determine the overall global impact of cloud changes; each cloud type and climate regime must be examined in particular. Moreover, alterations of cloud albedo, cloud emissivity, and cloud height can affect ERB even when cloud amount remains the same."

As Norris and Slingo described in some detail, measurements of clouds, whether from the ground or satellite, appear to suffer from various artifacts and are not trustworthy. GCMs suffer from an inability to represent clouds. As Norris and Slingo said:

"Many studies comparing simulated clouds with observed clouds have found that GCMs poorly represent clouds when evaluated on terms for which they were not explicitly tuned.

"Some GCMs suggest that the horizontal extent of low-level clouds over low-latitude oceans will increase with higher global temperature, whereas other GCMs suggest that low-level cloud amount will decrease."

Andronova *et al.* (2009) analyzed evolution of the tropical mean radiation budget at the top of the atmosphere since 1985 for the latitude range 20°S to 20°N. They found that since 1985, the Earth became less reflective to incoming short-wave radiation and more absorbent of OLR. Both effects produce heating of the atmosphere. Upon comparing with climate models, they found that "none of the models simulates the overall 'net radiative heating' signature of the Earth's radiative budget over the time period from 1985–2000". These changes occurred during a period when the Earth warmed. It is not at all clear whether these changes were induced primarily by rising CO_2 levels, or whether these changes were due to other factors such as El Niños, and it is difficult to assess the role of rising CO_2 levels. Dupont *et al.* (2009) also presented new data on the relation between cirrus cloud characteristics and the Earth's heat budget.

While considerable research has been carried out on clouds and their impact on the climate, we remain uncertain regarding several key aspects. We might hypothesize a normal period of years during which the CO_2 concentration is constant at $\sim$280 ppm with minimal volcanoes, El Niños, and La Niñas, and ask what is the long-term average cloud distribution across various regions of the Earth during such a period? A second question would be: How large is the annual statistical fluctuation about the mean cloud distribution? Accurate data on these two questions seems to be sadly lacking. There are some indications that annual variations are large and therefore, one requires data over many decades to establish averages and means. The problems with most attempts to characterize the effect of clouds are the brevity of the time period over which data are taken, and the effects of volcanoes, El Niños, and La Niñas can be significant. In particular, Dessler *et al.* (2008a) have used very short-term data in their analyses. However, Masters (2012) showed clearly that none of these short-term analyses has any validity. As Masters said: "Overall, there is little correlation between the changes in the CRF and surface temperatures on these timescales, suggesting that the net effect of clouds varies during this time period quite apart from global temperature changes."

6.5 THE RELATION BETWEEN ANCIENT CLIMATES AND CO_2 CONCENTRATION

6.5.1 Background

One of the most pressing issues of our time is the possibility that rising CO_2 concentrations in the atmosphere might lead to significant global warming in the future that could produce deleterious impacts on humankind. Hence, the relationship between CO_2 concentration and climate has become a very central and critical scientific issue. However, in addition to being a scientific issue, rising CO_2 has also become a political issue as well.

Thus we see that the current holy grail of climatology is to seek an estimate of how much the global average temperature will increase if the CO_2 concentration doubles from the pre-industrial value of about 280 ppm to 560 ppm. Attempts to estimate this directly from climate models are difficult due to uncertainties in secondary factors that accompany warming from increased CO_2 (humidity, cloudiness, winds, ocean currents, glaciers, ice sheets, etc.). Some climatologists have sought to estimate the dependence of the climate on CO_2 concentration by analyzing paleoclimatic data on climate and CO_2 concentration, with the intent of using the climate sensitivity derived from this to estimate the global average temperature increase if the CO_2 concentration doubles from the pre-industrial value of about 280 ppm.

Climatologists have attempted to estimate global average temperatures using proxies over geological time periods as long as hundreds of millions of years. Proxies are indirect indicators of past temperature based on some natural process that occurred in the past that was dependent on temperature. Many proxies have been proposed and utilized. The proxies that are of greatest value in estimating global temperatures over tens or hundred of millions of years are oxygen isotope ratios in benthic ocean sediments in which the ^{18}O concentration is an inverse measure of T_G. While the conversion of the direct signal $\delta^{18}O$ to temperature is highly approximate, the $\delta^{18}O$ measurements appear to be reliable and we have rough relative measures of how global average temperatures varied over the past 500 million years, even though absolute values are far less certain.

Over the past couple of million years in which we have had alternating Ice Ages and interglacials, there is good evidence that the trigger to set the cycles in motion is solar input to higher northern latitudes. But what does it mean to set the cycle in motion? It means that an albedo effect begins in the ~60°N latitude range as snow and ice accumulate, causing a regional cooling. As the ice sheet builds, and sea ice expands, and the ocean drops, other albedo effects occur and a greater regional cooling takes place. Dust is stirred up and this further amplifies the cooling. Then, as the cooling spreads, the CO_2 concentration in the atmosphere decreases, producing additional negative forcing worldwide, which lowers the worldwide temperature. However, changes in humidity and cloudiness are unknown and may be very large factors. The big question is: What is the effective climate sensitivity when all secondary factors are folded in?

A number of climatologists have attempted to circumvent the uncertainties in climate models by estimating the real-world climate sensitivity directly without the use of climate models. Hansen and Sato's procedure provides a good example. They claimed that "In contrast to climate models, which can only approximate the physical processes and may exclude important processes, the empirical result includes all processes that exist in the real world—and the physics is exact". They compared conditions at the LGM about 20,000 years ago with conditions in pre-industrial times (a few hundred years ago). These conditions include all the secondary effects. They estimated that the global average temperature was ~5°C colder at the LGM. Next, they estimate the forcing difference between the LGM and pre-industrial times. This includes changes in greenhouse gas concentrations, changes in Earth reflectivity due to changes in ice sheet extent, changes in sea ice, changes in vegetation, changes in land/sea ratio, and effects of dust in the atmosphere. Needless to say, such estimates are somewhat speculative. Nevertheless, the net result is that the total forcing according to Hansen and Sato is something like 6.5 W/m^2 and so they concluded that the real-world climate sensitivity (including all secondary effects) is 5/6.5 ~0.75°C per W/m^2. Hence their prediction is that doubling CO_2 in the 21st century will cause a temperature rise of 0.75×3.7 ~3°C. As I show later, when Hansen and Sato's model is modified with better data, it really suggests a climate sensitivity of around 0.6°C per W/m^2, resulting in a temperature rise of about 2.4°C. Another study by Chylek and Lohmann (2008) of the LGM-modern transition yielded a forcing of around 10 W/m^2 and a climate sensitivity of around 0.5°C per W/m^2, resulting in a temperature rise of only 2°C for doubling CO_2. Regardless of which values you prefer, all of these paleoclimatological analyses provide smaller temperature increases for doubling CO_2 in the 21st century than climate models.

Hansen and Sato (2011), Chylek and Lohmann (2008), and Kohler *et al.* (2009) independently estimated the forcing at the LGM (see Table 6.19). The contribution of the diminution of CO_2 at the LGM to the total cooling estimated by these studies was in the range 16% to 33%. While it seems likely that solar input to higher latitudes triggered the Ice Age—interglacial cycles, the variability of CO_2 concentration played a secondary part in determining the extremity of the temperature cycle that resulted from this trigger. The changes in CO_2 concentration between glacial maxima and interglacials (~180 ppm to ~280 ppm) are well documented in ice-core records, although no one seems to have a satisfactory explanation for why the CO_2 concentration changed this much (simple solubility in the oceans does not suffice). However, the estimates of forcings, particularly due to dust, vary considerably from investigator to investigator and it is difficult to pin down the climate sensitivity to CO_2 change. There are good estimates available of the global average temperature and the CO_2 concentration at the LGM 20,000 years ago and, if these data are compared with values in the pre-industrial era (a few hundred years ago), one can thereby estimate the sensitivity of the climate to CO_2 concentration over the range ~180 ppm to ~280 ppm. Using this estimated climate sensitivity, one can then estimate the global average temperature rise in going from 280 ppm to 560 ppm. The various investigators have come up with a range of projections. It is noteworthy that this range of estimates for the real-world ΔT due to doubling CO_2 from 280 ppm to 560 ppm is from ~1°C to ~3°C. However, these estimates do not take into account possible differences in humidity and cloudiness in Ice Age–interglacial cycles.

Over much longer time periods (up to 540 million years ago), the evidence from benthic sediments (and other geological evidence as well) is strong that there have been periods of great warmth with no glaciation at all on the Earth, with occasional periods when the Earth was heavily glaciated.

Believing that every effect has a cause or causes, climatologists have searched for possible causes of these long-term climate changes and inevitably, after eliminating all other candidates, they have assumed that variability of CO_2 concentration was the major factor that caused long-term climate changes over many millions of years:

> "The major transitions between climatic icehouse and greenhouse conditions are ultimately most probably driven by the deep Earth processes of plate tectonics, as a function of the long-term balance between CO_2 degassing at spreading centers and the conversion of atmospheric CO_2 to mineral carbon through long-term silicate weathering and oceanic carbonate formation." (NAS, 2011)

As Pierrehumbert (2009) said: "… the Urey reaction removes CO_2 from the atmosphere. When CO_2 dissolves in water, it forms a weak acid (carbonic acid), which reacts with silicate minerals (e.g. $CaSiO_3$) to form carbonate minerals (e.g. $CaCO_3$, or 'limestone'). The reaction takes place only in the presence of liquid water." The widely held view amongst geologists and climatologists alike, is that the primary cause of these climate changes was variability of CO_2 concentration due to long-term imbalances between CO_2 degassing at spreading centers and the conversion of atmospheric CO_2 to mineral carbon through long-term silicate weathering and oceanic carbonate formation. The argument goes (more or less): "If it wasn't CO_2, what else could it have been?" Foster *et al.* (2009) described this as the "accepted paradigm" that requires CO_2 to vary in unison with global temperature. So, paleoclimatologists have been trying for decades to establish a relationship between climate and CO_2 concentration over many millions of years. The more audacious of these have attempted to establish a quantitative relationship between climate and CO_2 concentration in order to estimate the Earth system climate sensitivity. Unfortunately, the proxy data for CO_2 over many millions of years are very widely scattered and the results are equivocal. There is a general tendency for warmer climates to be associated with higher CO_2 concentrations, but this mainly relates to very large temperature excursions and, even so, there are many exceptions. There is no direct one-to-one correspondence between CO_2 and climate. Evidently, factors other than CO_2 must also influence the climate. In his *Perspective Article*, Ruddiman (2010) emphasized that there is no present explanation for the fact that there was no significant drop in CO_2 concentration over the past 22 million years while the climate cooled substantially. Nevertheless, it is noteworthy that many paleoclimatologists are so convinced from the start that CO_2 is the main driver of long-term climate change that, even with the noisy data, they claim support for their theory. Royer (2010) began his commentary with the statement:

> "Global temperatures have covaried with atmospheric carbon dioxide (CO_2) over the last 450 million years of Earth's history."

It is noteworthy that, prior to 2004, a number of climatologists pointed out discrepancies between the geological records of climate and CO_2 over 500 million years, whereas, after 2004, most published papers emphasized the correlation of climate and CO_2 over that period. It is not clear whether new data made the difference, or whether it is necessary to support the orthodoxy in order to obtain research funding. Our conclusion here is that CO_2 is probably an important factor in long-term climate change, but other factors are also influential, such as the placement of the continents on the Earth, the functionality of ocean currents, worldwide distribution of clouds and aerosols, the past history of the climate, the orientation of the Earth's orbit relative to the Sun, the luminosity of the Sun, the presence of aerosols in the atmosphere, volcanic action, land clearing, biological evolution, etc. Hence, there is probably no single curve relating global average temperature to CO_2 concentration, but rather, a set of curves that depend on the above factors.

From the limited accuracy of the analyses of ancient climates conducted so far, we still cannot pin down the expected ΔT due to doubling CO_2 from 280 ppm to 560 ppm very precisely, but the rough indication from the LGM is that it may be in the range 1°C to 3°C. The merit of these estimates is highly questionable.

What we seek is a relationship between CO_2 concentration and the Earth's climate over long geological periods during which the CO_2 concentration varied over a wide range. We have already discussed this in Section 1.4. Here, we will refer to Section 1.4 in our discussion. It would be very nice if there were a single curve relating global average temperature (T_G) to CO_2 concentration such as that shown in Figure 1.9. In that case, if we could find several points on the curve, we could attempt to map out a good portion of the curve.

However, over long time periods, the variation of T_G with CO_2 concentration depends on various factors such as the placement of the continents on the Earth, the functionality of ocean currents, the past history of the climate, the orientation of the Earth's orbit relative to the Sun, the luminosity of the Sun, the presence of aerosols in the atmosphere, volcanic action, land clearing, biological evolution, etc. Hence, there is probably no single curve relating T_G to CO_2 concentration, but rather, a set of curves that depend on the above factors (see Figure 1.10).

Climatologists have mainly concentrated on the realm of CO_2 concentration between 280 ppm and 560 ppm, with some concern for higher concentrations up to ~900 ppm. In this regard, we can magnify the gray slice from Figure 1.11, and combine this with known values of T_G over the past ~120 years, as shown in Figure 1.12. Curves *1* to *4* show various estimates of the temperature rise that will be induced by further increases in CO_2 concentration.

A few comments of caution need to be made at this point. Use of the data and models from hundreds of millions of years ago may produce misleading results when extrapolated to conditions prevailing today. In addition, the Sun was some 6% reduced in intensity 500 million years ago. These, and other changes, add uncertainty as to whether such data can properly be extrapolated to the 21st century. Another comment is that proxies for T_G and CO_2 concentration over geologic time spans are not likely to be very accurate, and these should be critically reviewed before relying on them. As Zeebe (2011) said:

> "Unfortunately, palaeo data-derived climate sensitivities have large uncertainties. Errors can arise from issues such as dating, alteration of the climate signal after deposition, insufficient spatial and/or temporal coverage, and various uncertainties associated with the proxies for environmental variables such as temperature and past atmospheric CO_2 concentrations."

Climatologists use two different definitions of climate sensitivity. The *political definition* is the temperature rise (ΔT_G) in the 21st century resulting from doubling CO_2 from the pre-industrial level of ~280 ppm. The *scientific definition* is the temperature rise (ΔT_G) resulting from a forcing of 1 W/m^2 at the top of the atmosphere. The scientific definition is applicable for both short- and long-term periods.

It is common for paleoclimatologists to distinguish between two different types of *political climate sensitivity*: *fast feedback sensitivity* and *Earth system sensitivity* (Zeebe, 2011; Royer *et al.*, 2011). The former includes "water vapor, clouds, snow, and sea ice" operating on timescales of less than 100 years, whereas the latter includes these fast feedbacks as well as longer-term "changes in non-CO_2 greenhouse gases, vegetation, dust/aerosols, ice sheets, ocean circulation, marine productivity, weathering and more" (Zeebe, 2011). Climate models are aimed at estimating the temperature rise due to doubling CO_2 from the pre-industrial level of ~280 ppm during the 21st century, and thus deal with the *political fast feedback sensitivity*. However, consider this hypothetical scenario. Suppose climate models indicate that doubling CO_2 in the 21st century will produce an increase in T_G of X degrees (we need not specify X, except that it is a relatively large number). Suppose further (as some climatologists believe) that this temperature increase will gradually erode the ice sheets on Greenland and to some extent Antarctica, and produce other long-term effects that will be manifested well after the 21st century. These changes will add further warming leading to a higher *long-term political Earth system sensitivity* than the *short-term political fast feedback sensitivity*. However, the scientific climate sensitivity is the same in both eras. The only thing that changes from the short term to the long term is the magnitude of the forcing. Royer *et al.* (2011) provided a list of previous attempts to estimate the *long-term political Earth system sensitivity* from paleoclimatic data on climate and CO_2 levels. In the present article, I have attempted to provide insights and assessment of various models for paleoclimatic estimates of *scientific climate sensitivity*.

6.5.2 The transition from the LGM to the Pre-Industrial Era

About 20,000 years ago, the most recent Ice Age was at its maximum extent with gigantic ice sheets in the higher latitudes of the NH. There is reliable evidence from ice cores that the CO_2 concentration at that time was in the range 170 ppm to 180 ppm. The first question is what was T_G at the LGM?

Raymo (1992) estimated a swing in global average temperature from the LGM to the present of about 10°C. Dwyer *et al.* (1995) utilized the ratio of magnesium to calcium (Mg/Ca) in fossil ostracodes from Deep Sea Drilling Project Site 607 in the deep North Atlantic to infer that the change in bottom water temperature changed by ~4.5°C in going from the LGM to pre-industrial times. According to Leroux (2005), the difference in temperature between an Ice Age and an interglacial was about 10°C in the Antarctic and about 6°C globally. Broecker (2002) indicated that, at the LGM, Greenland was 16°C colder and the tropics were 4°C colder than today. Taylor *et al.* (2001) carried out an analysis in which they took into account the reduced CO_2 concentration and the extended ice sheets of the LGM in climate models to estimate the amount of cooling at the LGM compared to pre-industrial times. Using six different climate models, they obtained values of 3.5, 3.7, 3.8, 4.4, 5.2, and 5.9, for an average of 4.4°C. Crucifix (2006) provided a less optimistic view of the precision to which this is known: "The global temperature change is therefore is estimated to be comprised between 3°C and 9°C with 95% confidence." He also estimated that the tropical ocean SST decreased between 1.7°C and 2.7°C, and Antarctic surface air temperature decreased by 7°C to 11°C at the LGM compared to pre-industrial times.

Shakun and Carlson (2010) carried out an extensive review of the LGM–interglacial transition. They found, as expected, that the ΔT in this transition varied with latitude as shown in Figure 1.13. Their estimate of ΔT_G (the temperature at the LGM minus the pre-industrial temperature) was –4.5°C. If we couple the temperature during the LGM (14.3°C—4.5°C = 9.8°C) with an estimated CO_2 concentration of 170–180 ppm, we can plot a point representing the LGM, as shown in Figure 1.14.

Hansen and Sato (2011) also estimated ΔT_G. Hansen and Sato relied heavily on the paper by Zachos *et al.* (2001), but Zachos *et al.* mainly presented oxygen isotope data, and only obliquely and briefly tacked on temperature to their graph. Hansen *et al.* adopted their temperature scale, but no indication is provided by either Zachos *et al.* or Hansen and Sato as to how these temperatures were derived from the isotope data, or what the uncertainty was in the estimates. Zachos *et al.* interpreted the first part of the isotope ratio curve (from 500 million years ago to about 34 million years ago) as defining an equivalent deep-sea temperature. After that date, variable amounts of build-up of ice, initially in Antarctica, and most recently in Greenland,

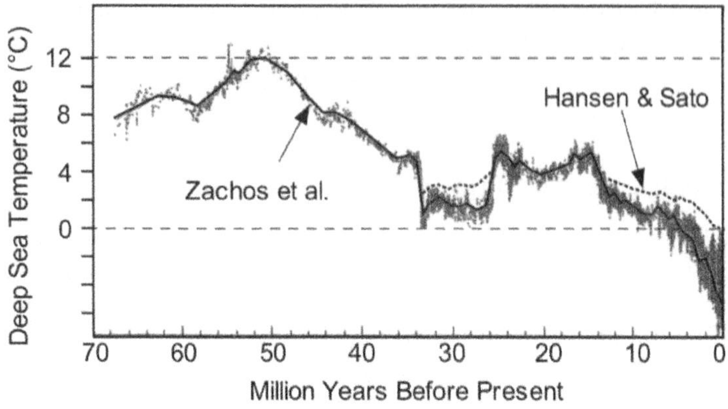

Figure 6.37. Deep-sea temperature over the past 60 million years (adapted from Hansen and Sato, 2011).

distorted the isotope ratios, and only part of the observed isotope ratio can be attributed to temperature change. Hansen and Sato pointed out that, when there is heavy glaciation at Antarctica and Greenland, the isotope ratios in sediments are partly due to deep-ocean temperature change and partly due to the isotope effect in evaporation as ice sheets form. Hansen and Sato assumed that half of the isotope ratio was due to temperature change during the last ~35 million years. They therefore modified the temperature vs. time curve of Zachos *et al.* as shown in Figure 6.37.

Hansen and Sato interpreted oxygen isotope data in deep-sea sediments to infer deep-ocean temperatures, and then asserted that these were representative of global average temperatures except as the deep-sea temperature approached the freezing point of water. They said that: "deep ocean temperature change becomes less representative of global surface temperature change as the ocean temperature approaches the freezing point of water, because the deep ocean temperature is limited by the freezing point while the global mean surface can continue to cool". They then asserted without proof that, over the past ~half million years, as deep-ocean temperatures approached the freezing point, "the amplitude of recent glacial-interglacial deep ocean temperature change is only about two-thirds the amplitude of global mean surface temperature change". It is not clear how they arrived at the 2/3-rule. However, conversion of oxygen isotope data to temperatures is not a simple matter—although Hansen and Sato seem to imply that it is trivial. Hansen and Sato concluded that the global average temperature change from the LGM to recent pre-industrial times was 5°C. However, it should be noted that Hansen and Sato's Figure (1c) indicates a deep-ocean temperature change from the LGM to recent pre-industrial times of ~3°C, which, after applying Hansen and Sato's 2/3 rule, would indicate a global average temperature change from the LGM to recent pre-industrial times of $3/2 \times 3 = 4.5$°C.

The reason why the LGM point lies so low is because important changes took place on the surface of the Earth as the ice sheets expanded. These changes go beyond the purely spectroscopic effect of less absorption of IR by CO_2 in the atmosphere as the CO_2 concentration was lowered to below 200 ppm. The growth of large ice sheets from recent pre-industrial times to the LGM resulted in an increase in the Earth's albedo across the ice sheets, as well as for mountain glaciers. In addition, the drop in sea level moved shorelines outward, converting ocean to land, thereby further increasing the Earth's albedo. As the climate got colder, biomass and vegetation grew less abundantly, increasing the Earth's albedo still further. Undoubtedly, there were other effects as well (humidity, cloudiness, dust, etc.).

In estimating the benchmark putative global average temperature rise expected in the 21st century due to doubling CO_2 from the pre-industrial level of 280 ppm, climatologists estimate forcings (W/m^2) and climate sensitivity (°C per W/m^2). If one considers only a doubling of CO_2 with no ancillary effects, the accepted forcing is ~3.7 W/m^2. The widely accepted best estimate for the climate sensitivity under this constraint is ~0.3 W/m^2. The product of the two values leads to a temperature rise of $0.3 \times 3.7 \sim 1.2$°C. As this temperature rise takes place, other changes occur in the Earth. Ice sheets shrink, glaciers retreat, humidity and clouds change, aerosols vary, etc. These secondary effects also contribute to temperature change.

According to Hansen and Sato (2011), based on Hansen *et al.* (2008), the transition between the LGM and pre-industrial times can be characterized by two major sources of forcing:

- The diminution of ice sheets from the LGM to recent pre-industrial times resulted in a decrease in the Earth's albedo across the ice sheets at high latitudes, as well as in mountain glaciers. In addition, the rise in sea level moved shorelines inward, converting land to ocean, thereby further decreasing the Earth's albedo. As the climate improved, more biomass and vegetation grew abundantly, further decreasing the Earth's albedo. Hansen and Sato estimated the net effect of these albedo changes as a forcing of 3.5 W/m^2, although they did not specify how they arrived at this estimate. Taylor *et al.* (2000, 2001) provided a range of estimates from 2 W/m^2 to 4 W/m^2. Crucifix (2006) estimated 4 W/m^2.
- The effect of rising greenhouse gases, predominantly CO_2, can be estimated as F_l from Figure 6.8. According to this figure, the effect of CO_2 alone would be about 3 W/m^2, but adding in the effects of rising methane and other trace gases, this would rise to roughly 3.7 W/m^2. (Hansen and Sato used 3 W/m^2 for the effect of all greenhouse gases.)

Their estimates for greenhouse gas forcings were 2.25 W/m^2 for CO_2 (185 ppm → 275 ppm), 0.43 W/m^2 for CH_4 (350 ppb → 675 ppb), and 0.32 W/m^2 for N_2O (200 ppb → 270 ppb) for a total greenhouse gas forcing of 3.0 W/m^2. They also estimated the forcing due to surface changes to be 3.5 W/m^2, but this estimate appears to be rather approximate. Nevertheless, they argued that a total negative forcing of 6.5 W/m^2 would bring about the LGM—pre-industrial transition. They assumed that the ΔT associated with this transition was 5°C. In that case, the Earth's *scientific* climate sensitivity would be:

$$\lambda = \Delta T/(\text{Forcing}) = 5.0/6.5 \sim 0.75\ ^\circ\text{C}/(\text{W/m}^2).$$

Hansen and Sato asserted that:

> "This empirical climate sensitivity incorporates all fast response feedbacks in the real-world climate system, including changes of water vapor, clouds, aerosols, aerosol effects on clouds, and sea ice. In contrast to climate models, which can only approximate the physical processes and may exclude important processes, the empirical result includes all processes that exist in the real world—and the physics is exact."

While this may be true in principle, the large uncertainties in the temperature change and the forcings make the quantitative estimate far from "exact". Obviously, Hansen and Sato, being alarmists, would prefer the climate sensitivity to be as high as possible, since the temperature rise due to increased CO_2 is proportional to the climate sensitivity. Hansen and Sato concluded that a net forcing of 6.5 W/m^2 occurred between the LGM and recent pre-industrial times producing a temperature change of 5°C. Hence the climate sensitivity was derived to be 5/6.5 = 0.75 $\pm$ 0.25 W/m^2. Crucifix (2006) argued that use of the climate sensitivity from LGM analysis to predict 21st-century temperatures due to increased CO_2 levels is not as straightforward as Hansen and Sato claim "for two reasons: (i) the forcing is not known accurately and (ii) the ratio between LGM and CO_2 feedback factors cannot be accurately estimated from current state-of-the-art coupled models".

There are several problems with this calculation. One problem is that the estimate of the forcing due to greenhouse gases appears to be a bit low. According to Figure 6.8, the forcing due to CO_2 is not 2.25 W/m^2 (as claimed), but 3 W/m^2, and the total forcing due to all greenhouse gases is not 3.0 W/m^2 (as claimed), but 3.7 W/m^2.

Another problem is that Hansen and Sato used ΔT=5.0°C, whereas 4.5°C appears to be a better choice.

In addition, Hansen and Sato did not appear to adequately consider the forcing due to high dust levels in the atmosphere during the LGM. One estimate is that dust would produce a forcing of about 1 W/m^2 (Crucifix, 2006). It is also noteworthy that Bielefeld (1997) estimated that, at the height of the last Ice Age (18,000 YBP) global radiation absorption was lower by 7%–10% than it is today. That would indicate a negative forcing of 24 W/m^2 to 34 W/m^2 which is far greater than other estimates. Another major concern is that no consideration was taken of possible changes in humidity or cloudiness.

If we modify Hansen and Sato's estimate by taking the forcing as 8.2 W/m^2 (instead of 6.5 W/m^2), and if we choose ΔT=4.5°C instead of 5.0°C, we obtain:

$$\lambda = \Delta T/(\text{Forcing}) = 4.5/8.2 \sim 0.55\ ^\circ\text{C}/(\text{W/m}^2).$$

Hansen and Sato used their value for the climate sensitivity (0.75 °C per W/m^2) in conjunction with the forcing due to doubling of the CO_2 concentration (from 280 ppm to 560 ppm): 3.7 W/m^2 (see Figure 6.8) to obtain $\Delta T_G \sim 3.0$°C for a doubling of the CO_2 concentration from the pre-industrial value of ~280 ppm. With the lower value of λ, they would have obtained 2.2°C.

Crucifix (2006) provided alternate estimates of the forcings:

- change in sea level and vegetation changes (~4 W/m^2);
- reduction in greenhouse gas concentrations (~2.85 W/m^2);
- other forcings, difficult to quantify, such as increased dust concentration (~1 W/m^2).

This sums to 7.85 W/m^2, but Crucifix added: "There is also a small contribution due to the surface being, on average, more elevated than today" that might bring the total close to the value 8.2 W/m^2 which was previously estimated. Crucifix felt that the value of λ could not be pinned down well, primarily because of uncertainty in ΔT. He attempted to use climate models to bridge this gap, but concluded that "the ratio between LGM and CO_2 feedback factors cannot be accurately estimated from current state-of-the-art coupled models".

Figure 6.44 shows the following:

A = Conditions at the LGM
B = Conditions in pre-industrial times
C = Projection by Hansen and Sato for doubling CO_2 from pre-industrial value.
D = Computer model projection for doubling CO_2 with no feedbacks
E = Present conditions
F = Range of climate model predictions including feedback
Curve 1 = Path we seem to be on
Curve 2 = Transition from LGM to pre-industrial times
Curve 3 = Predicted by climate models

Chylek and Lohmann (2008) (C&L) carried out an independent estimate of climate sensitivity by comparing the LGM to pre-industrial times. They asserted that "One of the uncertainties in the radiative forcing calculation during the LGM to the Holocene transition is the radiative forcing due to increased aerosol optical depth

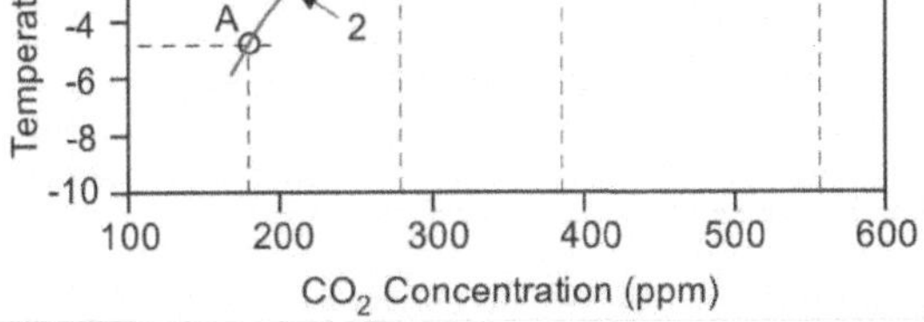

Figure 6.38. Dependence of global average temperature on CO_2 concentration according to several estimates.

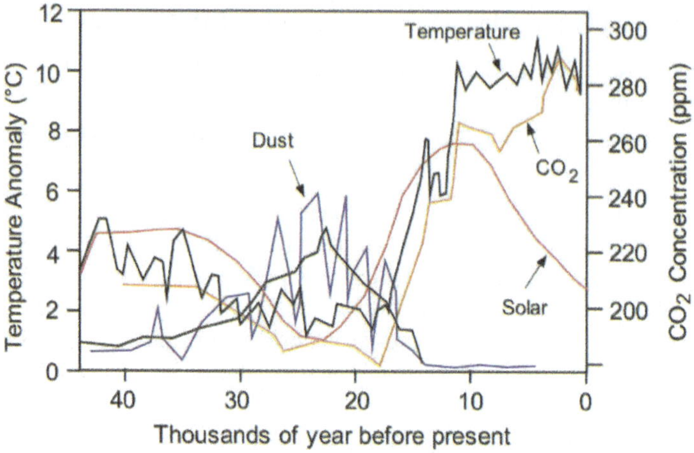

Figure 6.39. Smoothed data from Vostok ice core. The dust and solar scales are arbitrary. The solar curve represents midsummer solar intensity at 65°N (adapted from C&L).

during the peak of the last ice age". In their analysis, they used the LGM to pre-industrial transition and the cooling period between the warm period around 42,000 years before present and the LGM to deduce the change in aerosol radiative forcing and to estimate climate sensitivity. It was assumed that the climate sensitivity was the same for both periods.

C&L utilized data from the Vostok ice core for transitions between two time periods:

(1) warm period around 42,000 years ago → LGM (about 20,000 years ago);
(2) LGM → pre-industrial period (about 200 years ago).

Smoothed data from the Vostok ice core used by C&L are shown in Figure 6.39.

Based on these data, C&L estimated temperature differences and forcing due to greenhouse gases as shown in Table 6.18.

Table 6.18. Parameters used by C&L.

	CO_2 *(ppm)*	CH_4 *(ppb)*	ΔT *(°C)*	*Forcing via GHG* (W/m^2)
42K years ago	209	548		
LGM	182	340		
Pre-industrial	285	667		
42K → LGM			2.16 ± 0.23	0.93
LGM → pre-industrial			4.6 ± 0.5	2.67

The radiative forcing due to the surface albedo changes (extent of ice sheets, sea ice, and snow cover, exposure of a new land in a low-sea-level state, change in surface characteristics, and vegetation cover) for the LGM → pre-industrial transition was estimated to be roughly 3.5 W/m^2, but C&L used a range of values from 3.0 W/m^2 to 4.0 W/m^2.

C&L pointed out that the dust measurements in the Vostok ice core suggested that aerosol concentration differences from 42K to the LGM were about 53/58 as great as aerosol differences from pre-industrial time to the LGM. However, they were not able to attribute forcings to these changes *a priori*. They assumed that the forcing due to aerosols were 58X and 53X for the LGM → pre-industrial, and 42K years ago → LGM transitions, respectively, but X could not be specified *a priori*. In order to estimate the forcing due to aerosols, C&L carried out a comparison of the two transitions, assuming that the relation between ΔT and total forcing was the same for both transitions. Thus, they put:

$$\frac{\Delta T_1}{F_{GHG1} + F_{Alb1} + F_{Alb1}} = \frac{\Delta T_2}{F_{GHG2} + F_{Alb2} + F_{Alb2}}$$

in which transition (1) refers to LGM → pre-industrial, and transition (2) refers to 42K years ago → LGM. Their estimates for ΔT_1 and ΔT_2 and F_{GHG1} and F_{GHG2} are given in Table 6.19. Their estimate for F_{Alb1} was 3.5 W/m^2, but they did not seem to specify what they used for F_{Alb2}. If F_{Alb2} is known, and setting $F_{Aer1} = 58X$ and $F_{Aer2} = 53X$, the above equation provides a means to estimate X. C&L reported that their best estimate for X was 0.056 W/m^2. Working backwards, we may surmise that they must have used $F_{Alb2} = 1.58$ W/m^2. Using the above value for X, they estimated the total forcing for the LGM → pre-industrial to be $2.67 + 3.5 + 58 \times 0.056 = 9.4$ W/m^2, and with $\Delta T_1 = 4.6$°C, the climate sensitivity is $\lambda = 4.6/9.4 \sim 0.5$ °C per W/m^2). This implies that, when CO_2 goes from ~280 ppm to ~560 ppm, the expected temperature rise is 0.5 × 3.7 ~1.8°C. C&L also examined prior glacial to interglacial transitions and, from this, estimated slightly higher values for λ. However, they pointed out:

> "At this time it is not clear whether these higher sensitivities, compared to the climate sensitivity deduced from the LGM to Holocene transition, really reflect higher climate sensitivity at the time of the considered climate transitions or whether they are artifacts due to imperfect ice core data and uncertainties in the used approximations."

The main difference between the calculations of C&L and Hansen and Sato is the much higher values of aerosol forcing used by C&L. As in the case of Hansen and Sato, C&L did not consider changes in humidity or cloudiness.

Hargreaves and Annan (2009) (H&A) wrote a commentary on the paper by C&L. They pointed out (properly) that the data in Figure 6.39 are vacillating and, depending on exactly how one chooses the data points, one can derive different results. They provided two examples. In their first example, they chose to read the temperature curves such that ΔT_2 ~0.8°C instead of the value 2.16°C used by C&L.[34] In this case, however, the dust forcing turns out to be negative—the

[34] They did not actually provide the number 0.8°C but they did provide a graph and that was the value I read from their graph.

implication is that it was less dusty at the LGM. Hargreaves and Annan (2009) seem to imply that this is an equally good interpretation of the Vostok data. However, there are two things wrong with this. One is the choice of temperatures by H&A does not fit the data well in Figure 6.39. But, more importantly, the end result of a negative dust forcing at the LGM is contrary to our physical understanding and suggests that the figures chosen by H&A cannot be correct. In their second example, H&A claimed that they arrived at a dust forcing of 0.9 $\pm$ 1.2 W/m^2, as compared to the estimate by C&L of 58 $\times$ 0.056 = 3.25 W/m^2. This led to an estimate of ΔT_G $\sim$2.5 $\pm$ 0.7°C for a doubling of CO_2 from 280 ppm to 560 ppm. However, H&A did not specify which temperatures they used in this calculation, so it is impossible to reproduce what they did. H&A then extrapolated beyond science by asserting that an estimate ΔT $\sim$2.5 $\pm$ 0.7°C for a doubling of CO_2 from 280 ppm to 560 ppm "does not pose any significant challenge to the widely-held view that climate sensitivity is likely to lie in the range 2–4.5°C [3.25 $\pm$ 1.25]". H&A evidently desired to derive as high a climate sensitivity as they could from glacial–interglacial transitions, and the best they could do was 2.5 $\pm$ 0.7°C—which they said does not pose a challenge to 3.25 $\pm$ 1.25°C. If we consider that C&L (known skeptics) derived a value of ΔT_G = 1.8°C for a doubling of CO_2 from 280 ppm to 560 ppm, and H&A (defenders of the orthodoxy[35]) derived 2.5°C, it seem likely that perhaps the most credible value from this type of analysis is somewhere near 2.1°C.

None of these calculations takes into account potential changes in humidity and cloudiness during these transitions, which are likely to be as large as, or larger than, the forcings that were included.

Kohler *et al.* (2009) also performed an estimate of climate sensitivity based on glacial–interglacial cycles. They said: "Although water vapor is the most important GHG, the following compilation does not consider any changes in water vapor in the past due to missing constraints on its variability". In other words, they more or less said: *Water vapor may be the biggest factor, but since we have no data on it, we will neglect it*! Some of the data used by Kohler *et al.* (2009) are compared with data used by C&L and Hansen and Sato (2011) in Table 6.19.

If one were to simplistically take the result of Kohler *et al.* (2009) that a forcing of 12.43 W/m^2 produces a ΔT_G of 5.8°C, one might conclude that their estimate of climate sensitivity would be $\lambda = 5.8/12.43 = 0.47$°C per W/m^2) that agrees with the result of C&L, although the data are different in both cases. It appears likely that Kohler *et al.* made the most detailed analysis of the forcings, and it seems likely that their estimate of the total forcing (12.4 W/m^2) is likely to be the most reliable. However, great uncertainty remains regarding ΔT_G. If ΔT_G is as small as that estimated by C&L and Shakun and Carlson (2010), namely 4.5°C, the implied climate sensitivity would be $\lambda = 4.5/12.43 = 0.36$°C per W/m^2). This in turn would suggest a $\Delta T_G \sim 1.3$°C for doubling CO_2. However, Kohler *et al.* somehow arrived at a figure of 2.4°C by arguments that are difficult for this writer to comprehend.

[35] One can discern the attitude of these authors toward the orthodoxy regarding ΔT_G for doubling of CO_2 from their other publications. Furthermore, in the cited reference, H&A emphasize that the estimates by the IPCC are inviolable.

Table 6.19. Parameters for analyzing LGM—pre-industrial transitions. Forcings are in W/m^2. Blank elements are not available. Elements with dashes represent items that were not included.

	Chylek & Lohmann (2008)	*Kohler et al. (2009)*	*Hansen & Sato (2011)*
CO_2 forcing	2.4	2.1	2.25
CH_4 forcing	0.27	0.4	0.43
N_2O forcing	–	0.3	0.32
Total GHG forcing	2.67	2.8	3.0
Land cryosphere		4.54	
Land ice		3.17	
Sea ice		0.55	
Snow cover		0.82	
Sea ice		2.13	
Sea ice—north		0.42	
Sea ice—south		1.71	
Vegetation		1.09	
Total albedo	3.5	7.76	3.5
Dust/aerosols	3.2	1.88	
Water vapor, lapse rate, and clouds	–	–	–
Total forcing	9.4	12.43	6.5
ΔT_G (°C)	4.6	5.8[(36)]	5.0

In summary, we have the results shown in Figure 6.40. All of the estimates for the ΔT_G for doubling CO_2 are lower than the consensus value of 3.0°C based on climate models, except for the unmodified result of Hansen and Sato (2011). None of the estimates exceeds the consensus value.

All of these calculations suffer from a lack of understanding of changes in lapse rate, cloudiness, and humidity in the LGM → pre-industrial transition.

It is noteworthy that Hansen and Sato (2011), which came along three years after Chylek and Lohmann (2008) and two years after Kohler *et al.* (2009), did not refer to these prior papers that essentially carried out the same calculation but with different parameters and, in the case of Kohler *et al.*, with much greater detail.

Schwartz *et al.* (2010) carried out an analysis entitled "Why hasn't the Earth warmed as much as expected?" Of course, this depends on who did the expecting, and what they based their expectations on. The greenhouse forcing due to a doubling the CO_2 concentration from pre-industrial times has been estimated to be 3.7 W/m^2, and the forcing for a rise in CO_2 concentration from pre-industrial times to present levels is about 1.8 W/m^2. This increases to about 2.6 W/m^2 if the effect of other greenhouse gases is included. If one adopts the canonical prediction that doubling the CO_2

[36] Kohler *et al.* (2009) emphasized at considerable length that, although reasonable estimates can be made for the ΔT at Antarctica, the value of ΔT_G is far more elusive. They suggested that 5.8°C was perhaps one of the better estimates but emphasized that ΔT_G is not well pinned down.

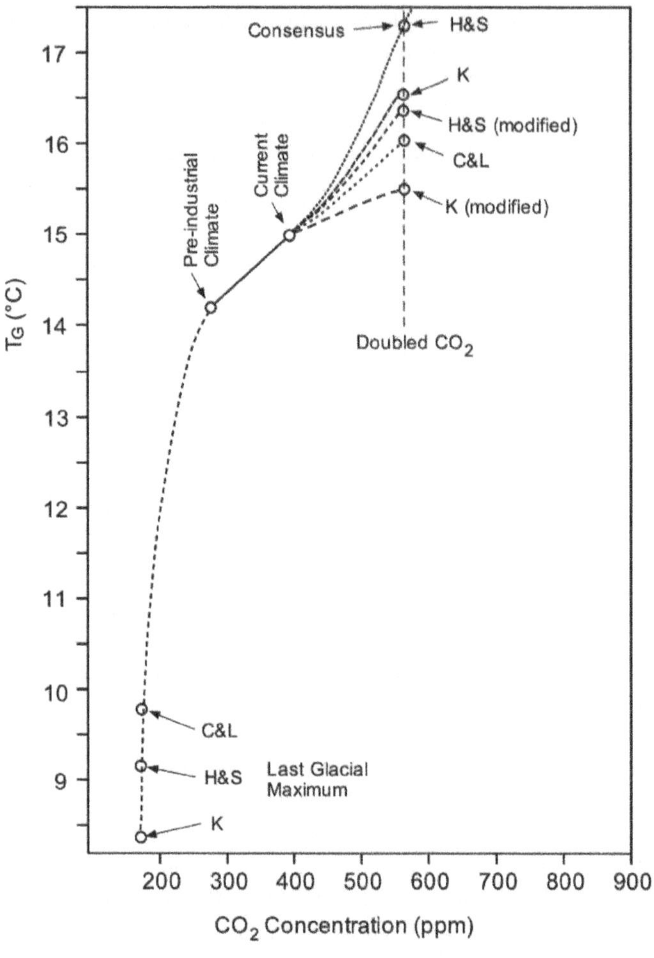

Figure 6.40. Summary of estimates from LGM → pre-industrial period transitions. C&L refers to Chylek and Lohmann (2008). H&S refers to Hansen and Sato (2011). K refers to Kohler *et al.* (2009). Modified values are produced herein as described in the text.

concentration from pre-industrial times will raise the Earth's temperature by ~3°C, then, based on CO_2 alone, the temperature rise from pre-industrial times should have been $(1.8/3.7) \times 3 = 1.5$°C. Based on all greenhouse gases, the rise in temperature would be predicted to be 2.1°C. The actual rise was ~0.8°C. Schwartz *et al.* (2010) considered:

> "... four major factors that might contribute to this discrepancy: (i) natural variation in global temperature over the industrial period, (ii) lack of attainment of equilibrium of the climate system to applied forcings over the industrial period, (iii) current estimates of climate sensitivity being too high, and (iv) countervailing forcings over the industrial period offsetting the warming forcings by incremental greenhouse gases."

They claimed "that relatively little of this warming discrepancy can be attributed to a countervailing natural cooling over this time period or to thermal lag of the climate system response to forcing" and argued " that this discrepancy is therefore due mainly to offsetting forcing by increased concentrations of atmospheric aerosols and/or to climate sensitivity being lower than current estimates; the discrepancy cannot be apportioned between these two causes primarily because of present uncertainty in aerosol forcing". While estimates of aerosol forcing tend to hover in the range –1.0 to –1.2 W/m^2, the uncertainty in these figures is large.

First, they analyzed natural variations over 150-year periods from year 1000 to 1850 to estimate how much natural variation in temperature might be expected to occur over a 150-year period. There are two problems with this. One is that they based the analysis of climate variability on the work by Juckes (2007) that resulted in very small climatic changes that are probably significant underestimates. The other problem is that, after 1850, the Earth came out of the LIA, and the Earth was warming to some extent due to natural causes. Therefore, their conclusion that natural cooling would only account for about 15% of the difference between predicted and observed temperatures is not credible. It seems more likely that there was natural warming, and that warming provides even less warming available to greenhouse gases as a cause of 20th-century warming. This is particularly true for warming that occurred prior to build-up of CO_2 after 1940.

Next, they analyzed lack of attainment of equilibrium of the climate system to applied forcings over the industrial period. Here, the issue is how much heat goes into the oceans vs. how much goes into the atmospheric climate system. After examining the relevant data on ocean heating, they concluded that the best estimate of ocean heating is 0.37 W/m^2, or only about 14% of the full greenhouse gas forcing.

They therefore concluded that most of the discrepancy between the high predicted temperature rise due to greenhouse gas forcing and the comparatively lower temperature rise of the 20th century was due to some combination of (3) overestimation of climate sensitivity by climate models and/or (4) countervailing forcings over the industrial period offsetting the warming forcings by incremental greenhouse gases.

Schwartz *et al.* (2010) then discussed estimates of climate sensitivity. They appraised the paleological approach, saying that "Although this paleological approach is thought by many to give reliable estimates of the Earth's climate sensitivity, the uncertainties are substantial" and they quoted estimates of uncertainty in carrying out such calculations (although their paper preceded Hansen and Sato (2011) and these authors claimed smaller uncertainty in their estimates). They were "further concerned" about "the applicability of the climate sensitivity inferred from such a large forcing and temperature change to the smaller anthropogenic perturbations associated with response to forcing by incremental GHGs". They also discussed estimates of climate sensitivity based on aftermath of volcanic eruptions but dismissed this as too inaccurate. They reviewed an empirical approach to "determine climate sensitivity from the known forcing and the increase in temperature over the industrial period". Such calculations suffer from uncertainties in parameters, leading to a wide swath in predicted climate sensitivities.

Hansen (2008a) claimed that "estimates of climate sensitivity based on the last 100 years of climate change are practically worthless, because we do not know the net climate forcing". Hence, those who estimate climate sensitivity from recent data claim the paleological approach is faulty, while those who use the paleological approach claim the approach based on recent data is faulty. It seems that none of these estimates can be trusted.

Schwartz *et al.* (2010) claimed that, if the effect of aerosols can be pinned down accurately, the remainder should provide a good estimate of climate sensitivity. However, because their estimate of natural variability is not credible, the climate sensitivity will be difficult to narrow down, even with a better estimate of the aerosol effect.

6.5.3 The early Pliocene: three to five million years ago

According to Haywood and Williams (2005):

> "Although the geography of our planet looked very similar to that of today three million years ago, the world was undergoing momentous changes everywhere, from the Americas to Tibet. At about this time, animals from South America first started to colonize North America, indicating that the Isthmus of Panama had finally risen above sea level. Along this trans-continental highway of migration, the armadillo was amongst the animals that migrated north, whilst dogs, cats, bears and many other animals headed south. In Africa, the spread of Savannah vegetation and retreat of forest habitats may have encouraged our primate ancestors to come down from the trees, colonizing the open plains of the rift valleys of east Africa and undergoing an evolutionary radiation into a number of 'graceful' and 'robust' australopithecines. Their fossil remains are found in modern Ethiopia and Tanzania. In Asia, the continued collision of India with the Eurasian land mass pushed the Himalayas still higher, intensifying the Asian monsoon. From the Americas, ancestral horses about the size of ponies migrated west along the Aleutian archipelago into Asia and Europe. In the Antarctic, the ice sheets and glaciers were not static, but fluctuated in size, influencing the global climate and sea level. In the oceans too, there were changes. The emerging Isthmus of Panama finally cut off the exit route for Atlantic water into the Pacific, and this contributed to a saltier Atlantic Ocean which may have encouraged the warm water current known as the 'Gulf Stream' in the North Atlantic to flow vigorously.
>
> "For much of the past three million years our planet's global climate has been cooler than today, particularly during the Ice Ages of the Pleistocene. However, during the mid-Pliocene there is strong evidence for a period lasting 300,000 years, when the global climate was warmer than it is today. On Antarctica, along the Trans-Antarctic Mountains, rocks of probable mid-Pliocene age yield fossils of southern beech plants suggesting that parts of Antarctica were ice-free. The question is, what caused this globally warmer climate and what relevance does it have to our understanding of current global warming? There are two key

questions. Did higher levels of greenhouse gases in the atmosphere cause Pliocene warmth? Or was this aided and abetted by other factors such as more intense ocean circulation transporting heat from the tropics to the higher latitudes?"

Schneider and Schneider (2010) reviewed the work of several investigators regarding the relationship of CO_2 concentration to climate in the early Pliocene (three to five million years ago). There is considerable evidence that, about three million years ago, the temperature was several degrees warmer than it is today. Haywood and Valdes (2004) provide numerous references to previous work on "sea surface temperatures (SSTs) reconstructed from planktonic foraminifera, ostracods, siliceous microfossil records, diatom records, terrestrial vegetation records and numerous records of higher than present sea levels". These investigators used "the alkenone CO_2 method to reconstruct Pleistocene Pliocene pCO_2 histories from six ocean localities". There was a wide diversity in inferred CO_2 concentration at the six sites, which lends some doubt as to their accuracy. Nevertheless, there were some common features. All of the sites indicated a significant decrease in CO_2 concentration from five million years ago toward ~one million years ago. According to Pagani *et al.* (2010), CO_2 concentrations were between 365 ppm and 415 ppm about 4.5 million years ago when temperatures were 3°C–4°C warmer than pre-industrial values. Seki *et al.* (2010) arrived at even lower CO_2 concentrations. If these estimates are correct, CO_2 concentrations were comparable to those of today, yet the Earth was considerably warmer. Alarmists such as Pagani *et al.* (2010) who assumed that CO_2 is the primary forcing for climate change, concluded that the longer-term Earth system climate sensitivity is much higher than the fast feedback climate sensitivity (using the political definition of climate sensitivity). In fact, Pagani *et al.* (2010) suggested values as high as 9.6°C per CO_2 doubling. It is not totally clear what they meant by this, but apparently they believe that, if we hold the CO_2 concentration at ~560 ppm and wait long enough, T_G will gradually rise by up to 9.6°C. Haywood and Valdes (2004) pointed out:

> "Numerous proposals exist within the literature to account for the relative climatic warmth of the middle Pliocene. These include increased concentrations of CO_2, enhanced thermohaline circulation, a more vigorous flow of surface ocean gyres, alterations in the outflow of Antarctic deep water, and changes in the elevations of mountain chains. All of these explanations have weaknesses when examined in detail and there may have been numerous contributing factors to middle Pliocene warmth. For example, it has been suggested that the warmth was generated through a combination of enhanced atmospheric CO_2 and an increase in thermohaline circulation."

Their modeling led them to conclude that the main forcing that produced higher temperatures during the Pliocene was reduced land ice cover, "but with strong positive feedbacks from clouds." Lunt *et al.* (2010) analyzed the Pliocene with a climate model and concluded that the Earth system climate sensitivity is 30–50% greater than the fast feedback sensitivity, which is considerably less extreme than the result of Pagani *et al.* By contrast, Brierley *et al.* (2009) claimed that "a vast poleward expansion of the ocean tropical warm pool" was responsible for Pliocene warmth. However, Haywood and Williams (2005) concluded from their study:

> "Our results suggest that mid-Pliocene warming was caused by more carbon dioxide in the atmosphere combined with climate feedbacks associated with smaller ice sheets. Since the pattern of sea temperature change reconstructed from alkenones, and predicted by our climate model, is not consistent with that produced through changes in ocean circulation/ocean heat transport, we also conclude that there was no major change in thermohaline circulation at that time."

Those who are devoted to the orthodoxy that CO_2 is the sole arbiter of climate will attribute the warming of the Pliocene entirely to CO_2 and will therefore conclude that, if we wait long enough at a fixed CO_2 concentration of ~395 ppm, the Earth will slowly approach Pliocene conditions and that, if we hold CO_2 at 560 ppm and wait long enough, T_G may rise by as much as 9.6°C (see Figure 6.41). Despite the many publications on the subject, what seems to be missing is this: We need a picture of the Earth at Pliocene conditions, particularly the extent of the ice sheets, as well as the ocean circulation, the degree of cloudiness, and the plant coverage of the Earth. Forcings need to be estimated for these factors (and more).

6.5.4 The past ~20 million years

Pearson and Palmer (2000) described "the boron-isotope ($\delta^{11}B$) approach to pCO_2 estimation that relies on the fact that a rise in the atmospheric concentration will cause more CO_2 to be dissolved in the surface ocean, causing a reduction in its pH". They were:

> "... able to estimate the pH of ancient sea water by measuring the boron-isotope composition of calcium carbonate ($\delta^{11}B_{CC}$) precipitated from it. This is because boron in aqueous solution occurs as two species, $B(OH)_3$ and $B(OH)_{4-}$, between which the equilibrium is strongly pH-dependent over the natural acidity range of sea water. Furthermore, there is a pronounced isotopic fractionation between the species ... so that the ($\delta^{11}B$) of each species is highly dependent on pH. Because boron incorporation into marine carbonates is predominantly from $B(OH)_{4-}$, ($\delta^{11}B_{CC}$) is a sensitive pH indicator. The pH of seawater is governed by the carbonate equilibria, such that, for a given pH value, it is possible to calculate the aqueous CO_2 concentration and thereby make quantitative estimates of atmospheric pCO_2. The pH and aqueous CO_2 concentration of the surface ocean vary spatially because of factors such as deep-water upwelling, local productivity regimes, and freshwater inflows. To arrive at pH estimates that most closely reflect atmospheric pCO_2, it is necessary to measure the ($\delta^{11}B$) of carbonates that were precipitated far from coastal influences and sources of upwelling. The ideal setting is in the low-latitude gyre systems, where a mixed layer of warm, low density, seawater in contact with the atmosphere generally overlies colder deep waters with little intermixing. Such environments support

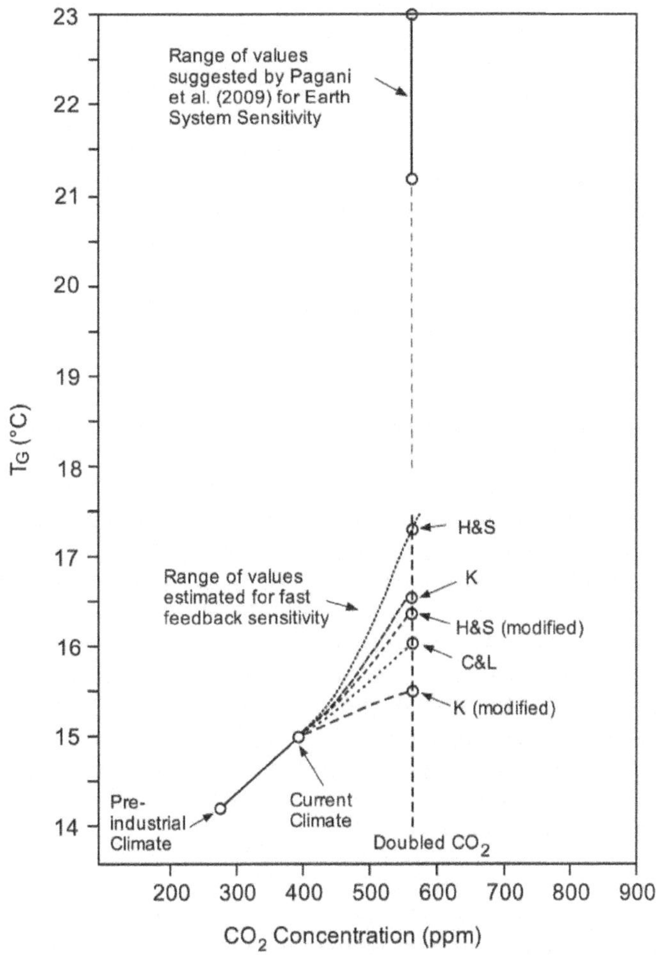

Figure 6.41. Range of values suggested by Pagani *et al.* (2010) for Earth System Sensitivity.

abundant planktonic foraminifera (a group of microscopic protists) that secrete calcite ($CaCO_3$) shells. The shells fall to the seafloor, from which a record of upper-ocean pH of many millions of years can be obtained."

Pearson and Palmer (2000) "analysed the (δ^{11}B) of monospecific sample splits of surface mixed-layer dwelling foraminifera from 32 sediment samples from the open tropical Pacific, spanning the past 60 Myr, augmenting data from six other previously studied samples". Their results for the past 25 million years are shown in Figure 6.42. Any putative relationship between CO_2 and climate is difficult to discern.

The use of proxies and climate models to infer relationships between climate and CO_2 concentration has been carried out by a number of investigators over various

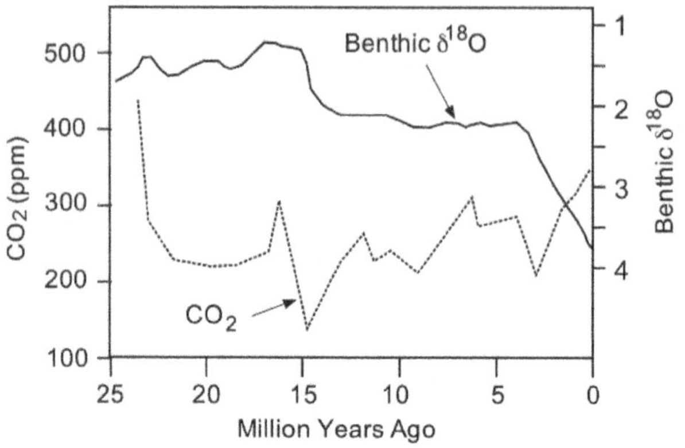

Figure 6.42. CO_2 concentration and Bethic $\delta^{18}O$ (inverse measure of temperature) over the past 25 million years (adapted from Pearson and Palmer, 2000).

timescales ranging up to hundreds of millions of years. In general, the results require distant extrapolations from short, recent calibration periods. Typically, there is much disagreement between different data sets, and considerable scatter within any particular set of data.

Van de Wal *et al.* (2011) developed an inverse modeling technique in an attempt to reconstruct a continuous high-resolution CO_2 record over the past 20 million years, by decomposing the global deep-sea benthic ^{18}O record into a mutually consistent temperature and sea-level record, using a set of 1-D models of the major NH and SH ice sheets. They compared their modeled temperature record to ice-core and proxy-derived CO_2 data to reconstruct a continuous CO_2 record over the past 20 million years. They reported:

> "Results show a gradual decline from 450 ppm around 15 million years ago to 280 ppm for pre-industrial conditions, coinciding with a gradual cooling of the Northern Hemisphere land temperatures by approximately 12°C."

There are several problems with this result. One problem is that the authors never defined exactly what they meant by "Northern Hemisphere". They used this term six times in the paper but never defined it. Normally, the "Northern Hemisphere" is defined as the entire region of the Earth north of the equator including the tropics north of the equator. However, it appears possible that, when the authors use the term "Northern Hemisphere" they might mean the land area north of 60°N. Furthermore, the authors claim that the ratio $\Delta T_{NH}/\Delta T_{global} = 2.5$, which seems incredibly high, unless the authors mean by "Northern Hemisphere" only the land area north of 60°N. This needs to be clarified. The authors plotted their estimate of ΔT_{NH} vs. CO_2 concentration as a steep upward line as shown in Figure 6.43. According to this result, when the CO_2 concentration in the atmosphere goes from 278 ppm to 390 ppm, the change in ΔT_{NH} is 14.8°C. Furthermore, if CO_2 is doubled to 556 ppm, Figure 6.43 would indicate that ΔT_{NH} would be 27.5°C. These estimates seem to be too high. The slope of the line in Figure 6.43 appears to be

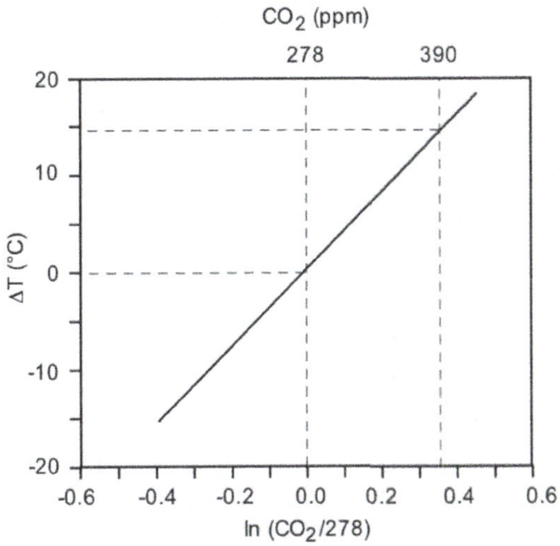

Figure 6.43. Dependence of ΔT_{NH} on CO_2 according to Van de Wal *et al.* (2011).

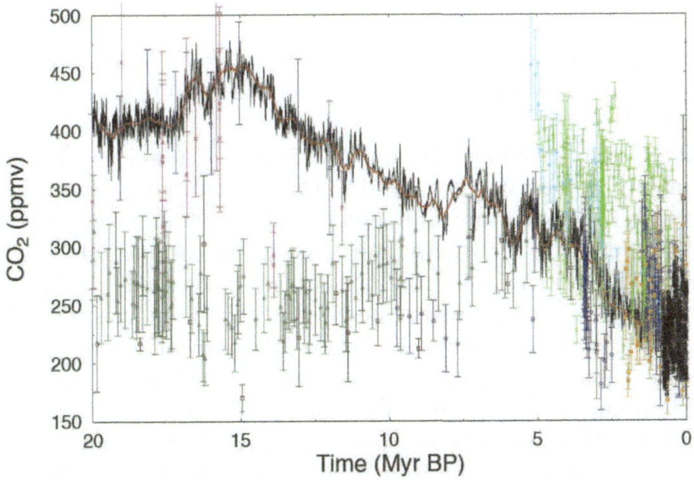

Figure 6.44. Estimates of CO_2 concentration over the past 20 million years (adapted from Kohler, 2011).

much too high. We have already reached 395 ppm of CO_2, and the measured temperatures are nowhere near what this model predicts.

Kohler (2011) carried out his own analysis partly built upon the work of van de Wal *et al.* His graph of estimates of CO_2 concentration over the past 20 million years is shown in Figure 6.44. The eight estimates listed in the upper right portion were provided by van de Wal *et al.* (2011) while Kohler's estimate is shown in black with a red 400-kyr running mean. Over the most recent 2.7 million years, CO_2 concentrations oscillated with the Ice Age–interglacial cycles. Note that Kohler's

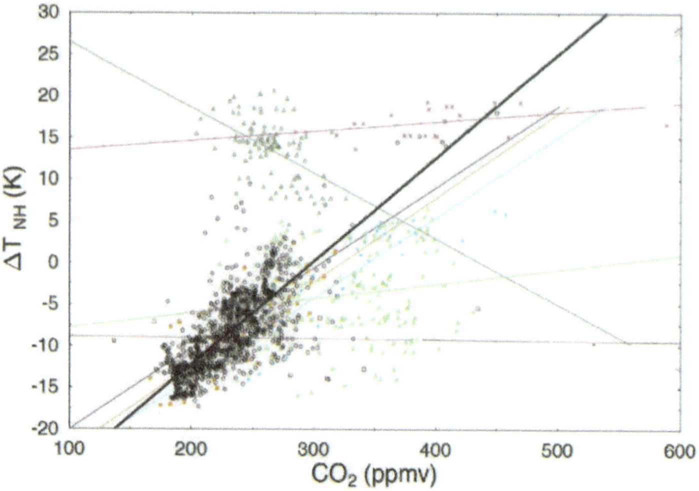

Figure 6.45. Dependence of (ΔT_{NH}) on CO_2 concentration as presented by Kohler (2011) based on van de Wal *et al.* (2011).

estimate for the early Pliocene was about 300 ppm, which is based on Seki *et al.* (2010), whereas Pagani *et al.* (2010) concluded that "CO_2 concentrations were between 365 and 415 ppm".

Kohler (2011) adapted a figure from van de Wal *et al.* (2011) as shown in Figure 6.45. Unfortunately, van de Wal *et al.* (2011) were not entirely clear on the meaning of "NH" in regard to temperature, although they did mention incidentally that "the reconstructed temperatures are strictly only valid in the continental areas where ice sheets develop in the NH (ΔT_{NH}), being mid- to sub-polar (NH) latitudes, implying that they are therefore not necessarily representative for the entire globe (ΔT_G)". In a personal communication to this writer, van de Wal indicated that (ΔT_{NH}) ~ 2.5 (ΔT_G). Evidently, some of the data were discounted, and the very wide scatter was not considered an impediment to drawing conclusions. The final result is the black line in Figure 6.45. This line passes through (ΔT_{NH})=0 at 300 ppm CO_2 and has slope 0.125 °C/ppm. Thus, in going from a pre-industrial level of ~280 ppm to the present level of ~395 ppm, van de Wal *et al.* (2011) would predict that (ΔT_{NH}) ~115 × 0.125 = 14.4°C and (ΔT_G) ~ 14.4/2.5 = 5.8°C. There are three possibilities: (1i) one possibility is that, if we hold CO_2 at 395 ppm and wait long enough, (ΔT_G) will approach 5.8°C; (2) the second possibility is that the climate is determined by factors other than CO_2; and (3) the third possibility is that the results of van de Wal *et al.* (2011) are inaccurate. This writer leans to the second and third possibilities. It seems likely that the variation of CO_2 and T_G over the past 20 million years has not been pinned down very accurately, but, even if it has, the putative slope of the black line in Figure 6.45 is based on the assumption that CO_2 is the sole determinant of climate change. Yet, the variability of CO_2 over the past 20 million years was moderate, and attributing all climate changes over that period to CO_2 leads to a

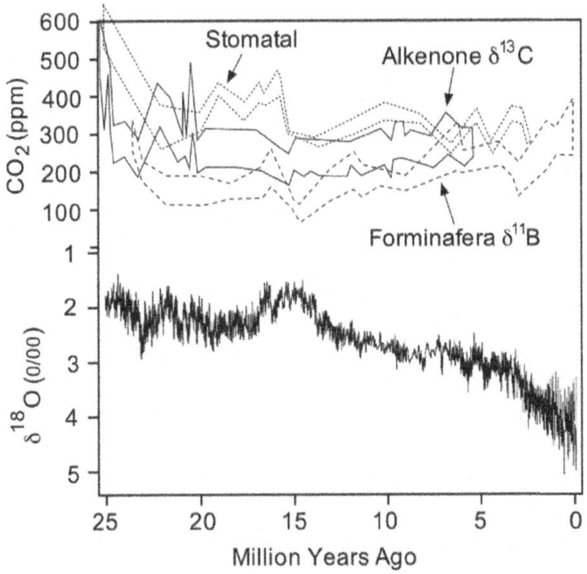

Figure 6.46. Comparison of $\delta^{18}O$ (an inverse measure of temperature) with CO_2 concentration over 25 million years. (adapted from Foster *et al.*, 2009).

severe overestimate of the importance of CO_2. It seems likely that there are more things than CO_2 in Heaven and Earth than are dreamt of in the philosophy of paleoclimatologists.

Foster *et al.* (2009) showed that, while the period from 25 to 5 million years ago was "a period of relative warmth" and only Antarctica was glaciated, "paradoxically" CO_2 concentrations were comparable to "pre-industrial values or even lower". "Records of ice rafted debris and the oxygen isotope composition of benthic foraminifera suggest that at several times over the last 25 million years substantial amounts of continental ice did build up in the Northern Hemisphere but none of these led to sustained glaciation." Foster *et al.* (2009) pointed out that the "accepted paradigm"[38] requires CO_2 to vary in unison with global temperature. However, they emphasized: "Reconstructing the concentration of atmospheric CO_2 beyond the reach of the Quaternary ice cores is, however, a notoriously difficult task. Nonetheless there is a growing consensus that pCO_2 did decline over the Cenozoic, but not exactly sympathetically with climate as the paradigm suggests" (see Figure 6.46). They also said: "This is likely because the pCO_2 records are not perfect and other phenomenon such as ocean circulation, continental configuration, and surface albedo (vegetation and ice coverage) also influence climate." They suggested that other geological factors could change the threshold for NH glaciation to occur. One

[37] The "accepted paradigm" is an almost religious belief that only CO_2 concentrations control climate change, and paleoclimatologists often interpret data and models with considerable bias toward that belief.

such factor is uplift of the North American Cordillera that "would have resulted in significant cooling of the Northern North American Continent. ... This suggests uplift of the North American Cordillera in the Late Miocene may have played an important role in priming the climate for the intensification of Northern Hemisphere glaciation in the Late Pliocene".

Tripati *et al.* (2009) said:

> "Although there is speculation about the role of the carbon cycle in driving these well-studied climate changes, there is surprisingly little direct evidence to support a coupling between pCO_2 and climate prior to the ice core record (i.e., before 0.8 Ma). Estimates of pCO_2 have been generated using several methods including the difference in the carbon isotopic composition ($\delta^{13}C$) of alkenones and co-occurring foraminifera, $\delta^{13}C$ of bulk carbon and of pedogenic carbonates, boron isotope composition ($\delta^{11}B$) of foraminifera, stomatal density on fossil leaves, and carbon cycle modeling. Most reconstructions support a decoupling between pCO_2 and climate during the Miocene and Late Pliocene, although very little pCO_2 data are available and the few published proxy reconstructions yield conflicting results. In addition, few pCO_2 proxies have replicated the ice core data of the past 0.8 Ma."

Their goal was "to test the hypothesis that CO_2 and climate were closely coupled across ... major transitions". They used boron/calcium ratios in foraminifera to estimate pCO_2 during major climate transitions of the past 20 million years. Their results are shown in Figure 6.47. They concluded:

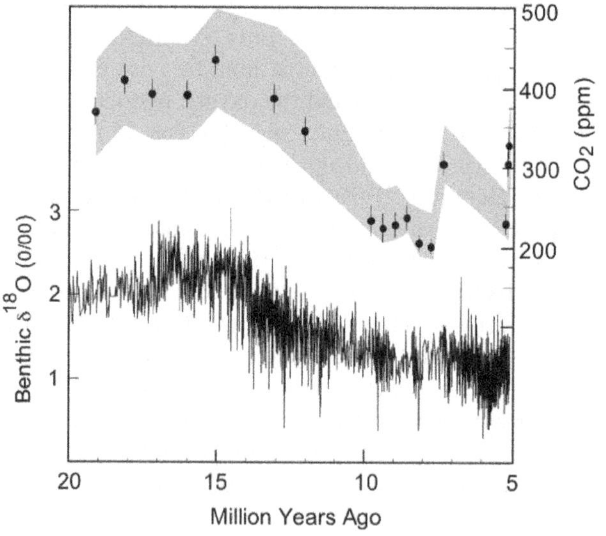

Figure 6.47. Estimated CO_2 vs. temperature over 20 million years (adapted from Tripati, *et al.*, 2009).

"These results show that changes in pCO_2 and climate have been coupled during major glacial transitions of the past 20 myr, ... supporting the hypothesis that greenhouse gas forcing was an important modulator of climate over this interval via direct and indirect effects."

However, they also said:

"During the Middle Miocene, when temperatures were ~3° to 6°C warmer and sea level was 25 to 40 meters higher than at present, pCO_2 appears to have been similar to modern levels."

One is left with this inference: assuming the results of Tripati are accurate, there appears to be a general tendency for pCO_2 to be higher during warmer periods, but as Foster *et al.* (2009) said the variation is "not exactly sympathetically with climate as the paradigm suggests". Nevertheless, as before, one is left with three possible interpretations similar to those reached in regard to van de Wal *et al.* (2011): (1) one possibility is that, if we hold CO_2 at 395 ppm and wait long enough, (ΔT_G) will approach 3°C to 6°C; (2) the second possibility is that the climate is determined by factors other than CO_2; and (3) the third possibility is that the results of Tripati *et al.* (2009) are inaccurate. This writer leans toward the second and third possibilities.

6.5.5 Initiation of Antarctic glaciation 34–33 million years ago

Liu *et al.* (2009) said:

"About 34 million years ago, Earth's climate shifted from a relatively ice-free world to one with glacial conditions on Antarctica characterized by substantial ice sheets. ... The abrupt shift to glacial conditions ... ~33.7 million years ago (Ma) is characterized by a ~ +1.5 per mil (‰) change in oxygen isotopic ($\delta^{18}O$) values of benthic foraminifera (1–3) in ~300,000 years, which is indicative of continental ice accumulation and high-latitude cooling. ... Proposed causes for this fundamental change in Earth's climate state include changes in ocean circulation due to the opening of Southern Ocean gateways, a decrease in atmospheric CO_2, and a minimum in solar insolation."

Liu *et al.* (2009) reported SST changes, which were determined from the alkenone unsaturation index and the tetrather index from 11 globally dispersed ocean localities. They estimated benthic cooling of 3°C to 5°C during the transition at 33.7 Ma.

Pearson *et al.* (2009) pointed out that the "principal geochemical fingerprint of the Eocene–Oligocene transition (EOT) is an approximately 11.5% 'shift' towards more positive values of the oxygen isotope ratio of deep-sea carbonates between 34.0 and 33.5 million years ago, the last part of which is a prominent 'step' of about 10.5% at about 33.5 Myr ago". They used "boron isotope ($\delta^{11}B$) analysis of the carbonate shells of upper-ocean planktonic foraminifera to establish palaeo-surface ocean pH" from which they inferred the dissolved CO_2 concentration, $[CO_2]_{aq}$, which they assumed was in approximate equilibrium with pCO_2 atm. The main

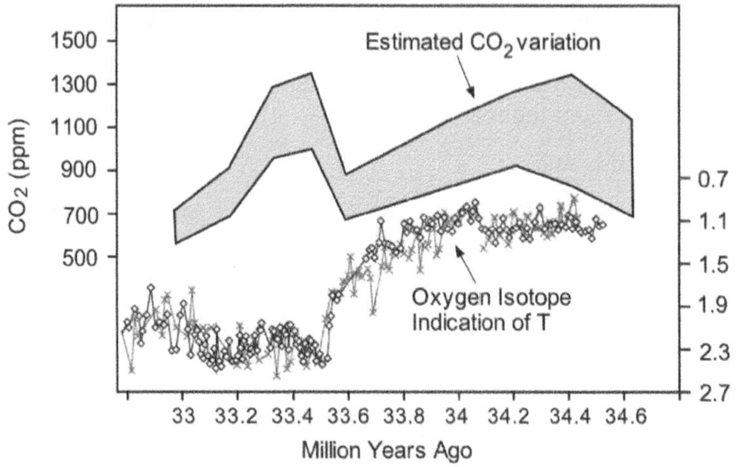

Figure 6.48. Comparison of CO_2 and temperature proxy across the Eocene–Oligocene boundary (adapted from Pearson *et al.*, 2009).

uncertainties were stated to be "the value for the boron isotope ratio of seawater ($\delta^{11}B_{sw}$), sea surface temperature, and the requirement to estimate one other parameter of the carbonate system (for example, total alkalinity)". Their results are shown in Figure 6.48.

Pearson *et al.* (2009) interpreted their results to:

> "... strongly suggest that the primary cause [for the transition to Antarctic glaciation] was a diminishing greenhouse effect. Although greenhouse gases other than CO_2 (for which there are no proxies) may have contributed, changing pCO_2 atm is likely to have had the greatest forcing. Ours is the first proxy-based study to confirm a substantial pCO_2 decline during the climate transition. We also find a sharp pCO_2 increase after maximum ice growth as the global carbon cycle adjusted to the presence of a large ice cap and there was a nonlinear hysteresis effect as the ice cap withstood this transient pCO_2 rise. This study reaffirms the links between cryosphere development and atmospheric carbon dioxide levels at the largest and most important climatic tipping point of the last 65 million years."

They also suggested that the threshold for initiation of Antarctic glaciation is in the range 700 ppm–850 ppm.

These conclusions seem to be influenced by adherence to the "accepted paradigm" that CO_2 is the main factor in climate change. While there was indeed a moderate decrease in CO_2 as the Earth approached the Eocene–Oligocene boundary, the so-called "hysteresis effect" cannot be brushed away so easily. CO_2 levels popped up to well above the threshold while temperatures remained low. While CO_2 is clearly a factor in climate change, once again, the comment by Foster *et al.* (2009) that CO_2 variations are "not exactly sympathetically with climate as the paradigm suggests" is appropriate.

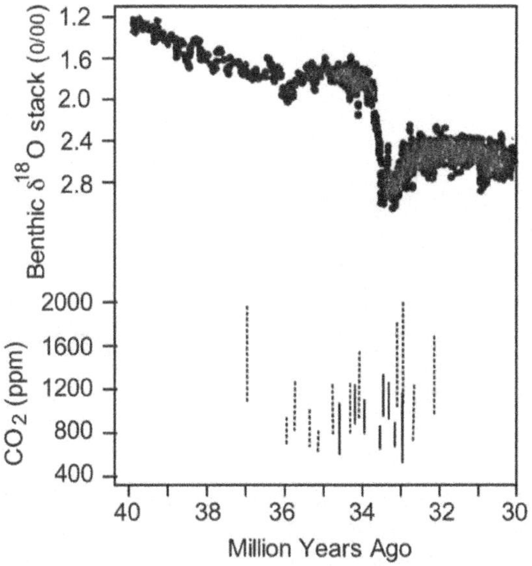

Figure 6.49. Comparison of CO_2 and temperature proxy across the Eocene–Oligocene boundary (adapted from Peters *et al.*, 2010).

Peters *et al.* (2010) used "an unusually well exposed coastal incised river-valley complex in the Western Desert of Egypt to show that eustatic sea level fell and then rose by ~40 m, 2 million years prior to establishment of a permanent Antarctic Ice Sheet".

They concluded that:

> "This fall in sea level is associated with a positive oxygen isotope excursion that records buildup of an Antarctic Ice Sheet with a volume ~70% of the present-day East Antarctic Ice Sheet. Both the sea-level fall and subsequent rise were coincident with a transient oscillation in atmospheric CO_2 concentration down to ~750 ppm, which climate models indicate may be a threshold for Southern Hemisphere glaciation. Because many of the carbon emission scenarios for the coming century predict that atmospheric CO_2 will rise above this same 750 ppm threshold, our results suggest that global climate could transition to a state not unlike the Late Eocene, when a large permanent Antarctic Ice Sheet was not sustainable."

The result presented by Peters *et al.* (2010) is shown in Figure 6.49. How in the world they reached their detailed conclusions from this mess of CO_2 data is beyond the ability of this writer to comprehend. As is the case in most paleoclimatological studies, they drew a dollar's worth of conclusions from a penny's worth of data.

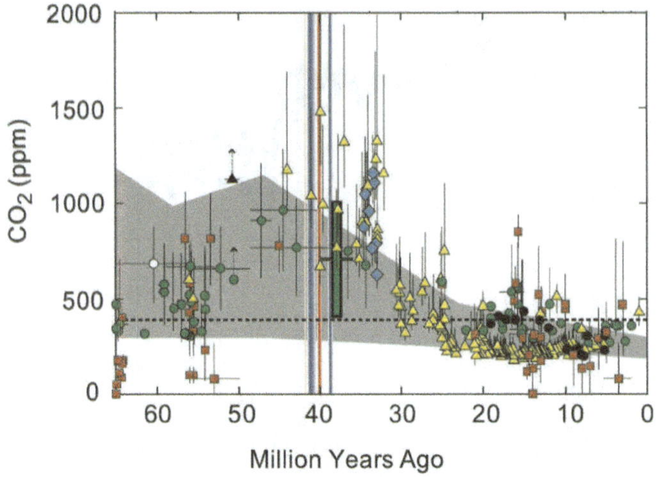

Figure 6.50. Estimates of CO_2 over the past 65 million years as provided by Doria *et al.* (2011). The gray area is modeled using "GEOCARB". The colored data points are measured by various techniques referred to by Doria *et al.* (2011). The vertical green bar was measured by Doria *et al.* (2011). The vertical red and blue lines are periods of relative warmth and cold.

6.5.6 Peak warming around 40 million years ago

Bohaty *et al.* (2009) described the so-called "Middle Eocene Climatic Optimum (MECO) as an enigmatic warming event that represents an abrupt reversal in long-term cooling through the Eocene". The event was centered on 40 million years ago with a duration of about half a million years. Their measurements of $\delta^{18}O$ at numerous sites "indicated that warming during the MECO event was globally ubiquitous". They found gradual warming prior to the event, and rather rapid cooling after the event. They also found a significant decrease in the mass accumulation rate of deep-sea carbonates during this period at some (but not all) sites. They therefore concluded that the event was tied to an increase in CO_2 concentration, although they had no direct evidence of this.

Doria *et al.* (2011) "estimated the concentration of atmospheric CO_2 during this critical interval using stomatal indices of fossil Metasequoia needles from ten levels in an exceptionally well-preserved core from the Giraffe kimberlite locality in northwestern Canada". They summarized estimates of CO_2 concentration as shown in Figure 6.50.

The connection between CO_2 and climate remains fuzzy to this writer based on Figure 6.50.

6.5.7 60 to 40 million years ago

Pearson and Palmer (2000) provided the results shown in Figure 6.51. There is a general tendency for higher CO_2 concentrations to be associated with higher T_G, although direct one-to-one correspondence is lacking.

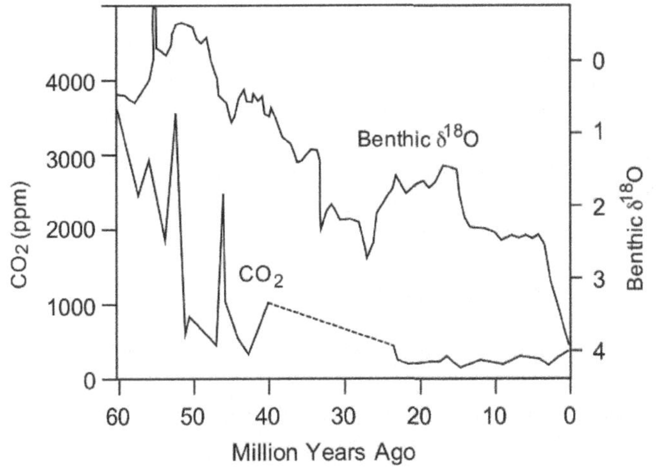

Figure 6.51. CO_2 concentration and Bethic $\delta^{18}O$ (inverse measure of temperature) over the past 60 million years (adapted from Pearson and Palmer, 2000).

Others have determined that CO_2 concentrations were relatively high about 50 million years ago. For example, Lowenstein and Demicco (2006) estimated that CO_2 was ~1000 ppm–3000 ppm about 50 million years ago. Pagani *et al.* (2005) pointed out that "the relation between the partial pressure of atmospheric carbon dioxide (pCO_2) and Paleogene climate is poorly resolved". They "used stable carbon isotopic values of di-unsaturated alkenones extracted from deep sea cores to reconstruct pCO_2 from the middle Eocene to the late Oligocene (~45 to 25 million years ago)". Their results indicated that pCO_2 ranged between 1,000 and 1,500 parts per million in the middle to late Eocene, then decreased in several steps during the Oligocene, and reached modern levels by the latest Oligocene.

Edwards *et al.* (2010) provided the result shown in Figure 6.52, which is similar in some ways to that of Pearson and Palmer (2000) in that the warm period from 60 to 40 million years ago is associated with generally higher values of the CO_2 concentration. However, Figure 6.52 shows very considerable scatter and, furthermore, there isn't much variation in CO_2 while temperatures changed considerably over the past 20 million years. These results seem to suggest that, on balance, the warmest climates are associated with higher CO_2 concentrations, but the wide scatter in estimates of CO_2 concentration preclude detailed comparisons between CO_2 and climate.

Kent and Muttoni (2008) suggested that:

> "... India's northward flight and collision with Asia was a major driver of atmospheric CO_2 concentration (pCO_2) and thus global climate in the late Cretaceous and Cenozoic. Subduction of Tethyan oceanic crust with a carpet of carbonate-rich pelagic sediments deposited during transit beneath the high productivity equatorial belt resulted in a component flux of CO_2 delivery to the atmosphere that maintained high pCO_2 levels and warm climate until the

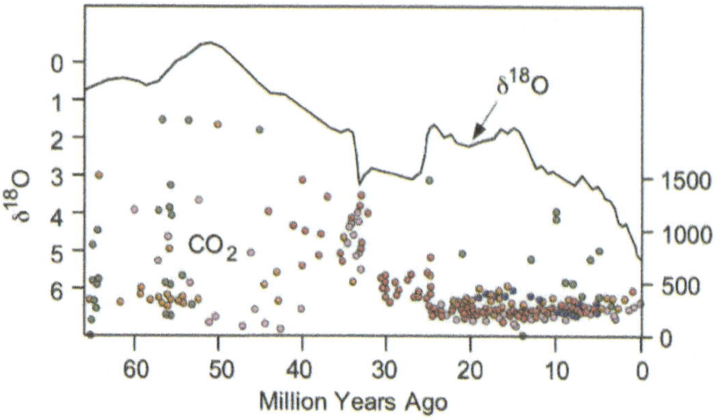

Figure 6.52. CO_2 concentration and Bethic $\delta^{18}O$ (inverse measure of temperature) over the past 60 million years (adapted from Edwards *et al.*, 2010).

decarbonation factory waned with the collision of Greater India with Asia at ~50 Ma, closely coinciding with the Early Eocene climatic optimum. At about this time, the India continent and the highly weatherable Deccan Traps drifted into the equatorial humid belt where uptake of CO_2 by silicate weathering further perturbed the equilibrium towards progressively lower pCO_2 levels and a cooling trend that eventually triggered the expansion of Antarctic ice sheets in the earliest Oligocene, even if global seafloor production rates remained steady."

This conclusion appears to be based on the supposition that climate is controlled by CO_2 and, for any time period in which climate changed, a geological model for a change in CO_2 must be developed to explain why the climate changed. The arguments in this study seem plausible, but it does not appear to this writer as clear-cut as Kent and Muttoni (2008) seem to think.

Cui *et al.* (2011) discussed the transient global-warming event known as the Palaeocene–Eocene Thermal Maximum that occurred about 55.9 million years ago. "The warming was accompanied by a rapid shift in the isotopic signature of sedimentary carbonates, suggesting that the event was triggered by a massive release of carbon to the ocean–atmosphere system." They claimed that "the source, rate of emission and total amount of carbon involved remain poorly constrained". They used "an expanded marine sedimentary section from Spitsbergen to reconstruct the carbon isotope excursion as recorded in marine organic matter [and found that] the total magnitude of the carbon isotope excursion in the ocean–atmosphere system was about 4". They used a climate model to infer that the peak rate of carbon addition was slower than the present rate of carbon emissions, although emissions were extended over a longer period.

Ruddiman (2010) wrote a "Perspective" article in *Science*: "A paleoclimatic enigma". In this paper, he emphasized that the Earth's climate had been cooling from pole to pole for 50 million years prior to the onset of alternating Ice Ages and interglacials about 2.7 million years ago. During this 50-million-year period:

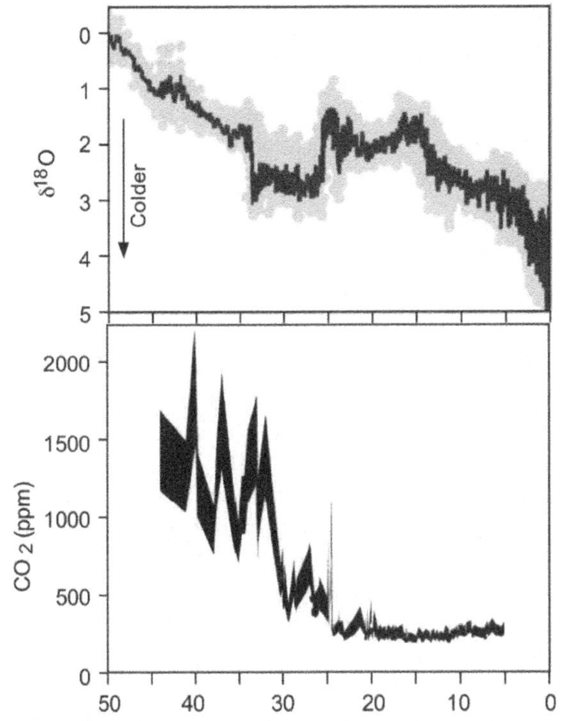

Figure 6.53. CO_2 concentration and Bethic $\delta^{18}O$ (inverse measure of temperature) over the past 60 million years. (adapted from Ruddiman, 2010).

"Arctic forests changed from frost-intolerant evergreens to temperate deciduous trees to cold-adapted spruce and larch and eventually to tundra. Antarctica was mostly ice-free until 34 million years ago; glaciers of varying size then existed on the continent until 14 million years ago, after which a large and relatively stable ice sheet formed. The gradual shift toward heavier $\delta^{18}O$ values in $CaCO_3$ shells of sea-floor foraminifera since 50 million years ago documents a combined deep-ocean cooling and increase in Antarctic ice" (see Figure 6.53).

Ruddiman (2010) went on to say:

"Until a decade ago, most paleoclimate modelers attributed this ongoing bipolar cooling to a gradual reduction in the CO_2 concentration in the atmosphere. This inferred CO_2 decrease was ascribed to a combination of reduced volcanic CO_2 input to the ocean and atmosphere because of a slowing rate of sea-floor spreading and increased CO_2 removal by enhanced chemical weathering in tectonically uplifting regions like Tibet.

"In a broad sense, this long-term CO_2 decrease provided some support for the idea that CO_2 has been the long-term driver of global cooling, but a closer look revealed major problems. By 22 million years ago, the alkenone and boron isotope data both showed that estimated CO_2 concentrations were already within the range typical of the glacial cycles of the past 800,000 years. If CO_2 concentrations of 180 to 300 ppm have played an integral role in allowing glacial cycles in the past 800,000 years, why did comparably low CO_2 values 22 million years ago not initiate glacial cycles? And if the average CO_2 trend has not fallen in the past 22 million years, what caused the substantial bipolar cooling during that time? Other proposed causes seem insufficient to explain large-scale cooling. Gradual plate motions and falling sea level have extended the northern margins of circum-Arctic continents into cooler near-polar latitudes, but models suggested that these factors were not enough to explain the major cooling observed."

The problem as Ruddiman explained is "persistently low CO_2 concentrations estimated for the past 22 million years" during which the climate cooled substantially. Although Ruddiman did not consider this, it appears that there are inconsistencies between CO_2 and climate prior to 22 million years ago as well, although to some extent, higher CO_2 was roughly associated with warmer climates. Ruddiman said: "Paleoclimatologists were left with three possibilities." These were:

(1) they "might have overlooked something crucial";
(2) effects other than CO_2 "could have had a much stronger effect than thought";
(3) "The proxy methods used to reconstruct CO_2 concentrations prior to ice-core records could be invalid."

Ruddiman discounted the first possibility and leaned toward the third. The argument seems to come down to this. Ruddiman doubts that there is a missing factor, and known factors do not seem to explain the variability of climate. Therefore, the only thing that he can think of that could be the cause of long-term climate change is variable CO_2. Since the data on CO_2 do not agree with this precept, the data must be wrong. This type of argument has been used a number of times recently in climatology. If the data do not agree with theory, throw the data out! However that seems antithetical to the scientific method.

One interesting event during this era was the so-called Paleocene-Eocene thermal maximum (PETM) that occurred about 55 million years ago. There is good evidence that T_G rose by at least several degrees (some estimates range from 4°C to 9°C) in as little as 10 to 30,000 years. It is widely believed that this could only result from a sudden massive input of greenhouse gases. However, Zeebe (2011):

> "... estimated the size of the PETM carbon input based on sediment records of deep-sea carbonate dissolution and showed that the subsequent rise in atmospheric CO_2 alone was insufficient to explain the full amplitude of global warming. We concluded that in addition to direct CO_2 forcing, other processes must have caused a portion of the PETM warming. ... Our study showed that there were processes in addition to CO_2 forcing that caused part of the warming, not that CO_2 was irrelevant. The processes are as yet unidentified—some may have operated independently, others as a response or feedback to the CO_2 release."

In contrast to Zeebe's indication of uncertainty regarding the PETM, Kump (2011) asserted that he understands the whole process. The initial release of CO_2 provided warming that added CH_4 to amplify the effects of CO_2. In fact, Kump (2011) provided a detailed description of the Earth during the PETM. Most of this seems to be subjective cloth woven from invisible thread. The methane hydrate hypothesis was discussed by Higgins and Schrag (2006), who concluded that analysis of the PETM leads to "a high climate sensitivity". Pagani *et al.* (2006) concluded that "the PETM either resulted from an enormous input of CO_2 that currently defies a mechanistic explanation, or climate sensitivity to CO_2 was extremely high".

As Royer *et al.* (2011) pointed out, "the PETM is considered a paleo-analog of present day climate change in terms of rate and magnitude of carbon release", although, as Kump (2011) emphasized, the annual release of carbon during the PETM was far less than today's, but it was sustained over a much longer time.

6.5.8 One hundred to 300 million years ago

Royer *et al.* (2011) compared crude estimates of CO_2 concentration to estimates of benthic $\delta^{18}O$ and tropical SSTs over the time range from 125 million years ago to 50 million years ago. The CO_2 and SST data show considerable scatter. There was a warm period at around 55 million years ago but it does not seem to have been accompanied by higher CO_2. In this regard, Royer *et al.* (2011) chose to ignore multiple CO_2 measurements near 55 million years ago that were low and, instead, accepted one outlier measurement that was four times higher. From this, they derived a high sensitivity of T_G to CO_2 concentration. This result does not seem credible to this writer.

6.5.9 Estimates of climate sensitivity based on CO_2 and climate in the Phanerozoic Eon

6.5.9.1 Introduction

During the Phanerozoic Eon (the past ~540 million years), the Earth experienced significant changes. These included redistribution of continents via continental drift, the emergence of vascular plants driving up oxygen content in the atmosphere, changing CO_2 concentrations (as high as 20 times current levels at some periods), and many other changes, as discussed by Berner (2004). One particular time period, the so-called Permo-Carboniferous period between about 330 and 280 million years ago, was marked by extensive world glaciation, low CO_2 levels, and high oxygen content (30%–35%). In addition, the brightness of the Sun increased with time across this eon.

As with almost every area of climatology, the data on the Phanerozoic climate and CO_2 concentrations are sparse and noisy, and the interpretation of the data in terms of climatological parameters requires complex models and a number of unverifiable assumptions. Various investigators have arrived at different interpretations regarding the connection between CO_2 concentrations and climate change during the Phanerozoic Eon. Some have concluded that changing CO_2 concentration was the main factor producing long-term climate change. Others claim that the effect of CO_2 was secondary and galactic cosmic ray variability was the important factor.

6.5.9.2 Climate during the Phanerozoic Eon

Veizer (2005) pointed out that "in the Phanerozoic, some organisms secreted their shells as the mineral calcite ($CaCO_3$), which often preserves the original oxygen isotope ratio, and this, in turn, reflects the ambient seawater temperature". In earlier work, Veizer and co-workers "generated a large database of several thousand well-preserved calcitic shells that cover this entire 545 million years timespan. Such

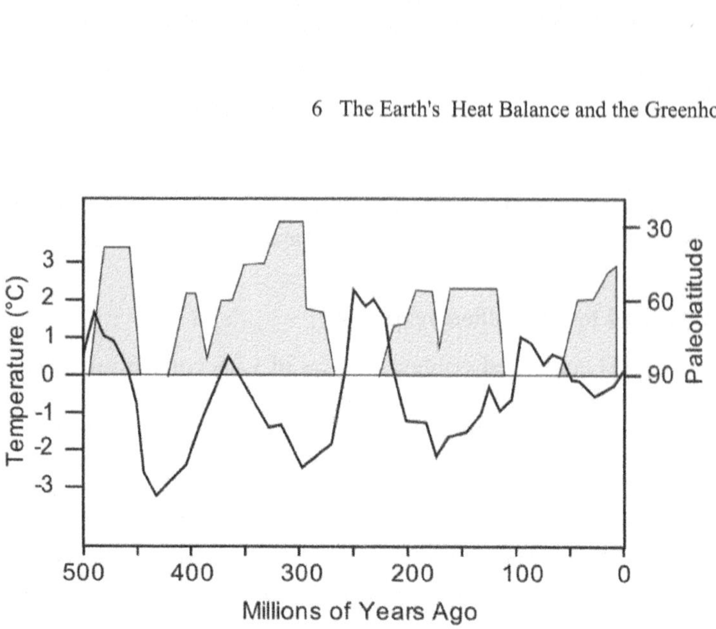

Figure 6.54. Estimated variations of tropical seawater temperatures during the Phanerozoic Eon (heavy line). Extent of glaciation (paleolatitude) shown as shaded areas (adapted from Veizer *et al.*, 2000; Veizer, 2005).

detrended isotope data correlate well with the climatic history of the planet, with tropical sea surface temperatures fluctuating by perhaps 5 to 9° C between the apexes of icehouse and greenhouse times, respectively".

Veizer (2005) indicated that "the record of climate variations during the Phanerozoic shows intervals of tens of millions of years duration characterized by predominantly colder or predominantly warmer episodes, called icehouses and greenhouses, respectively. Superimposed on these are higher order climate oscillations, such as the episodic waning and waxing of ice sheets". Veizer's estimates for Phanerozoic climates are shown in Figure 6.54.

Royer *et al.* (2004) corrected estimates of SST during the Phanerozoic Eon due to changes in pH of the oceans induced by changes in the CO_2 level of the atmosphere and changes in Ca concentrations and calcium carbonate saturation state in seawater. Their result is shown in Figure 6.55, in which the paleo SSTs are greatly increased when CO_2 concentrations are higher.

Subsequently, Shaviv and Veizer commented on the paper by Royer *et al.* (2004) and Royer *et al.* replied to their comment. According to Shaviv and Veizer:

> "The analysis of Royer *et al.* (2004) assumes an unrealistically high pH correction. First, it neglects the ice-volume effect, which changes the relation between $\delta^{18}O$ and ΔT. Second, this large pH correction implies high temperatures for seawater even during times of extensive glaciations. Moreover, the analysis of Royer *et al.* (2004) consists of bootstrapping, by introducing a correction to ΔT that is an implicit function of $R(CO_2)$. It is then not surprising that a correlation between ΔT and $R(CO_2)$ is obtained. This would be the case irrespective of the $R(CO_2)$ model utilized. A proper analysis, which avoids this bootstrapping and considers a more realistic pH correction, shows that the global temperature sensitivity to CO_2 is still relatively small. In summary, while

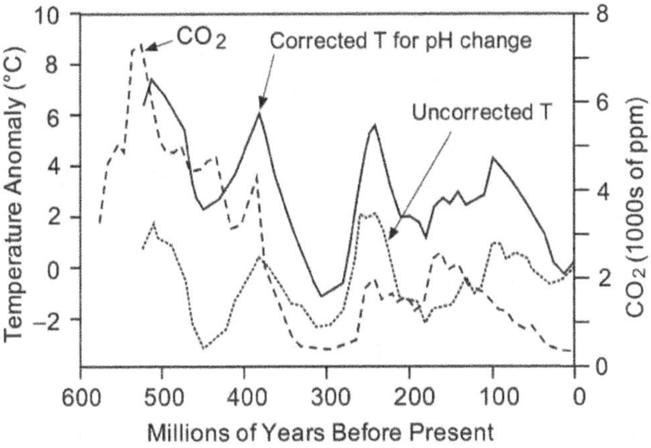

Figure 6.55. Corrected changes in tropical sea temperatures due to change in pH from changing CO_2 concentration (adapted from Royer *et al.*, 2004).

we acknowledge that the proposition of Royer *et al.* (2004) has some merit and likely will result in some modification of the $\delta^{18}O$ signal, the cosmic ray flux still remains the primary climate driver for any realistic pH correction. Even for the scenario that entirely disregards the ice-volume effect, the impact of cosmic ray flux would still be at par with that of CO_2."

$R(CO_2)$ is the ratio of CO_2 concentration to that prevailing in the pre-industrial period (280 ppm) ($R(CO_2) = CO_2/280$). Royer *et al.* disputed these conclusions. They argued that the claim of prolonged cold intervals during the periods (460–400 Ma) and (220–120 Ma), while $R(CO_2)$ was high are unjustified. They also disputed the validity of the ice-volume effect, particularly during periods with no glaciation. They concluded that:

"The correspondence between the Phanerozoic records of atmospheric CO_2 and glacial sediments, and the revision of the $\delta^{18}O$ paleo-temperature record toward values better matching the glacial sediment record, strongly implicate CO_2 as a primary driver of climate over these timescales. Cosmic ray flux is likely only of second-order significance."

Over roughly the same time period of the Shaviv/Veizer–Royer *et al.* controversy, Wallmann (2004) developed a box model that included pH corrections. He began with this introduction:

"Proxy data and box modeling demonstrate that pCO_2 has oscillated over the Phanerozoic due to changing rates of mantle degassing, weathering, organic carbon burial and carbonate turnover. The simulated changes in pCO_2 correspond roughly to paleoclimatic reconstructions supporting the view that Phanerozoic climate change has been driven mainly by changes in atmospheric pCO_2. A different picture emerges from the evaluation of $\delta^{18}O$ values in marine carbonate fossils. These data show regular oscillations with a shorter period indicating changes in surface temperatures at low latitudes that are consistent with some paleoclimatic reconstructions but not with surface temperatures derived from pCO_2 modeling. This difference may be regarded as evidence for the decoupling of pCO_2 and climate evolution. The latter view is supported by recent modeling of galactic cosmic radiation over the last billion years. These new data show a surprisingly strong correlation with $\delta^{18}O$-based temperature reconstructions suggesting that climate change has been driven mainly by cosmic radiation. Nevertheless, temperatures calculated from the marine $\delta^{18}O$ record are met with skepticism because the extremely low Jurassic surface temperatures derived from this proxy are not consistent with other observations."

Wallmann (2004) provided the estimate of the average global surface temperature shown in Figure 6.56.

Several Internet sites claim that Veizer *et al.* updated their global average Phanerozoic temperatures in 2000 based on oxygen isotope data on line in 2004 at *www.science.uottawa.ca/geology/isotope_data/*. However, that website is defunct. Nevertheless, Ziegler presented a graph that is claimed to be this update, at the following website *http://climaterealists.com/index.php?id=6680* (see Figure 6.57. This figure has been used by a number of websites and encyclopedias but it is not clear what the ultimate source is.

There is some considerable variance between various rough estimates of the Phanerozoic climate (see Figure 6.58). Nevertheless, the following salient points seem to be generally agreed upon:

(1) For much of the Phanerozoic Eon, global average temperatures were perhaps as high as 25°C as compared to present-day temperatures of about 14°C, showing that for much of the Protoerzoic Eon, the Earth was a veritable hothouse of warmth.

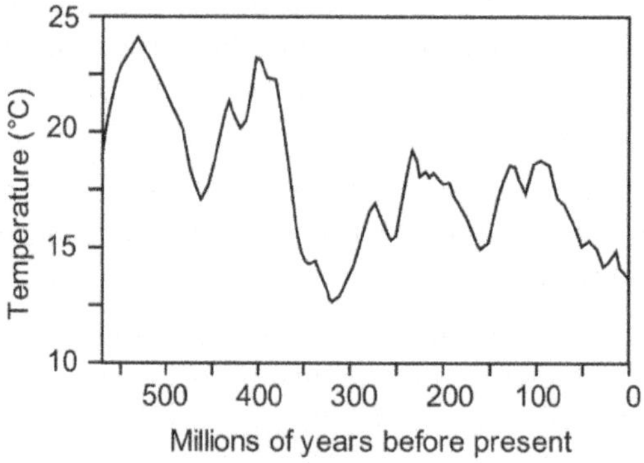

Figure 6.56. Estimate of average global temperature during the Phanerozoic eon (adapted from Wallmann, 2004).

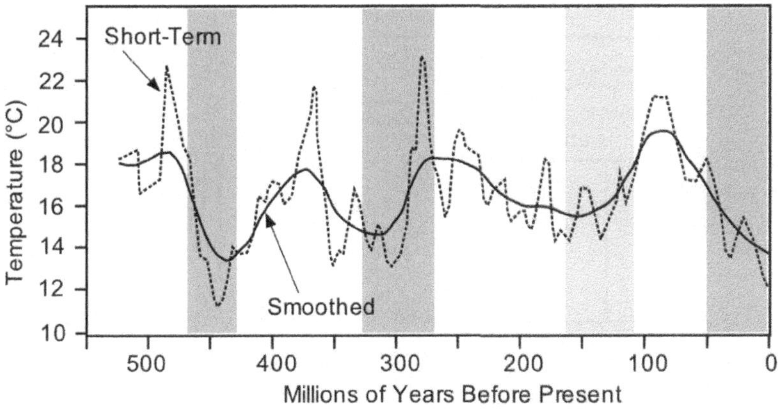

Figure 6.57. Temperatures derived from $\delta^{18}O$ values of calcitic shells for the Phanerozoic (adapted from Ziegler). Darker shading represents very cold glacial periods. Light shading represents a cool period.

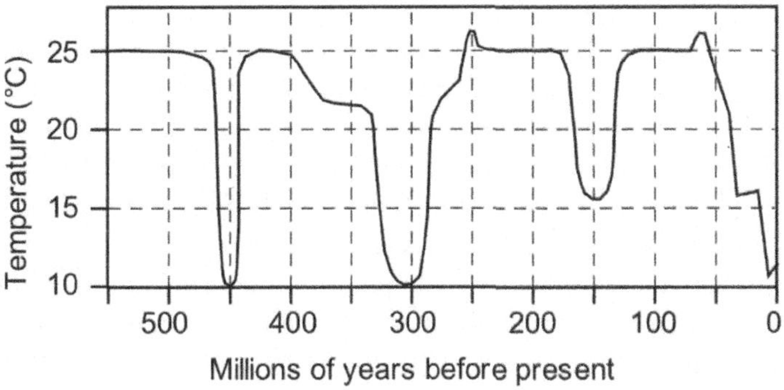

Figure 6.58. Estimate of global average temperature for the Phanerozoic (adapted from Scotese, 2002).

(2) There were two deep glacial cold periods embedded in the Phanerozoic Eon, during which evidence exists that glaciation may have reached latitudes down to 30°, and global temperatures dropped to perhaps as low as 10°C. These occurred over the rough ranges: 470–440 and 330–280 million years ago. The glacial period from 330 to 280 million years ago was the deepest and is referred to as the "Permo-Carboniferous period".

(3) There was an additional cold period centered on 150 million years ago when global average temperatures dropped by about 10°C to perhaps 15°C.

(4) Starting around 50 million years ago, the Earth entered into a period that was mainly a cooling trend.

6.5.9.3 CO_2 variability during the Phanerozoic Eon

In this Section, we discuss the evidence from the Phanerozoic Eon (the past 540

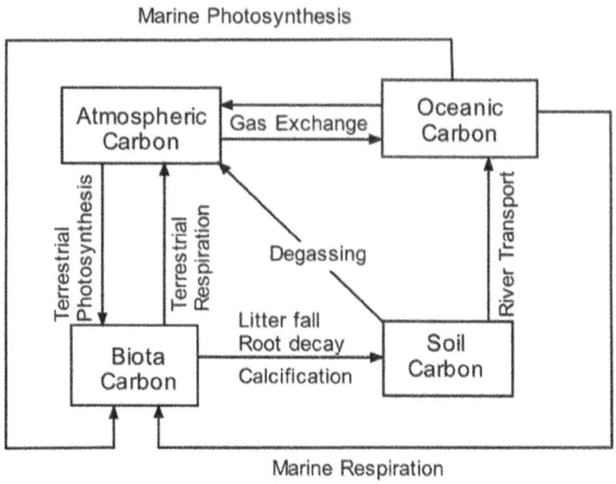

Figure 6.59. The short-term carbon cycle (adapted from Berner, 2004).

million years) regarding CO_2 concentrations. We begin with a brief summary of results provided by Berner (2004), which provides a good overview and introduction, and follow that with data from other sources. The "short-term carbon cycle," is described by Berner (2004) as shown in Figure 6.59. The word "short-term" is used for characteristic times for transferring carbon between reservoirs range from days to tens of thousands of years:

> "Carbon dioxide is taken up via photosynthesis by green plants on the continents or phytoplankton in the ocean. On land carbon is transferred to soils by the dropping of leaves, root growth, and respiration, the death of plants, and the development of soil biota. Land herbivores eat the plants, and carnivores eat the herbivores. In the oceans the phytoplankton are eaten by zooplankton that are in turn eaten by larger and larger organisms. The plants, plankton, and animals respire CO_2. Upon death the plants and animals are decomposed by microorganisms with the ultimate production of CO_2. Carbon dioxide is exchanged between the oceans and atmosphere, and dissolved organic matter is carried in solution by rivers from soils to the sea."

As the short-term cycle proceeds, concentrations of the two principal greenhouse gases, CO_2 and CH_4, can change as a result of perturbations of the cycle, resulting global warming and cooling over centuries and many millennia.

Over longer periods of time (millions of years), additional processes can add or remove CO_2. Because there is more than 1,000 times more carbon in rocks than there is in the oceans, atmosphere, biosphere, and soils combined, carbon transfers to and from rocks can result in significant changes in atmospheric CO_2 over long time periods. Berner (2004) discussed the processes whereby carbon is exchanged with rocks. Two opposing processes are involved. CO_2 is stored in rocks as calcium carbonate. Decarbonization via volcanism, metamorphism, and diagenesis releases

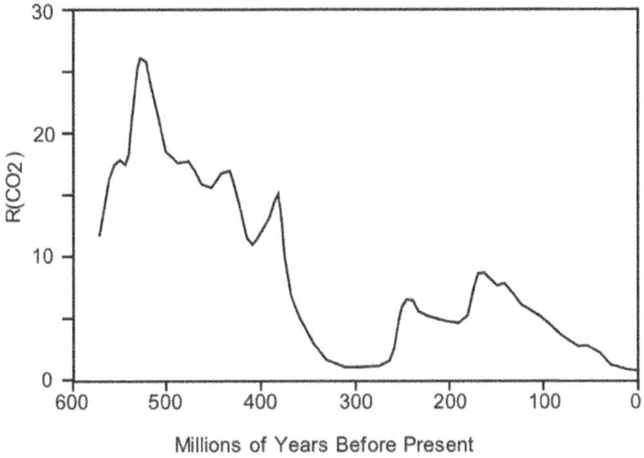

Figure 6.60. Plot of R(CO_2) versus time based on geological models (adapted from Berner, 2004).

CO_2 to the atmosphere while producing calcium ailicate. Berner (2004) described how these cycles operated over the past 550 million years. The details are extensive, and well beyond the scope of this review.

Using geological models, estimates have been made of the concentration of CO_2 during the Phanerozoic Eon, as shown for example in Figure 6.60. According to Berner (2004):

> "The most dramatic feature of the curve is the large drop in CO_2 occurring in the mid-Paleozoic (400–300 Ma). This drop is due mainly to a combination of changes brought about by the rise of large vascular land plants. The plants both accelerated weathering and provided biologically resistant organic remains for burial in sediments, causing a drop in CO_2."

Proxies have been used to estimate CO_2 levels during the Phanerozoic Eon. According to Berner (2004):

> "Methods include determining (1) the $\delta^{13}C$ of carbonates in paleosols; (2) the stomatal density of fossil leaves; (3) the degree of fractionation of carbon isotopes of specific compounds secreted by phytoplankton and preserved in sedimentary rocks; and (4) the boron isotopic composition of marine carbonate fossils. Each of the methods has its own problems, but if certain precautions are taken, they provide reasonable estimates of ancient CO_2 levels."

In addition, various proxies have yielded a curve of R(CO_2) vs. time as shown in Figure 6.61.

According to Berner (2004):

> "There are problems with all of the methods of CO_2 estimation, and the [R(CO_2)] curves are not intended to be used as an accurate CO_2 measure (as is

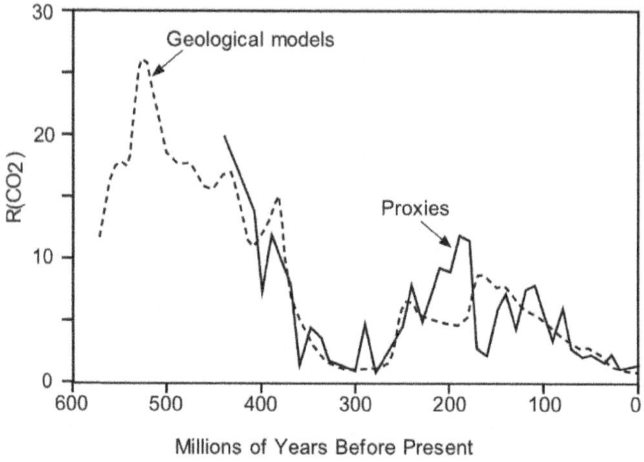

Figure 6.61. Plot of $R(CO_2)$ versus time based on proxies, compared with geological models (adapted from Berner, 2004).

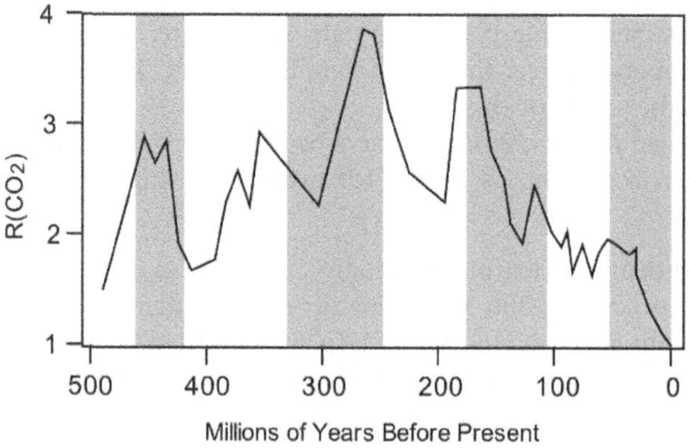

Figure 6.62. $R(CO_2)$ estimated by Rothman (2002). Gray areas are periods attributed to "relatively cool climates".

sometimes mistakenly done), but rather as a suggestion of how CO_2 has changed over the Phanerozoic. New advances ... will undoubtedly cause modifications. ..."

Rothman (2002) used the difference between the $\delta^{13}C$ of bulk organic matter and calcium carbonate ($\delta^{13}C$) to calculate the value of atmospheric CO_2 concentration during the Phanerozoic as shown in Figure 6.62. Rothman's estimates for $R(CO_2)$ are much smaller than those of Berner (2004). Rothman acknowedged this difference but felt that his estimates were justified. However, Berner criticized Rothman's approach, pointing out that "the calibrations based on the study of marine plankton do not apply generally to bulk material. Also, it has been shown that fractionation of carbon isotopes by plant-derived organic matter is not a simple function of CO_2, but rather a strong function of atmospheric O_2 that varies with time. ... These considerations indicate that the simple use of $\delta^{13}C$ to derive CO_2 values over the Phanerozoic is an inappropriate approach to the problem of deducing paleo-CO_2".

6.5.9.4 Comparison of Phanerozoic climate with CO_2 concentrations

Crowley and Berner (2001) said:

> "Geologists have long known that on time scales of tens of millions of years, intervals of continental glaciation were interspersed with times of little or no ice. The magnitude of warmth during these warm intervals is impressive. [About] 65 to 145 million years ago (Ma)], duck-billed dinosaurs [sometimes] roamed the northern slope of Alaska. Deep and bottom waters of the ocean, now near freezing, could reach a balmy 15°C. In the 1980s, a convergence of results from paleoclimatic data and geochemical and climate models suggested that such long-term variations in climate were strongly influenced by natural variations in the carbon dioxide (CO_2) content of the atmosphere. Lately, some geochemical results have raised concerns about the validity of this conclusion. CO_2 concentrations over the past 65 million years appear to have reached low levels well before the most recent phase (the past 3 million years) of Northern Hemisphere glaciation. This is especially true for times of elevated temperatures at about 50 to 60 Ma and 16 Ma, when CO_2 was apparently low. A study spanning the Phanerozoic also suggests some decoupling between times of predicted high CO_2 and some climate indices. In light of these results, it is important to reevaluate the validity of the assumed CO_2-climate link. Here we address this issue by comparing estimates of Phanerozoic CO_2 variations and net radiative forcing with the continental glaciation record and low-latitude temperature estimates."

Crowley and Berner (2001) then went on to compare the best available data on Phanerozoic temperatures from oxygen isotopic composition of fossils with levels of CO_2 based on geological modeling and proxy data (see Figure 6.63). The forcing represents a combination of two things. One is the effect of variable CO_2 due to the greenhouse effect. The other is the fact that the solar intensity of the Sun increased by about 6% during the Phanerozoic Eon. Early in the Phanerozoic, with the solar intensity 6% lower than at present, the solar forcing would have been about -14 W/m^2. This was counteracted by greenhouse forcing from CO_2 with R(CO_2) reaching values higher than 20. On balance, the forcing due to these two factors is shown as the middle curve in Figure 6.63. If CO_2 was the main driver of climate change, the tropical sea temperature should follow the forcing curve. For gray area *B* in Figure 6.69, there is a perfect consonance between low CO_2 concentration and low forcing. However, for gray areas *A* and *C*, the forcings are strongly positive but temperatures were low.

Crowley and Berner (2001) pointed out that:

> "There is a major discrepancy during the period 120 to 220 Ma between cold low-latitude temperatures and high levels of CO_2. ... The overall low

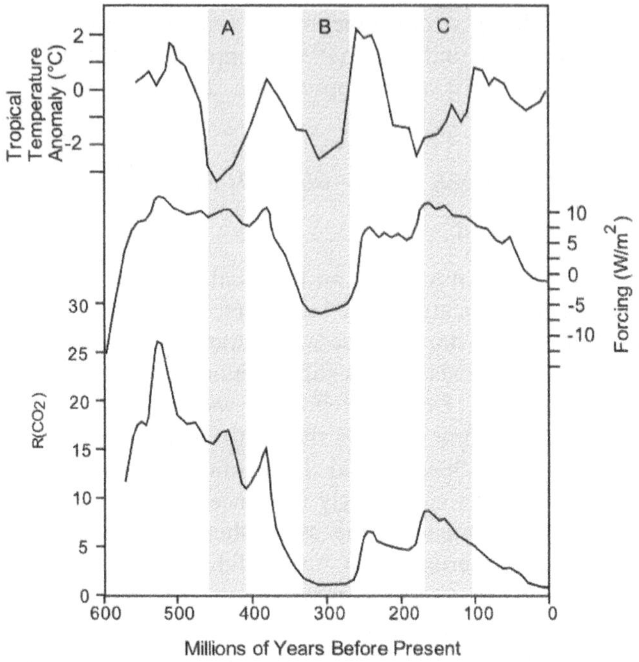

Figure 6.63. Comparison of $R(CO_2)$ with temperature changes during the Phanerozoic. Upper curve is tropical sea temperature anomaly. Middle curve is estimated forcing due to changing CO_2, taking into account the gradually strengthening Sun. Lower curve is estimated $R(CO_2)$. Gray areas *A* and *C* are time periods when CO_2 is disjoint with temperature, while gray area *B* has CO_2 and temperature in good agreement (adapted from Crowley and Berner, 2001).

correspondence between low-latitude $\delta^{18}O$ and net [CO_2 levels] begs for an explanation, especially because of the striking correspondence between low net [CO_2 levels] and major continental glaciation from 256 to 338 Ma."

They suggested that, in the case of the relatively short-lived glaciation of about 440 Ma, which occurred at a time of high $R(CO_2)$, "climate models suggest that the unusual continental configuration of … a large landmass tangent to the South Pole could result in conditions where high CO_2 and glaciation can co-exist". Nevertheless, they insisted that "the persistent Phanerozoic de-correlation between tropical $\delta^{18}O$ and $R(CO_2)$ demands a more comprehensive explanation". They suggested that one possibility is that the temperature estimates are erroneous. Another is that "climate change in the tropics can be largely decoupled from mid-high-latitude ice volume changes". While they insisted that "the first-order agreement between the CO_2 record and continental glaciation continues to support the conclusion that CO_2 has played an important role in long-term climate change", they nevertheless concluded that:

"Given the need for better confidence in some of the paleoclimate data and unanticipated complications arising from altered tectonic boundary conditions,

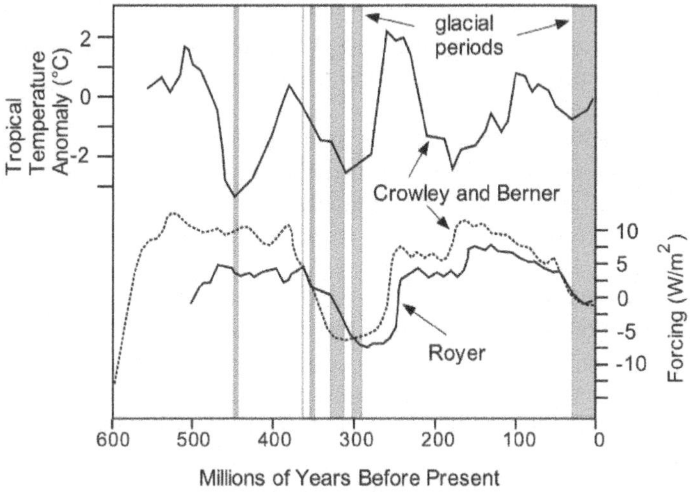

Figure 6.64. Royer's 2006 estimate of the net forcing due to CO_2 variability and a gradually strengthening Sun compared to that of Crowley and Berner (2001). The vertical gray bars are Royer's estimates of glacial periods, as compared to the curve of temperature given by Crowley and Berner.

it may be hazardous to infer that existing discrepancies between models and data cloud interpretations of future anthropogenic greenhouse gas projections."

Royer (2006) compared 490 published proxy records of CO_2 over the Phanerozoic with records of global cool events to evaluate the strength of CO_2-temperature coupling over the Phanerozoic. Figure 6.64 shows Royer's result in comparison with that of Crowley and Berner (2001). Royer found that the predominant glacial periods were between 350 and 290 million years ago, and the past 30 million years, when CO_2 concentrations were lower. The glacial period at 445 million years ago seem to be a contradiction of the CO_2-climate connection, but Royer (2006) argued that the glacial period was brief and the CO_2 level at that time is uncertain.

Royer (2006) concluded that:

"For periods with sufficient CO_2 coverage, all cool events are associated with CO_2 levels below 1000 ppm. A CO_2 threshold of below ~500 ppm is suggested for the initiation of widespread, continental glaciations, although this threshold was likely higher during the Paleozoic due to a lower solar luminosity at that time. Also, ... a CO_2 threshold of below ~1000 ppm is proposed for the initiation of cool non-glacial conditions. A pervasive, tight correlation between CO_2 and temperature is found both at coarse (10 million-year timescales) and fine resolutions up to the temporal limits of the data set (million-year timescales), indicating that CO_2, operating in combination with many other factors such as solar luminosity and paleogeography, has imparted strong control over global temperatures for much of the Phanerozoic."

With the passage of time since about 2004, the belief that CO_2 controls global temperature has become more widespread. Vaughn (2007) reviewed the field and concluded that CO_2 concentration is an important factor in major long-term changes in climate.

Came *et al.* (2007) claimed that they had developed an improved method for estimating paleotemperatures. They claimed that their results show a much better correlation of CO_2 variability and temperature change. However, they only estimated the temperature at two specific times during the Phanerozoic, and their results are not very convincing to this writer.

Fletcher *et al.* (2008) contributed new estimates of CO_2 concentrations from 200 million years ago to 60 million years ago. They concluded that there is a coupling of CO_2 and temperature.

Breecker *et al.* (2010) developed yet another estimate of paleo CO_2 concentrations over the past 400 million years. They also concluded that there is a coupling of CO_2 and temperature.

It is interesting that, since 2004, each successive published paper purports to find a tighter relationship between CO_2 and temperature in the Phanerozoic, yet the data have not changed much. One possibility is that there really is such a correlation, and better analyses have uncovered this. An alternate hypothesis is that only by emphasizing the role of CO_2 in climate change can one obtain research funding in the 21st century. As a result, more and more bias creeps into the published papers as investigators seek to ingratiate themselves with the orthodoxy. In all cases, the data are sparse, noisy, and difficult to interpret. It remains difficult to resolve the degree to which these two alternatives are involved.

6.5.9.5 Climate sensitivity assuming "the force" is with CO_2

Perhaps the most pervasive issue in modern climatology is the question of how much global warming (ΔT) in the 21st century would result from a doubling of the CO_2 concentration from the pre-industrial level of 280 ppm. While many estimates have been made, the canonical value often used is $\sim$3°C. Like the porridge in *The Three Bears*, this value is just right—not so great as to lack credibility, and not so small as to seem benign. Unfortunately, all of the estimates made to date by various procedures lack adequate data and require considerable speculation.

Assuming that variations in CO_2 concentration were the major cause of all historical climate change (which seems to be a widespread belief amongst paleoclimatologists), one can attempt to quantitatively estimate the relationship between changing CO_2 and global average temperature from the data. Unfortunately, most of the data are noisy and uncertain. According to Figures 6.56 and 6.58, the global average temperature dropped by at least 10°C prior to the great glaciation around 300 million years ago, while $R(CO_2)$ dropped from roughly 16 to 1. A factor of 16 represents four doublings. Hence, one might conclude that, going backward in time from 300 million years ago, each doubling of CO_2 produced a temperature rise of perhaps 10/4 = 2.5°C. Furthermore, the glacial period at 445 million years ago would seem to involve infinite climate sensitivity, since CO_2 did not appear to vary at all during that glacial period.

Royer *et al.* (2007) provided another estimate of climate sensitivity by fitting proxy data to a geological model for CO_2 during the Phanerozoic. Unfortunately, the data are sparse, noisy, and not completely reliable. Nevertheless, Royer *et al.* (2007) concluded that ΔT is greater than 1.5°C, and the best fit for the effect of doubling CO_2 is $\Delta T \sim 2.8$°C.

6.5.9.6 Correlation with galactic cosmic rays

Veizer *et al.* (2000) presented:

> "... a reconstruction of tropical sea surface temperatures throughout the Phanerozoic eon from [their] database of oxygen isotopes in calcite and aragonite shells. The data indicated large oscillations of tropical sea surface temperatures in phase with the cold and warm cycles, thus favoring the idea of climate variability as a global phenomenon."

But their data were not in consonance with reconstructed atmospheric carbon dioxide concentrations. They concluded that "The results can be reconciled if atmospheric carbon dioxide concentrations were not the principal driver of climate variability on geological timescales for at least one-third of the Phanerozoic eon, or if the reconstructed carbon dioxide concentrations are not reliable" (see Figure 6.65).

Shaviv and Veizer (2003) analyzed the "reconstructed seawater paleotemperature record for the Phanerozoic, and "compared it with the variable cosmic ray flux (CRF) reaching Earth" as well as "the reconstructed partial pressure of atmospheric CO_2 (pCO_2)". They found "that at least 66% of the variance in the paleotemperature trend could be attributed to CRF variations, likely due to solar system passages through the spiral arms of the galaxy. Assuming that the entire residual variance in temperature is due solely to the CO_2 greenhouse effect, [they proposed] a tentative upper limit to the long-term "equilibrium" warming effect of CO_2, one which is

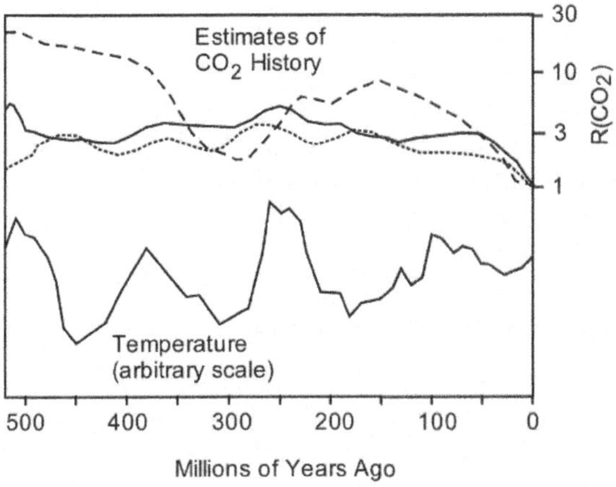

Figure 6.65. Comparison of estimates of CO_2 history with tropical sea temperature in the Phanerozoic (adapted from Veizer *et al.*, 2000).

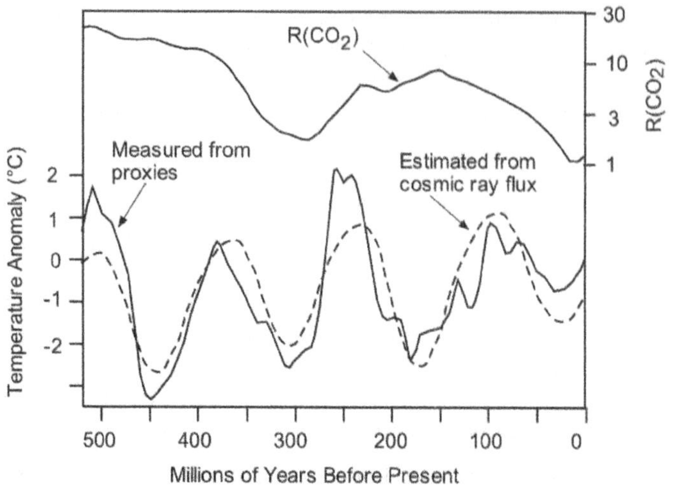

Figure 6.66. Comparison of $R(CO_2)$ and climatic effect of cosmic rays with estimated tropical sea-surface temperature anomalies from proxies during the Phanerozoic (adapted from Shaviv and Veizer, 2003).

potentially lower than that based on general circulation models". They used Berner's estimates for $p(CO_2)$, and $\delta^{18}O$ values of calcitic shells to estimate proxy-based paleo tropical SSTs. Cosmic ray activity indicators were based on ^{10}Be, ^{14}C isotopes. The results of Shaviv and Veizer (2003) are shown in Figure 6.66. Paleo SST anomalies were taken from estimates based on oxygen isotope proxies. The process by which Shaviv and Veizer arrived at temperature anomalies from variations in cosmic ray flux remains unclear to this writer. While Figure 6.66 is very impressive in favor of the argument made by Shaviv and Veizer that cosmic rays are more important than CO_2 in determining long-term climate change, the basis for this figure seems murky.

However, as we discussed previously, Royer *et al.* (2004) presented corrected estimates of SST during the Phanerozoic, due to changes in pH of the oceans induced by changes in the CO_2 level of the atmosphere and changes in Ca concentrations and calcium carbonate saturation state in seawater. With these changes, the variability of CO_2 was claimed to conform better with paleotemperatures, and the cosmic ray record does not. Shaviv and Veizer rebutted Royer's arguments, and the matter does not seem to be fully resolved.

6.5.9.7. Oxygen in the Phanerozoic atmosphere

It would be overly simplistic to treat the Phanerozoic Eon as if it were an extension of our current climate, but with variable CO_2 concentration. Continental drift produced significant changes in the distribution of landmasses on the Earth, producing variable feedback effects in response to changing CO_2. The Earth's atmosphere was quite different. Atmospheric O_2 concentration varied considerably over the Phanerozoic Eon. Oxygen levels reached as high as 30%–35% in so-called

Permo-Carboniferous between about 330 and 270 million years ago. During that period, "giant insects, including dragonflies reached wing spans up to 80 cm. Along with dragonflies, there are unusually large amphibians, mayflies, millipedes, hexapods, and arachnids confined to this same time span, and these organisms also metabolize by passive diffusion. Thus, animal fossils provide further evidence for the hypothesized high O_2 concentrations during the Permo-Carboniferous" (Berner, 2004). The principal cause of the high O_2 values was the rise of large vascular land plants that brought about increased O_2 production due to the increased global burial of microbially resistant, lignin-rich organic matter in sediments.

6.5.9.8. Phanerozoic summary

As with almost every area of climatology, the data on the Phanerozoic climate and CO_2 concentrations are sparse and noisy, and the interpretation of the data in terms of climatological parameters requires complex models and various assumptions. Various investigators have arrived at different interpretations regarding the connection between CO_2 concentrations and climate change during the Phanerozoic Eon. As Figure 6.72 shows, Shaviv and Veizer (2003) found a periodic variation in the climate that seemed to match variations in cosmic ray flux, and CO_2 variations did not seem to be highly relevant to climate change. However, a different interpretation of the data suggests that CO_2 variability was associated with climate change and cosmic rays play at most, a secondary role.

The evidence suggests that the Earth was relatively warm from about 550 to 400 million years ago, although the temperature may have varied considerably within that time frame. Around 400 million years ago, the Earth began to cool, and the cooling bottomed out with extensive glaciation for about 40 to 50 million years approximately 330 million years ago. Subsequently, the Earth warmed again, and finally cooled again during the most recent 100 million years. Variations of the climate on shorter timescales within that general scope are subject to considerable uncertainty. The data and models for CO_2 suggest that $R(CO_2)$ was very large prior to about 300 million years ago, peaking about 550 million years ago. $R(CO_2)$ declined slowly from 550 million years ago, and very rapidly from 400 to 350 million years ago. The cold period centered around 300 million years ago was associated with very low values of $R(CO_2)$. Temperature rose significantly after about 280 million years ago but CO_2 rose only very moderately. Over the last 100 million years, temperatures and $R(CO_2)$ both declined. There seems to be very little doubt that major changes in the Earth's climate are at least sometimes associated with large changes in $R(CO_2)$.

Boucota and Gray (2001) carried out a very detailed review of Phanerozoic climatic models and the relationship to the CO_2 content of the atmosphere. They concluded that:

> "... considerable disparity exists between the curves generated by the varied models. ... The wide disparities between the various published curves suggest that the presently published models are inadequate. Considerable disparity also

exists between all the models and the geological climatic evidence indicating changing climatic gradients through the Phanerozoic. This indicates, based on present knowledge of climates of the geological past, that there is no simple straightforward relation between levels of atmospheric CO_2, as estimated by the various modelers and changes in the global climatic gradient."

Although the review by Boucota and Gray (2001) was written prior to several of the papers cited herein, the conclusions seem to remain valid.

6.5.9.9 Concluding remarks

The widely held view amongst geologists and climatologists alike, is that the primary cause of long-term climate changes was variability of CO_2 concentration due to long-term imbalances between CO_2 degassing at spreading centers and the conversion of atmospheric CO_2 to mineral carbon through long-term silicate weathering and oceanic carbonate formation. The argument goes (more or less): "If it wasn't CO_2, what else could it have been?" Foster *et al.* (2009) described this as the "accepted paradigm" that requires CO_2 to vary in unison with global temperature. Thus, paleoclimatologists have been trying for decades to establish a relationship between climate and CO_2 concentration over many millions of years. There is some evidence that, over many millions of years, higher CO_2 concentrations are often, but not always, associated with warmer climates. However, there is a great deal of scatter in the CO_2 proxy data, and this relationship is difficult to pin down quantitatively. Royer (2010) began his commentary with the statement:

> "Global temperatures have covaried with atmospheric carbon dioxide (CO_2) over the last 450 million years of Earth's history. Critically, ancient greenhouse periods provide some of the most pertinent information for anticipating how the Earth will respond to the current anthropogenic loading of greenhouse gases. Paleo-CO_2 can be inferred either by proxy or by the modeling of the long-term carbon cycle. For much of the geologic past, estimates of CO_2 are consistent across methods."

This seems to be a rather optimistic view, considering the data from his paper (see Figure 6.67). This figure compares various estimates of tropical SST with estimates of CO_2 concentration over the time period 120 million years ago to 40 million years ago. Royer's point (I think) is that, throughout this period, SST was at least several degrees warmer than today and, even though there is much scatter in the CO_2 estimates, the general level of CO_2 concentration was much higher than today. This argument seems to make some sense from 120 million years ago to $\sim$90 million years ago. Yet, there are difficulties from 70 million years ago to 50 million years ago when SST remained high, yet the CO_2 concentration appears to have been much lower. In any event, the extreme scatter in the data in Figure 6.67 do not convey confidence that any valid conclusions can be drawn.

The best chance to use paleoclimatic data to infer climate sensitivity is probably the LGM some 20,000 years ago, when the total negative forcing produced a global average temperature decrease of roughly 4.5°C. There are good estimates available

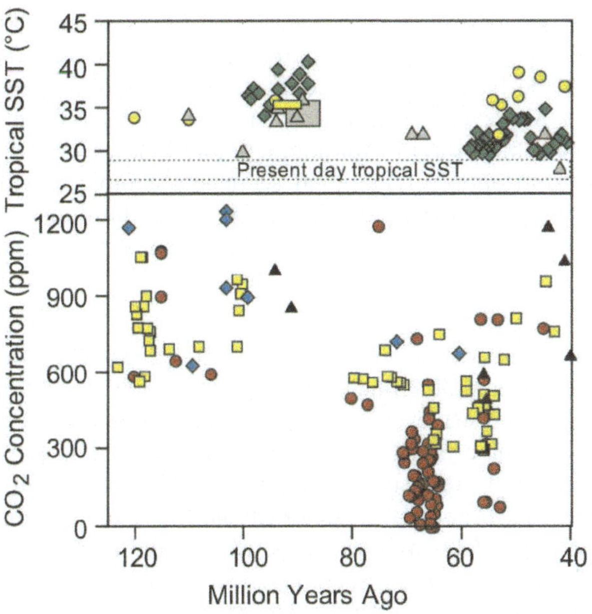

Figure 6.67. CO_2 and tropical sea-surface temperatures (adapted from Royer, 2010).

of the global average temperature and the CO_2 concentration at the LGM 20,000 years ago and, if these data are compared with values in the pre-industrial era (a few hundred years ago), one can thereby estimate the sensitivity of the climate to CO_2 concentration over the range $\sim$180 ppm to $\sim$280 ppm. This is discussed in Section 6.5.2. It is noteworthy that this range of estimates for the real-world ΔT due to doubling CO_2 from 280 ppm to 560 ppm is from $\sim$1°C to $\sim$3°C. However, these estimates do not take into account possible differences in humidity and cloudiness between Ice Ages and interglacials.

The data on temperature and CO_2 over hundreds of millions of years are far less reliable, and conclusions drawn from these time periods are dubious at best.

Our conclusion here is that CO_2 is probably one of several major factors in long-term climate change, but there are other factors, such as the placement of the continents on the Earth, the functionality of ocean currents, the past history of the climate, the orientation of the Earth's orbit relative to the Sun, the luminosity of the Sun, the presence of aerosols in the atmosphere, volcanic action, land clearing, biological evolution, etc. Hence, there is probably no single curve relating global average temperature to CO_2 concentration, but rather, a set of curves that depend on the above factors.

6.5.10 Relationship between sea level and climate forcing by CO_2 on geological timescales

Foster and Rohling (2013) recently published a paper in which they estimated CO_2

concentrations over geologic timescales using (1) gas bubbles trapped in ice cores, (2) carbon isotopic composition of sedimentary alkenones recovered from deep-sea sediments, and (3) the boron isotopic composition of planktic foraminifera from deep-sea sediments. Sea level over geologic time was estimated from: (1) changes in the oxygen isotopic composition of foraminifera and bulk carbonate from Red Sea sediments, (2) back-stripping of marginal sediments combined with estimates of paleo-water depth based on detailed litho-facies, ichnological, and benthic foraminiferal analyses, and (3) sea-level change reconstructed using Mg/Ca of foraminifera to isolate the ice-volume signal from foraminiferal $\delta^{18}O$. They concluded:

> "Here we use observations from five well-studied time slices covering the last 40 My to identify a well-defined and clearly sigmoidal relationship between atmospheric CO_2 and sea level on geological (near-equilibrium) timescales. This strongly supports the dominant role of CO_2 in determining Earth's climate on these timescales and suggests that other variables that influence long-term global climate (e.g., topography, ocean circulation) play a secondary role. The relationship between CO_2 and sea level we describe portrays the likely (68% probability) long-term sea-level response after Earth system adjustment over many centuries. Because it appears largely independent of other boundary condition changes, it also may provide useful long-range predictions of future sea level. For instance, with CO_2 stabilized at 400–450 ppm (as required for the frequently quoted "acceptable warming" of 2°C), or even at AD 2011 levels of 392 ppm, we infer a likely (68% confidence) long-term sea-level rise of more than 9 m above the present. Therefore, our results imply that to avoid significantly elevated sea level in the long term, atmospheric CO_2 should be reduced to levels similar to those of preindustrial times."

If the authors are correct, even the "moderate" guidepost of alarmists (accepting a ~2°C temperature rise from pre-industrial times) would lead to a 9 m rise in sea level! Hence, if these authors are correct, as they say, "atmospheric CO_2 should be reduced to levels similar to those of pre-industrial times [280 ppm]", which appears to be totally impossible considering we are already at 395 ppm and still heavily dependent on fossil fuels. We might then conclude that the damage has already been done and there is no return to the good old days of the LIA.

All of these proxies suffer from a variety of serious maladies, and therefore the data must be looked at with a jaundiced eye. Nevertheless, for the sake of argument, we might take the data at face value as presented and ask whether the data as reported support this sweeping conclusion. Their first analysis looks at the past 500,000 years in which they show (what has been discussed hundreds of times previously) a strong resemblance between the patterns of variation of CO_2 concentration and sea level (from ice volume in ice sheets) during Ice Age–interglacial transitions. That CO_2 rises and falls with global temperature is well known. But the rise and fall of CO_2 in these transitions occurs as a consequence of Ice Age–interglacial transitions, and are not a cause of Ice Age–interglacial transitions. The reported correlation coefficient of 0.68 occurs in these Ice Age–

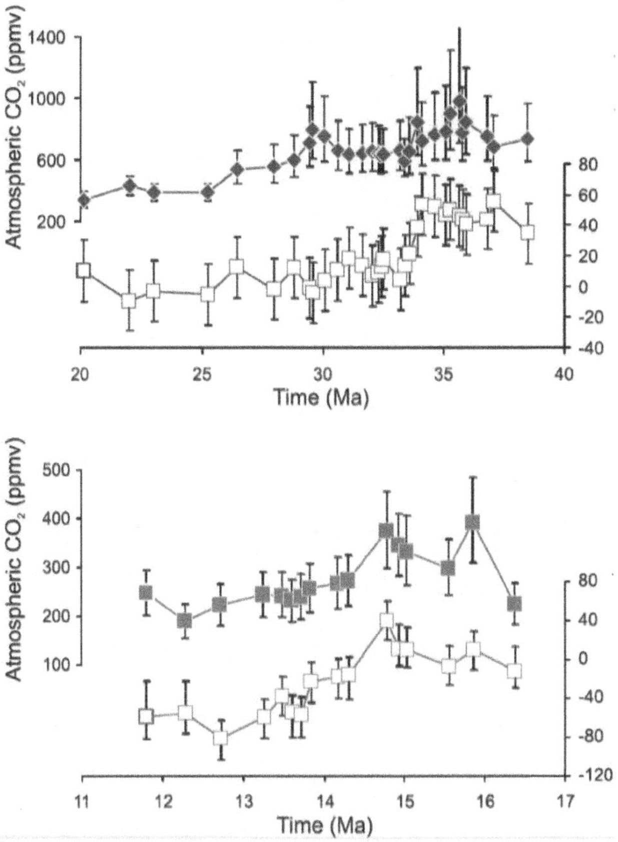

Figure 6.68. Comparison of estimated CO_2 and sea level over two geologic time periods (Foster and Rohling, 2013).

interglacial transitions and does not at all lead to the stated conclusion that "the dominant role of CO_2 in determining Earth's climate on these timescales and suggests that other variables that influence long-term global climate (e.g., topography, ocean circulation) play a secondary role", which has very little basis, if any.

Going further back in time, the data are suggestive that sea level is higher when CO_2 is higher but the correlation is very poor (see Figure 6.68). The estimates of sea level seem extreme, with changes in sea level of as much as 100 m associated with changes in CO_2 of 100 ppm to 300 ppm. As in the case of Ice Age–interglacial transitions of the past 500,000 years, there is a connection between CO_2 and sea level, but CO_2 is not the independent cause.

7

Anthropogenic influences on climate change

The fact that there has been an increase in global temperatures in the 20th century suggests that anthropogenic activity might have contributed significantly to climate change during this time period. Anthropogenic activity includes generation of effluents from power plants, cement manufacturing, other industrial operations (carbon dioxide, black soot, sulfate aerosols), land clearing and deforestation, and large-scale irrigation). The various contributing factors act as follows:

- Increased CO_2 (and other greenhouse gases) amplifies the absorption of outgoing infrared producing a heating effect at all latitudes.
- Deposition of black soot on high-latitude snow and ice increases solar absorptivity, thus producing a heating effect at higher latitudes.
- Sulfate aerosols in the atmosphere produce a cooling effect by reflecting incoming sunlight.
- Land clearing and utilizationn and deforestation increases the albedo of the Earth, producing a cooling effect.
- Water usage (irrigation, ground water depletion, etc.)

7.1 CO_2 CONCENTRATION PAST AND PRESENT

7.1.1 Introduction

One of the most pressing issues of our time is the possibility that rising CO_2 concentrations in the atmosphere might lead to significant global warming in the future that have been predicted to produce deleterious impacts on humankind. Hence, the relationship between CO_2 concentration and climate has become a very central and critical scientific issue. However, in addition to being a scientific issue, rising CO_2 has also become a political issue as well. This is due to several factors:

(1) In the process of consuming fossil fuels, cement production, and other industrial activities, the world produces large amounts of CO_2. Not all of this added CO_2 can be absorbed by the oceans and biosphere. If the world continues in a

D. Rapp, *Assessing Climate Change*, Springer Praxis Books,
DOI: 10.1007/978-3-319-00455-6_7

business-as-usual (BAU) scenario, CO_2 production will continue to rise in the 21st century, leading to higher CO_2 concentrations in the atmosphere. The potential cure for too much CO_2 requires a draconian change in the way that energy is generated and used by the world, and this change may not be technically feasible, and, even if it turns out to be technically feasible, it will likely be extremely costly. Indeed, it is possible that it may not be possible to provide the people of the world with energy to run the industrialized world if CO_2 emissions must be cut as dramatically as alarmists claim is needed.

(2) A significant number of climatologists believe that continuation of BAU energy policies in the 21st century will be disastrous to humankind. Many of them have voiced this viewpoint via the Internet, meetings, and media. Furthermore, this bias has crept into scientific publications published in peer-reviewed journals. A smaller number of climatologists have been skeptical of the certainty expressed by alarmists. Liberal politicians have been swayed by alarmists into enacting severe constraints on future CO_2 emissions. These constraints require that, by such and such a future year, we must emit considerably less CO_2. It is not clear that these constraints can be met technically or financially. Indeed, the benchmark used by several governments is an 80% reduction in CO_2 emissions by 2050—a goal that almost certainly cannot, and will not, be met. Conservative politicians tend to lean toward the skeptical view, more from a political perspective than from any scientific understanding. It should be noted that governments have not been clear whether this means an 80% reduction from present emission levels or an 80% reduction from that expected on the basis of a BAU scenario. If, as seems likely, it implies an 80% reduction from present levels, that is equivalent to approximately an 88% reduction from a future BAU scenario.

(3) Quite a few prominent climatologists in their zeal to save the world from overheating (and possibly to secure more funding for climate research) have engaged in unprofessional activities in an attempt to exclude the skeptics from science publications. They have also manipulated data to exaggerate the threat of rising CO_2 and they have presented their results in a biased and one-sided manner. Some have prevented others from checking their results by holding their data in secret. In many cases, they have drawn conclusions from sparse and noisy data, yet made bold claims of certainty in their conclusions. The exposure of these shenanigans has hurt their credibility in some quarters; nevertheless, the science questions remain regarding the impact of rising CO_2. The fact that they have acted unprofessionally does not make them wrong.

(4) Under auspices of the United Nations, the Intergovernmental Program on Climate Change (IPCC) has been co-opted by alarmists regarding the effect of CO_2 on climate, and they have widely promulgated the belief that "the debate is over" regarding the impact of rising CO_2, yet considerable uncertainty remains in all the issues.

As we showed in Chapter 6, there is considerable evidence that the CO_2 concentration in the atmosphere has varied over a very wide range over geological

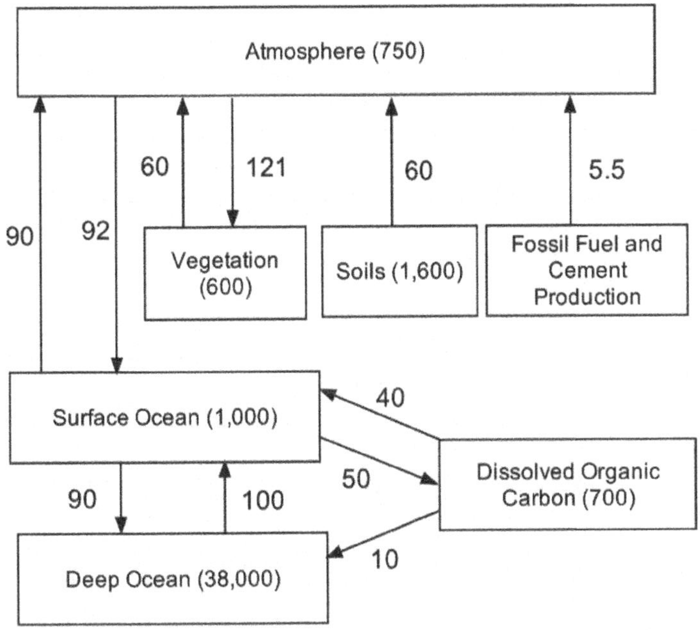

Figure 7.1. Rough estimates of carbon storage and annual carbon fluxes as of about year 2000. Storage is in gigatons (Gt) of carbon and fluxes are in Gt/yr of carbon. Ocean sediments are not shown, nor are many smaller contributors (adapted from Northrup, 2004). As of year 2012, the fossil-fuel and cement emissions are closer to 8.0 Gt/yr vs. 5.5 Gt/yr in the figure.

time. Nevertheless, despite this wide-ranging past, a seemingly repeatable pattern has emerged over the past few hundred thousand years. The Earth has alternated between Ice Ages and interglacial periods at roughly 100,000-year intervals. Ice cores reveal that, during interglacial periods, typical CO_2 concentrations were about 280 ppm while, at the height of glaciation during Ice Ages the CO_2 concentration dropped to roughly 190 ppm. About 20,000 years ago, the Earth was at its Last Glacial Maximum (LGM) and the CO_2 concentration was roughly 190 ppm. As the Ice Age waned, the CO_2 concentration rose, and reached a plateau meandering about 280 ppm, where it has remained for the past ~8,000 years. As a result, we are accustomed to thinking of 280 ppm as the "normal" pre-industrial level of CO_2 in the atmosphere.

Knowledge of carbon exchange between the atmosphere, land, and the oceans is important in understanding the rate of build-up of CO_2 in the atmosphere due to human intervention in the Earth environment. However, the Earth's carbon cycle is complex. The oceans, the biosphere, and land exchange very large amounts of CO_2 each year, and the contributions from fossil-fuel and cement production are relatively small. Nevertheless, the human contributions (from fossil-fuel and cement production) appear to be enough to upset the delicate Earth balance, leading to a build-up of CO_2 in the atmosphere. Figure 7.1 provides a very rough simplified version of the carbon content and annual carbon fluxes in the Earth's ecosystem.

7.1.2 Measurements and proxies

The atmospheric CO_2 measurements at Mauna Loa in Hawaii constitute the longest continuous record of atmospheric CO_2 concentrations available in the world. The Mauna Loa site is considered to be one of the most favorable locations for measuring undisturbed air because possible local influences of vegetation or human activities on atmospheric CO_2 concentrations are minimal and any influences from volcanic vents may be excluded from the records. The methods and equipment used to obtain these measurements have remained essentially unchanged during the 50-year monitoring program.

Because of the favorable site location, continuous monitoring, and careful selection and scrutiny of the data, the Mauna Loa record is considered to be a precise record and a reliable indicator of the regional trend in the concentrations of atmospheric CO_2 in the middle layers of the troposphere. Air samples at Mauna Loa are collected continuously from air intakes at the top of four 7 m towers and one 27 m tower. Four air samples are collected each hour for the purpose of determining the CO_2 concentration. Peaks occur every May and minima occur in September. During northern summer, the growth of plants removes CO_2 from the atmosphere while, in northern winter, decay of plants and oxidation produces extra CO_2. The Southern Hemisphere (SH) has less influence on CO_2 concentrations because it is dominated by water.

The Mauna Loa record shows a 19.4% increase in the mean annual concentration, from 316 ppm of dry air in 1959 to 377.4 ppm in 2004 and 395 ppm in 2012 (see Figure 7.2).

Humlum *et al.* (2012) analyzed in detail the annual variations in CO_2 as illustrated in Figure 7.3. While both CO_2 and temperature generally increased during this 31-year period, the rates of change varied significantly during the period. They showed that changes in CO_2 correlated somewhat with changes in sea-surface temperature (SST) but the CO_2 change lagged the SST change by about 11–12 months. They concluded

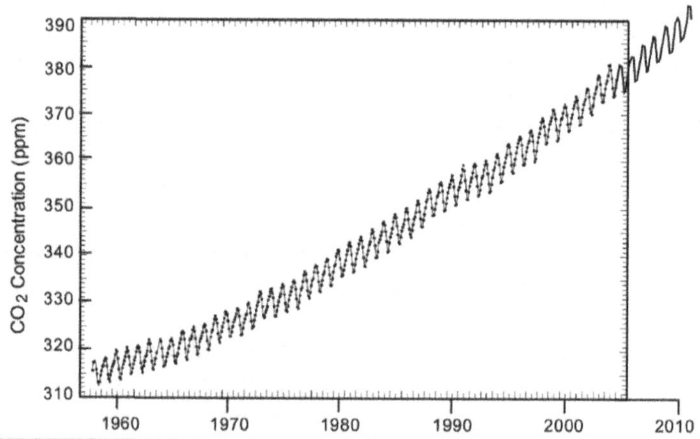

Figure 7.2. Measured CO_2 concentration at Mauna Loa.

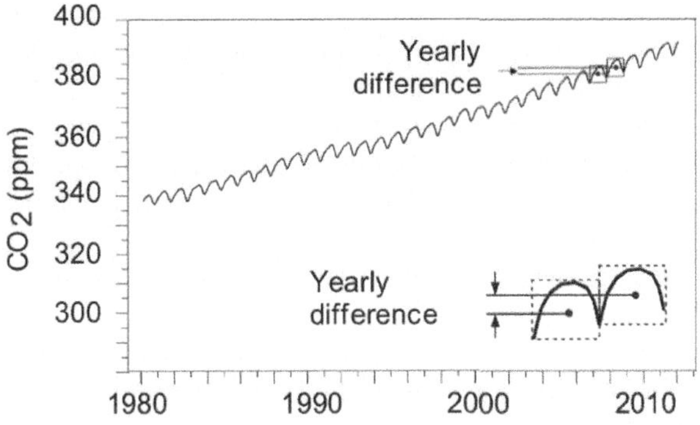

Figure 7.3. Annual and long-term variation of CO_2 concentration (Humlum *et al.*, 2012).

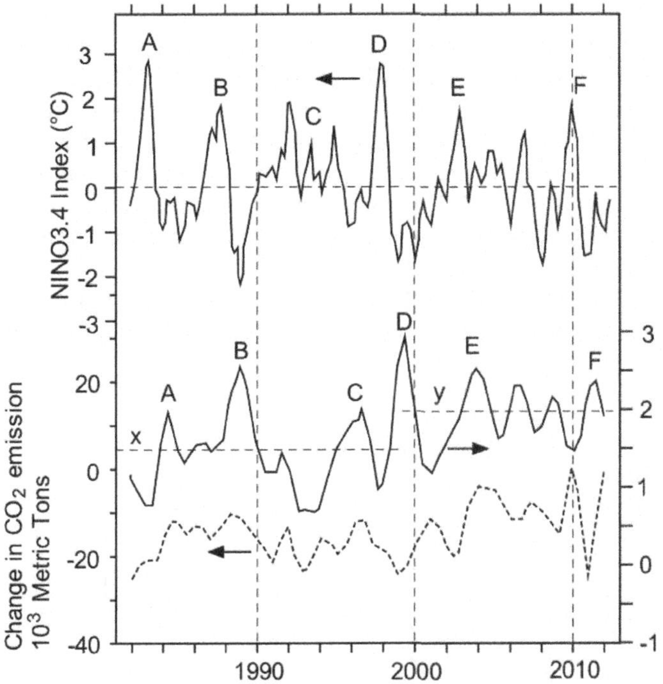

Figure 7.4. CO_2 emissions and concentration and El Niño index.

that "A main control on atmospheric CO_2 appears to be the ocean surface temperature". They mentioned a possible connection to the giant 1998 El Niño but did not elaborate on the relationship of the entire sequence of data to El Niño indices.

Consider Figure 7.4. The uppermost curve shows the NINO3.4 index from 1980 to 2011. Peak El Niños are labeled with letters *A* to *F*. The middle curve shows the change in CO_2 concentration per year plotted on a monthly basis. The peaks in this curve are also subjectively labeled *A* to *F*. The average change in CO_2 concentration per year can be interpreted either as a ramp or a step-function. Arbitrarily adopting the step-function, the average change in CO_2 concentration per year varied from year to year by about 1.5 ppm/yr prior to the 1998 El Niño, and varied from year to year by about 2.0 ppm/yr after the 1998 El Niño. These are depicted as horizontal dashed lines *x* and *y*.

The lowermost curve shows the annual change in anthropogenic CO_2 emissions plotted on a per-month basis. A rough rule of thumb is that each Gt of carbon (3.67 Gt of CO_2) produces the equivalent of about 0.5 ppm of CO_2 in the atmosphere if none of it is absorbed. Figure 7.4 below shows that annual variations in global emissions of carbon are typically about 2×10^4 metric tons per year which, if unabsorbed, would produce annual changes in CO_2 that are far too small to account for the observed variations in the average change in CO_2 concentration per year.

The point made by Humlum *et al.* is that the average change in CO_2 concentration per year lags the change in ocean temperature by about 11–12 months. As Tisdale (2012) showed in his book, El Niños leave behind them a pool of warm surface waters. As a result, the average change in CO_2 concentration per year tends to lag the NINO3.4 index by a bit more than a year. This correlation is far from perfect but it seems to have some validity, particularly for the major El Niño that started toward the end of 1997. The data suggest that the ability of the oceans to absorb CO_2 emitted by human activity responds to the state of the NINO3.4 index with a delay of slightly over a year.

Recent data on CO_2 emissions by country are shown in Figure 7.4a.

According to this website, these graphs of total CO_2 emission history show that up until 2012:

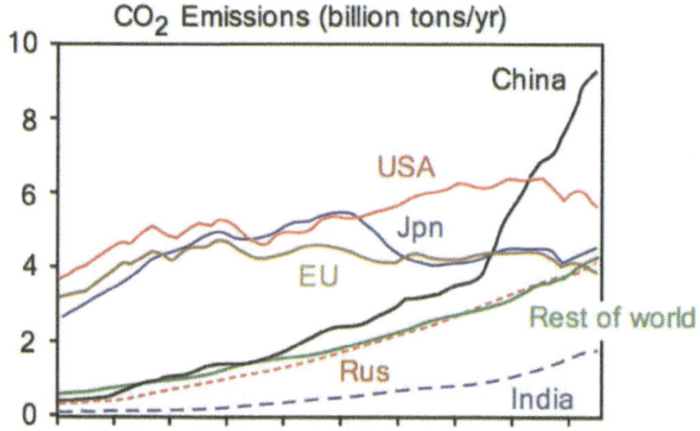

Figure 7.4a. CO_2 emissions by country (*http://wattsupwiththat.com/2013/07/17/global-warming-climate-change/#more-89993*).

- CO_2 emissions from the developing world as a whole overtook the developed world in 2007 and are now ~42% higher.
- There has been a very rapid escalation of Chinese CO_2 emissions since the year 2000.
- China overtook the U.S. CO_2 emissions in 2006, and, by 2012, Chinese emissions were already ~60% greater than the U.S., the escalation in Chinese CO_2 emissions is expected to continue.
- There is inexorable emissions growth from all the developing economies, from a low base.
- India has accelerating emissions, growing substantially, from a low base.
- There has been stabilization or reduction of emissions from developed economies.

However, probably more significant than the total CO_2 emissions output is the comparison of the actual emissions/head for the various populations.

- The E.U. even with active legal measures, have maintained a fairly level CO_2 emission rate but have managed to reduce their CO_2 emissions/head by ~29% since their peak in 1977. The recent downward trend is attributed to their declining economies.
- The U.S. has already reduced its CO_2 emissions/head by ~32% since its peak in 1970.
- Russia, Japan, and Canada reduced their emissions/head by ~24% since their peak in 1989.
- The eight rapidly developing nations have shown consistent growth from a low base in 1965 at 5.6 times. They exceeded the world average CO_2 emissions level in 1997.
- China's CO_2 emissions/head have grown a further 140% since 2000. China overtook the worldwide average in 2003 and surpassed the rapidly developing nations in 2005.
- India's CO_2 emissions have grown by 4.7 times over the period and are now showing recent modest acceleration. That increasing rate is likely to grow substantially.
- In the rest of the world (~160 nations), 36% of world population has grown CO_2 emissions consistently but only by 2.6 times in the period; this group will be the likely origin of major future growth.
- Overall average worldwide emissions/head have remained relatively steady but with early growth in the decade from 1965. It amounts to 1.6 times since 1965.

When the participating nations (particularly the EU) are compared with Chinese CO_2 emissions/head, an interesting picture arises:

- Chinese CO_2 emissions at 6.7 mt/head for its 1.3 billion population are already ~41% greater than the worldwide average. Those emissions are still growing fast.
- At 5.4 mt/head, France, with ~80% nuclear electricity generation, has the lowest CO_2 emission rates in the developed world and is at only ~12% above the world-wide average.

- China's CO_2 emissions/head exceeded France's CO_2 emissions/head in 2009.
- The U.K., at 7.2mt/head, is only ~50% higher than the worldwide average and only about ~12% higher than China.
- Germany, one of the largest CO_2 emitters in Europe, has emissions/head ~100% higher than the worldwide average and is still ~63% higher than China.

Human activity is presently emitting roughly 8 Gt/yr of carbon, which, if unabsorbed, would be sufficient to increase the atmospheric concentration of CO_2 by about 4 ppm per year. Over a period of years, we might assume that (very) roughly half of that CO_2 is absorbed by Earth systems (oceans, biosphere, etc.) and the other very rough half ends up in the atmosphere, raising the atmospheric concentration by about 2 ppm. However, on a year-by-year basis, the proportion of emitted CO_2 that is absorbed by the Earth systems varies considerably, mainly due to the presence of warm surface waters in the Pacific produced quasi-periodically by El Niños. According to the graphical data in Figure 7.4, the annual increase in CO_2 concentration can be as high as 3 ppm (following the 1998 El Niño) or as low as 1 ppm (between peaks *B* and *C*). During the most recent period after the 1998 El Niño, annual increases in CO_2 concentration seem to have varied roughly as 2 ± 0.5 ppm or $\pm 25\%$. These results suggest that, while roughly half of emissions might end up in the atmosphere over an extended period, annual variations in the distribution of emitted CO_2 between the atmosphere and the Earth system are significant, and strongly dependent on prevalence of El Niños.

Carbon dioxide measurements prior to the late 1950s derive primarily from ice cores. The conventional view is that the transition from snow to firn to ice occurs as the weight of overlying material causes the snow crystals to compress, deform, and recrystallize in more compact form. When firn is buried beneath subsequent snowfalls, the density is increased as air spaces are compressed due to mechanical

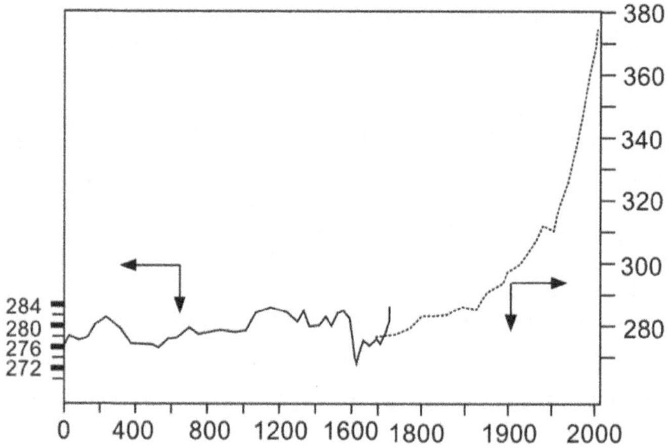

Figure 7.5. Atmospheric CO_2 concentration over the past 2,000 years (ppm) (adapted from Etheridge *et al.* (1996), and extended with recent data from NOAA).

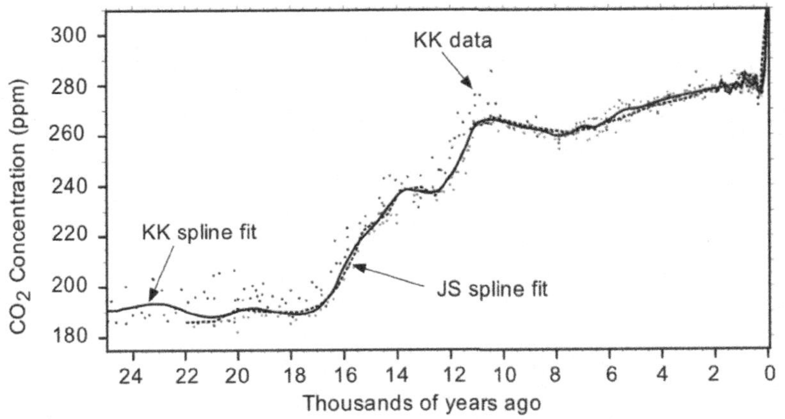

Figure 7.6. Variation of CO_2 concentration since the LGM. The KK data and spline fit were to 10 ice core results from Krumhardt and Kaplan (2010). The JS spline fit is to four ice core results (Joos and Sphani, 2008).

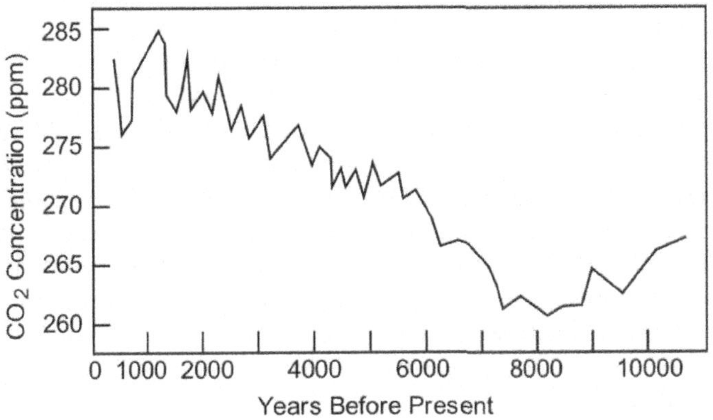

Figure 7.7. CO_2 concentration during Holocene from ice cores (Fliueckiger *et al.*, 2002).

packing as well as plastic deformation. Interconnected air passages may then be sealed and appear as individual air bubbles. This process is believed to take several hundred years, so the air entrapped within the ice is believed to be younger than the ice by a century or more. It seems likely that the trapped air within the ice represents a broad distribution of ages younger than the ice. Friedli *et al.* (1986) and Etheridge *et al.* (1996) provided ice core data from Antarctica. More recent estimates are available from Meure *et al.* (2006) and their results are shown in Figure 7.5.

As the LGM faded and the Earth warmed, the CO_2 concentration rose in response. Analysis of several ice cores led to the time dependence of CO_2 concentration over the past 25,000 years as shown in Figure 7.6.

During the Holocene, the reported CO_2 concentration was as shown in Figure 7.7.

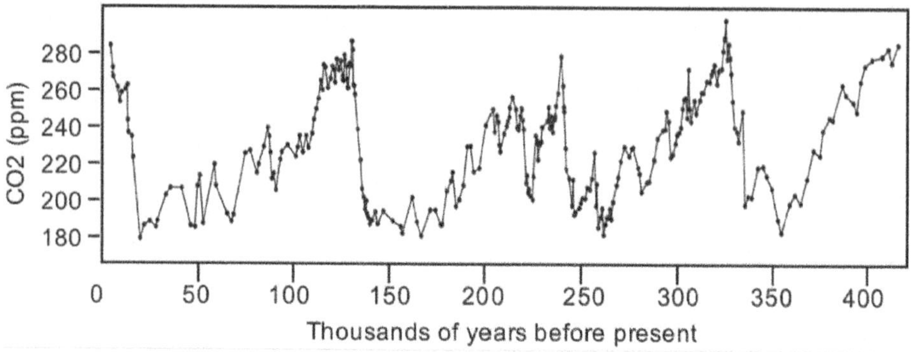

Figure 7.8. Historical variation of CO2 concentration during the past four decades (Petit *et al.*, 1999).

The reported CO_2 concentration during the past 400,000 years is shown in Figure 7.7. The CO_2 concentration peaked around 280 ppm–290 ppm at the height of interglacial eras, and decreased to less than 200 ppm during Ice Ages.

From a long-term point of view, the variation of CO_2 concentration in the atmosphere has varied widely across historical glacial cycles. Figure 7.8 shows the historical variation of CO_2 concentration during the past four glacial cycles. The peak CO_2 concentration in past interglacial periods was less than 300 ppm. The recent rise in CO_2 concentration from the pre-industrial level of roughly 280 ppm to the present value of 395 ppm appears to be primarily due to anthropogenic release of CO_2 and other greenhouse gases as well as the strong warming trend in the second half of the 20th century that releases CO_2 from the oceans. However, it is not clear what geological processes produce the changes in CO_2 concentration shown in Figure 7.8. Low CO_2 concentrations during periods of glaciation might be due, at least partially, to reduced deep-water ventilation associated with year-round Antarctic sea ice coverage.

Estimates of CO_2 concentration as far back as 500 million years are provided in Section 6.5.

7.1.3 Carbon cycle: CO_2 fluxes

Knowledge of carbon exchange between the atmosphere, land, and the oceans is important in understanding the rate of build-up of CO_2 in the atmosphere due to human intervention in the Earth environment. However, the Earth's carbon cycle is complex. The oceans, the biosphere, and land exchange very large amounts of CO_2 each year, and the contributions from fossil-fuel and cement production are relatively small. However, these contributions (from fossil-fuel and cement production) appear to be enough to upset the delicate Earth balance, leading to a build-up of CO_2 in the atmosphere. Figure 7.1 provides a very rough simplified version of the carbon content and annual carbon fluxes in the Earth's ecosystem.

Sabine *et al.* (2004) estimated the long-term (1800–1994) and recent (1980–1999)

Table 7.1. Exchange of CO_2 (Pg carbon) between biosphere, atmosphere and oceans (Sabine *et al.*, 2004).

	1800 to 1994	*1980 to 1999*
(1) Emissions from fossil fuels and cement production	244 ± 20	117 ± 5
(2) Storage in the atmosphere	165 ± 4	65 ± 1
(3) Uptake and storage in the ocean	118 ± 19	37 ± 8
(4) Net terrestrial balance = (2) + (3)—(1)	39 ± 28	15
(5) Emissions from land use	100 to 180	24 ± 12
(6) Terrestrial biosphere sink = (4)—(5)	–61 to –141	-39 ± 18

transfers of CO_2 between the biosphere, the atmosphere, and the oceans (measured as Pg of carbon).[1] Their results are summarized in Table 7.1.

Emissions from fossil-fuel burning and cement production were estimated from historical records. Storage in the atmosphere was estimated by converting measured ppm of CO_2 to Pg carbon, assuming 1 ppm CO_2 is equivalent to 2.11 Pg carbon. The uptake and storage in the oceans were estimated from ocean measurements described by Sabine *et al.* (2004). The net terrestrial balance is the net amount of carbon generated by the biosphere. As can be seen from Table 7.1, this quantity is >0, indicating that the biosphere has been a net contributor of CO_2 to the environment. This net contribution is the difference between two larger numbers (emissions from land use—the terrestrial biosphere sink). Although the biosphere absorbs a great deal of CO_2, it emits even more.

Of the historical net emissions from 1800 to 1994 (244 + 39 = 283 Pg carbon), Sabine *et al.* (2004) estimated that about 58% ended up in the atmosphere and 42% ended up in the oceans. It has been suggested that, over the past two decades, about half of net emissions remained in the atmosphere and half were taken up by the oceans. Douglass (2005) pointed out that the common approach is to treat nature as unchanging, and attribute all increases in atmospheric CO_2 concentration to anthropogenic sources. However, he argued that natural emissions of CO_2 increased in the latter part of the 20th century due to prevalence of El Niños that raised surface ocean temperatures, and some major volcanic eruptions that emitted large amounts of CO_2. When these factors were taken into account, he found that some of the increase of the late-twentieth-century atmospheric CO_2 was due to natural causes. After taking this into account, he concluded that the rate of increase of the late-twentieth-century atmospheric CO_2 concentration was approximately constant at around 44% of nominal anthropogenic emissions. In addition, he argued that massive coal fires in northern China and biomass burnings have added large amounts of CO_2 to the anthropogenic side of the ledger. Thus, his estimate of anthropogenic CO_2 emissions is greater than other estimates, and he concluded that only about 30% of late-twentieth-century CO_2 emissions from anthropogenic sources (other than biomass burnings) ended up in the atmosphere. This is

[1] 1 petagram (Pg) = 10^{15} grams.

considerably lower than the nominal figure 50% that is often used. A rough estimate is that, for each Gt of carbon[2] added to the atmosphere, the concentration of CO_2 rises by about 0.4 ppm to 0.5 ppm. Anthropogenic emissions of carbon have steadily increased since the 19th century, and currently run about 8 Gt/yr. If roughly half of them were to end up in the atmosphere, the concentration would rise at the rate of about 1.8 ppm per year.

Keeling (2005) suggested that the Sabine *et al.* (2004) estimates of ocean uptake of CO_2 should be slightly modified by including potential feedbacks due to ocean warming and increased stratification. For 1800 to 1994, Keeling estimated these to be −13 Pg carbon and +6 Pg carbon, for a net addition to the Sabine estimate of −7 Pg carbon. However, in a response to Keeling's comment, Sabine and Gruber claimed that such estimates are highly uncertain and, being small compared with the estimated error given in Table 7.1, need not be included until they can be resolved more accurately.

Schimel *et al.* (2001) provided an overview of the current state of knowledge of global and regional patterns of carbon exchange by terrestrial ecosystems somewhat different from that of Sabine *et al.* (2004). Schimel *et al.* concluded that the annual emissions from fossil-fuel burning and cement manufacture were roughly 6.3 Gt of carbon per year in the 1990s. According to Schimel *et al.*, about half of this ended up in the atmosphere, about 27% ended up in the oceans, and about 22% ended up in plants and soil. The increase of ~3.2 Gt of carbon per year in the atmosphere during the 1990s was the culmination of a growth of about 170 Gt of carbon above the pre-industrial figure of 580 Gt of carbon in the atmosphere. This 170/580 = 29% increase in atmospheric carbon increased the atmospheric CO_2 concentration from the pre-industrial level of ~280 ppm to about 385 ppm by 2008.

The description given above is the conventional one that is widely accepted. However, this picture is not without its difficulties. In this picture, the transport of huge amounts of CO_2 between the atmosphere, the oceans, and the land is so delicately balanced that the addition of a comparatively small annual anthropogenic CO_2 flux throws this system out of balance, resulting in growth of the CO_2 concentration in the atmosphere. There is still uncertainty as to whether biosphere systems absorb some of the additional CO_2 or whether the biosphere is a net emitter of CO_2.

Schuster and Watson (2007) utilized a time series of observations from merchant ships between the U. K. and the Caribbean to establish the variability of sea-surface pCO_2 and air-to-sea CO_2 flux from the mid-1990s to early 2000s. Two series of measurements were made: one for 15 months in 1994–1995 and one for three years in 2002–2005. However, the instrumentation was different in the two time periods. In the more recent data, they began with 180,000 data points but, because of the long lines connecting the equilibrator in the engine room to the detector some considerable distance away, they selected only 9,000 points for analysis chosen to be at the end of a line-flushing period. They measured pCO_2 in

[2] One Gt of carbon is equivalent to 44/12 = 3.67 Gt of CO_2.

the ocean (near-surface) and utilized other data for pCO_2 in the atmosphere. They found that, whereas pCO_2 in the ocean changed from ~328 μatm in 1994–1995 to ~365 μatm in 2002–2005, the atmospheric pCO_2 changed from ~354 μatm in 1994–1995 to ~370 μatm in 2002–2005.[3] If these data are correct, it would suggest that the driving force for absorption of CO_2 by the ocean (the difference between pCO_2 in the atmosphere and the ocean) decreased from 26 μatm to 5 μatm in 10 years, which hardly seems likely. Other investigators (e.g., see Schuster and Watson, 2007) have found less extreme diminution of the capability of the oceans to uptake CO_2.

Falkowski *et al.* (2000) emphasized that the current high concentrations of CO_2 in the atmosphere represent "unchartered waters" that have not been seen in the past four glaciation–interglaciation cycles. However, this paper was perhaps unduly pessimistic, implying that "ample fossil-fuel reserves" would lead to excessive growth of CO_2 in the 21st century (see Section 7.2.4 for a contrary view on fossil-fuel reserves). The total of dissolved inorganic carbon (DIC) in the oceans is 50 times that of the atmosphere, and atmospheric CO_2 continuously exchanges with oceanic CO_2 at the surface at the rate of ~90 Gt of carbon per year in each direction, leading to rapid equilibration of the atmosphere with the surface water (see Figure 7.1). On dissolution in water, CO_2 forms a weak acid that reacts with carbonate anions and water to form bicarbonate. The capacity of the oceanic carbonate system to buffer changes in CO_2 concentration depends on the addition of cations from the relatively slow weathering of rocks. Because the rate of anthropogenic CO_2 emissions is several orders of magnitude greater than the supply of mineral cations, the ability of the surface oceans to absorb CO_2 will inevitably decrease as the atmospheric concentration of the gas increases over timescales of millennia. The concentration of total DIC in the ocean increases markedly below about the upper 300 m, where it remains significantly above the surface ocean–atmosphere equilibrium value in all ocean basins. The higher concentration of inorganic carbon in the ocean interior results from two processes:

> "Since CO_2 is more soluble in cold, saline waters, sequestration of atmospheric CO_2 in the ocean interior is controlled by the formation of cold, dense water masses at high latitudes. As these water masses sink into the ocean interior and are transported laterally, CO_2 is effectively prevented from re-equilibrating with the atmosphere by a cap of lighter overlying waters. Re-equilibration occurs only when waters from the ocean interior are brought back to the surface, decades to several hundreds of years later. Coupled climate–ocean simulations suggest that CO_2-induced global warming will lead to increased stratification of the water column. If this occurs, the transport of carbon from the upper ocean to the deep ocean will be reduced, with a resulting decrease in the rate of sequestration of anthropogenic carbon in the ocean. The combined effects of progressive saturation of the buffering capacity and

[3] Here, we work with pCO_2 in units of μatm because we considered dissolved CO_2. For the atmospheric CO_2, 1 μatm is the same as 1 ppm.

increased stratification will weaken two important negative feedbacks in the carbon–climate system.

"Biological processes also contribute to the absorption of atmospheric CO_2 in the ocean. Phytoplankton photosynthesis lowers the partial pressure of CO_2 in the upper ocean and thereby promotes the absorption of CO_2 from the atmosphere. Approximately 25% of the carbon fixed in the upper ocean sinks into the interior, where it is oxidized through heterotrophic respiration, raising the concentration of dissolved inorganic carbon (DIC)." (Falkowski *et al.*, 2000).

Thus, on timescales of centuries, decreased solubility of CO_2 with increasing ocean temperature is counteracted by biological processes. According to Falkowski *et al.* (2000):

> "There are significant gaps in our knowledge that limit our ability to predict the magnitude of changes in oceanic uptake, but the likely changes in the biological pump are too small to counteract the projected CO_2 emissions in the coming century."

However, the "projected CO_2 emissions in the coming century" may be greater than fossil-fuel resources can supply (see Section 7.2.4). Nevertheless, Falkowski *et al.* (2000) concluded that:

> "If our current understanding of the ocean carbon cycle is borne out, the sink strength of the oceans will weaken, leaving a larger fraction of anthropogenically produced CO_2 in the atmosphere or to be absorbed by terrestrial ecosystems."

Terrestrial ecosystems remove CO_2 from the atmosphere through photosynthesis and organic-matter storage. It is returned to the atmosphere via a number of respiratory pathways as well as fire, in which large amounts of organic matter are oxidized in very short periods of time. Terrestrial carbon storage primarily occurs in forests. The turnover time of terrestrial carbon is in the order of decades. The activity of plants to remove CO_2 increases with the CO_2 concentration, but is believed by some to saturate between 800 ppm and 1,000 ppm CO_2. Falkowski *et al.* (2000) suggested that this "concentration will probably be reached early in the next century at the present emissions rate. Because the saturation function decreases as CO_2 increases, terrestrial plants will become less of a sink for CO_2 in coming decades".

However, as Section 7.2.4 indicates, it is unlikely that such high levels of CO_2 will be reached before 2100:

> "The combined effects of higher CO_2 concentrations, higher temperatures, and changes in disturbance and soil moisture regimes lead to considerable uncertainty about the ability of terrestrial ecosystems to mitigate against rising CO_2 in the coming decades. However, recent results from long-term soil warming experiments in a boreal forest contradict the idea that the projected rise in temperature is likely to lead to forests that are now carbon sinks becoming carbon sources in the foreseeable future. Again, as in the case of

marine ecosystems, we can predict that the negative feedback afforded by terrestrial ecosystems in removing anthropogenic CO_2 from atmosphere will continue; however, the sink strength will almost certainly weaken. The exact magnitude of the change in sink strength remains unclear." (Falkowski *et al.*, 2000).

7.1.4 CO_2 variations in glacial–interglacial cycles

Ice core records of CO_2 from Greenland reach back as far as 420,000 years. These records have been used to examine the changes in atmospheric CO_2 concentration as the Earth evolved through the past four Ice Ages interspersed with interglacial periods.

Ice core records have low time resolution. The gas occlusion process takes a considerable time, resulting in smearing of the time periods over which gases are trapped in the ice. According to Ehrelinger, Cerling, and Dearing (2005), "There are still open questions about the reliability of the CO_2 record". The difference in age between the trapped CO_2 and the ice in which it is trapped can be several hundred to several thousand years. The CO_2 levels in Greenland ice cores are consistently higher than those from Antarctica and, since atmospheric CO_2 is well mixed, at least one of these must be in error. Apparently, spurious variations in the Greenland data suggest that the Antarctic records provide the most reliable data (Ehrelinger *et al.*, 2005).

A number of studies of ice cores from Antarctica have shown that the CO_2 concentration rise (or fall) lags the temperature rise (or fall) that occurs during periods of increased glaciation or warming. The time lag was estimated to be $\sim$500 years by Roper (2006), 800 $\pm$ 200 years by Caillon *et al.* (2003), 1,300–5,000 years by Mudelsee (2001), "several thousand years" by Petit *et al.* (1999), 800 years by Monnin *et al.* (2001), and 400–1,000 years by Fischer *et al.* (1999). That would seem to imply that increased CO_2 is an effect—not a cause—of temperature change in Ice Age–interglacial transitions. Siegenthaler *et al.* (2005) reported on two deep ice cores from East Antarctica. One of the cores was the only ice core covering at least eight glacial cycles, four cycles longer than previously available from ice cores. This allowed them to reconstruct the record of the concentration of atmospheric CO_2 much further back in time than was possible before, over the interval between 390,000 and 650,000 YBP (years before present). Analyzing the air extracted from ice cores is the only way to directly determine atmospheric greenhouse gas concentrations for times before routine atmospheric measurements were begun. Antarctic ice cores are claimed to be very suitable for CO_2 measurements because of their low temperatures and low concentrations of impurities, which minimize the risk of artifacts, and it is asserted that Antarctic ice cores are reliable recorders of atmospheric CO_2. In addition, the concentration of deuterium was interpreted as a proxy for temperature. By shifting the timescales of the entire CO_2 and deuterium records between 390,000 and 650,000 YBP relative to each other, they obtained the best correlation for a lag of CO_2 of 1,900 years.

A number of skeptics seized on these observations with enthusiasm to argue against rising CO_2 concentration as a cause of global warming; instead, they claim that rising temperatures (from some other cause) produce increasing CO_2

concentrations. It is well known that the capacity of the oceans to hold CO_2 diminishes as the temperature increases. However, even though CO_2 lags the change in temperature, changing CO_2 concentrations will exert a forcing on the climate and act as an amplifier of trends originated from other sources. Ice sheets require several tens of thousands of years to build up, and the time lag for CO_2 evolution is not important in this regard.

Caillon *et al.* (2003) discussed the use of ice cores to provide detailed records of local temperature and atmospheric concentrations of greenhouse gases in the past. Analyses of the Vostok ice core in Antarctica show that concentrations of carbon dioxide correlate well with Antarctic temperature throughout the last four climatic cycles, with glacial–interglacial CO_2 increases of 80 ppmv to 100 ppmv. Determining the mechanisms that cause these variations is important for understanding climate change, but the explanation for the strong link between atmospheric CO_2 and Antarctic air temperature is still unclear. One reason for this uncertainty is that the relative timing of temperature and CO_2 changes is not easily resolved and thus obscures the phasing of gas variations with climate signals borne by the ice. An approach to circumvent this difficulty proposed by Caillon *et al.* (2003) used records of atmospheric CO_2 content and temperature (based on ^{40}Ar content) contained only in the trapped gases. Although they believed their procedure is valid, they admitted that they "don't clearly understand the underlying mechanisms". Nevertheless, their result was that CO_2 increases and peaks at a shallower depth in the ice core than ^{40}Ar. They concluded that, in a post-glacial warming period, the atmospheric CO_2 lags Antarctic warming by 800 ± 200 years.

Caillon *et al.* (2003) concluded that CO_2 is not the forcing that initially drives the climatic system during a deglaciation. Rather, deglaciation is probably initiated by another forcing (solar) with positive feedback that influences first the temperature change in Antarctica (and possibly in part of the SH and then the increase in CO_2 in the SH. According to Caillon *et al.* (2003), this sequence of events is still in full agreement with the idea that CO_2 plays, through its greenhouse effect, a key role in amplifying the initial orbital forcing that they suggest is solar-generated. First, the 800-year time lag is short in comparison with the total duration of the increases in temperature and CO_2 (5,000 years) in a post-glacial warming. Second, it is shown that the CO_2 increase clearly precedes the Northern Hemisphere (NH) deglaciation. However, the initiation of the deglaciation step may require a random occurrence of climate factors amplified by a positive feedback mechanism. The putative role of the Sun in this is speculative. Recently, Stott *et al.* (2007) found that the onset of deglacial warming throughout the SH occurred long before deglacial warming began in the tropical surface ocean. In a second paper (Timmermann *et al.*, 2008), this group carried out modeling that indicated that the likely cause of initiation of deglaciation after 20 KYBP (thousands of years before present) was the increase in insolation coupled with the sea ice–albedo feedback as sea ice went into retreat. As the CO_2 concentration rose, this added another warming feedback. They also claimed[4] that each of the last four major Ice Age

[4] Lowell Stott, personal communication, November 2008.

terminations were associated with increases in solar input to the far SH. The solar input to the far SH during the austral spring period when the ice pack is at a maximum appears to be a major factor in initiating deglaciation. It thus appears that CO_2 may be controlled in large part by the climate of the Southern Ocean. Although there is not yet unequivocal support for this assertion, a delay of about 800 years is claimed to be a reasonable time period to transform an initial Antarctic temperature increase into a CO_2 atmospheric increase through oceanic processes.

Ahn and Brook (2007) estimated a time lag for CO_2 to follow temperature changes of 720 $\pm$ 370 years.

In contrast to the previous discussion, Loulergue *et al.* (2007) claimed that:

> "The phase relationship between CO_2 and ice core temperature inferred at the start of the last deglaciation (lag of CO_2 by 800 $\pm$ 600 yr) is overestimated and that the CO_2 increase could well have been in phase or slightly leading the temperature increase at the EPICA Dome C."

Peacock, Lane, and Restrepo (2006) analyzed the variations of CO_2 concentration in glacial–interglacial cycles over the past 430,000 years. It was emphasized that there is no consensus on the underlying cause of the 80 ppm–100 ppm variation in the roughly 100 kyr glacial–interglacial cycles in CO_2 partial pressure (pCO_2). The inability of any proposed single mechanism to explain the observed cycles in pCO_2 (which show considerable similarity over the past 430,000 years) led them to consider a combination of mechanisms. They pointed out that physical changes (ocean circulation, temperature, mixing) can only explain part of the observed atmospheric pCO_2 variability, so they invoked changes in ocean chemistry to explain the remainder. In their model, they distinguish three phases in the glacial–interglacial cycles: glacial, intermediate, and interglacial. Physical changes in the ocean (mixing, temperature) account for interglacial-to-intermediate transitions. The transition from intermediate to glacial involves a small increase in mean ocean nutrient levels and mean ocean alkalinity, accomplished by falling sea level and subsequent erosion of organic-rich shelf sediments:

> "The first part of the transition out of full glacial conditions is achieved through increased temperature and increased mixing in the Southern Ocean. The final part of the atmospheric pCO_2 rise up to full interglacial conditions is accomplished through rising sea level and the subsequent change in mean ocean alkalinity and phosphate, and a rise in NH temperature and ocean mixing."

They claimed that they were able to explain the full magnitude of the glacial–interglacial cycle in atmospheric pCO_2. However, their explanation is not unique and represents only one possible rather contrived explanation, yet to be substantiated. This is a very lengthy, complex paper that only a specialist can follow in detail. It demonstrates how complex the processes are for CO_2 exchange between the oceans and the atmosphere during glacial–interglacial conditions.

There is no consensus on the causes of glacial–interglacial CO_2 changes. There are at least 11 hypotheses, which may be grouped into three basic themes: (1) physical/chemical "reorganization" of the oceans, (2) changes in the ocean carbonate system, and (3) changes in ocean nutrient inventories. However, many of these hypotheses are not mutually exclusive. The interactions between marine and terrestrial ecosystems, changes in ocean circulation, radiative forcing, and greenhouse gases all probably interact in a specific sequence to give rise to the natural cyclic atmospheric and climatic oscillations. These interactions are not adequately represented in detailed models of glacial–interglacial transitions. The scientific community has generally approached glacial–interglacial transitions from a disciplinary perspective. This approach has not produced completely satisfactory explanations for what is clearly a large natural perturbation in the global carbon cycle. Clearly, a systems approach is needed (Falkowski *et al.*, 2000).

Skinner (2006) provided a review of the subject of glacial–interglacial CO_2 cycles. He emphasized that, even though it is clear that changes in atmospheric CO_2 were tightly coupled to global climate change throughout the past 800,000 years, the mechanisms responsible for these changes in CO_2 concentration "remain a mystery". Archer *et al.* (2000) came to similar conclusions:

> "In spite of the clear importance of pCO_2 as an amplifier or even a primary driver of the glacial cycles, and the additional motivation provided by the threat of future climate change, we remain ignorant of the mechanisms responsible for the glacial/interglacial CO_2 cycles ... Fifteen years after the discovery of major glacial/interglacial cycles in the CO_2 concentration of the atmosphere, it seems that all of the simple mechanisms for lowering pCO_2 have been eliminated."

Sigman and Boyle (2000) echoed this sentiment: "... we have not yet identified the cause of these variations in CO_2."

A number of blogs on the Internet would have you believe that the explanation for the similarity of the CO_2 and T curves results simply from the difference in solubility of CO_2 in the oceans as a function of temperature. However, detailed analysis shows that this effect is insufficient to account for the change from about 180–200 ppm under full glacial conditions to about 280 ppm under full interglacial conditions.

Although most of the carbon on the Earth is incorporated into $CaCO_3$ in rocks, this carbon pool is too stable to account for pCO_2 changes over glacial cycles. Carbon in the terrestrial biosphere is available on shorter time frames but, in order to deplete pCO_2 by 100 ppm, the terrestrial biosphere and soil carbon reservoirs would have to approximately double in size over about 10,000 years. Instead, measurements of the $\delta^{13}C$ from deep-sea $CaCO_3$ suggest that the terrestrial biosphere released carbon during glacial times—the wrong direction to explain lower glacial pCO_2. The only remaining candidate driver for the atmospheric CO_2 change is the oceans, which can hold enough carbon to absorb the atmospheric decrease and can change on 1,000 to 10,000-year timescales (Archer *et al.*, 2000).

Archer *et al.* (2000) described two mechanisms that have been proposed to account for pCO_2 changes in glacial–interglacial CO_2 cycles (GICC). One proposed mechanism to lower glacial pCO_2 is based on an increased rate of biological productivity in surface waters of the oceans, leading to storage of carbon in the deep sea due to sinking particles. Either an increase in the ocean inventory of nutrients

(PO_4^{3-} and NO_3^-), or a change in the ratio of nutrient to C in phytoplankton, could have stimulated the ocean's "biological pump" in this way. Models of the ocean carbon cycle indicated that pCO_2 is extremely sensitive to the biological pump in high latitudes and relatively insensitive to low-latitude forcing. Since iron availability limits phytoplankton growth in remote parts of the ocean, a dustier, more iron-rich glacial climate would have intensified biological productivity in the surface waters of the oceans. A second mechanism to lower glacial pCO_2 is to change the pH of the whole ocean, converting seawater CO_2 into HCO_3^- and CO_3, which are unable to evaporate into the atmosphere. The pH of the ocean is controlled by any imbalance between the influx of dissolved $CaCO_3$ from chemical weathering on land and the removal of $CaCO_3$ by burial in the deep sea.

Skinner (2006) pointed out that the magnitude of the marine carbon reservoir and its interaction with atmospheric CO_2 suggests a major role for the oceans in GICC. While a simplistic model might suggest that the increased solubility of CO_2 in a colder glacial ocean would account for the reduction of CO_2 during Ice Ages, detailed models indicate that this would only amount to about 30 ppm of the total 80 ppm to 100 ppm reduction. Furthermore, even this moderate reduction in CO_2 would be counteracted by the reduced solubility of CO_2 as the oceans became saltier during Ice Ages, as well as by a large reduction in the terrestrial biosphere when land is covered by ice under glacial conditions. Thus, the net reduction in CO_2 during glacial conditions due to solubility, land changes, and salinity is probably more like ~10 ppm. Therefore, Skinner argued that "the bulk of the GICC remains to be explained by more complex inter-reservoir exchange mechanisms". He suggested that "the most viable proposals involve either the biological or physical 'carbon pumps' of the ocean," and, regardless of which mechanism is invoked, GICC involves changes in the sequestration of CO_2 in the deepest marine reservoirs.

Given the magnitude and dynamism of the deep marine carbon reservoir, it is almost certain that past glacial–interglacial fluctuations in atmospheric CO_2 have relied at least in part on changes in the carbon storage capacity of the deep sea. Skinner (2006) described three main types of conceptual models that have been offered to explain GICC:

(1) Biological pump—involving an increase in the export of organic carbon to the deep sea, either via increased nutrient availability at low latitudes or via increased efficiency of nutrient usage at high latitudes.
(2) Reduced ventilation of CO_2—of water exported to the deep Southern Ocean, either via sea ice "capping" or a change in ocean interior mixing efficiency.
(3) Changes in ocean chemistry and carbonate imbalance, possibly involving changes in the ratio of organic carbon and carbonate fluxes to the deep sea.

According to Skinner (2006), each of these approaches has difficulties individually in explaining the pattern and magnitude of past GICC, and it is likely that all three have participated to some extent. For example, Stephens and Keeling (2002) noted that outgassing of CO_2 from the oceans is enhanced when the partial pressure in the atmosphere is low, producing a high gradient between ocean and atmosphere. This acts in opposition to the innate increase in solubility when the oceans are colder. As a result, the sea ice cover of the Southern Ocean south of 55°S during glacial winters would have to be very high to produce a significant decrease in pCO_2 due to ice capping. They estimated that, if the sea ice cover of the Southern Ocean south of 55°S during glacial winters was as high as 99%, it could account for a 65 ppm reduction in pCO_2. However, Maqueda and Rahmstorf (2002) found the sea ice coverage to maximize at 92% using a sophisticated climate model, corresponding to a CO_2 decrease of only 35 ppm. They therefore concluded that the increase of sea ice in the Southern Ocean could explain only a moderate portion of the CO_2 decrease during glacial periods.

Skinner (2006) emphasized that one factor from these models that emerges as being fundamental is the competition between carbon export from the surface ocean to lower depths vs. carbon "reflux" by the overturning circulation of the ocean. This tug of war between these two processes, one biological and one physical, essentially determines the balance of carbon input into and output from the deepest marine reservoirs. This ultimately determines the magnitude of pCO_2 in the atmosphere. The thermohaline circulation of the oceans plays an important role in this process. The formation of North Atlantic deep water represents an efficient mechanism for mixing CO_2 deep into the ocean interior. The return flow of deep water to the surface occurs primarily in the Southern Ocean representing a net reflux of carbon to the atmosphere. Thus, the Southern Ocean plays a pivotal role in controlling the overall efficiency of the oceans' physical carbon pump. Any model to explain the GICC must provide scenarios for these oceanic processes.

Visser *et al.* (2003) found that SSTs decreased by larger amounts in the LGM than others had estimated, and thereby suggested that "a substantial portion of the atmospheric decrease in CO_2 during glacial periods" could be attributed to a simple difference in solubility. They based this on a previous estimate that "for a 1°C increase in temperature, the equilibrium partial pressure of CO_2 exerted by seawater increases by 11–16 ppm". However, Sigman *et al.* (2010) asserted that:

> "... the cause of the pCO_2 variation must be resolved if we are to understand its place in the causal succession that produces glacial cycles. ... The ocean is the largest reservoir of CO_2 that equilibrates with the atmosphere on the thousand-year timescale of glacial/interglacial changes in pCO_2, so the ocean must drive these changes. CO_2 was more soluble in the colder ice-age ocean, which should have lowered pCO_2 by ~30 ppm, but much of this appears to have been countered by other ocean changes (in salinity and volume) and a contraction in the terrestrial biosphere. The most promising explanations for the bulk of the pCO_2 decrease involve ocean biogeochemistry and its interaction with the ocean's physical circulation."

Sigman *et al.* (2010) concluded that:

> "Global climate and the atmospheric partial pressure of carbon dioxide (pCO_2) are correlated over recent glacial cycles, with lower pCO_2 during Ice Ages, but the causes of the pCO_2 changes are unknown. The modern Southern Ocean releases deeply sequestered CO_2 to the atmosphere. Growing evidence

suggests that the Southern Ocean CO_2 'leak' was stemmed during Ice Ages, increasing ocean CO_2 storage. Such a change would also have made the global ocean more alkaline, driving additional ocean CO_2 uptake. This explanation for lower ice-age pCO_2, if correct, has much to teach us about the controls on current ocean processes."

Studies have also been made over the more recent Holocene period. Kouwenberg and Ria (2005) and Kouwenberg *et al.* (2005) pointed out that:

> "An increasingly applied method to detect and quantify short-term fluctuations in Holocene CO_2 levels is the analysis of stomatal frequency of fossil leaves derived from peat and lake deposits. For a wide variety of tree species, there is observational and experimental evidence of an inverse relation between numbers of leaf stomata and ambient CO_2 concentration. Adjustment of stomatal frequency to changes in atmospheric CO_2 allows plants to retain the most profitable balance between carbon uptake for photosynthesis and loss of water through evaporation. Quantification of CO_2 responsiveness of individual tree species over the last century enables estimation of Holocene CO_2 levels by measuring stomatal frequency of fossil leaves. So far, stomata-based CO_2 reconstructions for the Holocene have mainly been derived from fossil leaves of broad-leaved trees and shrubs. However, because of the long-term dominance of conifers in temperate and boreal forest ecosystems, the use of fossil conifer needles could greatly improve the spatial and temporal coverage of such reconstructions."

Kouwenberg and Ria (2005) and Kouwenberg *et al.* (2005) then went on to say that conifer needles were largely neglected after initial studies had suggested that stomatal frequency of conifers would not adjust to CO_2 mixing ratios above 280 ppmv. However, it was indicated that this observation was due to inadequate stomata quantification methods because conifers exhibit a different (more complex) mode of leaf development and subsequent epidermal morphology than the broad-leaved trees commonly used for stomatal frequency analysis. They developed an improved quantification strategy, to enable conifer needle stomata to act as a viable proxy for CO_2 concentration, and thereby obtain high-resolution CO_2 reconstructions for the Late Holocene (their result is shown in Figure 7.10).

Kouwenberg and Ria (2005) and Kouwenberg *et al.* (2005) were concerned about the validity of their data on two counts: (1) disagreement with ice core data prior to 800, and (2) disagreement of their CO_2 data with *hockey stick* temperature profiles prior to 800. They argued that the ice core data tend to be smoother because they involve a much wider swath of times in each sample that tends to flatten out variations. However, they were unable to explain the significant differences prior to 800. They were also at a loss on how to relate their findings prior to 800 with *hockey stick* temperature profiles, and, in the end, they said:

> "Since the reconstructed enhanced CO_2 levels between 300 and 750 are incongruent with global climate changes, the extremely low stomatal frequency of T. heterophylla in this period is unlikely to reflect pronounced changes in the global atmospheric CO_2 regime."

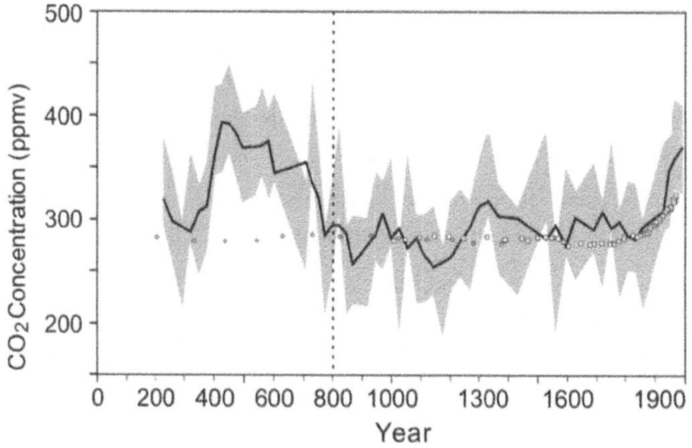

Figure 7.9. Reconstruction of paleo-atmospheric CO_2 levels based on stomatal frequency of fossil needles. The black line represents a three-point running average based on 35 needles per depth. The gray area indicates the uncertainty in the calibration. White diamonds and squares are measured data in ice cores (adapted from Kouwenberg and Ria (2005) and Kouwenberg *et al.* (2005)).

However, they were unaware of the inaccuracy of the *hockey stick* in this period, and perhaps should not have been so insecure about their results. The results shown in Figure 7.9 indicate diametrically opposite viewpoints. On the one hand, the ice core data suggest an essentially flat profile for 2,000 years; on the other hand, the data on stomatal frequency of conifers suggest considerable variation in CO_2 levels during that period. Neither record shows a significant decrease during the Little Ice Age (LIA). It is not clear which, if any, should be believed but the ice core data seem to be more credible.

7.1.5 CO_2 and global warming

It has been estimated that an increase of CO_2 concentration from the pre-industrial level of approximately 280 ppm to 560 ppm will produce a forcing of roughly 4 W/m^2 and the climate sensitivity neglecting feedbacks and secondary processes is roughly 0.3°C per W/m^2 (Hansen and Sato, 2011). Hence, in response to doubling CO_2, the expected global average temperature rise for a hypothetical Earth that did not change in any other way except to warm, would be about 1.2°C. Many studies have attempted to estimate the climate sensitivity taking into account additional feedbacks and secondary processes. Because of uncertainties in estimating these effects, climate models have produced a wide range of estimates of temperature rise from doubling CO_2, ranging from about 2°C to about 10°C. As we pointed out in the discussion of Figure 6.36, Roe and Baker (2007) showed that the predicted temperature rise is very sensitive to assumed feedback factors at the high end, leading to great uncertainty in the results. Hansen *et al.* (2008) emphasized the original work of Charney (1979), who came up with an estimate of 3°C $\pm$ 1.5°C. Amidst the great uncertainty in climate model predictions, Charney's figure has often been taken as a benchmark, not too great to lack credibility, but large enough to cause concern. Hansen *et al.* (2008) seem to take satisfaction in saying that "Climate models in the current IPCC assessment still agree with Charney's estimate". We have previously decribed this as a "Goldilocks approach" in which a very uncertain quantity is chosen not too big and not too small, so that it seems credible and yet has significant implications. This hides the fact that we really don't know the magnitude of the quantity.

Another approach for estimating the temperature rise caused by a doubling of CO_2 from the pre-industrial era utilizes paleoclimatic data, comparing climate parameters at the LGM with condition in the recent pre-industrial era. As we showed in Section 6.5.2, various estimates suggest a range from 2°C to 3°C. This would seem to erase the high-end tail from climate models.

Lacis *et al.* (2010) said (amongst other things) that:

> "Ample physical evidence shows that carbon dioxide (CO_2) is the single most important climate-relevant greenhouse gas in Earth's atmosphere. This is because CO_2, like ozone, N_2O, CH_4, and chlorofluorocarbons, does not condense and precipitate from the atmosphere at current climate temperatures, whereas water vapor can and does. Non-condensing greenhouse gases, which account for 25% of the total terrestrial greenhouse effect, thus serve to provide the stable temperature structure that sustains the current levels of atmospheric water vapor and clouds via feedback processes that account for the remaining 75% of the greenhouse effect. Without the radiative forcing supplied by CO_2 and the other non-condensing greenhouse gases, the terrestrial greenhouse would collapse, plunging the global climate into an icebound Earth state."

If you started with an Earth that had no greenhouse gases at all, it would be a very cold Earth indeed. Now, if you add greenhouse gases (water vapor, CO_2, CH_4, etc.) to that hypothetical Earth in proportion to how much we have today, you will get the percent contribution that each greenhouse gas makes to the present climate of the Earth. Lacis *et al.* (2010) carried out a study to estimate this. They found that water vapor accounts for 50%, clouds account for 25%, and CO_2 accounts for 20%, with all others amounting to 5% of the total greenhouse effect that produces the present climate, compared to how cold it would be if there were no greenhouse gases or clouds at all. There is no reason to doubt these results, even though at least some of the authors do have a track record of emphasizing the importance of CO_2.

Given that there is enough water vapor, clouds, and CO_2 in the atmosphere to account for 95% of the present greenhouse effect (present temperature compared to what it would be with no greenhouse gases), what were the drivers that caused the present distribution? If you had a cold Earth with low humidity and few clouds, permanent non-condensible greenhouse gases will nevertheless act to warm the atmosphere and, therefore, in some sense, CO_2 is more of a controlling factor than humidity and clouds. As CO_2 warms the atmosphere and more water evaporates and more clouds form, they produce an even greater greenhouse effect than CO_2, but CO_2 (via volcanic emissions and such) gets the ball rolling, so to speak. In that sense, one must agree with Lacis *et al.* (2010) that, as their press release states: "Water vapor and clouds are the major contributors to Earth's greenhouse effect, but a new atmosphere-ocean climate modeling study shows that the planet's temperature ultimately depends on the atmospheric level of carbon dioxide." But we did not need their study. It is just common sense. Humidity and clouds will not form on their own on a cold Earth. They will occur in response to some other factor warming the atmosphere. However, in addition to responding to CO_2-induced warming, they might also form as a consequence of other volcanic emissions, variability of total solar irradiance, variability of solar activity, variability of ocean currents, changes in ocean cycles (e.g., El Niño–La Niña), land development, soot deposition in polar areas, aerosol formation, etc.

Starting from the present (or perhaps from 100 years ago prior to build-up of CO_2 in the industrial era), how much will the global temperature rise if we raise the CO_2 level to, say, 560 ppm, and what fraction of this temperature rise will be due to rising CO_2? The answer is that, directly and indirectly, all of the temperature rise will be due to CO_2. The direct part is due to CO_2 absorption while the indirect part is from the consequent secondary effects such as increased humidity and cloud formation (as well as other effects) as the Earth warms. So, the issue comes down to what are the numbers? How much warming would be due to CO_2 alone? Hansen and Sato (2011) assert that this is 1.2°C. How much due to changes in humidity and clouds? In the case of clouds, is it warming or cooling? Climate models estimate that CO_2 is the prime mover, but that secondary effects due to humidity and clouds impact the putative 1.2°C temperature rise due to CO_2. It seems clear that CO_2 *is* a driver for climate change. Humidity and clouds tend to form as a consequence of other factors. But we didn't need a funded study to reach that conclusion. As Pielke, Sr. said on his website:[5]

> "My conclusion is that their paper does not present new scientific insight but is actually an op-ed presented in the guise of a research paper by Science magazine."

Roy Spencer said:[6]

> "Just because water vapor responds quickly to temperature change does not mean that there are no long-term water vapor changes (or cloud changes)—not due to temperature—that cause climate change. Asserting so is a *non sequitur*, and just leads to circular reasoning."

The whole point of the paper by Lacis *et al.* (2010) was an attempt to counteract the skeptics who insist that humidity and clouds are the most important greenhouse gases because they account for 75% of the greenhouse effect and therefore exert

[5] *http://pielkeclimatesci.wordpress.com/2010/10/15/comment-on-the-science-paper-atmospheric-co2-principal-control-knob-governing-earth%e2%80%99s-temperature-by-lacis-et-al-2010/.*

[6] *www.drroyspencer.com/2010/10/does-co2-drive-the-earths-climate-system-comments-on-the-latest-nasa-giss-paper/.*

more forcing than CO_2. The warmists say, yes that is true, but CO_2 is the underlying cause of it all, and thus CO_2 is *the* driver for climate change. The warmists are right that, without CO_2, the Earth would be a very cold place, and it is only through the warmth produced by CO_2 that we have enough humidity and clouds to produce even more warmth than the CO_2 is able to generate by itself. The warmists are also right that CO_2 is a driver for global warming; but the really important question is how much warming does one obtain for a given amount of CO_2 and we have no credible answer to that question.

7.2 PROJECTIONS OF FUTURE CO_2 CONCENTRATION BY CLIMATOLOGISTS

7.2.1 CO_2 emissions and build-up in the 21st century

A serious concern of global climate alarmists is that future fossil-fuel usage may generate a great deal more CO_2 than the Earth can absorb, leading to higher CO_2 levels in the atmosphere, increasing the greenhouse effect, and thereby raising world temperatures. The historical variation of CO_2 concentration in the atmosphere was shown in Figures 7.2 and 7.5. In the late 20th century, the CO_2 concentration has been advancing at the rate of roughly 2 ppm per year. Since the current CO_2 concentration is about 400 ppm, it would take about 80 years for the CO_2 concentration to double (from the pre-industrial level of ~280 ppm) if the rate remained unchanged. On the other hand, the rate of CO_2 production might change in the future. If one assumes that future growth in CO_2 concentration will not remain constant at ~2 ppm/year but will increase at its present percentage gain rate of 0.5% per year, then it would take about 65 years for the CO_2 concentration to double.

There are two elements in projecting future increases in atmospheric CO_2 concentration. One element is to predict future emissions. The other element is to estimate what fraction of these emissions accumulates in the atmosphere, and what fraction is absorbed by the Earth system (oceans and biosphere). Future emissions depend on growing world population, increased industrialization, changes in fuel mix, government regulations, and improvements in energy efficiency. All of these are affected by the state of the world economy. Future accumulation depends on the ability of the Earth system to absorb more CO_2, which can only be estimated roughly.

Various models to predict the future climate of the Earth have been based on widely different estimates of future CO_2 emissions in the 21st century, as shown in Figure 7.10. The resultant CO_2 concentrations in the atmosphere from emissions from various models are shown in Figure 7.11. The heavier curves show future CO_2 concentration if the future CO_2 concentration grows at the compounded rate of 0.5%/yr, or if future CO_2 grows at the fixed rate of 1.8 ppm/yr. It can be seen that some of the models assume much greater future growth of the CO_2 concentration.

The widely used IS92a projection (made by the IPCC in 1992 as a BAU scenario) leads to annual carbon emissions that continue to increase through the 21st

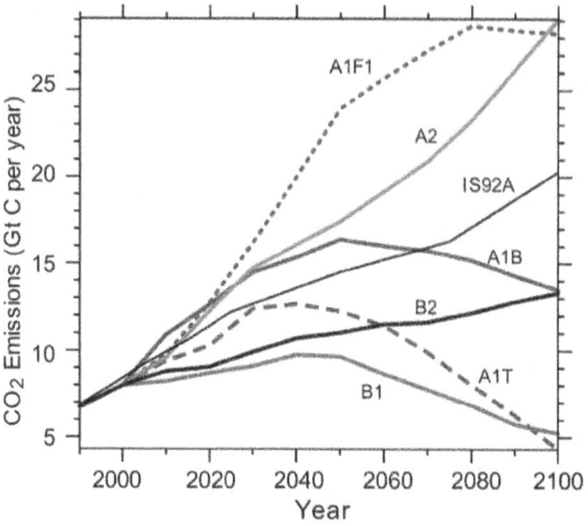

Figure 7.10. Range of projected future annual CO_2 emissions from various models (Gt/yr of carbon) (adapted from IPCC, 2001). Models A1F1 and A2 are variants of business as usual. Models A1T and B1 involve sever reductions in utilization of fossil fuels.

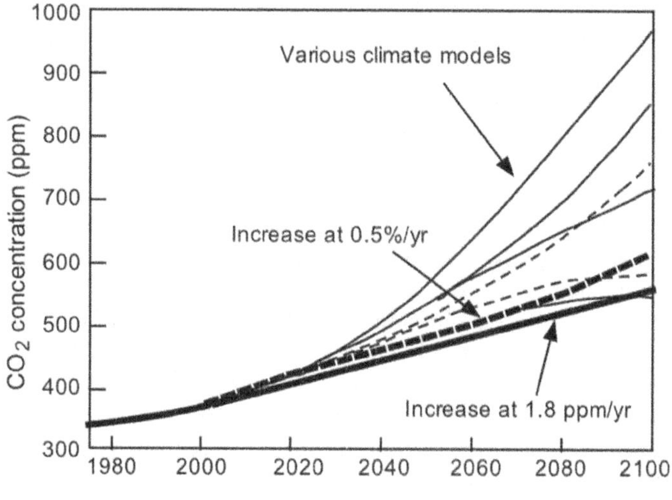

Figure 7.11. Various models for future CO_2 concentration in the 21st century (adapted from IPCC, 2001).

century from 6 Gt/yr in 1990 to 19 Gt/yr in 2100, even though the projection for carbon emitted per unit energy consumed included a significant decline over that period. This is due to the fact that the projection had a greater increase in total energy consumed (due to a burgeoning world population using increasingly more mechanization) than the decline in carbon emission per unit energy consumed. A feature of IS92a worth noting is that the share of carbon-intensive coal, relative to

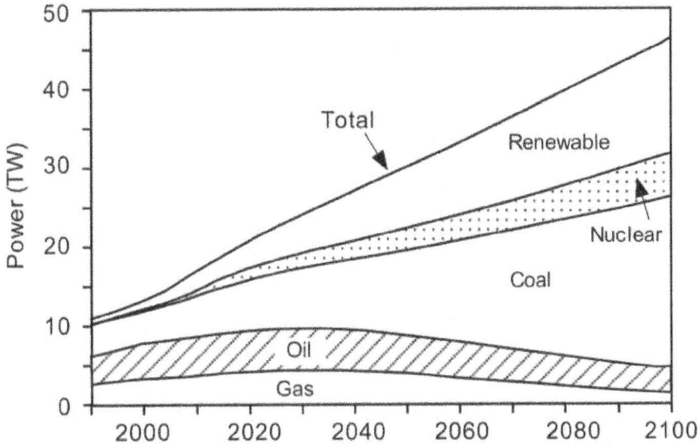

Figure 7.12. Energy mix for generation of electric power assumed by IS92a projection.

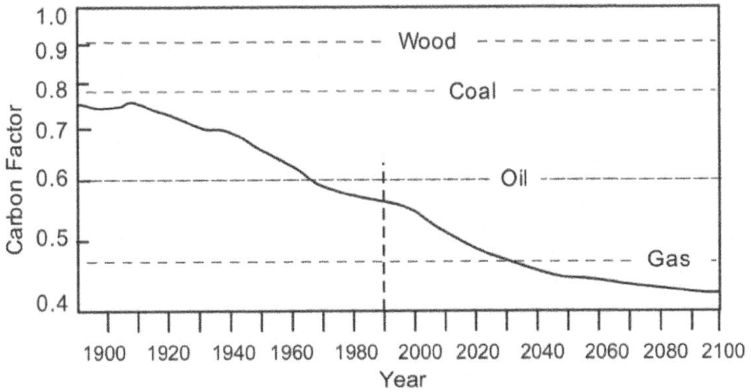

Figure 7.13. Carbon factor (kg of carbon emitted per Watt-year of power generated) assumed by IS92a projection.

less carbon-intensive natural gas and oil, rises after 2025, but the carbon emitted per unit energy consumed of the fuel mix declines overall, which implies a significant introduction of carbon-free energy sources and significant improvements in energy efficiency. Figures 7.12 and 7.13 illustrate some of the data inherent in the IS92a projection.

Webster *et al.* (2008) provided estimates of future carbon emissions correlated to ultimate CO_2 concentration stabilization targets as shown in Figure 7.14. They provided similar estimates for all greenhouse gas emissions with the results presented as gigatons of CO_2-equivalent as shown in Figure 7.15. The results in Figure 7.15, when converted to carbon (multiply data in Figure 7.15 by 14/44), are quite a bit higher than those in Figure 7.14. However, the tabulated data in Webster *et al.* (2009) do not seem to agree with the data in Figure 7.15. Their curves correspond to:

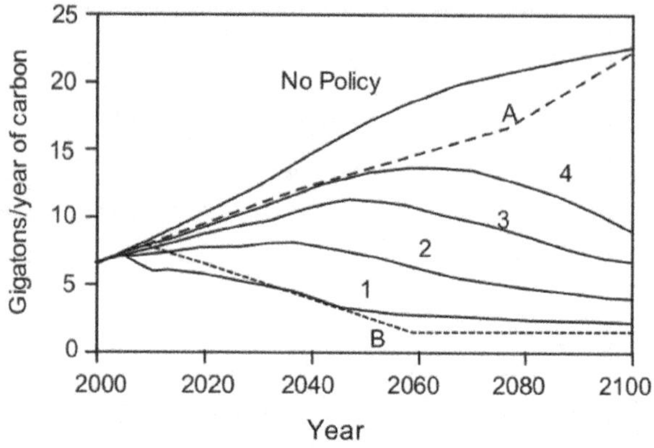

Figure 7.14. MIT estimates of carbon emissions for various levels of control (see text for details; Webster *et al.*, 2008).

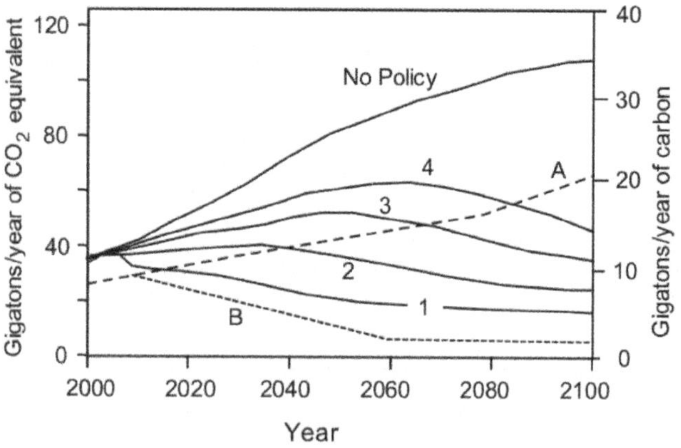

Figure 7.15. MIT estimates of all greenhouse emissions stated as CO_2 equivalent for various levels of control (see text for details.; Webster, *et al.* 2009).

- no policy for control of CO_2 emissions

1 = stabilization at 450 ppm
2 = stabilization at 550 ppm
3 = stabilization at 650 ppm
4 = stabilization at 750 ppm
A = IS92A
B = the "down-ramp" (dotted line).

The data are presented in Figure 7.16.

The common folklore is that about half the CO_2 emitted in the late 20th century was absorbed by the Earth system and the remaining half remained in the atmosphere. While this 50% assumption was representative of the past, it is not clear whether it will hold in the future. One model predicts that the 50% distribution will

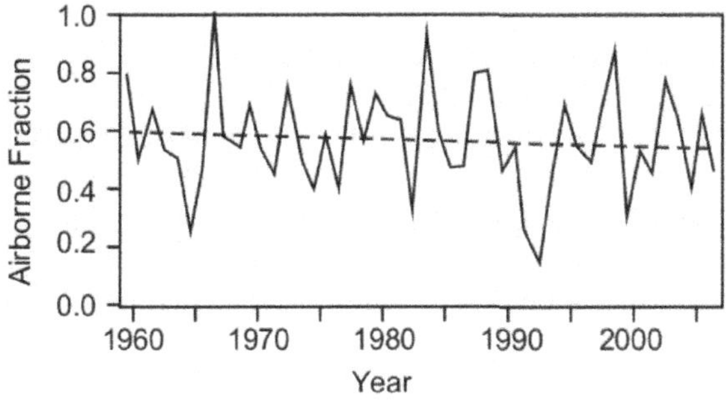

Figure 7.16. Absorbed fraction = 1 – (airborne fraction) of CO_2 (Curtin, 2009).

continue through at least 2040 (Mackenzie *et al.*, 2001). However, Sokolov *et al.* (2009) believe that the uptake by the Earth system will become saturated and reach a limit in the future. For any particular scenario for future CO_2 emissions, one must estimate the consequent rise in CO_2 concentration in the atmosphere. Knorr (2009) and Curtin (2009) reviewed data regarding the percentage of anthropogenically produced CO_2 that remains in the atmosphere as reported by studies such as that of Jones *et al.* (2005). The so-called "airborne fraction" of CO_2 remaining in the atmosphere "is known to have stayed remarkably constant over the past five decades" at "around 40%" (Knorr, 2009). Curtin (2009) estimated that:

> "... since 1958 on average 56 percent of total global emissions of CO_2 have been absorbed by the oceans (both by dissolving and by biotic uptakes) and by the terrestrial biosphere's vegetation, so that only 44 percent have remained 'aloft', and thereby increasing the atmospheric concentration of CO_2."

Several studies (e.g., Solomon *et al.*, 2009; Meinshausen *et al.*, 2009; Sokolov *et al.*, 2009; Schuster and Watson, 2007) claim that the capacities of the terrestrial and ocean sinks to uptake CO_2 are limited and will reach saturation levels of around 5 Gt C/year as CO_2 continues to be generated anthropogenically in the future. Sokolov *et al.* (2009) considered four levels of future CO_2 emissions and, in all four cases, the Earth system becomes saturated by around 2050. As a result, they claim that the percentage of emitted CO_2 taken up by the Earth system will decrease with time and, in the particular case of unrestricted future emissions, the percent of uptake by the Earth system will drop from the present value (estimated by them to be ~40%) to ~10% by 2100. If this proves to be correct, the projections of future CO_2 concentrations based on constant uptake at ~50% could be very low.

Curtin (2009) challenged this conclusion. He argued that this assumption was based on old data that suggested that, while plant growth rates increased with increasing CO_2 concentration at low to moderate CO_2 levels, plant growth rates would saturate when the CO_2 concentration becomes sufficiently high. The best data on world food production support the conclusion that there is no evidence of a slowing-down in the ability of the Earth system to absorb CO_2 as CO_2 increases. He pointed out that, over a 50-year period, the growth rate of atmospheric CO_2 has been slower than the growth rate of CO_2 emissions, while the growth rate of CO_2 emissions has increased significantly. He argued that if the Earth system continues to absorb CO_2 in the future at its present rate, and if CO_2 emissions are subjected to draconian reductions, this could result in a significant reduction in world food production—which is dependent on the present high CO_2 concentration.

Ballantyne *et al.* (2012) analyzed 50 years of global carbon dioxide measurements and found that the processes by which the planet's oceans and ecosystems absorb the greenhouse gas are not yet at capacity. They concluded that:

> "Globally, these carbon dioxide 'sinks' have roughly kept pace with emissions from human activities, continuing to draw about half of the emitted CO_2 back out of the atmosphere. However, we do not expect this to continue indefinitely."

This paper suggests that "we do not yet understand well enough the processes by which ecosystems of the world are removing CO_2 from the atmosphere, or the relative importance of possible sinks: regrowing forests on different continents, for example, or changing absorption of carbon dioxide by various ocean regions". Over the 50-year period 1960–2010, they found that cumulative emissions of carbon were 350 Gt. Of this, about 45% accumulated in the atmosphere and 55% was absorbed by the Earth's carbon sinks.

Hoffert *et al.* (1998) estimated the reductions in carbon emissions per year needed to stabilize the CO_2 concentration at various levels by 2100 as shown in Figure 7.17, based on the stabilization paths of Wigley, Richels, and Edmonds (1996). It is evident that, in their estimate, IS92a leads to a CO_2 concentration greater than 1,000 ppm toward the end of the 21st century. To achieve lower CO_2 concentrations, emissions must be significantly lower. Note, however, that, in Figure 7.19, this author's estimate for IS92a in 2100 is about 750 ppm. In order to stabilize the ultimate future CO_2 concentration in the atmosphere at some specific level, the annual emission rate must be controlled as shown in Figure 7.17.

The historical variation of CO_2 concentration in the atmosphere was shown in Figures 7.2 and 7.5. Based on these figures, it may appear that the CO_2

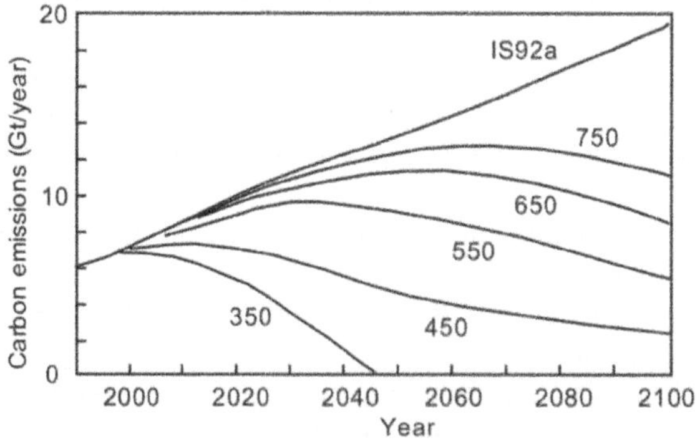

Figure 7.17. Carbon emissions per year needed to stabilize CO_2 concentration in the atmosphere at various levels of ppm (adapted from Hoffert *et al.*, 1998).

concentration is headed ever upward in the 21st century. Climatologists tend to be extrapolators. Taking trends from the past, they project forward into the future, often without regard to the constraints of finite resources. Figure 7.14 and 7.15 show the range of possible future CO_2 emissions that has been used by climatologists to predict future global warming.

A number of extrapolations involve huge future CO_2 emissions. Rutledge (2007) provided an even greater selection of 40 models, some of which yield carbon emissions rising as high as 36 Gt/yr, and 17 of which have higher emissions in 2100 than in 2005. It is not clear what assumptions (if any) were made regarding fossil-fuel resources in making these extrapolations. However, the sources of CO_2 in the atmosphere are mainly oil, gas, and coal, and all three of these have already been heavily exploited, leaving constrained resources remaining for future development.

Figure 7.18 shows five conceivable future scenarios for future emissions. One is the widely used middle-of-the-road BAU scenario from the IPCC known as "IS92a". In this scenario, annual CO_2 emissions continue to increase through the 21st century. However, it seems likely that a significant part of the predicted coal usage might be replaced by natural gas, which would reduce the level of future emissions. Hence, another scenario in which natural gas replaces part of the coal in IS92a is also shown as 2012-BAU. Three other hypothetical future scenarios are shown in this figure. In one scenario, the CO_2 emission rate is held constant at the 2010 rate (estimated to be about 8 Gt/yr of carbon) for the remainder of the 21st century. The 8 Gt/yr of carbon emissions consists of about 2 Gt/yr from land clearing and about 6 Gt/yr from fossil-fuel burning and cement production. The expectation in BAU is that the land-use figure will not change markedly but the fossil-fuel combustion will increase significantly in the BAU scenario. In the other scenarios, there are downward ramps to lower emission rates as the 21st century wears on. It should be noted that these

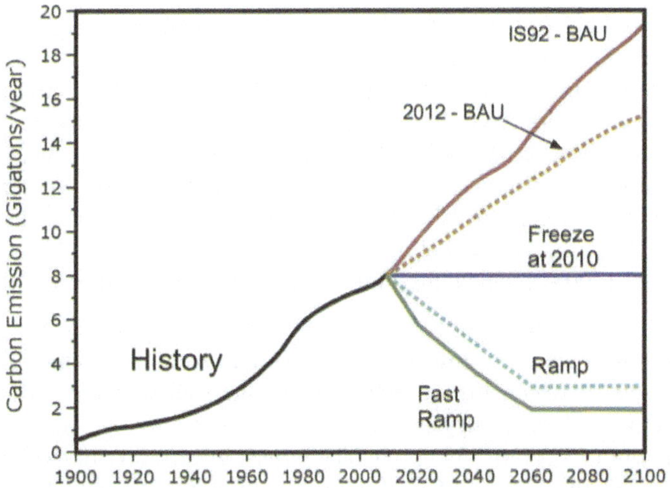

Figure 7.18. Annual emissions of carbon for five future scenarios in the 21st century.

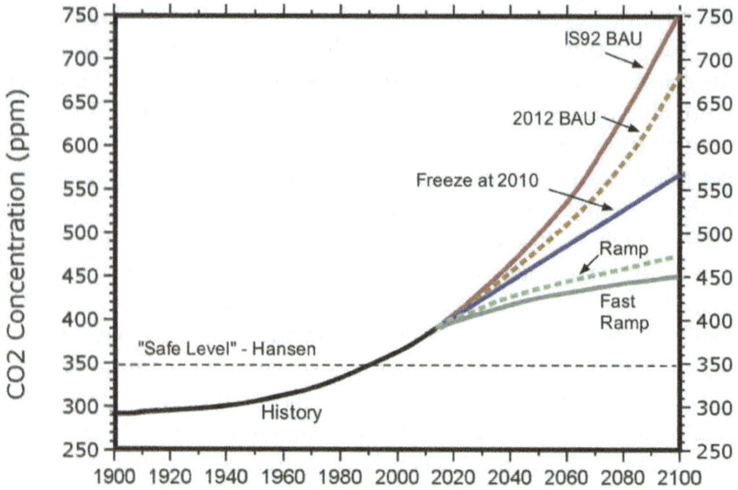

Figure 7.19. Build-up of CO_2 in the atmosphere corresponding to the five scenarios in Figure 7.18.

latter three scenarios require draconian modifications to the way that industrialized societies produce and consume energy.

For any arbitrary future scenario of emissions, one can roughly estimate how the CO_2 concentration will vary in the 21st century, assuming that 50% of the emitted CO_2 ends up as accumulation in the atmosphere. These five scenarios then lead to the build-ups of CO_2 in the atmosphere as shown in Figure 7.19, assuming that half of the CO_2 emitted ends up in the atmosphere. It was roughly estimated that each Gt of carbon emitted leads to a ~0.25 ppm increase in CO_2 concentration assuming half of the carbon ends up as atmospheric CO_2.

Krey and Clarke (2011) reported on 162 future scenarios based on 15 large-scale, energy-economic and integrated assessment models to deal with questions such as "What sorts of future levels of renewable energy deployment are consistent with different CO_2 concentration goals?" As it turns out, there is a wide range of possible future outcomes in these scenarios as shown in Figure 7.20. A very wide swath of scenarios corresponds to ultimate CO_2 concentrations in year 2000 greater than 600 ppm. According to scenarios, it will be necessary to achieve negative emissions (net absorption of CO_2) to keep CO_2 below 440 ppm in year 2100.

Hansen *et al.* (2008) pushed the alarmist agenda even farther off the charts. They argued that:

> "If humanity wishes to preserve a planet similar to that on which civilization developed and to which life on Earth is adapted, paleoclimate evidence and ongoing climate change suggest that CO_2 will need to be reduced from its current 385 ppm to at most 350 ppm. The largest uncertainty in the target arises from possible changes of non-CO_2 forcings. An initial 350 ppm CO_2 target may be achievable by phasing out coal use except where CO_2 is captured and

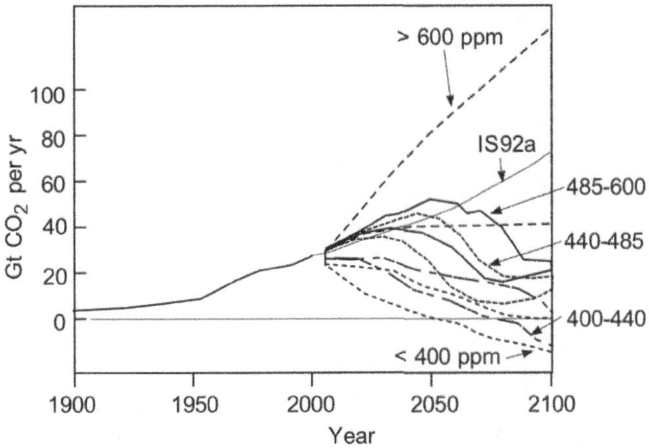

Figure 7.20. Range of future CO_2 emissions corresponding to each range of ultimate CO_2 concentration in year 2100. Note that 3.67 Gt of CO_2 are equivalent to 1 Gt of C (adapted from Krey and Clarke, 2011).

> adopting agricultural and forestry practices that sequester carbon. If the present overshoot of this target CO_2 is not brief, there is a possibility of seeding irreversible catastrophic effects."

But Hansen is not alone. Some climatologists recommend reducing the CO_2 level back to 300 ppm.[7]

7.2.2 Persistence of CO_2 beyond the 21st century

Several predictions indicate that the effects of increased CO_2 will persist for considerable time periods. Archer and Brovkin (2006) indicated that, even though most of the CO_2 released in the 21st century will be dissipated on a time scale of a century or so, a long tail will extend out to very long time periods. Montenegro *et al.* (2007) modeled the persistence of CO_2 in the atmosphere for very high emission rates (they assumed total emission of carbon to be 5,000 Gt whereas Figure 6.9a shows this to be a very unlikely scenario. Limitations on fossil-fuel availability, constraints imposed by governments, the rise of renewable energy sources, improvements in efficiency, and changes in consumption patterns will undoubtedly keep total emissions under 2,000 Gt. Nevertheless, Montenegro *et al.* (2007) found "about 75% of CO_2 emissions have an average perturbation lifetime of 1,800 years and 25% have lifetimes much longer than 5,000 years". They also found "significant surface ocean acidification, with pH decreasing from 8.16 to 7.46 units between years 2000 and 2300". Schmittner *et al.* (2008) adopted a similarly pessimistic scenario with CO_2 emission rising in an uncontrolled way until 2100, and then linearly decreasing to

[7] *http://sites.google.com/site/300orgsite/300-org—return-atmosphere-co2-to-300-ppm.*

2300, with total emissions of 5,100 Gt of carbon. They derived a doomsday prediction for the future:

> "Atmospheric CO_2 increases to a peak of more than 2000 ppm near year 2300 (that is an airborne fraction of 72% of the emissions) followed by a gradual decline to 1700 ppm at year 4000 (airborne fraction of 56%). … Global surface air warms by 10°C, sea ice melts back to 10% of its current area, and the circulation of the abyssal ocean collapses."

Solomon *et al.* (2009) estimated that, if the CO_2 concentration peaks at 500 ppm to 1,000 ppm around 2100 at 500 ppm to 1,000 ppm, and thereafter emissions stop altogether, high CO_2 concentrations will persist for thousands of years. However, these dire predictions are based on climate models whose ability to represent transfers of CO_2 within the ocean–atmosphere–land system are uncertain. There are many papers in the literature that underscore our lack of full understanding of transfers of CO_2 within the ocean–atmosphere–land system. For example, Ridgwell *et al.* (2007) said: "Our analysis highlights the importance of the prevailing uncertainties." Bousquet *et al.* (2000) emphasized uncertainty in "which regions and processes are responsible for interannual changes in the carbon balance of oceans and continents". Fung *et al.* (2005) emphasized uncertainties and "paucity of observations". Heimann (2007) said: "current models exhibit large differences—an indication of insufficient process knowledge." Sabine *et al.* (2004) emphasized uncertainties. Stauffer *et al.* (1998) said: "Determination of the relation between climate and global CO_2 concentration remains a challenge." Krakauer (2006) said:

> "The choice of parameters can significantly contribute to variations in CO_2 flux estimates obtained in inverse modeling studies. Methods such as generalized cross-validation that choose parameters systematically by optimizing a given objective function can improve inversion results. We recommend that uncertainty in inversion parameters be considered in future inverse modeling protocols and that formal parameter choice methods be used where appropriate."

Peacock, Lane, and Restrepo (2006) said:

> "There is not yet widespread agreement as to the underlying cause of the 80–100 ppmv roughly 100-kyr-duration glacial–interglacial cycles in atmospheric pCO_2. Most of the mechanisms which have been proposed to account for the observed pCO_2 variations appear to in some way violate interpretations of paleo proxy data. The inability of a single mechanism to explain the observed cycles in atmospheric CO_2 (which show amazing similarity over the past 430,000 years) is perplexing."

In addition, a number of studies have tried to understand what causes variations in CO_2 concentration in glacial–interglacial cycles and it is clear that understanding of these processes is fraught with uncertainty (e.g., Hogg, 2008; Sigman and Boyle, 2000; Maqueda *et al.*, 2002).

In the April 30, 2009 issue of *Nature*, three articles appeared that are essentially alarmist propaganda (Meinshausen *et al.*, 2009; Allen *et al.*, 2009; Schmidt and Archer, 2009). There is absolutely no doubt in these articles that CO_2 emissions were the prime cause of global warming in the 20th century. The only issue discussed is how rapidly CO_2 emissions must be reduced to save the world from disaster. The conclusion they reached is that, to save humanity, future CO_2 emissions must be reduced even more severely than that shown as the fast down-ramp in Figure 7.18—a recipe guaranteed to produce much more financial hardship in the world than putative global warming.

Predictions regarding CO_2 concentrations thousands of years in the future require further corroboration. The impacts of such accumulations of CO_2 are also uncertain. Nevertheless, it would seem prudent for the world to constrain future carbon emissions to the extent possible, while providing the people with energy and power, and avoiding global economic depression. How much constraint on future emissions is practical then becomes *the* crucial question in the climate change debate.

7.2.3 Practicality of reducing CO_2 build-up in the 21st century

Hoffert *et al.* (1998) analyzed the requirements to provide the world with needed energy while keeping a lid on ultimate CO_2 levels. In their formulation, the rate of emission of carbon to the atmosphere as CO_2 is the product of four terms:

(1) world population;
(2) gross domestic product (GDP) per person averaged for the world;
(3) energy required by the world per unit of GDP;
(4) mass of carbon emitted per unit energy consumed.

They began with a baseline of the "IS92a" projection made by the IPCC in 1992 as a BAU scenario, and then departed from there to various constraints on CO_2 production. The IS92a projection of world population is shown in Figure 7.21 with an estimated world population of 5.3 billion in 1990, a population of ~6.6 billion in 2006, and an eventual plateau of 11.4 billion at the end of the 21st century. The GDP per person was estimated by IS92a to be U.S.$4,100 in 1990 and was projected to increase at around 1.6% per year through the rest of the 21st century. The energy per unit GDP was estimated by IS92a to be 0.49 Watt-years per U.S.$ in 1990 and was (optimistically) projected to decrease at the rate of 1% per year through the remainder of the 21st century. Other aspects of IS92 are shown in Figures 7.11 and 7.12. This model evidently builds in an implicit departure from fossil fuels, as well as increases in efficiency, leading to a steadily decreasing carbon factor with time.

Hoffert *et al.* (1998) estimated the reductions in carbon emissions per year needed to stabilize the CO_2 concentration at various levels by 2100, as was shown in Figure 7.17. However, it seems evident from Figure 7.19 that IS92a leads to a CO_2 concentration of about 750 ppm in 2100, rather than 1000 ppm as given by Hoffert. To achieve lower CO_2 concentrations, emissions must be reduced significantly:

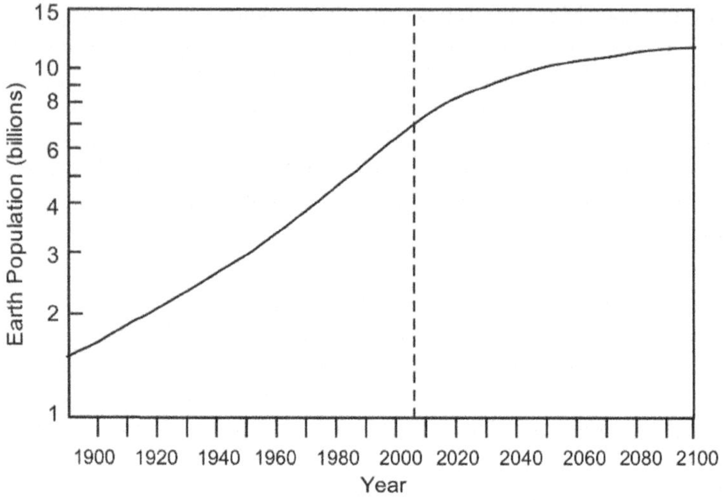

Figure 7.21. Projection of world population made by IS92a (adapted from Hoffert *et al.*, 1998).

> "Stabilizing atmospheric CO_2 at twice pre-industrial levels while meeting the economic assumptions of 'business as usual' implies a massive transition to carbon-free power, particularly in developing nations." (Hoffert *et al.*, 1998)

Lightfoot and Green (2002) estimated the required rate of world average annual energy intensity decline required to stabilize the level of CO_2 in the atmosphere at some level, such as 550 ppm in 2100 (about double the pre-industrial level). However, their projections of energy mix and energy efficiency for 2100 seem very optimistic.

The present world population of about seven billion may grow to as much as 10 to 11 billion by 2100. If prosperity expands, people will want to use more energy per capita. This burgeoning world population will seek a better life by consuming more energy per capita. Most of these people have a major goal in life to emulate the lifestyle of Americans, which entails using energy at a rate comparable with that of Americans. There is no way that they can be successful. The U.S. has about 5% of the world's population and uses about 20% of the world's energy. If the people of the world emulated Americans, world energy usage would increase by a factor of 4 even without an increase in world population. Increasing population would drive energy usage still higher. World primary power consumption in 2006 was 12 terrawatts (TW), of which ~85% was fossil-fueled. By 2050, world power demand may grow to as much as 30 TW. By 2100, it will likely be considerably higher. The main problem facing humankind in the 21st century is providing itself with these enormous energy requirements. If one were to assume that oil, gas, and coal resources are unlimited, and fossil fuels were produced in sufficient quantity to supply the impending demand during the 21st century with a mix typical of today, carbon emissions during the 21st century would be very high and the likelihood of global warming would be significant. If the role of coal increases in the 21st century, it will exacerbate the problem because coal has a higher carbon content than oil or gas per unit energy produced. However, the estimated fossil-fuel resources remaining are limited. Hydrocarbon production will likely top out before 2030 and coal production will likely top out around 2035. Assuming that greenhouse gas emissions are limited in the 21st century by the available fossil-fuel resources, emissions will peak by 2030–2035. The ultimate run-out to depletion of fossil resources would just be enough to double the CO_2 concentration, because energy production in the second half of the 21st century will have to undergo significant changes as fossil-fuel production (and consumption) inevitably pursues an increasingly downward trend.

The world might face a crisis some time around or after 2030. But that crisis will not be calamitous global warming. The crisis will likely be that, with oil, gas, and coal production going at full bore, the world will have difficulty supplying the energy that is demanded. This could lead to significantly higher energy costs, resulting in worldwide economic recession or depression. However, on the positive side, it provides incentive to develop renewable energy to become more competitive. Whether renewable energy can be developed and expanded rapidly enough to stave off economic collapse remains to be seen. Alternatively, Idso and Idso (2007) proposed a massive increase in the use of nuclear power, glossing over the problems inherent in such a strategy.

Brown *et al.* (2011) analyzed the relationship between energy use and economic growth. They concluded that:

> "Empirically, the central role of energy in modern human economies is demonstrated by the positive relationship between energy use and economic growth. ... To support a projected global population of 9.5 billion in 2050 with an average standard of living equivalent to the current US lifestyle would require about 268 terawatts, 16 times the current global energy use. Even maintaining this increased population at the more modest Chinese standard of living would require 2.5 times more energy than is used today. ... The bottom line is that an enormous increase in energy supply will be required to meet the demands of projected population growth and lift the developing world out of poverty without jeopardizing current standards of living in the most developed countries."

The problem with the over-emphasis on global warming by the warmists is that they have lost focus on the real problem facing humanity in the 21st century: providing the people of the world with energy as the developing countries gradually industrialize. While some naïve futurists think we can solve our problems with wind and solar energy, electric cars, and carbon capture and sequestration, it is difficult to see how we can make it through the current century without continued heavy use of fossil fuels. Even James Hansen, a leading warmist, said: "... suggesting that renewables will let us phase rapidly off fossil fuels in the United States, China, India, or the world as a whole, is almost the equivalent of believing in the Easter Bunny and Tooth Fairy." Meanwhile, the future looks pessimistic as the world population grows.

Hoffert *et al.* (2002) emphasized that the problem of stabilizing the CO_2 levels in the atmosphere relates to energy:

> "In the 20th century, the human population [of the Earth] quadrupled and primary power consumption increased 16-fold."

Creating a transition toward such stabilization will require reductions in energy consumption as well as development of primary energy sources that do not emit carbon dioxide to the atmosphere. Mid-century primary power requirements that are free of carbon dioxide emissions could be several times what we now derive from fossil fuels ($\sim 10^{13}$ watts), even with projected improvements in energy efficiency. Hoffert *et al.* (2002) surveyed potential future energy sources with emphasis on their capability to supply massive amounts of carbon emission-free energy. These included terrestrial solar and wind energy, solar power satellites, biomass, nuclear fission, nuclear fusion, fission–fusion hybrids, and fossil fuels from which carbon has been sequestered. They also studied non-primary power technologies that could contribute to climate stabilization including efficiency improvements, hydrogen production, storage and transport, superconducting global electric grids, and geo-engineering. They concluded that all of these approaches currently have severe deficiencies that limit their ability to stabilize the production of CO_2. Furthermore, they suggested that the IPCC (and many others) are overly optimistic regarding the potential for advanced energy technologies to quickly reduce CO_2 emissions significantly. Their conclusion was that we need a drastic expansion of research on renewable energy.

Anon. (H) provided a number of commentaries on Hoffert *et al.* (2002). Some of these came to the defense of their energy technologies (particularly solar-thermal and nuclear) but these appear to be self-serving to some degree. Several comments had to do with the role of energy conservation. Professor Albert Bartlett (2002) astutely said:

> "Even without the greenhouse problems, the obvious impossibility of continuing these [past energy] growth rates would lead rational people to say that … the world's first order of business should be to stop the growth of populations and the growth of *per capita* primary power consumption. Instead of advocating the obvious, the authors paint a picture of all manner of technological fixes that, at enormous expense, may provide some answers to the need to stop the growth in emissions of greenhouse gases that are associated with energy production. As is so often the case, technological fixes are offered without being reviewed in the light of Eric Sevareid's Law: 'The chief cause of problems is solutions.' One can be sure that each technological solution will create new problems that are not indicated by calculations, equations, and technical speculations."

G. W. Bush's administration placed major emphasis in their energy research on a hydrogen economy. Zubrin (2007) quotes a Secretary of Energy who said:

> "We envision a future economy in which hydrogen is America's clean energy choice—flexible, affordable, safe, domestically produced, used in all sectors of the economy, and in all regions of the country. … Environmental pollution will no longer be a concern. Every nation will have all the energy it needs available

> within its borders. ... The sources of hydrogen are abundant. The more you have of something relative to demand for that, the cheaper it's going to be, the less expensive it'll be for the consumer. ... One of the greatest results of using hydrogen power, of course, will be energy independence for this nation. ... If we develop hydrogen power to its full potential, we can reduce our demand for oil by over 11 million barrels per day by the year 2040."

In the words of Robert Zubrin: "Its all bunk." Zubrin then went on to describe alternative processes for producing, storing, and using hydrogen, and showed clearly that the whole concept of a hydrogen economy is impractical and costly (Zubrin, 2007).

Further analyses of the effects of various levels of future greenhouse gas emissions according to various scenarios continue to be made. Schewe *et al.* (2011) provide references for much of this work. Generally, these studies begin with various models for future emissions of greenhouse gases. From this, estimates are made of the future rise in CO_2 concentration. They then identify scenarios for future emissions that lead to various levels of CO_2 concentration in 2100 and beyond. From this, they derive estimates of future temperature rise from global climate models, assuming that these models are representative of reality (for which there is no proof). By selecting some level for future temperature growth as the ultimate acceptable limit, they identify which down-ramp for future emissions is the maximum tolerable. Until recently, the U. N. had concentrated on future emission scenarios designed to keep "global warming below 2°C", although it is not clear what baseline is taken at the end of the LIA to define "global warming". However, Schewe *et al.* (2011) indicated that the U. N. is now preoccupied with limiting "global warming" to 1.5°C "because climate change impacts associated with 2°C are considered to exceed tolerable limits for some regions, e.g. Small Island States".

Academics live in a cloistered environment dreaming up idealistic solutions to world problems. It is now common for academics, particularly in Europe, to map out strategies for supplying the world with energy in the future with sharply lower carbon emissions. For example, Held[8] accepted the results of climate models that predict a 2°C rise from pre-industrial times when CO_2 reaches 450 ppm. He then accepted that it would be impractical to reduce emissions fast enough to stay below 450 ppm by year 2100, and accepted the putative 2°C temperature rise by 2100 as unavoidable even with the best of efforts. Held claimed he could achieve the 2°C limit on temperature rise with the down-ramp for CO_2 emissions shown in Figure 7.22. In this graph, Held's proposed down-ramp and his version of BAU are shown as heavy lines. Also shown as light lines are the curves from Figure 7.18. It is evident that Held's down-ramp is similar to the down-ramps in Figure 7.18. It is also noteworthy that Held's version of BAU lies considerably higher than that given in Figure 7.18.

How does Held propose to achieve the fast-ramp? His answer is shown in Figure 7.23. Estimates for world energy consumption in 2100 vary from 500 to 1,500 EJ (Exajoules). Future world energy consumption will depend on the growth in world

[8] *www.newton.ac.uk/programmes/CLP/seminars/120916101.pdf.*

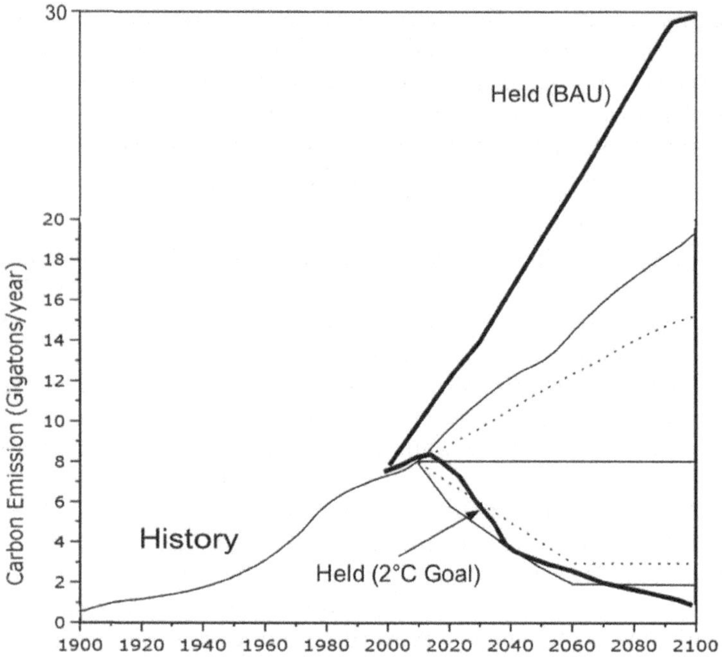

Figure 7.22. European comparison of BAU emissions with target for putative 2°C temperature rise by 2100 (compare with Figure 7.18; *www.newton.ac.uk/programmes/CLP/seminars/120916101.pdf*).

population, the degree of industrialization of developing nations, the state of the world economy, and the extent of new technology development for energy efficiency, and it remains difficult to predict. Held assumed about 1,000 EJ—a not unreasonable guess in the middle of the range.

There are several problems with Figure 7.23. One is that it is not exactly compatible with the Held down-ramp in Figure 7.22. If one considers only the period 2010–2050 in Figure 7.23, the sum of biomass, gas, oil, and coal actually rises from 2010 to 2020, and then very slowly decreases from 2020 to 2050. The sum in 2050 is not much different than the value in 2010. Hence, Figure 7.23 implicitly corresponds to a relatively flat rate of carbon emission at 8 Gt/yr from 2020 to 2050. This is quite different from the fast ramp proposed in Figure 7.22. Furthermore, according to Figure 7.23, the sum of coal, oil, gas and biomass in 2100 is about half of the sum in 2010. But the Held's fast ramp in Figure 7.22 shows carbon emissions in 2100 to be about 1/8 of the value in 2010.

Another problem is that, according to Figure 7.23, oil consumption in 2100 is almost the same as in 2010. As we point out in Section 7.2.4, it seems unlikely that the oil supply will be adequate for this continued rate of consumption. Countering this, it is possible that natural gas consumption will rise as oil becomes depleted. It is also indicated that nuclear power will greatly expand as the century wears on.

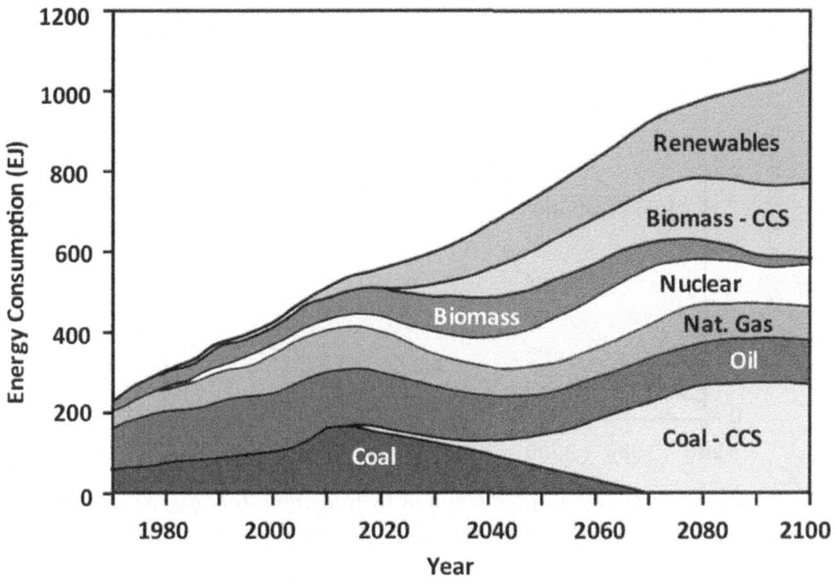

Figure 7.23. Held's projection of world energy supply to year 2100 (*www.newton.ac.uk/programmes/CLP/seminars/120916101.pdf*; CCS = carbon capture and storage).

Whether this can be achieved safely remains to be seen. Finally, there is great emphasis on carbon capture and sequestration (CCS) later in the century. Whether this technology will turn out to be technically feasible and affordable, remains to be seen. It seems likely that the cost of energy will go up significantly as these changes are implemented, which will impact the world economy, although it will also encourage technology development for energy efficiency.

7.2.4 Constraints on CO_2 production imposed by the limits of fossil fuels

The increase in CO_2 concentration in the atmosphere from the pre-industrial estimate of 280 ppm to the present value of 400 ppm shows that there has been an increase of about 120 ppm. In this period, the world burned a great deal of fossil fuel. Now we must ask how much fossil fuel remains and what is the probable limit to future CO_2 concentration in the atmosphere from burning the available fossil fuels in the 20th century?

Deffeyes (2001) provided an analysis of probable U.S. and world oil reserves. The best estimate was that cumulative world oil production will eventually reach about 2.1×10^{12} barrels, and the world has already produced roughly half that amount. Rogner (1997) provided a slightly more optimistic estimate. U.S. oil production was estimated at around 0.22×10^{12} barrels, and about 85% of that has already been consumed. Oil isn't the only fossil fuel; there are also natural gas, natural gas liquids, and coal. It was estimated that world production of total hydrocarbons (oil, gas, and gas liquids) may approach 4.5×10^{12} barrels of oil (equivalent), and the world has

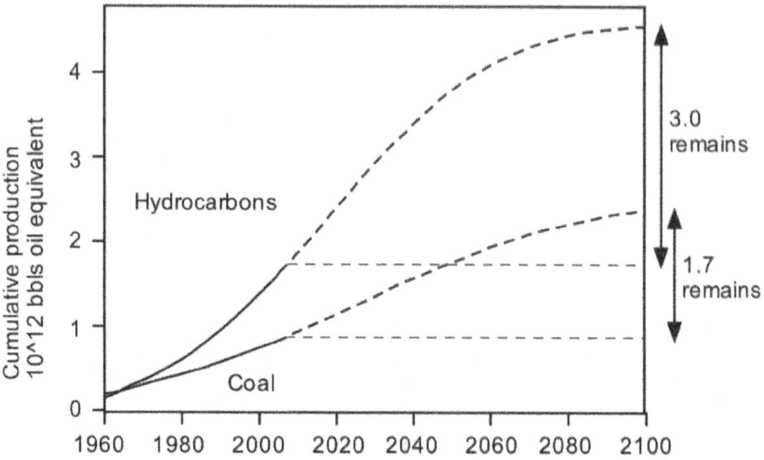

Figure 7.24. Estimated cumulative production of hydrocarbons (oil, gas, gas liquids) and coal to date, and projected ultimate production (adapted from Rutledge, 2007).

already produced roughly 38% of that amount. These resources are finite and significant fractions of their initial endowments in the Earth have already been exploited (Rutledge, 2007).

A simplistic analysis would suggest that, since burning ~38% of the world's fossil fuels has increased the CO_2 concentration in the atmosphere by about 100 ppm, burning the remainder is likely to add another ~150 ppm, bringing the ultimate level to perhaps 545 ppm by the end of the 21st century, at which time the world will have limited amounts of economically recoverable fossil fuels remaining. This suggests that future carbon emissions will be somewhat limited as fossil fuels become depleted. There may be barely enough fossil fuel to "double the CO_2 concentration in the atmosphere" (relative to pre-industrial levels), and all the climatologists who continue to deal with higher levels of CO_2 might be working in fantasy.

Estimation of future production of fossil fuels is a tricky business that requires extrapolation from past data. Rutledge (2007) has done a good job of analyzing the (admittedly sketchy) available information and projecting forward to the future. Figure 7.24 shows the expectation for future cumulative production of fossil fuels extrapolated from current levels. Figure 7.25 shows the annual production rates corresponding to the cumulative rates in Figure 7.24 (Figure 7.25 is just the slope of Figure 7.24).

Rutledge (2007) estimated CO_2 emissions based on a production rate limited by the finite resources as indicated in Figures 7.23 and 7.24. This is compared with the range of projections made by the IPCC in Figure 7.26. It can be seen that most projections of future CO_2 emissions are overly generous in their implied estimate of remaining fossil-fuel resources.

A recent discussion of fossil-fuel resources was provided by Rud Istvan on the website run by Judith Curry.[9] G. Maggio, and G. Cacciola wrote a paper entitled

[9] *http://judithcurry.com/2013/02/01/another-hockey-stick/#more-11053.*

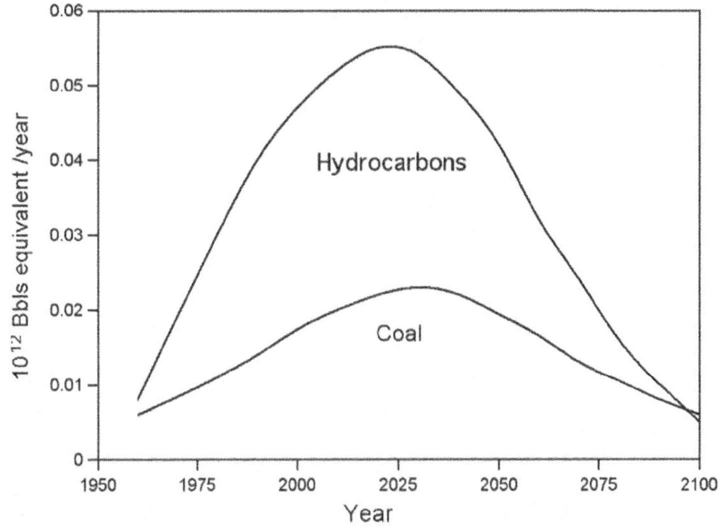

Figure 7.25. Estimated annual production of hydrocarbons (oil, natural gas, and natural gas liquids) and coal to date, and projected ultimate production by 2100 (from Rutledge, 2007).

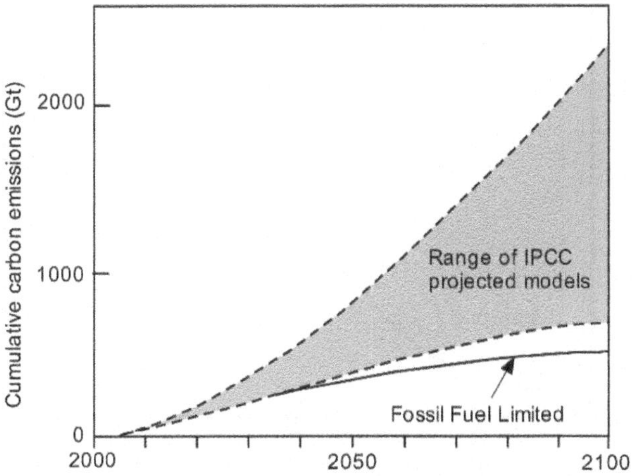

Figure 7.26. Comparison of cumulative CO_2 emissions based on fossil-fuel constraints with the range of IPCC projections for future CO_2 (from Rutledge, 2007).

"When will oil, natural gas, and coal peak?".[10] Their result is shown in Figure 7.27.

By contrast, the IPCC insists that there is no peak in any fossil-fuel production by 2100. This view is supported by Maugeri,[11] who said: "Contrary to what most people believe, oil supply capacity is growing worldwide at such an unprecedented

[10] *www.sciencedirect.com/science/article/pii/S001623611200230X.*

[11] *http://belfercenter.ksg.harvard.edu/publication/22144/oil.html.*

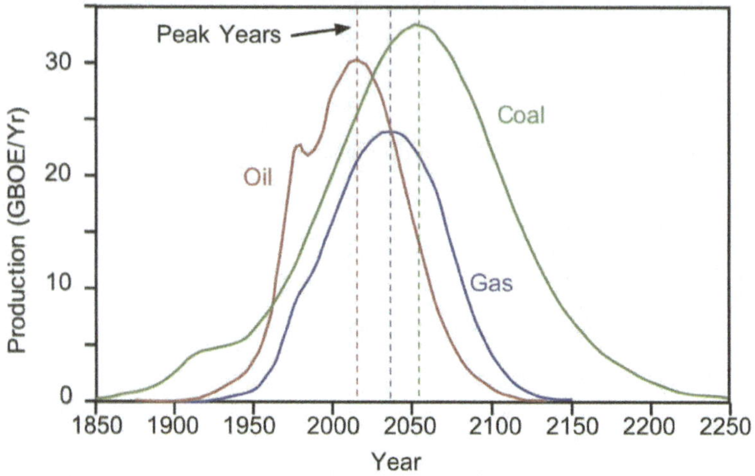

Figure 7.27. Estimated world production of oil, gas and coal showing peak years (gigabarrels of oil equivalent). (Maggio and Cacciola, 2009).

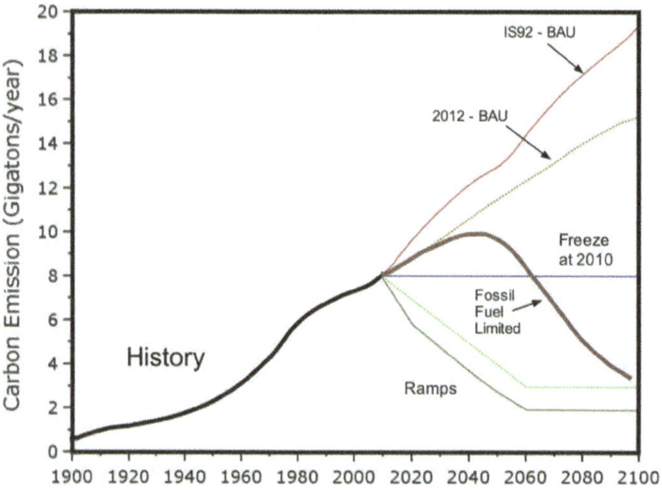

Figure 7.28. Annual emissions of carbon for five future scenarios in the 21st century compared with a fossil-fuel-limited scenario.

level that it might outpace consumption. This could lead to a glut of overproduction and a steep dip in oil prices." Istvan explains why he disagrees with Maugeri.

The history of estimating oil and gas reserves is reminiscent of "The Boy Who Cried Wolf". Ever since the early days of exploration, there have been repeated claims that we will soon run out of oil or gas. These claims have been disproved over and over again as new resources were discovered and new means of extraction were developed. Nevertheless, world resources of fossil fuels, though greater than heretofore estimated, will eventually run down. Peak production will be reached long

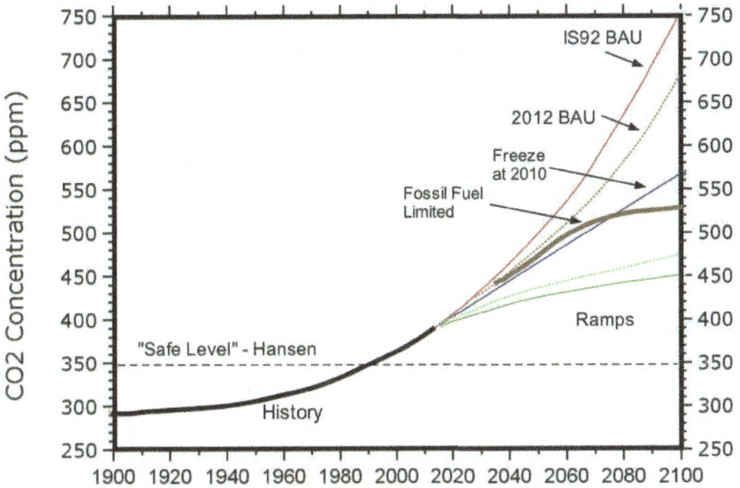

Figure 7.29. Build-up of CO_2 in the atmosphere corresponding to the six scenarios in Figure 7.28.

before the resources are depleted. The key issue for us here is the timing of peak production, which will determine whether finite resources of fossil fuels will limit growth of CO_2 concentration in the 21st century. There seems to be little doubt that global peak production of oil will be reached before 2025 and global peak production of gas will be reached before 2040. Coal remains a major unknown. As of early 2013, China continued to rapidly increase its use of coal while the remainder of the world flattened out its yearly usage. In fact, China uses almost as much coal as the rest of the world combined. Will there be sufficient fossil fuels to enable Figure 7.23 to be fulfilled? How will limitations on fossil fuels affect the curves in Figure 7.18? We don't know the answers to these questions. However, if the estimates of Maggio and Cacciola are taken at face value, then limitations of fossil-fuel production will impose limits on Figures 7.18 and 7.19 as shown in Figures 7.27 and 7.28.

7.3 BLACK CARBON AS A SOURCE OF 20TH-CENTURY GLOBAL WARMING

7.3.1 Warming early in the 20th century

There are aspects of Figure 3.21 that have created difficulties for climate models oriented toward greenhouse gases as the putative sole cause of climate change. One important factor is the rise in temperatures from 1910 to 1940 prior to large-scale CO_2 emissions. The increase in Arctic temperatures is particularly impressive, exceeding temperature changes in other regions by up to a factor of 5. As Nagashima *et al.* (2006) said:

> "The warming of the Earth's near-surface temperature in the latter part of the 20th century has been mainly attributed to increases in greenhouse gases, while the warming that occurred in the early 20th century has not yet been clearly attributed to any particular climate forcing agents."

The warming early in the century poses a major problem for the CO_2 advocates and they do not appear to have any answers.

As Bengtsson *et al.* (2004) said:

> "The huge warming of the Arctic that started in the early 1920s and lasted for almost two decades is one of the most spectacular climate events of the twentieth century. During the peak period 1930–40, the annually averaged temperature anomaly for the area 60°–90°N amounted to some 1.78°C. Whether this event is an example of an internal climate mode or is externally forced, such as by enhanced solar effects, is presently under debate. . . . It was a long-lasting event commencing in the early 1920s and reaching its maximum some 20 years later. The decades after were much colder, although not as cold as in the early years of the last century. It is interesting to note that the ongoing present warming has just reached the peak value of the 1940s, and this has underpinned some views that even the present Arctic warming is dominated by factors other than increasing greenhouse gases."

Tisdale (2009) provided the reconstructed SSTs shown in Figure 7.30. These data also show a sharp rise from about 1910 to 1940, a shallow dip from 1940 to

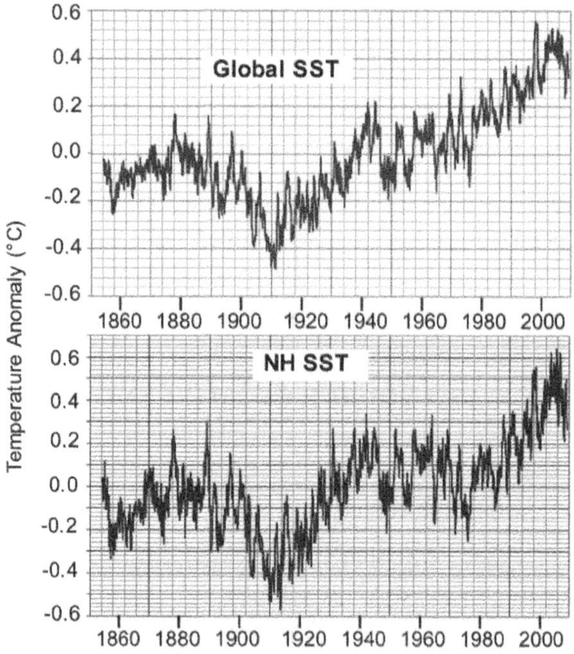

Figure 7.30. Reconstructed sea-surface temperatures (SSTs) based on ERSST.v3b as provided by Tisdale (2009).

about 1980, and a subsequent rise at roughly the same rate as that which prevailed from 1910 to 1940.

The sharp temperature rise in the North Polar region that occurred just prior to 1920 was remarkable. Bernaerts (2009) provided a number of historical references to this event. For example, in 1922, the *Washington Post* published a story entitled "Arctic Ocean getting warm; seals vanish and icebergs melt", stating that "ice conditions in the Northern North Atlantic were exceptional; in fact, so little ice has never before been noted". This article said: "The Arctic Ocean is warming up, icebergs are growing scarcer and in some places the seals are finding the water too hot." B.J. Birkeland in 1930 said the temperature rise was "probably the greatest yet known on Earth" and, a few years later, A.W. Ahlmann in 1946 called the event a "climatic revolution".

In 1935, Jules Schokalsky said: "The branch of the North Atlantic Current which enters it by way of the edge of the continental shelf round Spitsbergen has evidently been increasing in volume, and has introduced a body of warm water so great, that the surface layer of cold water which was 200 meters thick in Nansen's time, has now been reduced to less than 100 meters in thickness."

Bernaerts mentions an oral presentation at a symposium in 1956 in which Hesselberg and Johannessen said:

> "Of special interest are the data from Spitsbergen where the series of observations go back to 1912. During the first years the observations shows no conspicuous climatic change, but then comes a rapid rise of the temperature in the year 1917 to 1922. The increase of the mean temperatures in this period was about 7 degrees Celsius in the winter, 3 degrees in the spring, 3 degrees in the summer, 3 degrees in the autumn and 4 degrees for the whole year. After the year 1922 the temperature continued to rise ... but the rise was much slower. ... The rise of the temperature in Spitsbergen is large compared with the rise in other parts of the world (about five times as great as in Norway)."

Bernaerts (2009) pinpointed the sharp temperature rise in the Arctic region that occurred around 1920 by studying the recorded temperatures at Spitsbergen. Spitsbergen is a remote archipelago between the North Cape in Norway and the North Pole. It is an ideal site for climate research because it is located:

- between three huge water bodies in volume and size—the Norwegian/Greenland Sea; the Arctic Ocean; and the Barents Sea, with a modest volume (mean depth ~280 m) but considerable size;
- at the edge of sea ice, where, regardless of the time of season, at least a tiny space of the sea remains ice-free, which ensures a maritime-induced climatology, while a space covered with sea ice induces continental climatology;
- where the Sun does not rise above the horizon for the whole winter period, from October 26 to February 16.

As Bernaerts emphasized, the temperature rise of 1919:

- was greater in magnitude and more rapid than the recent one (1970s–2000s);

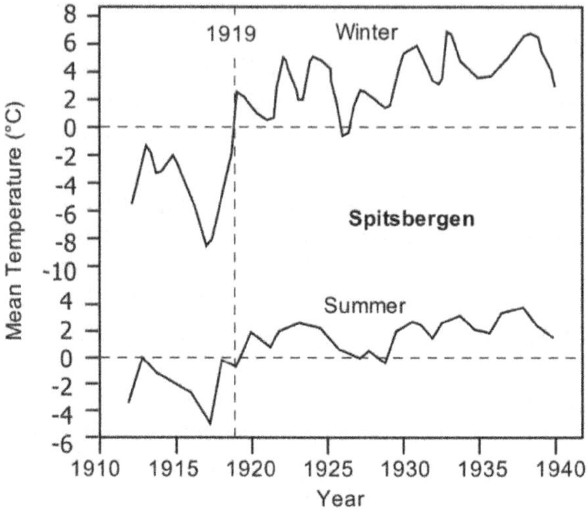

Figure 7.31. Measured summer and winter temperatures at Spitsbergen (Bernaerts, 2009).

- displayed exceptionally rapid winter warming;
- had no summer signal at all.

Thus, the temperature rise from 1910 to 1940 was dominated by a single-step rise around 1919. Bernaerts (2009) provided a good deal of evidence and argument that this sudden warming was due to an influx of warm ocean waters. However, he then gave rather murky (and incredible) arguments for why he believed that naval warfare in the First World War was responsible.

Figure 7.31 illustrates temperature variations at Spitsbergen.

Chylek *et al.* (2006) analyzed Greenland temperature records to compare the recent (1995–2005) warming period with the previous episode (1920–1930) of Greenland warming. They concluded that temperature increases in the two warming periods were of similar magnitude. However, the rate of warming in 1920–1930 was about 50% higher than that in 1995–2005.

Box *et al.* (2009) utilized a combination of meteorological station records and regional climate model output to develop a continuous 168-year (1840–2007) spatial reconstruction of monthly, seasonal, and annual mean Greenland ice sheet near-surface air temperatures. Their results are shown in Figure 7.32. These results also show a remarkable winter temperature rise just after 1920 while summer temperatures hardly changed. The post-1980 temperature increase is approaching the levels attained around 1940.

While particles suspended in the atmosphere produce a "dimming" effect, Flanner *et al.* (2008) found that the heating effect of settled black carbon (BC) is about six times greater than the dimming effect of carbon suspended in the air. The greatest warming effect occurs in spring, when there is still ample snow and ice on the surface but solar intensity is significant.

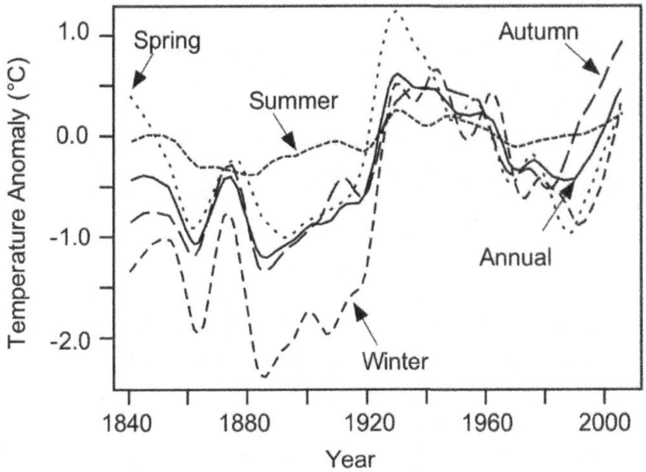

Figure 7.32. Reconstructed near-surface air temperatures over Greenland (Box *et al.*, 2009).

A variety of positive feedback mechanisms amplify the trends originated by these factors. Of particular importance are (1) the increase in humidity expected from a warming trend which adds water vapor greenhouse warming, and (2) the positive feedback of albedo changes as high-latitude snow and ice expands or contracts.

Estimates exist for the rate of emission of CO_2, BC, and sulfate aerosols in the 20th century. CO_2 emissions have increased steadily throughout the 20th century and are still increasing, even though emission per unit energy generated has decreased with time. BC emissions increased sharply toward the end of the 19th century and peaked in the 1920s. However, emissions continued through the remainder of the century, but the primary emission regions changed from the U.S., Europe, and the former U.S.S.R. in the first half of the century to developing nations (particularly China) in the second half of the century. Aerosol emissions rose throughout the 20th century but peaked in the early 1980s. Nevertheless, aerosol emissions remained significant after the peak.

In addition to these anthropogenic influences, there seem to be natural cycles in the climate that produce significant climate changes even in the absence of anthropogenic influences. Of particular interest is the behavior of the oceans which retain a great deal of heat energy. The Pacific Ocean underwent a major change in the mid-1970s that produced predominance of El Niño conditions for almost three decades. These warm surface waters undoubtedly contributed to the climate in the NH after the 1970s.

The task of resolving anthropogenic influences from natural cycles is rendered difficult by the variety of anthropogenic influences and the uncertainties in historical climate changes. Many climatologists have focused on the role of increased CO_2 as the putative driver of global warming via the greenhouse effect with amplification by increased water vapor. All climate models lead to significant amplification of temperature increases in polar areas, as discussed by Serreze *et al.* (2008) and Holland and Bitz (2003). While the majority of climatologists have focused on greenhouse gases as the putative source of 20th-century global warming with amplification in polar areas, it is also possible that a major source of global warming has been located in the Arctic with consequent spillover into neighboring latitudes.

Greenhouse gases cannot explain the sharp rise in Arctic temperatures prior to 1940 because carbon emissions were low during that period. A more likely explanation centers on the role of BC from the U.S., Europe, and the former U.S.S.R. in the first half of the century falling on Arctic snow and ice. By about 1940, BC emissions from mid-north latitudes diminished but sulfate aerosol emissions increased steeply after 1940. It seems likely that aerosols were dominant in producing a cooling effect from about 1940 to the mid-1970s. BC played a secondary role during this period, being overwhelmed by the sharp rise in sulfate aerosol emissions. It is not so clear what was responsible for the renewed global warming that began in the 1970s. While BC emissions began to rise again during this period, the region of predominant emission moved from more northerly locations to Asian regions at lower latitudes. It is not clear how efficiently such emissions can be transported to higher latitudes. CO_2 emissions rose sharply after the 1970s. But another factor is the previously mentioned state change of the Pacific Ocean in the 1970s. There is a close correspondence between surface air temperatures and El Niño indices over the past 30 years.

Bengtsson *et al.* (2004) suggested that "four possible mechanisms, individually or in combination, could have contributed to the early twentieth century warming: anthropogenic effects, increased solar irradiation, reduced volcanic activity, and internal variability of the climate system". They concluded: "It seems unlikely that anthropogenic forcing on its own could have caused the warming, since the change in greenhouse gas forcing in the early decades of the twentieth century was only some 20% of the present." However, in considering anthropogenic effects, they dealt only with greenhouse gases, and did not consider the deposition of BC. They pointed out the uncertainties in reconstructing past solar irradiance, and they dismissed volcanic activity as the cause of this warming. Therefore, they sought an answer in terms of the atmospheric flow pattern that drives ocean circulation and results in the advection of warm water into the north-eastern North Atlantic. Johannessen *et al.* (2004) concluded that the warming of the 1920s and 1930s was due to "natural fluctuations internal to the climate system". Reductions in albedo due to decreasing sea ice induced by wind changes were attributed as the cause of this early warming. However, they claimed that more recent warming in the 1980s and 1990s was due to greenhouse gases. However, Polyakov *et al.* (2003b) concluded that the Arctic is subject to natural oscillatory variations, the principal driver for climate change, and that "[greenhouse] warming alone cannot explain the retreat of Arctic ice observed in the 1980s–90s". Their final conclusion was:

> "The complicated nature of Arctic temperature and pressure variations makes understanding of possible causes of the variability, and evaluation of the anthropogenic warming effect most difficult."

In the present section, another possible factor in Arctic climate change is explored: deposition of BC on Arctic snow and ice.

Ramanathan and Carmichael (2008) presented a review of two opposing factors than emanate from "fossil fuel combustion (diesel and coal), open biomass burning (associated with deforestation and crop residue burning), and cooking with biofuels". Black carbon is often transported over long distances, mixing with other aerosols along the way. The aerosol mix can form transcontinental plumes of atmospheric brown clouds, with vertical extents of 3 km to 5 km. This can form "widespread atmospheric brown clouds in a mixture with other aerosols", which can produce a dimming effect for incoming solar irradiance, thus producing a cooling effect. At the same time, solar absorption of deposited BC is claimed to be the "second strongest contribution to current global warming, after carbon dioxide emissions". Painter *et al.* (2013) argue that the decline of the LIA was likely driven by black carbon deposits on alpine glaciers from expanding industrialization of Western Europe in the mid-to-late 19th century. This might have continued into the early 20th century as well.

7.3.2 Effect of carbon deposition on ice and snow

According to Jacobsen (2007) quoting the IPCC:

> "Soot is an amorphous-shaped particle emitted into the air during fossil-fuel combustion, biofuel combustion, and biomass burning. Soot particles contain black carbon, organic carbon, and smaller amounts of sulfur and other chemicals. ... Soot particles that fall to snow and sea ice surfaces, either on their own or within ice crystals or snow flakes, darken those surfaces, contributing to the melting of snow and ice and the warming of air above both. When soot particles age in the atmosphere, they become coated by relatively transparent or translucent chemicals, increasing their size and the probability that sunlight will hit and be absorbed by the particles. As such, aged, coated soot particles heat the air more than do new, uncoated soot particles. ... Calculations suggest a strong net global warming by fossil-fuel plus biofuel soot. Soot particles containing black carbon, from fossil-fuel and biofuel burning sources, have a strong probability of being the second-leading cause of global warming after carbon dioxide and ahead of methane. Because of the short lifetime of soot relative to greenhouse gases, control of soot emissions, particularly from fossil-fuel sources, is very likely to be the fastest method of slowing global warming for a specific period."

It is well known that carbon particles resulting from incomplete combustion that deposit on high-latitude snow and ice will increase the solar absorptivity of the snow and ice, producing a surface-warming effect. As the high-latitude region warms, the extent and depth of snow and ice begins to diminish. This, in turn, provides a positive feedback via reduced solar reflection, producing further warming. The effect of various levels of carbon deposition on absorptivity was estimated by Grenfell *et al.* Their results are shown in Figure 7.33.

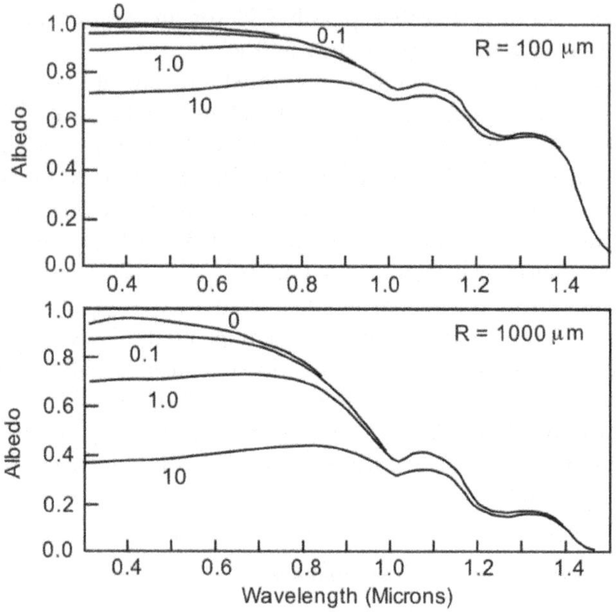

Figure 7.33. Effect of various concentrations of soot (ppm) on snow for two different grain sizes (Grenfell *et al.*, n.d.)

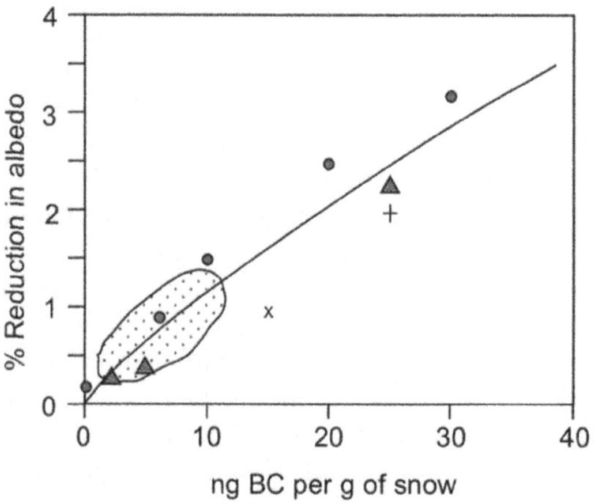

Figure 7.34. Estimates of reduction of albedo by deposition of BC. The data points are referenced in Hadley *et al.* (2010) and the irregular area encompasses the results of Hadley *et al.* (2010).

Hadley *et al.* (2010) measured the reduction in albedo due to deposition of BC on snow in the California and compared their results with other studies and models as shown in Figure 7.34. They concluded that BC deposition is a significant factor in reduction of the snowpack in the California mountains.

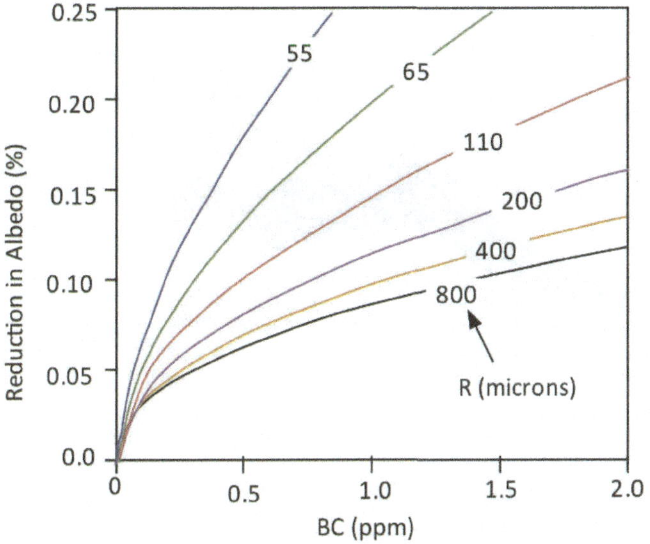

Figure 7.35. Snow-albedo reduction attributed to BC (Hadley and Kirchstetter, 2012).

Hadley and Kirchstetter (2012) "developed processes for making both pristine and BC-laden snow and techniques for measuring the morphology, albedo and BC content of snow. These methods have allowed [them] to quantify the snow-albedo reduction associated with increasing amounts of BC and as a function of snow grain size. Their results are summarized in Figure 7.35. These results suggest that warming due to BC could have been a major factor in early-20th-century warming.

7.3.3 Rate of emission of carbon in the 20th century

Jacobson (2002) provided a detailed summary of modeled and measured deposition rates for BC at various locations in the world covering a wide range of latitudes.

BC is a mix of elemental and organic carbon emitted by fossil-fuel combustion, biomass burning, and bio-fuel cooking (wood fires and cow dung) as soot. In the atmosphere, BC aerosols are mixed with sulfates and organics. Bond (2007) estimated BC emission rates by region since 1850, as shown in Figure 7.36. There was a sharp rise in BC emissions from 1900 to 1925, followed by a meandering period from 1925 to about 1970, and then a subsequent rise after 1970. The rise prior to 1925 derived mainly from the U.S. and Europe. After 1925, BC emissions from the U.S. dropped significantly and European emissions slowly diminished. Starting around 1970, emissions grew mainly from developing nations, particularly from China and other Asian nations.

More recently, Bond (2010) estimated that, in 2000, BC energy-related emissions (in ktons/year) were approximately as follows:

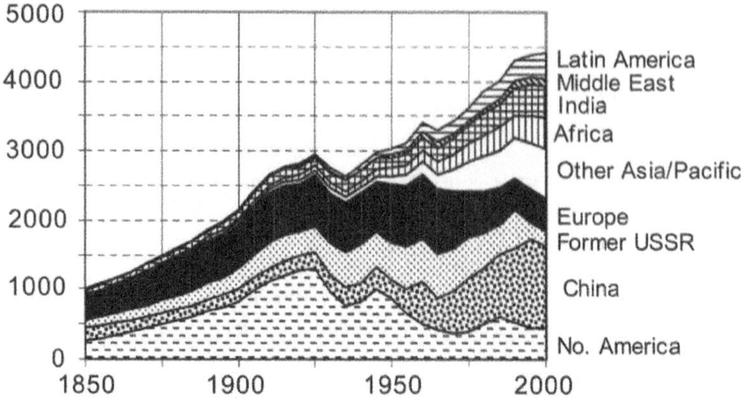

Figure 7.36. History of BC emissions by region (Bond, 2007).

China	1,300
India	500
Other Asia	800
Latin America	300
North America	500
Europe	500
Former U.S.S.R.	300
Middle East	100
Africa	500
Total	4,800

These figures do not include open burning, which would add another 55% to the total for energy-related emissions.

BC emissions prior to about 1940–1950 were primarily from the U.S., Europe, and the former U.S.S.R. These locations are proximate to high-latitude regions of the NH. Beginning around 1970, BC emissions from the U.S., Europe, and the former U.S.S.R. diminished and the increase in total BC emissions was due to Asian nations, somewhat removed from the high-latitude regions of the NH. Hence, it is to be expected that BC deposition in high-latitude regions of the NH would have peaked around 1925. As Bond (2010) showed, BC emissions in the U.S. peaked at around 1,200 ktons/yr in the late 1920s, and has now dropped to about 400 ktons/yr despite growth in energy usage, due to regulation of power generation and industry.

Graf *et al.* (2010) reported on sulfurous and black carbon emissions near Antarctica. Fahey *et al.* (2010) reported on aircraft observations of black carbon in northern Alaska.

7.3.4 Estimated role of BC in Arctic warming

Several studies have been conducted using sophisticated climate models to estimate the contribution of BC deposition on snow and ice to global warming. Hansen and

Nazarenko (2003) estimated that perhaps a quarter of the global warming of the 20th century was due to soot. These studies generally use models that were devised to estimate temperature rise due to increased greenhouse gas concentrations, particularly CO_2. The effect of BC can be taken into account by either adding the estimated forcing due to reduced albedo to the climate model, or by attributing the difference between the model and observed data to BC. In either case, questions remain as to how climate models utilized putative increased water vapor and clouds. Neither approach is likely to be very accurate, but the estimated order of magnitude of the BC effect may be more reliable. Although Jacobson (2002) continued to believe that greenhouse gases were the main cause of global warming, he concluded that "eliminating all fossil fuel BC + OM [organic matter] could eliminate 20–45% of net warming . . . within 3–5 years if no other change occurred". He also concluded that "emission reduction of fossil-fuel particulate BC plus associated organic matter (OM) may slow global warming more than any emission reduction of CO_2 or CH_4".

Flanner *et al.* (2007, 2008) pointed out that "of 22 climate models contributing to the IPCC Fourth Assessment Report, 21 under-predict the rapid warming (0.64°C/decade) observed over springtime Eurasia since 1979". They were therefore motivated to study the effects of BC + OM on climate change. To do this, they utilized sophisticated, comprehensive climate models to estimate the role of BC + OM in climate change. Their model requires a large number of parameters. This was the first global climate study that treated coupled snow aerosol heating and snow aging. The model required "plausible" ranges of present-day BC/snow forcing using combinations of BC emissions, BC optical properties, snow aging, meltwater scavenging of BC, and snow cover fraction. They estimated global annual mean BC/snow surface radiative forcings from all BC sources. While the authors clearly used the best values that could be estimated, it is difficult to resolve their reliability. There seems to be a considerable range of values, depending on the year chosen, as well as on the specific assumptions that were used. Nevertheless, the results are very interesting. These results should be read with the understanding that, in most cases, they were presented as global averages, and not as specific effects for the 66.5°N–90°N Arctic region. But, because all the effects of BC + OM are induced in the 66.5°N–90°N Arctic region, the estimated worldwide climate impact will depend on the model's ability to project regional forcings on global climate. The forcings produced in the Arctic region (W/m^2) at the snow level due to various factors was estimated to be as follows:

BC emissions	0.54–2.00
Snow aging	0.58–1.58
Melt scavenging	0.69–1.08
Optical properties	0.88–1.12
Snow cover fraction	0.83–1.08

These are very significant radiative forcings. However, most of their data were scaled to worldwide forcings that reduced the magnitudes significantly. Whereas they claim that the effect of inclusion of BC in snow was about +0.10°C to 0.15°C, the effect on the 66.5°N–90°N Arctic region was estimated to be 0.5°C to 1.6°C. Because of large Asian emissions of BC, large forcings were estimated over the Tibetan Plateau

(30–40°N, 80–100°E), averaging 1.5 W/m^2 over all land. During some spring months, forcing over snowy areas exceeds 10 W/m^2 and 20 W/m^2 over parts of eastern China and the Tibetan Plateau, respectively. In this context, they examined the recent, rapid springtime warming observed over Asia, where BC emissions rose from roughly 1.6 Tg/yr to 2.6 Tg/yr during 1980–2000. For this region, global warming due to BC + OM was estimated to be comparable with that due to greenhouse gases. They concluded that "on the global scale, positive surface and atmosphere forcings from carbonaceous particles drive significant reductions in springtime snow cover".

Shindell and Faluvegi (2009) utilized sophisticated climate models to estimate the historical temperature of the Earth by region. Unfortunately, using the usual forcings, they were unable to approximate the historical variation of temperature in the high NH latitudes. They postulated that deposition of BC on snow and ice in this region produced a much greater change in temperature than in other latitude regions. While they did not fully account for the dip from 1940 to 1970, they did point out that the rate of BC emissions slowed down after the 1920s and then turned up again in the 1970s. The total temperature rise in the 60°–90° latitude region of the NH in the 20th century was about 2.25°C. Shindell and Faluvegi (2009) estimated that somewhere between 0.5°C and 1.4°C of high-latitude NH warming was due to BC (22% to 62%). From 1976 to 2007, they estimated that the temperature increase was more than 50% due to BC. However, their method attributed the gap between climate theory and surface temperature observation to BC, but they did not invoke any direct measurements of BC levels or models to estimate forcing due to BC.

IPCC (2007) estimated the current forcing due to BC as +0.34 W/m^2 based on models of the global atmosphere. It can be compared with the forcing of carbon dioxide, which they estimated as +1.66 W/m^2. However, Bond (2010) suggested that these results neglected other effects, and she suggested that the forcing due to BC is about 0.5 W/m^2. She also said:

> "The emission rate of black carbon is another important factor in determining its forcing. Forcing is directly proportional to emission rate, so if emission estimates are doubled, the forcing estimate will double as well. Atmospheric measurements suggest that our current estimate of year 2000 emissions is too low in some regions. Forcing estimates as high as 1 W/m^2 have been published and are usually associated with models that assume more black carbon in the atmosphere than other models."

Wang (2002) modeled deposition of BC in Arctic areas. He was able to reproduce measured loadings in several regions. Surface warming was estimated in the range 0.6 W/m^2 to 0.9 W/m^2. However, BC aerosols suspended in the atmosphere can act in the opposite direction by reflecting incoming solar irradiance.

Menon *et al.* (2010) reported on "recent thinning of glaciers over the Himalayas (sometimes referred to as the third polar region)". They "quantified the impact of BC aerosols on snow cover and precipitation from 1990 to 2010 over the Indian subcontinental region". They found that "over the Himalayas, from 1990 to 2000, simulated snow/ice cover decreased by 0.9%. The contribution of enhanced Indian BC to this decline was estimated to be 36%, similar to that simulated for 2000 to 2010".

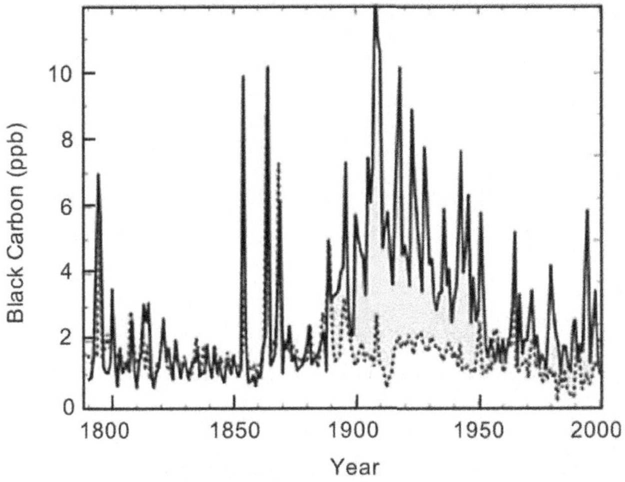

Figure 7.37. Annual average BC concentrations. The gray area is due to industrial emissions. The lower area is due to boreal fires (McConnell *et al.*, 2007).

7.3.5 Measurement of BC in Arctic ice cores

McConnell *et al.* (2007) used measurements of central Greenland ice cores to assess the origin and climate forcing of BC in snow during the past 215 years. Air mass back-trajectory modeling suggested that the eastern and northern U.S. and Canada were likely source regions prior to about 1950. They concluded that conifer combustion was the major source of BC in Greenland before 1850 and it remained a significant source during summer throughout the 215-year record. Their results are shown in Figure 7.37. They found that BC concentrations varied significantly during the past 215 years and were highly seasonal and erratic. They made rough estimates of surface radiative forcing. The vertical scale in Figure 7.37 can then be converted to forcing and the peak in Figure 7.37 would then correspond to a forcing of over 4 W/m^2. A five-year running mean of the forcing is shown in Figure 7.38. This was based on a rather crude calculation, but, if it holds up to further scrutiny, the forcing due to BC early in the 20th century would appear to be more than sufficient to account for global warming prior to 1940 without involving greenhouse gases at all.

It seems unlikely that BC alone would account for Arctic warming after 1976 because so much of the BC emissions of the past 30 years derive from Asian sources, further removed from the North Polar region.

Doherty *et al.* (2010) collected samples of Arctic snow:

> "... in Alaska, Canada, Greenland, Svalbard, Norway, Russia, and the Arctic Ocean during 2005–2009, on tundra, glaciers, ice caps, sea ice, frozen lakes, and in 10 boreal forests. Snow was collected mostly in spring, when the entire winter snowpack is accessible for sampling. Sampling was carried out in summer on the Greenland ice sheet and on the Arctic Ocean, of melting glacier snow and sea ice as well as cold snow."

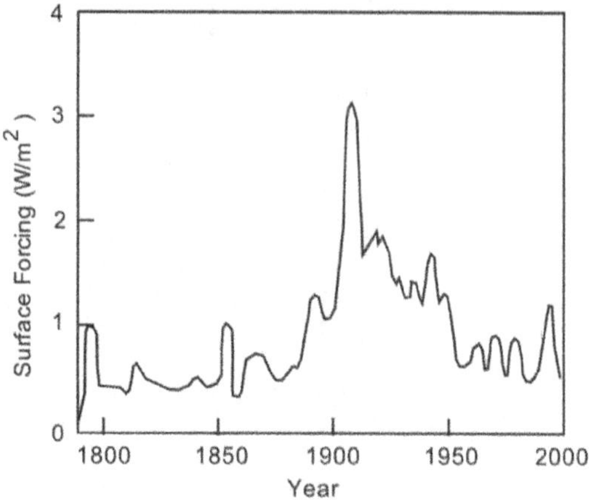

Figure 7.38. Five-year running mean of estimated surface radiative forcing for a seasonal snow cover derived using mean winter BC concentration measured in the core and early summer solar forcing conditions. Average early summer surface forcing was extrapolated throughout the Arctic region (McConnell *et al.*, 2007).

They concluded that "the reduction of snow albedo is primarily due to BC, but other impurities, principally brown (organic) carbon, are typically responsible for 40% of the visible and ultraviolet absorption". Median BC amounts in surface snow (ng per g snow) were: "Greenland 3, Arctic Ocean snow 7, melting sea ice 8, Arctic Canada 8, Subarctic Canada 14, Svalbard 13, Northern Norway 21, Western Arctic Russia 26, Northeastern Siberia 17."

7.4 ROLE OF SULFATE AEROSOLS

Figure 3.21 shows that, in the NH, average temperatures rose from about 1920 to 1940, dipped from 1940 to 1978, and then rose again from 1978 to 1998. Since 1998, NH average temperatures have meandered but have not increased. It has been argued that there was a build-up in sulfate aerosols as the number of power plants expanded in the mid-20th century, and these aerosols acted as a cooling force to counteract the build-up of greenhouse gases. The claim is that, as the 20th century wore on, greenhouse gases continued to build up while the developed nations began to control sulfur emissions and, therefore, late in the century, the temperature began to rise again. The reader is referred to Section 3.4.5 for a detailed review of the role of aerosols in changing climate.

The effect of aerosols on climate is not known precisely. According to Chin *et al.* (2009), "Clearly there are still large gaps in assessing the aerosol impacts on climate through modeling". These authors provide a detailed analysis of what is and what is not understood about aerosols and climate. Anderson *et al.* (2003) discussed two methods for incorporating aerosol forcing in climate models. In the "forward" approach, pertinent aerosol physics and chemistry are used to estimate the aerosol-induced forcing *a priori* and the climate model is based on this.

Ruckstuhl *et al.* (2008) reported:

> "... aerosol optical depth measurements from six specific locations and surface irradiance measurements from a large number of radiation sites in Northern Germany and Switzerland. The measurements show a decline in aerosol concentration of up to 60%, which have led to a statistically significant increase of solar irradiance under cloud-free skies since the 1980s. The measurements confirm solar brightening and show that the direct aerosol effect had an approximately five times larger impact on climate forcing than the indirect aerosol and other cloud effects. The overall aerosol and cloud induced surface climate forcing is + 1 W/m^2 per decade and has most probably strongly contributed to the recent rapid warming in Europe."

This was followed by another article (Philipona *et al.*, 2009) that pointed out that:

> "Mainland Europe's temperature rise of about 1°C since the 1980s is considerably larger than expected from anthropogenic greenhouse warming.
>
> "[They] analyzed the radiation and energy budgets using measured short-wave and empirically derived long-wave radiative forcings at the surface, and related them to observed temperature and humidity changes in Switzerland and Northern Germany."

They estimated roughly that perhaps two-thirds of the temperature rise since the 1980s was due to declining aerosol concentrations. This may be related to the finding by Pinker *et al.* (2005) that solar intensities at the surface in Europe underwent a "sustained increase after 1990".

Ghan and Schwartz (2007) and Chin *et al.* (2009) provided a comprehensive review of the status of modeling aerosols in various generations of global climate models, past, present, and future. Early approaches for incorporating the effects of aerosols simply adjusted the albedo as a crude approximation, while subsequent models used much more sophisticated representations of aerosols, but even these did not adequately account for interactions of aerosols with other climate elements:

> "It is now recognized that accurate representation of aerosol influences must take into account phenomena such as correlations of aerosol loading with meteorological variables and the influence of aerosol on clouds and precipitation, and hence that aerosol loading and those properties must be represented actively and interactively in climate models. It is this recognition that is driving much of the current effort to actively represent aerosol processes, properties, and effects in climate models." (Ghan and Schwartz, 2007).

The treatment of aerosols in future generations of climate models will rest on an improved understanding of the processes that control aerosol properties and the ways that they affect climate. Chin *et al.* (2009) provided a summary of unsolved issues remaining in regard to aerosols and climate. Ghan and Schwartz (2007) provided a plan for systematically improving our understanding of aerosols in the future.

7.5 LAND CLEARING AND SURFACE DEVELOPMENT

Several analyses of the Medieval Warm Period (MWP) were carried out with climate models. Unfortunately, we don't know much about differences in solar irradiance between the MWP and the LIA. Greenhouse gases were not an issue for the MWP and the LIA. Two forcings that have been modeled by several investigators are volcanic emissions and changes in land use (large-scale deforestation and conversion to farmland). While volcanic emissions provided several important short-term fluctuations, an important long-term difference between the MWP and the LIA that can be modeled is the change in land use. Cropland has a higher albedo than forest, and this can make a significant difference in the heat absorbed over a region. During the period after 1000, the conversion of forest to cropland accelerated, particularly in Europe.

According to Bauer, Claussen, and Brovkin (2003), global forest cover diminished over the period 1000 to 1992 by 30%, from 57 to 41.5×10^6 km^2. Before 1900, forests were mainly removed in the northern subtropical and temperate regions. In the second half of the 20th century, agriculture in these regions stopped expanding and even reversed while tropical deforestation was intensified. Figure 7.39 shows the deforestation model used by Bauer, Claussen, and Brovkin (2003).

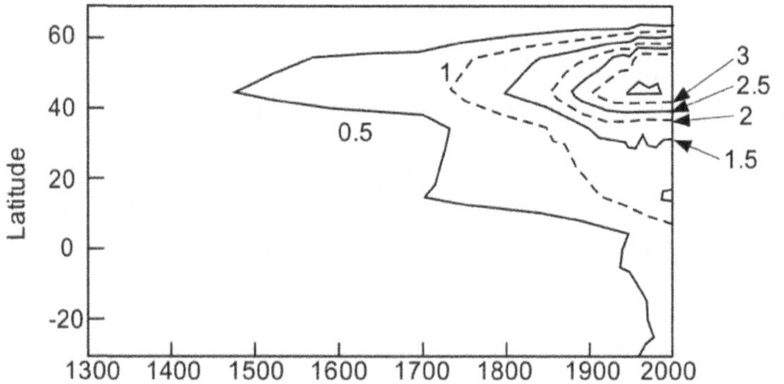

Figure 7.39. Deforested area shown as time-varying isolines in steps of units of 10^6 km^2 given for 10 latitude bands (adapted from Bauer *et al.*, 2003).

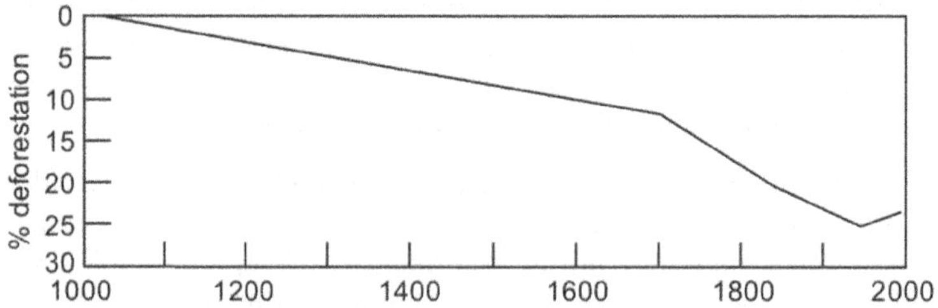

Figure 7.40. Estimate of deforestation (adapted from Goosse *et al.*, 2006).

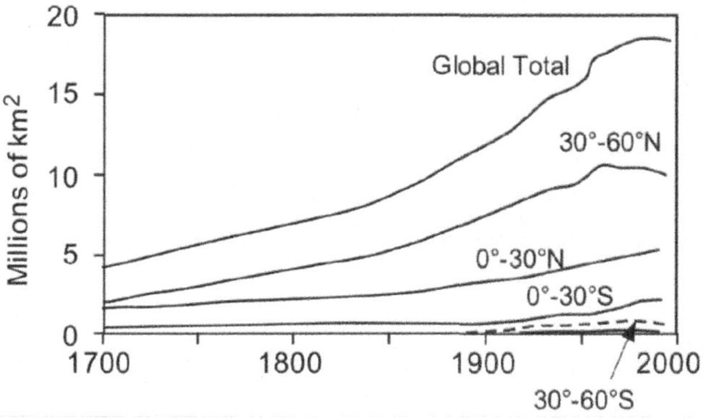

Figure 7.41. Cumulative area of land cleared by latitude and year (adapted from Brovkin *et al.*, 2006).

Goosse *et al.* (2006) assumed a linear increase of crop area from zero in 1000 to the value reconstructed for 1700. However, they note that, in a large number of regions of France, Belgium, the Netherlands, and Germany, deforestation was particularly intense between 1000 and 1250 and weaker during the two following centuries. Their estimate of deforestation is shown in Figure 7.40. The forcing due to changes in land use was applied through modifications in the surface albedo, which is the primary effect of land cover change.

Brovkin *et al.* (2006) modeled the effect of land cover changes and CO_2 emissions over the past millennium. The model used for land clearing is shown in Figure 7.41.

Bauer, Claussen, and Brovkin (2003) and Goosse *et al.* (2006) used global climate models to investigate the expected temperature variations over the past millennium, and compared the beginning of the millennium (the MWP) with the middle (the LIA) and later parts of the millennium (present day) to examine whether the putative MWP and LIA could be distinguished.

Starting in year 1000 as a baseline, Bauer, Claussen, and Brovkin (2003) examined the effect of various assumed forcings (solar, volcano, CO_2, and deforestation) on temperatures thereafter. Their model for solar irradiance was relatively flat with variances of about 0.1% from 1000 to 1900, and a greater increase after 1900. Therefore, the major forcing prior to the industrial era in their model was the change in land cover. A comparison of their estimated NH temperatures in the millennium based on all forcings (solar, volcano, CO_2, and deforestation) or only deforestation, is shown in Figure 7.42. According to this model, deforestation induced a slow and weak cooling in the NH from 1000 to 1850. The cooling accelerated after 1850 and reached about 0.35°C by the end of the millennium.

Goosse *et al.* (2006) also estimated millennium temperatures using models that accounted for deforestation. The results show a summer NH temperature drop of about 0.3°C from the MWP to the LIA, but it is not clear how much of this was due to the assumed pattern for solar irradiance, and how much was due to deforestation. Overall, from 1000 to 2000, temperatures dipped in the LIA and returned in 2000 to

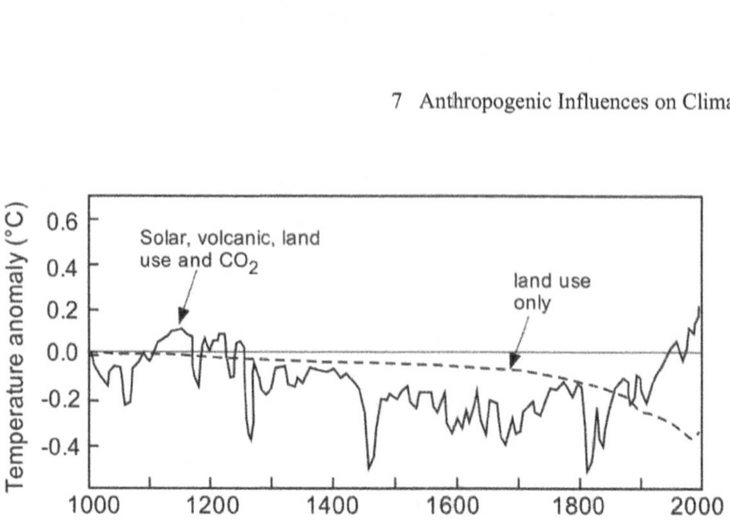

Figure 7.42. Modeled NH temperatures based on all forcings, or only deforestation (adapted from Bauer *et al.*, 2003).

levels near those of the MWP. Nevertheless, the effect of deforestation in the late 20th century was significant. As Goosse *et al.* (2006) said:

> "... there is no compelling evidence from either empirical proxy evidence or model simulation results that the European summer temperature during the last 25 years of the 20th century were the highest of the past millennium. This is largely due to the local negative radiative forcing caused by land-cover changes. The impact of this forcing at hemispheric scale has been underlined in recent studies. However, because of the large deforestation in Europe, land-use changes imply a larger negative temperature anomaly over Europe than on a global scale."

Unfortunately, our lack of knowledge of solar irradiance over the past millennium adds considerable uncertainty to any conclusion.

Brovkin *et al.* (2006) estimated global temperatures using climate models based on year 1000 as the starting point. When land clearing alone was used for forcing, the resultant global temperature decreased (relative to 1000) as shown in Table 7.2 (average of several models). Results are also shown for the case where land use and CO_2 forcing were both included.

Table 7.2. Estimated temperature change in the NH due to CO_2 and land clearing (Brovkin, *et al.*, 2006).

Year	*CO_2 (ppm)*	ΔT *(°C) (land-use forcing only)*	ΔT *(°C) (CO_2 forcing only)*	ΔT *(°C) (CO_2 and land-use forcing)*
1000	280	0.00	0.00	0.00
1600	280	–0.05	0.00	–0.05
1800	280	–0.10	0.00	–0.10
1900	300	–0.15	+0.20	+0.05
1990	355	–0.20	+0.40	+0.20

Matthews, *et al.* (2004) focused on the role of historical land cover change in forcing the climate of the last 300 years. In a detailed sensitivity analysis using different data sets of land cover change and varying model configurations and parameters, they found that the primary biogeophysical effect of historical land cover change was to increase local surface albedos, resulting in a cooling of 0.1°C to 0.3°C from 1700 to the present day, depending on the model. It was estimated that changes in land cover from 1700 to 1990 reduced the NH temperature by about 0.2°C, mitigating some global warming from other factors. However, local reductions were much greater, reaching ~0.8°C over the mid-western U.S. and Eastern Europe, where large-scale farming is widespread.

8

Impacts of global warming

8.1 ALARMISTS AND SKEPTICS

The 2001 Inter-governmental Panel on Climate Change (IPCC) Report provides not merely 1, but 19 separate chapters on the impacts of global warming, comprising over 1,000 pages of descriptions of impacts. It is difficult to summarize this huge body of material. Table 8.1 lists a number of potential climate disasters described in IPCC (2001). The other 999+ pages of the IPCC Report amplify this theme.

The Real Climate blog (*www.realclimate.org*) provides the alarmist view of global-warming impacts. This website is maintained by technically capable people but their objectivity seems doubtful.

On April 26, 2007, James E. Hansen (a well-known global-warming alarmist) gave testimony on the dangers of global warming to the Select Committee of Energy Independence and Global Warming of the U.S. House of Representatives. Hansen provided the case for alarmists in considerable detail. Only a few quotations are given here.

According to Hansen, the greatest near-term danger is sea level rise (SLR). He said that "sea level is already rising at a rate of 3.5 cm per decade and the rate is accelerating" due primarily to "ice sheet disintegration". He said that "there is increasing realization that SLR this century may be measured in meters if we follow business-as-usual fossil fuel emissions" and that "adaptation to a continually rising sea level is not possible". Hansen concluded that "increasingly rapid changes on West Antarctica and Greenland ... are truly alarming".

One of the major slow feedback processes that Hansen identified is "the effect of warming on emissions of long-lived greenhouse gases", caused by the "melting of tundra in North America and Eurasia", which "is observed to be causing increased ebullition of methane from methane hydrates". Hansen said that "continued business-as-usual greenhouse gas emissions threaten many ecosystems," and that "very little additional [climate] forcing is needed ... to cause the extermination of a large fraction of plant and animal species". He also said that "Earth's history shows that climate is remarkably sensitive to global forcings" and that "positive feedbacks predominate", causing "the entire planet to be whipsawed between climatic states".

D. Rapp, *Assessing Climate Change*, Springer Praxis Books,
DOI: 10.1007/978-3-319-00455-6_8

Table 8.1. Impacts of global warming according to IPCC (2001).

Projected changes during the 21st century in extreme climate phenomena and their likelihood	*Representative examples of projected impacts*
Higher maximum temperatures; more hot days and heat waves over nearly all land areas (very likely)	Increased incidence of death and serious illness in older age groups and urban poor Increased heat stress in livestock and wildlife Shift in tourist destinations Increased risk of damage to a number of crops Increased electric cooling demand and reduced energy supply reliability
Higher (increasing) minimum temperatures; fewer cold days, frost days, and cold waves over nearly all land areas (very likely)	Decreased cold-related human morbidity and mortality Decreased risk of damage to a number of crops, and increased risk to others Extended range and activity of some pest and disease vectors Reduced heating energy demand
More intense precipitation events (very likely over many areas)	Increased flood, landslide, avalanche, and mudslide damage Increased soil erosion Increased flood runoff could increase recharge of some flood plain aquifers Increased pressure on government and private flood insurance systems and disaster relief
Increased summer drying over most mid-latitude continental interiors and associated risk of drought (likely)	Decreased crop yields Increased damage to building foundations caused by ground shrinkage Decreased water resource quantity and quality Increased risk of forest fire
Increase in tropical cyclone peak wind intensities, mean and peak precipitation intensities (likely over some areas)	Increased risks to human life, risk of infectious disease epidemics, and many other risks Increased coastal erosion and damage to coastal buildings and infrastructure Increased damage to coastal ecosystems such as coral reefs and mangroves
Intensified droughts and floods associated with El Niño events in many different regions (likely)	Decreased agricultural and rangeland productivity in drought-and flood-prone regions Decreased hydro-power potential in drought-prone regions
Increased Asian summer monsoon precipitation variability (likely)	Increased flood and drought magnitude and damages in temperate and tropical Asia
Increased intensity of mid-latitude storms (little agreement between current models)	Increased risks to human life and health Increased property and infrastructure losses Increased damage to coastal ecosystems

Summarizing, Hansen said that "The dangerous level of CO_2 is at most 450 ppm, and it is probably less. ... Ignoring the climate problem at this time, for even another decade, would serve to lock in future catastrophic climatic change and impacts that will unfold during the remainder of this century and beyon.'.' The Earth "is close to dangerous climate change, to tipping points of the system with the potential for irreversible deleterious effects. ... The planet is on the verge of dramatic climate change". We "are forced to find a way to limit atmospheric CO_2 more stringently than has generally been assumed. ... We cannot shrink from our moral responsibilities ... to preserve the planet for future generations".

Idso and Idso (2007) reviewed this testimony and provided a skeptical commentary and critique.

Alarmists have found it rewarding to engage in a contest of "Can you top this?" by issuing a constant barrage of press releases about what supposedly "may", "might", or "could" happen in the future as a result of putative global warming. If you go to Google and punch in "global warming", you get thousands of responses predicting disaster from global warming. These include claims such as:

(1) the role of obese people in contributing to global warming by requiring extra resources;
(2) "climate change may be century's greatest health threat";
(3) "pets may be the latest victims";
(4) "climate change may halve South Africa";
(5) "increased incidence of tropical diseases, food shortages, natural disasters and heat waves threaten global ...";
(6) "climate change may drive refugees to Australia";
(7) "how climate change may be threatening national parks"; and
(8) "climate change will overload humanitarian system"

—and many more like this.

By contrast, the *www.co2science.org* blog provides a number of articles in the skeptics camp.

The main topics in Idso and Idso (2007) were the principal topics addressed by Hansen:

(1) ice sheet disintegration;
(2) sea level trends;
(3) atmospheric methane concentrations;
(4) climates of the past;
(5) predicted warming-induced extinctions of terrestrial plants and animals;
(6) the CO_2-induced preservation of terrestrial species and;
(7) predicted CO_2-induced extinctions of calcifying marine organisms.

In addition, Idso and Idso (2007) discussed a number of other topics that Hansen addressed in less detail, including:

(1) positive vs. negative climate feedbacks;
(2) effects of drought on agriculture in a CO_2-enriched world;

(3) SLR over the next 100 years;
(4) the adaptability of living organisms to rising sea levels;
(5) the "dangerous" level of atmospheric CO_2;
(6) the magnitude of climate forcing due to a doubling of the air's CO_2 content;
(7) empirical evaluations of the Earth's climate sensitivity;
(8) the ability of man to control global climate;
(9) the need to act now to reduce CO_2 emissions;
(10) the role of morality in the debate over what to do—or not do—about anthropogenic CO_2 emissions.

Idso and Idso (2007) provided point-by-point arguments in rebuttal to most of Hansen's claims. In regard to SLR, the data suggest a slower rate of rise than Hansen indicated, and the probability of disintegration of ice sheets in the 21st century appears to be exaggerated. While Idso and Idso dismissed Hansen's concerns out of hand, the consequences of such an eventuality are so profound that it must be considered seriously even if it is improbable.

While Hansen was concerned about release of methane from methane hydrates as polar areas warm up, Figure 6.9 shows that the methane concentration in the atmosphere reached a plateau in the 1990s and is increasing at a much slower rate. This figure suggests that the trend in atmospheric methane concentration, "has been decreasing for the last two decades until the present when it has reached near zero" and that "it is questionable whether human activities can cause methane concentrations to increase greatly in the future" (Khalil *et al.*, 2007). Idso and Idso (2007) capitalized on this short-term improvement to conclude that there is no long-term problem, but the long term remains uncertain. Alarmists claim that the present leveling-off of methane concentration is temporary.

Idso, Idso, and Idso (2003) and Idso and Idso (2007) provide rebuttals to Hansen on the claims of "extermination of a large fraction of plant and animal species". This material is beyond the scope of this book and the reader is referred to the Idso reports for details.

It is particularly noteworthy that Hansen was quoted as saying that "the dangerous level of CO_2 is at most 450 ppm, and it is probably less". If that were the case, humanity appears to be lost. With a current CO_2 concentration of $\sim$400 ppm, and annual increases of about 2 ppm per year, we will likely reach 450 ppm in roughly the next 35 years, and it seems most unlikely that any action by the world community can prevent this without economic disaster. More recently, Hansen *et al.* (2008) now claim that CO_2 concentrations greater than 350 ppm is the upper limit, making success even more unreachable.

The skeptical viewpoint is presented in its most authoritative form in Idso, Carter, and Singer (2011). This work comprises 432 pages with many illustrations and a very large number of references. As this book went to press in December 2013, the Heartland Institute, Chicago, IL, published a 1,000-page book sequel to the 2011 book, edited by C. Idso, R. Carter, and S.F. Singer, dealing with all aspects of climate change. This book provides a wealth of data and references. Although the discussions in this book are biased toward the skeptical viewpoint, they are no more biased than the IPCC Report, which is strongly biased toward the alarmist viewpoint. The two studies provide a good counterbalancce to one another. While most interpretations in this book are heavily slanted toward the skeptical viewpoint, this bias is no worse than the bias toward the alarmist view in the reports of the IPCC. The title of this work, Non-governmental International Panel on Climate Change (NIPCC), is a direct confrontation to the IPCC. This book is a good source of information about many aspects of climate change. It is particularly strong in regard to impacts of climate change—such as extreme weather, floods, impacts on flora and fauna, storms, hurricanes, human health effects, etc. Unfortunately, some of the conclusions in this book do not seem credible. For example, it is claimed that "... a doubling of the atmosphere's CO_2 concentration would result in only a 0.4°C or 0.5°C rise in temperature". While no one knows exactly how much of a rise in temperature would occur, it is very unlikely that it would be this low. The strange thing is that, if you read Section 1.3 of the NIPCC Report that deals with temperature, there does not seem to be any mention of the estimate that doubling CO_2 produces a temperature rise of less than 0.5°C. It seems likely that even most responsible skeptics would doubt this. Another problem with the NIPCC Report is that the authors seem willing to use almost any published paper to support their position, even if the published paper is clearly erroneous. For example, in discussing the inadequacies of climate models, the NIPCC utilizes the paper by Lean and Rind (2008), which claims that (as stated by the NIPCC): "contrary to recent assessments based on theoretical models (IPCC, 2007) the anthropogenic warming estimated directly from the historical observations is more pronounced between 45°S and 50°N than at higher latitudes, which finding, in their words,—is the approximate inverse of the model-simulated anthropogenic plus natural temperature trends ... which have minimum values in the tropics and increase steadily from 30 to 70°N". Unfortunately, Lean and Rind (2008) are wrong; there is ample evidence that the effect of greenhouse gases is to amplify the temperature rise toward higher northern latitudes (see Section 3.6).

8.2 FUTURE INCREASES IN GLOBAL TEMPERATURE

There have been many estimates of the putative rise in global average temperature that will result from increased CO_2 concentration above the pre-industrial level of roughly 280 ppm. Three key steps are involved in this process: (1) estimating the future increase in CO_2 concentration, (2) estimating the putative temperature increase due to the increase in CO_2, and (3) estimating the additional heating due to feedback, primarily from water vapor, clouds, aerosols, and other secondary processes. Many such models limited their study to a future doubling of CO_2 to approximately 560 ppm. It is of particularly great importance to understand how future temperature rise is distributed by latitude.

McIntyre[1] discussed a number of projections of future temperature rise due to

[1] *http://climateaudit.org/2013/03/02/mikes-agu-trick/.*

increasing CO_2 in the future. Back in 1988, Hansen projected three possible scenarios. Using 1960 as a base year, his projections for a temperature rise in year 2019 ranged from 1.5°C (high scenario), to 1.1°C (moderate scenario), to 0.6°C (low scenario). As we stand in year 2013, we seem to be following the low scenario fairly closely. As we pointed out in Section 3.4.3.2, the global average temperature has been flat from 1998 to 2013. In fact, most of the temperature rise after 1960 took place between 1976 and 1998 when El Niño events prevailed. According to McIntyre, Hansen has honestly described this in later publications, whereas Mann once again "hid the decline" and stopped plotting temperature data after it leveled off. There are many projections of future temperature rise in the 21st century by climate modelers. Only a very few are mentioned in the paragraphs that follow.

Nozawa *et al.* (2007) utilized three projections for future greenhouse gas concentrations in a coupled ocean–atmosphere general circulation model. One projection assumed 1% increases in CO_2 for the entire 21st century, leading to a concentration of over 800 ppm by 2100. The most moderate model had an asymptote of 540 ppm in 2100. The estimated temperature increases in the 21st century range from 4°C to 2.4°C.

Anon. (O) utilized several projections of future CO_2 concentrations that increase at 2100 over a range varying from 470 ppm to well over 1,000 ppm, leading to predicted temperature increases by 2100 ranging from 1.4°C to 5.8°C. They presented an ultimate *hockey stick* temperature profile with almost no temperature variation over the past millennium and a nearly vertical rise in the 20th century. Anon. (B) projected temperature increases by 2100 ranging from 2°C to 6°C with a similar exaggerated *hockey stick*.

At the heart of almost all alarmist positions are claims that we are already in an unprecedented, incredibly high-temperature cycle, and projections of significant future increases in global temperature will lead us still higher. For example, Mann, Bradley, and Hughes (1999) said:

> "... our results suggest that the latter 20th century is anomalous in the context of the last century. The 1990s was the warmest decade and 1998 the warmest year at moderately high levels of confidence."

Of course, this was the first MBH paper where the *hockey stick* was revealed, and these claims are exaggerated.

Singer and Avery (2007) quote a number of alarmist claims from various sources:

> "Nineteen ninety-nine was the most violent year in the modern history of weather. So was 1998. So was 1997. And 1996. ... A nine-hundred-year-long cooling trend has been suddenly and decisively reversed in the past fifty years ... Scientists predicted that the Earth will shortly be warmer than it has been in millions of years. A climatological nightmare is upon us. It is almost certainly the most dangerous thing that has ever happened in our history."

> "Climate extremes would trigger meteorological chaos-raging hurricanes such as we have never seen, capable of killing millions of people; uncommonly

long, record-breaking heat waves; and profound drought that could drive Africa and the entire Indian subcontinent over the edge into mass starvation."

"From sweltering heat to rising sea levels, global warming's effects have already begun. ... We know where most heat-trapping gases come from: power plants and vehicles. And we know how to limit their emissions."

"No matter if the science of global warming is all phony ... climate change [provides] the greatest opportunity to bring about justice and equality in the world."

Fortunately, Hansen (2004), written by a leading alarmist, provided a more balanced and credible picture:

"The IPCC scenarios may be unduly pessimistic, however. First, they ignore changes in emissions, some already under way, because of concerns about global warming. Second, they assume that true air pollution will continue to get worse, with ozone, methane and black carbon all greater in 2050 than in 2000. Third, they give short shrift to technology advances that can reduce emissions in the next 50 years."

Furthermore, it is likely that limitations on the availability of fossil fuels will have a dramatic effect as the 21st century wears onward. Hansen (2004) went on to say:

"Observed global carbon dioxide and methane trends for the past several years show that the real world is falling below all IPCC scenarios. It remains to be proved whether the smaller observed growth rates are a fluke, soon to return to IPCC rates, or are a meaningful difference."

Note that Figure 6.9 shows that increases in methane concentration have stagnated.

The projection made by Hansen (2004) was:

"... at the low end of the IPCC range of two to four watts per square meter. The IPCC four watts per square meter scenario requires 4 percent a year exponential growth of carbon dioxide emissions maintained for 50 years and large growth of air pollution; it is implausible."

One legitimate concern is that the Earth's climate appears to be precariously perched on the edge of a fence that can topple off one way or the other, once pushed by some "trigger" to get the process rolling. We are in a global-warming period, particularly at higher latitudes. As the ice and snow cover at higher latitudes diminishes, the heat input to the Earth increases and the warmth spreads. It is possible that we are entering such a positive feedback era in which global warming, regardless of greenhouse gases, has somehow gotten started and is likely to propagate by its own momentum.

8.3 SEA LEVEL RISE, THE ICE SHEETS, AND SEA ICE

Radic (2008) pointed out that:

> "In 1990, the near-coastal population ... was 1.2 billion people. ... Human settlements are also preferentially located close to the world's shoreline, including most of the largest cities, which means that the world's economy is also concentrated in the coastal zone. Thus, sea level rise has a major impact on coastal cities, deltaic lowlands, small islands, and coastal ecosystems. The potential threat has triggered studies on impacts and responses to sea-level rise which are focused on a range of direct and indirect socio-economic impacts such as loss of land and buildings, loss of tourist amenity, increasing flood risk, impact on variety of commercial infrastructure, coastal process plants and offshore oil and gas production."

8.3.1 Historical sea level rise since the last glacial maximum

Douglas and Peltier (2002) discussed many aspects of measurements of SLR and fall. At the height of the last glacial maximum (LGM), ~21,000 years ago, so much of the Earth's water was tied up in great high-latitude ice sheets that the oceans were about 120 m lower than they are today. Since then, sea level rose continuously as shown in Figure 8.1. Most of this increase in sea level was due to melting of the great ice sheets, although some rise was due to warming of the oceans. By about 5,000–6,000 years before present (YBP), the melting of the great high-latitude ice masses was essentially completed. Thereafter, the global sea level (GSL) rise was small, and appeared to have almost ceased by 3,000–4,000 YBP.

According to Singer and Avery (2007), when the great ice sheets began to melt at the end of the last period of glaciation, the initial rapid rise of sea level was about

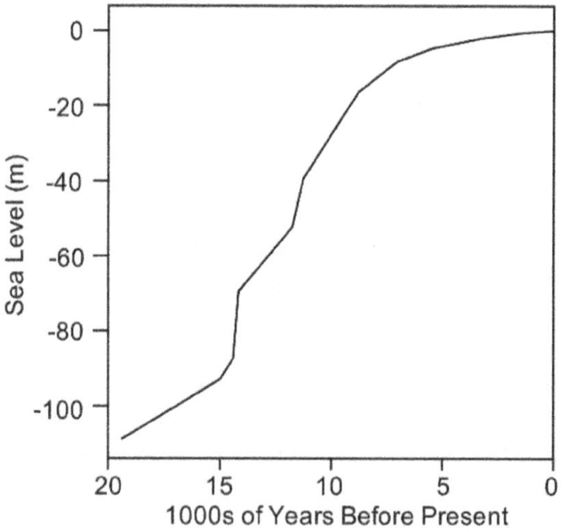

Figure 8.1. Rise in sea level since the LGM.

Table 8.2. Historical dependence of sea level on CO_2 concentration (Alley *et al.*, 2005).

Years before present	*CO_2 (ppm)*	*Eustatic sea level change (m)*
> 35 million	1,250	+ 73
~ 32 million	500	+ 45
21,000	185	–132
50	280	0

200 cm per century. This gradually changed to a slower rate of rise (15–20 cm per century) about 7,500 years ago, once the large ice masses covering North America and North Europe had melted away. But the slow melting of the West Antarctic Ice Sheet continued and will continue, barring another Ice Age, until it has melted away perhaps 6,000 years from now. This means that the world will continue to endure a SLR of about 18 cm per century (1.8 mm per year), just as it has been in previous centuries. And it is likely that there is nothing we can do about it. Thus, Singer and Avery (2007) attributed the continuing slow sea rise to the natural consequences of the post-glacial period but made no allowance for anthropogenic global heating adding to this rather bland picture.

Alley *et al.* (2005) provided some historical estimates of the relation between estimated atmospheric CO_2 concentration and the ice contribution to eustatic sea level referenced to modern (pre-Industrial Era) conditions (i.e., CO_2 ~280 ppmv, eustatic sea level 0.0 m) as shown in Table 8.2.

Huybrechts (2002) modeled the Greenland and Antarctica ice sheets from the last interglacial (120,000 YBP) through the LGM (20,000 YBP) to the present:

> "Together, their current volume contains enough ice to raise global sea level by almost 70 m, of which 61 m would derive from the Antarctic ice sheet and 7 m from the Greenland ice sheet. Major issues concern how much additional water was locked up in these ice sheets at the Last Glacial Maximum (LGM), when did this maximum occur, and over which period was the ice released back into the oceans? This problem bears directly on the amount of ice stored elsewhere on the globe, since the total eustatic sea-level depression is rather well constrained to have been between 125 and 135 m. The majority of ice at the LGM was contained in the ice sheets of Laurentia and Fennoscandia, but their combined estimated volume falls far short of the required ~130 m [sea level equivalent] in many assessments."

Huybrechts (2002) found that the sea level depressions at 21,000 YBP were between 13 m and 21 m for the Antarctic Ice Sheet and between 1.9 m and 3.5 m for the Greenland ice sheet (GIS). It was concluded:

> "It is hard to conceive that both ice sheets could have contributed more than ~25 m of equivalent sea level at the time of maximum sea-level depression. The implication is that the other northern hemisphere ice sheets [Laurentia and Fennoscandia] must have contained an equivalent sea-level volume of between

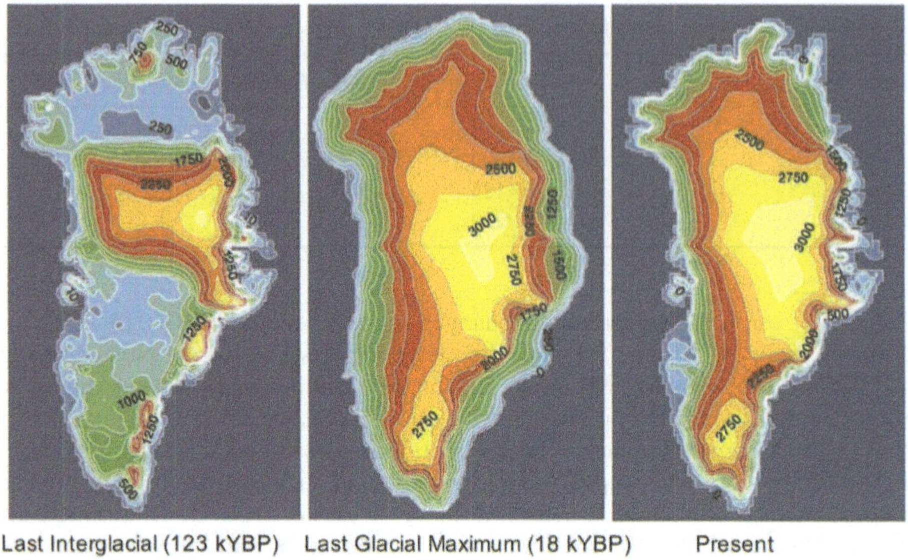

Figure 8.2. Modeled extent of Greenland ice at previous interglacial, last glacial maximum, and present (from Huybrechts (2002) by permission of Elsevier).

minimum 100 m and perhaps as much as 120 m at the LGM, substantially more than often assumed." (Huybrechts, 2002).

Figures 8.2 and 8.3 show the modeled results for Greenland and Antarctica (1) during the last interglacial, (2) near glacial maximum, and (3) at the present. For Greenland, it was estimated that the rise in sea level due to melting during the last interglacial was ~5.5 m, and the fall in sea level due to icing was ~2.7 m at the LGM, compared with the present. For Antarctica, it was estimated that the rise in sea level due to melting during the interglacial was ~1.4 m, and the fall in sea level due to icing was ~19.2 m at the LGM, compared with the present. These results suggest that a significant diminution of the GIS can occur during an interglacial. However, the calculations revealed that the Greenland minimum ice during the last interglacial is not very strongly constrained and is sensitive to assumptions. For plausible combinations of climatic conditions and only small shifts in the duration and magnitude of the peak warming, the ice sheet during the last interglacial could have varied from just a little smaller than today to only a small single dome covering only central–north Greenland. The major difference between the GIS during the LGM and today seems to be the much wider margins at the glacial maximum. If the model is correct, we might not yet have seen more than a small fraction of the ultimate disappearance of the GIS (Huybrechts, 2002). This might lead to a significant rise in sea level in the future if global warming occurs as predicted by alarmists. The model suggests that there is not much difference between Antarctica today and Antarctica during the last interglacial. The main difference during the glacial maximum was a significant spread of the margins. The model provides some hope that Antarctica may be less sensitive to global warming than Greenland.

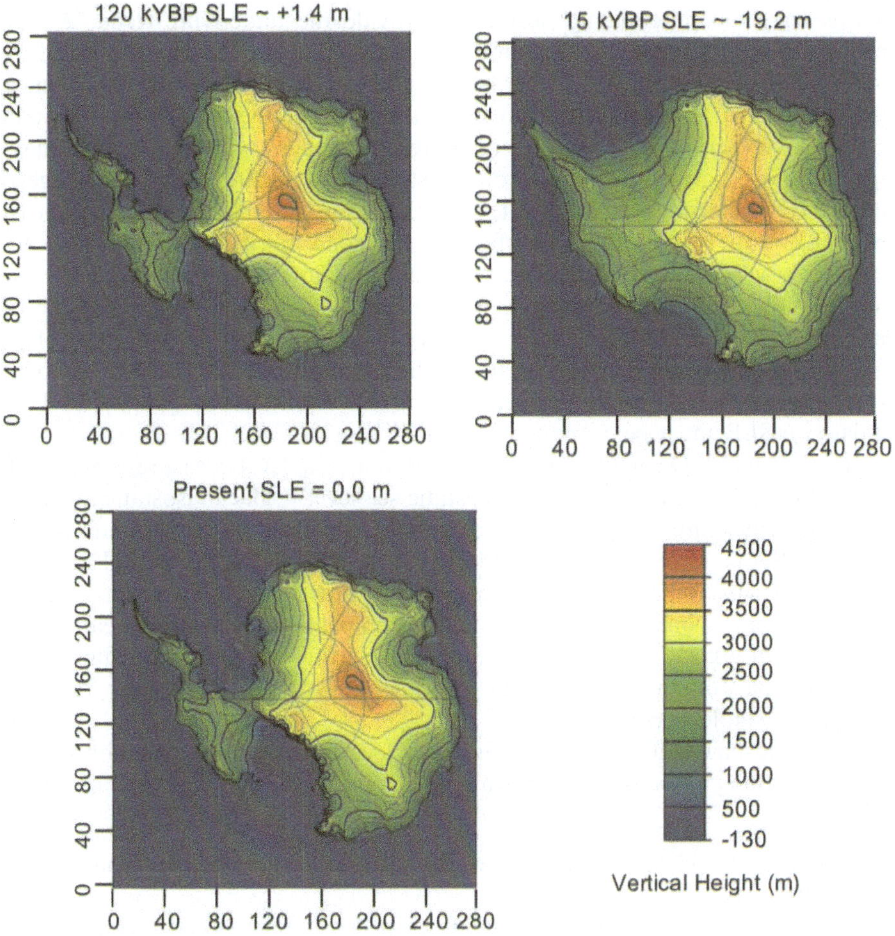

Figure 8.3. Modeled extent of Antarctica ice at previous interglacial, last glacial maximum, and present (from Huybrechts (2002) by permission of Elsevier).

8.3.2 Measurement of sea level

Sea level has been estimated by several methods. The greatest source of data is tide gauges, some of which date back well over 100 years. Over the past three decades or so, measurements were made from orbit using altimeters. The contribution of shrinking ice sheets in Greenland to rising sea level has been estimated by modeling, and, in the past decade, by satellite-based gravity detectors.

Most of our knowledge of historical variations in sea level is based on tidal gauges. Yet, interpretation of tide gauge data is problematic.

Douglas and Peltier (2002) emphasized that the measurements of relative sea level (RSL) at any location have large annual and decadal fluctuations that tend to obfuscate the long-term trends with noise. Consequently, only very long–term records have the potential to accurately provide the underlying trend. A CNES analysis showed that use of only four decades of data at the tide gauge sites led to an overestimate of the GSL rise. Other studies showed that the extreme dependence of trend on record length is real, and not an artifact of the tide gauge. To remain accurate over periods of a century or more, tide gauges, which over time may be repaired, moved, upgraded, and so on, must be kept consistent. But tide gauges, no matter how accurate and consistent, make local measurements. And they measure only RSL with respect to the surface of the solid Earth. Without independent estimates of vertical land movement, tide gauges cannot determine whether the water level is rising, the land is sinking, or both. Even though most estimates of the GSL rise in the past decade or so have used records that are as long as possible, the various estimates still differ significantly. The origin of the differences probably lies in the methods used to correct for vertical land movements at tide gauge sites. Hence, all data on SLR must be examined critically.

Houston and Dean (2011) pointed out that sea level measurements via tide gauges record the sum of worldwide eustatic sea level + glacial isostatic adjustment + local effects + noise. The glacial isostatic adjustment is the "rebound" from the weight of the massive ice sheets at the LGM. It is found to be an uplift of up to several meters per year at high latitudes, and a downtrend of up to about a meter per year in temperate zones. Local effects include compaction of sediments, earthquakes, withdrawal of ground fluids, and building heavy structures on weak sediments. The ideal location for tide gauge data provides very long continuous records in geologically stable regions.

As Fjeldskaar (2008) pointed out, sea level changes are commonly termed eustasy, meaning globally uniform sea level changes. However, sea level changes are caused by:

(1) changes in the ocean water volume caused by glaciations and deglaciations of ice sheets;
(2) variations in the ocean basin volume caused by sedimentation, changes in the volume of ocean ridge systems, and hydro-isostasy;
(3) variations in the Earth's gravity field as, for example, in mountain formation.

The geoid is defined an equipotential surface of the Earth's gravity field that would establish local sea level at equilibrium in the absence of atmospheric forces. Fjeldskaar (2008) also provided the following insights:

> "A mathematical figure representing the sea level surface with all irregularities removed is named the spheroid. The spheroid would be the sea level surface of an Earth with no lateral variations in density. The difference in elevation between the measured geoid and the spheroid is called the geoid anomaly. Some of the major geoid anomalies, like the geoid high over New Guinea (+70m) or the geoid low over India (–100 m) are probably related to mantle convection.
>
> "Any sea level change causes deflection of the ocean floor, hydro-isostasy, to attain isostatic equilibrium. The hydro-isostasy is approximately 1/3 of the sea

level change. Simultaneously the continents are deflected, with a mean magnitude over the continents twice the deflection of the ocean floor. This is due to the fact that the oceanic area is double the land area. An interesting implication of hydro-isostasy is the fact that the sea level history will differ between oceanic islands and continental margins. An island moving with the sea floor will record the full sea level change, while points near the continents record quite different sea level changes. Thus hydro-isostasy is an important factor in determining relative sea level fluctuations."

Evidently, eustasy is complex, and is certainly not globally uniform.

Relative sea level is the change in sea level relative to the surrounding land, taking into account changes in the elevation of the land due to factors such as glacial loading and rebound, and ocean floor subsidence. But Fjeldskaar (2008) emphasized that "relative sea level changes are not the same as eustatic changes" and "it is difficult to imagine where a eustatic change can be measured realistically".

The issues involved in contemplating the effect of global warming on SLR illustrate the complexity and uncertainty involved in predicting the future. Kolker and Hameed (2007) said:

> "Determining the rate of global sea level rise (GSLR) during the past century is critical to understanding recent changes to the global climate system. However, this is complicated by non-tidal, short-term, local sea-level variability that is orders of magnitude greater than the trend."

Of all the potential impacts of global warming, the potential rise of sea level is likely to be the most serious because of the possibility of much greater effects if the GIS is seriously impacted by modest temperature increases. However, it is not clear that much can be done about this.

The Gravity Recovery and Climate Experiment (GRACE) satellites are capable of measuring changes in the mass of the ocean on monthly time scales with an accuracy of a few millimeters (Boening *et al.*, 2012). Measurements have been made starting in 2005.

8.3.3 Measured sea level change in the 20th century

Jevrejeva *et al.* (2008) carried out a reconstruction of GSL since 1700 from tide gauge records. They performed a quadratic fit to the data, and concluded that sea level rise began accelerating at the end of the 18th century at the rate of about 0.01 mm/yr^2 (see Figure 8.4). According to them, sea level rose by 6 cm during the 19th century and 19 cm in the 20th century. Superimposed on the long-term acceleration were quasi-periodic fluctuations with a period of about 60 years. It was concluded that, if the conditions that established the acceleration continue into the future, then sea level would rise 34 cm over the 21st century. However, acceleration, like beauty, lies in the eye of the beholder. One could argue alternatively from Figure 8.4 that sea level oscillated about a linear trend from 1920 to 2000 with a slope of 2.0 mm/yr with no acceleration at all, and that the acceleration that did occur, took place in the 19th

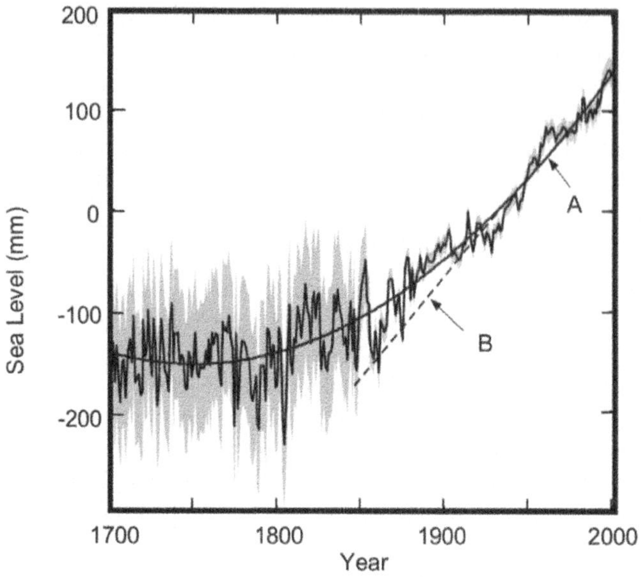

Figure 8.4. Reconstruction of relative sea level since 1700 according to Jevrejeva *et al.* (2008). Absolute numbers are meaningless. Curve "A" is a quadratic fit to the variable sea level. Line "B" is a linear fit to the period from 1920 to 2000. The gray area is an estimate of uncertainty.

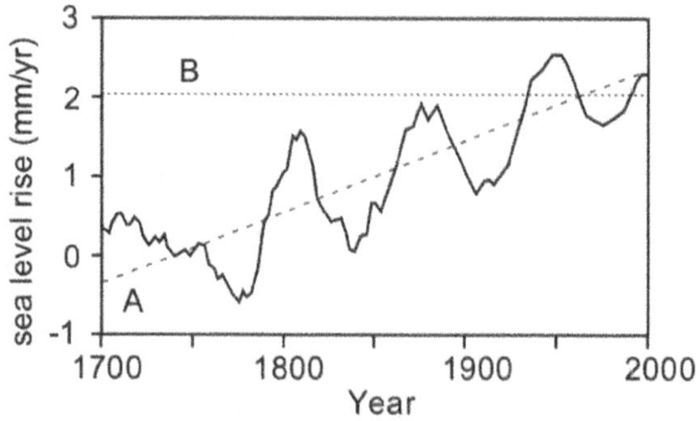

Figure 8.5. Reconstruction of rate of sea level rise since 1700 according to Jevrejeva *et al.* (2008). Line "A" represents constant acceleration at 0.009 mm/yr^2, while line "B" represents a constant rate of rise of 2 mm/yr since ~1920.

century as the Earth came out of the LIA. Jevrejeva *et al.* (2008) also produced Figure 8.5. One could interpret these data as a constant acceleration of 0.009 mm/yr^2 since 1700. Alternatively, one could argue that that sea level rose at an average constant rate of 2 mm/yr since 1920. In any event, there is no apparent correlation of the rate of sea level rise with the increase in CO_2 concentration in the 20th century.

Wopplemann *et al.* (2008) claimed that tide gauge records at Brest, France, were stable over the period 1889–2007. They found that the rate of sea level rise at Brest was constant over that period. This would suggest that rising CO_2 levels did not play a role in this case. They also found a "close matching of the Brest and Liverpool [U.K.] time series over more than 200 years." They found that both instrumental records "showed a roughly coincident increase in the rate of relative sea-level rise around the end of the 19th century"—well before the great increase in CO_2 concentration. This increase in the rate of sea level rise in the 19th century is also evident in the results of Jevrejeva *et al.* (2008) as shown in Figures 8.4 and 8.5.

Douglas and Peltier (2002) claimed that, although the rate of long-term average global sea level (GSL) rise for the past few millennia has been stable at a level near zero, the GSL abruptly began to rise around the mid-19th century. In the last dozen years, published values of 20th-century GSL rise ranged from 1.0 mm/yr to 2.4 mm/yr, even though all investigators used similar databases of tide gauge measurements. Douglas and Peltier (2002) made a significant distinction between estimates of a GSL rise of 2 mm/yr and 1 mm/yr. If the correct value of GSL rise was near 1 mm/yr, then it is argued that global warming provides an explanation: 1 mm/yr of the GSL rise corresponds to that expected from the thermal expansion of the oceans and the melting of small ice sheets and mountain glaciers caused by the 0.6°C increase in global surface temperature during the last 100 years. This explanation would further imply that melting of the great Greenland and Antarctic ice sheets is not currently contributing significantly to the GSL rise. But if, as Douglas and Peltier (2002) argued, the true rate of contemporary GSL rise was probably closer to 2 mm/yr, it is then likely that these ice sheets are contributing.

Douglas and Peltier concluded that the GSL rise for the past century is closer to 2 mm/yr than 1 mm/yr and it was suggested that this "must be seen as posing a definite geophysical puzzle". Current best estimates are that the thermal expansion of the oceans contributes 0.6 mm/yr to GSL rise and the melting of small ice sheets and glaciers also contributes about 0.3 mm/yr. Together, these two contributions amount to less than 1 mm/yr, leaving a deficit of the same order to be explained. Douglas and Peltier suggested three possible resolutions of this puzzle:

(1) The century-scale tide gauge records might be biased upward—perhaps as a consequence of a strong enhancement of the influence of thermal expansion at coastal locations.
(2) The true global rate of secular GSL rise might be closer to 2 mm/yr than to 1 mm/yr, implying that the great polar ice sheets on Antarctica and Greenland are losing mass at a net rate that contributes 1 mm/yr to the global value. However, Douglas and Peltier raised an objection to this possibility based on constraints on the present-day rate of polar ice mass loss provided by Earth rotation observations. They claimed that, if the best currently available result derived from satellite laser ranging is correct, it would be impossible to solve the puzzle by invoking the required rate of mass loss from Greenland, Antarctica, or both. If, however, it could be demonstrated that the laser-ranging result was in significant error, this solution would become possible.
(3) A third possible solution of the puzzle is that the current best estimate of the influence of thermal expansion is significantly biased downward as a

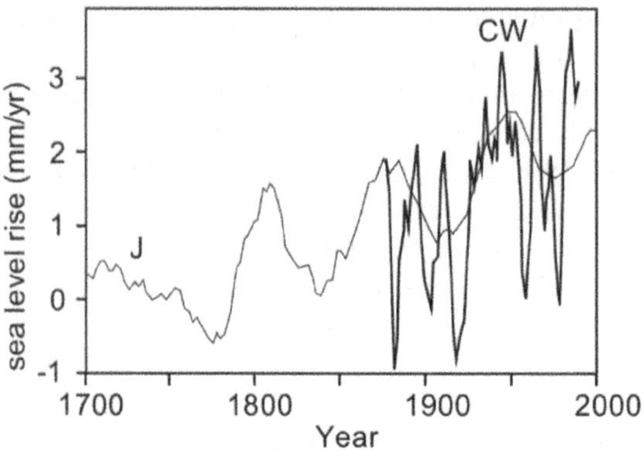

Figure 8.6. Rate of sea level rise. "J" = Jevrejeva *et al.* (2008). "CW" = Church and White (2006).

consequence of under-sampling the Southern Hemisphere ocean at all depths and the abyssal ocean in both hemispheres."

In the opinion of Douglas and Peltier (2002), none of these three individual possibilities could be entirely ruled out, nor can the possibility that the key may involve a mix of all three.

A detailed analysis of sea level rise was provided by Church and White (2006) (see Figure 8.6). They reconstructed GSL from a variety of data sources back to 1870. From 1870 to 2004 (135 years), the total GSL rise was 195 mm—an average of 1.44 mm/yr. For the 20th century, the rise was about 160 mm (1.7 $\pm$ 0.3 mm/yr) indicating a slight acceleration during the 20th century. As can be seen from Figure 8.6, there seems to have been a step-increase in the rate of sea level rise around 1920, and the rate of sea level rise oscillated wildly about 2 mm/yr since then.

Holgate and Woodworth (2004) estimated the sea level rise from 1952 to 1997 (45 years) based on 177 tide gauges divided into 13 regions with near-global coverage, and using a glacial isostatic adjustment model to correct for land movements. Sea level rise over these 45 years was estimated to have averaged 1.7 $\pm$ 0.2 mm/yr, although the curve showed periodic oscillations. Furthermore, the curve of sea level vs. time seemed to be accelerating upward in the 1990s. In a more recent study, Holgate (2007) chose nine long and nearly continuous sea level records from around the world to explore rates of change in sea level for 1904–2003. The lack of high-quality, long-life gauge records was circumvented by finding representative gauges that matched the data for 1952–1997 when more data were available. These records were found to capture the variability found in a larger number of stations over the last half-century studied in their 2004 paper. The addition of new data not only extended the time period back to 1904, but it also extended the time period forward to 2003. The new results indicated that the apparent acceleration noted in the 1990s tailed off and now appears to have just been another oscillation, while the extended curve indicated that the rate of rise of sea level was slightly higher early in the century than it was later in the century.

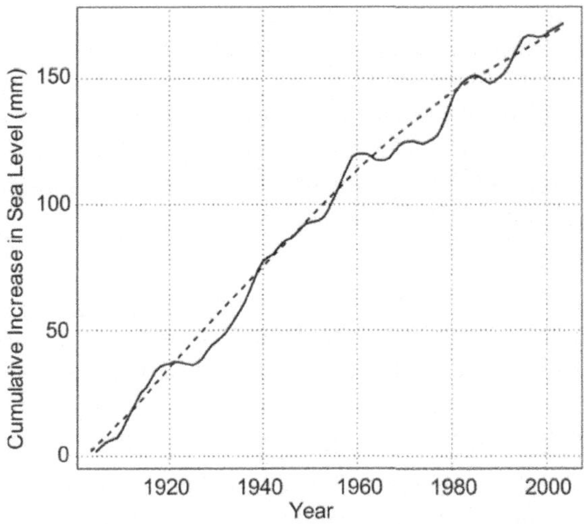

Figure 8.7. The mean sea level record from the nine tide gauges over the period 1904– 2003 based on the decadal trend values for 1907–1999 (adapted from Holgate and Woodsworth, 2004).

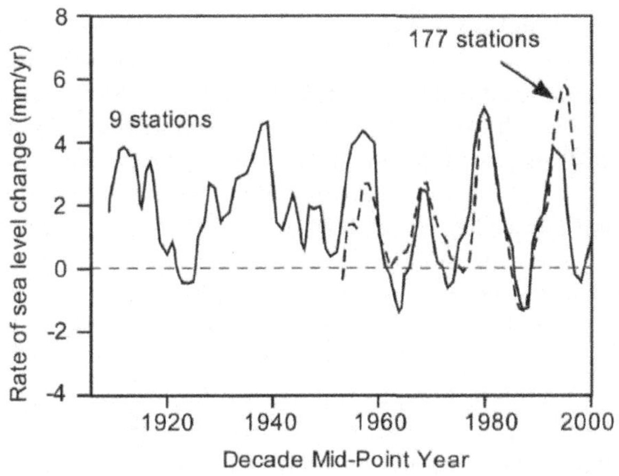

Figure 8.8. Decadal variation of sea level (Holgate, 2007).

According to Holgate (2007):

"The rate of sea level change was found to be larger in the early part of last century (2.03 ± 0.35 mm/yr 1904–1953), in comparison with the latter part (1.45 ± 0.34 mm/yr 1954–2003). The highest decadal rate of rise occurred in the decade centered on 1980 (5.31 mm/yr) with the lowest rate of rise occurring in the decade centered on 1964 (1.49 mm/yr). Over the entire century the mean rate of change was 1.74 ± 0.16 mm/yr."

For the 20th century as a whole, the average rate of rise was estimated by Holgate (2007) to be 1.7 mm/yr but, for the first part of the century, it was 2.0 mm/yr and, for the latter part of the century, it was 1.45 mm/yr. The cumulative curve is shown in Figure 8.7. According to this result, the rate of increase of sea level slowed in the latter part of the century. The detailed curves are shown in Figure 8.8.

Woppelmann *et al.* (2007) provided what appears to be a credible estimate of 20th-century sea level rise. As they discussed:

> "Two important problems arise when using tide gauges to estimate the rate of global sea-level rise. The first is the fact that tide gauges measure sea level relative to a point attached to the land that can move vertically at rates comparable to the long-term sea-level signal. The second problem is the spatial distribution of the tide gauges, in particular those with long records, which are restricted to the coastlines."

Corrections for land movement so far have included corrections for one of the many processes that can affect land stability, namely glacial–isostatic adjustment (GIA). However, different GIA models provide very different values in magnitude and sign. Moreover, GIA models do not account for other sources of vertical land motion. Woppelmann *et al.* (2007) utilized a dedicated GPS measurement system to estimate vertical land movement at 224 stations over a 7.7-year period.

Two important hypotheses were adopted for combining tide gauge and GPS results to derive "absolute" trends in sea level: (1) land motions are extremely low-frequency in character so that the current GPS vertical velocities can be applied for the last century, and (2) the vertical velocity observed at the GPS station applies to the tide gauge site. These assumptions were supported by other evidence.

The poor spatial distribution of historical gauges is problematic because of the evidence of regional variability of sea level trends, this being confirmed by satellite altimetry results. By selecting only tide gauges with long records (e.g., > 60 years), it is hoped that some of the errors might cancel out. The final result of Woppelmann *et al.* (2007) was that the global average sea level rise was estimated to be 1.3 mm/yr for the 20th century.

A comparison of estimates of sea level rise for the 20th century is given in Figure 8.9.

Domingues *et al.* (2008) pointed out that "Climate models ... do not reproduce the large decadal variability in globally averaged ocean heat content inferred from the sparse observational database, even when volcanic and other variable climate forcings are included". They claimed that they provided "improved estimates of near-global ocean heat content and thermal expansion for the upper 300 m and 700 m of the ocean for 1950–2003". They added their observational estimate of upper-ocean thermal expansion to other contributions to sea level rise and found that the sum of contributions from 1961 to 2003 was about 1.5 mm/yr. For the period from 1993 to 2003, they estimated sea level rise to be about 2.4 mm/yr.

Houston and Dean (2011) (H&D) provided a review of previous estimates of 20th-century sea level rise (see Table 8.3). The IPCC has projected a sea level rise of 180–590 mm from 1990 to 2100 based on rising temperatures. Melting ice sheets

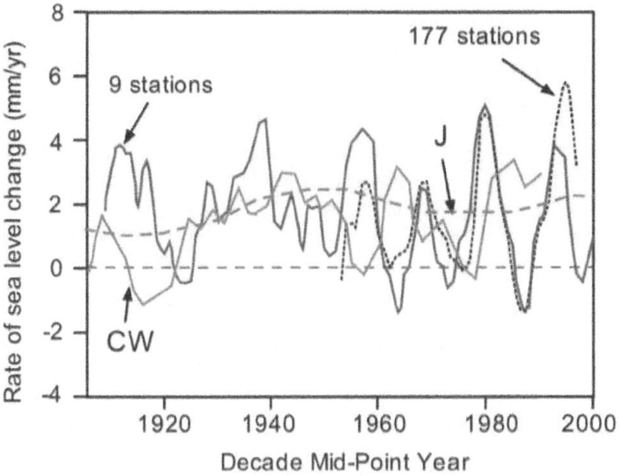

Figure 8.9. Comparison of estimates of sea level rise for the 20th century. The curves marked "9 stations" and "177 stations" are from Holgate (2007). "CW" is from Church and White (2006). "J" is from Jevrejeva *et al.* (2008).

might contribute another 200 mm. This wide range of predictions underlines the uncertainty in future sea level estimates. As H&D pointed out, continuation of the current rate of sea level rise of 1.7 mm/yr would lead to a total sea level rise by 2100 of only 190 mm. To achieve a total rise of 790 mm, acceleration of about 0.10 mm/yr^2 is required in the rate of sea level rise over the 110-year period. However, tide gauge records do not reveal such large accelerations. H&D pointed out the "lack of long-term tide gauge records and their concentration in the northern hemisphere, strong worldwide spatial variations of sea-level rise, vertical land movements, and seasonal-to-decadal temporal variations that can be large compared to sea-level trends and accelerations".

H&D analyzed tide gauge data at 44 U. S. sites and 7 long-term Florida sites that met their criteria for acceptability. They fitted a function:

$$\text{Sea Level} = a_0 + a_1 t + \tfrac{1}{2}\, a_2\, t^2$$

to the data. Here, t = time, a_0 is a constant that varies with location, $a_1 t$ represents a constant rate of rise of sea level, and $\frac{1}{2}\, a_2\, t^2$ represents a constant acceleration in the rate of rise of sea level. They found that many sites yielded a_1 = 1.7 mm/yr and $a_2 \sim 0$. The range of a_1 was from 1.25 mm/yr to 1.90 mm/yr. They concluded:

> "The results of all of our analyses are consistent—there is no indication of an overall world-wide sea level acceleration in the 20th Century data. Rather, it appears that a weak deceleration is present."

If sea level continues to rise at the rate of 1.7 mm/yr through the 21st century, the total rise from 2000 to 2100 would be 17 cm.

Watson (2011) concluded:

Table 8.3. Estimates of acceleration of the of sea level rise (Houston and Dean, 2011).

Reference	*Time period*	*Content*	*Acceleration (mm/yr^2)*
Woodworth (1990)	1870–1990	Oldest European tide gauges—a few gauges back to 1700s	+0.004
Jevrejeva *et al.* (2008)	1800–2000	Old European tide gauges	+0.01; greatest increase 1920–1950
Douglas (1992)	1905–1985 1850–1991	Worldwide tide gauges	–0.01 +0.001
Church *et al.* (2004)	1950–2000	9 yrs of ocean TOPography EXperiment (TOPEX) data + historical tide gauge data	No detectable acceleration (*)
Church and White (2006)	1870–2004	12 yrs of altimetry data + historical tide gauge data	+0.013
Woodworth *et al.* (2009)	1870–2004	Review paper	Slight acceleration; mostly before 1930
University of Colorado (2010)	1993–2010	Altimeter data	*Rate ~ 3 mm/yr; no acceleration given
Holgate (2007)	1910–2000	10-yr mean sea level trends	Rates varied widely from period to period (#)
Vermeer and Rahmsdorf (2009) Jevrejeva, *et al.* (2010) Grinsted, *et al.* (2010)	1990–2100	Semi-empirical models based on assumed future scenarios	+0.07 to +0.28 resulting in sea level rise in 2100 of 600–1,900 mm
Houston and Dean (2011)	20th century	Claimed consensus prior to their work	Rate = 1.7 to 1.8 mm/yr; acceleration uncertain but apparently small
Houston and Dean (2011)	1930–2010	Long-term U.S. tide gauges	Small deceleration (–0.001 to –0.01)

* H&D emphasized the "many uncertainties and sources of error in satellite-altimeter measurements.

H&D emphasized that Holgate (2007) and Church *et al.* (2004) showed that there have been several periods with rate >3 mm/yr even though there was no net acceleration due to the cyclic pattern of rate vs. year.

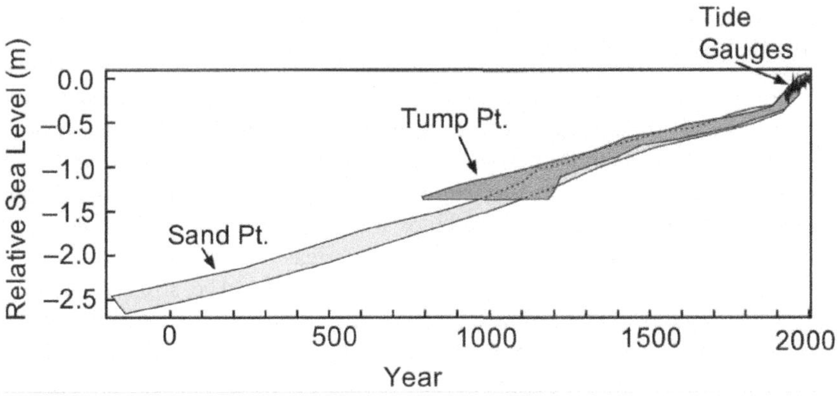

Figure 8.10. Calibration curve for sea level model based on salt-marsh sediments compared to measurements with tidal gauges. The overlap between model and measurements is 80 years (adapted from Kemp *et al.*, 2011).

> "The Australasian region has four very long, continuous tide gauge records which are invaluable for considering whether there is evidence that the rise in mean sea level is accelerating over the longer term at these locations in line with various global average sea level time-series reconstructions. These long records have been converted to relative 20-year moving average water level time series and fitted to second-order polynomial functions to consider trends of acceleration in mean sea level over time. The analysis reveals a consistent trend of weak deceleration at each of these gauge sites throughout Australasia over the period from 1940 to 2000. Short period trends of acceleration in mean sea level after 1990 are evident at each site, although these are not abnormal or higher than other short-term rates measured throughout the historical record."

Kemp *et al.* (2011) estimated sea level variations over the past 2,000 years based on salt-marsh sedimentary sequences from the North Carolina coast. It was claimed that "salt-marsh sediments and assemblages of foraminifera record former sea level because they are intrinsically linked to the frequency and duration of tidal inundation and keep pace with moderate rates of sea-level rise". They used a dataset of foraminifera (193 samples) from 10 salt marshes in North Carolina as proxies for sea level. Microfossils from sediment cores taken in the coastal salt water marshes of mainland North Carolina. These were then compared with North Carolina tidal gauge records from 1920 to 2000. The paper is very terse and does not provide details on how they made the connection between sea level and foraminifera in salt marshes. The comparison between modeled sea level and measured sea level is shown in Figure 8.10.

The calibration period was very short and the extrapolation of the model is very long. This calibration is highly suspect, for several reasons. One obviously is the short duration of overlap between model and measurements. A second is that, as we have amply demonstrated, tidal gauges are notoriously inaccurate. A third factor was provided by Ken Haapala:[2]

> "Environmentalists generally refer to these coastal saltwater marshes as 'fragile wetlands' and these wetlands have a number of interesting characteristics. They are broad, flat, generally marshy lands made of plants, silt, and sand that were formed by sediments from the long term erosion of the Appalachian Mountains and other uplands. ... These wetlands may stretch as far as 50 miles deep into main part of the state. As with most coastal areas built up by sediments, they are probably subject to subsidence. ... During the last Ice Age, streams and rivers cut channels through these sediments, but as the sea levels rose by about 120 meters after the last Ice Age, the channels became tidal estuaries resulting in wide rivers and bays. The areas are subject to erosion and accretion caused by the tides and storms such as hurricanes and northeasters. The areas are partially protected from ocean waves by a series of barrier islands made of sand that shift over the years. As the islands shift, they change the influence that tidal currents and storms have on these wetlands. To suggest that a model of global sea levels can be based on studies of such unstable lands is highly questionable."

In comparing their results to previous results based on tide gauges, Kemp *et al.* (2011) failed to refer to Alley *et al.* (2005), Huybrechts (2002), Wopplemann *et al.* (2007, 2008), Douglas and Peltier (2002), Holgate (2007), Domingues *et al.* (2008), Houston and Dean (2011), or Watson (2011). They did refer to Jevrejeva *et al.* (2008), and Church and White (2006).

It is worth noting that one of the authors of Kemp *et al.* (2011) (Mann) is the progenitor and advocate of the "*hockey stick*" and another (Rahmstorf) is a well-known alarmist (see Figure 8.14).

Boening *et al.* (2012) quoted Church and White (2011) as saying that "Observations from satellite altimeters, along with tide gauge data since the late 19th Century, reveal a fairly steady increase in global mean sea level (GMSL) of about 1.7 mm/ year, with a modest acceleration detectable over the 130 year record".

8.3.4 Recent sea level change

Lombard *et al.* (2005) found large oscillations in decadal changes in sea level due to the El Niño–Southern Oscillation and the Pacific Decadal Oscillation, and cautioned against extrapolating short-term sea level data from satellite measurements.

Jevrejeva *et al.* (2006) also found significant oscillations in sea level with periodicities ranging up to 30 years. For the period 1993–2000, they found a sea level rise of 2.4 ± 1.0 mm/yr, comparable with the value they estimated for 1920–1945. Since the period 1993–2000 was during an uptrend in the oscillatory pattern, the longer-term value will presumably be lower.

Wunsch *et al.* (2007) estimated regional patterns of GSL change from a 1°

[2] Haapala, K. (2011) "The week that was: June 25, 2011", *www.SEPP.org*.

horizontal resolution general circulation model based on about 100 million ocean observations and many more meteorological estimates during the period 1993–2004. Regional variability was found to be significant. They estimated a global mean of about 1.6 mm/yr, of which about 70% is from the addition of fresh water. They concluded, however, that "Useful estimation of the global averages is extremely difficult given the realities of space–time sampling and model approximations. Systematic errors are likely to dominate most estimates of global average change: published values and error bars should be used very cautiously".

Fjeldskaar (2008) presented a summary of a number of studies of sea level rise. Among the references that he cited are the following.

Monaghan *et al.* (2008) found statistically insignificant positive trends in sea level rise over most regions and months during 1960–2005. By contrast, 1970–2005 trends were weakly negative overall.

Harrison and Carson (2007) reported on subsurface temperature trends in the better-sampled parts of the oceans from 1950 to 2000. They found a large spatial variability with some regions showing cooling in excess of 3°C, and others warming of similar magnitude. They concluded that "The ocean neither cooled nor warmed systematically over the large parts of the ocean for the entire analysis period [1950–2000]".

It should be noted that a number of papers have referred to a TOPEX-POSEIDON result that the rise in sea level from 1993 to 2003 was 3.0 mm/yr (e.g., Shepherd and Wingham, 2007), and Hansen claims it is 3.5 mm/yr, but it now seems clear that, even if these measurements were accurate (which is still uncertain) this might be part of an upward cycle and must not be extrapolated. According to Radic (2008), "the error in the instrumental calibration dominates the error budget".

Since direct measurements of sea level appear to be problematic, estimates have been made of the amounts of ice contained in the great ice sheets on Greenland and Antarctica. If the volumes of ice at these sites diminish, and it is assumed that the lost ice appears in the oceans as liquid water, the sea rise resulting from any volume change in ice can be estimated (*360 Gt of ice is equivalent to ~1 mm of sea level*).

The contribution of melting ice sheets to sea level rise was estimated by Shepherd and Wingham (2007) based on 14 different satellite-based estimates of the imbalances of the polar ice sheets since 1998. These studies included standard mass budget analyses, altimetry measurements of ice sheet volume changes, and measurements of ice sheets' changing gravitational attraction. As might be expected, they have yielded a diversity of values, ranging from an implied sea level rise of 1.0 mm/yr to a sea level fall of 0.15 mm/yr. Based on their evaluation of these diverse findings, they estimated that the East Antarctica Ice Sheet (EAIS) is gaining some 25 Gt/yr, the West Antarctica Ice Sheet (WAIS) is losing about 50 Gt/yr, and the GIS is losing about 100 Gt/yr. These trends provide a modest contribution to sea level rise of about 0.35 mm/yr. However, these short-term results since 1998 should not be extrapolated because of the oscillatory behavior of ice sheet loss.

Alley *et al.* (2005) estimated that, for Greenland, between 1993/1994 and 1998/1999, the ice sheet lost 54 Gt/yr of ice, equivalent to a sea level rise of 0.15 mm/yr (the excess of melt water run-off over surface accumulation was about 32 Gt/yr, leaving ice flow acceleration responsible for a loss of 22 Gt/yr).

Rignot and Kanagaratnam (2006) used satellite radar interferometry observations of Greenland to detect widespread glacier acceleration below 66°N between 1996 and 2000, which rapidly expanded to 70°N in 2005. Accelerated ice discharge in the west and particularly in the east doubled the ice sheet mass deficit in the last decade from 90 km^3 to 220 km^3 per year. This provides a less optimistic outlook for Greenland.

Divine and Dick (2006) found evidence of persistent ice retreat in the Arctic since the second half of the 19th century. However, it was not clear whether this was a trend that will continue, or whether it was part of a cycle ("a similar shrinkage of ice cover was observed in the 1920s–1930s, during the previous warm phase of the LFO, when any anthropogenic influence is believed to have still been negligible").

Wingham *et al.* (2006), using radar altimetry, found that mass gains for East Antarctica slightly outweighed mass losses for West Antarctica, "exacerbating the difficulty of explaining twentieth century sea-level rise". Chen *et al.* (2006), using the GRACE satellite mission during its first 3.5 years (April 2002–November 2005), found that mass gains for East Antarctica roughly balanced mass losses for West Antarctica. Velicogna and Wahr (2006) found similar results. Davis *et al.* (2005) also found growth in East Antarctica, but of lesser magnitude.

Conway *et al.* (1999) found:

> "The history of deglaciation of the West Antarctic Ice Sheet (WAIS) gives clues about its future. Southward grounding-line migration was dated past three locations in the Ross Sea Embayment. Results indicate that most recession occurred during the middle to late Holocene in the absence of substantial sea level or climate forcing. Current grounding-line retreat may reflect ongoing ice recession that has been under way since the early Holocene. If so, the WAIS could continue to retreat even in the absence of further external forcing."

This would seem to suggest that the process of disintegration of the WAIS has been underway for some time, independently of anthropogenic influences. According to Dasgupta *et al.* (2007):

> "Until recently, studies of sea level rise (SLR) typically predicted a 0–1 meter rise during the 21st century. The three primary contributing factors have been cited as: (i) ocean thermal expansion; (ii) glacial melt from Greenland and Antarctica (plus a smaller contribution from other ice sheets); and (iii) change in terrestrial storage. Among these, ocean thermal expansion was expected to be the dominating factor behind the rise in sea level. However, new data on rates of deglaciation in Greenland and Antarctica suggest greater significance for glacial melt, and a possible revision of the upper-bound estimate for SLR in this century. Since the Greenland and Antarctic ice sheets contain enough water to raise the sea level by [7.2 m and 61.1 m, respectively] small changes in their volume would have a significant effect. Since the IPCC Report in 2001, there has

been an increased effort to improve measures of mass loss for the Greenland ice sheet and its contribution to SLR."

Dasgupta *et al.* (2007) claimed that satellite interferometry observations led to an estimate that the contribution of the GIS to SLR is double the rate assumed in the IPCC Report. They also indicated that the rate of loss of the WAIS is several times greater than that assumed in the IPCC Report.

Mörner (1973, 2004) provided a contrary view:

> "Sea level rose for glacial eustatic reasons up to about 5,000 years before present (YBP). After that, global sea level has been dominated by the redistribution of ocean water masses (and by ocean-stored heat). This redistribution of water masses is driven by the interchange of angular momentum between the solid Earth and the hydrosphere (in feedback coupling) primarily expressed as changes in the oceanic surface current systems. In view of this, it has been very hard to define any global eustatic signal. This is where and why a dialectic between models and observations enter the sea level debate. According to the glacial loading models, global sea level is now rising by 1.8 to 2.4 mm/year. The IPCC models have hypothesized a very rapid rise in the near future. ... Both the glacial loading models and the IPCC scenarios are strongly contradicted by observational data for the last 100–150 years that cannot have exceeded a mean rate of 1.0–1.1 mm/ year. In the last 300 years, sea level has oscillated close to the present with peak rates in the period 1890–1930. Sea level fell between 1930 and 1950. The late 20th century lacks any sign of acceleration. Satellite altimetry indicates virtually no changes in the last decade. Therefore, observationally based predictions of future sea level in the year 2100 give a value of ± 10 cm [in contradiction to] model outputs by IPCC as well as global loading models. In conclusion, there are firm observationally based reasons to free the world from the condemnation to become extensively flooded in the 21st century AD."

Mörner (2004) also concluded that sea level at the Maldive Islands had actually fallen between 1950 and 2001. However, Church, White, and Hunter (2006) contradicted this finding.

Rignot *et al.* (2011) reported that they resolved the disparities between two approaches to estimate ice sheet mass balance at Greenland and Antarctica over the past eight years. The surface mass balance (SMB) method utilizes "the sum of snowfall minus surface ablation reconstructed from regional atmospheric models with perimeter loss calculated from a time series of glacier velocity and ice thickness to deduce the rate of mass change". "The gravity method employs a monthly time series of time-variable gravity data from the GRACE to estimate the relative mass as a function of time. However, the details in their procedures are very complex and appear to require many assumptions of uncertain veracity. From this, they estimated mass loss rates from Greenland and Antarctica ice sheets over the past 18 years based on the SMB method. Over this period, the annual rate of mass loss has oscillated about a trend that accelerated at both sites. The annual loss in 2010 was about 250 Gt/yr of ice at each site, totaling about 500 Gt/yr for both sites. This corresponds to a rise in sea level of about 1.5 mm/yr—which seems a bit high when compared to independent estimates of the rise of sea level by direct measurement. Rignot *et al.* (2011) also noted that the annual rate of mass loss had accelerated over the 18-year period. They estimated that the acceleration for the sum of both sites was about 36 Gt/yr^2. If this acceleration persists into the future, it would imply that total sea level rise would be 15 cm by 2050 and 56 cm by 2100. However, there appears to be some alarmist chartsmanship in this conclusion. As Figure 8.11 shows, there are large oscillations in the data, and the mass gain in 1992–1993 skews the slope of the acceleration lines. Had the data begun in 1994, the slopes of the lines would be far less. Furthermore, the large positive loops in Antarctica since 2005 do not seem to be adequately considered. A strong case can be made for much flatter acceleration lines. In addition, measurements of sea level suggest that there are long-term oscillations, and extrapolations of short-term data are not justified.

One of the problems in estimating the rate and acceleration of sea level rise is that some satellite data and climate models indicate larger increases than we observe. If these estimates are correct, why haven't the oceans risen faster?

To add to our consternation, Wada *et al.* (2010) provided a global overview of groundwater depletion in sub-humid and arid areas. When more groundwater is removed than is replenished, most of that water ends up in the oceans. Between 1960 and 2000, depletion of groundwater increased by more than a factor of two. It was estimated that, in 2000, groundwater depletion added 0.8 mm/yr to the oceans. Yet, the measurements from tide gauges do not seem compatible with this result.

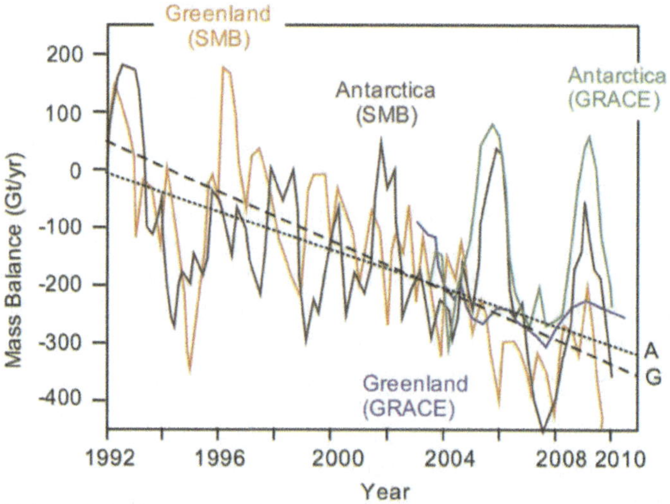

Figure 8.11. Ice sheet mass balance by surface mass balance and GRACE methods. The lines "A" and "G" represent the downward acceleration trends reported by Rignot *et al.* (2011).

According to Scott K. Johnson:[3]

"In many places, the water table is dropping as groundwater is depleted. When groundwater is pumped up for use, whether for drinking water or irrigation, some portion of it fails to infiltrate back down into the ground. (In drier regions, the portion that infiltrates approaches nil). Instead, the water evaporates into the atmosphere or ends up in surface streams. In either case, most of it eventually makes its way to the ocean. In many places, the amount of precipitation that infiltrates into the ground is too small to make up for that loss. And as the volume of groundwater decreases, sea level must rise in turn."

Although construction of dams on rivers creates large reservoirs (or lakes) behind them, increasing the storage of water on land, this has been estimated to be much smaller than the loss of groundwater to the oceans.

The 2007 IPCC Report assumed that dams and groundwater depletion roughly cancelled each other out.

Church *et al.* (2011) found the following results for the time period 1972 to 2008:

Source	*Rise in sea level (mm/yr)*
Ocean thermal expansion	0.8
Melting of glaciers and ice caps on Greenland and Antarctica	0.4
Melting of other glaciers and ice caps	0.7
Loss of ground water to oceans	0.3
Storage of water behind dams	−0.4
Total	1.8

These authors claimed that storage of water behind dams more than offset the loss of ground water to the oceans.

There are two basic methods to estimate loss of ground water contribution to sea level rise. One, used by Wada *et al.* (2010), used an indirect, flux-based water budget approach that assumed that groundwater depletion is equal to the difference between natural recharge and withdrawals—rather than an approach based on actual observations of groundwater conditions. According to Konikow (2011), while the flux-based method is global in nature, it does not take into account the fact that, as aquifers become depleted, they undergo reductions in natural discharge from the system (such as to springs and oases). Konikow also argued that the global modeling approach does not account for "non-natural" non-diffuse recharge, such as leakage from canals, sewers, or pipelines, or from artificial recharge. Hence, he argued that the flux-based water budget approach of Wada *et al.* (2010) "can substantially

[3] *http://arstechnica.com/science/2012/06/groundwater-responsible-for-nearly-half-of-sea-level-rise/.*

overestimate groundwater depletion". Konikow used a volumetric approach rather than a flux-based approach, in which he analyzed sequential changes in volume stored in various aquifers. Unfortunately, worldwide data are not widely available and Konikow had to extrapolate U.S. data to much of the rest of the world. He estimated that ground water depletion contributed about 6% to 7% of sea level rise from 1900 to 2008. However, he noted that there was a rapid acceleration in the contribution of ground water depletion to sea level rise in the latter half of the 20th century. Over the period 2000–2008, the contribution of ground water depletion to sea level was estimated to be 0.4 mm/yr.

Wada *et al.* (2012) defended the flux-based approach and pointed out that "volume-based assessments are only available for a limited number of aquifers and regions in the world, such that global estimates can be obtained only through extrapolation under assumptions, such as fixed depletion to abstraction ratios, that are difficult to verify".

Not only did Wada *et al.* (2012) estimate the contribution of ground water depletion to sea level rise in the 20th century, but they also estimated this quantity for the remainder of the 21st century using three possible future scenarios and three global climate models to project future climate (see Figure 8.12).

Pokhrel *et al.* (2012) "estimated sea-level change in response to human impacts on terrestrial water storage by using an integrated model that simulates global terrestrial water stocks and flows (exclusive of Greenland and Antarctica) and especially accounts for human activities such as reservoir operation and irrigation". They found that, "unsustainable groundwater use, artificial reservoir water impoundment, climate-driven changes in terrestrial water storage and the loss of water from closed basins have contributed a sea-level rise of about 0.77 mm/yr to sea level rise between 1961 and 2003, about 42% of the observed sea level rise". They noted that "of these components, the unsustainable use of groundwater represents the largest contribution". Rahmstorf, who previously predicted a sharp future rise in sea level, was "shocked" to find these results for groundwater.

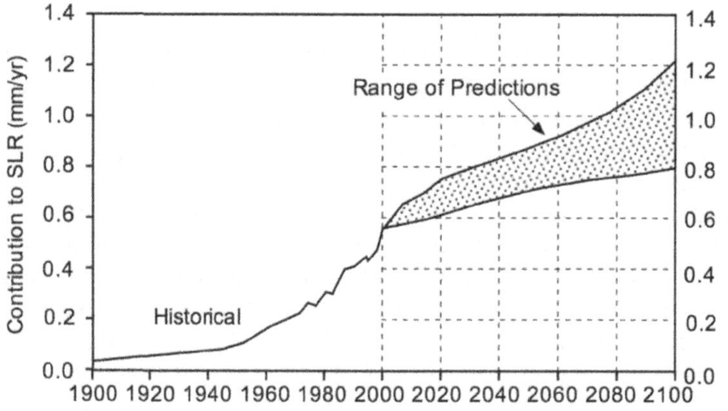

Figure 8.12. Projection of future contribution to sea level rise from ground water (Wada *et al.*, 2012).

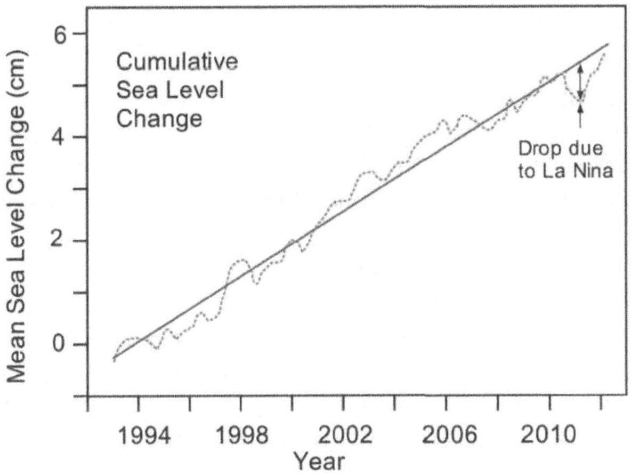

Figure 8.13. Global mean sea level from altimetry from 1992 to 2012 with annual and semi-annual variations removed and smoothed with a 60-day running mean filter (Boening *et al.*, 2012).

Using a combination of GRACE measurements since 2005, changes in ocean volume attributable to ocean temperature measurements, as well as inferences of ocean heat gain from energy flux measurements at the top of the atmosphere, Boening *et al.* (2012) derived a new estimate for growth of sea level shown in Figure 8.13. Their emphasis was on the fact that, during a major La Niña in 2010–2011, the ocean fell about 5 mm. They attributed this to "an ENSO-related transfer of mass between the oceans and the continents". However, precipitation is expected to diminish during La Niña events so it is not clear how or why this putative transfer took place. But the strangest thing about Figure 8.13 is that the slope of the line is about 3.2 mm/yr, which is considerably greater than previous estimates. Yet the authors admit that previous measurements indicated "a fairly steady increase in global mean sea level (GMSL) of about 1.7 mm/ year, with a modest acceleration detectable over the 130 year record" and they never discuss this discrepancy at all!

A recent article published while this book was in press (Baur, O., Kuhn, M., and Featherstone, W.E. (2013) "Continental mass change from GRACE over 2002–2011 and its impact on sea level", *Journal of Geodesy*, **87**, 117–125), reported new results from the GRACE satellite mission. Sea level rise occurs due to two main factors: melting of glacial ice, and steric expansion due to warming.

They quantified mass-change trends in 19 continental areas due to melting of glaciers:

> "During the 9-year period from May 2002 to April 2011, the average mass gain and loss contributed $-(0.7 \pm 0.4)$ mm/year of sea-level fall and $+(1.8 \pm 0.2)$ mm/year of sea level rise. The net effect was a sea level rise of $+(1.1 \pm 0.6)$ mm/year due to ice melt. Ice melting over Greenland, Iceland, Svalbard, the Canadian Arctic archipelago, Antarctica, Alaska and Patagonia was responsible for $+(1.4 \pm 0.2)$ mm/year of the total balance. Hence, land-water mass

accumulation compensated about 20% of the impact of ice-melt water influx to the oceans."

Modification of these data for geocenter correction and inclusion of regions outside the study increase the net sea level rise due to melting as 1.2 mm/yr.

Baur *et al.* reviewed estimates of the steric contribution and concluded that there is a wide range of estimates. They chose 0.5 mm/yr, bring total sea level rise from both components to 1.7 mm/yr.

Zhang and Church (2012) noted that "many sea level studies have an underlying purpose of detecting and quantifying sea level change due to anthropogenic climate change." However, "on a regional scale, such a signal is mixed with that due to natural climate variability," and "as a result, it is extremely difficult to separate the natural and anthropogenic signals, especially when they have comparable amplitudes and the available time series is short relative to the period of the natural variability".[4]

Zhang and Church (2012) used continuous near-global altimeter measurements since 1993 to attempt to separate inter-annual and decadal sea level variability in the Pacific from the long-term background sea level trend and, in doing so, they were able to show that "the decreasing regional sea level in the eastern equatorial Pacific is mainly associated with the Pacific Decadal Oscillation", while "in contrast, for those island countries in the western tropical Pacific and especially low-lying atolls, the high rate of sea level rise over the altimeter era has a significant component associated with natural variability".

In light of their illuminating findings, the two Australian researchers thus concluded that "it is tempting to use current-day altimeter-based regional sea level linear trends as a reference for future climate change projections". However, they say that "such practice needs to be treated with caution as regional sea level linear trends derived over the short altimeter era can be greatly affected by low-frequency climate variability".

8.3.5 Global warming and future sea level change

A number of climatologists have predicted the sea level rise for the remainder of the 21st century. This partly depends on projected scenarios for future CO_2 emissions, and the general benchmark question is what is the expected sea level rise if the CO_2 concentration rises from 280 ppm to 560 ppm?

Hansen (2004) said:

> "The dominant issue in global warming, in my opinion, is sea-level change and the question of how fast ice sheets can disintegrate."

According to Hoffman (1984):

> "Future global sea level will depend primarily on three factors: (1) the total quantity of water filling the oceans' basins; (2) the temperature of the oceans'

[4] Thanks to *www.nipccreport.org/articles/2013/feb/26feb2013a1.html* for this review.

layers, which determines the density and volume of their waters; and (3) the bathymetry (shape) of the ocean floor, which determines the water-holding capacity of the basins. A rise in global temperature can, by a variety of physical mechanisms, transfer snow and ice from land to the sea, increasing the quantity of water in the ocean basins, and can raise the oceans' temperatures, causing the thermal expansion of their volumes. Changes in the bathymetry of the oceans' floors occur independently of climate change. Because geological changes in the ocean floor could not raise or lower global sea level by more than a centimeter or two by 2100, this factor is not considered in constructing global scenarios. An evaluation of the impacts of sea level rise at specific coastal sites, however, will require consideration of local uplift or subsidence, which by 2100 could cause changes in land elevation that are large enough to be of significance to local planning. Projecting sea level rise requires the means to estimate future changes in atmospheric composition, to relate these changes to global warming, and then to determine how the warming can cause land-based snow and ice to enter the sea and the oceans to expand thermally."

Hoffman (1984) estimated the sea level rise expected from various degrees of global warming. Low, medium, and high scenarios were based on ultimate temperature rises of 1.5°C, 3°C, and 4.5°C by 2100, respectively, for a doubling of CO_2 concentration. The projected increases in sea level are summarized in Table 8.4.

Table 8.4. Projected rise in sea level (cm) (data from Hoffman, 1984).

Year	*Low* $\Delta T \sim 1.5°C$	*Medium* $\Delta T \sim 3°C$	*High* $\Delta T \sim 4.5°C$
1986	0		0
2000	5	11	17
2025	13	32	55
2050	24	66	117
2075	38	114	213
2100	56	180	345

According to Titus (1990):

"Since 1979, there has been a general consensus that a doubling of carbon dioxide would raise global temperatures 1.5 to 4.5°C, and that such a doubling is likely to occur over the next century. More recent assessments have pointed out that emissions of methane, nitrous oxide, and numerous other gases that absorb infrared radiation could further increase this warming, and that warmer temperatures may increase the rate of natural emissions of these gases. Although national policy makers are beginning to formulate strategies to slow global warming, there is an emerging consensus that at least a one or two degree warming is inevitable, due to past emissions and the time it will take to change production practices and retire existing machinery. In the late 1970s, some scientists suggested that the projected global warming might cause a 5 to 7 meter

Table 8.5. Predicted rise (cm) in sea level by 2100 (Titus, 1990).

Source	*Year*	*Low estimate*	*Mid–Low*	*Mid–High*	*High*
World Meteorological Organization	2050	20			170
Environmental Protection Agency	2100	70	160	210	340
National Research Council	2100	50			190

rise in sea level over the next few decades, due to a disintegration of the West Antarctic Ice Sheet. However . . . such a deglaciation would take at least 200–500 years. As a result, most recent assessments have focused on other contributors to future sea level rise: expansion of ocean water and the melting of mountain glaciers and parts of the ice sheet in Greenland."

Titus (1990) provided the estimates of future sea level rise shown in Table 8.5.

Alley *et al.* (2005) suggested that the GIS may melt entirely from future global warming, whereas the EAIS is likely to grow through increased accumulation (for warming less than 5°C). The future of the WAIS remains uncertain, with its marine-based configuration raising the possibility of important losses in the coming centuries. Alley *et al.* (2005) said:

> "Despite these uncertainties, the geologic record clearly indicates that past changes in atmospheric CO_2 were correlated with substantial changes in ice volume and global sea level. Recent observations of startling changes at the margins of the Greenland and Antarctic ice sheets indicate that dynamical responses to warming may play a much greater role in the future mass balance of ice sheets than previously considered. Models are just beginning to include these responses, but if they prove to be important, sea-level projections may need to be revised upward. Also, because sites of global deepwater formation occur immediately adjacent to the Greenland and Antarctic ice sheets, any notable increase in freshwater fluxes from these ice sheets may induce changes in ocean heat transport and thus climate."

IPCC (2001) said:

> "Disintegration of the West Antarctic Ice Sheet or melting of the Greenland Ice Sheet could raise global sea level up to 3 m each, over the next 1,000 years, submerge many islands, and inundate extensive coastal areas. . . . The projected sea-level rise of 5 mm/yr for the next 100 years would cause enhanced coastal erosion, loss of land and property, dislocation of people, increased risk from storm surges, reduced resilience of coastal ecosystems, saltwater intrusion into freshwater resources, and high resource costs to respond to and adapt to these changes (high confidence). . . . Many coastal areas will experience increased levels of flooding, accelerated erosion, loss of wetlands and mangroves, and

seawater intrusion into freshwater sources as a result of climate change. The extent and severity of storm impacts, including storm-surge floods and shore erosion, will increase."

The projection of 5 mm/yr is far higher than most other projections, and reflects the extreme alarmist views of the IPCC Report. Figure 8.14 shows the range of IPCC projections for the 21st century. Also shown is Rahmstorf's 2007 projection based on an assumed proportionality of 3.4 mm/yr per °C, and a hypothesized range of temperature increases of 1.5°C to 3.5°C. However, note that, for the 20th century, the proportionality was about 120 mm per 0.8°C temperature rise, or 1.5 mm/yr per °C increase.

As Fjeldskaar (2008) summarized, "There is no doubt that sea level is currently rising on a global scale ... but a key question is not whether sea level is actually rising, but rather, has there been any acceleration in its rise during recent decades?". Examining all the conflicting data, he reached "the first conclusion" that "we have not reached a consensus on the rate at which sea level rises".

According to Anon. (E):

"Over the past century, sea level has slowly been rising. This is in part due to the addition of water to the oceans through either the melting of or the 'calving' off of icebergs from the world's land ice. Many individual mountain glaciers and ice caps are known to have been retreating, contributing to the rising sea levels. It is uncertain, however, whether the world's two major ice sheets—Greenland and Antarctica—have been growing or diminishing. This is of particular importance because of the huge size of these ice sheets, with their great potential for changing sea level. Together, Greenland and Antarctica contain about 75% of the world's fresh water, enough to raise sea level by 70 m–75 m, if all the ice

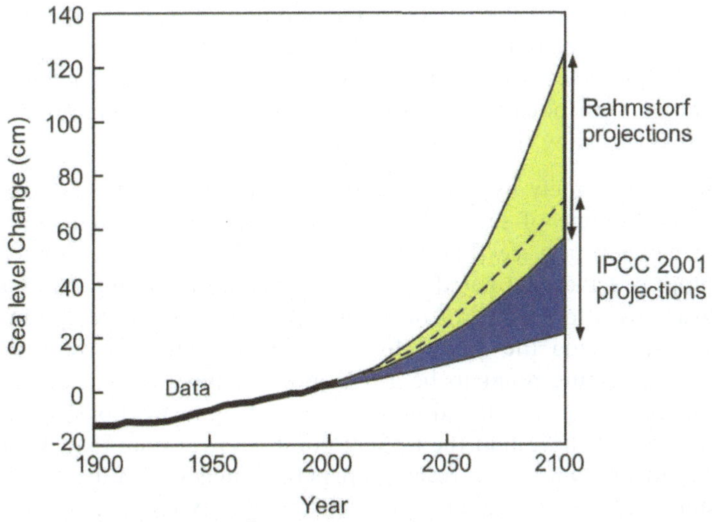

Figure 8.14. Projections of future sea level rise (IPCC, 2001; Rahmstorf, 2007).

were returned to the oceans. Measurements of ice elevations are now being made by satellite radar altimeters for portions of the polar ice sheets, and in the future measurements will be made by a laser altimeter as part of NASA's Earth Observing System (EOS). The laser altimeter will provide more accurate measurements over a wider area. The Greenland ice sheet is warmer than the Antarctic ice sheet and as a result, global warming could produce serious melting on Greenland while having less effect in the Antarctic. In the Antarctic, temperatures are far enough below freezing that even with some global warming, temperatures could remain sufficiently cold to prevent extensive surface melting. Where ice sheets extend outward to the ocean, the ice tends to move out over the surrounding water, forming ice shelves. There is concern that, with global warming, the water under the ice shelves would be warmer and cause them to break up more readily, forming very large icebergs. If the ice shelves of West Antarctica were to break up, this would release more inland ice in an irreversible process, possibly leading to sea level rises of several meters."

Alley *et al.* (2005) emphasized:

"Future sea-level rise is an important issue related to the continuing buildup of atmospheric greenhouse gas concentrations. The Greenland and Antarctic ice sheets, with the potential to raise sea level ~70–75 meters if completely melted, dominate uncertainties in projected sea-level change. Freshwater fluxes from these ice sheets also may affect oceanic circulation, contributing to climate change. Observational and modeling advances have reduced many uncertainties related to ice-sheet behavior, but recently detected, rapid ice-marginal changes contributing to sea-level rise may indicate greater ice-sheet sensitivity to warming than previously considered. Over the last century, sea level rose ~1.0 to 2.0 mm year, with water expansion from warming contributing 0.5 ± 0.2 mm (steric change) and the rest from the addition of water to the oceans (eustatic change) due mostly to melting of land ice. By the end of the 21st century, sea level is projected to rise by 0.5 ± 0.4 m in response to additional global warming, with potential contributions from the Greenland and Antarctic ice sheets dominating the uncertainty of that estimate."

It has been widely hypothesized that a warmer climate in Greenland would increase the volume of lubricating surface meltwater reaching the ice–bedrock interface, accelerating ice flow and increasing mass loss (Joughin *et al.*, 2008). Climatologists have postulated that there may be a tipping point in which the temperature rises to a point of no return, and an irreversible melting of the GIS would ensue, raising the oceans by some seven meters. Several alarmists have projected the tipping point to be a 3°C global temperature rise (of which about 0.8°C has already occurred), although recent prognostications by James Hansen would have you believe that the tipping point is lower than that. However, Carlin (2009) commented on several recent papers that suggest that the mass loss has been lower than predicted. (Joughlin *et al.*, 2008; van de Wal *et al.*, 2008). The paper by van de Wal *et al.* is particularly intriguing because the title "Large and rapid melt-induced velocity changes in the ablation zone of the Greenland ice sheet" belies the finding:

> "The overall picture obtained by averaging all stake measurements at all sites for individual years indicates a small but significant ($r = 0.79$, $P < 0.05$) decrease of 10% in the annual average velocity [of ice movement in western Greenland] over 17 years."

Most recently, Jonathan Bamber, an ice sheet expert at the University of Bristol, told the Copenhagen Climate Congress in 2009 that previous studies had misjudged the so-called Greenland tipping point, at which the ice sheet is certain to melt completely. He said the tipping point is closer to 6°C than 3°C. Murray *et al.* (2010) found that:

> "... the early 2000s speedup of SE Greenland tidewater outlet glaciers was followed by a widespread and synchronous slowdown event. ... Runoff lubrication of the glaciers does not provide the explanation for this speedup/ slowdown event. ..."

They indicated;

> "... that the speedup was the result of warm ocean waters coming into contact with the glaciers."

They suggested that:

> "... a negative feedback that currently mitigates against continued very fast loss of ice from the ice sheet in a warming climate ... namely the cold melt waters of the coastal Eastern Greenland Coastal Current act to stem further melting. Since these SE Greenland outlet glaciers have dominated recent changes in ice sheet mass loss, the negative feedback we identify will also help regulate Greenland's contribution to sea level rise against a background of increasing ocean temperatures. We should expect similar speedup and slowdown events of these glaciers in the future, which will make it difficult to elucidate any underlying trend in mass loss resulting from changes in this sector of the ice sheet."

Additional discussion and references are given in NIPCC (2011).

With respect to the melting of continental glaciers, Shepherd and Wingham (2007) reviewed what is known about sea level contributions arising from wastage of the Antarctic and GISs, focusing on the results of 14 different satellite-based estimates of the imbalances of the polar ice sheets that have been derived since 1998. The conclusion was that the current "best estimate" of the contribution of polar ice wastage to GSI change is a rise of 0.35 mm/yr.

Hansen (2005) claims that even this relatively benign scenario might nevertheless lead to destruction of the GIS. His arguments were based on a previous paper (Hansen, 2004) in which he pointed out that ice sheet growth is a slow, dry process, inherently limited by the snowfall rate, but disintegration is a wet process, driven by positive feedbacks, and, once well underway, it can be explosively rapid. Figure 8.15 shows a moulin on the GIS. The moulin, a near-vertical shaft worn in the ice by

Figure 8.15. A stream of snow melt cascades down a moulin on the Greenland ice sheet during a recent summer (adapted from Hansen (2004) and Zwally *et al.* (2002); original by Roger J. Brathwaite, University of Manchester, U.K., reproduced from *www.gisss.nasa.gov/research/briefs/grnitz_09/*).

surface water, carries water to the base of the ice sheet. There, the water acts as a lubricating fluid that speeds motion and disintegration of the ice sheet. Hansen reviewed the situation during past ice sheet disintegrations. He claimed that:

> "About 14,000 years ago, sea level rose about 20 m in approximately 400 years. That is an average of 1 m of sea level rise every 20 years. The nature of glacier disintegration required for delivery of that much water from the ice

sheets to the ocean would be spectacular (5 cm of sea level, the mean annual change, is about 15,000 cubic kilometers of water)." (Hansen, 2005).

Hansen believes that the Earth is "now out of energy balance by close to + 1 W/m^2, i.e., with that much more energy absorbed from sunlight than the energy emitted to space as thermal radiation" which is "due mainly to rapid growth of greenhouse gases, especially CO_2 and CH_4, and the thermal inertia of the ocean". The greenhouse gases produce a downward forcing while the thermal inertia of the oceans prevents a rapid temperature rise, thus limiting re-radiation from the Earth to space. Hansen pointed out that: "CO_2 and CH_4 amounts today are far outside the ranges that existed for hundreds of thousands of years" (Figure 8.16). However, that radiative imbalance has to show up in the oceans, and, so far, it has not. It seems likely that Hansen has overestimated the imbalance by a factor of two.

Most of this putative energy imbalance goes into warming the oceans. However, Hansen believes that mechanisms exist to transfer some of this energy to the ice sheets. He believes that a further 1°C temperature rise might be enough to do significant damage to the GIS. Such mechanisms are likely to have occurred during the rapid disintegration of the ice sheets after the last glaciation (14,000–11,000 years ago). According to Hansen (2005):

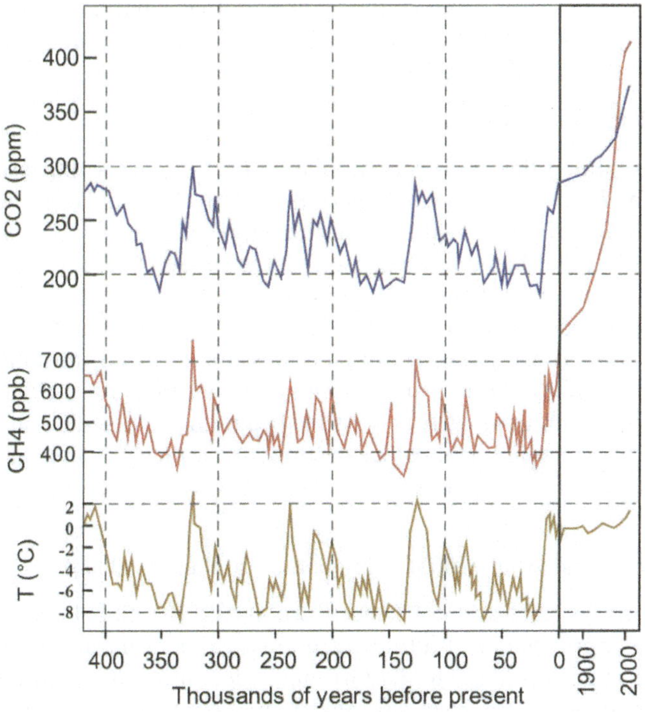

Figure 8.16. Record of atmospheric CO_2, CH_4, and temperature extracted from an Antarctic ice core (adapted from Hansen, 2005).

"The net effect of these processes, which eventually will include a positive feedback from lowering of the ice surface altitude, is the potential for a highly nonlinear response, a process that could run out of control, possibly to the ultimate demise of the entire south dome (64°N) of the Greenland ice sheet, if the strong planetary forcing is maintained long enough. The question is: how long is long enough?"

Hansen (2005) suggested that:

"Three time constants play critical roles in creating a slippery slope for human society: T_1, the time required for climate, specifically ocean surface temperature, to respond to a forced change of planetary energy balance; T_2, the time it would take human society to change its energy systems enough to reverse the growth of greenhouse gases; T_3, the time required for ice sheets to respond substantially to a large relentless positive planetary energy imbalance … T_1, the climate response time, is 50–100 years, as a result of the large thermal inertia of the ocean. T_2, the energy infrastructure time constant, also is perhaps 50–100 years … T_3, the ice sheet response time, is the time constant of issue.

"T_3 is of the order of centuries, not millennia, as commonly assumed. Growth of ice sheets requires millennia, as growth is a dry process limited by the snowfall rate. Ice sheet disintegration, on the other hand, is a wet process that can proceed more rapidly, as evidenced by the saw-toothed shape of glacial–interglacial temperature and sea level records."

Hansen (2005) summarized:

"The likelihood that T_3 is comparable to $T_1 + T_2$ has a staggering practical implication. $T_3 \gg T_1 + T_2$ would permit a relatively complacent 'wait and see' attitude toward ice sheet health. If, in the happy situation $T_3 \gg T_1 + T_2$, we should confirm that human forcings were large enough to eventually alter the ice sheets, we would have plenty of time to reverse human forcings before the ice sheets responded. Unfortunately, $T_3 \sim T_1 + T_2$ implies that once ice sheet changes pass a critical point, it will be impossible to avoid substantial ice sheet disintegration. If T_3 indeed is not very much larger than $T_1 + T_2$, it becomes of high priority to detect as early as possible beginnings of ice sheet disintegration. High precision measurements of ice motion and sea level change are needed for early detection of any acceleration in the global rates of ice movement and sea level rise."

Oppenheimer and Alley (2005) commented on Hansen (2005):

"If Hansen is right about ice sheet response to the global energy imbalance and if IPCC's projections of future greenhouse gas concentrations prove correct, [a 1°C rise might disintegrate the Greenland Ice Sheet] it would be too late to stem a catastrophic sea level rise, given the commitment to future warming already in the pipeline."

But Oppenheimer and Alley (2005) indicated that, based on evidence from the last interglacial, the ice sheets are likely to be stable to a 3°C temperature increase.

Graversen *et al.* (2010) applied a wide range of existing climate models to a detailed dynamical model of mass loss from Greenland expected in the 21st century. Unfortunately, the variations from model to model are so wide that the final result is a prediction that sea level will rise between 0 cm and 17 cm in the 21st century.

8.3.6 Evidence from previous deglaciations

Interglacial periods are of great interest because we are presently in an interglacial period, and there is widespread concern that rising CO_2, generated by human activity, may amplify and extend this climate period with negative consequences for civilization. It is therefore relevant to review data on past interglacials with particular emphasis on CO_2 levels and prevailing temperatures. It has been estimated that several past interglacials were warmer than the present interglacial, and therefore it has been hypothesized that such past interglacials might provide an analog for the future if global warming continues.

Holden *et al.* (2009) reported:

> "Ice core evidence indicates that even though atmospheric CO_2 concentrations did not exceed 300 ppm at any point during the last 800,000 years, East Antarctica was at least 3–4°C warmer than pre-industrial (CO_2 ~280 ppm) in each of the last four interglacials. During the previous three interglacials, this anomalous warming was short lived (~3,000 years) and apparently occurred before the completion of Northern Hemisphere deglaciation."

Holden *et al.* (2009) presented a speculative theory to explain this based on meltwater-forced slowdown of the Atlantic Meridional Overturning Circulation (AMOC) during glacial terminations.

Sime *et al.* (2009) presented oxygen isotope data from three sites in Antarctica showing that the change in isotope content was considerably greater in the previous three deglaciations than in the present deglaciation. They analyzed the relationship between isotope index and temperature at the three sites and concluded "that maximum interglacial temperatures over the past 340 kyr were between 6°C and 10°C above present-day values" and "there are serious deficiencies in our understanding of warmer than present day climates". A simplified version of their data is shown in Figure 8.17.

Dahl-Jensen *et al.* (2013) reconstructed the Eemian record from folded ice using globally homogeneous parameters known from dated Greenland and Antarctic ice core records. They estimated north Greenland surface temperatures after the onset of the Eemian (126,000 years ago) peaked at 8 ± 4°C above the mean of the past millennium. They estimated that, between 128,000 and 122,000 years ago, the thickness of the north-west GIS decreased by 400 ± 250 m, reaching surface elevations 122,000 years ago of 130 ± 300 m lower than the present. This would suggest that the temperature peak of 8 ± 4°C above the mean of the past millennium contributed only about 2 m of sea level rise due to melting of Greenland ice.

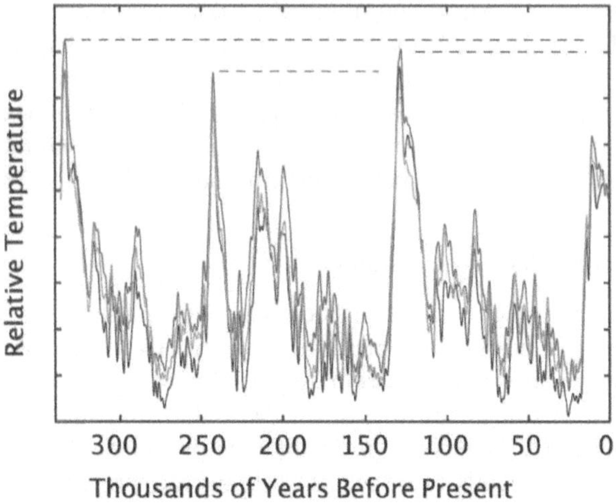

Figure 8.17. Relative temperatures of the last four interglacials. The previous three interglacials were significantly warmer than the present one, even though the CO_2 concentration was comparable (Sime *et al.*, 2009). The different curves refer to three different drilling sites in Antarctica.

Rohling *et al.* (2008) quoted previous studies that indicated that "The last interglacial period ... was characterized by global mean surface temperatures that were at least 2°C warmer than present [and] mean sea level stood 4–6 m higher than modern sea level, with an important contribution from a reduction of the Greenland ice sheet". They used "a combination of a continuous high-resolution sea level record, based on the stable oxygen isotopes of planktonic foraminifera from the central Red Sea and age constraints from coral data to estimate rates of sea level change during [the previous interglacial]". Rohling *et al.* (2008) estimated average rates of sea-level rise of 1.6 m per century. From this, they inferred that such a rate of sea level rise might occur in the next century. However, this conclusion is very misleading. Figure 8.18 shows their results. Two similar estimates were made from different sites. The rate of sea level rise in area "A" prior to about 123 KYBP was due to the melting of the ice sheet from the previous Ice Age. According to their result, the interglacial (shown as the shaded area "B") was a mere 1,500 years long, and was immediately followed by further glaciation in region "C". If these results are correct, the previous interglacial was very short-lived compared to the present interglacial, and is not directly comparable. By contrast, the sea level over the past 7,000 years in the Holocene has varied very slowly and has been almost constant for the past 3,000 years, as shown for example by Lambeck and Chappell (2001) and PALSEA (2010).

PALSEA (2010) provided data on sea level subsequent to the LGM as shown in Figure 8.19. It appears that there are major differences between the present interglacial and the previous one.

Clark and Huybers (2009) asked: "Why was sea level so much higher 125,000

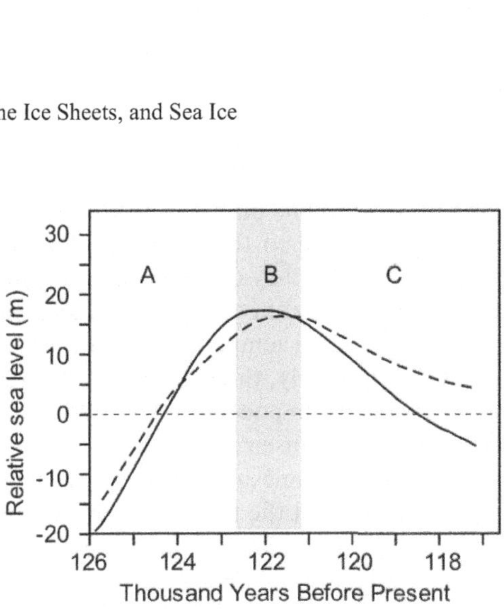

Figure 8.18. Variation of sea level near previous interglacial relative to present sea level (Rohling, *et al.*, 2008).

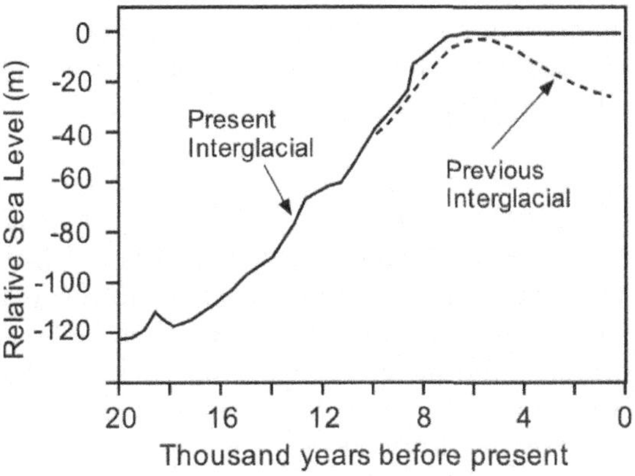

Figure 8.19. Sea level subsequent to the LGM (PALSEA, 2010).

years ago?" They suggested "one possibility is that ice sheets have multiple potential steady states for a given climate". They mention "the global temperature was apparently 1.5–2°C warmer than the pre-anthropogenic global average of the past 10,000 years despite there being essentially no difference in atmospheric greenhouse-gas concentrations". However, this conclusion, which seems amazing to me, did not seem to cause them much consternation. If the Earth can be up to 2°C warmer while the CO_2 concentration is unchanged, then how can CO_2 be the cause of all climate change as proposed by alarmists? Clark and Huybers went on to say "that the climate of the last interglacial might, by coincidence, provide a reasonable analog for establishing ice-sheet sensitivity to global warming" and this implies "that the equilibrium response of sea level to 1.5–2°C of global warming could be an increase of 7–9 meters". They didn't seem to see that there is a logical impasse here. If changes in CO_2 concentration accompany glacial–interglacial transitions, why indeed did the CO_2 concentration not rise sharply above 300 ppm during warmer interglacials? And if rising CO_2 toward 400 ppm has produced the global warming of the past century, how is that connected to the warming of past interglacials when CO_2 remained below 300 ppm? Nor did they consider the possibility that the present interglacial has not yet reached its maximum temperature independently of CO_2, and perhaps (who knows?) is now extending temperatures upward to emulate previous interglacials regardless of CO_2. Ultimately, the relationship between CO_2 concentration and global temperature seems to be poorly understood.

As we discussed in Section 6.5.2, Hansen and Sato (2011) analyzed paleoclimatic data to infer climatological properties relevant to projected climate change in the 21st century. Hansen and Sato estimated the forcings over the past ~ 800,000 years including about eight Ice Ages and interglacials using the measured CO_2 and CH_4 concentrations in ice cores, and estimates of sea level during that period. These fit the oxygen isotope data from the ice cores quite well. But this should not be a surprise. The variation of CO_2 and CH_4 over the past 800,000 years is known to have very nearly the same shape as the isotope curve. Hansen and Sato assigned temperatures to isotope ratios both from the ice cores as well as ocean sediments and claimed good agreement. According to Hansen and Sato, variation of global average temperature from glacial maxima to interglacials was estimated to be typically about 3°C about 800,000 years ago, and this slowly increased to about 5°C in the most recent Ice Age–Interglacial transition.

Hansen and Sato embarked on a discussion of the temperatures reached in the past several interglacial periods as compared to our own interglacial of the past ~ 10,000 years. In particular, ice core data indicate that the interglacials that peaked about 125,000 and 400,000 years ago were warmer than today. Hansen and Sato (2011) concluded that the peak warm periods were "less than 1°C warmer than peak Holocene global temperature" and therefore "were also less than 1°C warmer than global temperature in year 2000". It is not clear how they reached this conclusion, considering that ice core measurements clearly show that past interglacials were warmer than 1°C above our interglacial. It was necessary for Hansen and Sato, as global warming activists, to insist that previous interglacials were not much warmer than our own interglacial, since that would suggest that even a moderate warming compared to today's climate would result in potentially catastrophic increases in sea level. However, as we pointed out, the data show that previous interglacials were several degrees warmer than our own.

Sime *et al.* (2009) presented oxygen isotope data from three sites in Antarctica showing that the change in isotope content was considerably greater in the previous three deglaciations than in the present deglaciation. They analyzed the relationship between isotope index and temperature at the three sites and concluded "that maximum interglacial temperatures over the past 340 KYBP were between 6°C and 10°C above present-day values" and "there are serious deficiencies in our understanding of warmer than present day climates". Applying Hansen and Sato's "factor of two" rule for global average temperature change vs. Antarctic temperature change, this would imply that global average temperatures were 3°C to 5°C higher in previous interglacials.

Kopp *et al.* (2009) concluded that sea level during the previous interglacial was

6–9 m higher than during the present interglacial. This was presumably due to the warmer temperatures. (They estimated that polar temperatures were 3°C to 5°C warmer in the previous interglacial). Kopp *et al.* projected that such a warming today could raise the oceans by a similar amount: 6–9 m. Clark and Huybers (2009) echoed this theme. However, it is not clear why previous interglacials had higher temperatures with about the same concentration of CO_2, and therefore it is not clear at all that increased CO_2 should produce conditions like that of the past interglacial.

Another aspect of the rhetoric by Hansen and Sato (2011) regarding climate alarmism is their need to claim that today, after a global temperature rise of ~0.7°C in the past 120 years, "global temperature in year 2000 had returned, at least, to approximately the Holocene maximum". The data are not accurate enough to be certain whether this is true, and several investigators find the contrary result. Nevertheless, Hansen and Sato (2011) would like to conclude that a temperature rise of another 1°C from today's climate might raise the ocean level by 7–9 m.

A related argument has to do with the global average temperature some five million years ago and the associated sea level. Hansen and Sato (2011) asserted that the global average temperature five million years ago was a mere 1°C to 2°C warmer than that prevailing in the 19th century prior to the industrial era, or only 1°C warmer than today. They also asserted that sea level was some 25 m higher five million years ago. Therefore, another global average temperature rise of ~1°C from today's climate might produce a huge rise in sea level. However, there is extensive evidence that indicates that the global average temperature five million years ago was far warmer than 1°C above current temperatures. Robinson *et al.* (2008) said that, about three million years ago, the global average temperature was 2°C to 3°C warmer than today. However, "global warmth was distributed differently". Three million years ago, "temperatures at high northern latitudes, above 70°N, were as much as 10°–20°C higher than today, but tropical temperatures were near the same". This points out the limitation of using a single global average temperature to characterize climate.

Fedorov *et al.* (2006) discussed the so-called Pliocene Paradox: "During the early Pliocene, 5 to 3 million years ago, globally averaged temperatures were significantly higher than they are today even though the external factors that determine climate were essentially the same".

> "The early Pliocene was both similar to, and also very different from the world of today. The intensity of sunlight incident on the Earth, the global geography, and the atmospheric concentration of carbon dioxide, were close to what they are today, but surface temperatures in polar regions were so much higher that continental glaciers were absent from the northern hemisphere, and sea level was approximately 25 m higher than today. ... Conditions today, and those during the early Pliocene, are two different climate states in response to practically the same external forcing".

They provided evidence that a permanent El Niño existed between five and three million years ago:

> "Persistent El Niño conditions would have had a huge impact on the global climate given that, today, even brief El Niño episodes can have a large influence. The reasons are [due to] a remarkably high correlation between tropical sea surface temperature and rainfall patterns. Tall, rain-bearing, convective clouds cover the warmest waters but highly reflective stratus decks that produce little rain cover the cold waters. During El Niño, the warming of the eastern equatorial Pacific reduces the area covered by stratus clouds thus decreasing the albedo of the planet, while the atmospheric concentration of the powerful greenhouse gas, water vapor increases. Calculations with a General Circulation Model of the atmosphere indicate that this happened during the early Pliocene and contributed significantly to the warm conditions at that time.'

They concluded that:

> "A major factor in the warmth of the early Pliocene was the persistence of El Niño in the Pacific; it contributed to global warming by causing the absence of stratus clouds from the eastern equatorial Pacific, thus lowering the planetary albedo, and by increasing the atmospheric concentration of water vapor, a powerful greenhouse gas."

Ballantyne, *et al.* (2010) said:

> "The consensus among proxies suggests that Arctic temperatures were ~19°C warmer during the Pliocene than at present, while atmospheric CO_2 concentrations were ~390 ppm. These elevated Arctic Pliocene temperatures result in a greatly reduced and asymmetrical latitudinal temperature gradient that is probably the result of increased poleward heat transport and decreased albedo."

Pagani *et al.* (2010) found evidence that CO_2 concentrations ramped linearly downward from between ~390 ppm five million years ago to about 280 ppm in recent pre-industrial times. They concluded that "the Earth-system climate sensitivity has been significantly higher over the past five million years than estimated from fast feedbacks alone". But this presupposes that CO_2 is the sole driving force for climate change. As we have seen, climate is not regulated by CO_2 alone.

Hansen and Sato (2011) utilized the data of Velicogna (2009) that indicated that the rate of ice sheet mass loss from Greenland and Antarctica accelerated over the period 2002–2008. Velicogna found that the rate of mass loss from Greenland increased from ~200 Gt/yr to ~300 Gt/yr over the interval 2003–2008. The rate of mass loss from Antarctica increased from ~150 Gt/yr to ~200 Gt/yr over the interval 2003–2008. Using the rule of thumb that adding 100 Gt of liquid water to the oceans raises sea level by about 0.3 mm, these rates correspond to sea level rise of a bit over 1 mm/yr. Yet, Hansen and Sato (2011) have interpreted these data by extrapolation to a projected sea level rise of 5 m to 25 m by year 2100. This was based on their estimate that previous interglacials were a mere ~1°C warmer than ours, while sea level was several meters higher.

8.3.7 Impact of sea level rise

According to the IPCC:

"Coasts are experiencing the adverse consequences of hazards related to climate and sea level (very high confidence). Coasts are highly vulnerable to extreme events, such as storms, which impose substantial costs on coastal societies. Annually, about 120 million people are exposed to tropical cyclone hazards, which killed 250,000 people from 1980 to 2000. Through the 20th century, global rise of sea level contributed to increased coastal inundation, erosion and ecosystem losses, but with considerable local and regional variation due to other factors. Late 20th century effects of rising temperature include loss of sea ice, thawing of permafrost and associated coastal retreat, and more frequent coral bleaching and mortality."
[Comment: While it is true that coasts are indeed "highly vulnerable to extreme events", there is little evidence that the incidence of such extreme events increases due to global warming. "Increased coastal inundation, erosion and ecosystem losses" will result from global warming, but the issue comes down to the cost of such effects vs. the cost of draconian reduction of carbon emissions.]

"Coasts will be exposed to increasing risks, including coastal erosion, over coming decades due to climate change and sea-level rise."
[Comment: This is very likely. The question now comes down to how much rise, and what can we do about it? Evidently, we should take all reasonable steps to reduce carbon emissions in the years ahead. But what is reasonable? What can be done technically? What are the costs?]

"Anticipated climate-related changes include: an accelerated rise in sea level of up to 60 cm or more by 2100; a further rise in sea surface temperatures by up to 3°C; an intensification of tropical and extra-tropical cyclones; larger extreme waves and storm surges; altered precipitation/run-off; and ocean acidification. These phenomena will vary considerably at regional and local scales, but the impacts are virtually certain to be overwhelmingly negative."
[Comment: The future rise of sea level is uncertain. Currently, sea level is rising at 2 mm/yr or 3 mm/yr, depending on who one believes. This suggests that, by year 2100, the cumulative rise might be 18 cm to 27 cm if there is no further acceleration. Rising sea level is a threat. As we have stated many times, the question is what can we afford to do about it? As we discussed in Section 7.1.5. in regard to the Goldilocks principle, the choice of "a further rise to 3°C" was made to create a serious impression, yet not outside the realm of credibility. Climatologists do not know what this quantity is. However, 3°C is possible. That is a serious concern. Yet, there is little evidence that this will bring about putative "intensification of tropical and extra-tropical cyclones; larger extreme waves and storm surges; altered precipitation/run-off". The oceans will become less alkaline. The impacts do appear to be "overwhelmingly negative" even though we cannot yet quantify them. As we have stated many times, the question is what can we afford to do about it?]

"Corals are vulnerable to thermal stress and have low adaptive capacity. Increases in sea surface temperature of about 1 to 3°C are projected to result in more frequent coral bleaching events and widespread mortality, unless there is thermal adaptation or acclimating by corals."
[Comment: This is a legitimate concern.]

"Coastal wetland ecosystems, such as salt marshes and mangroves, are especially threatened where they are sediment starved or constrained on their landward margin."
[Comment: This is probably true. However, as sea level rises, such formations might be simply pushed inland.]

"Degradation of coastal ecosystems, especially wetlands and coral reefs, has serious implications for the well-being of societies dependent on the coastal ecosystems for goods and services. Increased flooding and the degradation of freshwater, fisheries and other resources could impact hundreds of millions of people, and socio-economic costs on coasts will escalate as a result of climate change. The impact of climate change on coasts is exacerbated by increasing human-induced pressures. Utilization of the coast increased dramatically during the 20th century and this trend is virtually certain to continue through the 21st century. Under the SRES scenarios, the coastal population could grow from 1.2 billion people (in 1990) to 1.8 to 5.2 billion people by the 2080s, depending on assumptions about migration. Increasing numbers of people and assets at risk at the coast are subject to additional stresses due to land-use and hydrological changes in catchments, including dams that reduce sediment supply to the coast. Populated deltas (especially Asian mega-deltas), low-lying coastal urban areas and atolls are key societal hotspots of coastal vulnerability, occurring where the stresses on natural systems coincide with low human adaptive capacity and high exposure. Regionally, South, Southeast and East Asia, Africa and small islands are most vulnerable. Climate change therefore reinforces the desirability of managing coasts in an integrated manner."
[Comment: These concerns are probably justified. However, one thing is striking here. With all of these risks and impacts in coastal areas, why isn't the IPCC arguing more vociferously against increases in coastal population, which is probably a great deal more practical than draconian reductions in carbon emissions? The IPCC seems to accept that billions of people will move to the coasts despite continuous degradation of conditions there, and their response is to cut off the world's energy supply.]

"Adaptation for the coasts of developing countries will be more challenging than for coasts of developed countries, due to constraints on adaptive capacity (high confidence)."
[Comment: This is probably correct.]

"Adaptation costs for vulnerable coasts are much less than the costs of inaction."

[Comment: This is probably correct. But the best action is not to deprive the world of energy, but rather to limit coastal development.]

"Sea-level rise has substantial inertia and will continue beyond 2100 for many centuries. Irreversible breakdown of the West Antarctica and/or Greenland ice sheets, if triggered by rising temperatures, would make this long-term rise significantly larger, ultimately questioning the viability of many coastal settlements across the globe. The issue is reinforced by the increasing human use of the coastal zone. Settlement patterns also have substantial inertia, and this issue presents a challenge for long-term coastal spatial planning. Stabilization of climate could reduce the risks of ice sheet breakdown, and reduce but not stop sea-level rise due to thermal expansion. Hence, it is now [even] more apparent that the most appropriate response to sea-level rise for coastal areas is a combination of adaptation to deal with the inevitable rise, and mitigation to limit the long-term rise to a manageable level."

[Comment: It is probably correct that sea level rise will continue beyond 2100. In fact, even extreme skeptics such as Singer and Avery (2007) claim that sea level has been rising through the Holocene and will continue to rise (although they ascribe this to evolution from the last Ice Age rather than increase in greenhouse gases). It is also correct that the West Antarctica and/or GIS might be vulnerable to breakdown and this would drive up sea level a great deal. The probability of this occurring over the next few hundred years is not understood but it is certainly not zero. Stabilization of climate is indeed a consummation devoutly to be wished. But how to do that while providing the world with energy and power?]

8.3.8 Sea ice extent

Goody (1980) described sea ice as a "bizarre substance". He went on to say that:

"The minimum extent differs greatly in the two hemispheres. In the north more than half of the sea ice survives the summer; in the south less than one-eighth. ... The ice margin is also subject to year-to-year changes and to large secular changes, although the geological record indicates that sea ice has never completely disappeared from the Arctic Ocean during the last million years.

"The thickness of sea ice does not differ greatly in the two polar regions, and averages ~2 m. Ice can grow almost to this thickness in one year. Thereafter, it stabilizes at ~3 m with a seasonal variation of 1 m. The ice is broken up by leads and open areas or polynyas and also by long pressure ridges up to 25 m high, which, like icebergs, are indicators of much larger changes at the bottom surface. Sea ice first forms from a slush which ultimately aggregates into a matrix of pure ice with brine inclusions. The physical properties of the sea ice are governed by the phase equilibrium between ice and brine giving rise to a substance with no definite melting point and with strongly temperature-dependent physical properties such as specific heat and conductivity. The freezing of the ocean surface dramatically changes both its thermal and mechanical properties.

"The presence or absence of leads is important in at least four major respects. First, during the winter a lead replaces a solid surface at –30°C with a liquid surface at –1.9°C and both radiative and turbulent heat fluxes increase dramatically. Second, leads allow the surface to drain during the summer, a process that, as we shall see, controls the ice thickness. Third, cracking changes the mechanical properties of the sea ice permitting the ice to flow and to export negative latent heat to other latitudes. Finally, leads are precursors' to pressure ridges that alter the character of both atmospheric and oceanic boundary layers.

" [It was estimated that] the heat flux (radiation, sensible and latent heat together) through 3 m of ice is only ~1.5% of that from the open ocean. Leads form 1% or so of even old sea ice; they, therefore, compete in thermal importance with the rest of the ice surface."

Humlum (2011) described sea ice as follows:

"Sea ice occupies about 7% of the surface area of planet Earth. The sea ice thickness, its spatial extent, and the fraction of open water within the ice pack can vary rapidly and profoundly in response to weather and climate. Sea ice typically covers about 14 to 16 million square kilometers in late winter in the Arctic and 17 to 20 million square kilometers in the Southern Ocean around the Antarctic. The seasonal decrease is much larger around the Antarctic, with only about three to four million square kilometers remaining at summer's end, compared to approximately seven to nine million square kilometers in the Arctic. The main reason for this difference is that the Arctic Ocean is centered on the Pole, while the Southern Ocean is not.

"Sea ice variations have recently attracted much public interest. Part of the reason for this is the high albedo (c. 80%) of sea ice, which reflects much of the incoming solar short-wave radiation during the summer time. If not reflected, this radiation may instead be consumed by warming ocean water, thereby initiating a positive feedback, leading to more warming. This simple analysis however ignores that evaporation will increase from the ocean when the total sea ice cover are reduced in size. Increased evaporation usually results in an increased cloud cover and increased reflectance of incoming solar radiation, which tend to counteract the above process. The decrease or increase of sea ice has no effect on the global sea level."

Two important parameters are the extent of Arctic sea ice coverage and the average thickness of Arctic sea ice. Serreze *et al.* (2007) showed that Arctic sea ice extent declined in every year from 1979 to 2006. The most pronounced rate of loss occurred in September of each year. However, they said that "evidence for accompanying reductions in ice thickness is inconclusive". While there are some indications that the sea ice has thinned, "sparse sampling complicates interpretation". According to Serreze *et al.* (2007):

> "The observed decline in ice extent reflects a conflation of thermodynamic and dynamic processes. Thermodynamic processes involve changes in surface air temperature, radiative fluxes, and ocean conditions. Dynamic processes involve changes in ice circulation in response to winds and ocean currents. These include changes in the strength and location of the Beaufort Gyre (a mean annual clockwise motion in the western Arctic Ocean) and characteristics of the Transpolar Drift Stream (a motion of ice that progresses from the coast of Siberia, across the pole, and into the North Atlantic via the Fram Strait). Nearly all of the ice export from the Arctic to the Atlantic occurs through this narrow strait between northern Greenland and Svalbard."

Divine and Dick (2006) found evidence of persistent ice retreat in the Arctic since the second half of the 19th century. However, it was not clear whether this was a trend that will continue, or whether it was part of a cycle ("a similar shrinkage of ice cover was observed in the 1920s–1930s, during the previous warm phase of the LFO, when any anthropogenic influence is believed to have still been negligible").

Koberle and Gerdes (2003) explored the time variability of Arctic sea ice properties in a coupled ocean–sea ice model of the Arctic and the northern North Atlantic driven by 50-year NCEP–NCAR reanalysis data No appreciable trend in sea ice volume was found for the period 1948–1998, although rather long sub-periods (e.g., 1965–1995) exhibited a large decline in sea ice volume. Their analysis concluded that wind forcing significantly contributes to the decadal variability in the Arctic ice volume. They concluded that "these results make connecting 'global warming' to Arctic ice thinning very difficult for two reasons." The two reasons are (1) large decadal and longer-term variability masks any trend, and (2) there might be long-term trends in wind stress that tend to either increase or decrease ice volume that would mask the effects of temperature change.

Lindsay and Zhang (2005) pointed out that recent observations of summer Arctic sea ice over the satellite era show that record or near-record lows for the ice extent occurred in the years 2002–2005. It was hypothesized that:

> "The thinning since 1988 is due to preconditioning, a trigger, and positive feedbacks: 1) the fall, winter, and spring air temperatures over the Arctic Ocean have gradually increased over the last 50 years, leading to reduced thickness of first-year ice at the start of summer; 2) a temporary shift, starting in 1989, of two principal climate indexes (the Arctic Oscillation and Pacific Decadal Oscillation) caused a flushing of some of the older, thicker ice out of the basin and an increase in the summer open water extent; and 3) the increasing amounts of summer open water allow for increasing absorption of solar radiation, which melts the ice, warms the water, and promotes creation of thinner first-year ice, ice that often entirely melts by the end of the subsequent summer. Internal thermodynamic changes related to the positive ice–albedo feedback, not external forcing, dominate the thinning processes over the last 16 years. This feedback continues to drive the thinning after the climate indexes return to near-normal conditions in the late 1990s. The late 1980s and early 1990s could be considered a tipping point during which the ice–ocean system began to enter a

new era of thinning ice and increasing summer open water because of positive feedbacks. It remains to be seen if this era will persist or if a sustained cooling period can reverse the processes."

By contrast, Winsor (2001) found that "there was no trend towards a thinning ice cover during the 1990s". He also concluded: "that the mean ice thickness has remained on a near-constant level around the North Pole from 1986 to 1997".

The website *http://arctic.atmos.uiuc.edu/cryosphere/* provides extensive data on Arctic and Antarctic sea ice extent since 1980. This website provided Figures 8.20 and 8.21. These figures show that, since 1980, global sea ice extent has followed a repetitive pattern with some significant negative anomalies during the past five years. Over this time period, Antarctic sea ice extent has held steady but Arctic sea ice has retreated.

Perovich *et al.* (2011) said that:

"... observed changes [in sea ice extent] are intricately linked to sea-ice dynamics and thermodynamics and are driven by atmosphere and ocean forcing. Several factors have been established as contributors to the decline in the ice cover: a general warming, changes in atmospheric circulation patterns, changes in cloudiness, advected ocean heat from lower latitudes, increased ice export from the Fram Strait, and increased solar heating of the upper ocean."

Serreze *et al.* (2007) indicated that surface air temperatures in the Arctic increased from 1979 to 1997, and they increased as well during the period 2000 to 2006 relative to 1979 to 1999. While many climatologists assume automatically that this was due to the effect of greenhouse gases, Serreze *et al.* (2007) pointed out that:

"... at least part of the recent cold-season warming ... is itself driven by the loss of ice, because this loss allows for stronger heat fluxes from the ocean to the atmosphere. The warmer atmosphere will then promote a stronger long wave flux to the surface."

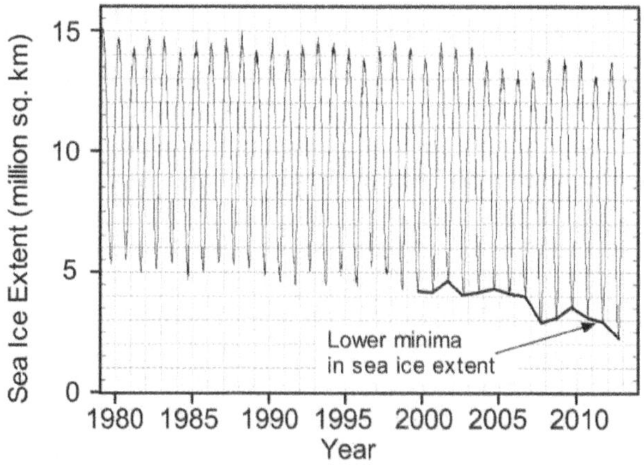

Figure 8.20. Global sea ice extent since 1980 (*http://arctic. atmos.uiuc.edu/cryosphere/*).

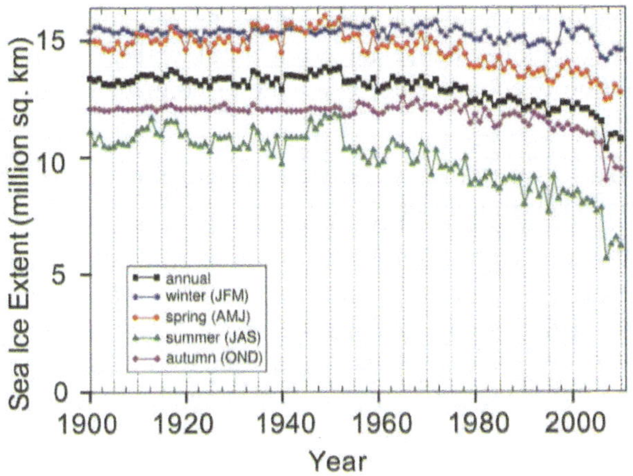

Figure 8.21. Northern Hemisphere seasonal sea ice extent (*http://arctic.atmos.uiuc.edu/cryosphere/*).

Serreze *et al.* (2007) said that "links have been established between ice loss and changes in ice circulation associated with the behavior of the North Atlantic Oscillation (NAO), Northern Annular Mode (NAM), and other atmospheric patterns". ... Over the period 1970 through the mid-1990s, winter indices of the NAO-NAM shifted from negative to strongly positive". While changes in circulation might explain sea ice loss over that period, Serreze *et al.* (2007) asserted that "these processes cannot readily explain the extreme September sea-ice minima of recent years". However, Serreze *et al.* (2007) seem to have concentrated on the post-1970s period. There is some evidence that changes in surface air temperatures during the 20th century were cyclic with an upswing in the cycle after 1970 (Frolov *et al.*, 2009). There was a significant upswing from 1910 to 1940 while CO_2 levels were much lower. Serreze *et al.* (2007) was published prior to Frolov *et al.* (2009) or Chylek *et al.* (2009). Chylek *et al.* (2009) said that:

> "Temperature trend reversals in 1940 and 1970 separate two Arctic warming periods (1910–1940 and 1970–2008) by a significant 1940–1970 cooling period. Analyzing temperature records of the Arctic meteorological stations we find that (a) the Arctic amplification (ratio of the Arctic to global temperature trends) is not a constant but varies in time on a multi-decadal time scale, (b) the Arctic warming from 1910–1940 proceeded at a significantly faster rate than the current 1970–2008 warming, and (c) the Arctic temperature changes are highly correlated with the Atlantic Multi-decadal Oscillation (AMO) suggesting the Atlantic Ocean thermohaline circulation is linked to the Arctic temperature variability on a multi-decadal time scale."

Serreze *et al.* (2007) had acknowledged this, saying:

> "To further complicate the picture, it appears that changes in ocean heat transport have played a role. ... To summarize, the observed sea-ice loss can in

part be connected to arctic warming over the past several decades. Although this warming is part of a global signal suggesting a link with greenhouse gas (GHG) loading, attribution is complicated by a suite of contributing atmospheric and oceanic forcings."

Arctic sea ice loss was very high in the past decade. Kay *et al.* (2008) compared years 2006 and 2007. They found that:

> "Over the Western Arctic Ocean, total summertime cloud cover estimated from space-borne radar and lidar data decreased by 16% from 2006 to 2007. The clearer skies led to down-welling shortwave (long-wave) radiative fluxes increases of +32 W/m^2 (–4 W/m^2) from 2006 to 2007. ... Longer-term observations show that the 2007 cloudiness is anomalous in the recent past, but is not unprecedented."

This suggests that variable cloudiness is a potent force for Arctic climate variability, but it does not reveal what causes the change in cloudiness. However, more recently, Screen and Simmonds (2010) said that:

> "Arctic warming is strongest at the surface during most of the year and is primarily consistent with reductions in sea ice cover. Changes in cloud cover, in contrast, have not contributed strongly to recent warming. Increases in atmospheric water vapor content, partly in response to reduced sea ice cover, may have enhanced warming in the lower part of the atmosphere during summer and early autumn. We conclude that diminishing sea ice has had a leading role in recent Arctic temperature amplification."

Francis and Hunter (2007) asserted that "while increases in greenhouse gases are believed to be the underlying cause of the melting, interactions among the Arctic's changing thermodynamic and dynamic processes driving ice loss are poorly understood". They found that, in areas dominated by low clouds containing liquid water the emission of infrared radiation from the atmosphere to the surface during spring around the periphery of the Arctic Ocean down-welling long-wave flux is driven primarily by increasing cloud fraction and more abundant water vapor. In ice-cloud-dominated regions, they found that changing water vapor is more important than changing cloud fraction.

According to Winton (2006), "it is difficult to associate the Arctic-global differences with specific features of the atmosphere's CO_2 response" because multiple processes contribute feedbacks and forcing. He claimed that "there are reasons to expect significant contributions to the Arctic-global long-wave feedback difference from cloud, water vapor and temperature feedbacks".

Polyakov *et al.* (2003b) provided data on sea ice in various regions of the Arctic dating back to 1900. Figure 8.22 shows the annual change in extent of sea ice summed over four Arctic areas. The long-term trend was downward (shown as a straight line in the figure) but oscillations about this trend were huge. Polyakov *et al.* (2003b) also provided data on fast sea ice thickness at five Arctic locations as shown in Figure 8.23. Here, the trend is less clear, but, after 1990, it was predominantly

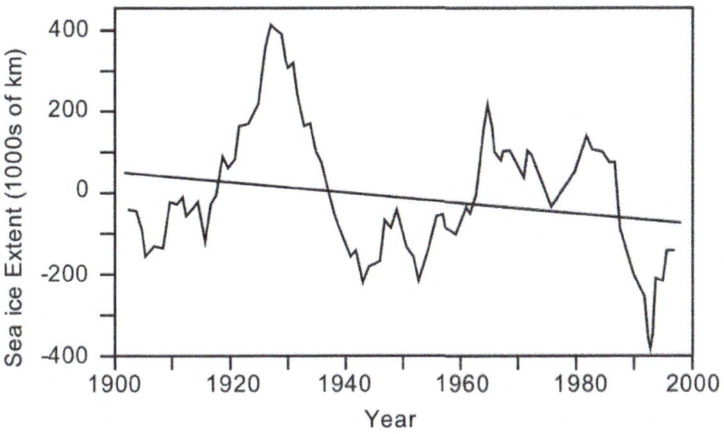

Figure 8.22. Six-year running mean of sum of changes in sea ice extent in four Arctic seas (Kara, Laptey, East Siberian, and Chukchi) (adapted from Polyyakov *et al.*, 2003a).

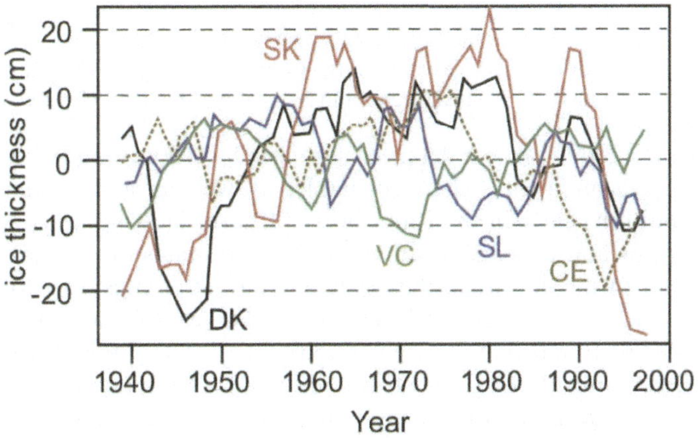

Figure 8.23. Six-year running mean of fast sea ice thickness in five Arctic locations (SK = Sterelegova (Kara Sea), DK = Dikson (Kara Sea), SL = Sannikova (Laptev Sea), CE = Chetirekhstolbovii (East Siberian Sea), and VC = Vrangelya (Chukchi Sea) (adapted from Polyyakov *et al.*, 2003a).

downward. In addition, Vinje (2001) provided data showing a gradual withdrawal of sea ice extending back to 1860.

Polyak *et al.* (2010) compared estimates of historical sea ice extent in the Nordic Seas based on ice core and tree-ring data with more recent measurements of maximum sea ice extent in the Arctic-wide area. They claimed that this showed an anomalous reduction in sea ice extent in the 20th century. However, the estimates of past sea ice extent based on ice cores and tree rings appears somewhat fragile. Furthermore, the analysis only went back as far as year 1200. Had it been carried back through the MWP to, say, year 850, it is possible that the recent decrease in sea ice extent might not appear so anomalous.

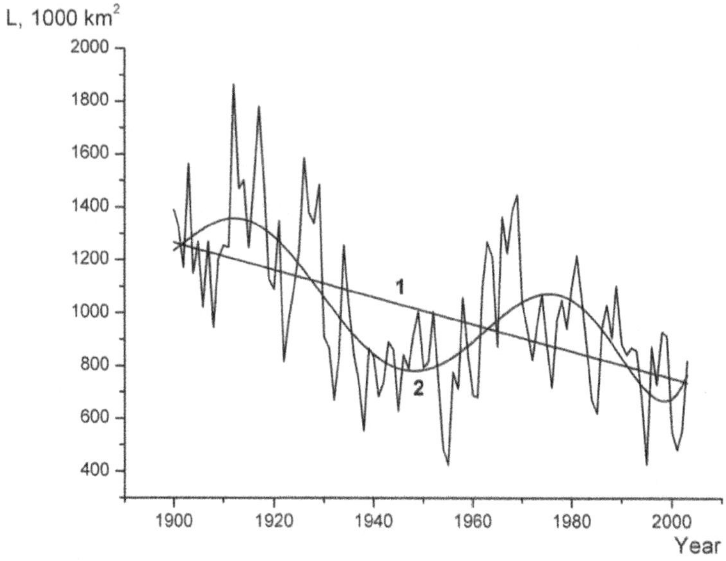

Figure 8.24. Variability of the total ice extent in the Greenland, Barents, and Kara Seas for the period 1900–2003 (August): 1 = linear trend and 2 = polynomial fit. (Frolov *et al.*, 2009).

Frolov *et al.* (2009) wrote a detailed monograph on Eurasian Arctic seas. They utilized direct measurements where available, as well as anecdotal data on ice navigation conditions early in the 20th century. Charts showing ice conditions with the routes of ships and ice edge locations, with coordinates or orientation markers were used to fill in missing quantitative data. They were thus able to reconstruct Arctic climate and sea ice extent over the past ~100 years for a vast area of about 5×10^6 km^2. They found that sea ice variability was quite different for two Arctic regions: (1) Greenland, Barents, and Kara Seas; and (2) Laptev, East Siberian, and Chukchi Seas. Their results for these two regions are shown in Figures 8.24 and 8.25. It can be seen that, in the Greenland, Barents, and Kara Sea region, there has been an overall descending extent of sea ice throughout the 20th century, although there have been significant cycles about the main trend line. The trend for the Laptev, East Siberian, and Chukchi Sea region has been predominantly flat with significant variations about this trend line.

Frolov *et al.* (2009) carried out spectral analyses of these data and found several frequencies that stood out. For the Greenland, Barents, and Kara Sea region, an important frequency corresponded to a ~60-year period. The Laptev, East Siberian, and Chukchi Sea region was dominated by higher frequencies. They interpreted the data as if the Arctic sea ice extent followed natural cycles and dismissed global warming via the greenhouse effect as a factor in these variations. Clearly, warming of the Arctic region is tied to changes in sea ice in the Greenland, Barents, and Kara Sea region. Frolov *et al.* (2009) attributed these changes to changes in solar activity. However, other interpretations are probably more credible. By extrapolating these figures into the future according to this belief, they predicted significant increases in

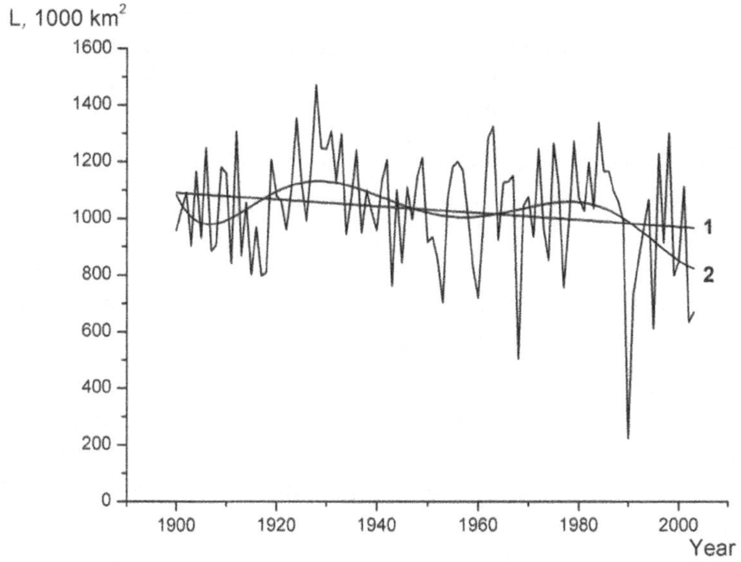

Figure 8.25. Variability of total ice extent in the Laptev, East Siberian, and Chukchi Seas for the period 1900–2003 (August): 1 = linear fit and 2 = polynomial fit. (Frolov *et al.*, 2009).

Arctic sea ice extent in the second and third decades of the 21st century. Time will reveal whether this unlikely speculation holds true.

Over the past decade or so, a number of climatologists have raised alarms about the dire state of declining Arctic sea ice. For example, according to Wang and Overland (2009):

> "Summers in the Arctic may be nearly ice-free in as few as 30 years, not at the end of the century as previously expected. The updated forecast is the result of a new analysis of computer models coupled with the most recent summer ice measurements."

Polyak *et al.* (2010), in their *History of Sea Ice in the Arctic*, said that "Arctic sea-ice extent and volume are declining rapidly. Several studies project that the Arctic Ocean may become seasonally ice-free by the year 2040 or even earlier".

The Internet is full of similar dire predictions. On the other hand, Frolov *et al.* (2009) argue, perhaps overly optimistically, that we are in the midst of the down leg of a natural cycle and all we need to do is wait for the next expansion of sea ice. Similarly, according to Holloway and Sou (2000):

> "Previous reports that Arctic sea ice volume decreased nearly by half in recent decades have been widely cited in popular media and scientific considerations, e.g., IPCC 2001. We find instead ... that Arctic sea ice volume has decreased more slowly. Misleading inferences of rapid ice loss were a result of variable wind stress forcing a natural component of sea ice variability. In particular a dominant mode of variability moves ice between the central Arctic

and the Canadian sector, suggestive of the modes of Arctic ice motion. ... We conclude that observations and model results together suggest only modest reduction of ice volume, like the modest decline in ice areal extent. Previously inferred rapid loss of ice volume is unlikely."

A more recent report (Lindsay *et al.*, 2009) concluded that:

> "[although] the minimum of Arctic sea ice extent in the summer of 2007 was unprecedented in the historical record ... the loss in total ice mass was not. Rather, the 2007 ice mass loss is largely consistent with a steady decrease in ice thickness that began in 1987. Since then, the simulated mean September ice thickness within the Arctic Ocean has declined from 3.7 to 2.6 m at a rate of 0.57 m per decade. Both the area coverage of thin ice at the beginning of the melt season and the total volume of ice lost in the summer have been steadily increasing. The combined impact of these two trends caused a large reduction in the September mean ice concentration in the Arctic Ocean. This created conditions during the summer of 2007 that allowed persistent winds to push the remaining ice from the Pacific side to the Atlantic side of the basin and more than usual into the Greenland Sea. This exposed large areas of open water, resulting in the record ice extent anomaly."

Arctic sea ice extent passes through an annual curve with maximum extent around mid-March and minimum extent around mid-September. Arctic sea ice extent bottomed out from 2004 to 2007, but more recent data than that reported by Lindsay *et al.* (2009) show that sea ice extent expanded in 2008 and 2009 to levels about 5% greater than in 2007. While the prospects for Arctic sea ice remain gloomy, alarmists might have exaggerated the threat. More to the point, they uniformly blame Arctic sea ice diminution on greenhouse gases and rarely, if ever, mention black carbon. On the other hand, those who argue in favor of natural cycles have no credible explanation for the source of these putative cycles. Similarly, those who argue for the rise in CO_2 as the sole cause cannot explain the changes that occurred earlier in the 20th century. Furthermore, Sedlacek and Mysak (2008) pointed out that "The wind-driven changes in sea-ice area are about twice as large as those due to thermodynamic (i.e., radiative) forcing".

By contrast, Antarctic sea ice has been expanding. Ozsoy-Cicek *et al.* (2009) reported that "Antarctic sea ice cover has shown a slight increase ($<1\%$/decade) in overall observed ice extent as derived from satellite mapping from 1979 to 2008, contrary to the decline observed in the Arctic regions". It was claimed that this increase in Antarctic sea ice extent resulted from the ozone hole that strengthened surface winds around Antarctica and deepened the storms in the South Pacific area of the Southern Ocean that surrounds the continent. This resulted in greater flow of cold air over the Ross Sea (West Antarctica) leading to more ice production in this region (Turner *et al.*, 2009). Turner *et al.* said that:

> "The only thing we can conclude at this point in time, therefore, is that for some still-unproven reason, and in spite of the supposedly unprecedented increases in mean global air temperature and CO_2 concentration that the planet

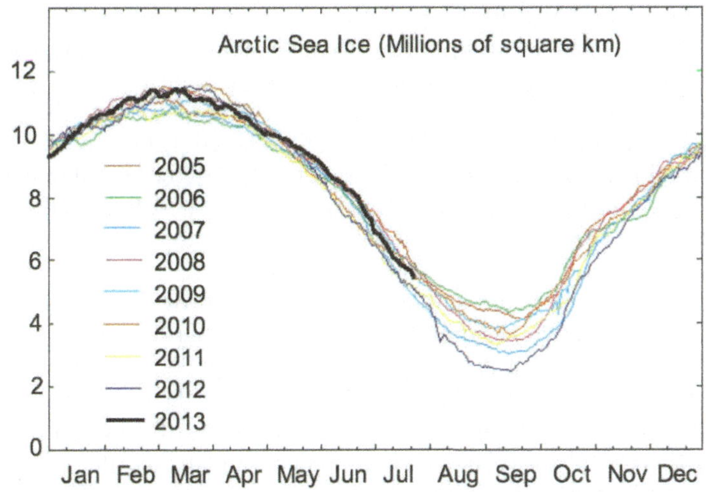

Figure 8.25a. Variation of extent of Arctic sea ice during the year since 2005 (*wattsupwiththat.com/*).

has experienced since the late 1970s, Antarctica sea ice extent has stubbornly continued to just keep on growing."

Figure 8.25a shows the variation of Arctic sea ice during the course of a year for 2005 to 2013. The curves seemed to be headed endlessly downward until there was a substantial recovery in 2013.

8.3.9 Summary

Sea level rise is the most credible potential global problem that could be caused by continued global warming. Hansen (2004) said that "The dominant issue in global warming, in my opinion, is sea-level change and the question of how fast ice sheets can disintegrate".

As the oceans warm, they expand, and sea level rises. Coming out of the LGM, sea level rose roughly 120 m over the past 15,000 years, with the rate of rise gradually diminishing with time from an initial rate of 200 cm per century. Historical tide gauge data suggest that, as we evolved out the LIA in the late 19th century, sea level rose at a fairly steady average rate of ~1.7 mm/yr, with significant oscillations about this average, but essentially no acceleration of the average rate. Tide gauge data are difficult to interpret properly, requiring corrections for local uplift or subsidence, and consideration of ENSO and other oceanic effects. Nevertheless, there appears to be adequate tide gauge data to support the conclusion that, at least until recently, sea level rose at an average rate of ~1.7 mm/yr. About 70% of this might have been due to addition of fresh water, with the remainder due to thermal expansion.

More recently, satellite observations have been used to infer changes in sea level. The TOPEX-POSEIDON observations indicated that the rise in sea level from 1993 to 2003 was 3.0 mm/yr; the accuracy of these measurements is suspect, and it is possible that this was part of an upward cycle and should not be extrapolated.

Instead of measuring sea level directly, one may try to measure the extent of ice sheets and infer from their diminution what the rise in sea level must be (360 Gt of melted ice is equivalent to ~1 mm of sea level). The contribution of melting ice sheets to sea level rise has been estimated based on 14 different satellite-based mass budget analyses, altimetry measurements of ice sheet volume changes, and measurements of ice sheets' changing gravitational attraction. These measurements seem to vary widely and appear to depend on many fragile assumptions. In any event, they are all short-term measurements that do not provide a good estimate of the average rate of change. Rignot *et al.* (2011) used gravity measurements to estimate the change in mass of the GIS. They found a significant acceleration in the mass loss over an 18-year period. However, the details in their procedures are very complex and appear to require many assumptions of uncertain veracity. Furthermore, there appears to be some chartsmanship in interpreting their data that accentuates the recent rate of increase.

The relationship between sea level and rising CO_2 concentration remains uncertain. The major uncertainty is how much warming will occur and, more importantly, how much melting of ice sheets will take place as the Earth warms. Several alarmists predict sea level rise in the range 20 cm to 120 cm by year 2100. Hansen and Sato (2011) projected a sea level rise of 5 m to 25 m by year 2100. Ultimately, this may depend on which curve in Figure 6.39 represents reality.

The bottom line on sea level rise is that there is no bottom line. The data for the past and present are uncertain, but seem to indicate a steady rise since the 19th century of about 1.5 mm/yr to 2.0 mm/yr. While satellite measurements indicate an increased rate toward the end of the 20th century, the evidence in favor of acceleration of sea level rise is fragmentary. Projections for the future vary widely. In the worst-case scenario, sea level rise could be a significant problem toward the end of the 21st century.

The National Oceanic and Atmospheric Administration (NOAA) Climate Program Office summed it up succinctly:

> "Global sea level rise has been a persistent trend for decades. It is expected to continue beyond the end of this century, which will cause significant impacts in the United States. Scientists have very high confidence (greater than 90% chance) that global mean sea level will rise at least 8 inches (0.2 meter) and no more than 6.6 feet (2.0 meters) by 2100."

8.4 CHANGES IN PRECIPITATION: FLOODS, DROUGHT, AND STORMS

8.4.1 Drought

A number of climatologists of the alarmist persuasion have made dire predictions regarding the occurrence of drought, floods, and intense storms. In a sort of "Can you top this?" mode, each new report hypothesizes more and more dire outcomes. For example, recently, Andrew Dessler has suggested that the 2012 drought in Texas was caused by greenhouse gas global warming.

In contrast, Idso (2008) wrote a 133-page rebuttal on the putative relationship between global warming and extreme weather. Although his approach is clearly one-sided, he provides many references to support his claim that (1) many climate models are unable to predict recent African Sahel drought, and (2) severe drought in Africa is not unique to the 20th century; there is evidence for severe drought dating back many hundreds of years. Hence, the current drought is not necessarily a by-product of global warming, and predictions of future drought are based on models that don't work in the recent past. Idso (2008) also provided evidence from Asia, North America, and Europe that global warming does not necessarily lead to the occurrence of more frequent or more severe drought.

The particular case of the continental U.S. is interesting. As Idso (2008) pointed out:

> "Andreadis and Lettenmaier (2006) examined 20th-century trends in soil moisture, runoff and drought over the conterminous United States with a hydro-climatological model forced by real-world measurements of precipitation, air temperature and wind speed over the period 1915–2003. This work revealed, in their words, that 'droughts have, for the most part, become shorter, less frequent, less severe, and cover a smaller portion of the country over the last century' … Van der Schrier *et al.* (2006) constructed maps of summer moisture availability across a large portion of North America and concluded that over the area as a whole, 'the 1930s and 1950s stand out as times of persistent and exceptionally dry conditions, whereas the 1970s and the 1990s were generally wet'."

They also said that "no statistically significant trend was found in the mean summer drought severity index over the 1901–2002 period".

Fye *et al.* (2003) used tree-ring proxies to infer that periods of sustained U.S. drought have been recurrent phenomena over hundreds of years. Both Fye *et al.* (2003) and Stahle *et al.* (2000) concluded that drought in the 16th century was far worse than the severe drought of the "dust bowl" in the 1930s. Other papers are cited by Idso (2008) that indicate that occurrences of periodic U.S. drought date back 1,000 years or more.

NIPCC (2011) updated Idso's analysis with additional references that indicate that periodic droughts have occurred in the past and there is nothing unique about the 20th century. The reader is referred to NIPCC (2011) for details.

Sheffield *et al.* (2012) pointed out that the alarmist viewpoint is that "drought is expected to increase in frequency and severity in the future as a result of climate change, mainly as a consequence of decreases in regional precipitation but also because of increasing evaporation driven by global warming":

> "Previous assessments of historic changes in drought over the late twentieth and early twenty-first centuries indicate that this may already be happening

globally. In particular, calculations of the Palmer Drought Severity Index (PDSI) show a decrease in moisture globally since the 1970s with a commensurate increase in the area in drought that is attributed, in part, to global warming."

Sheffield *et al.* (2012) showed:

"... that the previously reported increase in global drought was overestimated because the PDSI uses a simplified model of potential evaporation that responds only to changes in temperature and thus responds incorrectly to global warming in recent decades. More realistic calculations, based on the underlying physical principles that take into account changes in available energy, humidity and wind speed, suggest that there has been little change in drought over the past 60 years."

8.4.2 Floods

If alarmists are correct in claiming that global warming is producing increased flooding in the world, the data should reflect this. Idso (2008) examined data for flooding during the 20th century when global warming took place. He examined data from the literature for Asia, Europe, and North America. In general, the data show no enhanced flooding in the 20th century compared with the past, and periodic flooding has occurred for centuries. As Idso pointed out, every time there is a flood somewhere, alarmists immediately blame it on global warming and a hue and cry is raised to politically constrain emissions. But Idso quotes from many articles (e.g., Mudelsee *et al.*, 2003) that conclude there was no increase in the occurrence of floods during the 20th century.

More recently, NIPCC (2011) provided an extensive discussion of evidence regarding flooding based on references as recent as 2010. They began by pointing out that it is difficult to derive accurate trends for the effect of climate on flooding because there has been a multitude of modifications to navigable rivers such as dams, dikes, weirs, bridges, etc., nevertheless, examining the data such as they are, there is no evidence that flooding was unusually frequent or severe as the Earth warmed in the 20th century. The reader is referred to Idso (2008) and NIPCC (2011) for details.

8.4.3 Storms

8.4.3.1 Tropical hurricanes

Idso (2008) reviewed hurricane activity in the Atlantic, Pacific, and Indian Ocean basins, as well as for the globe as a whole, beginning with Atlantic Basin hurricanes. He provided a very large number of references. However, his review appears to be biased toward the skeptical persuasion.

Climate alarmists contend that global warming increases the frequency and intensity of hurricanes. Al Gore's film *An Inconvenient Truth* makes the contention that global warming produces more fierce storms. His film shows whistling winds, flooded cities, and people in despair from storm damage. In fact, Al Gore specifically claims that globally warmed gulf waters were responsible for the increase in Katrina's strength after it left Florida. The implication of the film is that the Earth is besieged by waves of storms of unprecedented intensity.

But Idso (2008) and Pielke *et al.* (2005) made the point that, even if storm damage does increase due to global warming, the gross damage to the biosphere in the 21st century will be dictated far more by population growth and industrialization than it will by intense storms. Yet these topics are often ignored while global warming commands attention from the media.

There is a huge literature on the question of the degree to which increased sea-surface temperatures (SSTs) produce more frequent and more intense hurricanes. Closely related to this are the questions of whether such increases in SST are due generally to anthropogenic activity, and specifically to the greenhouse effect from CO_2 emissions.

As is the case for many areas of climatology related to global warming, there is a diversity of opinion and one must be careful in reading papers to discern biased views, whether of the alarmist persuasion or the skeptical persuasion. Knutson (2008) provided a review that appears to be comprehensive and even-handed. The intensity of tropical cyclones is measured by the power dissipation index (PDI) that includes contributions by frequency and intensity of storms. Emanuel (2005, 2007) studied the occurrence of tropical cyclones and compared the historical record of SST with the PDI from 1870 to 2005. His results are shown in Figure 8.26. He found a significant correlation between SST and PDI over this time span but craftily only quoted the correlation coefficient after 1970. Emanuel (2005) claimed there was a 90% increase in the PDI, due to a 67% increase in the frequency and a 19% increase in intensity from 1870 to 2004. Actually, however, as Figure 8.26 shows, this is a biased view. The PDI varied around 6 prior to 1930, and jumped to about 8 from 1930 to 2000. The spike in the curves after 2000 may represent a fluctuation (as in 1950), swayed by Katrina and a few other hurricanes. And, indeed, more recent data show that the PDI dropped sharply after 2005. Nevertheless, Knutson (2008) concluded that observed records of Atlantic hurricane activity show a strong correlation, on multi-year time scales, between local tropical SSTs and the PDI. He said that "Both Atlantic SSTs and PDI have risen sharply since the 1970s, and there is some evidence that PDI levels in recent years are higher than in the previous active Atlantic hurricane era in the 1950s and 60s". He went on to say that, if the projected temperature increases in the 21st century (made by alarmists) are used to infer corresponding increases in SSTs, such a correlation between SST and PDI would be of grave concern.

Michaels *et al.* (2006) compared SST with wind speed of storms for Atlantic Basin tropical systems from 1982 to 2005. They found that a SST higher than a threshold value of 28.25°C is a virtual necessity for attaining category-3 or higher winds. However, they also found that an SST greater than 28.25°C does not act to further increase the intensity of tropical cyclones. In other words, they found a bimodal distribution:

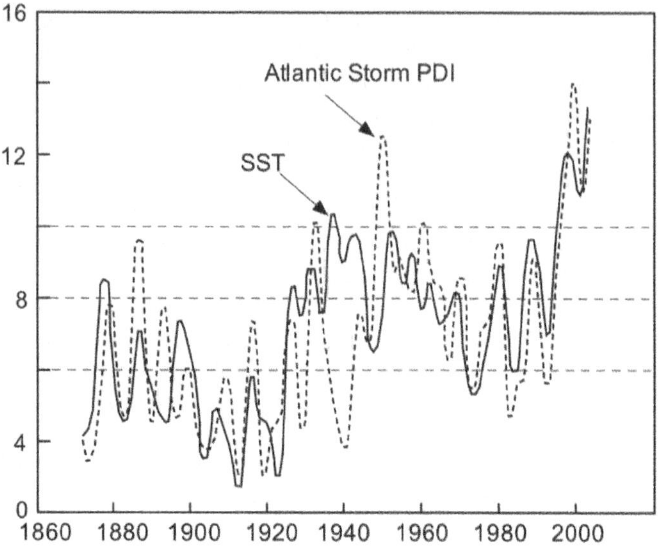

Figure 8.26. Scaled smoothed curves for August–October SST and storm lifetime maximum PDI for the North Atlantic (Emanuel, 2005).

(1) below 28.25°C, very few strong hurricanes occur;
(2) above 28.25°C, strong hurricanes do occur but the frequency is roughly independent of temperature above 28.25°C (at least up to 30.5°C, where measurements ended).

This is in contrast to conclusions reached by some alarmists that the frequency of strong hurricanes increases continuously with increasing SST. Michaels *et al.* (2006) also concluded that the actual rise in SST was far too small to explain the post-1994 changes in tropical cyclone characteristics relative to prior decades. Other factors must cause the increase in hurricane intensity.

Vecchi, Swanson, and Soden (2008) distinguished between two correlations of PDI and SST. In one case, the so-called "absolute SST" is taken as "the SST in the main development region of Atlantic hurricanes". Another index is "relative SST", taken as "the SST in the tropical Atlantic main development region relative to the tropical mean SST that controls fluctuations in Atlantic hurricane activity". Using global climate models, projections suggest that, if global warming proceeds as predicted by alarmists, the absolute SST will increase enough to produce PDI levels comparable with that of 2005 (see Figure 8.26). However, these models predict little change in relative SST, suggesting little change in future PDI. Hence, Knutson (2008) raised the question as to "which of the two future Atlantic hurricane scenarios ... is more likely?".

To gain more insight on this problem, Knutson (2008) attempted to analyze much longer (> 100-year) records of Atlantic hurricane activity to determine at the century scale any increase in global and tropical Atlantic SSTs accompanied by a long-term rising trend in PDI. Although existing records of past Atlantic tropical storm numbers since 1878 show a pronounced upward trend, correlated with rising SSTs, it is believed that the reporting of storms was incomplete in the historical era. After adjusting for an estimated number of missing storms, there is a small nominally positive upward trend in tropical storm occurrence from 1878 to 2006 that is not significantly distinguishable from zero. Thus, the historical tropical storm count record does not provide compelling evidence for a greenhouse warming-induced long-term increase in PDI.

When Knutson (2008) analyzed Atlantic Basin hurricanes, rather than all Atlantic tropical storms, the result was similar. The evidence for an upward trend was even weaker for U.S. land-falling hurricanes. Knutson's conclusions were:

> "In summary, neither our model projections for the 21st century nor our analyses of trends in Atlantic hurricane and tropical storm counts over the past 120+ yr support the notion that greenhouse gas-induced warming leads to large increases in either tropical storm or hurricane numbers in the Atlantic.
>
> "Therefore, we conclude that despite statistical correlations between SST and Atlantic hurricane activity in recent decades, it is premature to conclude that human activity—and particularly greenhouse warming—has already had a discernible impact on Atlantic hurricane activity.
>
> "Similarly, efforts to project future levels of Atlantic hurricane activity using observed SST–PDI statistical relations derived from recent decades should be treated as highly speculative at this stage."

Some other relevant papers are mentioned below.

Webster *et al.* (2005) found a doubling of the number of category-4 and category-5 hurricanes in the 15-year period 1990–2004 as compared with 1975–1989. However, this article pointed out that they:

> "... deliberately limited this study to the satellite era because of the known biases before this period, which means that a comprehensive analysis of longer-period oscillations and trends has not been attempted. There is evidence of a minimum of intense cyclones occurring in the 1970s, which could indicate that our observed trend toward more intense cyclones is a reflection of a long-period oscillation."

Klotzbach (2006) found only a 10% growth in global category-4 and category-5 hurricanes from 1986–1995 to 1996–2005, of which most was in the Southern Hemisphere. In another publication, Klotzbach said:

> "These findings indicate that there has been very little trend in global tropical cyclone activity over the past twenty years, and therefore, that a large portion of the dramatic increasing trend found by Webster and Emanuel is likely due to the diminished quality of the datasets before the middle 1980s. One would expect that if the results of Webster *et al.* and Emanuel were accurate reflections of what is going on in the climate system, then a similar trend would be found over the past twenty years, especially since SSTs have warmed considerably (about 0.2°C– 0.4°C) during this time period."

Klotzbach (2011) showed that, over the period 1900 to 2009, tropical cyclone activity in the Atlantic is reduced by roughly a factor of 2 in El Niño years.

Some alarmists have pointed out that dollar losses in the U.S. from hurricanes rose sharply in the 20th century and particularly since 1990. However, Lomborg (2007) pointed out that "just comparing costs over long periods of time does not make sense without taking into account the change in population patterns and demography as well as economic prosperity. [Today] there are many more people, residing in much more vulnerable areas, with many more assets to lose". He shows that, if one evaluates past hurricanes assuming the population distribution of today, one finds that hurricane damage would not have changed in the 20th century. Furthermore, the great 1926 Miami hurricane would have created far more devastation than Katrina. Ackerman (2008) presented a lengthy criticism of Lomborg's work, and indeed he found a number of errors and imperfections in Lomborg's analysis, as well as a bias toward skeptics' publications. However, Lomborg's conclusions remain valid: (1) the connection between human activity and global warming is uncertain, (2) the effects of global warming have been grossly exaggerated, (3) the economic problems faced by the world are severe, and a severe worldwide reduction in carbon emissions would add measurably to these problems, (4) there are bigger problems than global warming and we are not dealing with them effectively.

Pielke, Jr. *et al.* (2008) studied the total economic damage related to hurricane landfalls along the U.S. Gulf and Atlantic coasts from 1900 to 2005. They *normalized* the data by estimating damage that historical storms would have caused had they made landfall under contemporary levels of societal development by adjusting historical damages by three factors: inflation, wealth, and population. They said that "As people continue to flock to the nation's coasts and bring with them ever more personal wealth, losses will continue to increase". They concluded "that while 2004 and 2005 were exceptional from the standpoint of the number of very damaging storms, there is no long-term trend of increasing damage over the time period covered by this analysis. Even Hurricane Katrina is not outside the range of normalized estimates for past storms". Neumayer and Barthel (2011) pointed out that, in assessing the change of hurricane damage over many years, "affected areas become wealthier over time and rational individuals and governments undertake defensive mitigation measures, which requires normalizing economic losses if one wishes to analyze trends in economic loss from natural disasters for detecting a potential climate change signal". They argued "that the conventional methodology for normalizing economic loss is problematic since it normalizes for changes in wealth over time, but fails to normalize for differences in wealth across space at any given point of time". They introduced "an alternative methodology that overcomes this problem in theory, but faces many more problems in its empirical application". In general, they found "no significant upward trends in normalized disaster damage over the period 1980–2009 globally, regionally, for specific disasters or for specific disasters in specific regions".

Bender *et al.* (2010b) produced a model that predicts "nearly a doubling of the frequency of category 4 and 5 storms by the end of the 21st century". They also estimated that it would take 60 years to discern the putative anthropogenic influence above and beyond normal fluctuations. However, Knutson *et al.* (2010) reviewed the data and models regarding the possible anthropogenic influence on tropical cyclones. They said that:

> "Whether the characteristics of tropical cyclones have changed or will change in a warming climate—and if so, how—has been the subject of considerable investigation, often with conflicting results. Large amplitude fluctuations in the frequency and intensity of tropical cyclones greatly complicate both the detection of long-term trends and their attribution to rising levels of atmospheric greenhouse gases. Trend detection is further impeded by substantial limitations in the availability and quality of global historical records of tropical cyclones. Therefore, it remains uncertain whether past changes in tropical cyclone activity have exceeded the variability expected from natural causes."

They pointed out that models predict that greenhouse warming will produce fewer storms (6% to 34% less) but stronger storms (2% to 11% greater). However, the wide range of modeling results suggests that the issue is not well understood. Crompton *et al.* (2011) revised the analysis of Bender *et al.* (2010b) and concluded that it would take between 120 years and 550 years to discern the putative anthropogenic influence.

Landsea *et al.* (2010) said that:

> "Records of Atlantic basin tropical cyclones (TCs) since the late nineteenth century indicate a very large upward trend in storm frequency. This increase in documented TCs has been previously interpreted as resulting from anthropogenic climate change. However, improvements in observing and recording practices provide an alternative interpretation for these changes: recent studies suggest that the number of potentially missed TCs [in the past] is sufficient to explain a large part of the [apparent] recorded increase in TC counts."

Their study examined TC duration, using a widely used Atlantic hurricane database (HURDAT). It was found that "the occurrence of short-lived storms (duration of 2 days or less) in the database has increased dramatically, from less than one per year in the late nineteenth–early twentieth century to about five per year since about 2000, while medium- to long-lived storms have increased little, if at all". The authors went on to say:

> "While it is possible that the recorded increase in short-duration TCs represents a real climate signal, ... it is more plausible that the increase arises primarily from improvements in the quantity and quality of observations, along with enhanced interpretation techniques. These have allowed National Hurricane Center forecasters to better monitor and detect initial TC formation, and thus incorporate increasing numbers of very short-lived systems into the TC database."

The authors made "quantitative estimates of the frequency of missed TCs, focusing just on the moderate to long-lived systems with durations exceeding 2 days in the raw HURDAT. Upon adding the estimated numbers of missed TCs, the time series of moderate to long-lived Atlantic TCs showed substantial multidecadal variability, but neither time series exhibited a significant trend since the late nineteenth century".

It is revealing that Kerry Emanuel, a leading exponent of the interpretation that increased storm activity derived from global warming, backed off this position quite sharply in Emanuel (2010). He said that:

> "When applied to the 1908–1958 reanalysis, the derived global frequency of tropical cyclones shows no significant trend over the period, while the frequency of events in the southern hemisphere shows a statistically significant decline and that of the northern hemisphere shows a marginally significant increase. There are statistically significant increases in frequency over the period in the North Atlantic, eastern North Pacific, and northern Indian Oceans, while frequency declines in the western North Pacific. … Finally, while it is tempting to believe that specification of sea surface temperature is sufficient for capturing most aspects of the general state of the atmosphere relevant to tropical cyclones, we show, using simple arguments, that failure to account for changing radiative properties of the atmosphere can distort the response of tropical cyclone activity to changing distributions of sea surface temperature; moreover, models appear to systematically underestimate the response of near-tropopause temperatures to changing surface temperature, and this too can affect the response of potential intensity."

Smith *et al.* (2010) claimed some progress in predicting hurricane activity in the tropical Atlantic. Klotzbach and Gray (2011) developed a methodology for predicting hurricane activity in the next season. In this report, they have a section entitled "Why CO_2 increases are not responsible for Atlantic SST and hurricane activity increases". Amongst other things, they said that:

> "Theoretical considerations do not support a close relationship between SSTs and hurricane intensity. … We have no plausible physical reasons for believing that Atlantic hurricane frequency or intensity will significantly change if global or Atlantic Ocean temperatures were to rise by 1–2°C. Without corresponding changes in many other basic features, such as vertical wind shear or mid-level moisture, little or no additional tropical cyclone activity should occur with SST increases."

They went on to say that:

> "Any potential CO_2 influence on tropical cyclone activity is deeply buried as turbulence within the tropical atmospheres' many other energy components. It is possible that future higher atmospheric CO_2 levels may cause a small influence on global tropical cyclone activity. But any such potential influence would likely never be able to be detected, given that our current measurement capabilities only allow us to assess tropical cyclone intensity to within about 5 mph."

Villarini *et al.* (2011a) re-examined the historical record of tropical storm observations from the Atlantic Ocean since 1880, dividing storms into those that lasted less than two days and those that lasted longer than two days. They found no increase in the longer storms but there was a significant increase in the number of shorter storms. They attributed most of this increase to better observing techniques and questionable data, rather than an actual increase in short-storm frequency.

Villarini *et al.* (2011b) began with the introductory statement:

> "The impact of future anthropogenic forcing on the frequency of tropical storms in the North Atlantic basin has been the subject of intensive investigation. However, whether the number of North Atlantic tropical storms will increase or decrease in a warmer climate is still heavily debated and a consensus has yet to be reached."

To shed light on this issue, Villarini *et al.* (2011b) "used a recently developed statistical model, in which the frequency of North Atlantic tropical storms was modeled by a ... rate of occurrence parameter that is a function of tropical Atlantic and mean tropical sea surface temperatures (SSTs)". Their "results do not support the notion of large (~200%) increases in tropical storm frequency in the North Atlantic basin over the twenty-first century in response to increasing greenhouse gases (GHGs)". The authors also modeled "projected changes in U.S. land-falling tropical storm activity under a variety of different climate change scenarios and climate models. These results were similar to those for the overall number of North Atlantic tropical storms, and do not point to a large increase in U.S. land-falling tropical storms over the twenty-first century in response to increasing GHGs".

NIPCC (2011) provided a more general discussion of storms, going beyond tropical hurricanes to include storminess in the Arctic, Europe, Australia, New Zealand, and China. As in other work, there is no evidence of more intense storminess in the late 20th century than at other times.

Weinkle *et al.* (2012) reviewed historical global tropical cyclone (TC) landfalls through 2010, and Pielke Jr.[5] updated the data to include 2011 and 2012. Pielke Jr. summarized the findings as follows:

> "Anyone who'd like to argue that the world is experiencing a 'new normal' with respect to tropical cyclones is simply mistaken. Over the past 4 years, the world is actually in the midst of a very low period in tropical cyclone landfalls—at least as measured over the past 43 years.
>
> "There is even evidence in our paper ... that the period before 1970 saw more intense hurricane landfalls than the period since. Older data ... indicates that landfalling intense hurricanes ... occurred at a 40% higher rate from 1950–1969 than 1970–2010. There were 9 intense landfalls in 1964 and 1965 in just these two basins, which equals the global record for all basins post-1970.
>
> "What we can glean from this data is that in terms of U.S. and global damage, things will get much worse when the statistics return to the 'old normal' (and this is independent of whether you think such a return is due to natural

[5] *http://rogerpielkejr.blogspot.com/2012/12/global-tropical-cyclone-landfalls-2012.html.*

variability, human-caused climate change or the prophesies of Nostradamus). It will happen—and you can take that to the bank."

Weinkle *et al.* (2012) pointed out that "in recent decades, economic damage from tropical cyclones (TCs) around the world has increased dramatically". They went on to say that "Scientific literature published to date finds that the increase in losses can be explained entirely by societal changes (such as increasing wealth, structures, population, etc.) in locations prone to tropical cyclone landfalls, rather than by changes in annual storm frequency or intensity". They concluded that:

> "Using historical TC best-track records, a global database focused on hurricane-force strength landfalls was constructed. The analysis does not indicate significant long-period global or individual basin trends in the frequency or intensity of land-falling TCs of minor or major hurricane strength. The evidence in this study provides strong support for the conclusion that increasing damage around the world during the past several decades can be explained entirely by increasing wealth in locations prone to TC landfalls, which adds confidence to the fidelity of economic normalization analyses."

Van Soelen *et al.* (2013) found evidence in peat fields that "suggests that on average tropical cyclone frequency did not change during the past 200 years".

8.4.3.2 Tornadoes

Donald Prothero (2011) provided a short article on tornadoes. Tornadoes form when:

> "… warm moist air mass rising from the Gulf of Mexico that moves north and meets cooler, drier air from the northern Plains and the Rockies. When these collide, a strong front develops which causes a big horizontal cylindrical vortex to form. The warm air slides beneath the cold air and thunderheads grow. If there is also strong shear from the jet stream, the horizontal cylindrical spiral of air will tilt into a vertical funnel. If it continues to grow, it will touch the ground and become a tornado.
>
> "The U.S. has by far most of the world's tornadoes due to its favorable geography. … The only other country in the world with significant tornadoes is Bangladesh."

The U.S. tornado season got off to "a rip-roaring start in 2011 … with more than 322 deaths in the April 27 outbreak and the more than 144 deaths in the May 22 outbreak". However, … these numbers don't yet approach the deadliest tornadoes in U.S. history, including over 700 killed and 2,027 injured by the Tri-State Tornado of March 18, 1925.

So, the question arises: Has the incidence and severity of tornadoes increased as the U. S. warmed in the 20th century? Unfortunately, this is very difficult, and may even be impossible, to answer. According to Brooks (2006):

> "The historical records of the occurrence of and losses from severe thunderstorms and tornadoes present significant challenges in attempting to

establish trends, if they exist. ... It is likely that the highest-quality dataset of significant length is the tornado dataset of the United States, which began in the early 1950s. Even these data have serious problems with consistency."

The data, such as they are, indicate that "the number of tornadoes reported in the United States per year has been increasing steadily (~14 per year) over the past half century. However, they state that determining whether this is a robust trend in tornado occurrence is difficult because the historical record is both relatively short and non-uniform in space and time. In addition, the increase in yearly tornado numbers runs parallel with the concurrent increase in the country's population, which makes for that much better geographical coverage and more complete (i.e., numerous) observations".[6] On the other hand, the data indicate the number of severe tornadoes per year has decreased over the past few decades. Attempts have been made to estimate changes in tornado occurrence from data on damages, but these are obfuscated by changes in population, increases in wealth, and changes in reporting.

In his "Concluding Remarks", Brooks said:

> "Problems in the reporting databases mean that it is extremely unlikely that climate change will be detectable in severe thunderstorm and tornado reports, even if there is a physical effect. ... There is no evidence to date to suggest that changes in damage are related to anything other than changes in wealth in the U.S. ... In the absence of databases of at least reasonable quality, it is extremely difficult to say much of substance."

8.4.3.3 Extreme weather

The *UKCIP Climate Digest*[7] recently raised the question: "Has human-induced global warming increased the risk of heavy rainfall and flooding in the UK?" They answered this question by claiming that "two new papers[8] published in *Nature* add to the body of evidence that suggests that human-induced global warming is leading to an increased risk of heavy rainfall and flooding. They show that human activity has contributed to an increased intensity of heavy rainfall events in the Northern hemisphere over the second half of the 20th century and substantially increased the risk of flood occurrence in England and Wales in autumn 2000". This report concludes that "These results show that human activity has contributed to an increased intensity of heavy rainfall events in the northern hemisphere over the second half of the 20th century. ... This implies that extreme precipitation events may strengthen more quickly in the future than projected and therefore have more severe impacts than previously estimated". However, the papers by Min *et al.* (2011) and Pall *et al.* (2011) are utterly unconvincing to this writer.

[6] NIPCC Report (2011).

[7] *www.ukcip.org.uk/climate-digest/cd-february-2011/*.

[8] Min *et al.* (2011) and Pall *et al.* (2011).

John Christy discussed extreme events.[9] He reviewed a number of recent extreme climate events and compared the event with the historical record. He showed that recent flooding in Australia is not out of character with events in the 19th century and, to a lesser extent, an event earlier in the 20th century. In regard to flooding in England, he said:

> "For the Thames River, there has been no trend in floods since records began in 1880, though [there was] a lull in flooding events from 1965 to 1990. … Flooding events on the Thames since 1990 are similar to, but generally slightly less than those experienced prior to 1940. One wonders that if there are no long-term increases in flood events in England, how could a single event (Fall 2000) be pinned on human causation as in Pall *et al.* (2011), while previous, similar events obviously could not? Indeed, on a remarkable point of fact, Pall *et al.* 2011 did not even examine the actual history of flood data in England to understand where the 2000 event might have fit."

He also showed that the 2010 heat wave in Russia, and high snowfall in the U. S. in 2009–2010 and 2010–2011 can be explained by natural variability. Then he went on to show why occurrence of extreme events is not a good measure of climate change, using an example of extreme high and low temperatures as recorded by the states of the U.S.:

> "For each of the 50 states, there are records kept for the extreme high and low temperatures back to the late 19th century. In examining the years in which these extremes occurred (and depending on how one deals with "repeats" of events) we find about 80 percent of the states recorded their hottest temperature prior to 1955. And, about 60 percent of the states experienced their record cold temperatures prior to that date too. One could conclude, if they were so inclined, that the climate of the U. S. is becoming less extreme because the occurrence of state extremes of hot and cold has diminished dramatically since 1955. … Then, one might look at the more recent record of extremes and learn that no state has achieved a record high temperature in the last 15 years (though one state has tied theirs.) However, five states have observed their all-time record low temperature in these past 15 years (plus one tie.) This includes last month's record low of 31°F below zero in Oklahoma, breaking their previous record by a rather remarkable 4°F. If one were so inclined, one could conclude that the weather that people worry about (extreme cold) is getting worse in the U. S. (Note: this lowering of absolute cold temperature records is nowhere forecast in climate model projections, nor is a significant drop in the occurrence of extreme high temperature records.)"

[9] Written Statement of John R. Christy, The University of Alabama in Huntsville Subcommittee on Energy and Power Committee on Energy and Commerce, March 8, 2011, *republicans.energycommerce.house.gov/Media/file/Hearings/.../Christy.pdf.*

8.5 SPECIES EXTINCTION

The media are replete with alarmists' claims that global warming has produced or will produce extensive extinction of species. In recent years, the number of voices raised in this regard has increased significantly.

The CO_2-induced global-warming extinction hypothesis claims that, as the world warms in response to the ongoing rise in the air's CO_2 content, many species of plants and animals will not be able to migrate either poleward in latitude or upward in elevation fast enough to avoid extinction as they try to escape the stress imposed by the rising temperature.

Anon. (D) claims that "climate change has led to some 25 per cent of the world's mammals and 12 per cent of birds being at significant risk of extinction" and "that climate warming scenarios for 2050 could lead to extinction of approximately 18% to 35% of species" for low, mid, and high scenarios. Anon. (D) provides references to a number of recent papers that make such dire predictions. However, these all seem to utilize terms such as "at risk of", "could lead to", and "could possibly lead to".

Idso, Idso, and Idso (2003) argued, however, that:

> "... as long as the atmosphere's CO_2 concentration rises in tandem with its temperature, most species will not 'feel the heat,' as their physiology will change in ways that make them better adapted to warmer conditions. Hence, although Earth's plants will likely spread poleward and upward at the cold-limited boundaries of their ranges in response to a warming-induced opportunity to do so, their heat-limited boundaries will probably remain pretty much as they are now or shift only slightly. Consequently, in a world of rising atmospheric CO_2 concentration, the ranges of most of Earth's plants will likely expand if the planet continues to warm, making plant extinctions even less likely than they are currently. Animals should react much the same way. In response to concurrent increases in atmospheric temperature and CO_2 concentration, they will likely migrate poleward and upward, where cold temperatures prevented them from going in the past, as they follow Earth's plants. ... A goodly portion of Earth's plants and animals should actually expand their ranges and gain a stronger foothold on the planet as the atmosphere's temperature and CO_2 concentration continue to rise."

NIPCC (2011) provided an extensive discussion of the impact of global warming on species but their detailed analysis is beyond the scope of this book. Only a short discussion based on NIPCC (2011) is given here. One approach for estimating the impact of global warming on species is to determine the "climate envelopes" of over 1,000 species. "Each of these envelopes represented the current climatic conditions under which a given species was found in nature. Then, after seeing how the habitat area of each of the studied species would be expected to change in response to an increase in temperature, they used an empirical power-law relationship that relates species number to habitat area size to determine extinction probability calculations." However, in a warmer climate, the potential habitat of many species would be expanded (e.g., the climate would be milder at higher altitudes) which would open up new living space for species presently threatened with extinction by non-climatic factors, "due to increasing habitat loss attributable to expanding urbanization and agricultural activities; while it may help other species that are threatened with extinction by habitat fragmentation to cross geographical barriers that were previously insurmountable obstacles to them". In general, as NIPCC (2011) pointed out in considerable detail, analysts seem to be too ready to pounce on any and all extinctions of amphibians, birds, butterflies, insects, lizards, and animals, with the claim that global warming was the cause. NIPCC (2011) refuted these claims with references to studies that show that these were either caused by non-climatological human impact on the habitats or changes in climatological parameters other than temperature (winds, precipitation, etc.). In some cases, such extinctions were counterbalanced by evolution of new species.

The main point here seems to be that habitat degradation is indeed occurring on a global scale, and most of this is due to human expansion, extension, development, and globalization. Attributing all of this to warming is ridiculous.

It seems likely, in general, that, given enough time, the species on the Earth reach some approximation of equilibrium with the prevailing climate. Those species that are attuned to any locality can prosper there. When the climate changes rapidly, the species might not have enough time to readjust to changing conditions, resulting in some extinctions. With the expansion of human population and human activity over the past 200 years, we have had a devastating impact on the biota of the world. Russian scientists have been particularly concerned with the impacts of human development on the environment and the biota, and have concluded that, in order to prevent global catastrophe, fundamental limits must be imposed on growth and development.

Krapivin and Varotsos (2007) said:

> "A characteristic of the present global ecological situation is increasing instability or put another way—a crisis in the civilization system, the global scale of which is expressed through a deterioration of human and animal habitats. The most substantial features of global ecodynamics of the late 20th and early 21st centuries include the rapid increase in world population (mainly in developing countries), increase in the size of the urban population (considerable growth in the number of megalopolises), and increase in the scales of such dangerous diseases as HIV/AIDS, hepatitis, tuberculosis, etc. With growing population size, the problems of providing people with food and improving their living conditions in many regions will not only not be resolved but will become even more urgent. Any possible benefit from decrease in per capita consumption as a result of increased efficiency of technologies will be outweighed by the impact of such a growth in population size. Despite the predominant increase of population in developing countries, their contribution to the impact on the environment will not necessarily exceed that of developed countries. ...
>
> "As civilization has developed, so the problem of predicting the scale of

expected climate change and associated change in human habitats has become more urgent. The matter primarily concerns the origin and propagation of natural phenomena causing the death of many creatures and large-scale economic damage. In the historical past, natural anomalies at various spatio-temporal scales are known to have played a certain role in the evolution of nature, causing and activating mechanisms for natural system regulation. With the development of industry and growing population density, these mechanisms have suffered considerable changes and acquired a life-threatening character. This is primarily connected with the growth and propagation in the level of anthropogenic disturbances in the environment. Numerous studies of resultant problems carried out in recent decades have shown that the frequency of catastrophic phenomena in nature and their scale have been continuously growing, leading to a growing risk to human life, financial losses, and a breakdown of the social infrastructure."

Danilov-Danil'yan *et al.* (2009) organized their book with the following summary:

"*Part I: Civilization in crisis*: On the edge of an abyss begins with an overview of the ecological, demographic, and other aspects of the global crisis. The global environmental situation involves unprecedented atmospheric change. Destruction of natural ecosystems is shown to be the primary cause of the crisis. With the chemicalization of the Earth's biosphere (48 tons of waste per capita) the global cost of local environmental cleanup is huge. The planet is critically overpopulated. Demographic growth is discussed in the light of biological constraints on the numbers of species. The energy and resource components of demographic growth are explored. Anthropogenic pressure is viewed through the prism of human ecological equivalents.

"*Part II: Civilization teetering over the abyss of crisis*: The social dimensions of the crisis are explored, and it is concluded that poverty is a cause of pollution. The problems of hunger and malnutrition, and the harmful environment of the megalopolis are discussed. The 'contribution' of the centralized economy and the pros and cons of the market economy are discussed, as well as the market's unreceptiveness to long-term strategies. The limitations placed on the free market by the environment are derived. Ultimately, it is the spiritual crisis of man that is the primary cause of the challenge to the environment.

"*Part III: The world community: Politicians and scientists in search of a solution* provides a history of the first ecology forums and the establishment of a systemic approach to the study of the biosphere. The theory of biotic regulation and stabilization of the environment is the next step in the development of a systemic approach. This part ... suggests approaches toward a systemic understanding of the biosphere and the concept of biotic regulation as a theoretical basis for sustainability. In this discussion, the following topics are covered: (1) the universal role of the biota in the maintenance of chemical and physical parameters suitable for life, (2) the biotic mechanisms of compensating for disruptions, (3) the destruction of the biotic regulation mechanism under the influence of anthropogenic pressure, and (4) the conservation of ecosystems still in existence as well as the partial reconstruction of destroyed ecosystems as humanity's key goal.

"*Part IV: Sustainable development*: Between complacency and reality deals with the basis of sustainability in nature and in civilization. Two concepts are interpreted: (1) sustainability and sustainable development, and (2) principles of evolutionary strategy in nature vs. growth and development in society. National programs aimed at moving toward sustainable development are inadequate. Finally, the benefits of the nature-society co-evolution concept are discussed. In the process, the evolution of the biosphere 'toward humanity' is contrasted with the evolution of humanity 'toward the biosphere', as the only alternative.

"*Part V: On the scale of a scientific approach*: deals with sustainable development in the context of the biosphere's carrying capacity. Among the topics covered are: (1) distribution of energy streams in biota among groups of organisms, (2) energy quota of man as a large mammal, and (3) the law of energy stream distribution and the biosphere's carrying capacity. In addition, the starting conditions of sustainable development and the safety of ecosystems are given by country and continent. Specific topics include: (a) world centers of environmental stabilization and destabilization, (b) absorption of anthropogenic carbon by the World Ocean and ecosystems on dry land, (c) the rebirth of destroyed ecosystems: a path toward the stabilization of atmospheric carbon, and (d) transition to sustainable development and starting conditions in various countries of the world with respect to the degree of natural ecosystem preservation. Finally, sailing directions and the compass are provided: indicators of sustainable development.

"*Part VI: 'Is there enough community, responsibility, discipline and love?'* is the final part of this book. It deals with the barricades of backward thinking and dwindling chances for the future as well as psychological obstacles on the way to grasping the ecological threat. It summarizes what the market economy can and cannot do, and outlines the creative and destructive potential of the market economy. It discusses sustainable development and the actual condition of man. Sustainable development is put forward as a world idea. The social prerequisites of sustainable development and the relationship to globalization are summarized."

Evidently, the world faces a crisis from the impacts of overpopulation and global development. Global warming is only one of many factors, and possibly not the most important. Ultimately, growing world population is by far, the biggest and most pervasive problem from which all other problems evolve. History shows that as regional populations grow more prosperous, birth rates decrease sharply. In a sense, therefore, an even more fundamental problem for the world is poverty. Allowing the world population to increase while depriving them of fossil fuels will only increase poverty, thus accentuating the problem.

8.6 VEGETATION

NIPCC (2011) discussed the impacts of climate change on terrestrial plants and soils in great detail with many references. This included: plant growth responses to atmospheric CO_2 enrichment, below-ground biotic responses to atmospheric CO_2 enrichment, transpiration and water use efficiency, ecosystem responses to elevated temperature, responses of plants under stress to atmospheric CO_2 enrichment, ecosystem biodiversity, soil carbon sequestration, extinction, evolution, food production, greening of the Earth, nitrogen, dating of arrival of spring, and range expansion. In general, the effect of increased CO_2 was beneficial.

Idso (2012) pointed out that the IPCC claims the following:

> "Current levels of temperature and changing precipitation patterns are beginning to stress Earth's natural and agro-ecosystems now by reducing plant growth and development. And looking to the future, they claim that unless drastic steps are taken to reduce the ongoing rise in the air's CO_2 content (e.g., scaling back on the use of fossil fuels that, when consumed, produce CO_2), the situation will only get worse—that crops will fail, food shortages will become commonplace, and many species of plants (and the animals that depend on them for food) will be driven to extinction."

However, Idso (2012) argued that such concerns "are not justified". In his lengthy report, he "reviewed the peer-reviewed scientific literature, examining how the productivities of Earth's plants have responded to the 20th and now 21st century rise in global temperature and atmospheric CO_2". He concluded that:

> "The productivity of the planet's terrestrial biosphere, on the whole, has been increasing with time. ... There is no empirical evidence to support the model-based claim that future carbon uptake by plants will diminish on a global scale due to rising temperatures. In fact, just the opposite situation has been observed in the real world. ... Over the past 50 years, global carbon uptake has doubled from 2.4 ± 0.8 billion tons in 1960 to 5.0 ± 0.9 billion tons in 2010. ... There is compelling evidence that the atmosphere's rising CO_2 content is most likely the primary cause of the observed greening trends."

Evidently, Idso is optimistic about the state of the global biosphere, in contrast to the pessimistic views of Danilov-Danil'yan *et al.* and Krapivin and Varotsos. Perhaps Idso is right that the biosphere has responded to the increase in atmospheric CO_2 by absorbing more carbon, but that does not negate the serious threats to the biosphere posed by growing population and development. Absorption of carbon is not the sole arbiter of the state of the biosphere.

8.7 CORAL REEFS

Silverman *et al.* (2009) pointed out that the recent anthropogenic increase in atmospheric CO_2 is increasing the acidity of the surface oceans. Models predict that doubling CO_2 from the pre-industrial value of ~280 ppm will reduce the ocean pH by about 0.5 units. They claim that this may produce "a severe global decline of coral reef abundance but not a complete extinction".

NIPCC (2011) said that:

> "According to the IPCC, CO_2-induced global warming is increasing the temperatures of Earth's oceans and seas and lowering their pH values, a process called acidification. Both processes, according to the IPCC, are likely to harm aquatic life. The IPCC said: 'Many studies incontrovertibly link coral bleaching to warmer sea surface temperature ... and mass bleaching and coral mortality often results beyond key temperature thresholds' Modeling, the IPCC goes on to say, 'predicts a phase switch to algal dominance on the Great Barrier Reef and Caribbean reefs in 2030 to 2050'. The IPCC further claims that "coral reefs will also be affected by rising atmospheric CO_2 concentrations ... resulting in declining calcification."

NIPCC (2011) took issue with these claims. According to the NIPCC:

> "While some corals exhibit a propensity to bleach and die when sea temperatures rise, others exhibit a positive relationship between calcification, or growth, and temperature. Such variable bleaching susceptibility implies that there is a considerable variation in the extent to which coral species are adapted to local environmental conditions. ... The latest research suggests corals have effective adaptive responses to climate change, ... that allow reefs in some areas to flourish despite or even because of rising temperatures. Coral reefs have been able to recover quickly from bleaching events as well as damage from cyclones. ... Bleaching and other signs of coral distress attributed to global warming are often due to other things, including rising levels of nutrients and toxins in coastal waters caused by runoff from agricultural activities on land and associated increases in sediment delivery. ... The IPCC expresses concern that rising atmospheric CO_2 concentrations are lowering the pH values of oceans and seas, a process called acidification, and that this could harm aquatic life. But the drop in pH values that could be attributed to CO_2 is tiny compared to natural variations occurring in some ocean basins as a result of seasonal variability, and even day-to-day variations in many areas. Recent estimates also cut in half the projected pH reduction of ocean waters by the year 2100."

Loss of coral reefs is a legitimate concern that provides incentives for reduction of carbon emissions.

8.8 FOOD PRODUCTION

In the spring of 2011, several reports appeared claiming that consequences of human-caused global warming are more significant than previously projected in the 2007 IPCC Report. These new claims were quickly repeated by many news organizations. In one of these papers, Lobell *et al.* (2011) claimed that:

"... global average temperatures have risen by roughly 0.13°C per decade since 1950, yet the impact this has had on agriculture is not well understood. An even faster pace of roughly 0.2°C decade of global warming is expected over the next 2–3 decades, with substantially larger trends likely for cultivated land areas. Understanding the impacts of past trends can help to gauge the importance of near-term climate change for supply of key food commodities. In addition, identifying which particular crops and regions have been most impacted by recent trends would assist efforts to measure and analyze ongoing efforts to adapt."

Ken Haapala[10] discussed the paper by Lobell *et al.* As Haapala pointed out:

"Lobell *et al.* (2011) claim that global warming is restricting world food production. They analyze four major crops: maize (corn), wheat, rice and soybeans, which they state account for 75% of the world's human caloric consumption, either directly or indirectly. To estimate reduction in food production, the authors created a model estimating ... production ... without global warming (climate change) and compared it with actual production. They concluded that global wheat production is 5.5% below and maize production is 3.8% below what they would be without global warming. The main question is: does the model actually measure what the authors claim or are there significant confounding variables that are not identified in the study?"

Haapala made two key points:

(1) 'The claim that "global average temperatures have risen by roughly 0.13°C decade since 1950" is mistaken. This would imply that world temperatures have increased by 0.91°C from 1950 to 2010, which is considerably greater than the widely accepted range of 0.5°C to 0.6°C. (Furthermore, as we discussed in Section 3.4.3.2, global average temperature did not rise from 1998 to 2012.)
(2) The lower 48 states of the U.S. have experienced a cooling from 1980 to 2008 and since the U.S. experienced a cooling, it is exempt from the conclusions of the study even though the U.S. is the world's major producer of two of the four food commodities studied, maize and soybeans, accounting for some 40% of world production.

Haapala also discussed "the production of wheat and rice in China and India, the two largest producers and consumers of these staples, accounting for approximately 29% of world wheat production and 48% of world rice production. Historically, these countries were noted for widespread famines, often due to changing weather patterns (climate change)". Haapala pointed out that:

"From 1960 to 2008, in China, wheat production went up by 437% and milled rice production went up by 221%, or an average annual increase through the period of 9% and 5%, respectively. In India, wheat production went up by

[10] Haapala, K., Executive Vice President, Science and Environmental Policy Project (SEPP) The Week That Was: 2011-05-14 (May 14, 2011), *www.sepp.org/the-week-that-was.cfm.*

661% and rice production by 186%, or an average annual increase through the period of 14% and 4%, respectively.

"These remarkable increases in production occurred during a period of global warming, including the great climate shift of the late 1970s. Those inclined to hasty generalizations, without consideration of confounding variables, may conclude that global warming has been a boon to agriculture production in China and India. However, much of the increase is due to the green revolution, carbon dioxide enhancement, and changes in government policies.

"Also during this period both China and India rapidly increased maize production becoming the world's second and fourth largest producers, respectively. Chinese maize production increased more than 10-fold. . . .

"Are these reductions in the spectacular increases in production due to global warming? No! An analysis of the complete data suggest otherwise, showing that, generally, production increases began to taper off in the 1990s. Basically, China and India became self-sufficient in grain production. Famines are no longer an issue and grain imports in 2008 were less than 0.5% of domestic production. [Lobell *et al.*, 2011] mistakenly attribute to warming the reductions in production increases due to market stabilization from the green revolution, carbon dioxide enhancement, and changes in government policies.

"The authors claim that any excess would be available for export. However, export of grains requires an integrated system for such purposes that China and India do not have. Low-cost producers, such as the U.S. and Canada, have such systems, and dominate the world markets. There is no incentive for farmers in China and India to produce more than what they can sell in domestic markets.

"Hunger remains a major problem in much of the world, especially in sub-Sahara Africa, parts of which are subject to incessant warfare and political turmoil, resulting in low production of foodstuffs. To falsely attribute this hunger to global warming ignores the real causes and is a disservice to science and humanity."

NIPCC (2011) presented considerable evidence that food production has been higher historically during warm periods.

9

Global climate change and public policy

9.1 U.S. GOVERNMENT POLICY ACTIONS AND INACTIONS

Over the past two or three decades, climatologists issued repeated warnings in press releases and other media emphasizing the putative dangers associated with global warming from increasing greenhouse gas (GHG) concentrations. Al Gore's film *An Inconvenient Truth* (see Appendix I) had a major public relations effect. The U.N., through the Inter-governmental Panel on Climate Change (IPCC) as well as several other of its agencies, issued many reports and other media to support the alarmist viewpoint. Published materials, including some rather bizarre forms of the *hockey stick* were circulated to hundreds of millions of schoolchildren. The U.N. took the lead in organizing the international community to take action in setting goals and standards for reducing carbon emissions. Some of this is discussed in later sections of this chapter. In the U.S., perhaps predictably, Democrats reacted favorably and introduced legislation to control future emissions, while Republicans remained skeptical and opposed stringent measures proposed by Democrats.

Wikipedia[1] provides a summary of U.S. federal actions in response to this media blitz on climate change. Some of the material below is taken from this website, However, the Wikipedia article portrays a fictitious conspiracy by special interests to subvert the good intentions of climate alarmists, whereas the material below has a different spin.

In 1997, the U.N. passed the Kyoto Protocol to reduce future emissions, although it excepted developing nations. The U.S., although a signatory to the Kyoto Protocol, neither ratified nor withdrew from the protocol. The signature alone is symbolic, as the Kyoto Protocol is non-binding on the U.S. unless ratified. The U.S. was, as of 2005, the largest single emitter of carbon dioxide from the burning of fossil fuels. China has since taken over this honor in the 2011 time frame. China was given a free hand to emit under the Kyoto Protocol.

On July 25, 1997, before the Kyoto Protocol was finalized (although it had been fully negotiated, and a penultimate draft was finished), the U.S. Senate unanimously

[1] *http://en.wikipedia.org/wiki/Climate_change_policy_of_the_United_States.*

D. Rapp, *Assessing Climate Change*, Springer Praxis Books,
DOI: 10.1007/978-3-319-00455-6_9

passed (by a 95:0 vote) the Byrd–Hagel Resolution, which stated that the sense of the Senate was that the U.S. should not be a signatory to any protocol that did not include binding targets and timetables for developing as well as industrialized nations or this "would result in serious harm to the economy of the U.S.". On November 12, 1998, Vice President Al Gore symbolically signed the protocol. Both Gore and Senator Joseph Lieberman indicated that the protocol would not be acted upon in the Senate until there was participation by the developing nations. The Clinton Administration never submitted the protocol to the Senate for ratification.

President George W. Bush also did not submit the treaty for ratification, partly because of the exemption granted to China. Bush also opposed the treaty because of the strain he believed the treaty would put on the economy; he emphasized the uncertainties that are present in the climate change issue:

> "In October 2003 and again in June 2005, the *McCain-Lieberman Climate Stewardship Act* failed a vote in the US Senate. In the 2005 vote, Republicans opposed the Bill 49-6, while Democrats supported it 37–10."

The Global Warming Pollution Reduction Act of 2007 was introduced in the U.S. Senate. However, the Bill died in committee.

The American Clean Energy and Security Act of 2009 was approved by the House of Representatives by a vote of 219:212, but died in the Senate.

The Environmental Protection Agency (EPA) did a good job of cleaning up America in the latter decades of the 20th century. As the 21st century began, the need for an EPA diminished. A cynic might conclude that the EPA seized upon control of emissions of GHGs as a new *cause célèbre* to justify its continued existence. Republicans naturally opposed this. In March 2011, the Republicans submitted a Bill to the U.S. Congress that would prohibit the EPA from regulating GHGs as pollutants. However the EPA continues to oversee regulation of GHG emissions under the Clean Air Act. The U.S. Supreme Court unwisely affirmed the notion that CO_2 is a pollutant and therefore emissions can be legally regulated by the EPA.

In 2001 and 2002, the Bush Administration continued previous U.S. policy of not implementing the Kyoto Protocol, but proposed a much more limited plan to reduce emissions by providing tax credits to businesses that use renewable energy sources.

In 2006, Governor Arnold Schwarzenegger said: "We simply must do everything we can in our power to slow down global warming before it is too late. ... The science is clear. The global warming debate is over." This seems to confirm his nickname as "The Terminator".

In October 2007, Chairman Dingell's Energy and Commerce Committee of the U.S. Congress released its first White Paper in a series on "Meeting the Climate Change Challenge".[2] This was claimed to be "the next step in the legislative process leading to enactment of a mandatory, economy-wide climate change program". The essential basis for this program is the belief that:

[2] *http://energycommerce.house.gov/Climate_Change/White_Paper.100307.pdf.*

> "The United States should reduce its greenhouse gas emissions by between 60 and 80 percent by 2050 to contribute to global efforts to address climate change. To do so, the United States should adopt an economy-wide, mandatory greenhouse gas reduction program. … The central component of this program should be a cap-and-trade program. The cap-and-trade program will have increasingly stringent caps on greenhouse gas emissions, eventually reaching a level that reduces emissions by 60 to 80 percent in 2050. The Government will distribute allowances equal to the level of allowed greenhouse gas emissions. Allowances can then be bought and sold."

It is not clear from the White Paper whether the Committee means a 60%–80% reduction from projected levels in 2050 based on a business-as-usual (BAU) scenario, or a 60%–80% reduction from present levels. However, the wording seems to suggest a reduction from present levels. Considering that, under a BAU scenario, energy usage in the U.S. is likely to grow between now and 2050, that would imply an even greater percentage reduction from projected levels in 2050 under BAU. Assuming that energy usage in the U.S. in 2050 under BAU would increase by, say, 40% compared with today, that would imply that the required reduction would be between 72% and 86% compared with energy usage in the U.S. in 2050 under a BAU scenario.

In 2008, President Obama also proposed an 80% reduction in emissions.[3] As part of his plan, he would increase the use of coal and use a cap-and-trade system for emissions. The cap-and-trade system would allow those who can afford it to continue to emit. Thus, the percentage reduction for those not buying the right to emit would be even higher than 80%. Obama planned the use of "clean coal". But, coal pollutes in the mines, in the runoff from the mines, in the desecration left behind, in the railroads that transport the coal, in the power plants that burn the coal, in the emissions from the power plants, and in the ash left over. Coal produces a lot more CO_2 per unit energy produced than petroleum or natural gas. In the process of cleaning up coal for combustion, a considerable amount of CO_2 is emitted. The economic impact of such policies will be measured in many trillions of dollars, and the technical and economic challenges in implementing such policies have generally been underestimated (Pielke *et al.*, 2008). Thus, if one accepts the alarmist view that continued use of fossil fuels will produce unacceptable global warming, humanity is caught between the proverbial "rock and a hard place". According to this belief, we cannot accept the consequences of continuing BAU; however, we have neither technical nor economic capability to do otherwise without creating great financial and operational dislocations. It is noteworthy that, in 2012, the EPA (under orders from President Obama) passed stringent new policies on coal-fired power plants that might kill the entire coal industry in the U.S. Mr. Obama now seeks to kill coal altogether, rather than clean it.

Senator John McCain authored S. 2191, a Bill that required a reduction of CO_2 emissions 70% below current levels by 2050 (about 80% below levels in 2050 in a BAU scenario). Senator Barbara Boxer advocated a 90% reduction (Michaels and

[3] *Los Angeles Times*, November 19, 2008.

Balling, 2009). House Leader Nancy Pelosi said that fossil fuels should be replaced by natural gas.

In November 2008, President-elect Barack Obama said that the U.S. should enter a cap-and-trade system to limit global warming. The American Clean Energy and Security Act, a cap-and-trade Bill was passed in 2009 in the House of Representatives, but was not passed by the Senate.

In 2009, Secretary of State Clinton said that "... the United States will be energetic, focused, strategic and serious about addressing global climate change and the corollary issue of clean energy".

In 2009, President Barack Obama said "that if the international community would not act swiftly to deal with climate change that we risk consigning future generations to an irreversible catastrophe. ... The security and stability of each nation and all peoples—our prosperity, our health, and our safety—are in jeopardy, and the time we have to reverse this tide is running out".

In late June 2009, the U.S. House of Representatives passed a Bill requiring reduction in carbon emissions of 17% by 2020 and by 83% by 2050, and establishment of a cap-and-trade system for exchanging rights to emit CO_2. Apparently, none of the members of the House read the 1,500-page Bill. President Obama's views were expressed during his candidacy (widely quoted on the Internet):

> "So if somebody wants to build a coal-powered plant, they can; it's just that it will bankrupt them because they're going to be charged a huge sum for all that greenhouse gas that's being emitted.
>
> "You know, when I was asked earlier about the issue of coal, uh, you know, under my plan of a cap-and-trade system, electricity rates will necessarily skyrocket. Even regardless of what I say about whether coal is good or bad. Because I'm capping greenhouse gases, coal power plants, you know, natural gas, you name it—whatever the plants were, whatever the industry was, uh, they would have to retrofit their operations. That will cost money. They will pass that money on to consumers."

In 2010, President Obama said that it was time for the United States "to aggressively accelerate" its transition from oil to alternative sources of energy and vowed to push for quick action on climate change legislation.

In his second inaugural address, President Obama said "We will respond to the threat of climate change, knowing that the failure to do so would betray our children and future generations. Some may still deny the overwhelming judgment of science, but none can avoid the devastating impact of raging fires and crippling drought and more powerful storms". He claimed that "Heat waves, droughts, wildfires, and floods - all are now more frequent and intense". Clearly, these statements have no scientific basis at all, and constitute nonsense from the mouth of the president.

In the 1920s, H.L. Mencken summarized the role of legislators:

> "The whole aim of practical politics is to keep the populous alarmed, and hence clamorous to be lead to safety, by menacing it with an endless series of hobgoblins, all of them imaginary."

9.2 THE KYOTO PROTOCOL

This section is based to a considerable extent on Anon. (G).

9.2.1 Description of the Kyoto Protocol

The Kyoto Protocol was an agreement made under the U.N. Framework Convention on Climate Change (UNFCCC). Countries that ratify this protocol commit to reduce their emissions of carbon dioxide and five other GHGs, or engage in emissions trading if they maintain or increase emissions of these gases.

The Kyoto Protocol covered more than 160 countries globally and over 55% of global GHG emissions. At its heart, the Kyoto Protocol establishes the following principles:

(1) Kyoto is underwritten by governments and is governed by global legislation enacted under the U.N.'s aegis.
(2) Governments are separated into two general categories:
 - developed countries, referred to as "Annex I countries" (who have accepted GHG emission reduction obligations and must submit an annual greenhouse gas inventory); and
 - developing countries, referred to as "Non-Annex I countries" (who have no GHG emission reduction obligations but may sell emission rights if they do reduce emissions).

The treaty was negotiated in Kyoto, Japan, in December 1997, and signed by most countries by March 15, 1999. The treaty could not take effect until it included enough Annex I countries to account for at least 55% of emissions from Annex I countries. The agreement came into force on February 16, 2005, following ratification by Russia on November 18, 2004, that drove the total to over 55%. As of December 2006, a total of 169 countries and other governmental entities have ratified the agreement (representing over 61.6% of emissions from Annex I countries). Emissions from Non-Annex I countries were not stated. Notable exceptions include the U.S. and Australia. Other countries, like India and China, being Annex II countries, have ratified the protocol but are not required to reduce carbon emissions under the present agreement.

The Kyoto Protocol was an agreement under which industrialized countries will reduce their collective emissions of GHGs by 5.2% compared with the year 1990 (but this represents a 29% drop from what it would be in 2010 without Kyoto). The goal is to lower overall emissions of six GHGs (carbon dioxide, methane, nitrous oxide, sulfur hexafluoride, hydrofluorocarbons (HFCs), and perfluorocarbons (PFCs)) calculated as an average over the five-year period of 2008–2012. National limitations range from 8% reductions for the E.U. and some others to 7% for the U.S., 6% for Japan, 0% for Russia, and permitted increases of 8% for Australia and 10% for Iceland. This agreement was negotiated as an amendment to the UNFCCC, which was adopted at the Earth Summit in Rio de Janeiro in 1992.

China, India, and other developing countries were exempt from the requirements of the Kyoto Protocol on the specious grounds that they were not the main contributors to the GHG emissions during the industrialization period that are believed by alarmists to be causing today's putative runaway climate change.

Between 2008 and 2012, Annex I countries were required to reduce their GHG emissions by an average of ~5% below their 1990 levels (for many countries, such as the E.U. member states, this corresponds to some 15% below their expected GHG emissions in 2008). While the average emissions reduction is 5%, specific national limitations vary widely (e.g., 8% reductions for the E.U. and 10% emissions increase for Iceland), but, since the E.U. intended to meet its target by distributing different rates among its member states, much larger increases (up to 27%) were allowed for some of the less developed E.U. countries. Reduction requirements expires at the end of 2012.

However, critics of Kyoto argued that China, India, and other developing countries will soon be the top emitters of GHGs (In 2012, China was the world's greatest emitter and emits almost as much as the rest of the world combined.).More importantly, without Kyoto restrictions on these countries, industries in developed countries would be driven toward these non-restricted countries, and thus there would be no net reduction in carbon. Furthermore, if the effects of CO_2 are anywhere near what the U.N. thinks they are, most of the global warming still lies ahead of us and China will be the main contributor to it.

If the global climate models are correct, and if CO_2 emissions continue to rise in the 21st century, the global warming in the 21st century will far surpass that of the 20th century. The Kyoto agreement exempts the main contributor to this putative 21st-century warming: China. As more and more industrialization is shifted from the U.S. and Europe to China, China's output of GHGs will swell, and there are no constraints on these emissions!

Kyoto includes "flexible mechanisms" which allow Annex I economies to meet their GHG emission limitation by purchasing GHG emission reductions from elsewhere. What this means in practice is that Non-Annex I nations have no GHG emission restrictions so they can produce as many GHG emissions as they prefer but, if they do go about implementing GHG reductions, they can sell the rights to these reductions to Annex I nations, thus enabling them to generate the CO_2 instead.

The Kyoto linking mechanisms were in place for two main reasons:

(1) The cost of complying with Kyoto is prohibitive for many Annex I countries (especially those countries, such as Japan or the Netherlands, with highly efficient, low GHG-polluting industries, and high prevailing environmental standards). Kyoto therefore allows these countries to purchase Carbon Credits instead of reducing GHG emissions domestically. Hence, those countries that have already done a good job are punished for doing a good job but can bail themselves out by purchasing credits from others. Instead of exempting these countries for their good past performance, they are required to pay for it.

(2) This is seen as a means of encouraging Non-Annex I developing economies to reduce GHG emissions since, by doing so, they can profit from the sale of Carbon Credits. It is not clear to this writer why developing economies are

merely "encouraged" while responsible developed countries are punished with unfair requirements.

The protocol also reaffirms the principle that developed countries will have to pay billions of dollars, and supply technology to developing countries for climate-related studies and projects. This was originally agreed to in the UNFCCC. Critics claim that the Kyoto Protocol is a cleverly disguised plan to transfer assets from developed countries to developing countries.

If the Enforcement Branch determines that an Annex I country is not in compliance with its emissions limitation, then that country is required to make up the difference plus an additional 30%. In addition, that country will be suspended from making transfers under an emissions-trading program. However, it is not clear what pressures can be brought to bear on a country that flouts the agreement.

9.2.2 Commentary on Kyoto Protocol

The U.N. has supported studies of global climate change for a number of years through its IPCC. IPCC (2001) provides not 1, but 19 separate chapters on the impacts of global warming, totaling over 1,000 pages.

Against this backdrop of alarmism regarding global warming, the Kyoto Protocol evolved under the auspices of the U.N.

Some public policy experts who are skeptical of claims of insidious effects of global warming see Kyoto as a scheme to either slow the growth of the world's industrial democracies or to transfer wealth to the Third World in what they claim is a global socialism initiative. Others claim that the costs of the Kyoto Protocol outweigh the benefits, and the standards that Kyoto sets are too optimistic. Others see a highly inequitable and inefficient agreement that would do little to curb GHG emissions, and they are right. While the conspiracy theories of Jaworowski (2007) seem to be rather paranoid, his quotations by a high-level U.N. diplomat:

> "We may get to the point where the only way of saving the world will be for industrial civilization to collapse";

and by U.S. State Department officials:

> "We have got to ride the global warming issue. Even if the theory of global warming is wrong, we will be doing right thing in terms of economic policy and environmental policy."

> "A global warming treaty must be implemented even if there is no scientific evidence to back the [enhanced] greenhouse effect".

make one wonder about the motivations of government officials.

There are idealists in the U.N. who want to save the world from calamity, real or imagined. And there are politicians everywhere, willing to "go with the flow" if it is politically expedient.

Nevertheless, if action were taken to reduce GHG emissions in the 21st century, the number one culprit is clearly China. With its rapid growth in GHG emissions,

China will soon become not merely the world's greatest generator of GHGs, but the world's leader in emissions by a wide margin. Exemption of China from any controls at all under the Kyoto Protocol defies logic.

But there are other forms of pollution besides GHGs, and China is clearly the greatest source of pollution in the world. The Kyoto Protocol does nothing about that.[4]

It is noteworthy that even the *Los Angeles Times*, a fervent believer in the theory that greenhouse gases cause catastrophic global warming, believes that the Kyoto Protocol is fatally flawed. As a *Los Angeles Times* editorial[5] pointed out:

> "The choice of 1990 as a base year simply rewards countries whose economies have shrunk since then and punishes growth. Russia, Eastern Europe, Germany and Britain are strong backers of Kyoto, and if one looks at the costs and benefits of the pact, it's no wonder. Today, these countries emit either less than they did in 1990 or just a little bit more. In Britain, that's because the privatization of the coal industry led to a decline in coal-fired power plants in favor of natural gas; elsewhere, it's because the collapse of the Soviet Union was followed by the closing of filthy Soviet-era industrial plants, while economies in Russia and much of Eastern Europe stagnated. The U.S. economy, meanwhile, has grown significantly since 1990, with a corresponding rise in power demand that ... has caused carbon dioxide emissions to jump 20.4%. What a global carbon-trading scheme boils down to, then, is a massive wealth transfer from the U.S. to Russia.
>
> "U.S. polluters would pay billions of dollars to buy carbon credits from other countries mostly Russia, because it would have the most to sell. Why should we inject huge sums into a country with a rotten human rights record, rampant corruption and opposing geopolitical views? And what did Russia do to earn the cash, other than shrink?
>
> "Further, because there is no world body that polices greenhouse gas emissions, countries and polluters are on the honor system—we have to trust them to be honest about how much they're polluting. Governments in Russia or Ukraine aren't capable of monitoring emissions from every pollution source even if they wanted to, and under Kyoto, there's no reason for them to want to. After all, if Ukraine claims to be cleaner than it really is, rich countries such as the U.S. and Japan will shower it with money for carbon credits. And corrupt governments will tend to distribute credits unfairly, using them to reward political supporters and reducing the market's effectiveness.
>
> "India and China are left out. ... If China keeps growing at its current rate, its per capita income is expected to reach U.S. levels within 25 years. Once that

[4] It is noteworthy that China already is the world's number one polluter. The satellite picture of Chinese pollution at *http://visibleearth.nasa.gov/view_rec.php?id=1036* demonstrates the extent of pollution from Chinese coal burning. But, with all the hoopla about global warming, hardly anyone worries about conventional pollution anymore.

[5] *Los Angeles Times* editorial, June 11, 2007.

> happens ... there might be 1.1 billion vehicles in China. Currently, there are only 800 million vehicles in the entire world. Meanwhile, China builds a new coal-fired power plant every week to stoke its growth, and ... in a quarter of a century, its CO_2 emissions will be double those of the other industrialized nations combined. India, the world's fourth-biggest polluter, is also growing at a blistering pace.
>
> "Unless these two countries can be persuaded to embrace green power, they will render Kyoto and every other attempt to reduce greenhouse gases moot. Unfortunately, they're not eager to change their ways. Last Monday, China unveiled its climate-change plan in response to pressure from the G-8; it made no commitments to any quantifiable carbon reductions and rejected international efforts to impose them. India also refuses to consider anything that might slow its economic development."

The *Los Angeles Times* then went on to propose

> "... a new, improved version of Kyoto that brings India and China onboard and commits them to 'grow green,' but still leaves the tougher cuts up to those nations better able to make them, such as the U.S., Canada, Japan and Europe."

It is not clear to this writer why China, the world's worst polluter (including CO_2), should get any favoritism. The *Los Angeles Times* also said:

> "A better treaty would scrap the unworkable carbon-trading scheme and instead impose new taxes on carbon-based fuels."

In a review of the Kyoto Protocols, Adam (2008) reported on figures released by the U.N. suggesting that the world is on track to meet its Kyoto targets for GHGs. It was claimed that emissions by the 40 industrialized nations that agreed to binding cuts in pollution are down 5% compared with 1990 levels. But it turns out that the drop had little to do with climate policies: the bulk of the decline was due to the collapse of the Soviet Union and the subsequent economic decline in eastern Europe in the 1990s. Without these so-called "economies in transition", GHG emissions have grown by almost 10% since 1990. Furthermore, emissions are once again rising in eastern Europe. In fact, the most efficient way to reduce emissions is to have an economic recession.

Adam (2008) reported that, among industrialized nations, 16 were on target to meet their Kyoto obligations, including France, the U.K., Greece, and Hungary, while 20 countries are off-course, including Canada, Germany, Ireland, Italy, Japan, New Zealand, and Spain. Nations that missed their Kyoto target in 2012 will have incurred a penalty of an additional third added to whatever cut they agree under a new treaty being drafted in Copenhagen.

According to Yvo de Boer, executive secretary of the U.N. Climate Secretariat, Kyoto has been successful at establishing an architecture for future reductions, but has not been successful at actually reducing emissions significantly. Anon. (N) provided data on current emissions and socio-economic factors as shown in Tables

9.1 and 9.2. The Climate Change Performance Index (CCPI) world rank is subjective and not worth discussing. Table 9.1 shows that China has now surpassed the U.S. in absolute emissions per year. Table 9.2 (column "A/B") shows that all nations except Brazil are comparable when emissions are compared with energy consumed. Column "A/C" shows that Russia and China are exceptionally high emitters on the basis of comparison with their values of gross domestic product (GDP). In terms of population, China is exceptionally low, while the U.S. and Canada are high.

Table 9.1. Emissions and socio-economic quantities as national percentages of world total (Anon. (N)).

Country	*CCPI Rank*		*A = % share of global CO_2 emissions*	*B = % share of global primary energy supply*	*C = % share of global GDP*	*D = % share of global population*
	2012	*2013*				
Germany	6	8	2.34	2.56	3.99	1.19
India	18	24	4.94	5.42	5.49	17.15
Brazil	14	33	4.19	2.08	2.86	2.85
Indonesia	32	36	2.33	1.62	1.36	3.51
U.S.	50	43	16.26	17.36	19.02	4.54
Japan	42	47	3.52	3.89	5.69	1.86
Korea	44	51	1.73	1.95	1.93	0.71
China	55	54	21.42	19.34	13.76	19.71
Russian Federation	54	56	4.84	5.49	2.93	2.07
Canada	57	58	1.65	1.97	1.75	0.50
Total			64.96	61.49	62.91	50.10

Table 9.2. National emissions as ratios of various national socio-economic quantities (from Table 9.1) (Anon. (N)).

Country	A/B	A/C	A/D
Germany	0.91	0.59	1.97
India	0.91	0.90	0.29
Brazil	2.01	1.47	1.47
Indonesia	1.44	1.71	0.66
U.S.	0.94	0.85	3.58
Japan	0.90	0.62	1.89
Korea	0.89	0.90	2.44
China	1.11	1.56	1.09
Russian Federation	0.88	1.65	2.34
Canada	0.84	0.94	3.30

9.2.3 Future of the Kyoto Protocol

This section is based on the website: *http://knowledge.allianz.com/environment/climate_change/?1741/what-future-kyoto-protocol*:

> "The Durban climate talks at the end of 2011 "saved" the Kyoto Protocol from oblivion, ensuring that there would not be a total collapse of legally-binding [emission control] commitments after the first phase of the Protocol expires at the end of 2012."

This maintains the Kyoto Protocol's accounting rules, mechanisms, and markets that will help smooth the way for a global treaty on emission controls by 2015:

> "However, the sole international agreement to cut greenhouse gas emissions has been severely weakened. Canada withdrew from the treaty while Japan and Russia said they would not commit themselves beyond 2012. The United States remains aloof, while emerging economies like Brazil, China and India are exempt.
>
> "That leaves 34 industrialized countries—Australia, New Zealand and European nations—who have agreed a second commitment period of the Kyoto Protocol from January 1, 2013. Significantly, they will represent just 15 percent of global man-made CO_2 emissions. In the 1990s, Kyoto Protocol countries accounted for 33 percent of world CO_2 emissions.
>
> "So far, only European countries have pledged targets for the second commitment period, with the European Union reiterating its longstanding policy of reducing emissions by 20 percent by 2020 and by 30 percent if other major polluters commit to curbing their emissions.
>
> "Kyoto countries will also have to decide this year whether the Protocol's second phase should last until 2017 or until 2020—the longer the timeframe, the easier it is to hit targets, or just postpone action.
>
> "The question is: will the rest of the world, responsible for 85 percent of global CO_2 emissions, pay much attention? After all, it's debatable how effective the Kyoto Protocol has been. It has clearly failed to influence global emissions as they have reached record highs, driven upward overwhelmingly by non-Kyoto countries.
>
> "And it is far from certain that all Kyoto Protocol countries will meet their relatively modest targets for phase one. The most recent data from the UN Climate Change Secretariat, up to 2009, shows that performance has been very mixed.
>
> "France, Germany, Russia and the U.K. were comfortably ahead of target, Japan and Italy were not far off, while Australia and Canada were lagging way behind—Canada quit Kyoto largely because it has no chance of making its targets and wants to avoid the resulting penalties.
>
> "Since then, we've seen another economic downturn, the Eurozone crisis, and the Fukushima meltdown, all of which could, depending on the country, impact emissions positively or negatively during this period. Slowing economies tend to reduce emissions while replacing nuclear power production with imported gas or coal will increase emissions.

> "Critics say that the Kyoto Protocol is increasingly marginal, if not irrelevant, to the future trend of man-made greenhouse gas emissions and the pace of global warming. Many argue it is simply the wrong way to go about fighting the complexities of climate change and therefore other models of climate action, such as focusing on green technology investment, should be prioritized instead.
>
> "Supporters counter that Kyoto, despite recent history, can still set a positive example for sustainable development and green growth, and lay the groundwork for a global climate treaty.
>
> "Events in 2012 will help decide whether Kyoto Part II is merely an uninspiring rerun of the original or whether it is a more effective tool to combat climate change and a more persuasive call to action."

9.3 ECONOMICS: WILL IT COST MORE TO DO NOTHING?

9.3.1 The Stern Report

The Stern Report (CCSP, 2007) is a lengthy study (about 600 pages) on the economics of global climate change. The report presupposed that:

(1) drastic climate impacts to the world will occur if no action is taken;
(2) we have the means to take action to prevent climate change; and
(3) global warming is the biggest threat faced by humanity in the 21st and 22nd centuries that supersedes all other problems facing humanity.

Based on this, the economic study reached the conclusion that it will cost less to invest in climate control now, than to pay later for the problems that will be created by future climate change. In this regard, climate control is assumed to be achieved through draconian reductions in CO_2 emissions. The study was conducted as a cost–benefit analysis for the 200-year period ending around 2200 that assessed the likely cost of climate impacts in a BAU scenario without active control of GHG emissions, and compared this with the cost and benefits of various levels of control of GHGs.

However, the Stern Report is based on a number of scientific and technical assertions that at best are quite uncertain, and at worst, are simply wrong. The Stern Report estimated that the current CO_2 equivalent level ("CO_{2e}") is about 430 ppm (375 ppm of CO_2 + about 55 ppm equivalent from other GHGs). A fundamental basis of the Stern Report is stated as:

> "The level of 550 ppm CO_{2e} could be reached as early as 2035. At this level there is at least a 77% chance—and perhaps up to a 99% chance, depending on the climate model used—of a global average temperature rise exceeding 2°C. … On current trends, average global temperatures will rise by 2–3°C within the next fifty years or so."

At the rate of 2 ppm/yr, the likely CO_2 concentration would be 430 ppm and CO_{2e} would be about 490 ppm. The prediction of 550 ppm is exaggerated.

The Stern Report delineates the putative impacts of global warming in great detail: floods, reduced water supplies for one-sixth of the world's population, declining crop yields, leaving hundreds of millions without the ability to produce or purchase sufficient food, spread of disease, displacement of 200 million people by rising sea levels, extinction of 15% to 40% of species, destruction of the Amazon rainforest, etc. Various large-scale discontinuities are hypothesized. The Stern Report predicts a temperature rise of 3.9°C to 4.3°C by 2100 and 7.4°C to 8.6°C by 2200 in the BAU scenario, assuming that the world supply of fossil fuels is limitless, that there are no economic and social limitations to continued expansion of fossil fuel consumption, and that climate models are essentially correct. This is, of course, an absurd scenario because the world will take action long before such temperature increases are approached. There are fundamental flaws in the underlying basis for the Stern Report. A fundamental problem facing humanity in the 21st century is how to provide the burgeoning population with energy. Oil supplies, then gas, will gradually run down by mid-century. Coal is problematic for many reasons. There is no presently known technical fix for the energy problem, but we can be fairly sure of one thing: if there is a fix, it will have to involve ramping down use of fossil fuels as the century wears on—not so much because of fear of global warming, but because fossil fuel supplies are finite. Hence, the principal we face in the coming century is not global warming, but rather, providing the people of the world with affordable energy. In the short run, plans to urgently reduce carbon emissions will exacerbate the bigger problem of providing energy. Fossil fuels (and carbon emissions) must be viewed as a bridge to a new energy paradigm (as yet unknown), but premature curtailing of fossil fuel use can only bring on economic ruin. The Stern Report promulgates the false belief widely held by economists (but not by energy specialists) that technology exists for replacing fossil fuels by non-carbon-emitting technologies, and all we need to do is implement such technologies at an affordable cost.

Limitations on the availability of fossil fuels will prevent CO_{2e} from reaching extreme levels, and it is far from certain how large a temperature increase will result from increased CO_{2e}. In the past, an increase in CO_2 (not equivalent) of 100 ppm from the pre-industrial era to the early 21st century used up about one-third of original world fossil fuel resources. The temperature rise during that period was about 0.7°C, but only part (possibly a small part) of that rise can be attributed to GHGs. Why should we believe that further increases in CO_2 will produce a 2°C to 3°C temperature rise in 50 years, and an 8°C rise in 200 years?

According to the Stern Report:

> "… the poorest countries and people will suffer earliest and most. And if and when the damages appear, it will be too late to reverse the process. Thus we are forced to look a long way ahead. … Climate change may initially have small positive effects for a few developed countries, but is likely to be very damaging for the much higher temperature increases expected by mid-to late-century."

However, the Stern Report admits that "there is much to learn about these risks" but insists that "the temperatures that may result from unabated climate change will take the world outside the range of human experience". Since the human experience includes Ice Ages, this is certainly an exaggeration.

There are three problems with the impacts listed by the Stern Report: (1) we don't know what the future CO_{2e} will rise to, (2) we don't know what temperature changes will result from a putative increase in CO_{2e} and how they will vary regionally, and (3) the speculations on how any given temperature rise will affect the world economies are quite subjective.

The Stern Report carried out detailed academic exercises predicting future reductions in GDPs due to global warming. However, the Stern Report cautions:

> "Economic forecasting over just a few years is a difficult and imprecise task. The analysis of climate change requires, by its nature, that we look out over 50, 100, 200 years and more. Any such modeling requires caution and humility, and the results are specific to the model and its assumptions. They should not be endowed with a precision and certainty that is simply impossible to achieve. Further, some of the big uncertainties in the science and the economics concern the areas we know least about (for example, the impacts of very high temperatures), and for good reason—this is unknown territory."

But the Stern Report maintains that, despite these uncertainties:

> "The main message from these models is that when we try to take due account of the upside risks and uncertainties, the probability-weighted costs look very large. Much (but not all) of the risk can be reduced through a strong mitigation policy, and we argue that this can be achieved at a far lower cost than those calculated for the impacts. In this sense, mitigation is a highly productive investment."

Until recently, North America and Europe have produced around 70% of all the CO_2 emissions due to energy production, while developing countries have accounted for less than one-quarter. However, most future emissions growth will come from today's developing countries and China is now the world's leading emitter.

The Stern Report insists: "the world does not need to choose between averting climate change and promoting growth and development". It argues for "decarbonization" of energy technologies to achieve climate stabilization. This requires that annual emissions be brought down to the level that balances the Earth's natural capacity to remove GHGs from the atmosphere. The Stern Report indicates that the rate of CO_{2e} emission[6] in 2000 was about 41 Gt/yr (equivalent to $41/3.67 = 11.2$ Gt/yr of carbon). However, Figure 7.18 suggests a lower figure. The Stern Report postulates a number of future scenarios in which annual CO_{2e} emissions rise in the near term (as high as ~63 Gt/yr) and then fall back after

[6] It should be noted that CO_{2e} emissions include CO_2 and all other GHG emissions weighted by their relative effectiveness. The relation between CO_2 emission and carbon emission rate is that CO_2 emission is $(44/12) = 3.67$ times the carbon emission rate.

2040, dropping to about 20 Gt/yr by 2100. It indicates that if the peak CO_{2e} emission rate were less than 50 Gt/yr around 2030 to 2040, and it dropped to 30 Gt/yr in 2050 and 20 Gt/yr by 2100, stabilization of CO_{2e} at around 550 ppm might be achievable. However, these cuts will have to be made in the context of a world economy in 2050 that may require three to four times more energy than today, so the emissions per unit of GDP would need to be about 20% to 25% of current levels by 2050.

Note that the Stern Report deals with CO_2 emissions. One mass unit of carbon is equivalent to (44/12) = 3.67 mass units of carbon dioxide. Thus, when the Stern Report postulates a peak CO_2 emission rate of 50 Gt/yr, that is equivalent to 13.6 Gt/yr of carbon. The emission scenarios postulated by the Stern Report would more or less follow IS92a (Figure 7.18) out to 2030 to 2040 and then ramp down to about 5.4 Gt/yr by 2100.

Achieving deep cuts in emissions will have a cost. GHG emissions can be cut in four ways:

(1) reducing demand for emissions-intensive goods and services;
(2) increasing efficiency;
(3) action on non-energy emissions, such as avoiding deforestation;
(4) switching to lower carbon technologies for power, heat, and transport.

The Stern Report estimates the annual costs of stabilization at 550 ppm CO_{2e} to be around 1% of GDP by 2050 (3%–4% of present GDP), a level that is claimed to be "significant but manageable". It is assumed that this can be achieved technically while providing the world with adequate power and energy.

The Stern Report claims that, by 2050, energy efficiency has the potential to be the biggest single source of emissions savings in the energy sector. Non-energy emissions make up one-third of total GHG emission. A substantial body of evidence suggests that action to prevent further deforestation could be relatively cheap compared with other types of mitigation. Large-scale uptake of a range of clean power, heat, and transport technologies is required for radical emission cuts in the medium to long term. The power sector around the world would have to be decarbonized by at least 60%, and perhaps as much as 75%, by 2050 to stabilize at or below 550 ppm CO_{2e}. Deep cuts in the transport sector are likely to be more difficult in the shorter term, but will ultimately be needed.

The Stern Report says that the shift to a low-carbon global economy can take place despite "an abundant supply of fossil fuels". It says that stocks of hydrocarbons that are profitable to extract (under current policies) are more than enough to take the world to levels of GHG concentrations well beyond 750 ppm CO_{2e}, which is a great deal more optimistic than the estimates given in Figures 7.28 and 7.29. It is postulated that:

> "Even with very strong expansion of the use of renewable energy and other low-carbon energy sources, hydrocarbons may still make over half of global energy supply in 2050. Extensive carbon capture and storage would allow this continued use of fossil fuels without damage to the atmosphere."

Here it is assumed that carbon capture and storage are technically and economically feasible on a large scale. It is not clear in the Stern Report how the GDP impacts of this policy vary with country. Discussions of GDP seem to be limited to a world average. The argument seems to be that the consequences of BAU in the 21st century are disastrous for the whole world, and the cost of stabilization at say 550 ppm are less than the cost of global-warming impacts. To achieve such stabilization, the Stern Report provides hundreds of pages on policy responses for mitigation and adaptation, and international collective action (*à la* Kyoto). That would certainly be a boon for bureaucracy. The Stern Report appears to be based on unfounded expectations of disaster and the economic analyses of consequences are of dubious credibility. When impending shortages of fossil fuel are taken into account, it is likely that many of the proposed mitigations will take place through market forces without micromanagement by the U.N.

9.3.2 Nordhaus's review of the Stern Report

Nordhaus (2006) wrote a review of the Stern Report. On the positive side, Nordhaus said that (1) the Stern Report is an impressive document, (2) he suspects that the results are fundamentally correct in sign if not in size, and (3) the Stern Report argues correctly that it is critical to have a harmonized carbon tax or similar regulatory device both to provide incentives to individual firms and households, and to stimulate research and development in low-carbon technologies.

On the negative side, Nordhaus claimed that the Stern Report's use of a near-zero (0.1%) social discount rate is unfounded and leads to very unrealistic conclusions. Most of Nordhaus' review was concerned with this aspect.

The social discount rate has to do with social time preference; it measures the importance of the welfare of future generations relative to the present:

> "It is calculated in percent per year, like an interest rate, but refers to the discount in future utility or welfare, not future goods or dollars. A zero social discount rate means that future generations into the indefinite future are treated equally with present generations; a positive social discount rate means that the welfare of future generations is reduced or discounted compared to nearer generations. Philosophers and economists have conducted vigorous debates about how to apply social discount rates in areas as diverse as economic growth, climate change, energy policy, nuclear waste, major infrastructure programs such as levees, and reparations for slavery." (Anon. (K)

On a far smaller scale, each individual has a similar problem in deciding how much of their current income to sequester into 401K, 403b, and other tax shelter plans for future retirement vs. taking current income for current expenses in living. Achieving a balance between current needs and preparing for the future is a problem that confronts individuals, institutions, and governments. It is desirable to avoid the extremes of dying as the "richest man in the cemetery" vs. living in poverty in one's old age. As Nordhaus (and others) have discussed, there are many possible philosophies on how to value future gains in the present. Economists have their

Greek letters (*deltas* and *etas*) that they put into formulas, but it all comes down to the question of how to value the future against the present.

As we stand in 2013, the world economy is in desperate straits and there exists a notable shortage of income and jobs, with a tremendous excess of borrowing against the future. The approach used by world governments to deal with the credit crisis produced by excessive debt is to borrow more. Against this backdrop, the proposal to expend several percent of the world GDP per year to head off a hypothetical problem that might occur 100 years hence would exacerbate the world economic problem to the point where it might cause a world economic collapse.

On the other hand, if one really believes the threat of global warming is a blight on the lives of our grandchildren, the moral imperative would be to spend the money now to prevent future disaster, at whatever the cost to us in the present. In times of war, when we are fighting for survival, we spend what is necessary, regardless of the cost. But, can economists really be certain of the future? Should they take the views of alarmists verbatim? The answer seems to be that we must face the real problem of the 21st century: providing the people with energy, while reducing carbon emissions is a secondary goal. In the process, carbon emissions will gradually diminish, but not as quickly as the alarmists would prefer.

Nordhaus showed that the basis for the extreme economic impacts in the Stern Report rests on "selectively chosen studies that emphasized high damage estimates, some of which are highly speculative". More importantly, Nordhaus demonstrated that the Stern Report used considerable *legerdemain* in reporting losses from global warming. Nordhaus asked:

> "How do damages, which average around 5 percent of output over the next two centuries turn into a [claimed] 14.4 percent reduction in consumption now and forever? The answer lies in the way that near-zero discounting magnifies distant impacts. With near-zero discounting, the low damages in the next two centuries get overwhelmed by the long-term average over many centuries."

Nordhaus posed this illustration:

> "Suppose that scientists discover that a wrinkle in the climatic system will cause damages equal to 0.01 percent of output starting in 2200 and continuing at that rate thereafter. How large a one-time investment would be justified *today* to remove the wrinkle starting *after two centuries*? The answer is that a payment of 15 percent of world consumption today (approximately $7 trillion) would pass the Stern Review's cost–benefit test. This seems completely absurd. The bizarre result arises because the value of the future consumption stream is so high with near-zero discounting that we would trade off a large fraction of today's income to increase a far-future income stream by a very tiny fraction. ... Hence, the damage puzzle is resolved. The large [reported] damages from global warming reflect large and speculative damages in the far-distant future; the impacts now, as in today, are small; and ... the [proposed] 20 percent cut in consumption from global-warming might be reduced by an order of magnitude if alternative assumptions about discounting are used."

Nordhaus went on to say:

> "A further unattractive feature of the Review's near-zero social discount rate is that it puts present decisions on a hair-trigger in response to far-future contingencies. Under conventional discounting, contingencies many centuries ahead have a tiny weight in today's decisions. Decisions focus on the near future. With the Review's discounting procedure, by contrast, present decisions become extremely sensitive to uncertain events in the distant future."

Nordhaus provides further examples of absurd requirements for present investment to cover unlikely small impacts in the distant future if a zero social discount rate is used.

In his "summary verdict", Nordhaus raised the questions:

> "How much and how fast should the globe reduce greenhouse-gas emissions? How should nations balance the costs of the reductions against the damages and dangers of climate change?"

He pointed out that:

> "The Stern Review answers these questions clearly and unambiguously: we need urgent, sharp, and immediate reductions in greenhouse-gas emissions."

Nordhaus then mentioned that economists are always saying: "On the one hand this and on the other hand that." He noted that Harry Truman wanted a one-handed economist. Nordhaus concluded that:

> "The Stern Review is a Prime Minister's dream come true. It provides decisive and compelling answers instead of the dreaded conjectures, contingencies, and qualifications. However, a closer look reveals that there is indeed another hand to these answers. The radical revision of the economics of climate change proposed by the Review ... depends decisively on the assumption of a near-zero social discount rate. The Review's unambiguous conclusions about the need for extreme immediate action will not survive the substitution of discounting assumptions that are consistent with today's market place. So the central questions about global-warming policy—how much, how fast, and how costly remain open. The Review ... does not answer these fundamental questions."

9.3.3 Other critiques of the Stern Report

A number of other critiques of the Stern Report have been published, most notably that of Weitzman (2007). In response, there have been a number of defenses of the Stern Report by economists, most notably one by Cole. Quiggan (2006) also wrote a defense of the Stern Report. Quiggan (an economist) provides us with assurance that the credibility of the doubters regarding the connection between global warming and rising CO_2 "was on the verge of collapse" by 2006, due to the success of the documentary *An Inconvenient Truth*, and the fact that the "scientific controversies have now been resolved". In his view, the Stern Report:

"... outflanked the remaining skeptics. They could either continue denying the results of scientific analysis, or try to salvage the fallback position, undermined by the Stern Report, that although global warming is real, the costs of doing anything significant about it exceed the benefits, at least in the short term."

He then proceeded to fill about a dozen pages with economic hash, full of sound and fury, signifying very little. It is nice to hear from an economist that the "scientific controversies have now been resolved" but, considering the ineptitude of economists at their own trade, it would be better if they did not venture into unknown territories.

The "Dual Critique" (Part I: Science—Carter *et al.*, 2006 and Part II: Economics—Byatt *et al.*, 2006)) provided a serious independent assessment of the Stern Report. Part I began by taking issue with the quotation:

"... what is not in doubt is that the scientific evidence of global warming caused by greenhouse gas emissions is now overwhelming ... [and] ... that if the science is right, the consequences for our planet are literally disastrous ... what the Stern Review shows is how the economic benefits of strong early action easily outweigh any costs."

As Part I emphasized, the Stern Report:

"... presumes without question that moderate further increases in atmospheric CO_2 levels will give rise to major climatic changes and that these are likely to be seriously damaging; that the climatic changes observed over recent decades can be reliably blamed on emissions of 'greenhouse gases' in general, and CO_2 in particular; and that climate model projections and forecasts present a sufficiently accurate view of the future at relevant geographic and temporal scales to form a basis for major policy decision."

Part I argues against the conclusion expressed in the Stern Report that warming in the 20th century was unprecedented "for at least the last 1,000 years". This of course implies acceptance by the Stern Report of the *hockey stick* and all the claims of alarmists regarding the late 1900s and early 2000s being the hottest since the last Ice Age. Part I mentions the Wegman Report as contradicting this finding. But the Stern Report insists that the *hockey stick* is "only one of a number of lines of evidence", although these other lines of evidence are not presented.

The confidence expressed in the Stern Review appears to derive heavily from a single published paper (Stott, *et al.*, 2000) that utilizes a global climate model in an attempt to separate natural variations from those induced by human generation of CO_2. Unfortunately, the model predicts a much greater temperature rise in the 20th century than was observed and the modelers had to invoke a significant cooling due to aerosols *ad hoc* to reduce the heating produced by the model.

Part I goes on to discuss other aspects of the Stern Review's technical basis for alarmism, as well as the putative impacts of predicted global warming, much of which we have already discussed in this book.

Part II deals with economics issues. The first point that is made is that the lack of clarity in the Stern Report makes it difficult to determine precisely which procedures were used. Part II is divided into six elements: (1) economic impacts of global warming, (2) costs of mitigation, (3) discounting the future, (4) choice of policy instruments, (5) major omissions from the Stern Review, and (6) a summary and conclusions.

In regard to the economic impacts of global warming, Part II pointed out that 80% to 90% of the proposed impacts are subjective, being attributed to "non-market impacts" and "catastrophes with little further definition provided". These impacts are further amplified by the fact that BAU as defined in the Stern Report does not take into account economic pressures for conservation, and adoption of new technologies because they are profitable. Because the Stern Review deals with the long run, such factors will change even in a BAU scenario.

Just as the Stern Report is pessimistic in regard to costs of global warming, it is grossly optimistic in regard to the costs of mitigation. One topic discussed in Part II was revenue recycling in which "some emission pricing policies (taxes, auctioned permits) generate revenue for the government, and this added revenue could be used to finance a cut in other tax rates". Only economists could think that this is a benefit. Wealth is created by efficiently producing things. Capturing CO_2 and storing it provides no wealth and only adds to our costs for living. The fact that revenues are raised by taxing emissions merely transfers the tax burden from one group to another; it creates no wealth.

Part II also discusses the discount rate. However, their discussion is not as clear as that of Nordhaus.

A number of comments were published in response to the "Dual Critique". These were rebutted. The reader is directed to Volume 8 of *World Economics* (2007).

Pirilä provided an independent commentary on the Stern Review.[7] He said:

> "The Stern Review presents a quantitative cost-benefit analysis (CBA) comparing the discounted damages from climate change to the costs of mitigation. The costs of damages are very high, because they are summed over a very long period. Having a social discount rate of 0.1% corresponds for a typical time span of 1000 years. Changing that to 0.2% would halve the time span and changing it to 0.05% would double it. Choosing the value as 0.1% is totally arbitrary; there is certainly no good argument to support that value rather than any other between 0% and 0.5%. How sensitive the final cost estimates are to this value depends on the other choices done in the calculation. Emphasizing the worst outcomes in the spirit of risk aversion and making the uncertainties large with a possibility for very serious consequences turned out to be important in their analysis. Again this contained several parameters, which cannot be determined with any accuracy from our present knowledge. Other choices made in the calculation have received less emphasis, but they are equally important in determining the final results."

[7] *http://pirila.fi/energy/*.

Pirilä went on to say:

> "The analysis assumes that we can calculate the consequences of near-term decisions to very distant future. How impossible that is can be envisioned by thinking, how decisions of the 19th or 18th century influence the present. Some decisions of those periods may have a great influence, but how can we decide, what would be the alternative counter-factual history, and how much worse or better its results would have been."

Pirilä argued that the decisions we make, even very important ones such as going to war, typically only impact humanity for years to decades:

> "For many present issues a history of hundreds or thousands or years can be identified, but even in these cases the influence of any single original decision had hardly significant influence for long, when the comparison is done to potential counter-factual histories. It's mostly a fallacy to say that we are here with the present state of the world, because of some specific decision of distant history, and even in those very few cases, we don't know, how the alternative would be different.
>
> "It may be argued that influencing the global resources and the global environment is different. There is some truth in that, as fossil fuels can be burned only once and using them soon means less fossil fuels for the later future. Similarly carbon dioxide added to the atmosphere today disappears from there slowly."

Pirilä emphasized the importance of coal in the world's future. As oil, and eventually gas supplies wind down in the 21st century, expanded use of coal seems to be the only way to maintain industrial economies, particularly in view of prospects for expansion in China and India. The main alternatives of continued or expanded use of coal include widespread introduction of alternative energy production technologies (renewable, nuclear), or reduction of consumption via more efficient energy systems, and/or changes in consumption patterns. Alternatively, carbon sequestration may permit greater use of coal without great build-up of CO_2. Nevertheless, we don't know what limit on CO_2 concentration can be achieved in a practical sense while "offering ... social and economic development of emerging economies like China and India. It cannot present politically unrealistic requirements for the industrial economies either. The present optimism on possibilities expressed by many European states and organizations is not based on solid arguments."

In further comments on the Stern Review, Pirilä emphasized that the:

> "... conclusion of the Review was that the cost of climate change is very much higher than the cost of mitigation. The analysis was based on calculating the present value of the expected damages using a very low pure time preference of 0.1%/annum. The total discount rate was 1.4 % since the economic growth rate was assumed to be 1.3% and the elasticity of marginal utility was taken as one corresponding to a logarithmic utility function. ... The low discount rate

alone would not have led to the very high damage estimate without ... the possible damages including a non-negligible possibility of extreme scenarios. ... These damage scenarios dominate the expectation value, which was found to be 10% of the total discounted world wealth. The result is very sensitive to the selected parameters."

As Pirilä pointed out, a number of economists:

"... criticized the ... Stern Review claiming both that the properly discounted damages are much less and possibly also that the mitigation is likely to be much more costly than that estimated in Stern Review where the cost is estimated as 1–2% of world GDP. On the other hand, [others] that the damages may be even larger when the expected future development of relative prices is taken into account."

He concluded that:

"But is this the whole picture? Does this approach succeed in determining an answer that we should use to guide policy decisions? Not necessarily as all these analyses may be beside the point. They may all answer a wrong question, or at least we may have an alternative question of more immediate importance that has a little better hope of having well defined answers. The problem that the above analysis has, is trying to know too much about the future. The correct posterior answers known in the future may differ to a very substantial extent from the best estimates that we can produce right now. We should not forget that the future will be different from all scenarios that we can create now and that we can generate now extremely different plausible scenarios for the distant future—say 30 years from the present and beyond. We are really incapable of guessing, what the 22nd century will be like. Trying to calculate discounted sums over 200 or 300 years just doesn't make sense. This doesn't mean that we should forget intergenerational justice, but it means that we should come back to the Brundtland Commission definition of the sustainable development. Sustainable development is development that meets the needs of the present without compromising the ability of future generations to meet their own needs."

Judith Curry ran a posting on her website by Michael Cunningham entitled: "The costs of tackling or not tackling anthropogenic global warming".[8] This posting, in turn, was based on a review of the Stern Report written by U.K. MP Peter Lilley. Lilley provided a critical analysis of many aspects of the Stern Report.

One problem that Lilley did not seem to emphasize is that the consensus seems to believe that GHG concentrations of $\sim$550 ppm, which are permitted by the Stern analysis on the grounds that it is impractical to seek a smaller bound, would still be disastrous.

[8] *http://judithcurry.com/2012/09/12/the-costs-of-tackling-or-not-tackling-anthropogenic-global-warming/*.

9.3.4 Investment opportunities in climate change

With the passage of time, the literature on impacts of global warming has expanded considerably. Almost any academic can further his or her career by writing on the dangers and impacts of global warming, always adopting the view that such warming is entirely produced by GHGs, CO_2 in particular. The advent of global warming as an integrative theme for academia has provided new opportunities, new platforms for publication and, what is most important of all, new funding. A representative example (of many) is the 39-page document by the European Climate Forum (Anon. (P)).

The Internet is full of reports by organizations such as the *Wall Street Journal* and (the now defunct) Goldman-Sachs emphasizing the opportunities for investment in climate change.

The Deutsche Bank Group (2008) has been a leader in arguing for investment in mitigation of global warming (presuming that carbon emissions are the sole culprit). They said that "the growing investment opportunities in climate change ... [will] continue into the foreseeable future" and went on to say that:

> "In the energy sector alone, the International Energy Agency estimates that about $45 trillion will be needed to develop and deploy new, clean technologies between now and 2050. This represents nothing less than a low-carbon Industrial Revolution. Writers and policymakers from across the political and intellectual spectrum have recognized the potential this holds for long-term job growth and industry creation. The debate around climate change is shifting away from cost and risk towards the question of how to capitalize on exciting opportunities."

In their view, in seeking new investment capital to renew the world economy, they "believe that for investors, climate change has a built-in advantage over most other sectors. Its regulated markets hold the promise of enormous secular growth. In the long-term, *the earnings of companies and projects that are supported by governments for policy reasons are more trustworthy*." (emphasis added).

The Deutsche Bank distributed a 142-page report entitled "Investing in climate change 2011" dealing with how to profit from investments based on climate change policies.

Scientists recently discussed planetary engineering as a climate change mitigation strategy. Some of the proposals include:

- fertilizing the oceans with iron, as this is the limiting nutrient in some ocean areas, to encourage blooms of planktonic algae that draw carbon dioxide out of the atmosphere;
- recycling carbon dioxide from the atmosphere into fuel, by reacting it with hydrogen;
- ejecting carbon dioxide from the atmosphere at the Earth's poles, using the planet's magnetic field;
- reflecting sunlight back to outer space to cool the planet. This is proposed by increasing the amount of pollution in the atmosphere so that particles reflect sunlight back to space;

- spraying clouds with seawater, resulting in particles of salt forming through evaporation. These act as nuclei around which droplets of water can condense, creating more sunlight-reflecting clouds.

9.4 ECONOMIC ANALYSES

The first law of economics says that, if you have two economists, they have diametrically opposite opinions. The second law says they are both wrong.

Starting with the Stern Report, economists have taken over the business of estimating costs for implementing draconian reductions in carbon emissions. The MIT Joint Program on the Science and Policy of Global Change developed a series of funded studies on the economics of carbon emission reduction. One report of note is Paltsev *et al.* (2009). This report considered three scenarios: (1) constant emissions rate from 2008 to 2050 totaling 287 billion metric tons of CO_2-e (CO_2 equivalent in all GHGs), (2) linear reduction in emissions from 2008 to 2050 down to a 50% reduction in 2050 totaling 203 billion metric tons of CO_2-e, and (3) linear reduction in emissions from 2008 to 2050 down to an 80% reduction in 2050 totaling 167 billion metric tons of CO_2-e. It is noteworthy that they mention that, because of the current recession, they downgraded their estimates of future GNP, resulting in a reduction in emissions of 20% compounded over 40 years. This suggests that the best (and perhaps only) way to meet the 80% reduction target is to have a permanent recession. The "reference scenario" used by Paltsev *et al.* (2009) is in some ways a rosy picture. It assumes that with no policy at all, the annual U.S. emissions of GHGs in units of CO_2 equivalent will slowly rise between 2008 and 2050 from about 7 to 11 billion tons per year (similar to the 2012 BAU curve in Figure 7.18). Over this same period, the U.S. GNP is estimated to rise from $12 trillion to $37 trillion.

The problems for politicians in planning policy center around uncertainty. The MIT Group published a series of reports dealing with climate change policy under uncertainty (e.g., Webster *et al.*, 2008, 2009). They pointed out that:

> "Though the climate policy challenge is essentially one of risk management, requiring an understanding of uncertainty, most analyses of the emissions implications of these various policy targets have been deterministic, applying [specific] scenarios of emissions and reference (or at best median) values of parameters that represent aspects of the climate system response, and the cost of emissions control. ... These efforts provide insight to the nature of the human-climate relationship, but necessarily they fail to represent the effects of uncertainty in emissions, or to reflect the interacting uncertainties in the natural cycles of CO_2 and other gases or the response of the climate system to these gases."

While these authors utilized a range of possible future emission scenarios, they used specific model results for the ultimate concentrations of GHGs, forcings, and temperature increases resulting from any emission scenario. In other words, they accounted for uncertainty in the emission scenario, but not for uncertainty in the climate impact of such emissions. Their final results are sets of probability distributions for the global mean temperature rise from the average for 1981–2000 to the decadal average for 2091–2100. If there is no policy to restrict carbon emissions (BAU), they end up with a most probable temperature rise of 5°C and a range of about 3.5-7°C based on a most probable CO_2 concentration of about 900 ppm in the last decade of the 21st century. (Note that this is much higher than the 2012 BAU curve in Figure 7.19.) They postulate four possible levels of carbon emission control. The least stringent emission scenario leads to a most probable temperature rise of 3.2°C and a range of about 2°C to 5°C based on a most probable CO_2 concentration of about 700 ppm in the last decade of the 21st century, and the most stringent emission scenario leads to a most probable temperature rise of 1.8°C and a range of about 1°C to 3°C based on a most probable CO_2 concentration of about 480 ppm in the last decade of the 21st century.

Unfortunately, the definition of GNP used by economists includes almost any kind of activity; yet the thing that we are really interested in is activity that produces wealth. Wealth is produced by activity that efficiently produces goods and services that better the quality of life of the people. Consider the hypothetical case of employing 200,000 bean counters to monitor CO_2 emissions at factories and power plants across the country. The government would take credit for creating 200,000 jobs, and the GNP would value this activity at something roughly like $300,000 per person, adding up to $60 billion. But these people would produce nothing of value for the population and would not add to the wealth of the nation. On the contrary, they would be a drag on the wealth of the nation because taxes would be needed to raise the $60 billion needed to pay for the bean counters. Of course, the proponents of CO_2 reduction would argue that these activities would improve the quality of life by stemming global warming, and if the global climate models are correct, and if the alarmist estimates of impacts of global warming are correct, they would have a point. At the same time, the projections of increasing GNP over the next 40 years assure that, as the cost of reducing CO_2 emissions builds up, the cost will appear more moderate when written as a percentage of GNP. According to Paltsev *et al.* (2009) the "welfare change" associated with the 80% reduction scenario reaches about 2.5% for the decade 2040 to 2050. Multiplying by an average GNP of 40 trillion dollars during that period implies a cost of a trillion dollars a year. Paltsev *et al.* (2009) also estimated that, during the 2040 to 2050 decade, the cost of CO_2 removal is about $200/ton, which, when multiplied by ~70% of 10 billion tons per year, amounts to $1.4 trillion dollars per year. However, actual costs always seem to be much higher than those predicted by economists.

Prior to the late 1990s, one might have winced at the thought of taking on a charge of a trillion dollars a year. However, since then, trillions of dollars are acquired by borrowing. There is an economic nirvana. The U.S. government can spend as much as it pleases, and, when it does not have the tax revenues to cover these expenditures, it simply borrows. As long as there is a more or less permanent recession, the demand for money is low, interest rates are equally low, and the government can spend trillions more than it takes in. With the prospect of permanent recession facing us, this system should work for the foreseeable future.

Roe and Bauman (2011) described the possibility that the climate sensitivity might be at the upper end of predictions based on climate models as the "fat tail" of the probability distribution function for climate sensitivity (see the sharp rise at the right of Figure 6.36). They emphasized that recent economic analyses suggested that a "rational policy strategy" should be based on this worst-case scenario "if the damages associated with such high temperatures are large enough". Some have claimed that damages rise non-linearly with temperature and therefore the worst-case scenario should dictate government policy. Roe and Baker showed that the "fat tail" is a necessary consequence of uncertainty in parameters that enter into climate models. But Roe and Bauman were "skeptical of this approach" because they claimed that two factors prevent the high temperatures in the fat tail from being reached for many centuries. One factor is:

> "... the enormous thermal inertia represented by the deep ocean. The whole climate system cannot reach a new equilibrium until the deep ocean has also reached equilibrium. In response to a positive climate forcing (i.e., a warming tendency), the deep ocean draws heat away from the surface ocean, and so buffers the surface temperature changes, making them less than they would otherwise be. The deep ocean is capable of absorbing enormous amounts of heat, and not until this reservoir has been exhausted can the surface temperatures attain their full, equilibrium values."

As Roe and Bauman pointed out, "a second key [factor] is the inherent relationship between feedbacks and adjustment time scales in physical systems". If the climate sensitivity is high, that is because there are strong positive feedbacks:

> "A positive feedback reflects a tendency to retain energy within the system, inhibiting its ultimate emission to space, and therefore requiring a larger temperature response in order to achieve energy equilibrium. Moreover, it is generally true that, all else being equal, an inefficient system takes longer to adjust than an efficient one. A useful rule-of-thumb is that the relevant response time of the climate system is given by the effective thermal inertia of the deep ocean multiplied by the climate sensitivity parameter. ... As time progresses, more and more of the ocean abyssal waters become involved in the warming, and so the effective thermal inertia of the climate system increases. Hansen *et al.* (1985) solved a simple representation of this effect and showed that the adjustment time of climate is proportional to the square of climate sensitivity. In other words, if it takes 50 yrs to equilibrate with a climate sensitivity of 1.5°C, it would take 100 times longer, or 5,000 yrs to equilibrate if the climate sensitivity is 15°C."

Roe and Bauman (2011) then launched into a discussion of the economics of future damage from global warming based on Weitzman (2009 and 2011), taking into account the point that the higher the ultimate temperature rise from GHGs, the longer it will take to reach that temperature. The Weitzman papers are framed in mathematical expressions for economic models, but there are three unknowns for the next few hundred years: (1) the future rise in temperatures due to GHGs, (2) the impacts of this rise on human welfare, and (3) the economic health of the world.

Since none of these can be estimated with any reliability, all the economic mathematics in the world cannot help us make policy decisions. Humans always face the challenge of how much to spend now to provide for the future but the problem here is the future is far out, and the need for various degrees of remediation is highly uncertain.

9.5 RENEWABLE ENERGY

As fossil fuel resources dwindle down in the 21st century, it is desirable to rely to a greater extent on renewable energy sources (solar, wind, biomass, etc.). At the same time, use of such resources will reduce CO_2 emissions, which are likely to provide benefits in reducing the rate of climate change. Federal and state governments have provided funding for development of these resources as well as policies to provide financial incentives for the public to use these resources while they are more expensive than conventional energy. Since these energy sources are intermittent, it remains unclear as to what percentage of our total energy consumption can ultimately be supplied by renewable energy. As in the case of reducing fossil fuel emissions of GHGs, governments act by setting quotas for the future, regardless of the technical feasibility of achieving such quotas.

Krey and Clarke (2011) discussed the potential the role of renewable energy in climate change mitigation by reviewing 162 recent medium- to long-term scenarios from 15 large-scale, energy-economic, and integrated assessment models (see Figure 7.20). It seems clear that significant increases in the use of renewable energy are required to reduce CO_2 emissions but, as they pointed out:

> "One cannot say with certainty today whether a future heavily reliant on renewable energy will be extraordinarily costly or whether the costs will only be modest. The scenarios in this study demonstrate no meaningful correlation between carbon prices and renewable energy production. Indeed, this sort of variability in indicators of mitigation cost is common in multi-model scenario analyses."

They also concluded that the high level of uncertainty that pervades future scenarios is "unsatisfactory" and that unpacking this uncertainty is a "very challenging task".

The IPCC (2011) published a 1,544-page report[9] on the relationship between expanded use of renewable resources and climate change mitigation that was heavily based on the scenarios reviewed by Krey and Clarke (2011). One important parameter of interest is the percentage of total world energy consumed that is generated by renewable energy. Despite the 1,544 pages in the IPCC Report, very

[9] Intergovernmental Panel on Climate Change, Working Group 111, Mitigation of Climate Change, Special Report on Renewable Energy Sources and Climate Mitigation, June, 2011, *www.ipcc.ch/news_and_events/docs/ipcc33/SRREN_FD_SPM_final.pdf*.

little mention of this parameter is provided. However, it is briefly mentioned in passing in Chapter 10 of the report that:

> "... the global primary energy supply share of renewable energy (RE) differs substantially among the scenarios. More than half of the scenarios show a contribution of RE in excess of a 17% share of primary energy supply in 2030, rising to more than 27% in 2050. The scenarios with the highest RE shares reach approximately 43% in 2030 and 77% in 2050. RE deployment levels in 2100 are substantially larger than these, reflecting continued growth throughout the century."

In order to estimate the percent of total energy supplied by RE, one must first estimate the total energy consumption. According to IPCC (2011), the various scenarios for future world energy use range from 450 to 1,000 EJ in year 2050, and 500 to 1,500 EJ in year 2100.[10] There is also a wide variation of estimates of RE corresponding to each estimate of total energy. Hence, predicting future RE percentage amounts to "any number can play". It is interesting that the IPCC issued a press release prior to releasing the IPCC (2011) report. The press release claimed (amongst other things) that:

> "Close to 80 percent of the world's energy supply could be met by renewables by mid-century (2050) if backed by the right enabling public policies a new report shows."

McIntyre[11] pointed out that this was highly misleading. As he said:

> "The report does NOT show that 'close to 80 percent of the world's energy supply could be met by renewables by mid-century if backed by the right enabling public policies'. It does list a scenario from Greenpeace in which 77% of world energy is supplied by renewables, but the report itself did not conduct any independent assessment of the validity of the Greenpeace scenario and did not 'show' that the claim in the press release was true."

McIntyre also pointed out that the scenario that led to 77% renewable energy by 2050 was the most extremely optimistic of 164 different scenarios. The least optimistic scenario provides for 15% RE by 2050, up from 12.9% in 2008 (see Figure 9.1). A large number of blog enthusiasts responded to McIntyre's report, most of which added little to McIntyre's presentation.

Trainer (2011) wrote a critical review of the 2011 *IPCC Report on Renewable Energy*. The IPCC Report purported to confirm that:

> ",,, the widespread belief that renewable energy can replace about 80% of fossil fuels and more or less meet world energy demand by 2050. It is more than 1000 pages long, has 38 lead authors and input ... from over 120 leading experts from all over the world ..., reports on 164 studies, and digests 24,766 comments

[10] 1 exajoule (EJ) = 10^{18} J.

[11] *http://climateaudit.org/2011/06/16/responses-from-ipcc-srren/#comments*.

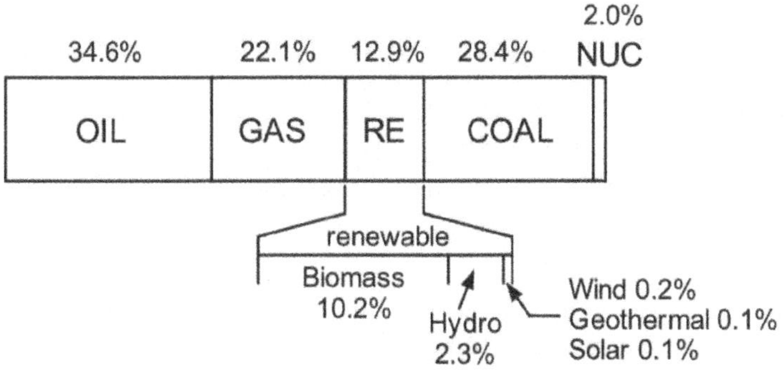

Figure 9.1. Distribution of sources of world energy use in 2008.

... from more than 350 expert reviewers and government and international authorities".

Trainer's points included:

(1) *No case is made.* "The report does not show that renewable sources can meet future energy demand, or a large fraction of it. ... It does not attempt to show what proportion of demand could be met by renewables. It presents much evidence relevant to the issue, but this is not put together into a case that sets out reasoning leading to the conclusion that the necessary quantities could be provided, how they could be provided, and that the difficulties could be overcome. The report merely presents the results of some studies that state conclusions about renewable energy's potential, without attempting to assess their worth."

(2) *No critical review.* "There is no critical examination of the 164 studies [upon which it is based]. There is [not even] a list of the studies enabling their examination. ... In other words the IPCC has not carried out an evaluation of literature in the field; it has only summarized the conclusions of (a select number of) studies, with no apparent effort to check on their validity."

(3) *The case for biomass is greatly overestimated.* Trainer provided a variety of cogent reasons why the potential for biomass energy is far lower than the Report indicates.

(4) *Variability, winter, peak demand, etc.* Trainer pointed out that the Report glosses over problems of intermittency or "the crucial problem of meeting demand in mid winter". The report utilizes "annual and/or average demand and supply, whereas what matters much more are the figures for maximum demand, e.g., peak quantities, when they coincide with minimum renewable resource availability". As a result of this mismatch, particularly in winter, reserve backup capacity will be very large, adding to costs. There will be occasional long gaps of several days in a row with little or no wind and sun. On the other hand, there will be considerable overcapacity during some periods when sun and wind are available in quantity.

(5) *Integration limits.* Trainer estimated realistically that, as an extreme upper limit, "wind plus PV [photovoltaics] might contribute at best only 55% of electricity, i.e., only 14% of all energy. The Report does not deal with the question of from which sources the other 86% is to come, apart from biomass". The Report makes it appear as if viable energy storage technology is "just around the corner" whereas, in reality, energy storage remains a major hurdle for renewable energy.

Trainer also provided additional arguments regarding overly optimistic cost estimates and several other aspects.

In dealing with beliefs regarding the related questions of whether rising CO_2 is a very serious problem, and the extent to which renewable energy can replace fossil fuels over the next 40 years (or more), we can form analogous categories:

(1) You think that rising CO_2 is a very serious problem, and renewable energy can replace ~80% of fossil fuels by 40 years from now.
(2) You think that rising CO_2 is a very serious problem, but renewable energy cannot replace more than ~20% of fossil fuels by 40 years from now.
(3) You think that rising CO_2 is not a very serious problem; nevertheless, renewable energy can replace ~80% of fossil fuels by 40 years from now.
(4) You think that rising CO_2 is not a very serious problem, and renewable energy cannot replace more than ~20% of fossil fuels by 40 years from now.

Most of us probably fit into one category or another. The IPCC and the whole community of alarmists tend to be of Type (1). Most of those who think that CO_2 is a major problem also think that renewable energy provides the solution (they are Type (1)). Most of those who doubt that CO_2 is a major problem also doubt that renewable energy could solve the problem even if it were a problem (they are Type (4)). I am a hybrid between Type (2) and Type (4). I am not sure about the danger presented by CO_2 but I am pretty sure that renewable energy could not solve the problem if the alarmists are right.

It is interesting that Trainer evidently subscribes to the extreme alarmist view that rising CO_2 poses an extreme danger to mankind, and that immediate draconian reductions in CO_2 emissions are needed. He is also an advocate for renewable energy development. However, unlike most advocates for renewable energy (who are Type (1)) he analyzed the potential for renewable energy in detail, and he concluded that it is quite limited—far less than that needed to replace most of the fossil energy we expect to need by 2050. Hence, as he said:

> "So what's the solution? The point is that there isn't one. . . . Global problems are basically due to the commitment to grossly unsustainable levels of consumption and to limitless economic growth. The problems cannot be solved on the supply side, i.e., by trying to provide the quantities of energy that a consumer-capitalist society for 10 billion would require. That kind of society is generating other major problems in addition to energy and climate, including the poverty of billions, the destruction of the ecosystems of the planet, resource conflicts, and deteriorating social cohesion. These problems cannot be solved

unless there is vast and radical transition to a *Simpler Way* of some kind. This IPCC WG3 Report reinforces the dominant faith that there is no need to think about this perspective on our global situation."

In his book, he describes the *Simpler Way*:

- far less affluent living standards;
- very different economic system based on minimal production and consumption with a much lower GDP and no growth;
- shift in values away from competition, individualism, and acquisition to frugality, self-sufficiency, cooperation, participation, and non-material satisfaction.

It appears that Trainer's review of the IPCC Report is on target. Certainly, we should push renewable energy for all it is worth but, as he says, it is not worth nearly as much as enthusiasts claim. Rising CO_2 over the next few decades may create some problems, but none that is insuperable. We'll probably muddle through, although continuing economic problems will plague us. Indeed, they are already upon us to a limited degree. Around 2040–2050, it will get worse. With rising CO_2, population growth, worldwide industrialization, and declining fossil fuel availability, the world is in for some severe dislocations. While renewable energy will gradually take on a greater share of our energy consumption, Trainer is probably right: there is no solution on the horizon.

9.6 PIRILÄ'S SUMMARY

This section is based on various essays written by Pekka Pirilä.[12] Pirilä wrote:

> "The world is facing two related developments: running out of high quality resources of oil and gas, and adding CO_2 to the atmosphere. The most straightforward answer to the first problem, switching gradually to coal and perhaps shale oil, is at the same time the worst path concerning the second problem. The oil crisis of 1970's led to a rapid increase in funding of alternative energy technologies. It's certainly possible to find some signs of success from that research, but in my general judgment it has produced pitifully little. We have learned that it's very difficult to develop solutions that have a major effect on the scale of the overall energy use. Some of the most promising alternatives, most notably new technological alternatives of nuclear energy production, have been out of favor, but mostly the lack of success reflects rather the difficulty of the task than lack of trying."

According to Pirilä, "the economics of wind power has improved slowly, and has led to competiveness under [the] most favorable conditions". However, on a worldwide basis, "wind power cannot produce enough to solve more than a small

[12] *http://pirila.fi/energy/*.

fraction of the energy [requirements]" and "solar electricity is significantly behind". Regions with high wind or high solar availability "could perhaps reach acceptable economics fairly soon", especially when "combined with a load peaking". However, "expanding from that to acceptable costs on a wider basis appears still to be a very distant goal". He also says that the net environmental of biofuels is "questionable", and he raises the question of whether government support "has given any positive results at all, or whether the money spent is used to deteriorate also the state of the environment".

What, then, should we do about energy policy? As Pirilä said, "Pointing out failures of present policies is not enough, but neither is it useful to continue these policies in spite of their failures". He argued that:

> "We know far too little on, how different incentives actually influence decision-making. The lack of proper knowledge leads at least in Europe continuously to choosing policies that seem nice and make us feel that we have done something rather than policies that influence efficiently the future development leading to the results envisioned and avoiding unnecessary and excessive costs or collateral damage to environment and social structures. … Presently I do not see better alternatives than emphasizing research on a wide range of alternatives, and also research on policies for efficient advancement of technological change and on economic incentives of environmental and climate policies."

Finally, he concluded that "Making wise decisions is extremely difficult, because both the severity of the threat and the value of the proposed policies are highly uncertain. An accurate uncontroversial quantitative cost–benefit analysis is certainly beyond our capabilities".

10

Final remarks

10.1 CONCLUSIONS

As we pointed out at the beginning of Chapter 2, *weather* represents the short-term variability of temperature, wind, precipitation, humidity, cloudiness, etc. and *climate* represents the long-term variability of average weather. How long a period is needed to differentiate between weather fluctuations and climate change is not clear. While the popular folklore says 30 years is the dividing line between weather and climate, strong arguments can be made that this should be extended to at least 50 years, and perhaps as much as 100 years. It is not clear how to define the climate over any time period by averaging the weather at myriad points. Indeed, the whole concept that there even is a global climate may not have much meaning, except in discussing the extremes such as Ice Ages and hothouse Earth. Only for such large changes do worldwide parameters tend to conform to a considerable degree. For small variations in global climate (less than about 1°C), the dividing line between weather variability and climate variability may be obscure. Characterization of the climate of the Earth by a single average temperature appears to have a low signal-to-noise ratio except for glacial–interglacial cycles.

The climate of the Earth has undergone large changes over long time periods. For example, as we discussed in Section 6.5, there is evidence that the Earth was much warmer than today during much of the Phanerozoic Eon (the past 540 million years). Some estimates indicate that the global average temperature may have been up to about 10°C warmer than today. The poles were free of ice. However, there were extreme glacial periods during this Eon as well. In seeking a cause for this, variability of CO_2 concentration has been a prime suspect. There is evidence that the CO_2 concentration was 10 to 20 times higher than today's concentration when temperatures were highest during the Phanerozoic Eon, and CO_2 concentrations plunged during the period from around 330 to 280 million years ago when the world was heavily glacial. However, attempts to correlate CO_2 with climate across the entire Phanerozoic Eon are difficult to resolve because of noisy data.

As we discussed in Section 6.5, the climate of the Earth cooled over the past

D. Rapp, *Assessing Climate Change*, Springer Praxis Books,
DOI: 10.1007/978-3-319-00455-6_10

50 million years, and decreasing CO_2 is widely suspected as being at heavily involved as a cause, although the data remain fuzzy.

Starting about 2.7 million years ago, the climate of the Earth began to oscillate between glacial and interglacial periods. About one million years ago, the variations took the form of long, extended Ice Ages interspersed by relatively short interglacial periods. The last Ice Age peaked around 20,000 YBP and the current interglacial (Holocene) commenced about 10,000 YBP. As the global average temperature rose and fell in these Ice Age—interglacial transitions, CO_2 concentrations rose and fell in unison. However, there is no known physical reason why the CO_2 concentration should have gone through such undulations independently; hence the changing CO_2 concentrations in this instance are effects of temperature change rather than causes, although changing CO_2 concentrations would have acted in a secondary role to amplify temperature changes due to the primary cause. The prevailing theory is that quasi-periodic variations in the Earth's orbit about the Sun induce changes in solar input to higher latitudes, which produces the initial forcing that causes these climate changes. A variety of feedback effects amplify this forcing (Rapp, 2012).

As Figure 6.8 shows, variable CO_2 concentrations exert a forcing of the climate. Other factors being equal, higher CO_2 concentrations will produce higher temperatures, and vice versa. The question is how much higher, especially since other factors are seldom equal.

Climatologists have attempted to characterize the Earth's climate during the past century or so, and during the past millennium or two. This is important in:

(1) establishing trends that may have evolved in the past century or so from industrialization and urbanization of the Earth; and
(2) defining the range of past fluctuations in climate prior to industrialization as a conjectural baseline of expected variations, independently of human intervention.

If it were possible to unambiguously extract (1) and (2) from the data, anthropogenic-induced changes could then be compared with expected fluctuations from natural causes. If such anthropogenic changes were found to be considerably greater in frequency and amplitude than natural fluctuations, that would provide an some evidence for the belief that human impacts on climate are real.

One important factor is the question of how temperatures on the Earth have varied over the past century (or more in some cases) as measured by monitoring stations dispersed around the world. While the scientists who process such data have made affirmative claims for the reliability of these data, the fact remains that the network for monitoring world temperatures suffers from a number of maladies including uneven spatial and temporal representation of large areas, poor maintenance and recording at many stations, effects of urban heating and land use of stations, and uncertainties in the measurement of sea temperatures (see Chapter 3). In addition, the number of stations reporting data decreased sharply earlier in the 20th century. Climatologists typically work in terms of a single global average temperature, or hemispheric average temperatures. In order to estimate how such averages have varied with time during the past century or so, a space–time grid is typically created from station temperature data. Unfortunately, the sparseness of spatial and temporal coverage creates considerable uncertainty in these averages—more so than the purveyors will typically admit to. Furthermore, such averages (1) have little physical or thermodynamic significance, (2) tend to average out regional variations that convey more incisive information, and (3) can be constructed in various ways to lead to different results. Nevertheless, it seems evident that the global average temperature has increased by roughly 0.8°C over the past 120 years. While this warming has been far from uniform, and one-third of measurements stations reported an increase in temperature over this interval, there is no doubt that, on balance, the Earth is warmer today than it was 120 years ago. But 120 years ago, the Earth was emerging from the Little Ice Age (LIA). There is considerable evidence that the Earth began pulling out of the LIA in the 19th century, well before the build-up of CO_2 concentrations in the second half of the 20th century. For example, mountain glaciers typically began their retreat well before the build-up of CO_2 in the atmosphere in the late 20th century.

There is also evidence that the temperature rise of the 20th century occurred in two steps: one from 1900 to about 1940, and the other from about 1976 to the present. The initial rise occurred prior to massive build-up of CO_2 in the atmosphere and deposition of black carbon on Arctic snow and ice was likely to be one significant contributor. Warming was greatest by far in the Arctic region. The dip from 1940 to 1976 has been attributed to build-up of aerosols in the atmosphere. The sharp rise in temperature after 1976 coincides with a change in the Pacific Ocean in which the usual upwelling of deep cold waters seems to have suddenly diminished (see Section 3.4.4). In addition, the regions of greatest temperature rise appear to be near the greatest concentration of urban centers. There has been some divergence of opinion regarding the importance of the urban heating effect on measured temperatures. However, recent measurements indicate that this effect was much larger than previously thought (see Section 3.1.3).

In attempting to extend temperature estimates backward in time prior to the advent of surface measurements, climatologists have relied on a variety of temperature proxies that leave remnants from an earlier time that were formed via a temperature-dependent process, from which (in principle) the temperatures at the times of formation can be extracted. Unfortunately, all proxies are subject to confounding influences of one type or another that add noise to the signal and introduce considerable uncertainty in the veracity of the derived temperatures. In the past decade, a number of climatologists attempted to integrate the results of a large number of proxies with varying spatial and temporal coverage in an effort to extract a historical global average (or hemispheric average) temperature over the past millennium or two. The accuracy of such procedures is limited by the sparse spatial and temporal coverage provided by the proxies, especially for dates prior to about 1600 (see Table 2.1). However, an even more serious problem with these models has emerged. The use of principal components analysis to reveal the principal trends in the integrated data was not carried out properly, and the resultant temperature profile derived from these models had the characteristic *hockey stick* form in which the temperature of the Earth was estimated to be almost flat with little variation for

1,000 years (or more) prior to the 20th century, followed by a steep rise in the 20th century. This result was interpreted to mean that the human intervention of the 20th century produced an unprecedented, alarming rate of temperature rise, and this led to a number of extravagant claims that the last few years of the 20th century were the hottest in at least several millennia, with some making even wilder claims that they were the hottest years dating back millions of years. This became a rallying point for global climate alarmists and the *hockey stick* has been widely promulgated by the U.N., Al Gore, and numerous other organizations, institutions and governments. McIntyre and McKitrick (M&M), Wegman, and others pointed out this error, but the climatologists ignored the criticism and stubbornly defended their incorrect procedures. A close-knit group of co-proposing climatologists control manuscript publication in journals, and this *paleoclimatological cabal* has managed to exclude the major criticisms from the journals and, as a result, criticisms have been mainly relegated to blogs.

Even if the statistical processing of the data did not contain this serious error, there is another fundamental problem with summing up large numbers of proxies. As Section 2.4 shows, the sum of multiple noisy proxies is noise. If one adds up noisy proxies for 1,000 years one obtains 1,000 years of noise. When an exaggerated version of the measured temperature rise in the 20th century is added to a flat long-term profile from noisy proxy data, the inevitable result is the *hockey stick*.

There is anecdotal and proxy evidence to suggest that a relatively warm Medieval Warn Period (MWP) occurred in the time frame of about 900–1000 and a relatively cold LIA occurred from perhaps 1600 to 1850. However, the magnitude and extent of these climate fluctuations remain uncertain.

Thus, we have a controversy. If the *hockey stick* picture were correct, that would provide support for the belief that human intervention has drastically altered a pattern of a flat temperature profile for 1,000 years (or more). However, we know that the *hockey stick* result is based on inadequate data and incorrect data processing. On the other hand, if the fluctuations in the MWP and the LIA were of comparable magnitude with the temperature rise of the 20th century, one might infer that the current warming is within the range of natural fluctuations. However, we don't have enough data to pin down the temperature changes in the MWP and the LIA. We do know that the temperatures around 1880 represented the final vestiges of the LIA, and, since the temperature around 1880 was about 0.8°C lower than today, we can be fairly sure that, at the depth of the LIA in the 18th century, temperatures were at least 1°C cooler than they are today. The fact that the LIA was colder than the present climate may reflect unusual cold during the LIA rather than unusual warmth today. The real issue is how current temperatures compare with those in the MWP, for this would provide insight into how unusual the current warmth is. Unfortunately, the data are not accurate enough to answer this question.

Data taken over the past 50 years show that the CO_2 concentration has risen continuously and now well exceeds the levels found in ice cores from past interglacial periods. The pre-industrial level appears to be about 280 ppm, and it is presently about 400 ppm. Ice core data suggest that the rise in CO_2 began in the late 1800s, at about the time that temperatures started rising. The coincidence between the CO_2

increase and the temperature rise has suggested that the rise in CO_2 (together with other greenhouse gases) produced the observed rise in temperatures via the greenhouse gas effect. In order to examine this hypothesis, another genre of climatologists developed global climate models in which the Earth is mathematically divided into cells that interact with nearest neighbors and evolve in time according to equations that attempt to represent all the physical processes that take place. The goal of these models has typically been to estimate the rise in global average temperature that will occur due to a future doubling of CO_2 concentration from the pre-industrial level of 280 ppm, although some models have also considered higher CO_2 concentrations. The basic greenhouse heating effect of CO_2 is limited. Typical models predict that a doubling of CO_2 would produce (by itself, without feedback effects) a temperature rise of perhaps 1°C or slightly higher. However, the warming that results from the CO_2 greenhouse effect would increase evaporation of water, and the increased water vapor content in the atmosphere would tend to amplify the warming via a water vapor greenhouse effect. Some models predict that this would add 2°C, bringing the total rise to about 3°C. However, the modeling of the entire water vapor cycle is primitive. As Lindzen and co-workers have shown, the effect of an increase in global average water vapor can have vastly different impacts on global temperature, depending on the regional distribution of the increase in water vapor. Furthermore, the effect of clouds and aerosols are poorly understood. Because of uncertainties, different models predict a net temperature rise from a doubling of CO_2 that vary by a factor of 3 from one model to another. Unfortunately, it is difficult to test these models against actual data. Some investigators have attempted to run their models backward in time to see whether they reproduce the known temperature variations of the 20th century. In this respect, they had to cope with the fact that warming has not been monotonic, and there was a temperature dip from about 1940 to 1978 (see Section 3.4.5). It seems likely that increased CO_2 produces some global heating, but how much remains very uncertain.

A number of modelers have modeled future growth of CO_2 concentration in the 21st century and used their climate models to thereby estimate the future temperature increase in the 21st century. Many of these assumed future emission rates of CO_2 in the business-as-usual (BAU) mode that are far beyond what seems to be possible from estimated fossil fuel resources. Even the least aggressive of these use CO_2 emissions in the 21st century in the BAU mode that are unlikely because of fossil fuel limitations, likely controls on greenhouse gas emissions, improvements in energy efficiency, and the advent of renewable energy resources. Equally important is the fact that, even if projections of future CO_2 emissions are correct, the climate models lack fundamental understanding of many factors, particularly effects of changing humidity, clouds, ocean currents, aerosols, dust, black carbon, etc. As a result, the climate models are not trustworthy.

A major source of uncertainty in understanding the Earth's climate, past and future, is the possibility of variation in the irradiance emitted by the Sun. Solar irradiance can only be measured above the Earth's atmosphere and we only have data since 1980. Nevertheless, we do know quite a bit about the Sun. We have observations of the solar cycle and the relevance of sunspots and other surface markings to this cycle. Sunspot data go back several hundred years. Our data indicate that sunspot activity has varied considerably over the past 200 years, and we have anecdotal data to suggest that sunspots disappeared altogether from about 1645 to 1715. A number of investigators have attempted to model past solar irradiance based on sunspot indices, length of the solar cycle, comparison with Sun-like stars, and other phenomena. None of these is very credible because we only have absolute measurements in space for the past 30 years, and even these are somewhat equivocal. As a result, we simply do not know how much the solar irradiance has varied in the past or how much it might vary in the future. This adds further uncertainty to climate models.

The thing that distinguishes climatology from some other sciences is that the data are almost always sparse and noisy. Whereas climate by its very nature requires long-term data (typically ≫ 100 years) to infer trends and relationship between cause and effect, the data are usually too short in duration to be adequate for this purpose. While climatologists have invented many creative proxies for past data, the calibration periods are almost always short and require lengthy extrapolations into past realms that might differ in some important ways from the calibration period. Furthermore, in many cases, the agreement between proxy and data during the calibration period is not impressive. This might be why the calibration curves are rarely included in publications. The thing that distinguishes climatologists from some other scientists is that climatologists draw firm conclusions, make bold assertions and go beyond the realm of peer-reviewed publications with multiple press releases and media blitzes to promulgate these conclusions and assertions, all based on flimsy data with inadequate duration and excess noise.

John R. Christy, a noted climatologist, published a commentary in the November 1, 2007, *Wall Street Journal* in which he said:

> "It is my turn to cringe when I hear overstated-confidence from those who describe the projected evolution of global weather patterns over the next 100 years, especially when I consider how difficult it is to accurately predict that system's behavior over the next five days. Mother Nature simply operates at a level of complexity that is, at this point, beyond the mastery of … the tools available to us. … I see jump-to-conclusions advocates and, unfortunately, some scientists who see in every weather anomaly the specter of a global-warming apocalypse. Explaining each successive phenomenon as a result of human action gives them comfort and an easy answer. Others of us scratch our heads and try to understand the real causes behind what we see. We discount the possibility that everything is caused by human actions, because everything we've seen the climate do has happened before."

The neat thing about predicting the climate 100 years from now is that no one can prove you wrong! In the field of climatology, a little humility would go a long way. However, humility does not go with the culture.

Martin (1979) wrote an interesting report in which he described the biases that inevitably creep into scientific research and reporting. According to Martin, scientists "do not disinterestedly look at the available evidence, do not make a balanced analysis, and do not present results in a neutral manner". Instead, he suggested that "from the beginning [they] support or favor a particular conclusion, and in a number of ways organize their scientific work so as to selectively support this conclusion". He labels this as "pushing the argument". He went on to say:

> "A scientist in developing an argument to support an hypothesis draws evidence from a number of sources. In presenting evidence one must always be selective—all the evidence and arguments cannot be presented. Often different authorities support different viewpoints, present different 'facts', and offer different interpretations of evidence. Depending on the field, a scientist may draw sound support for many points of view and find some support for nearly any view. Therefore it is easy for a scientist, knowingly or unknowingly, to push an argument by selective choice and use of available evidence."

Most of Martin's treatise was framed in terms of the debate during the early 1970s as to whether emissions from high-flying supersonic transports (SSTs) would destroy the ozone layer and thereby endanger the Earth's population by exposure to excessive radiation. For purposes of discussion, he presented detailed summaries and reviews of two prime scientific papers in the field with contrasting approaches. One paper was said to contain "the built-in assumption that the burden of proof lies with those who claim that SSTs are safe: that all that he must demonstrate is that there is at least some small possibility of danger". By contrast, the other paper used "the [implied] assumption that the burden of proof lies with those who claim that SSTs are dangerous to ozone: that all [they needed to] demonstrate was that the likelihood of significant danger was small".

Above all, Martin emphasized that scientists are motivated by various forces and factors in their lives. According to Martin:

> "People tend to selectively observe and interpret information in a way that supports their preconceived ideas. Because of this, the personal commitments of individual scientists can help to explain the link between the scientists' presuppositions and their pushing of arguments. ... In a scientist, this process might operate as follows. The scientist starts with an original idea or hypothesis, perhaps arrived at as a creative solution to a certain problem. In testing or validating the idea, the scientist will tend to notice and use supporting evidence and arguments. Data that seems mainly supportive will be studied, analyzed and applied so that every possible advantage can be drawn from it. Seemingly irrelevant or inconclusive items will be filtered from advantageous components, or interpreted in a way that promotes the argument. Evidence that seems mainly to contradict or challenge the argument at hand may be ignored completely or explained away or reinterpreted and twisted into support for the argument.
>
> "Some of the ways in which a person may deal with a challenging item of information are (1) flat denial of the item; (2) skepticism about the source of the item; (3) ascription of a motive to the source of the item; (4) isolation of the item from the context of one's attitude; (5) minimization of the importance of the

item; (6) interpretation of the item to suit one's purpose; (7) misunderstanding of the item; and (8) thinking away or just forgetting the item."

According to Martin, one may often detect a deep-rooted personal commitment or bias of a scientist by examining a series of published papers and detecting a constancy of attitude that repeats itself from year to year. He argued:

> "The idea that scientists are often strongly committed or biased is quite compatible with the fact that scientists are human beings. ... That is, they are subject to motivations and failings similar to those of other people. They may strive for money, power and prestige; they may work for the satisfaction of a job well done or for revenge or to relieve boredom; they may make terrible blunders as well as have brilliant insights. It is sometimes said or suggested that scientists, at least when it comes to their work, live on a higher moral plane than other mortals. Don't believe it!"

In examining the literature on SST emissions and their impact on the ozone layer, Martin concluded:

> "From my point of view, the authors do not disinterestedly look at the available evidence, do not make a balanced analysis, and do not present results in a neutral manner. Rather, it appears to me that the authors from the beginning support or favor a particular conclusion, and in a number of ways organize their scientific work so as to selectively support this conclusion."

These claims made by Martin (1979) are backed up by lengthy and detailed discussions and analyses that seem quite credible to this writer.

In the 30 years that have passed since Martin wrote this report, several major changes have taken place in the way that scientific information is distributed. With the advent of the Internet, the monopoly of scientific journals has been weakened. Other cultural changes have taken place. Of some relevance is the fact that scientists are now far more prone to issue press releases on their work prior to publication, and these tend to find their way onto many websites. Other scientists, disagreeing with the orthodoxy of an established consensus, have difficulty getting published in the journals. A number of so-called web blogs dealing with climate change have emerged over the past several years, and these have become foci for discussions and commentary. Most blogs are rabidly one-sided and present forums for either alarmists or skeptics to agree with one another. Any moron can voice his or her opinion. Two blogs that stand out above the others are *climateaudit.org*, which has become a universal watchdog for reviewing statistical analysis of large data sets, and *judithcurry.com* which provides even-handed intelligent postings on climate. Unfortunately, the responses by the public on these blogs have become so numerous that the small amount of wheat gets lost in the large amount of chaff.

Starting in the 1990s, and building up with time, concern has grown amongst many climatologists that putative global warming produced by greenhouse gas emissions presents a grave danger to humanity. With the advent of Al Gore's film *An Inconvenient Truth* in 2006, this concern has been conveyed to the public as well as politicians. Concern about human-induced climate change has escalated to become one of the major defining issues of our time. Governments are contemplating policies for extreme reduction of carbon emissions that are likely to cost many trillions of dollars and could produce global economic woe. Climatology, which used to be a minor science that was widely ignored, became thrust into the limelight. Prominent climatologists responded by issuing many repeated warnings of impending disaster, and now receive much attention and adulation (and funding) as a result. It appears that many climatologists routinely bias their results in such a way as to exaggerate the threat of rising CO_2. Others outside the field of climatology strive to relate almost any distant phenomenon to greenhouse gases—often to an absurd degree. The in-group has operated in a heavy-handed way to shut out opposing views and avoid criticism of their own work. Funding for climate research and analysis has become the goose that laid a golden egg, and, human nature being what it is, many have succumbed to the temptation to seek a share of the lucre.

The subject of global climate change seems to have bifurcated into two groups, each opposed to one another, each absolutely certain that they are correct, and each predisposed to interpret everything from a one-sided viewpoint. One group, the alarmists, believe that insidious global warming will cause great havoc and suffering in the 21st century if we don't move quickly and draconically to suppress greenhouse gas emissions. The great majority of climatologists are alarmists. Unfortunately, their insistence on use of the *hockey stick* result, their extravagant predictions of high future CO_2 emissions, their excessive predictions of catastrophe, and their inordinate belief in climate models, provides the opposition with considerable ammunition for attack. Many alarmists are motivated by high ideals; they want to protect the planet. But it appears that the underpinnings of many are based on a similar situation that occurs in earthquake science with their mantra: "the big one is coming". If the climate is not a threat, who would want to fund their work? The opposition, the skeptics, tend to be equally one-sided in the opposite direction.

The Inter-governmental Panel on Climate Change (IPCC) has taken an alarmist position. In its reports, it has devoted 1,000 pages to descriptions of the negative impacts of climate change. Much of this is grossly exaggerated. The only tangible possibility for a serious consequence in the next 100 years is a rise in sea level. Nevertheless, the consequence of the IPCC activities has led to the Kyoto Protocol in which the countries of the world have banded together to constrain future greenhouse gas emissions. Unfortunately, the protocol is severely flawed in its uneven treatment of developed vs. developing countries, its unrealistic requirements in the short run, and its creation of a world market for emission rights. Quite properly, the U.S. has refused to sign it.

Curry and Webster[1] wrote a review of the notion of consensus in science, with particular regard to climate change. The report by Curry and Webster follows two paths. One is the "consensus findings" of the IPCC regarding the role of greenhouse

[1] Curry, J.A., and Webster, P.J. (2012), "Climate change: no consensus on consensus", *http://judithcurry.com/2012/10/28/climate-change-no-consensus-on-consensus/*.

gases on warming over the past century, and the other is a review of philosophical views of consensus in science. They reviewed a number of philosophical questions regarding the role of consensus in science. They concluded that:

> "Arguing from consensus to enforce conclusions does not work with the extended peer community. What is needed are serious attempts to engage the extended peer community with the modes of expert reasoning used to reach those conclusions."

There seems to be quite a bit of confusion in the world of climate science as to what the consensus is consenting to. For example, Curry and Webster focused on a IPCC conclusion that warming in the 20th century was primarily caused by anthropogenic generation of greenhouse gases. There might be considerable variability in the extent of widespread consensus on several beliefs as shown below (degree of consensus shown in brackets: 5 = greatest consensus; 1 = least consensus):

(1) The Earth's climate would have been steady and constant, were it not for the impact of anthropogenic activity on the climate system. [4]
(2) The climate of the mid-19th century was ideal. [3]
(3) The global average of Earth temperatures rose over the past ~ 120 years. [5]
(4) Rising concentrations of greenhouse gases warm the Earth's climate. [5]
(5) Since the climate warmed over the past ~ 120 years, it must have been due to emissions of greenhouse gases by human activity. [4]
(6) Continued emission of greenhouse gases in the future will lead to further warming. [5]
(7) Global climate models, while not perfect, are good enough to predict future global temperature rise due to future increases in greenhouse gas concentrations. [2]
(8) The impacts of future temperature rise at any level are well understood. [3]
(9) The impacts of future temperature rise will be disastrous to the world unless we immediately drastically reduce the world rate of carbon emissions. [3]
(10) We can immediately drastically reduce the world rate of carbon emissions by a combination of introducing green energy, sequestration, and energy conservation at a more rapid rate, without severely impacting the world economy. [1]

There is no accurate way to estimate the degree of consensus on each of these relevant issues. Subjectively, my impression from reading papers, blogs, various press releases, and conversations with individuals is as follows. On a scale from 1 to 5, where 5 is the most widespread consensus of belief, and 1 is the least widespread consensus of belief, my impression is as given in brackets in the above list.

Let us consider these issues one by one.

(1) *The Earth's climate would have been steady and constant, were it not for the impact of anthropogenic activity on the climate system.* [4] There seems to be a fairly widespread belief in this postulate that the Earth's climate is steady unless acted upon by an outside influence. This is parallel to Newton's law of

motion that a body remains at rest or in uniform motion unless acted upon by an outside force. This does not allow for changes in the Sun–Earth relationship that can produce major climate changes due to innate variability of the Sun, as well as variations in the Earth's orbit relative to the Sun. A problem with this belief is that the Earth is a very complex system with many feedback effects. Weather, ocean currents, and other aspects of the Earth system can vary widely, triggering feedbacks that produce longer-term variations (i.e., internally generated climate change). The historical record shows that, over hundreds of thousands of years we have had Ice Ages and interglacials without human influence. More to the point, we have had small but significant climate fluctuations within our current interglacial period over the past couple of millennia prior to large-scale human influence. The alarmists have tried unsuccessfully to dispute this via the *hockey stick* picture of millennial climates. This widely held belief that the climate would not change were it not for the impact of anthropogenic activity is probably justified over periods of one generation (30 years) but, over longer periods of time, it does not necessarily hold up. Over very long periods of time, it is patently false.

(2) *The climate of the mid-19th century was ideal.* [3] I don't think that most people have really thought about this very much and the degree to which there might be consensus on this point is uncertain. But there seems to be widespread concern that we are presently warmer than they were the mid-19th century, which seems to imply that they wish we were back in the midst of the LIA. We can either argue that it is warmer now, or it was colder then. The majority seem to favor that we are warmer now. But, actually, a strong case can be made that averaged over the whole world population, the 2013 climate is more benign than the 1850 climate.

(3) *The global average of Earth temperatures rose over the past ~120 years.* [5] This is very widely believed and it is a matter of fact. However, what is not widely known is that this temperature rise took place in two steps: one prior to build-up of CO_2 and one after. In addition, one-third of land measurement stations recorded a decrease in temperature over this time span, so this increase was not uniform geographically or in time.

(4) *Rising concentrations of greenhouse gases warm the Earth's climate.* [5] This is very widely believed and it is a matter of fact. However, the quantitative relationship between CO_2 concentration and Earth temperature remains elusive.

(5) *Since the climate warmed over the past ~120 years, it must have been due to emissions of greenhouse gases by human activity.* [4] This is a corollary to *The Earth's climate would have been steady and constant, were it not for the impact of anthropogenic activity on the climate system.* It is a fairly widespread belief. But it is too digital. In retrospect, it seems likely that rising greenhouse gas concentrations contributed to warming over the past 120 years, but it also seems likely that natural fluctuations in climate (e.g., pulling out of the LIA) may have contributed as well. The influence of changes in the Pacific Ocean might be very significant. The degree to which various factors that influenced climate change is unknown.

(6) *Continued emission of greenhouse gases in the future will lead to further warming.* [5] This is very widely believed and it is almost certainly correct. However, the quantitative relationship between CO_2 concentration and Earth temperature remains elusive.
(7) *Global climate models, while not perfect, are good enough to predict future global temperature rise due to future increases in greenhouse gas concentrations.* [2] I am under the impression that this is not widely believed, although those with a religious fervor of alarmism seem to believe it almost as a biblical pronouncement. However, it is probably widely believed that there is a reasonable chance that the models could be in the right ballpark, and the models at least provide us with a worst-case scenario. But, 99% of believers know nothing about climate models, and place their trust in the claims of the climatologists.
(8) *The impacts of future temperature rise at any level are well understood.* [3] I am under the impression that this is believed to the extent of perhaps 50%. However, it is probably widely believed that the analyses at least provide us with a worst-case scenario.
(9) *The impacts of future temperature rise will be disastrous to the world unless we immediately drastically reduce the world rate of carbon emissions.* [3] I am under the impression that this is believed to the extent of perhaps 50%. But, 99% of believers know nothing about impact studies, and place their trust in the claims of the climatologists.
(10) *We can immediately drastically reduce the world rate of carbon emissions by a combination of introducing green energy, sequestration and energy conservation at a more rapid rate, without severely impacting the world economy.* [1] I think that this is only believed by a limited number of rabid "greenies". I am under the impression that the overwhelming majority wish that this were so, but are very uncertain of its practicality.

10.2 THE NINE QUESTIONS

In the Foreword of this book, nine essential questions were raised. Here, I provide short summary responses to these questions.

(1) How well has the world monitored near-surface temperatures of the 30% land and 70% ocean areas on the Earth during the past 120 years or more, and how well can we characterize the changes in climate over that time span?

✓ While the purveyors of these temperature data networks seem to be well satisfied with their data, others have pointed out their inadequacies. Station data are suspect because, at many sites, there were (1) changes in the observing time during the station's history, (2) changes in instrumentation, (3) station relocations, (4) poor siting due to nearby reflectors or shades, and (5) bias caused by urbanization. More importantly, spatial and temporal coverage over many regions was very sparse. In recent years, new techniques have been added (space observation of the troposphere, ocean temperature network), but their histories have been too short to provide needed long-term trends. Nevertheless, we can safely conclude that the global average Earth surface temperature rose by roughly 0.7°C over that time period, with most warming at far northern latitudes and a much weaker warming in the tropics and Southern Hemisphere.

(2) What is the utility and significance of a single global average temperature?

✓ Use of a single average global or hemispheric temperature tends to cancel out regional variations, and one ends up with only small apparent net changes. Yet, the single average global or hemispheric temperature is derived from local and regional measurements, which provide much better insight into the climate changes occurring on the Earth. The utility of such a global average temperature is limited and the Earth's climate is better described in terms of regional variations.

(3) How has the Earth's climate varied over the past Ice Ages, the Holocene, the last millennium, and the past century, and what can we infer about the "natural" variability of the climate prior to industrialization by humankind?

✓ The Earth's climate has gone through wild gyrations from cold to warm over the past few hundred thousand years. The Holocene (last ~ 10,000 years) has primarily been a relatively benign period with much smaller fluctuations. The last millennium experienced an MWP and an LIA, but the extent and amplitude of these variations are difficult to resolve. In the last century, we have emerged from the LIA, and global temperatures have risen by about 0.8°C. However, this rise has not been uniform spatially, and there have been notable downturns imbedded in this overall uptrend. We remain unable to distinguish the current trend as clearly distinct from natural fluctuations, although it is likely that the greenhouse effect contributed to some extent.

(4) How reliable are proxies for historical temperatures? What do we really know about past temperature variations? Is the hockey stick version of millennium temperatures credible, in which temperatures were relatively flat for 2,000 years prior to a sudden rise in the 20th century?

✓ There are many challenges in using proxies to infer historical temperatures. Proxies suffer from a variety of ills. In combining many proxies to generate historical global average temperatures, principal component analysis has been misapplied, leading to an invalid *hockey stick* result. Climatologists attempted to cover up] the failure of proxies late in the 20th century by "hiding the decline". We simply do not know how much the temperature has varied quantitatively over the past 1,000 years, but we do know that it was relatively warm around 1000 (MWP) and relatively cold from about 1600 to 1880 (LIA).

(5) How does the current global-warming trend compare with past fluctuations in the Earth's climate, and what is the likelihood that the warming trend we are experiencing now is primarily just another in a series of natural climate fluctuations as opposed to a direct result of human production of greenhouse gases?

✓ We don't know whether the current warming trend is greater than that experienced in the MWP or further back in time, in various interglacial periods. Most of the warming occurred in high latitudes of the Northern Hemisphere. The acceleration in warming that occurred after 1976 may be tied to the change in the Southern Oscillation Index that occurred at that time, but it is unclear which is the cause and which is the effect. The strong correlation between temperatures and El Niño–Southern Oscillation (ENSO) indices since 1980 suggests that global warming since 1980 was driven more by oceanic effects than by the greenhouse effect. Nevertheless, despite all these arguments, it seems likely that rising greenhouse gas concentrations contributed to rising temperatures in the 20th century.

(6) How credible are the global climate models that claim that greenhouse gases produced most of the temperature rise of the 20th century, and forecast much greater impacts in the 21st century?

✓ Global climate models suffer from a number of maladies and uncertainties, particularly the uncertain treatment of water vapor, clouds, and aerosols, and uncertainty about variations in solar irradiance over multi-decade time periods. The wide range in results testifies to the lack of precision in these models. The models are heavily focused on greenhouse gas effects, and they all (to a greater or lesser degree) claim that the temperature rise of the past century is mainly due to greenhouse gas effects. However, about half of that rise was early in the century, prior to major greenhouse gas build-up. As Lindzen has pointed out, the treatment of water vapor by climate models is highly suspect. The climate models will undoubtedly improve in the future, but, right now, they are not nearly as credible as their developers claim.

(7) How good were the "good old days?" Was the climate of the Little Ice Age ideal, should we abhor warming from that baseline, and do we want to return to the climate of the 19th century?

✓ Without necessarily realizing it, most alarmists implicitly assume that the climate of the mid-19th century was ideal, and regard the warming of the 20th century as highly detrimental. However, there are numerous references in the literature to climatic hardships from the cold of the 19th century.[2] Balling, Vose, and Weber (1998) found that the coldest period in Europe since 1751 was around 1890. But this is just the period that most climatologists use as a baseline period against which to compare global warming in the 20th century. As Balling, Vose, and Weber (1998) said:

> "... it is entirely possible that the warming in the record of the past century has been caused by an unusually cool period 100 years ago as opposed to an unusually warm period in recent decades."

[2] For example, *http://kclibrary.nhmccd.edu/19thcentury1820.htm*: "The bitterly cold winters of 1825–26 and 1826–27 caused great hardship in that country [Germany] and motivated many Germans to leave their homeland." Hardships endured due to the cold climate of the LIA are documented in *Climate of Fear*, by Thomas Gale Moore, Cato Institute, 1998.

(8) How will limits on fossil energy supplies constrain future CO_2 production and climate change, even if the climate models are accurate?

✓ Most temperature projections for the 21st century using climate models with the BAU assumption assume an annual rate of production of CO_2 that is very unlikely to be sustainable far into the century as fossil fuel resources fall short of energy demand by a growing world population that desires a more modern technological existence. While future fossil fuel resources remain debatable, the preponderance of evidence suggests that, as world energy demand grows (by at least a factor of 5 by 2100), fossil fuel resources will not be able to keep up with demand.

(9) How can the world provide itself with the energy needed for a burgeoning population that will demand more and more energy in the future, considering the finite limits on fossil fuel resources?

✓ No one knows the answer to this question, but it is clear that the world has not even begun to face up to this serious problem in any significant way. The prospects for renewable energy remain uncertain, and nuclear energy introduces many problems. Lack of availability of water is a growing problem that rivals energy. All of this is driven by the impending rise in population and industrialization in developing countries. With projections of world energy demand rising by at least a factor of 5 by 2100, providing the people of the world looms as a far greater challenge than dealing with presumed global warming.

10.3 ABOUT CLIMATOLOGY AND CLIMATOLOGISTS

The field of climatology, by its nature, deals with phenomena and data spread across the globe in vastly different environments. Much of these data are complex, and it is difficult to resolve cause–effect relationships because of so many confusing cross-factors. Furthermore, climate data need to be long-term, typically 100 years or more, and most direct data are far shorter. Proxies are beset with a variety of problems. In this situation, it has come to pass that climatologists tend to draw definitive conclusions from this checkerboard of inadequate, noisy data. A subtle compact has evolved whereby climatologists accept the results of one another, regardless of how flimsy the support base might be. Members of the club preserve this outward aura of scientific rigor, which in most cases is not justified. With the advent of climate alarmism over the past couple of decades, climatologists seem to have taken on a more strident tone. Assertions today are presented as fact regardless of the weak technical underpinnings. This has created a great market for selling climate research, and climatologists have banded together to minimize and snuff out those who would raise the specter of "the emperor has no clothes".

One of the conclusions I have drawn in this book is that, in many instances, we simply do not know enough about climate change, past and future, to draw firm conclusions. I have referred to Wunsch several times, who emphasized that, sometimes we might have to await more and better data. But we find that in science in general, and in climatology in particular, scientists are prone to hypothesizing scenarios from sparse data, and these hypotheses often gel into a consensus that gradually becomes established as an orthodoxy. Those of the alarmist persuasion argue that, despite uncertainties in predictions of future climate change, we know enough to be sure that we must immediately take action toward reducing greenhouse gas emissions.

For example, according to Anon. (F):

> "Special interest groups and policymakers opposed to legislative action to reduce human emissions of CO_2 and other greenhouse gases often cite 'uncertainty' in climate change science to justify their position. While there is much uncertainty in climate science (and there always will be), many researchers in the field insist that this uncertainty does not justify the lack of a policy response. In fact, scientists know a great deal about climate change, and there is a strong scientific consensus that the Earth is warming significantly, primarily due to human activities."

What is interesting here is that the International Center for Technology Assessment (ICTA) has posed the issue as a confrontation between "the special interests" or "advocates for industries that produce the bulk of greenhouse gas emissions" on the one hand, and idealistic "researchers" on the other. Yet, many of those of the alarmist persuasion have a special interest in the alarmist view (funding, recognition, positions) while many of those who emphasize uncertainty are independent, unfunded researchers. Furthermore, emissions control with cap and trade is likely to become a multi-trillion-dollar business.

For example, the Deutsche Bank Group (2008) has been a leader in arguing for investment in mitigation of global warming (presuming that carbon emissions are the sole culprit) and they said that "the growing investment opportunities in climate change ... [will] continue into the foreseeable future". They went on to say that:

> "In the energy sector alone, the International Energy Agency estimates that about $45 trillion will be needed to develop and deploy new, clean technologies between now and 2050. This represents nothing less than a low-carbon Industrial Revolution. Writers and policymakers from across the political and intellectual spectrum have recognized the potential this holds for long-term job growth and industry creation. The debate around climate change is shifting away from cost and risk towards the question of how to capitalize on exciting opportunities."

In their view, in seeking new investment capital to renew the world economy, they:

> "... believe that for investors, climate change has a built-in advantage over most other sectors. Its regulated markets hold the promise of enormous secular growth. In the long-term, *the earnings of companies and projects that are supported by governments for policy reasons are more trustworthy*." (Emphasis added)

According to Goldman–Sachs:

> "Global warming and the steps to prevent it are increasingly on the minds of policymakers, pundits, executives and investors. 'Mitigation' and 'adaptation' are the watchwords of the Stern Report. Both offer significant opportunities for research and *development, for investment and for financial innovation.*" (Emphasis added.)

In addition, the ugly head of "consensus" rears itself once again in the ICTA argument.

While the ICTA said that "Unfortunately, some special interests opposed to addressing the problem of climate change have picked up on the wait-to-learn concept and have begun trumpeting the uncertainties in climate models and projections". Yet, almost all of the "trumpeting" one hears today is from Al Gore, James Hansen, the IPCC, and many others of the alarmist persuasion.

The ICTA argues that "many of the uncertainties about future climate change concern the rate at which people will emit greenhouse gases in the coming decades". Yet, that is the least of the uncertainties about future climate change. We do have reasonable projections for future world population growth. We can reasonably assume that the world will continue to industrialize in the 21st century, requiring increased energy, even with new conservation technologies and practices. While we cannot accurately predict the future usage of nuclear and renewable energy, we can be fairly certain that coal will remain an important, probably growing energy source, while oil and gas will very slowly fade downward in usage. These assumptions are not ironclad. The state of the world economy and many other factors will affect ultimate energy consumption. But the uncertainty in future projections of energy usage are miniscule compared with projections of future temperature rise putatively caused by energy usage, and even smaller compared with projections of the global economic impacts of future temperature rise.

The ICTA also places incredible faith in global climate models, claiming that "they can accurately simulate the climate over the past century", which is simply not true. Furthermore, the ICTA (like many of the alarmist persuasion) equates "human activity" with greenhouse effects and does not consider deposition of black carbon on snow and ice, which appears to be another important effect of "human activity".

The ICTA then goes on to suggest various ways that climatologists can provide convincing arguments that, even though some uncertainty remains, there is enough known to make probabilistic predictions about future climate change. In doing this, they rely on the spectrum of various predictions by climate models that span a wide range of future scenarios, none of which may be correct. The problem with all this is that the uncertainties inherent in the estimates of uncertainty are large enough to vitiate the whole procedure. As the saying goes, "you can't make a silk purse out of a sow's ear".

Ha-Duong *et al.* (2006) discussed the evolution of viewpoints on dealing with uncertainty in the IPCC. In early reports, the IPCC found it useful to adopt the phrase "the balance of evidence suggests . . ." and, in later reports, "predicting" was replaced by "projecting" climatic changes on the basis of a set of scenarios. These

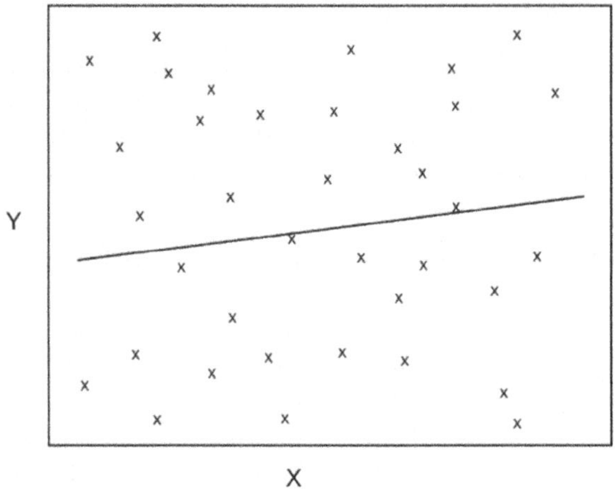

Figure 10.1. Cartoon showing derivation of trend line from noisy data.

authors distinguish between what they call "risk" and "uncertainty". They define "risk" as an uncertain outcome that can be categorized by quantitative probabilities of specific outcomes. Thus, if outcome A is 30% probable and outcome B is 70% probable, one does not know specifically which outcome will occur, but the risk can be estimated fairly precisely. They define "uncertainty" as a situation where information is too imprecise to meaningfully use a probability distribution. In addition, there is "ignorance", where "we don't know what we don't know". Evidently, the *alarmist cabal* believes that they can estimate risk probabilities but the naysayers think otherwise.

About 10 years ago, as a relatively new entrant in the field of climatology, I was immediately struck by the following characteristics of this field that differed from other scientific fields I have worked in:

(1) In most cases, data are very sparse, noisy, and difficult to unravel. This is particularly true in regard to climate proxies. Measured data span too short a period to be convincing.
(2) Systems are highly complex and subject to frequent chaotic meandering, making it difficult to perceive underlying trends.
(3) Models require data in order to set parameters. Data are typically grossly inadequate. For example, as the Earth warms, how do clouds change, and do they produce heating or cooling? And is there a repeatable pattern to this?

Despite these difficulties, climatologists have forged ahead with ingenious methods to extract information from murky sources. Nevertheless, there are usually significant uncertainties, and, where the data are lacking, climatologists tend to make unsupportable assumptions, or just guess. Figure 10.1 shows a cartoon of how a typical climatologist would interpret data. In determining the dependence of some variable Y on X, the data are as shown. A least-squares fit to the data results in the trend line as shown. Now, a typical climatologist might assert Y depends on X

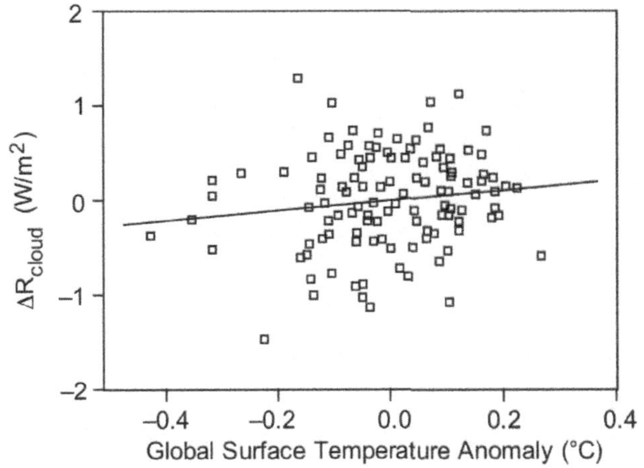

Figure 10.2. Measured cloud feedback vs. Earth surface temperature (Dessler, 2010).

exactly as the trend line. Yet *Y* might depend on several other variables, and its dependence on *X* might not be discernible from data! It is very likely that *Y* hardly depends on *X* at all.

As a result, much of the substance of climatology is built upon very shaky foundations. Yet these tenuous hypotheses quickly gel and harden into orthodoxies. We hear repeatedly that "the debate is over". But the debate isn't over, the facts are not in, and the processes are not well understood.

A case in point is the vital issue of cloud feedback. When the Earth warms due to increases in greenhouse gas concentrations, does this warming produce deterministic changes in cloud cover that produce a feedback, and is the feedback positive (amplifying the greenhouse warming) or negative (opposing the greenhouse warming)? McIntyre discusses the debate on this issue at some length.[3] The effect of a change in greenhouse gas concentration is expressed as an equivalent forcing at the top of the atmosphere in W/m^2. Similarly, the feedback from cloud cover changes is also expressed as an equivalent forcing ΔR_{cloud} also in W/m^2. The debate between Spencer and Braswell (2010, 2011) and Dessler (2010) centers about whether ΔR_{cloud} is positive or negative. The raw data provided by Dessler (2010) are shown in Figure 10.2. Unfortunately, as Dessler admits, these global cloud feedback data were taken "in response to short term climate fluctuations" where "the primary source of climate variations [was] the El Niño–Southern Oscillation (ENSO)". The data suffer from two lacks: (1) the data are extremely short-term, and (2) the data are not based on greenhouse gas warming. While Dessler derived the straight-line fit shown in the figure, indicating a weak positive feedback, it seems evident that cloud cover was not driven by Earth surface temperature at all, and varied chaotically during this short period due to unknown factors.

As McIntyre, pointed out, an argument could be made that there is a time lag of

[3] *http://climateaudit.org/2011/09/06/the-stone-in-trenberths-shoe/.*

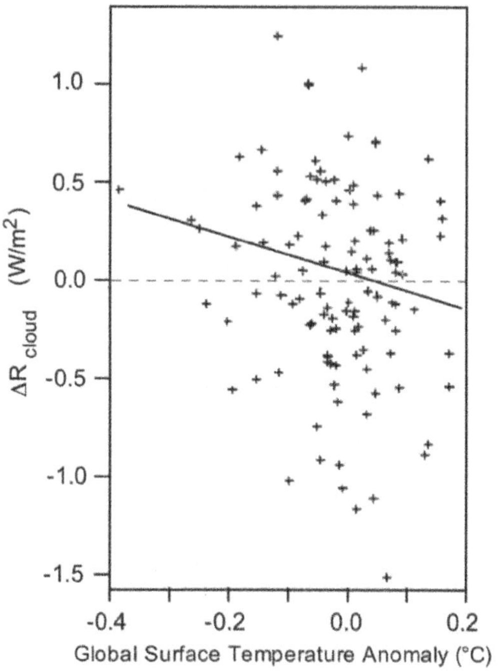

Figure 10.3. Cloud feedback plot with four-month time lag. (*http://climateaudit.org/2011/09/06/the-stone-in-trenberths-shoe/*).

several months between a change in temperature and a change in cloud cover, so he replotted the data with a four-month time lag and obtained Figure 10.3. While McIntyre now obtained a negative feedback, the scatter in the data is even greater than before. Only a climatologist or a statistician could believe there is any information contained in this plot.

Studies indicate that there is a huge statistical inversion between the personalities of climate scientists vs. the public in that climate scientists greatly lean toward intuition whereas the public heavily leans toward sensing. This implies the climate scientists "focus on theories" and "follow hunches to reach conclusions" whereas the public tends to "focus on experience" and "build carefully and logically towards conclusions". Climate scientists tend to "prefer to make decisions quickly, come to closure and move on". This is clearly evident in the many papers in climatology that utilize a penny's worth of data to draw a dollar's worth of conclusions.

10.4 LOGICAL FALLACIES USED IN CLIMATE DISCUSSIONS

In discussing criticisms of the hypothesis that increased greenhouse gases have increased the frequency of intense tropical storms, Curry *et al.* (2006) prepared an excellent list of fallacious arguments typically used in discussions of climate. While this paper focused on use of these arguments by "greenhouse deniers", it is clear that alarmists use these arguments even more frequently. Their list of fallacies is reproduced below:

- ***An ad hominem fallacy*** consists of asserting that someone's argument is wrong because of something discreditable/not authoritative about the person or persons cited by them rather than addressing the soundness of the argument itself.
  One example of this is the disparaging of works that use references in the so-called "gray literature"—those not published in peer-reviewed journals. In a review of an earlier edition of this book, a reviewer dismissed the book entirely because it had references to the gray literature. However, 95% of the references were to peer-reviewed journal articles. In any event, the reviewer found nothing technically wrong in the book. He merely looked down his very long nose and sneered at the fact that 5% of the references were to non-peer-reviewed articles. The *alarmist cabal* have repeatedly charged myself and others with plagiarism in regard to well-known background material, in an attempt to divert attention from the real content of published work.
- ***Appeal to authority*** cites a person or organization that is an authority in the relevant field and therefore should carry more weight.
  But almost all "authorities" in climatology are only authoritative within their narrow disciplines and typically do not have synoptic expertise. Furthermore, authorities make mistakes and may be biased. Ultimately, it is not the position of the proposer that matters, but rather, the strength of the data supporting their arguments. For example, as we showed in Section 4.2.3, Oreskes has lately made a career out of appealing to authorities in the climate debate. Similarly, those who would muzzle journalists to only present the alarmist persuasion appeal to authorities (Section 4.4).
- ***Appeal to motive*** is a pattern of argument that consists of challenging a thesis by calling into question the motives of its proposer. While the motives of some involved in climate debates (on both sides) can be questioned, this again should be put aside in favor of the argument itself. As we pointed out in Section 4.5, it is claimed on the Internet that Exxon funded the National Center for Policy Analysis and the Heritage Foundation to publicize skeptics' viewpoints. However, the claimed amounts of funding involved ($75,000 and $50,000, respectively) were miniscule compared to the hundreds of millions doled out by government agencies to support the alarmist agenda. We also pointed out in Section 10.3 that the ICTA posed the issue as a confrontation between "the special interests" or "advocates for industries that produce the bulk of greenhouse gas emissions" on the one hand, and idealistic "researchers" on the other.
- ***An unrepresentative sample*** is one that is falsely taken to be typical of a population from which it is drawn.
  Here, we might point to the attempt by Andrew Dessler to infer a deterministic relationship between surface temperature and cloud cover based on very short periods of data (less than two years) involving El Niños.
- ***Begging the question*** is a fallacy occurring in deductive reasoning in which the

proposition to be proved is assumed implicitly or explicitly in one of the premises. This fallacy is used by most climate debaters on both sides of the question.

- ***Correlation implies causation*** is a logical fallacy by which two events that occur together are claimed to be cause and effect.
  This fallacy is quite common on both sides of the climate debate. The most common example is that temperature and CO_2 both increased in the 20th century; therefore, rising CO_2 caused the temperature rise. Another is the rise and fall of CO_2 with temperature in Ice Age–Interglacial transitions. There is sometimes confusion on which is the cause and which is the effect.
- ***A fallacy of distribution*** occurs when an argument assumes that what is true of members is true of the class (composition), or what is true of the class is true of its members (division).
  Because some notable skeptics are members of right-wing conservative groups, this attribute is often cast upon other skeptics who are actually politically liberal. Because several outspoken skeptics do not have extensive lists of peer-reviewed articles in climatology, skeptics are typically described as amateurs.
- ***Hasty generalization*** is the logical fallacy of reaching an inductive generalization based on too little evidence.
  An example of this is the previously mentioned attempt by Dessler to infer a deterministic relationship between surface temperature and cloud cover based on very short periods of data (less than two years) involving El Niños. But, more generally, the entire field of climatology suffers from this malady. In almost all cases, climatological data are inadequately distributed spatially, too short in duration, and too affected by too many random variables to draw the kinds of generalizations that are typically drawn by climatologists.
- ***Statistical special pleading*** occurs when the interpretation of the relevant statistic is "massaged" by looking for ways to reclassify or re-quantify data from one portion of results, but not applying the same scrutiny to other categories.
  An example of this was the treatment of tree-ring proxies by Mann, Jones, and co-workers in which they omitted the down-trending tree-ring data in the late 20th century, and used Jones's "trick" of replacing these data (which showed the inadequacy of tree-ring proxies) with surface temperature data. More generally, throughout the realm of climatology, one rarely encounters the raw data; instead, one sees only massaged data that are processed by one means or another. There have been several notable revisions in this connection, and they always seem to be in a direction to support the alarmist position.
- ***Fallacy of the single cause*** occurs when it is assumed that there is one simple cause of an outcome when, in reality, it may have been caused by a number of only jointly sufficient causes.
  This is the most common and pervasive fallacy committed by alarmists who have a *idée fixe* attributing all climate change to CO_2.

These various fallacies permeate the climatology literature, both peer-reviewed, and on the Internet. My final words to anyone reading anything on climatology (including my book) are: *caveat lector*.

Appendix I

Review of the film *An Inconvenient Truth*

Al Gore's film: *An Inconvenient Truth*, has been widely hailed and, to a lesser extent, criticized. The website: *www.sourcewatch.org/index.php?title=An_Inconvenient_-Truth* provides links to more than 50 reports and reviews of the film. More often than not, those on the political left (e.g., Arianna Huffington, Carl Pope) embrace Gore's findings, and those on the political right (e.g., Glenn Beck, Jonah Goldberg) reject them.

Lewis (2007) provided a very detailed and thorough rebuttal to Al Gore's film (and book): *An Inconvenient Truth*, (AIT), although he presented a very one-sided view, just as Al Gore provided a very one-sided view from the opposite vantage point. In this Appendix, only the film is reviewed. Lewis's review was very helpful in preparing this Appendix.

Al Gore's film is full of innuendos and implications in which he occasionally flashes graphs and data but, by the time your eyes focus on them, they are gone from the screen. The film jumps around and flits from one point to another in a dizzying sequence of sound bites made for an audience with an attention span of 10 seconds or less. While it is all very impressive visually, the actual transfer of data to the audience is minimal. Occasionally, Al Gore leaves the topic of global warming for a moment and provides some introspection into his own life. The defeat for the presidency and the accident to his son stand out. In doing this, he provides a satisfying connection to Al Gore, the human being, as opposed to Al Gore the advocate.

Scene 1. The film opens with scenes of factories belching out black smoke. Such scenes recur several times in the film. But, as Marlo Lewis said:

> "The 'smoke' is probably steam, but it looks dark and ominous against the inferno colors of a fading sunset. Thus, film viewers are set up to believe they are literally seeing CO_2 spew out of smokestacks, even though CO_2 is as invisible as oxygen. Pictorially, AIT presents CO_2 as an air pollutant, anticipating Gore's later oft-repeated description of CO_2 as 'global warming pollution'. This iconic and rhetorical depiction of CO_2 as pollution is inaccurate and manipulative."

D. Rapp, *Assessing Climate Change*, Springer Praxis Books,
DOI: 10.1007/978-3-319-00455-6

I don't know why, but I am reminded of the opening scene of Charlie Chaplin's *Modern Times*. Charlie sees a truck laden with timber roar around the corner, which drops its red warning flag. He picks up the flag but the truck is gone. Just then, a Communist throng swings around another corner, sweeping Charlie in the vanguard waving a red flag. ...

Scene 2. The scene switches from belching smoke to a placid stream with trees and flowers and grass. The message is: "You can either have nature or pollution; choose one!" This was followed by pictures of the Earth from space, showing pollution filling up the atmosphere. As before, this is misleading because CO_2 is colorless and, furthermore, since the recent emphasis on global warming, few people seem to worry about conventional pollution anymore.

Scene 3. Next, we see quick flashes of melting ice (which ice? where? when?) and hurricanes blowing. A graph is flashed but it cannot be read. No data are transferred, only fear is communicated.

Scene 4. Then Al Gore asserts (correctly) that we should not assume that mankind is too puny to impact the Earth.

Scene 5. Al Gore goes on to explain a simplified version of the greenhouse effect reasonably well, pointing out that it is a good thing that we have a greenhouse effect or we would be very cold. However, he again illustrates the argument improperly with belching black smoke, mixing metaphors of "global warming" and "pollution". For those who found this description too difficult to comprehend, he provides a cartoon of melting ice cream but, unfortunately, he describes the greenhouse effect in terms of little green monsters that repel reflected sunlight from the Earth, whereas the green monsters actually absorb infrared (IR) radiant energy and re-emit it.

Scene 6. Next, he shows the jagged curve of rising CO_2 concentration, although the scene switches around so fast that all you get is a vague sense of a saw-tooth rising to the right. He provides a nice, concise explanation for why there is a yearly saw-tooth pattern.

Scene 7. The next topic is mountain glaciers. He begins with photos of Kilimanjaro before and after global warming. Unfortunately, as we showed in Section 3.5, and Lewis discusses, the depletion of glaciers on Kilimanjaro has more to do with changing precipitation patterns than temperature change. Indeed, the rate of African mountain glacier retreat was roughly constant since 1880. However, Al Gore is correct that many mountain glaciers are retreating. The Earth is warming and there is no better example of its effect than the contracting mountain glaciers. However, Lewis makes the point that these retreats began prior to the major build-up of CO_2.

The critical issue here is what is normal? The key mistake that Al Gore makes over and over again is that he uses the depths of the LIA as his baseline for comparison, and, of course, we have had significant warming since then. If we treat the LIA as the "normal" climate then, compared to this baseline, we are undergoing significant warming. The extreme expansion of mountain glaciers experienced during the LIA was followed by a retreat as the LIA waned at the end of the 19th century.

Much of this occurred prior to build-up of CO_2 concentrations in the atmosphere. Yes, we have had significant warming since the LIA, and this is evidenced in retreat of mountain glaciers, but can we attribute this to a CO_2-induced greenhouse effect? Retreat of glaciers in the later stages might be partly due to a greenhouse gas effect, but the strong retreat prior to ~1930 can hardly be due to greenhouse gases.

Should we treat the climate of the LIA as "normal"? I hope not. Today's climate is far more temperate and pleasant for many people in the world, with occasional exceptions, than the climate of the LIA. I would rather think of the LIA as an aberration instead of a "normal" baseline.

Scene 8. Al Gore goes on to provide a very brief, sketchy sound bite on the history of Earth temperatures over the last millennium based on the work of his friend, Lonnie Thompson, and colleagues (1998). As Chapter 2 of this book amply demonstrates, the history of Earth temperatures over the last millennium is a complex subject that has been studied by many investigators, of which Thompson's work is one minor element. Al Gore claims that the MWP did not exist, and he treats the LIA as a period of normalcy. He is a subscriber to the *hockey stick* that we have adequately disposed of in Section 2.4. As always, when the LIA is used as a baseline, temperatures do indeed rise during the 20th century. But Al Gore does not explain why temperatures rose early in the century prior to the main build-up of CO_2, and does not discuss the dip from 1940 to 1978, nor does he discuss the inadequacies of the temperature measurement network or the problems in defining a global average temperature.

Scene 9. Al Gore discusses Ice Ages of the past 650,000 years, and shows a strong correlation of CO_2 with temperature. However, he does not discuss the phasing of these, and implies that rising CO_2 produces higher temperatures, rather than vice versa. However, the rise in CO_2 typically lagged the rise in temperature by up to 1,000 years.

As we discussed in this book, there are quantitative problems in understanding the effects of CO_2 concentration on temperature (and vice versa). If the difference between an Ice Age and an interglacial is a change in CO_2 concentration from ~200 ppm to ~280 ppm, why isn't the Earth burning up at 395 ppm?

Nevertheless, Al Gore is right that the present CO_2 concentration of ~395 ppm is alarmingly high, and provides cause for concern. While Lewis and other skeptics seem to be unconcerned about this CO_2 level, and have even suggested it may be beneficial to plant growth, sober reflection suggests that it is important to understand the effects of this rise (and future increases) in CO_2.

Scene 10. Al Gore deplores the lack of action in Congress in the late 1980s, and then goes on to describe his own personal agony when his son was hit by a car in 1989. This made him more conscious of the possibility of losing what is precious, and he then transfers this thought to the environment.

Scene 11. At this point, Al Gore goes into a diatribe about recent hot years setting "all-time records". Indeed, we have set some records—when compared to the LIA as a baseline. In a period of warming after the LIA, it is only natural that, with continued warming, some records will be set. However, that does not prove that, prior to the LIA, as for example during the MWP, or earlier in the Holocene, temperatures were not higher than they are today. It all depends on how far back you want to look. There is one glaring error in the Gore presentation. The European heat wave of 2003 that was so devastating is portrayed by Al Gore as a product of global warming, whereas, as Lewis points out (with several references to various detailed studies), this was an unusual fluctuation in air mass circulation, unrelated to global warming. However, it is true that global warming seems to have accelerated after 1976, and some have attributed this rise in temperatures to a change in the Pacific Ocean with warm surface waters prevailing (see Section 3.4.3.3).

Scene 12. The contention is made that warmer temperatures produce more fierce storms such as Katrina. The screen is aglow with whistling winds, flooded cities, and people in despair from storm damage. In fact, Al Gore specifically claims that globally warmed gulf waters were responsible for the increase in Katrina's strength after it left Florida. The implication of the film is that the Earth is besieged by waves of storms of unprecedented intensity. In Section 8.4.3.1, we discussed the lack of evidence regarding the claim that global warming of the 20th century produced an increase in frequency and intensity of hurricanes.

Scene 13. In a personal note, Al Gore describes his disappointment at losing the presidency, but mentions that this led him to re-devote himself to environmental issues.

Scene 14. In the next scene, all the vagaries of climate fluctuation such as heavy rains and drought in any and all regions are blamed on global warming. This section begins with a bar chart, but it was not clear what is plotted. Floods in India and China and desertification in Africa; you name it—all were attributed to rising CO_2. Striking photos of Lake Chad in Africa (before and after) show boats lined up at the shore of a dry lake. However, Lewis concluded that "Lake Chad's decline probably has nothing to do with global warming" based on a scientific study that attributed the Lake's condition to "a combination of regional climate variability and societal factors such as population increase and overgrazing". As Idso (2008) showed, drought in Africa has a long history that goes back well beyond the industrial era.

Scene 15. Al Gore returned to the scenes of his youth, his double life in a Washington apartment, and a Midwest farm. He describes his idyllic life on the farm in a climate that "was unchanged since the last ice age". This, of course, is misleading, when one considers the LIA, the MWP, and the various fluctuations in the Holocene.

Scene 16. Next, he discusses the Arctic. He is correct that warming has been most severe in the Arctic compared to other parts of the globe. There is evidence of a significant reduction in the Arctic ice pack but, as we pointed out in Section 8.3.8, wind patterns and factors other than temperature may be playing a significant role in this regard. But there is nothing new here. There has always been an Arctic amplification factor, as we showed in Figures 3.21 and 3.22. Furthermore, during the mid-century pause in warming, Arctic temperature plummeted compared to the rest of the globe (see Figure 3.21).

Scene 17. He presents a likely explanation for some of the sudden climate changes experienced at the transition from the last Ice Age to the Holocene. However, he states this as firm fact, rather than speculation—which it is.

Scene 18. We see Ronald Reagan, George Bush Sr., and Senator Imhof making idiotic statements arguing against protection of the environment, although global warming and conventional pollution seem not to be distinguishable here. The conclusion Gore reaches is that the public (and its leaders) are not yet convinced of what he perceives as proven fact.

Scene 19. Here, Al Gore blames most of the afflictions of mankind (beetles, mosquitoes, vermin, flu, tuberculosis, West Nile virus, etc.) on global warming.

Scene 20. The devastation of coral reefs is presented. Lewis discusses this issue in some detail. Arguments can be made on either side (see Section 8.7).

Scene 21. Al Gore makes it seem as if the entire Antarctica ice sheet is almost ready to disintegrate. He seems to repeat the phrase many times: "Scientists were astonished!" Yet, if scientists were astonished, that only goes to show that their models were inadequate to predict events. Furthermore, as we showed in Sections 3.6.1, most of the endangered part of Antarctica is the peninsula, and the East Antarctic Ice Sheet may be growing almost as quickly as the West Antarctic Ice Sheet is diminishing. He shows "moulins" in Greenland (see Figure 8.15). He then discusses sea level rise and describes the effects of flooding from a putative 20- to 40-foot rise in sea level. China, India, Holland, and New York would be inundated.

Sea level rise is discussed in Section 8.3. Sea level rise is the most credible potential impact of global warming, and future reductions in the Greenland ice sheet are of greatest concern. Fortunately, there is some evidence that as losses occur at the margins of this ice sheet, more ice is being added (via snow) to the center of Greenland. Nevertheless, this bears careful monitoring. While the effect of future global warming on the Greenland ice sheet is a legitimate concern, the connection of this warming to the CO_2 greenhouse effect remains uncertain.

Scene 22. Al Gore then visits China with its "huge coal resources". He points out that China is rapidly becoming a leading CO_2 generator. In the process, he once again shows black smoke billowing from Chinese power plants, which, as we have shown previously, is not CO_2. One must wonder why Al Gore supports Kyoto, which absolves China from any controls on CO_2 emissions. Al Gore tries to mitigate the Chinese emissions by presenting them on a per capita basis but, as of 2013, China has become by far the world's leading CO_2 emitter. In addition, China is the world leader in conventional pollution.[1]

[1] With current emphasis on greenhouse gases, we seem to have forgotten about conventional pollution (nitrogen and sulfur oxides, soot, ozone depletion, etc.). The satellite picture of Chinese pollution at *http://visibleearth.nasa.gov/view_rec.php?id=1036* demonstrates the extent of conventional pollution from China. In Graham Greene's book, *Our Man in Havana*, the War Office is delighted to find a new super weapon because "it will make the atomic bomb a conventional weapon," and "nobody worries about conventional weapons".

Scene 23. Al Gore traces the growth of world population during his lifetime from 2 billion to 6.6 billion with the prospect of it reaching 9 billion by the time he dies. He rightly emphasizes that this population growth has strained the Earth's capacity to provide food, water, natural resources, and energy to the people of the world. However, he makes no proposal to place international constraints on family size—which is likely to be a critical need in mitigating future global warming.

Scene 24. Here, Al Gore diverts to the subject of tobacco, and his sister's untimely death from lung cancer. He uses this as an illustration of the fact that human nature often requires a long time to deal with a problem but, in the case of global warming, he asserts that we don't have time.

Scene 25. Al Gore asserts that, essentially, the entire science community agrees with his precepts regarding global warming as evidenced by a survey of 928 scientific articles. This argument is akin to "Fifty Million Frenchmen Can't Be Wrong".[2] While there are number of contrarians, Al Gore is correct that the majority of climatologists are concerned about global warming, and most of them believe the CO_2-induced greenhouse effect is a major problem facing humanity. Solar physicists are less prone to endorsing CO_2 and tend to lean toward solar variations as at least a contributor to climate change. Some climatologists have emphasized changes in the oceans as a major factor in climate change. However, climate modelers have been fairly unanimous in emphasizing temperature growth since the LIA and the role of the greenhouse effect in producing this warming. Yet, truth in science is not a matter of voting. As we pointed out in Section 4.2, the emergence of an orthodoxy and a consensus does not prove anything. Furthermore, there are many climatologists and scientists who do not subscribe to this persuasion.

Al Gore argues that the oil and gas lobby opposes all efforts to reduce carbon emissions and mentions an egregious case of a Bush appointee in "cahoots" with the oil and gas lobby. He also points out the actions of NASA and Environmental Protection Agency (EPA) administrators in changing publications by scientists to conform to Republican political prejudices. In this, he is certainly correct. In fact, it is most strange that the far right in American politics seems to uniformly disbelieve global warming, but the connection to their other agenda (Christianization of America, anti-abortion, anti-gay, and low taxes for the wealthy) is difficult to understand.

However, Al Gore does not mention that:

[2] *Fifty Million Frenchmen* is a musical comedy with a book by Herbert Fields and music and lyrics by Cole Porter. It opened on Broadway in 1929 and was adapted for a film two years later. The title is a reference to the hit 1927 song "Fifty Million Frenchmen Can't Be Wrong" by Willie Raskin, Billy Rose, and Fred Fisher, which compared free attitudes in 1920s Paris with censorship and prohibition in the U.S. The musical's plot is consistent with the standard boy-meets-girl plots of musical comedies of the first half of the twentieth century (*http//en.wikipedia.org/wiki/Fifty_Million_Frenchmen*).

(1) climatologists need funding for their work and they are unlikely to obtain funding if they don't tie their work to a calamitous problem facing humanity;
(2) like earthquake analysts who are always predicting the "big one is coming", climatologists tend to be alarmists to call attention to the importance of their work;
(3) the climatologists who use proxies to infer historical temperatures and derive global average temperatures are close-knit and have often co-published with one another; this *paleo-climatological cabal* has prevented valid criticism of their work from being published (see Section 2.4.11);
(4) those who generate global climate models use the world's biggest computers for which funding can only be justified if there is a major global crisis;
(5) most of the criticism of global warming alarmism appears on Internet blogs which is typically ignored by the elite climatological science community; yet, some of this criticism is valid.

Scene 26. We are treated to a view of more billowing black smoke. This is followed by a picture of scales with gold bars on one side, and the Earth on the other side. We are asked to choose between making money and saving the Earth. However, we are also assured that, by saving the the Earth we will also create wealth, although the details seem muddled.

He then suggests approaches to reduce emissions by increasing the efficiency of vehicles, buildings, appliances, etc., carbon sequestration, expansion of renewable energy. He points out the disparity of mileage requirements for American cars vs. Europe and Asia. In this respect, he is right. American gasoline is too cheap, and cars are too big and heavy and too powerful. Even if you don't believe in global warming, the balance of payments and America's dependence on Organization of Petroleum Exporting Countries (OPEC) dictate that the U.S. needs to downsize vehicles and raise the price of gasoline. Increasing the efficiency of systems and renewable energy is vital and requires expanded tax incentives. Carbon sequestration needs further study. All of these steps are good for America and the world. Here, Al Gore is on target, independently of greenhouse gases.

Unfortunately, Al Gore concludes this scene with a strong endorsement of the Kyoto accords, which, as we showed in Section 9.2, is ill-conceived, impractical, and grossly unfair.

Scene 27. Al Gore closes with a pep talk. He mentions all of the crises that have faced our country, including the revolutionary war in 1776, the Civil War, women's suffrage, the second World War, going to the Moon, conquering polio, and restoring the ozone layer. He says that, if we put our minds and hearts to it, we can deal with global warming. However, in all the examples given, there was a clear-cut problem of known dimensions, whereas, in global warming, everything is much more nebulous.

Appendix II

Warming of oceans by increases in greenhouse gas concentrations in the atmosphere

In the sections that follow, we attempt to clarify (to the extent possible) the role of forcing due to increased CO_2 in the atmosphere on ocean warming.

A2.1 BRIEF OVERVIEW OF THE MODEL

In our model, we start with an ocean in equilibrium with the air above it. The profile of temperatures below the surface at a tropical location is shown schematically in Figure A2.1. Initially, there is a temperature profile (*A*) for the equilibrium state before applying a forcing to the surface. The short-term curve (*B*) shows the initial response to the forcing. The initial temperature rise at the surface is ΔT_i. As a result, the mixed layer of the ocean begins warming. As it warms, there is a progression of temperature profiles until a new equilibrium is established at a new ΔT. Figure A2.2 shows the same information in greater detail, with curve *B* representing an intermediate stage in the passage from no forcing to ultimately a new equilibrium under forcing with curve *D*. The ultimate temperature rise of the mixed layer of the ocean is ΔT in this figure.

The ultimate new equilibrium established after passage of sufficient time is shown as curve (*c*). This graph is not to scale. The difference in temperatures between the mixed layer and the ocean surface was exaggerated to make the graphic clearer.

As these changes take place, changes are likely to occur in the air above the ocean surface. In some zero'th order models, it has been assumed that the air remains unchanged in temperature and humidity, even as the ocean surface warms. At the other extreme, one can assume that the air temperature tracks the ocean surface temperature and the relative humidity in the air remains constant (the absolute humidity increases). Reality probably lies between these extremes.

We begin with a model for the initial response to an increase in long-wave back radiation to the surface. This radiant flux is absorbed in the surface, resulting in a temperature increase at the surface.

D. Rapp, *Assessing Climate Change*, Springer Praxis Books,
DOI: 10.1007/978-3-319-00455-6

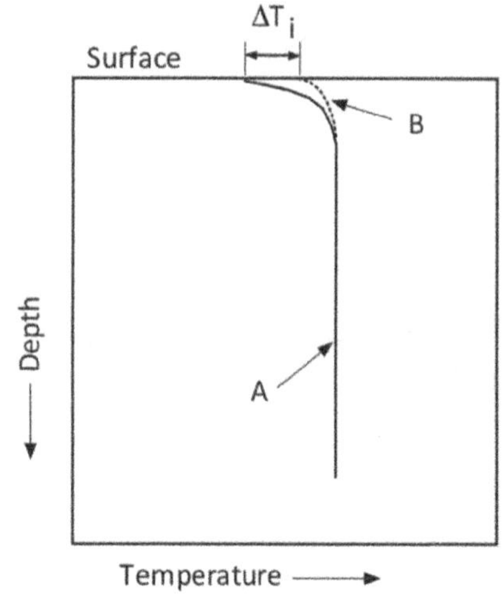

Figure A2.1. Schematic temperature profile in a tropical ocean. The initial curve (*A*) refers to equilibrium before applying a forcing to the surface. The short-term curve (*B*) shows the initial response to the forcing. This graph is not to scale. The difference in temperatures between the mixed layer and the ocean surface was exaggerated to make the graphic clearer.

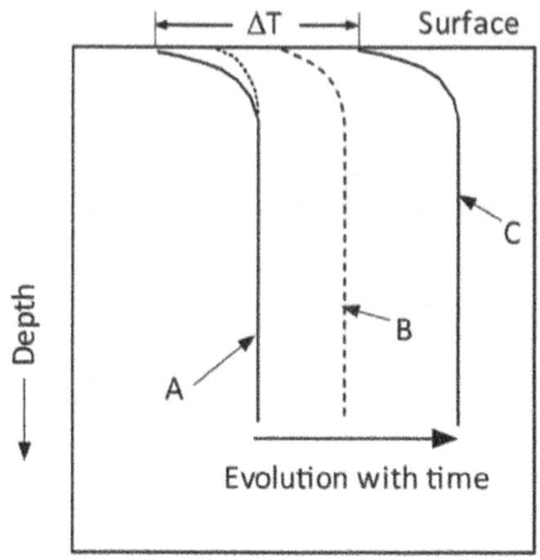

Figure A2.2. Schematic temperature profile in a tropical ocean. The initial curve (*A*) refers to equilibrium before applying a forcing to the surface. The response (*B*) shows an intermediate stage in the process as time progresses.

A2.2 THE INITIAL RESPONSE TO FORCING AT THE OCEAN SURFACE

Professor Anthony Mills (private communication) carried out an energy balance about the ocean surface, in which he modeled the ocean in two layers. The upper layer is a thin surface layer at T_S below which is a mixed layer at T_L. As Mills pointed out, on average, T_L always exceeds T_S by a small amount due to absorption of short-wave radiation in the mixed layer below the ocean surface while the surface is losing energy to the air. When back radiation is increased, say by 1 W/m^2, the initial effect will be an increase in T_S while T_L will hardly change. As a consequence of the rise in T_S, the convective, latent, and radiative losses to the atmosphere will increase. But, in addition, the ongoing heat transfer from the mixed ocean layer to the ocean surface will be reduced because T_L—T_S will be reduced compared to what it was before the forcing was applied to the surface. As time progresses, T_L will slowly rise and eventually establish a new equilibrium value, slightly greater than the increased value of T_S. This is illustrated in Figures A2.1 and A2.2. As a result, the mixed ocean layer will gradually warm due to the influence of an increase in long-wave back radiation impinging on the ocean.

As we shall see, the effect of an increase in back radiation due to increased CO_2

results in an increase in the surface temperature T_S that reduces the convective heat loss from the well-mixed water layer to the surface. In this sense, the effect of increased back radiation due to increased CO_2 is to heat the ocean because the ocean will lose heat to the atmosphere less effectively, resulting in an increase in the bulk temperature of the ocean.[1]

An energy flux balance about the ocean–atmosphere interface, prior to application of a forcing is:

$$(1 - \alpha)\, Q_S + Q_L - Q_{LH} - Q_{SH} - Q_B = 0$$

in which:

Q_L = rate of heat gain by the ocean mixed layer per unit area
Q_{LH} = rate of latent heat loss from surface to air per unit area
Q_{SH} = rate of sensible heat loss from surface to air per unit area
Q_B = rate of net back radiation from surface to air per unit area
Q_S = solar intensity falling on ocean surface
α = reflectivity of ocean surface
F = increase in back radiation from the sky to the ocean surface due to increasing CO_2 per unit area.

At night, we can discard the solar intensity term. The flux balance at night is:

$$Q_L - Q_{LH} - Q_{SH} - Q_B = 0.$$

When we apply a forcing F to the ocean surface at night, this equation becomes:

$$F + q_L - q_{LH} - q_{SH} - q_B = 0$$

in which the lower-case q's represent values after the forcing is turned on. As we pointed out above, T_S will increase initially while T_L will lag. The increase in T_S will cause q_{LH}, q_{SH}, and q_B to increase, while q_L will decrease. Subtracting the two flux balance equations, we obtain:

$$F = -\Delta Q_L + \Delta Q_{LH} + \Delta Q_{SH} + \Delta Q_B,$$

where each $\Delta Q_i = q_i - Q_i$. The T_S values before and after forcing are designated as T_{S1} and T_{S2} (same for T_A).

Approximations for Q_{LH}, Q_{SH}, and Q_B were given by Newell and Dopplick (1979). From these, we find:

Latent Heat Loss = $Q_{LH} = 6.08\,(h_S - h_A)\,w$
Sensible Heat Loss = $Q_{SH} = 2.51\,(T_S - T_A)\,w$
Net Back Radiation = $Q_B = 0.94\,[\sigma\, T_S^4\,(0.56 - 0.08\, e_A^{1/2})]$.

It should be noted that there are quite a number of empirical formulas for net back

[1] This concept was presented on the *realclimate.org* website at *www.realclimate.org/index.php/archives/2006/09/why-greenhouse-gases-heat-the-ocean/*. However, no quantitative estimates were given.

radiation. The formula used by Newell and Dopplick (1979) is one of several such formulas that are similar.

e_A = water vapor pressure in the air (product of relative humidity and saturation water vapor pressure) (mb)
R = relative humidity of air (%)
h_S = saturation specific humidity (g/kg) at T_S
h_A = $R\, h_S$ at T_A
R = relative humidity at T_A
T_A = air temperature (K)
T_S = ocean surface temperature (K)
w = wind speed (m/s)

The expressions for the changes in energy flux after applying the forcing are:

$\Delta Q_{LH} = 6.08\ (3)\ (0.25)\ [h_S(T_{S2}) - h_S(T_{S1})]\ \mathrm{W/m^2}$
$\Delta Q_{SH} = 22.59\ [(T_{S2} - T_{A2}) - (T_{S1} - T_{A1})]\ \mathrm{W/m^2} = 22.59\ (T_{S2} - T_{S1})\ \mathrm{W/m^2}$
$\Delta Q_B = 5.33 \times 10^{-8}\ \{T_{S2}^{\,4}\ (0.56 - 0.08 \times [e_A(T_{S2})]^{0.5}) - T_{S1}^{\,4}\ (0.56 - 0.08 \times [e_A(T_{S1})]^{0.5})\}\ \mathrm{W/m^2}$

where $e_A(T_{S2})$ means e_A at T_{S2}, the relative humidity was fixed at 75%, and the wind speed was taken as 3 m/s.

In the zero'th approximation, one assumes that the air temperature does not change and the humidity of the air does not change.

At night, $Q_L = h_L(T_L - T_S)$, where h_L (W/m^2-K) is a convective heat transfer coefficient controlling heat transfer from the well-mixed layer at temperature T_L, to the interface at temperature T_S. The heat transfer coefficients were estimated as shown below:

Ocean condition	h_L *W/m²-K*
(1) Still—natural convection	216
(2) Wind speed 3 m/s	420
(3) Wind speed 5 m/s	1,050
(4) Wind speed 10 m/s	3,600

If one assumes that, initially, T_L is unchanged by the forcing, it follows that:

$$\Delta Q_L = h_L(T_{S1} - T_{S2}).$$

This term is negative. The coefficient h_L is very large. As a result, the dominant term is $-\Delta Q_L$ in the expression:

$$F = -\Delta Q_L + \Delta Q_{LH} + \Delta Q_{SH} + \Delta Q_B.$$

For example, consider a case where T_{S1} = 294 K and F = 1 W/m^2. We can use trial and error, varying a guessed value of T_{S2} until F is equal to $-\Delta Q_L + \Delta Q_{LH} + \Delta Q_{SH} + \Delta Q_B$.

When this procedure is carried out, we find that a temperature rise ΔT_S equal to about 0.0045°C produces the following changes in flux (assuming T_A does not change):

$$\Delta Q_{LH} = 0.08\ \mathrm{W/m^2}$$
$$\Delta Q_{SH} = 0.04\ \mathrm{W/m^2}$$
$$\Delta Q_B = 0.02\ \mathrm{W/m^2}$$
$$\Delta Q_L = 0.986\ \mathrm{W/m^2}.$$

With a higher wind speed, ΔQ_L would be amplified. Clearly, the overwhelming effect of a rise in T_S is to reduce the heat flux from the mixed layer to the ocean surface. Therefore, most of the effect of an increase in back radiation is to heat the mixed layer. This conclusion is independent of any assumptions regarding changes in the atmosphere that result from warming of the ocean surface. If we repeated the calculation assuming that the air temperature tracks the ocean surface temperature and the relative humidity of the air remains constant, that would decrease energy loss from the ocean surface to the air, and amplify the importance of energy loss from the mixed layer to the ocean surface.

A2.3 ESTABLISHMENT OF A NEW EQUILIBRIUM

In the previous section, we found that the initial response of the ocean to a forcing applied to the ocean surface is a rise in the surface temperature and a decrease in energy flux from the mixed layer to the ocean surface. This is illustrated as curves *B* in Figures A2.1 and A2.2.

As time progresses, the mixed layer continues to lose energy at a lower rate than prior to application of the forcing to the surface. As a result, the mixed layer gradually warms (as shown in curve *C* in Figure A2.2). However, as the mixed layer warms, the rate of energy transfer from the mixed layer to the surface increases, and thus the rate of warming of the mixed layer decreases as time progresses. Eventually, the rate of energy transfer from the mixed layer increases enough to establish a new equilibrium at a higher mixed layer temperature (see curve *C* in Figure A2.1 or curve *D* in Figure A2.2). When this new equilibrium is established, we can treat the upper ocean as a mixed layer and ignore the small difference in temperature between the mixed layer and the surface. We then carry out an energy balance about the ocean surface.

A2.4 ZERO'TH ORDER MODEL

Newell and Dopplick (1979) used an idealized model to estimate the effect of increased back radiation on the ocean surface. They started with an atmosphere and ocean in equilibrium. In their model, they kept all parameters of the atmosphere constant (e.g., temperature, absolute humidity) but allowed the temperature of the upper mixed layer of the ocean to rise as a consequence of the additional back radiation. The back radiation is absorbed into the top few microns of the ocean surface. This additional energy flux to the ocean surface can end up in four possible responses:

(1) warming of the mixed layer of ocean as expressed in an increase in $T_S \approx T_L$;
(2) sensible heating of the atmosphere—if the ocean warms and the atmosphere is unchanged, the ocean will transfer some heat to the atmosphere; sensible heat transfer depends on wind speed.
(3) latent heat loss—if evaporation from the ocean surface increases, some energy will be transferred from the ocean to the atmosphere in the form of water vapor; latent heat loss depends very sharply on wind speed.
(4) radiant heat loss—if the ocean surface warms, it will radiate more energy back into the atmosphere; since we assume that the air does not change as the ocean warms, the back radiation from the air does not increase, and therefore the net back radiation from the ocean to the air increases.

The temperature of the mixed layer, $T_S = T_L$, will rise until the sum of sensible, latent heat and radiative losses increases by an amount equal to the increase in back radiation forcing due to increasing CO_2. As we pointed out previously, Newell and Dopplick (1979) used the following equations:

Latent Heat Loss = $Q_{LH} = 6.08\ (h_S - h_A)\ w$
Sensible Heat Loss = $Q_{SH} = 2.51\ (T_S - T_A)\ w$
Net Back Radiation = $Q_B = 0.94\ [\sigma\ T_S^4\ (0.56 - 0.08\ e_A^{1/2})]$.

Prior to forcing, an energy balance on the mixed layer is:

$$(1 - \alpha)\ Q_S - Q_{LH} - Q_{SH} - Q_B = 0$$

in which heat transfer to the deep ocean is ignored. Since IR is absorbed in the top few microns of the ocean surface, we assume that any forcing (X) W/m^2 injects (X) W/m^2 into the ocean surface. Since convective heat transfer between the surface and the top mixed layer of ocean is very efficient, this warms the upper ocean layer. As the ocean warms, energy losses to the atmosphere via convection, radiation, and latent heat take place. Ultimately, a new equilibrium is reached at a higher value of T_S. In the zero'th order model used by Newell and Dopplick, the atmosphere does not change as the ocean warms. To proceed numerically, we add an arbitrarily chosen increment of temperature to T_S. For this value of ΔT_S, we calculate the terms in the energy balance and determine the increase in total heat loss to the atmosphere. We can then repeat this by trial and error until the increase in heat loss is equal to the forcing, and this increase in T_S is due to the forcing.

As an illustration, let $T_S = T_L = T_A$. Further, assume that $R = 75\%$, and choose a wind speed of 3 m/s. For example, if we arbitrarily choose $T_S = T_L = T_A$ = 294 K, and use the saturation vapor pressure of water as 21°C = 24.79 mb, we obtain prior to forcing:

Latent Heat Loss = $Q_{LH} = 6.08\ (h_S - h_A)\ w = 6.08\ (15.70 - 11.78)\ 3 = 6.08\ (3.92)\ (3) = 71.59$ W/m^2

Sensible Heat Loss = Q_{SH} = 2.51 ($T_S - T_A$) 3 = 0
Net Back Radiation = Q_B = 0.94 [σ T_S^4 (0.56 – 0.08 $e_A^{1/2}$)] = 0.94 × (5.67 × 10^{-8}) × $(294)^4$ × (0.56 – 0.08 × $[0.75 \times 24.79]^{0.5}$) = 85.64 W/m^2

According to this simplistic model for the unforced case, $T_S = T_L = T_A$ will adjust to provide a total energy flux from the ocean surface to match an incoming energy flux of 157.23 W/m^2. Had we used a lower relative humidity for the air, the latent heat loss would have been higher, and vice versa.

When forcing is introduced in this simple model, $T_S = T_L$ will increase while we assume for simplicity that T_A remains unchanged. With forcing:

Latent Heat Loss = Q_{LH} = 6.08 (h_S – 11.78) 3 = 18.24 (h_S – 11.78) W/m^2
Sensible Heat Loss = Q_{SH} = 2.51 (T_S – 294) 3 W/m^2 = 7.53 (T_S – 294) W/m^2
Net Back Radiation = Q_B = 0.94 [σ T_S^4 (0.56 – 0.08 $e_A^{1/2}$)] = 0.94 × (5.67 × 10^{-8}) × $(T_S)^4$ × (0.56 – 0.08 × $[0.75 \times 24.79]^{0.5}$) W/m^2
Total Heat Loss = Latent Heat Loss + Sensible Heat Loss + Net Back Radiation.

We can now calculate these values as a function of $T_S = T_L$. For each assumed value of $T_S = T_L$, we subtract 157.23 W/m^2 from the calculated total heat loss and attribute this difference to the forcing that produced the increase in $T_S = T_L$. As it turns out, the total heat loss is a very sensitive function of $T_S = T_L$. Increasing $T_S = T_L$ by 0.02°C increases the total heat loss by about 0.39 W/m^2. Hence, according to this simple model, adding, say, 0.36 W/m^2 of forcing to the back radiation will only increase $T_S = T_L$ by about 0.02°C. Adding 1.0 W/m^2 of forcing to the back radiation will only increase $T_S = T_L$ by about 0.05°C. Some calculations are shown in Table A2.1.

As we showed in Figure 3.34, the forcing at the surface resulting from a doubling of CO_2 from the pre-industrial value, is about 1 W/m^2. From Table A2.1, it follows that a surface forcing of about 1 W/m^2 will only warm the oceans by about 0.05°C in this simple model in which it is assumed that the atmosphere does not change as the ocean warms. Had we used a relative humidity of, say, 60% instead of the value of 75% used for Table A2.1, only small changes in the numbers would result.

Table A2.1. Change in total heat loss as a function of T_S when the atmosphere is unchanged (W/m^2).

T_S (K)	*Latent Heat Loss*	*Sensible Heat Loss*	*Net Back Radiation*	*Total Loss*	*Change in Total Loss*
294	71.59	0.00	85.64	157.23	0.00
294.01	71.77	0.08	85.65	157.43	0.20
294.02	71.96	0.15	85.66	157.62	0.39
294.03	72.14	0.23	85.68	157.82	0.59
294.04	72.32	0.30	85.69	158.01	0.78
294.05	72.50	0.38	85.70	158.20	0.97
294.06	72.69	0.45	85.71	158.40	1.17

A2.5 FIRST ORDER MODEL

Watts (1980) commented on the paper by Newell and Dopplick (1979), to be denoted ND1. Watts pointed out that:

> "... under the assumption that the temperature and humidity ratio of the atmospheric air over the sea remain constant as the radiative flux downward at the ocean surface increases. If the ocean and atmosphere temperatures change together, the change in sensible heat flux will be zero according to the author's equation. The temperature of the atmosphere will increase, of course, because the latent heat that leaves the ocean must end up in the atmosphere after condensation occurs. The specific humidity of the atmosphere will also increase."

He went on to say "Estimating how the latent heat, sensible heat and back radiation change in response to long-term changes in the ocean temperature is a difficult problem indeed". Watts (1980) suggested that there is a "constant value of relative humidity when the temperature changes". However, as we discussed in Section 6.1.7, although this is a widely used approximation, the basis for it is actually quite weak.

Newell and Dopplick (1981) replied to Watts. One point they made was that "a good fraction of the additional energy received by the surface air from the warmer sea could be radiated away by the additional water vapor". However, they admitted that "Clearly, we should go further than our 'first iteration' and additional computations [to] include consideration of the variation of sea surface temperature with air temperature and moisture". They also said: "We agree that the additional energy used in evaporation will eventually appear in the free atmosphere above the surface layer. It is by no means obvious that it will cause an increase in temperature. First of all part of the increase may be offset by additional cooling to space from the increased water vapor. Second, there are a myriad of water-cloud-circulation feedback processes that may come into play."

Watts (1982) provided a rebuttal to this note. He claimed that the method used in ND1 is merely a first iteration. In this simple model, one fixes atmospheric conditions to the initial value and calculates the forcing corresponding to any change in $T_S \approx T_L$. According to Watts, one must take into account changes in the atmosphere above the ocean that result from an increase in $T_S \approx T_L$, and use these (with the increased $T_S \approx T_L$) as the starting point for a second iteration. The problem is how to assess the changes in atmospheric parameters resulting from an increase in $T_S \approx T_L$. Watts (1982) made the assumption that T_A remains $\approx T_L$ as $T_S \approx T_L$ rises, and furthermore the relative humidity in the atmosphere remains constant as T_A rises. These assumptions are qualitatively in the right direction. Surely, one expects T_A to rise as $T_S \approx T_L$ increases, but why should it remain equal to $T_S \approx T_L$? One also expects the absolute humidity to rise, but why should the relative humidity remain constant? In fact, we don't really know much about the relative humidity. It was assumed to be 75% for purposes of calculating Table A2.1, but that was merely a wild guess. Nevertheless, if we make the assumptions suggested by Watts (1982) and set $T_A \approx T_S \approx T_L$ and assume constant relative humidity = 75% as T_A rises, for purposes of a second iteration, we must use:

Latent Heat Loss = Q_{LH} = 6.08 [h_S – 0.75 h_S] 3 = 18.2 [0.25 h_S] W/m^2
Sensible Heat Loss = Q_{SH} = 2.51 (T_S – T_A) 3 W/m^2 = 0
Net Back Radiation = Q_B = 0.94 [σ T_S^4 (0.56 – 0.08 $e_A^{1/2}$)] = 0.94 × (5.67 × 10^{-8}) × $(T_S)^4$ × (0.56 – 0.08 × $[e_A]^{0.5}$) W/m^2 = 5.33 × 10^{-8} T_S^4 (0.56 – 0.08 × $[e_A]^{0.5}$) W/m^2
Total Heat Loss = Latent Heat Loss + Sensible Heat Loss + Net Back Radiation

We obtain the results shown in Table A2.2.

In this approximation, since $T_A \approx T_S \approx T_L$, there is no sensible heat loss. The latent heat loss increases as $T_S \approx T_L$ increases. However, the net back radiation decreases significantly as $T_S \approx T_L$ increases. The ultimate result is that, if one assumes $T_A \approx T_S \approx T_L$ and the relative humidity remains constant, the modeled temperature rise increases significantly compared to the model of Newell and Dopplick. For example, $T_S \approx T_L$ rises by about 0.6°C due to a forcing of 1 W/m^2. For a forcing of 0.4 W/m^2, characteristic of the average for the past 55 years, the temperature gain is estimated to be about 0.25°C.

Table A2.2. Change in total heat loss as a function of T_S when we set $T_A \approx T_S \approx T_L$ (W/m^2).

T_S (K)	*Saturation Humidity (mb)*	*Sensible Heat Loss*	*Net Back Radiation*	*Latent Heat Loss*	*Total Loss*	*Change in Total Loss*
294.0	24.79	0.00	85.64	71.59	157.23	0.00
294.1	24.92	0.00	85.38	72.05	157.43	0.20
294.2	25.07	0.00	85.08	72.50	157.59	0.36
294.3	25.22	0.00	84.79	72.96	157.75	0.52
294.4	25.37	0.00	84.49	73.42	157.91	0.68
294.5	25.51	0.00	84.23	73.87	158.10	0.87
294.6	25.66	0.00	83.94	74.33	158.26	1.03
294.7	25.80	0.00	83.64	74.78	158.42	1.19
294.8	25.95	0.00	83.35	75.24	158.59	1.36
294.9	26.10	0.00	83.05	75.70	158.75	1.52
295.0	26.25	0.00	82.76	76.15	158.91	1.68
295.1	26.41	0.00	82.42	76.61	159.03	1.80
295.2	26.56	0.00	82.13	77.06	159.19	1.96
295.3	26.72	0.00	81.81	77.52	159.33	2.10
295.4	26.88	0.00	81.50	77.98	159.47	2.24
295.5	27.03	0.00	81.19	78.43	159.62	2.39
295.6	27.19	0.00	80.87	78.89	159.76	2.53

However, Watts (1982) pointed out that:

"... the surface energy-balance model gives results that are extremely sensitive to slight model-parameter changes."

Small changes in the assumed value of the humidity in the atmosphere produce large changes in the temperature rise. For example, if Table A2.2 is repeated with a wind speeds of 5 m/s, the temperature rise for 1 w/m^2 forcing is reduced to 0.25°C. If the wind speed is reduced to 2 m/s, one finds the calculation to be unstable. As Watts emphasized, "The surface heat flux calculation is very, very sensitive to small changes in the components of the heat balance". He pointed out that a heat balance at the tropopause does not suffer from this problem. However, a heat balance at the tropopause does not provide us with direct insight as to the effect of back radiation on the oceans.

It is evident from these calculations that a crucial unknown is the change in the air above the ocean as the ocean surface warms. In the zero'th order model, it was assumed that the air did not change as the ocean surface warmed. As a result, when the ocean surface warms in this model, the net back radiation increases extremely slowly, while the latent heat loss and sensible heat loss increase at significant rates. This allows the ocean surface to rid itself of excess energy that would have accumulated due to a reduction in flux from the mixed layer to the ocean surface. Hence, a new equilibrium is achieved with a rather small increase in ocean temperature. In the first order model, the air temperature is set equal to the ocean temperature and the relative humidity is assumed to be constant. Thus, the sensible heat loss is zero. The latent heat loss increases with temperature as before. However, the net back radiation now decreases sharply as the ocean and temperatures increase, due to the increased humidity in the air. Thus, the ocean surface is less able to lose energy to the air, and the temperature rise is greater. The problem is that the total heat loss from the ocean depends on three terms, one of which is assumed to be zero, a second decreases with temperature, and the third increases with temperature. The term that decreases with increasing temperature depends critically on the humidity in the air above the ocean. The term that increases with temperature depends critically on wind speed. Depending on assumptions made about these variables, the final equilibrium temperature of the mixed layer of the ocean can be almost any value.

A2.6 ANALYSIS BY RAMANATHAN (1981)

Ramanathan (1981) argued that, as a consequence of doubling CO_2 in the atmosphere, not only is there an increase in back radiation at the surface of about 1.2 W/m^2, but there are two additional feedback fluxes that affect the surface.

One additional factor is direct radiative heating warming of the troposphere that increases the IR emission by the radiatively active constituents of the tropospere (clouds, H_2O, CO_2, O_3, and other trace gases). This amplifies the back radiation flux. However, the transmission efficiency for the tropospheric emission to penetrate to the surface is unclear from Ramanathan's paper.

The second process "concerns the interactions between ocean surface temperature, the hydrological cycle and tropospheric convective adjustment processes". According to Ramanathan (1981):

"The surface warming due to [increased back radiation and radiation from the warm troposphere] enhances H_2O evaporation into the troposphere, which indirectly amplifies the surface warming in two ways: (i) The latent heat released within the troposphere (resulting from the enhanced evaporation) warms the troposphere, thus enhancing tropospheric IR emission; and (ii) the enhancement in the evaporation also increases the absolute humidity of the troposphere which, of course, increases tropospheric IR emission. A fraction of the increase in tropospheric IR is emitted upward to space and the remainder is emitted downward to the surface. The downward fraction of the enhanced emission amplifies the surface warming by [increased back radiation and radiation from the warm troposphere]. This feedback between temperature, H_2O evaporation and IR emission is primarily controlled by ocean-atmosphere interactions since the world oceans are the primary source for atmospheric H_2O. The magnitude of the amplification is strongly determined by tropospheric convective adjustment processes and its subsequent effect on tropospheric lapse rates. This dependence arises because the partitioning of the enhanced IR emission between upward and downward components is controlled by lapse rate changes."

However, the latent heat release seems to have been taken into account in the first process, and one wonders whether this effect was counted twice. Furthermore, it is not clear how much effect a supposed increase in tropospheric humidity would have on back radiation at the surface, considering that air above the ocean is already quite humid and clouds act as a barrier to IR transmission. Ramanathan seems to have assumed a low opacity for the atmosphere.

According to Ramanathan, a back radiation flux of 1.2 W/m^2 leads to increases of 2.3 W/m^2 due to increased radiation from the warming troposphere, and 12.0 W/m^2 from increases in radiation from increases in atmospheric humidity. Thus, he argued that the total back flux is about 15.5 W/m^2, rather than 1.2 W/m^2. He claimed that "Newell and Dopplick's approach underestimates surface warming by about a factor of 30-40". As stated above, this writer suspects that the feedback is inhibited by clouds and low-lying humidity of air over the ocean, so Ramanathan's estimates seem grossly exaggerated. It should be noted that Trenberth *et al.* (2009) provided a wide range of estimates of variable sign and this value has considerable uncertainty.

Using a multi-layer global climate model in one dimension, Ramanathan estimated the effect of the three factors mentioned above on ultimate equilibrium ocean temperature due to a doubling of CO_2:

	Increased back radiation	*Warming of atmosphere*	*Increased water vapor*	*Total*
Surface flux (W/m^2)	1.2	2.3	12.2	15.5
ΔT_S (°C)	0.17	0.33	1.7	2.2

There are several aspects of Ramanathan's treatment that are difficult to comprehend. One is that, when he compared his calculation (omitting feedbacks) with that of Newell and Dopplick, he agreed that a 1.2 W/m^2 forcing produces a temperature rise without feedbacks is about 0.04°C. Yet, in the above table, he indicates a rise of 0.17°C without feedback.

Ramanathan's estimates for the effects of warming of the atmosphere and increased humidity depend on assumptions regarding changes to the atmosphere (temperature and humidity) that result from increased ocean temperature and, apparently, he assumed a relatively transparent atmosphere. One wonders whether he has properly taken into account clouds over the oceans. Despite the length of his article, it is written in such a confusing manner that it is difficult to determine exactly what he assumed for these changes.

Nevertheless, one thing stands out from Ramanathan's calculation. Even with his seemingly exaggerated large feedback terms, he found that the oceans only warm by about 2.2°C. Since ocean temperatures will ultimately control the Earth's climate, this would seem to limit the future rise in global temperature due to a doubling of CO_2 from the pre-industrial value.

A2.7 SUMMARY OF MODELS

In our model, the initial response of the ocean to an increase in back radiation flux is an increase in the ocean surface temperature, T_S, while the mixed layer of ocean remains at its original temperature, T_L. Although the increase in T_S tends to increase heat loss from the surface, this increase is far less than the increase in back radiation flux. But the effect of an increase in T_S is a decrease in $T_L - T_S$, resulting in a large reduction in energy flux transported from the mixed layer to the surface. The new (increased) value of T_S is that obtained when the sum of increased heat losses upward and decreased ocean energy flux to the surface balances the increase in back radiation flux. As time progresses, the energy flux from the mixed layer to the surface remains less than it was before the forcing was applied, and therefore it warms (T_L increases with time). As T_L increases, the ocean energy flux to the surface gradually increases. After passage of sufficient time, a new equilibrium is eventually established in which T_S and T_L are both higher than before the forcing was applied. To model this new equilibrium, it is sufficient to use a lumped ocean model for the mixed layer and surface by assuming $T_S \approx T_L$ because the difference between T_S and T_L is expected to be smaller than the temperature change resulting from the forcing. One then takes an energy flux balance about the ocean surface, and varies $T_S \approx T_L$ until the calculated heat loss from the surface equals the forcing. This is the new equilibrium value of $T_S \approx T_L$.

In order to utilize the model quantitatively, we need to estimate the increase in back radiation flux at the ocean surface due to increased CO_2, and we also need to estimate the changes that occur in the air above the ocean (in order to estimate heat losses from the ocean to the air).

The basic forcing at the ocean surface due to increased CO_2 is well understood, as described in Figure 3.34. The forcing at the surface due to doubling of CO_2 from the pre-industrial value is about 1 W/m^2 to 1.2 W/m^2. What is not clear is to what extent this basic forcing is augmented by radiation from a warmer troposphere. Ramanathan (1981) concluded that this effect is very large but his results seem exaggerated.

The effect of increased ocean surface temperature on the air above can only be conjectured. If radiation from a warmer troposphere is neglected, and one makes the zero'th order assumption that the atmosphere remains unchanged as the ocean warms, one finds a very small increase in ocean temperature ($\sim$0.05°C) due to an increase in back radiation of 1 W/m^2. On the other hand, if radiation from a warmer troposphere is neglected, and one assumes that the air temperature remains equal to the ocean surface temperature and the relative humidity of the air remains constant, the calculated increase in ocean temperature due to an increase in back radiation of 1 W/m^2 is about 0.6°C. Unfortunately, this calculation is very sensitive to assumptions made about the condition of the air, and even more sensitive to assumptions about the wind velocity. Hence, it is not possible to obtain precise quantitative estimates of the ultimate equilibrium temperature of the mixed ocean layer when subjected to a forcing at the surface.

One thing we can assert, however, is that warming of the oceans by an increase in back radiation to the surface is an efficient process, and the initial rate of heat loss by the mixed layer is only slightly less than the magnitude of the imposed forcing. As shown in Section 3.4.3.4, the average surface forcing due to increased CO_2 over the past 55 years was roughly 0.4 W/m^2. It is therefore not unreasonable to expect that the upper mixed level of the ocean would have warmed by an input of roughly this amount over that time period. The equilibrium increase in temperature due to this forcing depends on the level of additional radiation from a warmer troposphere at the surface, as well as the characteristics of the air above the oceans (temperature, humidity, wind speed, cloud cover).

A2.8 EFFECTIVE BACK RADIATION AT THE OCEAN SURFACE

Sverdrup *et al.* (1942) discussed the effective back radiation over the ocean surface. According to these authors:

> "The sea surface emits long-wave heat radiation, radiating nearly like a black body, the energy of the outgoing radiation being proportional to the fourth power of the absolute temperature of the surface. At the same time the sea surface receives long-wave radiation from the atmosphere, mainly from the water vapor. A small part of this incoming long-wave radiation is reflected from the sea surface, but the greater portion is absorbed in a small fraction of a centimeter of water, because the absorption coefficients are enormous at long wave lengths. The effective back radiation from the sea surface is represented by the difference between the 'temperature radiation' of the surface and the long-wave radiation from the atmosphere, and this effective radiation depends mainly upon the temperature of the sea surface and the water-vapor content of the atmosphere [a few meters above the surface]. ... According to Ångström

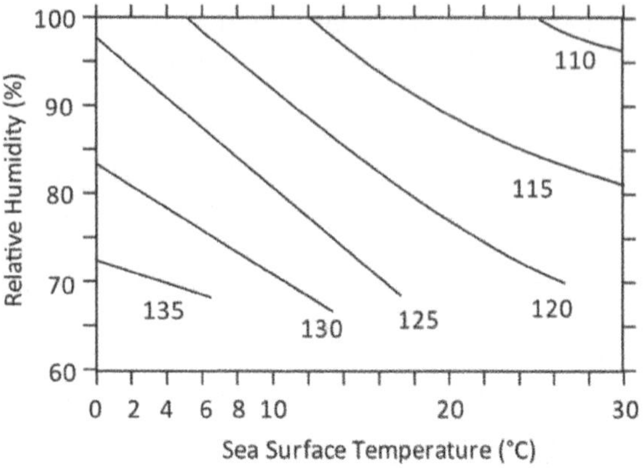

Figure A2.3. Effective back radiation in W/m^2 from the sea surface to a clear sky as a function of sea-surface temperature and relative humidity of the air at a height of a few meters (Sverdrup *et al.*, 1942).

(1920), the latter is proportional to the local vapor pressure, which can be computed from the relative humidity if the air temperature is known. Over the oceans, the air temperature deviates so little from the sea-surface temperature that the vapor pressure can be obtained with sufficient accuracy from the sea-surface temperature and the relative humidity of the air at a short distance above the surface."

Estimates of the back radiation as a function of temperature and cloud cover date back to fairly early in the 20th century. Several empirical formulas were proposed in the first half of the 20th century and these were widely used to analyze ocean–atmosphere interactions. As the years went by, remembrance of the basis for the formulas gradually faded while use of the formulas expanded.

Ångström (1920) presented the graph shown in Figure A2.3.

Sverdrup *et al.* went on to say:

"In the presence of clouds the effective back radiation is cut down because the radiation from the atmosphere is increased. The empirical relation can be written

$$Q = Q_o\,(1 - 0.083\ C)$$

where Q_o is the back radiation for a clear sky and where C is the cloudiness on the scale 1 to 10. A diurnal or annual variation in the cloudiness will lead to a corresponding variation in the effective back radiation. On an average, the diurnal variation of cloudiness over the oceans is very small and can be neglected, but the annual variation is in some regions considerable. The above equation is applicable to average conditions only, because the reduction of the effective back radiation due to clouds depends upon the altitude and the density of the clouds. Thus, if the sky is completely covered by cirrus, alto-stratus, or stratocumulus clouds, the effective radiation is about 0.75 Q_o, 0.4 Q_o, and 0.1 Q_o, respectively."

Assuming cloud cover over the ocean averages roughly 0.6, the back radiation would be roughly half of the values given in Figure 1 for clear skies.

Sverdrup *et al.* cited "Brunt's empirical formula":

$$Q_B = Q\,(1 - 0.44 - 0.08\,e_A^{0.5}),$$

where Q is the radiation of a black body having the temperature of the sea surface and e_A is the vapor pressure of water vapor in the air in millibars.

These formulas were used by a number of investigators (e.g., Koto (1966) and Hasse (1971) used the graph provided by Sverdup *et al.* together with the correction for cloudiness given above).

Baldwin (1970) used results of Anderson (1954) and James (1966) to estimate the effective back radiation. He used Anderson's equation:

$$Q_B = (4.75 \times 10^{-9})\, T_S^4\, (1 - a + b\, e_A),$$

where

$$a = 0.74 + 0.025\, C \exp(-0.0584\ \mathrm{h})$$
$$b = 0.0049 - 0.00054\, C \exp(-0.060\, h)$$

in which:

e_A = water vapor pressure in the air at $\mathrm{T_S}$ (mb)
C = cloud amount in tenths
h = average cloud height in thousands of feet.

Haney (1971) estimated the net upward flux of long-wave radiation, using an empirical relationship due to Brunt, which was also used by others:

$$Q_B = Q^*\, \sigma\, T_S^4,$$

where

$$Q^* = 0.985\,[0.39 - 0.05\,(e_A)^{0.5}](1 - 0.6\, C^2).$$

Dorman (1974) used:

$$Q_B = [\sigma\, T_S^4\, (a - b\, e_A^{1/2})](1 - c\, C) + d\, T_S\, (T_S - T_A)$$

with constants given in a book by Wyrtki.

Lane (1989) provided an empirical equation for the effective back radiation "that takes into account the complex atmospheric absorption and radiation":

$$Q_B = 0.96\, \sigma\, T_A^4\, (11.7 - 0.23\, e_A)(1 - c\, C) + 3.84\, \sigma\, T_A^3\, (T_S - T_A)$$

in which

C = fractional cloud cover
c = a coefficient that varies with latitude and is roughly 0.5 at tropical latitudes ranging to 0.6 at 30° latitude.

Gill (1982) suggested use of the formula:

$$Q_B = 0.985\ [\sigma\ T_S^4\ (0.39 - 0.05\ e_A^{1/2})(1 - 0.6\ C^2).$$

Maughan (1966) measured the outgoing and incoming IR levels over a body of water and subtracted these to obtain the back radiation flux.

Huang and Park (1975) attempted to use insolation measurements from ocean buoys to infer cloud cover, based on correlations of dependence of back radiation on cloud cover in the mid-latitude North Pacific Ocean (43°N). They referenced a number of earlier publications that utilized the so-called Berliand formula:

$$Q_B = Q_o\ (1 - \mathrm{a}\ C - \mathrm{b}\ C^2),$$

where C is cloud cover, Q_o is the clear sky back radiation, and a and b are constants. At the latitude of measurements, a ~ b ~ 0.38. They mention a formula attributed to Berliand and Berliand in 1952 as follows:

$$Q_B = 0.985\ [\sigma\ T_S^4\ (0.39 - 0.05\ e_A^{1/2})(1 - \mathrm{k}\ C^2) + 2.91\ \sigma\ T_A^3\ (T_S - T_A).$$

Sopkin (2008) used a similar formula attributed to Berliand and Berliand in 1952.

Kraus and Rooth (1961) attributed the following formula to Brunt in 1944 and Dorsey in 1940:

$$Q_B = 0.985\ \sigma\ T_S^4\ (0.39 - 0.0504\ e_A^{1/2})(1 - k\ C^2).$$

Siegel and Dickey (1986) measured the net long-wave radiation at the sea surface over the eastern North Pacific Ocean for 22 days during the fall of 1982. They referred to a number of previous attempts to correlate simple formulas to measurements of net back radiation, and then went on to compare their measurements with a number of these formulas (Table A2.3). The mean value of T_S was 297.8 K.

Table A2.3. Clear sky formulas for back radiation (Siegel and Dickey, 1986).

Reference	*Formula for back radiation*	*(W/m²)*
Berliand and Berliand (1952)	$Q_o = \varepsilon\sigma\ T_S^4\ (0.39 - 0.05\ e_A^{1/2}) + 4\varepsilon\sigma T_A^3\ (T_S - T_A)$	85.2
Brunt (1932)	$Q_o = \varepsilon\sigma\ T_S^4\ (0.39 - 0.05\ e_A^{1/2})$	81.1
Efimova (1961)	$Q_o = \varepsilon\sigma\ T_S^4\ (0.254 - 0.00495\ e_A)$	73.0
Bunker (1976)	$Q_o = \varepsilon\sigma\ T_S^4\ (0.257 - 0.005\ e_A) + 4\varepsilon\sigma\ T_A^3\ (T_S - T_A)$	78.9
Anderson (1952)	$Q_o = \varepsilon\sigma\ [T_S^4 - T_A^4\ (0.74 + 0.0049 e_A)]$	81.0
Swinbank (1963)	$Q_o = \varepsilon\sigma\ [\mathrm{T_S}^4 - 9.36 \times 10^{-6}\ T_A^6]$	88.5
Clarke *et al.* (1974)	$Q_o = \varepsilon\sigma\ T_S^4\ (0.39 - 0.05\ e_A^{1/2}) + 4\varepsilon\sigma\ T_S^3\ (T_S - T_A)$	86.5

The various formulas accounted for clouds using a formula of the type:

$$Q_B = Q_o\ (1 - \mathrm{B}\ C^{1/\mathrm{N}}),$$

where is the clear sky formula and B and N are constants. The predictions of the formulas for C ~ 0.72 are given in Table A2.4. The measured mean effective back radiation was 52.0 W/m^2.

Table A2.4. Simple formulas for back radiation including clouds.

	N	B	Q_B
Berliand and Berliand (1952)	1	0.55	52.3
	2	0.63	55.3
Brunt (1932)	1	0.49	51.4
	2	0.57	53.6
Efirnova (1961)	1	0.44	50.2
	2	0.52	51.3
Bunker (1976)	1	0.50	51.3
	2	0.57	53.5
Anderson (1952)	1	0.52	51.6
	2	0.60	53.9
Swinbank (1963)	1	0.58	52.0
	2	0.66	54.8
Clark *et al.* (1974)	1	0.56	52.6
	2	0.64	55.8

Newell and Dopplick (1979) claimed that estimates of Q_B were "discussed extensively in the literature" and, based on this, they used:

$$Q_B = 0.94\ [\sigma\ T_S^4\ (0.56 - 0.08\ e_A^{1/2})],$$

which presumably includes clouds and pertains to tropical latitudes. Their estimate at 297.8 K is 47.5 W/m^2.

Zapadka *et al.* (2007) simultaneously measured data regarding the long-wave radiation of the sea surface and its contiguous air layer, the water vapor pressure in the air above the water, and the cloud cover. These data were gathered during numerous research cruises in the Baltic in 2000–2003 and were supplemented by satellite data characterizing the cloud cover over the whole Baltic. From this, they derived an improved formula for the back radiation:

$$Q_B = 0.985\ \sigma\ T_S^4 - \sigma\ T_A^4\ (0.685 - 0.00452\ e_A)(1 - B\ C^N).$$

They provide values of B and N for various types of clouds.

For purposes of estimating the temperature rise when the oceans are exposed to a long-term increase in back radiation, the absolute value of the back radiation is not very important. What is important is the temperature dependence of the back radiation. The temperature dependence of the back radiation depends on two terms: T_S and e_A. Using the Brunt formula (for example), we have:

$$dQ_o/dT_S = 4\varepsilon\sigma T_S^3\ (0.39 - 0.025\ e_A^{-1/2}\ de_A/dT_S).$$

References

Abe-Ouchi, A., Segawa, T., and Saito, F. (2007), "Climatic conditions for modelling the Northern Hemisphere ice sheets throughout the ice age cycle", *Climate of the Past Discussions*, **3**, 301–336.

Ackerman, F. (2008), "Hot, it's not: Reflections on *Cool It!*, by Bjorn Lomborg" (*ase.tufts.edu/gdae/Pubs/rp/Ackerman_CoolIt.pdf*).

Adam, D. (2008), "Analysis: Has the Kyoto Protocol worked?", *The Guardian*, 8 December, (*www.guardian.co.uk/environment/2008/dec/08/kyoto-poznan-environment-emissions-carbon*).

Agee, E.M., Cornett, E., and Gleason, K. (2010), "An extended solar cycle 23 with deep minimum transition to cycle 24: Assessments and climatic ramifications", *Journal of Climate*, **23**, 6110–6114.

Agee, E.M., Kiefer, K., and Cornett, E. (2011), "Relationship of lower troposphere cloud cover and cosmic rays: An updated perspective", *Journal of Climate* (*http://journals.ametsoc.org/doi/abs/10.1175/JCLI-D-11-00169.1*).

Ahn, J., and Brook, E.J. (2007), "Atmospheric CO_2 and climate from 65 to 30 ka B.P.", *Geophysical Research Letters*, **34**, L10703.

Allen, M.R., Frame, D.J., Huntingford, C., Jones, C.D., Lowe, J.A., Meinshausen, M., and Meinshausen, N. (2009), "Warming caused by cumulative carbon emissions towards the trillionth tonne", *Nature*, **458**, 1163–1166.

Alley, R.B., Clark, P.U., Huybrechts, P., and Joughin, I. (2005), "Ice-sheet and sea-level changes", *Science*, **310**, 456.

Ammann, C.M., Joos, F., Schimel, D.S., Otto-Bliesner, B.L., and Tomas, R.A. (2007), "Solar influence on climate during the past millennium: Results from transient simulations with the NCAR Climate System Model", *Proceedings of the National Academy of Sciences (PNAS)*, **104**, 3713–3718.

Anderson, E.R. (1952), "Energy budget studies", *U.S. Geological Survey Circular*, **229**, 71–119.

Anderson, E.R. (1954), "Energy budget studies", in *Technical Report*, United States Geological Survey Professional Paper 269, Water Loss Investigations—Lake Hefner Studies, pp. 71–117.

Anderson, T.L., Charlson, R.J., Schwartz, S.E., Knutti, R., Boucher, O., Rodhe, H.,

D. Rapp, *Assessing Climate Change*, Springer Praxis Books,
DOI: 10.1007/978-3-319-00455-6

and Heintzenberg, J. (2003), "Climate forcing by aerosols: A hazy picture", *Science*, **300**, 1103–1104.

Andreadis, K.M., and Lettenmaier, D.P. (2006), "Trends in 20th century drought over the continental United States", *Geophysical Research Letters*, **33**, L10403.

Andronova, N., Penner, J.E., and Wong, T. (2009), "Observed and modeled evolution of the tropical mean radiation budget at the top of the atmosphere since 1985", *Journal of Geophysical Research*, **114**, D14106.

Ångström, A.K. (1920), "Applications of heat radiation measurements to problems of evaporation from lakes and heat convection at their surfaces", *Gerlands Beitrage Geografiska Annaler*, **2**, 237.

Anon. (A), "Antarctic connection" (*www.antarcticconnection.com/antarctic/weather/index.shtml*).

Anon. (B), *Climate Change and the Greenhouse Effect*, Hadley Center, U.K., Department of Meteorology (*www.reefrelief.org/scientificstudies/climate_greenhouse.pdf*).

Anon. (C) (2005), "Correcting temperature data sets", comments by several groups, *Science*, **310**, 972–973.

Anon. (D), "Hot topics in climate change science, species extinctions will increase due to global warming", Australian Government (*www.climatechange.gov.au/science/hottopics/pubs/topic11.pdf*).

Anon. (E), "Ice over the Poles" (*http://earthobservatory.nasa.gov/Library/Polar Ice/*).

Anon. (F) (2008), "International Center for Technology Assessment: Communicating uncertainty in the science of climate change" (*www.icta.org/doc/Uncertainty %_20in%20science-9-04.pdf*).

Anon. (G), "Kyoto Protocol", Wikipedia (*http://en.wikipedia.org/wiki/Kyoto_Protocol*).

Anon. (H), "Published e-letter responses to Hoffert et al. (2002)" (*http://intl.sciencemag.org/cgi/eletters/298/5595/981?ck=nck—676*).

Anon. (I), "Satellite finds warming relative to humidity", Goddard Earth Science and Technology Center, NASA press release: 04-090 (*http://gest.umbc.edu/science_hilt/highlights.html*).

Anon. (J) (1994), *Solar Influences on Global Climate Change*, National Academy Press, Washington, DC, U.S.

Anon. (K), "Spotless Sun: Blankest year of the Space Age" (*http://science.nasa.gov/headlines/y2008/30sep_blankyear.htm*).

Anon. (L), *Stern Review on the Economics of Climate Change*, HM Treasury (*www.hm-treasury.gov.uk./independent_reviews/stern_review_economics_climate_change/stern_review_report.cfm*).

Anon. (M) (2006), "Surface temperature reconstructions for the last 2,000 years", Committee on Surface Temperature Reconstructions for the Last 2,000 Years, National Research Council (*www.nap.edu/catalog/11676.html*).

Anon. (N), "The Climate Change Performance Index" (*http://germanwatch.org/en/5698*).

Anon. (O), *The Greenhouse Effect and Climate Change*, Australian Government

Department of Meteorology (*www.bom.gov.au/info/GreenhouseEffectAndClimate Change.pdf*).

Anon. (P) (2004), *What Is Dangerous Climate Change? Initial Results of a Symposium on Key Vulnerable Regions Climate Change and Article 2 of the UNFCCC*, Buenos Aires, 14 December.

Anon. (Q), "Windows to the Universe" (*www.windows.ucar.edu/*).

Archer, D., and Brovkin, V. (2006), "Millennial atmospheric lifetime of anthropogenic CO_2" (*www.pik-potsdam.de/~victor/archer.subm.clim.change.pdf*).

Archer, D., Winguth, A., Lea, D., and Mahowald, N. (2000), "What caused the glacial/interglacial atmospheric PCO_2 cycles?", *Reviews of Geophysics*, **38**, 159–189.

Archibald, D.C. (2009), "Solar cycle 24: Expectation sand implications", *Energy and Environment*, **20**, 1–10.

Armour, K.C., and Roe, G.H. (2011), "Climate commitment in an uncertain world", *Geophysical Research Letters*, **38**, L01707.

Badalyan, O.G., Obridko, V.N., and Sykora, J. (2001), "Brightness of the coronal green line and prediction for activity cycles", *Solar Physics*, **199**, 421–435.

Baldwin, R.C. (1970), "A dispersion model for heated effluent from an ocean outfall", thesis, United States Naval Postgraduate School (*www.dtic.mil/dtic/tr/fulltext/u2/710730.pdf*).

Baker, M.B., and Roe, G.H. (2009), "The shape of things to come: Why is climate change so predictable?", *Journal of Climate*, **22**, 4574–4589.

Ball, T. (2007), "The science isn't settled: The limitations of global climate models", Marshall Institute (*www.marshall.org/article.php?id=524*).

Ballantyne, A.P., Greenwood, D.R., Sinninghe Damsté, J.S., Csank, A.Z., Eberle, J.J., and Rybczynski, N. (2010), "Significantly warmer Arctic surface temperatures during the Pliocene indicated by multiple independent proxies", *Geology*, **38**, 603–606.

Balling, R.C., Jr., Vose, R.S., and Weber, G.-R. (1998), "Analysis of long-term European temperature records: 1751–1995", *Climate Research*, **10**, 193–200.

Ban-Weiss, G.A., Bala, G., Cao, L, Pongratz, J., and Caldeira, K. (2011), "Climate forcing and response to idealized changes in surface latent and sensible heat", *Environmental Research Letters*, **6**, 034032.

Barclay, D.J., Wiles, G.C., and Calkin, P.E. (2009), "Tree-ring crossdates for a First Millennium AD advance of ebenkof Glacier, southern Alaska", *Quaternary Research*, **71**, 22–26.

Bard, E., Raisbeck, G., Yiou, F., and Jouzel, J. (2000), "Solar irradiance during the last 1200 years based on cosmogenic nuclides", *Tellus*, **B52**, 985–992.

Barnett, T., Zwiers, F., Hegerl, G., Allen, M., Crowley, T., Gillett, N., Hasselmann, K., Jones, P., Santer, B., Schnur, R., *et al.* (2005a), "Detecting and attributing external influences on the climate system: A review of recent advances", *LLRL Report UCRL-JRNL-209353*.

Barnett, T.P., Pierce, D.W., AchutaRao, K.M., Gleckler, P.J., Santer, B.D., Gregory, J.M., and Washington, W.M. (2005b), "Penetration of human-induced warming into the world's oceans", *Science*, **309**, 284–287.

Barron, E.J. (1984), "Ancient climates: Investigation with climate models (1984)", *Reports of Progress in Physics*, **47**, 1563–1599.

Barron, E.J., Thompson, S.L., and Hay, W.W. (1984), "Continental distribution as a forcing factor for global-scale temperature", *Nature*, **310**, 574–575.

Bartlett, A. (2002), "Slowing population growth", *Science*, 24 April, e-letter response to Hoffert *et al.* (2002).

Bauer, E., Claussen, M., and Brovkin, V. (2003), "Assessing climate forcings of the Earth system for the past millennium", *Geophysical Research Letters*, **30**, 1276.

Bauer, M., Del Genio, A.D., and Lanzante, J.R. (2002), "Observed and simulated temperature–humidity relationships: Sensitivity to sampling and analysis", *Journal of Climate*, **15**, 203–215.

Beckman, J.E., and Mahoney, T.J. (1998), "The Maunder Minimum and climate change: Have historical records aided current research?", Library and Information Services in Astronomy III ASP Conference Series 153.

Beer, J., Mende, W., and Stellmacher, R. (2000), "The role of the sun in climate forcing", *Quaternary Science Reviews*, **19**, 403–415.

Bejan, A., and Heitor Reis, A. (2005), "Thermodynamic optimization of global circulation and climate", *International Journal of Energy Research*, **29**, 303–316.

Bender, F.A.-M., Ekman, A.M.L., and Rodhe, H. (2010a), "Response to the eruption of Mount Pinatubo in relation to climate sensitivity in the CMIP3 models", *Climate Dynamics* (*www.springerlink.com/index/a75134944633kt74.pdf*).

Bender, M.A., Knutson, T.R., Tuleya, R.E., Sirutis, J.J., Vecchi, G.A., Garner, S.T., and Held, I.M. (2010b), "Modeled impact of anthropogenic warming on the frequency of intense Atlantic hurricanes", *Science*, **327**, 454–458.

Benestad, R.E. (2005), *Solar Activity and Earth's Climate*, 2nd edn, Praxis Publishing, Chichester, U.K.

Bengtsson, L., Semenov, V.A., and Johannessen, O.M. (2004), "The early twentieth-century warming in the Arctic: A possible mechanism", *Journal of Climate*, **17**, 4045–4057.

Berliand, M.E., and Berliand, T.G. (1952), "Determination of effective radiation of the Earth as influenced by cloud cover", *Izvestiia Akademii Nauk S.S.S.R.*, Seriia Geofiziceskaya 1, 64–78.

Berner, R.A. (2004), *The Phanerozoic Carbon Cycle: CO_2 and O_2*, Oxford University Press, Oxford, U.K.

Bernaerts, A. (2009), "Arctic heats up: Spitzbergen 1919–1939", iUniverse.com, Bloomington, IN, U.S.

Bhatt, N.J., Jain, R., and Aggarwal, M. (2009), "Predicting maximum sunspot number in Solar Cycle 24", *Journal of Astrophysics and Astronomy*, **30**, 71–77.

Bielefeld, B. (1997), "Investigation into albedo-controlled energy loss during the last glaciation", *Geojournal*, **42**, 329–336.

Bischof, J. (2000), *Ice Drift, Ocean Circulation and Climate Change*, Springer/Praxis, Heidelberg, Germany/Chichester, U.K.

Boening, C., Willis, J.K., Landerer, F.W., Nerem, R.S., and Fasullo, J. (2012), "The 2011 La Niña: So strong, the oceans fell", *Geophysical Research Letters*, **39**, L19602–L19606.

Bohaty, S.M., Zachos, J.C., Florindo, F., and Delaney, M.L. (2009), "Coupled greenhouse warming and deep-sea acidification in the middle Eocene", *Paleoceanography*, **24**, PA2207.

Bonan, G.B. (1997), "Effects of land use on the climate of the United States", *Climatic Change*, **37**, 449–486 (*http://sheridan.geog.kent.edu/geog41066/bonan.pdf*).

Bonan, G.B. (2002), *Ecological Climatology*, 1st edn (*www.cgd.ucar.edu/tss/aboutus/staff/bonan/ecoclim/index.htm*).

Bond, G., Kromer, B., Beer, J., Muscheler, R., Evans, M.N., Showers, W., Hoffmann, S, Lotti-Bond, R., Hajdas, I., and Bonani, G. (2001), "Persistent solar influence on North Atlantic climate during the Holocene", *Science*, **294**, 2130–2136.

Bond, T.C. (2007), Testimony for the Hearing on Black Carbon and Global Warming House Committee on Oversight and Government Reform United States House of Representatives, The Honorable Henry A. Waxman, Chair, 18 October.

Bond, T.C. (2010), Testimony for "Clearing the Smoke: Black Carbon Pollution" House Committee on Energy Independence and Global Warming United States House of Representatives The Honorable Edward Markey, Chair, 16 March.

Bony, S., Colman, R., Kattsov, V.M., Allan, R.P., Bretherton, C.S., Dufresne, J.-L., Hall, A., Hallegatte, S., Holland, M.M., Ingram, W., *et al.* (2006), "How well do we understand and evaluate climate change feedback processes?", *Journal of Climate*, **19**, 3446–3483.

Booker, C., and North, R. (2007), *Scared to Death: From BSE to GlobalWarming: Why Scares Are Costing Us the Earth*, Continuum, London, U.K.

Boucota, A.J., and Gray, J. (2001), "A critique of Phanerozoic climatic models involving changes in the CO_2 content of the atmosphere", *Earth Science Reviews*, **56**, 1–159.

Bounoua, L., Collatz, G.J., Los, S.O., Sellers, P.J., Dazlich, D.A., Tucker, C.J., and Randall, D.A. (2000), "Sensitivity of climate to changes in NDVI", *Journal of Climate*, **13**, 2277–2292.

Bousquet, P., Peylin, P., Ciais, P., Le Quere, C., Friedlingstein, P., and Tans, P.P. (2000), "Regional changes in carbon dioxide fluxes of land and oceans since 1980", *Science*, **290**, 1342–1346.

Bouville, M. (2008), "Plagiarism: Words and ideas", *Science and Engineering Ethics*, **14**, 311–322.

Box, J.E., Yang, L, Bromwich, D.H., and Bai, L.-S. (2009), "Greenland ice sheet surface air temperature variability: 1840–2007", *Journal of Climate* (submitted, *http://adsabs.harvard.edu/abs/2008AGUFM.C21D..04B*).

Boykoff, M.T., and Boykoff, J.M. (2004), "Balance as bias: Global warming and the US prestige press", *Global Environmental Change Human and Policy Dimensions*, **14**, 125–136.

Braganza, K., Karoly, D.J., and Arblaster, J.M. (2004), "Diurnal temperature range as an index of global climate change during the twentieth century", *Geophysical Research Letters*, **31**, L13217–L13220.

Braithwaite, R.J. (2009), "After six decades of monitoring glacier mass balance we still need data but it should be richer data", *Annals of Glaciology*, **50**, 191.

Breecker, D.O., Sharp, Z.D., and McFadden, L.D. (2010), "Atmospheric CO_2 concentrations during ancient greenhouse climates were similar to those predicted for A.D. 2100", *Proceedings of the National Academy of Sciences (PNAS)*, **107**, 576–580.

Brierley, C.M., Fedorov, A.V., Liu, Z., Herbert, T.D., Lawrence, K.T., and LaRiviere, J.P. (2009), "Greatly expanded tropical warm pool and weakened Hadley circulation in the Early Pliocene", *Science*, **323**, 1714–1718.

Briffa, K.R., and Osborn, T.J. (1999), "Seeing the wood from the trees", *Science*, **284**, 926–927.

Briffa, K.R., Osborn, T.J., Schweingruber, F.H., Harris, I.C., Jones, P.D., Shiyatov, S.G., and Vaganov, E.A. (2001), "Low-frequency temperature variations from a northern tree-ring-density network", *Journal of Geophysical Research*, **106**, 2929–2941.

Briffa, K.R., Schweingruber, F.H., Jones, P.D., Osborn, T.J., Shiyatov, S.G., and Vaganov, E.A. (1998), "Reduced sensitivity of recent tree-growth to temperature at high northern latitudes", *Nature*, **391**, 678–682.

Brillinger, D., Curry, J., Jacobsen, R., Muller, E., Muller, R., Perlmutter, S., Rohde, R., Rosenfeld, A., Wickham, C., and Wurtele, J. (2011), "Berkeley Earth surface temperature analysis" (*www.berkeleyearth.org/*).

Broecker, W.S. (1995), "Chaotic climate", *Scientific American*, November.

Broecker, W.S. (2002), *The Glacial World According to Wally*, Eldigo Press, New York, U.S.

Brohan, P., Kennedy, J.J., Harris, I., Tett, S.F.B., and Jones, P.D. (2006), "Uncertainty estimates in regional and global observed temperature changes: A new dataset from 1850", *Journal of Geophysical Research*, **111**, D12106–D12140.

Brooks, H.E. (2006), *Tornado and Severe Thunderstorm Damage*, NOAA/National Severe Storms Laboratory, Norman, Oklahoma, U.S.

Brovkin, V., Claussen, M., Driesschaert, E., Fichefet, T., Kicklighter, D., Loutre, M.F., Matthews, H.D., Ramankutty, N., Schaeffer, M., and Sokolov, A. (2006), "Biogeophysical effects of historical land cover changes simulated by six Earth system models of intermediate complexity", *Climate Dynamics*, **26**, 587–600.

Brown, J.H., Burnside, W.R., Davidson, A.D., DeLong, J.P., Dunn, W.C., Hamilton, M.J., Mercado-Silva, N., Nekola, J.C., Okie, J.G., Woodruff, W.H., and Zuo, W. (2011), "Energetic limits to economic growth", *BioScience*, **61**, 19–26.

Brunetti, M. (2003), "Solar signals in instrumental historical series of meteorological parameters", *Memorie della Società Astronomica Italiana*, **74**, 778–785.

Brunt, D. (1932), "Notes on radiation in the atmosphere", *Quarterly Journal of the Royal Meteorological Society*, **58**, 389–420.

Bunker, A.F. (1976), "Computation of surface energy flux and annual air–sea cycle of the North Atlantic Ocean", *Monthly Weather Review*, **104**, 1122–1140.

Burger, G., and Cubasch, U. (2005), "Are multiproxy climate reconstructions robust?", *Geophysical Research Letters*, **32**, L23711.

Burrett, C.F. (1982), "Phanerzoic land–sea and albedo variations as climate controls", *Nature*, **296**, 54–56.

Byatt, I., Castles, I., Goklany, I.M., Henderson, D., Lawson, N., McKitrick, R., Morris, J., Peacock, A., Robinson, C., and Skidelsky, R. (2006), "The Stern Review: A dual critique. Part II: Economic aspects", *World Economics*, **7**, 199–230.

Caillon, N., Severinghaus, J.P., Jouzel, J., Barnola, J.-M., Kang, J., and Lipenkov, V.Y. (2003), "Timing of atmospheric CO_2 and Antarctic temperature changes across Termination III", *Science*, **299**, 1728–1747.

Callebaut, D.K. (2008), 'Approach of a deep minimum in Cycle 26 and effect on climate" (*mearim.cu.edu.eg/new/41-Mearin-Callebaut-Approach.pdf*).

Came, R.E., Eiler, J.M., Veizer, J., Azmy, K., Brand, U., and Weidman, C.R. (2007), "Coupling of surface temperatures and atmospheric CO_2 concentrations during the Palaeozoic era", *Nature*, **449**, 198–202.

Canadell, J.G., Le Quere, C., Raupacha, M.R., Fielde, C.B., Buitenhuis, E.T., Ciais, P., Conway, T.J., Gillett, N.P., Houghton, R.A., and Marlandi, G. (2007), "Contributions to accelerating atmospheric CO_2 growth from economic activity, carbon intensity and efficiency of natural sinks", *Proceedings of the National Academy of Sciences*, **104**, 18866–18870.

Cane, M.A. (1983), "Oceanographic events during El Nino", *Science*, **222**, 1189–1195.

Cane, M.A., and Zebiak, S.E. (1985), "A theory for El Nino and the Southern Oscillation", *Science*, **228**, 1085–1087.

Carlin, A. (2009), "Comments on Draft Technical Support Document for Endangerment Analysis for Greenhouse Gas Emissions under the Clean Air Act, March 16, 2009" (*http://wattsupwiththat.com/2009/06/27/released-the-censored-epa-document-final-report/*).

Carter, R.M., de Freitas, C.R., Goklany, I.M., Holland, D., and Lindzen, R.S. (2006), "The Stern Review: A dual critique. Part I: The science", *World Economics*, **7**, 165–198.

CCSP (2005), "Temperature trends in the lower atmosphere", U.S. Government Climate Change Science Program, 17 November (*www.climatescience.gov/Library/sap/sap1-1/public-review-draft/sap1-1prd-all.pdf*).

CCSP (2007), "CCSP Synthesis and Assessment Report 3.1. Climate models: An assessment of strengths and limitations for user applications", Public Review Draft, U.S. Government Climate Change Science Program, 15 May (*www.climatescience.gov/Library/sap/sap3-1/public-review-draft/sap3-1prd-cover.pdf*).

Changnon, S.A. (ed.) (2000), *El Niño 1997–1998: The Climate Event of the Century*, Oxford University Press, New York, U.S.

Chapman, W.L., and Walsh, J.E. (2007), "A synthesis of Antarctic temperatures", *Journal of Climate*, **20**, 4096–4117.

Charbit, S., Ritz, C., Philippon, G., Peyaud, V., and Kageyama, M. (2007), "Numerical reconstructions of the Northern Hemisphere ice sheets through the last glacial–interglacial cycle", *Climate of the Past*, **3**, 15–37.

Charlson, R., Francisco, J., Valero, P.J., and Seinfeld, J.H. (2005), "In search of balance", *Science*, **308**, 806–807.

Charney, J. (1979), *Carbon Dioxide and Climate: A Scientific Assessment*, National Academy of Sciences Press, Washington, DC, U.S.

Chen, J., Carlson, B.E., and Del Genio, A.D. (2002), "Evidence for strengthening of the tropical general circulation in the 1990s", *Science*, **295**, 838–841.
Chen, J.L., Wilson, C.R., Blankenship, D.D., and Tapely, B.D. (2006), "Antarctic mass rates from GRACE", *Geophysical Research Letters*, **33**, L11502.
Chilingar, G.V., Khilyuk, L.F., and Sorokhtin, O.G. (2008), "Cooling of the atmosphere due to CO_2 emission", *Energy Sources*, **A30**, 1–9.
Chin, M., Kahn, R.A., Remer, L.A., Yu, H., Rind, D., Feingold, G., Quinn, P.K., Schwartz, S.E., Streets, D.G., DeCola, P., and Halthore, R. (2009), "Atmospheric aerosol properties and climate impacts", U.S. Climate Change Science Program, Synthesis and Assessment Product 2.3, January.
Chou, M.-D., Lindzen, R.S., and Hou, A.Y. (2002), "Comments on 'The Iris Hypothesis: A Negative or Positive Cloud Feedback?'", *Journal of Climate*, **15**, 2713–2715.
Choudhuri, A.R., Chatterjee, P., and Jiang, J. (2007), "Predicting Solar Cycle 24 with a solar dynamo model", *Physical Review Letters*, **98**, 131101.
Christy, J.R. (2012), Written Statement of John R. Christy, The University of Alabama in Huntsville, Subcommittee Energy and Power, U.S. House of Representatives, 20 September (*http://energycommerce.house.gov/sites/republicans.energycommerce.house.gov/files/Hearings/EP/20120920/HHRG-112-IF03-WState-ChristyJ-20120920.pdf*).
Christy, J.R., Herman, B., Pielke, R., Sr., Klotzbach, P., McNider, R.T., Hnilo, J.J., Spencer, R.W., Chase, T., and Douglass, D. (2010), "What do observational datasets say about modeled tropospheric temperature trends since 1979?", *Remote Sensing*, **2**, 2148–2169.
Christy, J.R., Norris, W.B., Spencer, R.W., and Hnilo, J.J. (2007), "Tropospheric temperature change since 1979 from tropical radiosonde and satellite measurements", *Journal of Geophysical Research*, **112**, D06102.
Christy, J.R., Spencer, R.W., and Braswell, W.D. (2000), "MSU tropospheric temperatures: Dataset construction and radiosonde comparisons", *Journal of Atmospheric and Ocean Technology*, **17**, 1153–1170.
Church, J.A., and White, N.J. (2006), "20th century acceleration in global sea-level rise", *Geophysical Research Letters*, **33**, L01602.
Church, J.A., and White, N.J. (2011), "Sea level rise from the late 19th to the early 21st century", *Surveys in Geophysics*, **32**, 585–602.
Church, J.A., White, N.J., and Hunter, J.R. (2006), "Sea-level rise at tropical Pacific and Indian Ocean islands", *Global and Planetary Change*, **53**, 155–168 (*http://timesonline.typepad.com/times_tokyo_weblog/files/john_church_document.pdf*).
Church, J.A., White, N.J., Coleman, R., Lambeck, K., and Mitrovica, J.X. (2004), "Estimates of the regional distribution of sea-level rise over the 1950 to 2000 period", *Journal of Climate*, **17**, 2609–2625.
Church, J.A., White, N.J., Konikow, L.F., Domingues, C.M., Cogley, J.G., Rignot, E., Gregory, J.M., van den Broeke, M.R., Monaghan, A.J., and Velicogna, I. (2011), "Revisiting the Earth's sea level and energy budgets from 1961 to 2008", *Geophysical Research Letters*, **38**, L18601.
Chylek, P., and Lohmann, U. (2008), "Aerosol radiative forcing and climate

sensitivity deduced from the Last Glacial Maximum to Holocene transition", *Geophysical Research Letters*, **35**, L04804.
Chylek, P., Box, J.E., and Lesins, G. (2004), "Global warming and the Greenland ice sheet", *Climatic Change*, **63**, 201–221.
Chylek, P., Dubey, M.K., and Lesins, G. (2006), "Greenland warming of 1920–1930 and 1995- 2005", *Geophysical Research Letters*, **33**, L11707.
Chylek, P., Folland, C.K., Lesins, G., Dubey, M.K., and Wang, M. (2009), "Arctic air temperature change amplification and the Atlantic Multidecadal Oscillation", *Geophysical Research Letters*, **36**, L14801.
Chylek, P., Lohmann, U., Dubey, M., Mishchenko, M., Kahn, R., and Ohmura, A. (2007), "Limits on climate sensitivity derived from recent satellite and surface observations", *Journal of Geophysical Research*, **112**, D24S04–D24S12.
Clark, N.E., Eber, L., Laurs, R.M., Renner, J.A., and Saur, J.F.T. (1974), "Heat exchange between ocean and atmosphere in the eastern North Pacific for 1961–71", NOAA Tech. Rep. NMFS SSRF–682, U.S. Department of Commerce, Washington, DC, U.S.
Clark, P.U., and Huybers, P. (2009), "Interglacial and future sea level", *Nature*, **462**, 856.
Clilverd, M.A., Clarke, E., Ulich, T., Rishbeth, H., and Jarvis, M.J. (2006), "Predicting Solar Cycle 24 and beyond", *Space Weather*, **4**, S09005.
Collins, W.D., Ramaswamy, V., Schwarzkopf, M.D., Sun, Y., Portmann, R.W., Fu, Q., Casanova, S.E.B., Dufresne, J.-L., Fillmore, D.W., Forster, P.M.D., *et al.* (2006), "Radiative forcing by well-mixed greenhouse gases: Estimates from climate models in the Intergovernmental Panel on Climate Change (IPCC) Fourth Assessment Report (AR4)", *Journal of Geophysical Research*, **111**, D14317.
Comiso, J.C. (2000), "Variability and trends in Antarctic surface temperatures from *in situ* and satellite infrared measurements", *Journal of Climate*, **13**, 1674–1697.
Conway, H., Hall, B.L., Denton, G.H., Gades, A.M., and Waddington, E.D. (1999), "Past and future grounding-line retreat of the West Antarctic Ice Sheet", *Science*, **286**, 280–283.
Crichton, M. (2003), "Aliens cause global warming" (*www.crichton-official.com/speech-alienscauseglobalwarming.html*).
Crompton, R.P., Pielke, R.A., Jr., and McAneney, K.J. (2011), "Emergence timescales for detection of anthropogenic climate change in US tropical cyclone loss data", *Environmental Research Letters*, **6**, 014003.
Cronin, T.M., Hayo, K., Thunell, R.C., Dwyer, G.S., Saenger, C., and Willard, D.A. (2010), "The medieval climate anomaly and Little Ice Age in Chesapeake Bay and the North Atlantic Ocean", *Palaeogeography, Palaeoclimatology, Palaeoecology*, **297** 299–310.
Crowley, T.J. (2002), "Cycles, cycles everywhere", *Science*, **295**, 1473–1474.
Crowley, T.J., and Berner, R.A. (2001), "Paleoclimate: Enhanced CO_2 and climate change", *Science*, **292**, 870–872.
Crowley, T.J., and Lowery, T.S. (2000), "How warm was the *Medieval Warm Period*?", *Ambio*, **29**, 51–54.

Crucifix, M. (2006), "Does the Last Glacial Maximum constrain climate sensitivity?", *Geophysical Research Letters*, **33**, L18701.

Cubasch, U., Zorita, E., Gonzales-Rouco, F., von Storch, H., and Fast, I. (2002), "Cool stars, the sun and climate variability: Is there a connection?" (*www.hs.unihamburg.de/cs13/day1/02_Cubasch.ppt*).

Cui, Y., Kump, L.R., Ridgwell, A.J., Charles, A.J., Junium, C.K., Diefendorf, A.F., Freeman, K.H., Urban, N.M., and Hardin, I.C. (2011), "Slow release of fossil carbon during the Palaeocene–Eocene Thermal Maximum", *Nature Geoscience*, **4**, 481–485.

Curry, J.A., Webster, P.J., and Holland, G.J. (2006), "Mixing politics and science in testing the hypothesis that greenhouse warming is causing a global increase in hurricane intensity", *Bulletin of the American Meteorological Society (BAMS)*, August, 1025–1037.

Curtin, T. (2009), "Climate change and food production", *Energy and Environment*, **20**, 1098–1116.

D'Arrigo, R., Wilson, R., and Jacoby, G. (2006a), "Further discussion on: Tree-ring temperature reconstructions for the past millennium" (*www.ldeo.columbia.edu/res/fac/trl/.../D'AWilsJac.nrc.followup.pdf*).

D'Arrigo, R., Wilson, R. and Jacoby, G. (2006b), "On the long-term context for late twentieth century warming", *Journal of Geophysical Research*, **111**, D03103.

Dahl-Jensen, D., and the North Greenland Eemian Ice Drilling (NEEM) community members (2013), "Eemian interglacial reconstructed from a Greenland folded ice core", *Nature*, **493**, 489–494.

Dahl-Jensen, D., Mosegaard, K., Gundestrup, N., Clow, G.D., Johnsen, S.J., Hansen, A.W., and Balling, N. (1998), "Past temperatures directly from the Greenland ice sheet", *Science*, **282**, 268–271.

Dai, A., Karl, T.R., Sun, B., and Trenberth, K.E. (2006), "Recent trends in cloudiness over the United States: A tale of monitoring inadequacies", *Bulletin of the American Meteorological Society (BAMS)*, May, 597–606.

Danilov-Danil'yan, V., Losev, K.S., and Reyf, I.E. (2009), *Sustainable Development and the Limitation of Growth*, Springer-Praxis Publishing, Chichester, U.K.

Dansgaard, W. (2005), *Frozen Annals*, Niels Bohr Institute, Copenhagen, Denmark.

Dasgupta, S., Laplante, B., Meisner, C., Wheeler, D., and Yan, J. (2007), *The Impact of Sea Level Rise on Developing Countries: A Comparative Analysis*, World Bank Policy Research Working Paper 4136, February (*www-wds.worldbank.org/external/default/WDSContentServer/IW3P/IB/2007/02/09/000016406_20070209161430/Rendered/PDF/wps4136.pdf*).

Davey, C.A., and Pielke, R.A., Sr. (2005), "Microclimate exposures of surface-based weather stations implications for the assessment of long-term temperature trends", *Bulletin of the American Meteorological Society*, **86,** 497–506.

Davis, C.H., Li, Y., McConnell, J.R., Frey, M.M., and Hanna, E. (2005), "Snowfall-driven growth in East Antarctic ice sheet mitigates recent sea-level rise", *Science*, **308**, 1898–1901.

de Jager, C. (2008), "Solar activity and its influence on climate", *Netherlands Journal of Geosciences—Geologie en Mijnbouw*, **87**, 207–213.

de Laat, A.T.J., and Maurellis, A.N. (2004), "Industrial CO_2 emissions as a proxy for anthropogenic influence on lower tropospheric temperature trends", *Geophysical Research Letters*, **31**, L05204.

de Laat, A.T.J., and Maurellis, A.N. (2006), "Evidence for influence of anthropogenic surface processes on lower tropospheric and surface temperature trends", *International Journal of Climatology*, **26**, 897–913.

de Silva, S.L. (2003), "Volcanic eruptions and their impact on the Earth's climate" (*www.space.edu/papers/Volcanoclimate.pdf*).

Deffeyes, K.S. (2001), *Hubbert's Peak: The Impending World Oil Shortage*, Princeton University Press, Princeton, NJ, U.S.

Degaetano, A.T., and Allen, R.J. (2002), "Trends in twentieth-century temperature extremes across the United States", *Journal of Climate*, **15**, 3188–3206.

Del Genio, A.D. (2002), "The dust settles on water vapor feedback", *Science*, **296**, 665–666.

Desler, C., Alexander, M.A., and Timlin, M.S. (1996), "Upper-ocean thermal variations in the North Pacific during 1970–1991", *Journal of Climate*, **9**, 1840–1855.

Dessler, A.E. (2010), "A determination of the cloud feedback from climate variations over the past decade", *Science*, **330**, 1523–1527.

Dessler, A.E. (2011), "Cloud variations and the Earth's energy budget", *Geophysical Research Letters* (*geotest.tamu.edu/userfiles/216/Dessler2011.pdf*).

Dessler, A.E., and Davis, S.M. (2010), "Trends in tropospheric humidity from reanalysis systems", *Journal of Geophysical Research*, **115**, D19127.

Dessler, A.E., Yang, P., Lee, J., Solbrig, J., Zhang, Z., and Minschwaner, K. (2008a), "An analysis of the dependence of clear-sky top-of-atmosphere outgoing long-wave radiation on atmospheric temperature and water vapor", *Journal of Geophysical Research*, **113**, D17102–D17111.

Dessler, A.E., Zhang, Z., and Yang, P. (2008b), "Water-vapor climate feedback inferred from climate fluctuations, 2003–2008", *Geophysical Research Letters*, **35**, L20704–L20707.

Deutsche Bank Group (2008), "Investing in climate change 2009: Necessity and opportunity in turbulent times" (*http://dbadvisors.com/climatechange*).

Dietz, R.S., and Holden, J.C. (1970), "Reconstruction of Pangaea: Breakup and dispersion of continents, permian to present", *Journal of Geophysical Research*, **75**, 4939–4956.

Dikpati, M., de Toma, G., and Gilman, P.A. (2006), "Predicting the strength of Solar Cycle 24 using a flux-transport dynamo-based tool", *Geophysical Research Letters*, **33**, L05102–L05105.

DiLorenzo, E., Schneider, N., Cobb, K.M., Franks, P.J.S., Chhak, K., Miller, A.J., McWilliams, J.C., Bograd, S.J., Arango, H., Curchitser, E., *et al.* (2007), "North Pacific Gyre Oscillation links ocean climate and ecosystem change", *Geophysical Research Letters*, submitted.

Divine, D.V., and Dick, C. (2006), "Historical variability of sea ice edge position in the nordic seas", *Journal of Geophysical Research*, **111**, C01001.

Doherty, S.J., Warren, S.G., Grenfell, T.C., Clarke, A.D., and Brandt, R.E. (2010),

"Light-absorbing impurities in Arctic snow", *Atmospheric Chemistry and Physics*, **10**, 18807–18878.

Domingo, V., Ermolli, I., Fox, P., Fröhlich, C, Haberreiter, M., Krivova, N., Kopp, G., Schmutz, W., Solanki, S.K., Spruit, H.C., *et al.* (2009), "Solar surface magnetism and irradiance on time scales from days to the 11-year cycle", *Space Science Reviews*, **145**, 337–380.

Domingues, C.M., Church, J.A., White, N.J., Gleckler, P.J., Wijffels, S.E., Barke, P.M., and Dunn, J.R. (2008), "Improved estimates of upper-ocean warming and multi-decadal sea-level rise", *Nature*, **453**, 1090–1093.

Doran, P.T., Priscu, J.C., Berry Lyons, W., Walsh, J.E., Fountain, A.G., McKnight, D.M., Moorhead, D.L., Virginia, R.A., Wall, D.H., Clow, G.D., *et al.* (2002), "Antarctic climate cooling and terrestrial ecosystem response", *Nature*, **415**, 517–520.

Doria, G., Royer, D.L., Wolfe, A.P., Fox, A., Westgate, J.A., and Beerling, D.J. (2011), "Declining atmospheric CO_2 during the late Middle Eocene climate transition", *American Journal of Science*, **311**, 63–75.

Dorman, C.E. (1974), "Analysis of meteorological and oceanographic data from Ocean Station Vessel N (30N 140W)", Ph.D. thesis.

Douglas, B.C. (1992), "Global sea level acceleration", *Journal of Geophysical Research*, **97**, 12699–12706.

Douglas, B.C., and Peltier, W.R. (2002), "The puzzle of global sea-level rise", *Physics Today*, March (*www.aip.org/web2/aiphome/pt/vol-55/iss-3/p35.html*).

Douglass, D.H. (2005), "Observational climate data and comparison with models", in R. Ragaini (ed.), *The Science and Culture Series Nuclear Strategy and Peace Technology*, World Scientific Publishing Co. Ltd.

Douglass, D.H. (2010), "El Niño southern oscillation: Magnitudes and asymmetry", *Journal of Geophysical Research*, **115**, D15111.

Douglass, D.H., and Christy, J.R. (2009), "Limits on CO_2 climate forcing from recent temperature data of Earth", *Energy and Environment*, **20**, 177–189.

Douglass, D.H., and Knox, R.S. (2005), "Climate forcing by the volcanic eruption of Mount Pinatubo", *Geophysical Research Letters*, **32**, L05710 (revised version at *http://arxiv.org/abs/physics/0509166*).

Douglass, D.H., Christy, J.R., Pearson, B.D., and Singer, S.F. (2007), "A comparison of tropical temperature trends with model predictions", *International Journal of Climatology*, 10.1002/joc.1651.

Dupont, J.C., Haeffelin, M., and Long, C.N. (2009), "Cirrus cloud radiative effect on surface-level shortwave and longwave irradiances at regional and global scale", *Atmospheric Chemistry and Physics*, **9**, 26777–26832.

Dwyer, G.S., Cronin, T.M., Baker, P.A., Raymo, M.E., Buzas, J.S., and Corrège, T. (1995), "North Atlantic deepwater temperature change during late pliocene and late quaternary climatic cycles", *Science*, **270**, 1347–1351.

Dyke, A.S., Andrews, J.T., Clark, P.U., England, J.H., Miller, G.H., Shaw, J., and Veillette, J.J. (2002), "The Laurentide and Innuitian ice sheets during the Last Glacial Maximum", *Quaternary Science Reviews*, **21**, 9–31.

Dyurgerov, M.B., and Meier, M.F. (1999), "Twentieth century climate change:

Evidence from small glaciers", *Proceedings of the National Academy Sciences*, **97**, 1406–1411.

Eastman, R., Warren, S.G., and Hahn, C.J. (2011), "Variations in cloud cover and cloud types over the ocean from surface observations, 1954–2008", *Journal of Climate*, **24**, 5914–5934.

Eddy, J. (1976), "The Maunder Minimum", *Science*, **192**, 1189–1202.

Eddy, J. (2004), "The Sun–climate connection" (*www.ams.confex.com/ams/pdfpapers/74103.pdf*).

Edwards, E.J., Osborne, C.P., Strömberg, C.A.E., Smith, S.A., and C4 Grasses Consortium (2010), "The Origins of C4 Grasslands: Integrating evolutionary and ecosystem science", *Science*, **328**, 587–591.

Efimova, N.A. (1961), 'On methods of calculating monthly values of net long-wave radiation", *Meteorol Gidrol*, **10**, 28–33 (in Russian).

Ehrelinger, J.R., Cerling, T.E., and Dearing, M.D. (2005), *A History of Atmospheric CO_2 and Its Effects on Plants, Animals and Ecosystems*, Springer-Verlag, New York, U.S.

Emanuel, K. (2005), "Increasing destructiveness of tropical cyclones over the past 30 years", *Nature* **436**, 686–688.

Emanuel, K. (2010), "Tropical cyclone activity downscaled from NOAA-CIRES reanalysis, 1908–1958", *Journal of Advances in Modeling Earth Systems* , **2**, 1–12.

Emanuel, K.A. (2007), "Environmental factors affecting tropical cyclone power dissipation", *Journal of Climate*, **20**, 5497–5510.

Esper, J., Büntgen, U., Timonen, M., and Frank, D.C. (2012b), "Variability and extremes of northern Scandinavian summer temperatures over the past two millennia", *Global and Planetary Change*, **88–89**, 1–9.

Esper, J., Cook, E.R., and Schweingruber, F.H. (2002a), "Low-frequency signals in long tree-ring chronologies for reconstructing past temperature variability", *Science*, **295**, 2249–2252.

Esper, J., Frank, D.C., Timonen, M., Zorita, E., Wilson, R.J.S., Luterbacher, J., Holzkämper, S., Fischer, N., Wagner, S., Nievergelt, D., Verstege, A., and Büntgen, U. (2012a), "Orbital forcing of tree-ring data", *Nature Climate Change Online*, 8 July.

Esper, J., Schweingruber, F.H., and Winiger, M. (2002b), "1300 years of climatic history for Western Central Asia inferred from tree-rings", *The Holocene*, **12**, 267–277.

Esper, J., Wilson, R.J.S., Frank, D.C., Moberg, A., Wanner, H., and Luterbacher, J. (2005), "Climate: Past ranges and future changes", *Quaternary Science Reviews*, **24**, 2164–2166.

Essex, C., McKitrick, R., and Andresen, B. (2007), "Does a global temperature exist?", *Journal of Non-Equilibrium Thermodynamics*, **32**, 1–27.

Etheridge, D.M., Steele, L.P., Langenfels, R.I., Francey, R.J., Barnola, J.M., and Morgan, V.I. (1996), "Natural and anthropogenic changes in atmospheric CO_2 over the last 1000 years from air in Antarctic ice and firn", *Journal of Geophysical Research*, **101**, 4115–4128.

Evan, A.T., Heidinger, A.K., and Vimont, D.J. (2007), "Arguments against a

physical long-term trend in global ISCCP cloud amounts", *Geophysical Research Letters*, **34**, L04701.
Fagan, B. (2000), *The Little Ice Age*, Basic Books, New York, U.S.
Fahey, D.W., Holloway, J.S., Ryerson, T.B., Peischl, J., and Brock, C.A. (2010), "Aircraft observations of enhancement and depletion of black carbon mass in the springtime Arctic", *Atmospheric Chemistry and Physics*, **10**, 15167–15196.
Falkowski, P., Scholes, R.J., Boyle, E., Canadell, J., Canfield, D., Elser, J., Gruber, N., Hibbard, K., Hogberg, P., Linder, S., *et al.* (2000), "The global carbon cycle: A test of our knowledge of Earth as a system", *Science*, **290**, 291–296.
Fall, S., Watts, A., Nielsen-Gammon, J., Jones, E., Niyogi, D., Christy, J.R., and Pielke, R.A., Sr. (2011), "Analysis of the impacts of station exposure on the U.S.: Historical Climatology Network temperatures and temperature trends", *Journal of Geophysical Research*, **116**, D14–D27.
Fedorov, A.V., and Philander, S.G. (2000), "Is El Nino Changing?", *Science*, **288**, 1997–2003.
Fedorov, A.V., Dekens, P.S., McCarthy, M., Ravelo, A.C., deMenocal, P.B., Barreiro, M., Pacanowski, R.C., and Philander, S.G. (2006), "The Pliocene Paradox (mechanisms for a permanent El Niño)", *Science*, **312**, 1485–1489.
Feulner, G., and Rahmstorf, S. (2010), "On the effect of a new grand minimum of solar activity on the future climate on Earth", *Geophysical Research Letters*, **37**, L05707.
Fischer, H., Wahlen, M., Smith, J., Mastroianni, D., and Deck, B. (1999), "Ice core records of atmospheric CO_2 around the last three glacial terminations", *Science*, **283**, 1712–1714.
Fjeldskaar, W. (2008), "Web commentaries on sea-level rise, stiger virkelig havnivaet mer enn før?" (*www.geoportalen.no/planetenjorden/klima/sealevel/*, *www.geo365.no/planetenjorden/klima/sealevel/*).
Flanner, M.G., Zender, C.S, Hess, P.G., Mahowald, N.M., Painter, T.H., Ramanathan, V., and Rasch, P.J. (2008), "Springtime warming and reduced snow cover from carbonaceous particles", *Atmospheric Chemistry and Physics*, **8**, 19819–19859.
Flanner, M.G., Zender, C.S., Randerson, J.T., and Rasch, P.J. (2007), "Present-day climate forcing and response from black carbon in snow", *Journal of Geophysical Research*, **112**, D11202–D11219.
Fletcher, B.J., Brentnall, S.J., Anderson, C.W., Berner, R.A., and Beerling, D.J. (2008), "Atmospheric carbon dioxide linked with Mesozoic and early Cenozoic climate change", *Nature Geoscience*, **1**, 43–48.
Fligge, M., Solanki, S., Meunier, N., and Unruh, Y. (2000), "Solar surface magnetism and the increase of solar irradiance between activity minimum and maximum", Proceedings of the First Solar and Space Weather Euroconference, 25–29 September, Santa Cruz de Tenerife, Tenerife, Spain.
Fligge, M., Solanki, S.K., Unruh, Y.C, Fröhlich, C., and Wehrli, C. (1998), "A model of solar total and spectral irradiance variations", *Astronomy and Astrophysics*, **335**, 709–718.
Flückiger, J., Monnin, E., Stauffer, B., Schwander, J., Stocker, T.F., Chappellaz, J.,

Raynaud, D., and Barnola, J.-M. (2002), "High-resolution Holocene N_2O ice core record and its relationship with CH_4 and CO_2", *Journal of Geophysical Research*, **16**, 1010 (*www.ncdc.noaa.gov/paleo/pubs/flueckiger2002/flueckiger2002.html*).

Foster, G., Annan, J.D., Jones, P.D., Mann, M.E., Renwick, J., Salinger, J., Schmidt, G.A., and Trenberth, K.E. (2010), "Comment on 'Influence of the southern oscillation on tropospheric temperature' by J. D. McLean, C. R. de Freitas and R. M. Carter", *Journal of Geophysical Research*, **115**, D09110.

Foster, G., Annan, J.D., Schmidt, G.A., and Mann, M.E. (2008), "Comment on 'Heat capacity, time constant and sensitivity of Earth's climate system' by S. E. Schwartz", *Journal of Geophysical Research*, **113**, D15102.

Foster, G.L., and Rohling, E.J. (2013), "Relationship between sea level and climate forcing by CO_2 on geological timescales", *Proceedings of the National Academy of Sciences (PNAS)*, **110**, 1209–1214.

Foster, G.L., Lunt, D.J., and Parrish, R.R. (2009), "Mountain uplift and the threshold for sustained Northern Hemisphere Glaciation", *Climate of the Past Discussions*, **5**, 2439–2464.

Foukal, P., and Lean, J. (1990), "An empirical model of TSI variation between 1874 and 1988", *Science*, **247**, 556–559.

Foukal, P., Frohlich, C., Spruit, H., and Wigley, T.M.L. (2006), "Variations in solar luminosity and their effect on the Earth's climate", *Nature*, **443**, 161–166.

Foukal, P., North, G., and Wigley, T. (2004), "A stellar view on solar variations and climate", *Science*, **306**, 68–69.

Frakes, L.A., and Kemp, E.M. (1972), "Influence of continental positions on early Tertiary climates", *Nature*, **240**, 97–100.

Francis, J.A., and Hunter, E. (2007), "Changes in the fabric of the Arctic's greenhouse blanket", *Environmental Research Letters*, **2**, 045011.

Frauenfeld, O.W. (2005), "Predictive skill of the ENSO and related teleconnections", in P.J. Michaels (ed.), *Shattered Consensus*, Rowman and Littlefield, New York, U.S.

Frauenfeld, O.W., Davis, R.E., and Mann, M.E. (2005), "A distinctly interdecadal signal of Pacific Ocean atmosphere interaction", *Journal of Climate*, **18**, 1709–1718.

Friedli, H., Lotscher, H., Oeschger, H., Siegenthaler, U., and Stauffer, B. (1986), "Ice core record of the $^{13}C/^{12}C$ ratio of atmospheric CO_2 in the past two centuries", *Nature*, **324**, 237–238.

Friis-Christensen, E., and Lassen, K. (1991), "Length of the solar cycle: An indicator of solar activity closely associated with climate", *Nature*, **254**, 698–700.

Frohlich, C. (2009), "Evidence of a long-term trend in total solar irradiance", *Astronomy and Astrophysics*, **501**, L27–L30.

Frolov, I., Gudkovich, Z., Karklin, V., Kovalev, E., and Smolyanitsky, V. (2009), *Climate Change in Eurasian Arctic Shelf Seas: Centennial Ice Cover Observations*, Praxis Publishing, Chichester, U.K.

Fu, Q., Baker, M., and Hartmann, D.L. (2002), "Tropical cirrus and water vapor: An effective Earth infrared iris feedback?", *Atmospheric Chemistry and Physics*, **2**, 31–37.

Fung, I.Y., Doney, S.C., Lindsay, K., and John, J. (2005), "Evolution of carbon sinks in a changing climate", *Proceedings of the National Academy of Sciences (PNAS)*, **102**, 11201–11206.

Fye, F.K., Stahle, D.W., and Cook, E.R. (2003), "Paleoclimatic analogs to twentieth-century moisture regimes across the United States", *Bulletin of the American Meteorological Society*, **84**, 901–909.

Ganapolski, A., Rahmstorf, S., Petoukhov, V., and Claussen, M. (1998), "Simulation of modern and glacial climates with a coupled global model of intermediate complexity", *Nature*, **391**, 351–356.

Ge, Q.-S., Zheng, J.Y., Hao, Z.-X., Shao, X.-M., Wang, W.-C., and Luterbacher, J. (2010), "Temperature variation through 2000 years in China: An uncertainty analysis of reconstruction and regional difference", *Geophysical Research Letters*, **37**, L03703.

Gerard, J.C., and Hauglustaine, D.A. (1991), "Transient climate response to solar irradiance: Reconstruction for the last 120 years", *Climate Research*, **1**, 161–167.

Gergis, J.L., and Fowler, A.M. (2006), "How unusual was late 20th century El Niño-Southern Oscillation (ENSO)? Assessing evidence from tree-ring, coral, ice-core and documentary palaeoarchives, A.D. 1525–2002", *Advances in Geosciences*, **6**, 173–179.

Gergis, J.L., and Fowler, A.M. (2008), "A history of ENSO events since A.D. 1525: Implications for future climate change", *Journal of Climatic Change*, **92**, 343–387.

Gerhard, L.C., and Harrison, W.E. (2001), "Distribution of oceans and continents: A geological constraint on global climate variability, 2001", in L.C. Gerhard, W.E. Harrison, and B.M. Hanson (eds), *Geological Perspectives of Global Climate Change*, American Association of Petroleum Geologists Studies in Geology #47, Tulsa, OK, U.S., pp. 35–49.

Gerlich, G., and Tscheuschner, R.D. (2007), "Falsification of the atmospheric CO_2 greenhouse effects within the frame of physics", Version 2.0 (24 July) (*http://arxiv.org/pdf/0707.1161, http://arxiv.org/pdf/0707.1161*).

Gettleman, A., and Fu, Q. (2008), "Observed and simulated upper tropospheric water vapor feedback", *Journal of Climate*, **21**, 3282.

Ghan, S.J., and Schwartz, S.E. (2007), "Aerosol properties and processes: A path from field and laboratory measurements to global climate", *Bulletin of the American Meteorological Society (BAMS)*, **88**, 1059–1083.

Gill, A. (1982), *Atmosphere–Ocean Dynamics*, Academic Press, London, U.K.

GISS (2007), "GISS surface temperature analysis" (*http://data.giss.nasa.gov/gistemp/graphs/*).

Goody, R. (1980), "Polar process and world climate (a brief overview)", *Monthly Weather Review*, **108** (December).

Goosse, H., Arzel, O., Luterbacher, J., Mann, M.E., Renssen, H., Riedwyl, N., Timmermann, A., Xoplaki, E., and Wanner, H. (2006), "The origin of the European *Medieval Warm Period*", *Climate of the Past*, **2**, 99–113.

Goosse, H., Renssen, H., Timmermann, A., and Bradley, R.S. (2005), "Internal and forced climate variability during the last millennium: A model–data comparison using ensemble simulations", *Quaternary Science Reviews*, **24**, 1345–1360.

Gorodetskaya, I.V., Cane, M.A., Tremblay, L.B., and Kaplan, A. (2006), "The effects of sea-ice and land-snow concentrations on planetary albedo from the Earth Radiation Budget Experiment", *Atmosphere–Ocean*, **44**, 195–205.

Graf, H.-F., Shirsat, S.V., Oppenheimer, C., Jarvis, M.J., Podzun, R., and Jacob, D. (2010), "Continental scale Antarctic deposition of sulphur and black carbon from anthropogenic and volcanic sources", *Atmospheric Chemistry and Physics*, **10**, 2457–2465.

Graversen, R.G., Drijfhout, S., Hazeleger, W., Bintanja, R., van de Wal, R., and Helsen, M. (2010), "Greenland's contribution to global sea-level rise by the end of the 21st century", *Climate Dynamics* (*www.knmi.nl/publications/fulltexts/gravesen.pdf*).

Gray, L.J., Beer, J., Geller, M., Haigh, J.D., Lockwood, M., Matthes, K., Cubasch, U., Fleitmann, D., Harrison, G., Hood, L., *et al.* (2010), "Solar influences on climate", *Reviews in Geophysics*, **42**, RG2002.

Gregory, J. (2004), "Review of concepts and methods relating to climate sensitivity", IPCC Workshop in Paris on Climate Sensitivity, July (*www. met.reading.ac.uk/jonathan/talks/paris0407.pdf*).

Gregory, J.M., Ingram, W.J., Palmer, M.A., Jones, G.S., Stott, P.A, Thorpe, R.B., Lowe, J.A., Johns, T.C., and Williams, K.D. (2004), "A new method for diagnosing radiative forcing and climate sensitivity", *Geophysical Research Letters*, **31**, L03205.

Grenfell, T., Warren, S., Clarke, T., and Radionov, V. (n.d.), "Black carbon in Arctic snow: Concentrations and effect on surface albedo" (*http://niflheim.nilu.no/spac/nilu-workshop/summary/grenfell-bc-and-arctic-snow*).

Grinsted, A., Moore, J.C., and Jevrejeva, S. (2010), "Reconstructing sea level from paleo and projected temperatures 200 to 2100 AD", *Climate Dynamics*, **34**, 461–472.

Grove, J.M. (1988), *The Little Ice Age*, Routledge, London, U.K.

Grove, J.M. (2001), "The initiation of the '*Little Ice Age*' in regions around the North Atlantic", *Climatic Change*, **48**, 53–82.

Guilderson, T.P., and Schrag, D.P. (1998), "Abrupt shift in subsurface temperatures in the tropical Pacific associated with changes in El Nino", *Science*, **281**, 240–243.

Ha-Duong, M., Swart, R., Bernstein, L., and Petersen, A. (2006), "Uncertainty management in the IPCC: Agreeing to disagree" (*ideas.repec.org/p/hal/journl/halshs-00117303_v1.html*).

Hadley, O.L., and Kirchstetter, T.W. (2012), "Black-carbon reduction of snow albedo", *Nature Climate Change* (*www.nature.com/nclimate/journal/vaop/ncurrent/full/nclimate1433.html*).

Hadley, O.L., Corrigan, C.E., Kirchstetter, T.W., Cliff, S.S., and Ramanathan, V. (2010), "Measured black carbon deposition on the Sierra Nevada snow pack and implication for snow pack retreat", *Atmospheric Chemistry and Physics*, **10**, 7505–7513.

Haerter, J.O., Roeckner, E., Tomassini, L., and von Storch, J.-S. (2009), "Parametric uncertainty effects on aerosol radiative forcing", *Geophysical Research Letters*, **36**, L15707.

Hale, R.C., Gallo, K.P., Owen, T.W., and Loveland, T.R. (2006), "Land use/land cover change effects on temperature trends at U.S. Climate Normals Stations", *Geophysical Research Letters*, **33**, L11703.

Hall, A., and Manabe, S. (1999), "The role of water vapor feedback in unperturbed climate variability and global warming", *Journal of Climate*, **12**, 2327–2346.

Hall, B.L., Koffman, T., and Denton, G.H. (2010), "Reduced ice extent on the western Antarctic Peninsula at 700–907 cal. yr B.P.", *Geology*, **3**, 635–638.

Hall, J.C., Henry, G.W., and Lockwood, G.W. (2007), "The Sun-like activity of the solar twin 18 SCORPII", *The Astronomical Journal*, **133**, 2206–2208.

Hall, J.C., Henry, G.W., Lockwood, G.W., Skiff, B.A., and Saar, S.H. (2009), "The activity and variability of the Sun and Sun-like stars. II. Contemporaneous photometry and spectroscopy of bright solar analogs", *The Astronomical Journal*, **138**, 312–322.

Hall, M.H.P., and Fagre, D.B. (2003), "Modeled climate-induced glacier change in Glacier National Park, 1850–2100", *Bioscience*, **53**, 131–140.

Hamon, M., Reverdin, G., and Le Traon, P.-Y. (2012), "Empirical correction of XBT data", *Journal of Atmospheric and Oceanic Technology*, **29**, 960–973.

Haney, R.L. (1971), "Surface thermal boundary condition for ocean circulation models", *Journal of Physical Oceanography*, **1**, 241–248.

Hanna, E., and Cappelen, J. (2003), "Recent cooling in coastal southern Greenland and relation with the North Atlantic Oscillation", *Geophysical Research Letters*, **30**, 1132.

Hansen, J., and Nazarenko, L. (2003), "Soot climate forcing via snow and ice albedos", *Proceedings of the National Academy of Sciences (PNAS)*, **101**, 423–428.

Hansen, J., and Sato, M. (2011), "Paleoclimate Implications for human-made climate change" (*www.columbia.edu/˜jeh1/mailings/2011/20110118_Milankovic Paper.pdf*).

Hansen, J., Ruedy, R., Sato, M., and Lo, K. (2010), "Global surface temperature change", *Reviews of Geophysics*, **48**, RG4004 (*http://data.giss.nasa.gov/gistemp/paper/gistemp2010_draft0601.pdf*).

Hansen, J., Russell, G., Lacis, A., Fung, I., Rind, D., and Stone, P. (1985), "Climate response times: Dependence on climate sensitivity and ocean mixing", *Science*, **229**, 857–859.

Hansen, J., Sato, M., and Ruedy, R. (2012), "Perception of climate change", *Proceedings of the National Academy of Sciences (PNAS)*, published online (*www.pnas.org/content/early/2012/07/30/1205276109.abstract*).

Hansen, J., Sato, M., Kharecha, P., Beerling, D., Berner, R., Masson-Delmotte, V., Pagani, M., Raymo, M., Royer, D.L., and Zachos, J.C. (2008), "Target atmospheric CO_2: Where should humanity aim?", *Open Atmosheric Science Journal*, **2**, 217–231.

Hansen, J., Sato, M., Ruedy, R., Kharecha, P., Lacis, A., Miller, R., Nazarenko, L., Lo, K., Schmidt, G.A., Russell, G., *et al.* (2007), "Climate simulations for 1880–2003 with GISS model", *Climate Dynamics*, **29**, 661–696.

Hansen, J., Sato, M., Ruedy, R., Nazarenko, L., Lacis, A., Schmidt, G.A., Russell,

G., Aleinov, I., Bauer, M., Bauer, S., *et al.* (2005a), "Efficacy of climate forcings", *Journal of Geophysical Research*, **110**, D18104.

Hansen, J.E. (2004), "Defusing the global warming time bomb", *Scientific American*, March, 69– 77.

Hansen, J.E. (2005), "A slippery slope: How much global warming constitutes 'dangerous anthropogenic interference'? An editorial essay", *Climatic Change*, **68**, 269–279 (*www.columbia.edu/jeh1/hansen_slippery.pdf*).

Hansen, J.E. (2008a), "Threat to the planet: Dark and bright sides of global warming", *EOS, Transactions, American Geophysical Union*, **89** (Fall).

Hansen, J.E., Nazarenko, L., Ruedy, R., Sato, M., Willis, J., Del Genio, A., Koch, D., Lacis, A., Lo, K., Menon, S., *et al.* (2005b), "Earth's energy imbalance: Confirmation and implications", *Science*, **308**, 1431–1434.

Hansen, J.E., Ruedy, R., Glascoe, J., and Sato, M. (1999), "GISS analysis of surface temperature change", *Journal of Geophysical Research*, **104**, 30997–31022.

Hansen, J.E., Ruedy, R., Sato, M., Imhoff, M., Lawrence, W., Easterling, D., Peterson, T., and Karl, T. (2001), "A closer look at United States and global surface temperature change", *Journal of Geophysical Research*, **106**, 23947–23964.

Hansen, J.E., Sato, M., Ruedy, R., Lacis, A., and Oinas, V. (2000), "Global warming in the twenty-first century: An alternative scenario", *Proceedings of the National Academy of Sciences (PNAS)*, **97**, 9875–9880.

Hare, S.R., and Mantua, N.J. (2000), "Empirical evidence for North Pacific regime shifts in 1977 and 1989", *Progress in Oceanography*, **47**, 103–145.

Hargreaves, C., and Annan, J.D. (2009), "Comment on 'Aerosol radiative forcing and climate sensitivity deduced from the Last Glacial Maximum to Holocene transition', *Geophysical Research Letters*', *Climate of the Past*, **5**, 143–145.

Hargreaves, J.C., and Annan, J.D. (2008), "Comment on 'Aerosol radiative forcing and climate sensitivity deduced from the Last Glacial Maximum to Holocene transition', by P. Chylek and U. Lohmann, *Geophysical Research Letters*, 2008", *Climate of the Past Discussions*, **4**, 1319–1326.

Harrison, D.E., and Carson, M. (2007), "Is the world ocean warming? Upper-ocean temperature trends: 1950– 2000", *Journal of Physical Oceanography*, **37**, 174–187.

Hasse, L. (1971), "The sea surface temperature deviation and the heat flow at the sea–air interface", *Boundary-Layer Meteorology*, **1**, 368–379.

Hathaway, D.H., Wilson, R.M., and Reichmann, E.J. (2002), "Group sunspot numbers: Sunspot cycle characteristics", *Solar Physics*, **211**, 357–370.

Hausfather, Z., Menne, M.J., Williams, C.N., Masters, T., Broberg, R., and Jones, D. (2013), "Quantifying the effect of urbanization on U.S. historical climatology network temperature records", *Journal of Geophysical Research—Atmospheres*, **118**, 481–494.

Haywood, A., and Williams, M. (2005), "The climate of the future: Clues from three million years ago", *Geology Today*, **21**, 138–143.

Haywood, A.M., and Valdes, P.J. (2004), "Modelling Pliocene warmth: Contribution of atmosphere, oceans and cryosphere", *Earth and Planetary Science Letters*, **218**, 363–377.

Hegerl, G.C., Crowley, T.J., Allen, M., Hyde, W.T., Pollack, H.N., Smerdon, J., and Zorita, E. (2007), "Detection of human influence on a new, validated 1500-year temperature reconstruction", *Journal of Climate*, **20**, 650–666.

Heimann, M. (2007), "The global carbon cycle in the Earth system" (*www.bgc-jena.mpg.de/bgc-systems/Department_presentations_publications/talks/martin.heimann/heimann_veldhoven_04_2007.pdf*).

Held, I.M., and Soden, B.J. (2000), "Water vapor feedback and global warming", *Annual Review of Energy and the Environment*, **25**, 441–475.

Hester, R.E., and Harrison, R.M. (eds) (2002), *Global Environmental Change*, Royal Society of Chemistry, Cambridge, U.K.

Higgins, J.A., and Schrag, D.P. (2006), "Beyond methane: Towards a theory for the Paleocene–Eocene thermal maximum", *Earth and Planetary Science Letters*, **245**, 523–537.

Hoffert, M.I., Caldeira, K., Benford, G., Criswell, D.R., Green, C., Herzog, H., Jain, A.K., Kheshgi, H.S., Lackner, K.S., Lewis, J.S., *et al.* (2002), "Advanced technology paths to global climate stability: Energy for a greenhouse planet", *Science*, **298**, 981.

Hoffert, M.I., Caldeira, K., Jain, A.K., Haites, E.F., Harvey, L.D., Potter, S.D., Schlesinger, M.E., Schneider, S.H., Watts, R.G., Wigley, T.M.L., *et al.* (1998), "Energy implications of future stabilization of atmospheric CO_2 content", *Nature*, **395**, 881–884.

Hoffman, J.S. (1984), *Estimates of Future Sea Level Rise* (*www.epa.gov/climatechange/effects/coastal/SLRChallenge.html*).

Hogg, A.McC. (2008), "Glacial cycles and carbon dioxide: A conceptual model", *Geophysical Research Letters*, **35**, L01701–L10705.

Holden, P.B., Edwards, N.R., Wolff, E.W., Lang, N.J., Singarayer, J.S., Valdes, P.J., and Stocker, T.F. (2009), "Interhemispheric coupling and warm Antarctic interglacials", *Climate of the Past Discussions*, **5**, 2555–2575.

Holgate, S.J. (2007), "On the decadal rates of sea level changes during the twentieth century", *Geophysical Research Letters*, **34**, L01602.

Holgate, S.J., and Woodworth, P.L. (2004), "Evidence for enhanced coastal sea level rise during the 1990s", *Geophysical Research Letters*, **31**, L07305-9.

Holland, M.M., and Bitz, C.M. (2003), "Polar amplification of climate change in coupled models", *Climate Dynamics*, **21**, 221–232.

Holloway, G., and Sou, T. (2000), "Is Arctic sea ice rapidly thinning?", *Journal of Climate*, **15**, 1691–1701 (*http://adaptation.nrcan.gc.ca/projdb/pdf/73_e.pdf*).

Holzhauser, H., Magny, M., and Zumbuhl, H.J. (2005), "Glacier and lake-level variations in west-central Europe over the last 3500 years", *The Holocene*, **15**, 789–801.

Houghton, J. (2004), *Global Warming*, 3rd edn, Cambridge University Press, Cambridge.

Houston, J.R., and Dean, R.G. (2011), "Sea-level acceleration based on U.S. tide gauges and extensions of previous global-gauge analyses", *Journal of Coastal Research*, **27**, 409–417 (*www.jcronline.org/doi/pdf/10.2112/JCOASTRES-D-10-00157.1*), "Sea level acceleration characteristics in the 20th century and extrapolation to 2100" (*www.fsbpa.com/2011TechPresentations/Dean%20 Robert.pdf*).

Howat, I.M., Joughin, I., and Scambos, T.A. (2007), "Rapid changes in ice discharge from Greenland outlet glaciers", *Science*, **315**, 1559–1561.

Hoyt, D. (2006), *A Critical Examination of Climate Change* (*www.warwickhughes.com/hoyt/climate-change.htm*).

Hoyt, D.V., and Schatten, K.H. (1993), "A discussion of plausible solar irradiance variations 1700–1992", *Journal of Geophysical Research*, **98**, 18895–18906.

Hoyt, D.V., and Schatten, K.H. (1998), "Group sunspot numbers: A new solar activity reconstruction", *Solar Physics*, **179**, 189–219, Reprinted with figures in *Solar Physics*, **181**, 491.

Huang, J.C.K., and Park, J.M. (1975), "Effective cloudiness derived from ocean buoy data", *Journal of Applied Meteorology*, **14**, 240–245.

Huang, S., Pollack, H.N., and Shen, P.Y. (1997), "Late Quaternary temperature changes seen in worldwide continental heat flow measurements", *Geophysical Research Letters*, **24**, 1947–1950.

Huang, S., Pollack, H.N., and Shen, P.Y. (2000), "Temperature trends over the past five centuries reconstructed from borehole temperatures", *Nature*, **403**, 756–758.

Huang, S.P., Pollack, H.N., and Shen, P.-Y. (2008), "A late Quaternary climate reconstruction based on borehole heat flux data, borehole temperature data and the instrumental record", *Geophysical Research Letters*, **35**, L13703.

Humlum, O. (2009), (*www.climate4you.com/*).

Humlum, O. (2011), (*www.climate4you.com/*).

Hurrell, J.W., and Dickson, R.R. (2004), "Climate variability over the North Atlantic", in N.C. Stenseth, G. Ottersen, J.W. Hurrell, and A. Belgrano (eds), *Marine Ecosystems and Climate Variation: The North Atlantic*, Oxford University Press, Oxford, U.K., pp. 15–31.

Huss, M., Jouvet, G., Farinotti, D., and Bauder, A. (2010), "Future high-mountain hydrology: A new parameterization of glacier retreat", *Hydrology and Earth System Sciences*, **14**, 815–829.

Huybers, P. (2010), "Compensation between model feedbacks and curtailment of climate sensitivity", *Journal of Climate*, **23**, 3009–3018.

Huybrechts, P. (2002), "Sea-level changes at the LGM from ice-dynamic reconstructions of the Greenland and Antarctic ice sheets during the glacial cycles", *Quaternary Science Reviews*, **21**, 203–231.

Iacono, M.J., Delamere, J.S., Mlawer, E.J., Shephard, M.W., Clough, S.A., and Collins, W.D. (2008), "Radiative forcing by long-lived greenhouse gases: Calculations with the AER radiative transfer models", *Journal of Geophysical Research*, **113**, D13103.

Idso, C., Carter, R., and Singer, S.F. (2011), *Climate Change Reconsidered*, NIPCC Report, Heartland Institute, Chicago, IL.

Idso, C.D. (2008), "Public comment in response to EPA's proposed rulemaking on regulating greenhouse gas emissions under the Clean Air Act", 24 November (*www.co2science.org/*).

Idso, C.D. (2012), *The State of Earth's Terrestrial Biosphere: How Is It Responding to*

Rising Atmospheric CO_2 and Warmer Temperatures?, Center for the Study of Carbon Dioxide and Global Change, Tempe, AZ.

Idso, S.B., and Idso, C.D. (2007), *Carbon Dioxide and Global Change: Separating Scientific Fact from Personal Opinion*, a critique of the 26 April 2007 testimony of James E. Hansen made to the Select Committee of Energy Independence and Global Warming of the United States House of Representatives entitled "Dangerous human-made interference with climate", Center for the Study of Carbon Dioxide and Global Change, 6 June (*www.co2science.org*).

Idso, S.B., Idso, C.D., and Idso, K.E. (2003), *The Specter of Species Extinction* (*www.co2science.org*).

Imhoff, M.L., Zhang, P., Wolfe, R.E., and Bounoua, L. (2010), "Remote sensing of the urban heat island effect across biomes in the continental USA", *Remote Sensing of Environment*, **114**, 504–513.

Indermuehle, A., Stocker, T.F., Joos, F., Fischer, H., Smith, H.J., Wahlen, M., Deck, B., Mastroianni, D., Tschumi, J., Blunier, T., *et al.* (1999), "Holocene carbon-cycle dynamics based on CO_2 trapped in ice at Taylor Dome, Antarctica", *Nature*, **398**, 121–126.

IPCC (2001), *Third Assessment Report*, Intergovernmental Panel on Climate Change (*www.ipcc.ch/*).

IPCC (2007), *Fourth Assessment Report*, Intergovernmental Panel on Climate Change.

IPCC (2011), *IPCC Special Report on Renewable Energy Sources and Climate Change Mitigation*, Intergovernmental Panel on Climate Change.

Ishii, M., and Kimoto, M. (2009), "Reevaluation of historical ocean heat content variations with time varying XBT and MBT depth bias corrections", *Journal of Oceanography*, **65**, 287–299.

Jacobson, M.Z. (2002), "Control of fossil-fuel particulate black carbon and organic matter, possibly the most effective method of slowing global warming", *Journal of Geophysical Research*, **107**, 4410–4431.

Jacobson, M.Z. (2007), Testimony for the hearing on black carbon and global warming, House Committee on Oversight and Government Reform United States House of Representatives, The Honorable Henry A. Waxman, Chair, 18 October.

Jacoby, G.C., Lovelius, N.V., Shumilov, O.I., Raspopov, O.M., Karbainov, J.M., and Frank, D.C. (2000), "Long-term temperature trends and tree growth in the Taymir region of northern Siberia", *Quaternary Research*, **53**, 312–318.

James, R.W. (1966), *Ocean Thermal Structures Forecasting*, U.S. Naval Oceanographic Office.

Jaworowski, Z. (2007), "CO_2: The greatest scientific scandal of our time", EIR *Science* (*www.21stcenturysciencetech.com/Articles 2007/20_1-2_CO2_ Scandal.pdf*).

Jevrejeva, S., Grinsted, A., Moore, J.C., and Holgate, S. (2006), "Nonlinear trends and multiyear cycles in sea level records", *Journal of Geophysical Research*, **111**, C09012–C09022.

Jevrejeva, S., Moore, J.C., and Grinsted, A. (2010), "How will sea level respond to changes in natural and anthropogenic forcings by 2100?", *Geophysical Research Letters*, **37**, L07703.

Jevrejeva, S., Moore, J.C., Grinsted, A., and Woodworth, P.L. (2008), "Recent global sea level acceleration started over 200 years ago?", *Geophysical Research Letters*, **35**, L08715.

Joerin, U.E., Stocker, T.F., and Schluchter, C. (2006), "Multicentury glacier fluctuations in the Swiss Alps during the Holocene", *The Holocene*, **16**, 697–704.

Johannessen, O.M., Bengtsson, L., Miles, M.W., Kuzmina, S.I., Semenov, V.A., Alekseev, G.V., Nagurnyi, A.P., Zakharov, V.F., Bobylev, L.P., Pettersson, L.H., *et al.* (2004), "Arctic climate change: Observed and modelled temperature and sea-ice variability", *Tellus*, **56A**, 328–341.

Jones, C.D., Cox, P.M., Simmonds, P., and Manning, A. (2005), "On the significance of atmospheric CO_2 growth rate anomalies in 2002–2003", *Geophysical Research Letters*, **32**, L14816.

Jones, P.D., Briffa, K.R., Barnett, T.P., and Tett, S.F.B. (1998), "High-resolution palaeoclimatic records for the last millennium: Interpretation, integration and comparison with General Circulation Model control-run temperatures", *The Holocene*, **8**, 455–471.

Jones, P.D., Briffa, K.R., Osborn, T.J., Lough, J.M., van Ommen, T.D., Vinther, B.M., Luterbacher, J., Wahl, E.R., Zwiers, F.W., Mann, M.E., *et al.* (2009), "High-resolution palaeoclimatology of the last millennium: A review of current status and future prospects", *The Holocene*, **19**, 3–49.

Jones, P.D., New, M., Parker, D.E., Martin, S., and Rigor, I.G. (1999), "Surface air temperature and its changes over the past 150 years", *Reviews of Geophysics*, **37**, 173–199.

Jones, P.D., Osborn, T.J., and Briffa, K.R. (2001), "Evolution of climate over the last millennium", *Science*, **292**, 662–668.

Joos, F., and Spahni, R. (2008), "Rates of change in natural and anthropogenic radiative forcing over the past 20,000 years", *Proceedings of the National Academy of Sciences (PNAS)*, **105**, 1425–1430.

Joughin, I., Das, S.B., King, M.A., Smith, B.E., Howat, I.M., and Moon, T. (2008), "Seasonal speedup along the western flank of the Greenland Ice Sheet", *Science*, **320**, 781–783.

Juckes, M.N. (2007), "Millennial temperature reconstruction intercomparison and evaluation", *Climate of the Past*, **3**, 591–609.

Juckes, M.N., Allen, M.R., Briffa, K.R., Esper, J., Hegerl, G.C., Moberg, A., Osborn, T.J., Weber, S.L., and Zorita, E. (2006), "Millennial temperature reconstruction intercomparison and evaluation", *Climate of the Past Discussions*, **2**, 1001–1049.

Kalnay, E., and Cai, M. (2003), "Impact of urbanization and land-use change on climate", *Nature*, **423**, 528–531.

Kane, R.P. (2008), "Prediction of Solar Cycle 24 based on the Gnevyshev–Ohl–Kopecky rule and the three-cycle periodicity scheme", *Annuls of Geophysics*, **26**, 3329–3339.

Kaufman, D.S., Ager, T.A., Anderson, N.J., Anderson, P.M., Andrews, J.T., Bartlein, P.J., Brubaker, L.B., Coats, L.L., Cwynar, L.C., Duvall, M.L., *et al.* (2004), "Holocene thermal maximum in the western Arctic (0–180W)", *Quaternary Science Reviews*, **23**, 529–560.

Kay, J.E., L'Ecuyer, T., Gettelman, A., Stephens, G., and O'Dell, C. (2008), "The contribution of cloud and radiation anomalies to the 2007 Arctic sea ice extent minimum", *Geophysical Research Letters*, **35**, L08503.

Keeling, R.F. (2005), "Comment on 'The Ocean Sink for Anthropogenic CO_2'", *Science*, **308**, 1743, and response by C. Sabine and N. Gruber, same issue.

Kemp, A.C., Horton, B.P., Donnelly, J.P., Mann, M.E., Vermeer, M., and Rahmstorf, S. (2011), "Climate related sea-level variations over the past two millennia", *Proceedings of the National Academy of Sciences (PNAS)* (*www.pnas.org/content/early/2011/06/13/1015619108.abstract*).

Kennett, J.P. (1977), "Cenozoic evolution of Antarctic glaciation, the circum-Antarctic oceans and their impact on global paleoceanography", *Journal of Geophysical Research*, **82**, 3843–3859.

Kent, D.V., and Muttoni, G. (2008), "Equatorial convergence of India and changes in atmospheric CO_2 and climate during the Cenozoic", *Proceedings of the National Academy of Sciences (PNAS)*, **105**, 16065–16070.

Kernthaler, S.C., Toumi, R., and Haigh, J.D. (1999), "Some doubts concerning a link between cosmic ray fluxes and global cloudiness", *Geophysical Research Letters*, **26**, 863–865.

Khalil, M.A.K., Butenhoff, C.L., and Rasmussen, R.A. (2007), "Atmospheric methane: Trends and cycles of sources and sinks", *Environmental Science and Technology*, **41**, 2131–2137.

Kiehl, J.T., and Trenberth, K.E. (1997), "Earth's annual global mean energy budget", *Bulletin of the American Meteorological Society (BAMS)*, **78**, 197–208.

Kilty, K.T. (1997), *Past Climate and Borehole Temperatures* (*www.kilty. com/pdfs/t-d.pdf*).

Kim, H.-J., and Miller, A.J. (2007), "Did the thermocline deepen in the California current after the 1976/77 climate regime shift?", *Journal of Physical Oceanography*, **37**, 1733–1739.

Kirkby, J. (2008), "Cosmic rays and climate", *Surveys in Geophysics*, **28**, 333–375.

Kirkby, J., Mangini, A., and Muller, R.A. (2004), *The Glacial Cycles and Cosmic Rays* (*http://arxiv.org/pdf/physics/0407005*).

Kirschvink, J., as reported by Hoffman, P.F., and Schrag, D.P. (2002), "The snowball Earth hypothesis: Testing the limits of global change", *Terra Nova*, **14**, 129–155.

Klotzbach, P.J. (2006), "Trends in global tropical cyclone activity over the past twenty years (1986–2005)", *Geophysical Research Letters*, **33**, L010805.

Klotzbach, P.J. (2011), "El Niño—southern oscillation's impact on Atlantic basin hurricanes and U. S. landfalls", *Journal of Climate*, **24**, 1252–1263.

Klotzbach, P.J., and Gray, W.M. (2011), "Extended range forecast of Atlantic seasonal hurricane activity and landfall strike probability for 2011" (*http://hurricane.atmos.colostate.edu/Forecasts*).

Klotzbach, P.J., Pielke, R.A., Sr., Pielke, R.A., Jr., Christy, J.R., and McNider, R.T. (2009), "An alternative explanation for differential temperature trends at the surface and in the lower troposphere", *Journal of Geophysical Research*, **114**, D21102.

Knaack, R., Fligge, M., Solanki, S.K., and Unruh, Y.C. (2001), "The influence of an inclined rotation axis on solar irradiance variations", *Astronomy and Astrophysics*, **376**, 1080–1089.

Knight, J.R., Allan, R.J., Folland, C.K., Vellinga, M., and Mann, M.E. (2005), "A signature of persistent natural thermohaline circulation cycles in observed climate", *Geophysical Research Letters*, **32**, L20708–L20711.

Kniveton, D.R., and Todd, M.C. (2001), "On the relationship of cosmic ray flux and precipitation", *Geophysical Research Letters*, **28**, 1527–1530.

Knorr, W. (2009)), "Is the airborne fraction of anthropogenic CO_2 emissions increasing?", *Geophysical Research Letters*, **36**, L21710.

Knox, R.S., and Douglass, D.H. (2010), "Recent energy balance of Earth", *International Journal of Geosciences*, **1**, 99–101.

Knutson, T.R. (2008), "Global warming and hurricanes: An overview of current research results' (*www.gfdl.noaa.gov/˜tk/glob_warm_hurr_webpage.html*).

Knutson, T.R., McBride, J.L., Chan, J., Emanuel, K., Holland, G., Landsea, C., Held, I., Kossin, J.P., Srivastava, A.K., and Sugi, M. (2010), "Tropical cyclones and climate change", *Nature Geoscience*, **3**, 157–163.

Knutti, R. (2008), "Should we believe model predictions of future climate change?", *Philosophical Transactions of the Royal Society*, **A366**, 4647–4664.

Kobashi, T., Severinghaus, J.P., Barnola, J.-M., Kawamura, K., Carter, T., and Nakaegawa, T. (2010), "Persistent multi-decadal Greenland temperature fluctuation through the last millennium", *Climatic Change*, **100**, 733–756.

Koberle, C., and Gerdes, R. (2003), "Mechanisms determining the variability of Arctic sea ice conditions and export", *Journal of Climate*, **16**, 2843–2858.

Kohler, P. (2011), "Measuring paleoclimate sensitivity" (*epic.awi.de/Publications/Khl2011c.pdf*).

Kohler, P., Bintanja, R., Fischer, H., Joos, F., Knutti, R., Lohmann, G., and Masson-Delmotte, V. (2009), "What caused Earth's temperature variations during the last 800,000 years? Data-based evidence on radiative forcing and constraints on climate sensitivity", *Quaternary Science Reviews*, **29**, 129–145.

Kolker, A.S., and Hameed, S. (2007), "Meteorologically driven trends in sea level rise", *Geophysical Research Letters*, **34**, L23616.

Kondratyev, K.Y. (2003), "Uncertainties of climate observation data and simulation modelling" (*www.nwicpc.ru/documents/Uncertaintiesofgcc-eng.doc*).

Konikow, L.F. (2011), "Contribution of global groundwater depletion since 1900 to sea level rise", *Geophysical Research Letters*, **38**, L17401.

Kopp, G., and Lean, J.L. (2011), "A new, lower value of total solar irradiance: Evidence and climate significance", *Geophysical Research Letters*, **38**, L01706.

Kopp, R.E., Simons, F.J., Maloof, A.C., and Oppenheimer, M. (2009), "Probabilistic assessment of sea level during the last interglacial stage", *Nature*, **462**, 863–867.

Kotlyakov, V.M. (ed.) (1996), *Variations of Snow and Ice in the Past and at Present on a Global and Regional Scale*, UNESCO, Paris, SC-96/WS-13.

Koto, H. (1966), "Stagnation, mixing and renewal of the water of the Funka Bay", *Memoirs of the Faculty of Fisheries Hokkaido University*, **13**, 65–78.

Kouwenberg, L., and Ria, L. (2005), "Application of conifer needles in the reconstruction of Holocene CO_2 levels', Ph.D. thesis, University of Utrecht, The Netherlands.

Kouwenberg, L., Wagner, R., Kurschner, W., and Visscher, H. (2005), "Atmospheric CO_2 fluctuations during the last millennium reconstructed by stomatal frequency analysis of Tsuga heterophylla needles", *Geology*, **33**, 33–36.

Krakauer, N.Y. (2006), "Characterizing carbon-dioxide fluxes from oceans and terrestrial ecosystems", thesis, California Institute of Technology Pasadena, California.

Krapivin, V.F., and Varotsos, C.A. (2007), *Globalization and Sustainable Development*, Springer-Praxis Publishing, Chichester, UK.

Kraus, E.B., and Rooth, C. (1961), "Temperature and steady state vertical heat flux in the ocean surface layers", *Tellus*, **13**, 231–238 (*www.tellusb.net/index.php/tellusb/article/download/12986/14753*).

Krey, V., and Clarke, L. (2011), "Role of renewable energy in climate mitigation: A synthesis of recent scenarios", *Climate Policy*, **11**, 1–28.

Krivova, N.A., Solankia, S.K., and Unruh, Y.C. (2009), "Towards a long-term record of solar total and spectral irradiance", *Journal of Atmospheric and Solar-Terrestrial Physics*, **73**, 223–234.

Krumhardt, K.M., and Kaplan, J.O. (2010), "A spline fit to atmospheric CO_2 records from Antarctic ice cores and measured concentrations for the last 25,000 years", ARVE Technical Report #2, 1ARVE Group, Environmental Engineering Institute, Ecole Polytechnique Fédérale de Lausanne, Station 2, 1015 Lausanne, Switzerland (*ecospriv4.epfl.ch/pub/ARVE_tech_report2_co2spline.pdf*).

Kump, L.R. (2011), "The last great global warming", *Scientific American*, July, 56–61.

Kunkel K.E., Bromirski, P.D., Brooks, H.E., Cavazos, T., Douglas, A.V., Easterling, D.R., Emanuel, K.A., Groisman, P.Y., Holland, G.J., Knutson, T.R., *et al.* (2008), "Observed changes in weather and climate extremes", in T.R. Karl, G.A. Meehl, C.D. Miller, S.J. Hassol, A.M. Waple, and W.L. Murray (eds), *Weather and Climate Extremes in a Changing Climate: Regions of Focus: North America, Hawaii, Caribbean and US Pacific Islands*, a report by the US Climate Change Science Programme and the Sub-committee on Global Change Research, Washington, DC.

Lacis, A., Schmidt, G.A., Ruedy, R., and Miller, R.L. (2010), "Atmospheric CO_2: Principal control knob governing Earth's temperature", *Science*, **330**, 356–359.

Lambeck, K., and Chappell, J. (2001), "Sea level change through the last glacial cycle", *Science*, **292**, 679–686.

Landsea, C.W., Vecchi, G.A., Bengtsson, L., and Knutson, T.T. (2010), "Impact of duration thresholds on Atlantic tropical cyclone counts", *Journal of Climate*, **23**, 2508–2518.

Lane, A. (1989), "The heat balance of the North Sea", Report No. 8, Proudman Oceanographic Laboratory (*http://nora.nerc.ac.uk/3872/*).

Lane, C.S., Horn, S.P., Orvis, K.H., and Thomason, J.M. (2011), "Oxygen isotope evidence of Little Ice Age aridity on the Caribbean slope of the Cordillera Central, Dominican Republic", *Quaternary Research*, **75**, 461–470.

Lea, D.W. (2004), "The 100 000-yr cycle in tropical SST, greenhouse forcing and climate sensitivity", *Journal of Climate*, **17**, 2170–2179.
Lean, J. (2000), "Short term, direct indices of solar variability", *Space Science Reviews*, **94**, 39–51.
Lean, J., and Foukal, P. (1988), "A model of solar luminosity modulation by magnetic activity between 1954 and 1984", *Science*, **240**, 906.
Lean, J., and Rind, D. (1998), "Climate forcing by changing solar radiation", *Journal of Climate*, **11**, 3069–3094.
Lean, J., Beer, J., and Bradley, R. (1995), "Reconstruction of solar irradiance since 1610: Implications for climate change", *Geophysical Research Letters*, **22**, 3195–3198.
Lean, J., Rottman, G.J., Kyle, H.L., Woods, T.N., Hickey, J.R., and Puga, L.C. (1997), "Detection and parameterization of variations in solar mid-and near-ultraviolet radiation (200–400 nm)", *Journal of Geophysical Research*, **102**, 29939–29956.
Lean, J., Skumanich, A., and White, O. (1992), "Estimating the Sun's radiative output during the Maunder Minimum", *Geophysical Research Letters*, **19**, 1595–1598.
Lean, J.L. (2010), "Cycles and trends in solar irradiance and climate", *wires.wiley.com/climatechange*, **1**, 111–122.
Lean, J.L., and Rind, D.H. (2008), "How natural and anthropogenic influences alter global and regional surface temperatures: 1889 to 2006", *Geophysical Research Letters*, **35**, L18701–L18706.
Lean, J.L., Cook, J., Marquette, W., and Johannesson, A. (1998), "Magnetic sources of the solar irradiance cycle", *Astrophysical Journal*, **492**, 390–401.
Lean, J.L., Wang, Y.-M., and Sheeley, N.R., Jr. (2002), "The effect of increasing solar activity on the Sun's total and open magnetic flux during multiple cycles: Implications for solar forcing of climate", *Geophysical Research Letters*, **29**, 2224–2227.
Leathers, D.J., Palecki, M.A., Robinson, D.A., and Dewey, K.F. (1998), "Climatology of the daily temperature range annual cycle in the United States", *Climate Research*, **9**, 197–211.
Lefohn, A.S., Husar, J.D., and Husar, R.B. (1999), "Estimating historical anthropogenic global sulfur emission patterns for the period 1850–1990", *Atmospheric Environment*, **33**, 3435–3444.
Leroux, M. (2005), *Global Warming: Myth or Reality*, Springer/Praxis, New York, U.S./Chichester, U.K.
Leroy, M. (2010), "Siting classification for surface observing stations on land, climate and upper-air observations", JMA/WMO Workshop on Quality Management in Surface, Tokyo, Japan 27–30 July.
Levitus, S. (2010), "Ocean climate laboratory", *Earth System Monitor*, **18**, 10.
Levitus, S., Antonov, J., and Boyer, T. (2005), "Warming of the world ocean, 1955–2003", *Geophysical Research Letters*, **32**, L02604.
Levitus, S., Antonov, J.I., Boyer, T.P., Baranova, O.K., Garcia, H.E., Locarnini, R.A., Mishonov, A.V., Reagan, J.R., Seidov, D., Yarosh, E.S., and Zweng,

M.M. (2012), "World ocean heat content and thermosteric sea level change (0–2000 m), 1955–2010", *Geophysical Research Letters*, **39**, L10603.

Levitus, S., Antonov, J.I., Boyer, T.P., Locarnini, R.A., Garcia, H.E., and Mishonov, A.V. (2009), "Global ocean heat content 1955–2008 in light of recently revealed instrumentation problems", *Geophysical Research Letters*, **36**, L07608.

Lewis, M. (2007), "A skeptic's guide to 'An inconvenient truth'" (*www.CEI.org*).

Li, B., Nychka, D.W., and Ammann, C.M. (2010), "The value of multiproxy reconstruction of past climate", *Journal of the American Statistical Association*, **105**, 883–911.

Lightfoot, H.D., and Green, C. (2002), *Energy Intensity Decline Implications for Stabilization of Atmospheric CO_2 Content*, Department of Economics, McGill University, McGill Center for Climate and Global Change Research (C2GCR) Report No. 2001-7, October 2001 (January 2002) (*http://people.mcgill.ca/files/christopher.green/energyintensity decline.pdf*).

Lim, Y.-K., Cai, M., Kalnay, E., and Zhou, L. (2005), "Observational evidence of sensitivity of surface climate changes to land types and urbanization", *Geophysical Research Letters*, **32**, L22712.

Lin, B.T.W., Wielicki, B.A., and Hu, Y. (2004), "Examination of the Decadal Tropical Mean ERBS Nonscanner radiation data for the Iris Hypothesis", *Journal of Climate*, **17**, 1239–1246.

Lindsay, R.W., and Zhang, J. (2005), "The thinning of Arctic sea ice, 1988–2003: Have we passed a tipping point?", *Journal of Climate*, **18**, 4879–4894.

Lindsay, R.W., Zhang, J., Schweiger, A., Steele, M., and Stern, H. (2009), "Arctic sea ice retreat in 2007 follows thinning trend", *Journal of Climate*, **22**, 165–176.

Lindzen, R.S. (1997), "Can increasing carbon dioxide cause climate change?", *Proceedings of the National Academy of Sciences (PNAS)*, **94**, 8335–8342.

Lindzen, R.S. (2007), "Taking greenhouse warming seriously", *Energy and Environment*, **18**, 937–950.

Lindzen, R.S. (2008), "Climate Science: Is it currently designed to answer questions?", Creativity and Creative Inspiration in Mathematics, Science and Engineering: Developing a Vision for the Future, San Marino, 29–31 August.

Lindzen, R.S., and Choi, Y.-S. (2009), "On the determination of climate feedbacks from ERBE data", *Geophysical Research Letters*, **36**, L16705.

Lindzen, R.S., and Choi, Y.-S. (2011), "On the observational determination of climate sensitivity and its implications", *Asia-Pacific Journal of Atmospheric Sciences*, **47**, 377–390 (*www.pensee-unique.fr/Erice-ERBEpaper-2009finalL.pdf*).

Lindzen, R.S., Chou, M.-D., and Hou, A.Y. (2001), "Does the Earth have an adaptive infrared iris?", *Bulletin of the American Meteorological Society*, **82**, 417–432.

Lindzen, R.S., Hou, A.Y., and Farrell, B.F. (1982), "The role of convective model choice in calculating the climate impact of doubling CO_2", *Journal of the Atmospheric Sciences*, **39**, 1189–2005.

Liu, J., von Storch, H., Zorita, E., Chen, X., and Wang, S. (2005), "Simulated and reconstructed temperature in China since 1550 AD", *History of Meteorology*, **2**, 87–104.

Liu, Z., Pagani, M., Zinniker, D., DeConto, R., Huber, M., Brinkhuis, H., Shah, S.R., Leckie, R.M., and Pearson, A. (2009), "Global cooling during the Eocene–Oligocene climate transition", *Science*, **323**, 1187–1190.

Livingston, W., and Penn, M. (2008), "Sunspots may vanish by 2015" (*wattsupwiththat.files.wordpress.com/2008/06/livingston-penn_sunspots2.pdf*).

Ljungqvist, F.C. (2011), "A regional approach to the medieval warm period and the little ice age" (*www.intechopen.com/download/pdf/pdfs_id/11433*), also see Ljungqvist, F.C. (2010), "A new reconstruction of temperature variability in the extra-tropical Northern Hemisphere during the last two millennia", *Geografiska Annaler*, **92A**, 339–351.

Lobell, D.B., Schlenker, W., and Costa-Roberts, J. (2011), "Climate trends and global crop production since 1980" (*www.sciencemag.org/content/early/2011/05/04/science.1204531*).

Lockwood, G.W., Skiff, B.A., and Radick, R.R. (1997a), "The photometric variability of Sun-like stars: Observations and results, 1984–1995", *Astrophysical Journal*, **485**, 789–811.

Lockwood, G.W., Skiff, B.A., Henry, G.W., and Radick, R.R. (1997b), "Patterns of photometric and chromospheric variation among Sun-like Stars: A 20 year perspective" (*http://arxiv.org/pdf/astro-ph/0703408*).

Lockwood, G.W., Skiff, B.A., Henry, G.W., Henry, S., Radick, R.R., Baliunas, S.L., Donahue, R.A., and Soon, W. (2007), "Patterns of photometric and chromospheric variation among Sun-like Stars: A 20 year perspective", *Astrophysical Journal Supplement Series*, **171**, 260–303.

Lockwood, M. (2002), "An evaluation of the correlation between open solar flux and total solar irradiance", *Astronomy and Astrophysics*, **382**, 678–687.

Lockwood, M., and Fröhlich, C. (2007), "Recent oppositely directed trends in solar climate forcings and the global mean surface air temperature", *Proceedings of the Royal Society A*, **463**, 2447–2460.

Lockwood, M., and Stamper, R. (1999), "Long-term drift in the coronal source magnetic flux and the total solar irradiance", *Geophysical Research Letters*, **26**, 2461–2465.

Loeb, N.G., Lyman, J.M., Johnson, G.C., Allan, R.P., Doelling, D.R., Wong, T., Soden, B.J., and Stephens, G.L. (2012), "Observed changes in top-of-the-atmosphere radiation and upper-ocean heating consistent within uncertainty", *Nature Geoscience* (*www.nature.com/ngeo/journal/vaop/ncurrent/full/ngeo1375.html*).

Loehle, C. (2007), "A 2000-year global temperature reconstruction based on non-tree ring proxies", *Energy and Environment*, **18**, 1049–1058.

Lombard, A., Cazenave, A., Le Traon, P.-Y., and Ishii, M. (2005), "Contribution of thermal expansion to present-day sea-level change revisited", *Global and Planetary Change*, **47**, 116.

Lomborg, B. (2007), "Perspective on climate change" (*www.climatechangefacts.info/ClimateChangeDocuments/lomborg_testimony.pdf*), see also Lomborg, B. (2001), *The Skeptical Environmentalist*, Cambridge University Press and Lomborg, B. (2004), *Global Crises, Global Solutions*, Cambridge University Press.

Loulergue, L., Parrenin, F., Blunier, T., Barnola, J.-M., Spahni, R., Schilt, A.,

Raisbeck, G., and Chappellaz, J. (2007), "New constraints on the gas age–ice age difference along the EPICA ice cores, 0–50 kyr", *Climate of the Past Discussions*, **3**, 435–467.

Lovejoy, S. (2013), "What is climate?", *Eos, Transactions American Geophysical Union*, **94**, 1–2.

Lowenstein, T.K., and Demicco, R.V. (2006), "Elevated eocene atmospheric CO_2 and its subsequent decline", *Science*, **313**, 1928.

Lüdecke, H.-J. (2011), "Long-term instrumental and reconstructed temperature records contradict anthropogenic global warming", *Energy and Environment*, **22**, 723–745.

Lüdecke, H.-J., Link, R., and Ewert, F.-K. (2011), "How natural is the recent centennial warming? An analysis of 2249 surface temperature records", *International Journal of Modern Physics C*, **22**, 1139–1159.

Lunt, D.J., Haywood, A.M., Schmidt, G.A., Salzmann, U., Valdes, P.J., and Dowsett, H.J. (2010), "Earth system sensitivity inferred from Pliocene modeling and data", *Nature Geoscience*, **3**, 60–64.

Lyman, J.M., Good, S.A., Gouretski, V.V., Ishii, M., Johnson, G.C., Palmer, M.D., Smith, D.M., and Willis, J.K. (2010), "Robust warming of the global upper ocean", *Nature*, 465, 334–337.

Maasch, K.A. (2009), "El Niño and interannual variability of climate in the western hemisphere" (*http://climatechange.umaine.edu/people/profile/kirk_allen_maasch*).

MacFarling Meure, C., Etheridge, D., Trudinger, C., Steele, P., Langenfelds, R., van Ommen, T., Smith, A., and Elkins, J. (2006), "Law dome CO_2, CH_4 and N_2O ice core records extended to 2000 years BP", *Geophysical Research Letters*, **33**, L14810.

Mackenzie, F.T., Lerman, A., and Ver, L.M.B. (2001), "Recent past and future of the global carbon cycle", in L.C. Gerhard, W.E. Harrison, and B.M. Hanson (eds), *Geological Perspectives of Global Climate Change*, American Association of Petroleum Geologists, Special Publication, Tulsa, OK, pp. 51–82.

Mackey, R. (2007), "The climate dynamics of total solar variability" (*www.coastal conference.com/2007/papers2007/Richard%20Mackey.doc*).

Maggio, G., and Cacciola, G.(2009), "A variant of the Hubbert curve for world oil production forecasts", *Energy Policy*, **37**, 4761–4770.

Mahmood, R., Quintanar, A.I., Conner, G., Leeper, R., Dobler, S., Pielke, R.A., Sr., Beltran-Przekurat, A., Hubbard, K.G., Niyogi, D., Lawrence, P., *et al.* (2010), "Impacts of land use/land cover change on climate and future research priorities", *Bulletin of the American Meteorological Society*, 10 January, 37–46.

Makowski, K., Wild, M., and Ohmura, A. (2008), "Diurnal temperature range over Europe between 1950 and 2005", *Atmospheric Chemistry and Physics*, **8**, 6483–6498.

Mann, M.E., and Jones, P.D. (2003), "Global surface temperatures over the past two millennia", *Geophysical Research Letters*, **30**, 1820.

Mann, M.E., Bradley, R.S., and Hughes, M.K. (1998), "Global-scale temperature patterns and climate forcing over the past six centuries", *Nature*, **392**, 779–807.

Mann, M.E., Bradley, R.S., and Hughes, M.K. (1999), "Northern Hemisphere temperatures during the past millennium: Inferences, uncertainties and limitations", *Geophysical Research Letters*, **26**, 759–762.

Mann, M.E., Bradley, R.S., and Hughes, M.K. (2004), "Corrigendum: Global-scale temperature patterns and climate forcing over the past six centuries", *Nature*, **430**, 105.

Mann, M.E., Rutherford, S., Wahl, E., and Ammann, C. (2005), "Testing the fidelity of methods used in proxy-based reconstructions of past climate", *Journal of Climate*, **18**, 4097–4107.

Mann, M.E., Rutherford, S., Wahl, E., and Ammann, C. (2007), "Robustness of proxy-based climate field reconstruction methods", *Journal of Geophysical Research*, **112**, D12109.

Mann, M.E., Zhang, Z., Hughes, M.K., Bradley, R.S., Miller, S.K., Rutherford, S., and Ni, F. (2008), "Proxy-based reconstructions of hemispheric and global surface temperature variations over the past two millennia", *Proceedings of the National Academy of Sciences (PNAS)*, **105**, 13252–13257.

Marsh, G.E. (2002), *Global Warming Primer, National Center for Public Policy Research*, Report No. 420 (July) (*www.nationalcenter.org/NPA420.pdf*).

Marsh, N.D., and Svensmark, H. (2000), "Low cloud properties influenced by cosmic rays", *Physical Review Letters*, **85**, 5004–5007.

Marshall, J., Kushnir, Y., Battisti, D., Chang, P., Czaja, A., Dickson, R., Hurrell, J., McCartney, M., Saravanan, R., and Visbeck, M. (2001), "North Atlantic climate variability: Phenomena, impacts and mechanisms", *International Journal of Climatology*, **21**, 1863–1898.

Martin, B. (1979), "The bias of science" (*www.uow.edu.au/arts/sts/bmartin/pubs/79bias/index.html*).

Masters, T. (2012), "On the determination of the global cloud feedback from satellite measurements", *Earth System Dynamics*, **3**, 97–107.

Matthews, H.D., Weaver, A.J., Meissner, K.J., Gillett, N.P., and Eby, M. (2004), "Natural and anthropogenic climate change: Incorporating historical land cover change, vegetation dynamics and the global carbon cycle", *Climate Dynamics*, **22**, 461–479.

Matthews, J.A., and Briffa, K.R. (2005), "The '*Little Ice Age*': Re-evaluation of an evolving concept", *Geografiska Annaler*, **87A**, 17–36.

Maughan, P.M. (1966), "Measurement of radiant energy over mixed water body", Ph.D. thesis, Oregon State University.

Maunder, E.J (1922), "The prolonged sunspot minimum 1645–1715", *Journal of the British Astronomical Association*, **32**, 140.

McConnell, J.R., Edwards, R., Kok, G.L., Flanner, M.G., Zender, C.S., Saltzman, E.S., Ryan Banta, J., Pasteris, D.R., Carter, M.M., and Kahl, J.D.W. (2007), "20th-century industrial black carbon emissions altered Arctic climate forcing", *Science*, **317**, 1381–1384.

McIntyre, S. (2007), *Climate Audit Website* (*www.climateaudit.org*).

McIntyre, S., and McKitrick, R. (2003), "Corrections to the Mann et al. (1998), 'Proxy data based and Northern Hemispheric average temperature series'", *Energy and Environment*, **14**, 751–771.

McIntyre, S., and McKitrick, R. (2005), "Hockey sticks, principal components and spurious significance", *Geophysical Research Letters*, **32**, L03710.
McIntyre, S., and McKitrick, R. (2006), "Surface temperature reconstructions for the past 1,000–2,000 years", presentation to the National Academy of Sciences Expert Panel, 2 March, Washington, D.C.
McIntyre, S., and McKitrick, R. (2007), *The M&M Critique of the MBH98 Northern Hemisphere Climate Index: Update and Implications*, informal report (*www.uoguelph.ca/rmckitri/research/trc.html*).
McKitrick, R. (2005a), "What is the 'hockey stick' debate about?" (*www.climatechangeissues.com/files/PDF/conf05mckitrick.pdf*).
McKitrick, R. (2008), "Atmospheric oscillations do not explain the temperature–industrialization correlation", 21 July (*http://ssrn.com/abstract=1166424*).
McKitrick, R.T. (2005b), *The Mann* et al. *Northern Hemisphere Climate Index*, edited by P.J. Michaels, Rowman and Littlefield, New York.
McKitrick, R.T., and Michaels, P.J. (2007), "Quantifying the influence of anthropogenic surface processes and inhomogeneities on gridded global climate data", *Journal of Geophysical Research*, **112**, D24S09–D24S22.
McLean, J. (2007a), "Fallacies about global warming", September (*http://mclean.ch/climate/global_warming.htm*).
McLean, J. (2007b), "Ignoring a natural event to blame humans", October (*http://mclean.ch/climate/global_warming.htm*).
McLean, J.D. (2010), "Global warming issues" (*http://mclean.ch/climate/ENSO_paper.htm, http://mclean.ch/climate/global_warming.htm*).
McLean, J.D., de Freitas, C.R., and Carter, R.M. (2009), "Influence of the southern oscillation on tropospheric temperature", *Journal of Geophysical Research*, **114**, D14104.
McNider, R.T., Steeneveld, G.J., Holtslag, B., Pielke, R., Sr., Mackaro, S., Pour Biazar, A., Walters, J.T., Nair, U.S., and Christy, J.R. (2012), "Response and sensitivity of the nocturnal boundary layer over land to added longwave radiative forcing", *Journal of Geophysical Research*, **117**, D14106.
McShane, B.B., and Wyner, A.J. (2011), "A statistical analysis of multiple temperature proxies: Are reconstructions of surface temperatures over the last 1,000 years reliable?", *Annals of Applied Statistics*, **5**, 5–44.
McWilliams, J.C. (2007), "Irreducible imprecision in atmospheric and oceanic simulations", *Proceedings of the National Academy Sciences (PNAS)*, **104**, 8709–8713.
Mears, C.A., Schabel, M.C., and Wentz, F.J. (2003), "A reanalysis of the MSU Channel 2 tropospheric temperature record", *Journal of Climate*, **16**, 3650–3664.
Meehl, G.A., and Washington, W.M. (1996), "El Nino-like climate change in a model with increased atmospheric CO_2 concentrations", *Nature*, **382**, 56–60.
Meehl, G.A., Covey, C., Delworth, T., Latif, M., McAvaney, B., Mitchell, J.F.B., Stouffer, R.J., and Taylor, K.E. (2007), "The WCRP CMIP3 multimodel dataset: A new era in climate change research", *Bulletin of the American Meteorological Society (BAMS)*, **88**, 1383–1394.
Meehl, G.A., Washington, W.M., Wigley, T.M.L., Arblaster, J.M., and Dai, A.

(2002), "Solar and greenhouse gas forcing and climate response in the twentieth century", *Journal of Climate*, **16**, 426–444.

Meinshausen, M., Meinshausen, N., Hare, W., Raper, S.C.B., Frieler, K., Knutti, R., Frame, D.J., and Allen, M.R. (2009), "Greenhouse-gas emission targets for limiting global warming to 2C", *Nature*, **458**, 1158–1163.

Menne, M.J., Williams, C.N., Jr., and Vose, R.S. (2009), "The United States Historical Climatology Network Monthly Temperature Data—Version 2", *Bulletin of the American Meteorological Society (BAMS)* (*http://ams.allenpress.com/perlserv/?request=get-abstract&doi=10.1175%2F2008BAMS2613.1*).

Menon, S., Koch, D., Beig, G., Sahu, S., Fasullo, J., and Orlikowski, D. (2010), "Black carbon aerosols and the third polar ice cap", *Atmospheric Chemistry and Physics*, **10**, 4559–4571.

Michaels, P.J., and Balling, R.C., Jr. (2009), *Climate of Extremes*, Cato Institute, Washington, D.C.

Michaels, P.J., Knappenberger, P.C., and Davis, R.E. (2006), "Sea-surface temperatures and tropical cyclonesin the Atlantic basin", *Geophysical Research Letters*, **33**, L09708–L09711.

Miller, G.H., Brigham-Grette, J., and Alley, R.B. (2010a), "Arctic amplification: Can the past constrain the future?", *Quaternary Science Reviews*, **29**, 1779–1790.

Miller, G.H., Brigham-Grette, J., Alley, R.B., Anderson, L., Bauch, H.A., Douglas, M.S.V., Edwards, M.E., Elias, S.A., Finney, B.P., Fitzpatrick, J.J., *et al.* (2010b), "Temperature and precipitation history of the Arctic", *Quaternary Science Reviews*, **29**, 1679–1715.

Min, S.-K., Zhang, X., Zwiers, F.W., and Hegerl, G.C. (2011), "Human contribution to more-intense precipitation extremes", *Nature*, **470**, 378–381.

Minnis, P., Kirk Ayers, J., Palikonda, R., and Phan, D. (2004), "Contrails, cirrus trends and climate", *Journal of Climate*, **17**, 1671–1685.

Minschwaner, K., and Dessler, A.E. (2004), "Water vapor feedback in the tropical Upper Troposphere: Model results and observations", *Journal of Climate*, **17**, 1272–1282.

Minschwaner, K., Dessler, A.E., and Sawaengphokhai, P. (2006), "Multimodel analysis of the water vapor feedback in the tropical Upper Troposphere", *Journal of Climate*, **19**, 5455–5464.

Miskolczi, F.M. (2007), "Greenhouse effect in semi-transparent planetary atmospheres", *IDŐJÁRÁS Quarterly Journal of the Hungarian Meteorological Service*, **111**, 1–40.

Miskolczi, F.M. (2010), "The stable stationary value of the Earth's global average atmospheric planck-weighted greenhouse-gas optical thickness", *Energy and Environment*, **21**, 243–263.

Mitchell, J.F.B., and Johns, T.C. (1997), "On modification of global warming by sulfate aerosols", *Journal of Climate*, **10**, 245–267.

Moberg, A., Sonechkin, D.M., Holmgren, K., Datsenko, N.M., Karlen, W., and Lauritzen, S.E. (2005), "Highly variable Northern Hemisphere temperatures reconstructed from low- and high-resolution proxy data", *Nature*, **433**, 613–617.

Monaghan, A.J., Bromwich, D.H., Chapman, W., and Comiso, J.C. (2008), "Recent

variability and trends of Antarctic near-surface temperature", *Journal of Geophysical Research Atmospheres*, **113**, D04105.
Monnin, E., Indermuhle, A., Dällenbach, A., Fluckiger, J., Stauffer, B., Stocker, T.F., Raynaud, D., and Barnola, J.-M. (2001), "Atmospheric CO_2 concentrations over the last glacial termination", *Science*, **291**, 112–114.
Montandon, L., Fall, S., Pielke, R.A., Sr., and Niyogi, D. (2011), "Distribution of landscape types in the global historical climatology network", *Earth Interactions*, **15**, 1–24.
Montenegro, A., Brovkin, V., Eby, M., Archer, D., and Weaver, A.J. (2007), "Long term fate of anthropogenic carbon", *Geophysical Research Letters*, **34**, L19707.
Montford, A.W. (2010), *The Hockey Stick Illusion: Climategate and the Corruption of Science*, Stacey International, London, U.K.
Morales Maqueda, A., and Rahmstorf, S. (2002), "Did Antarctic sea-ice expansion cause glacial CO_2 decline?", *Geophysical Research Letters*, **29**, 1011–1013.
Morice, C.P., Kennedy, J.J., Rayner, N.A., and Jones, P.D. (2012), "Quantifying uncertainties in global and regional temperature change using an ensemble of observational estimates: The HadCRUT4 data set", *Journal of Geophysical Research*, **117**, D08101–D08122.
Mörner, N.-A. (1973), "Eustatic changes during the last 300 years", *Palaeogeography, Palaeoclimatology, Palaeoecology*, **9**, 153–181.
Mörner, N.-A. (2004), "Estimating future sea level changes from past records", *Global and Planetary Change*, **40**, 49–54 (*http://gsa.confex.com/gsa/inqu/finalprogram/abstract_ 54461.htm*).
Mosher, S., and Fuller, T.W. (2010), *Climategate: The CRUTape Letters*, self-published.
Mote, P.W., and Kaser, G. (2007), "The shrinking glaciers of Kilimanjaro: Can global warming be blamed?", *The American Scientist*, **95**, 318–326.
Mudelsee, M. (2001), "The phase relations among atmospheric CO_2 content, temperature and global ice volume over the past 420 ka", *Quaternary Science Reviews*, **20**, 583–589.
Mudelsee, M., Borngen, M., Tetzlaff, G., and Grunewald, U. (2003), "No upward trends in the occurrence of extreme floods in Central Europe", *Nature*, **425**, 166–169.
Muller, R.A., Curry, J., Groom, D., Jacobsen, R., Perlmutter, S., Rohde, R., Rosenfeld, A., Wickham, C., and Wurtele, J. (2011a), "Decadal variations in the global atmospheric land temperatures" (*www.berkeleyearth.org/*).
Muller, R.A., Curry, J., Groom, D., Jacobsen, R., Perlmutter, S., Rohde, R., Rosenfeld, A., Wickham, C., and Wurtele, J. (2011b), "Earth atmospheric land surface temperature and station quality in the United States" (*www.berkeley earth.org/*).
Murphy, D.M., Solomon, S., Portmann, R.W., Rosenlof, K.H., Forster, P.M., and Wong, T. (2009), "An observationally based energy balance for the Earth since 1950", *Journal of Geophysical Research*, **114**, D17107.
Murray, T., Scharrer, K., James, T.D., Dye, S.R., Hanna, E., Booth, A.D., Selmes, N., Luckman, A., Hughes, A.L.C., Cook, S., and Huybrechts, P. (2010), "Ocean

regulation hypothesis for glacier dynamics in southeast Greenland and implications for ice sheet mass changes", *Journal of Geophysical Research*, **115**, F03026.

Muscheler, R., Joos, F., Müller, S.A., and Snowball, I. (2005), "How unusual is today's solar activity?", *Nature*, **436**, E3–E4.

Nagashima, T., Shiogama, H., Yokohata, T., Crooks, S.A., and Nozawa, T. (2006), "The effect of carbonaceous aerosols on surface temperature in the mid twentieth century", *Geophysical Research Letters*, **33**, L04702.

National Academy of Sciences (NAS) (2011), *Understanding Earth's Deep Past: Lessons for Our Climate Future*, National Academies Press.

Neumayer, E., and Barthel, F. (2011), "Normalizing economic loss from natural disasters: A global analysis", *Global Environmental Change*, **21** 13–24.

Newell, R.E., and Dopplick, T.G. (1979), "Questions concerning the possible influence of anthropogenic CO_2 on atmospheric temperature", *Journal of Applied Meteorology*, **18**, 822–825.

Newell, R.E., and Dopplick, T.G. (1981), "Reply to Robert G. Watts' Discussion of questions concerning the possible influence of anthropogenic CO_2 on atmospheric temperature", *Journal of Applied Meteorology*, **20**, 114–117.

NIPCC (2011), *Climate Change Reconsidered, Center for the Study of Carbon Dioxide and Global Change*, Tempe, AZ, U.S.

Nordhaus, W. (2006), "Review of Stern Review", 17 November (*www.carbontax.org/wp-content/uploads/2006/12/nordhaus-_-stern-review-_-dec-2006.pdf*).

Norris, J.R. (2005), "Multidecadal changes in near-global cloud cover and estimated cloud cover radiative forcing", *Journal of Geophysical Research*, **110**, D08206,

Norris, J.R., and Slingo, A. (2009), "Trends in observed cloudiness and Earth's radiation budget: What do we not know and what do we need to know?" (*meteora.ucsd.edu/˜jnorris/.../02_Norris_and_Slingo.pdf*).

Northrup, A. (2004), *The Global Carbon Cycle in the Ocean and the Threat of Climate Change Because of Anthropogenic Carbon Emissions*, BISC 419, 22 April (*http://bioweb.usc.edu/courses/2004-spring/documents/bisc419-carboncycle.pdf*).

Nozawa, T., Nagashima, T., Ogura, T., Yokohata, T., Okada, N., and Shiogama, H. (2007), *Climate Change Simulations with a Coupled Ocean–Atmosphere GCM (MIROC)*, Center for Global Environmental Research, Japan, Report No. CGER 1073.

Nozawa, T., Nagashima, T., Shiogama, H., and Crooks, S.A. (2005), "Detecting natural influence on surface air temperature change in the early twentieth century", *Geophysical Research Letters*, **32**, L20719.

NRC (2011), "Frontiers in understanding climate change and polar ecosystems: Summary of a workshop", National Research Council (*www.nap.edu/catalog.php?record_id=13132*).

O'Donnell, R., Lewis, N., McIntyre, S., and Condon, J. (2011), "Improved methods for PCA-based reconstructions: Case study using the Steig et al. (2009), Antarctic temperature reconstruction", *Journal of Climate*, **24**, 2099–2115.

Ogilvie, A.E.J., and Jonsson, T. (2001), "*Little Ice Age* research: A perspective from Iceland", *Climatic Change*, **48**, 9–52.

Oke, T.R. (1973), "City size and the urban heat island", *Atmospheric Environment*, **7**, 769–779.

Oppenheimer, M., and Alley, R.B. (2005), "Ice sheets, global warming and Article 2 of the UNFCCC: An editorial essay", *Climatic Change*, **68**, 257–267 (*www.princeton.edu/step/people/Oppenheimer%and%Alley%II%published.pdf*).

Oreskes, N. (2004), "The scientific consensus on climate change", *Science*, **306**, 1686.

Oreskes, N. (2007a), "The long consensus on climate change", *Washington Post*, 1 February (*www.washingtonpost.com/wp-dyn/content/article/2007/01/31/AR2007 01301808_pf.html*).

Oreskes, N. (2007b), *The Science of Global Warming: How Do We Know We're Not Wrong*? AMS Environmental Science Seminar Series, American Meteorological Society, Washington, DC, U.S. (*www.lpl.arizona.edu/resources/globalwarming/ documents/oreskeson-science-consensus.pdf*).

Oreskes, N., Conway, E.M, and Shindell, M. (2008), "From Chicken Little to Dr. Pangloss: William Nierenberg, global warming and the social deconstruction of scientific knowledge" (*www.lse.ac.uk/collections/CPNSS/projects/Contingency DissentInScience/DP/DPOreskesetalChickenLittleOnlinev2.pdf*).

Ou, H.-W. (2001), "Possible bounds on the Earth's surface temperature: From the perspective of a conceptual global-mean model", *Journal of Climate*, **14**, 2976–2988.

Overpeck, J., Hughen, K., Hardy, D., Bradley, R., Case, R., Douglas, M., Finney, B., Gajewski, K., Jacoby, G., Jennings, A., *et al.* (1997), "Arctic environment change of the last four centuries", *Science*, **278**, 1251.

Ozsoy-Cicek, B., Xie, H., Ackley, S.F., and Ye, K. (2009), "Antarctic summer sea ice concentration and extent: Comparison of ODEN 2006 ship observations, satellite passive microwave and NIC sea ice charts", *The Cryosphere*, **3**, 1–9.

Pagani, M., Caldeira, K., Archer, D., and Zachos, J.C. (2006), "An ancient carbon mystery", *Science*, **314**, 1556–1557.

Pagani, M., Liu, Z., LaRiviere, J., and Ravelo, A.C. (2010), "High Earth-system climate sensitivity determined from Pliocene carbon dioxide concentrations", *Nature Geoscience*, **3**, 27–30.

Pagani, M., Zachos, J.C., Freeman, K.H., Tipple, B., and Bohaty, S. (2005), "Marked decline in atmospheric carbon dioxide concentrations during the paleogene", *Science*, **309**, 600–604.

Painter, T.H., Flanner, M.G., Kaser, G., Marzeion, B., Van Curen, R.A., and Abdalati, W. (2013), "End of the Little Ice Age in the Alps forced by industrial black carbon", *Proceedings of the National Academy of Sciences (PNAS)*, **110**, 15216–15221.

Pall, P., Aina, T., Stone, D.A., Stott, P.A., Nozawa, T., Hilberts, A.G.J., Lohmann, D., and Allen, M.R. (2011), "Anthropogenic greenhouse gas contribution to flood risk in England and Wales in autumn 2000", *Nature*, **470**, 382–386.

Palmer, M.D., and Haines, K. (2009), "Estimating oceanic heat content change using isotherms", *Journal of Climate*, **22**, 4953–4969.

PALSEA (2010), "The sea-level conundrum: Case studies from palaeo-archives", *Journal of Quaternary Science*, **25**, 19–25.

Paltridge, G., Arking, A., and Pook, M. (2009), "Trends in middle- and upper-level tropospheric humidity from NCEP reanalysis data", *Theoretical and Applied Climatology*, **98**, 351–359.

Paltsev, S., Reilly, J.M., Jacoby, H.D., and Morris, J.F. (2009), *The Cost of Climate Policy in the United States*, MIT Joint Program on the Science and Policy of Global Change Report No. 173, April.

Parker, D.E. (2004), "Climate: Large-scale warming is not urban", *Nature*, **432**, 290.

Parker, D.E. (2010), "Urban heat island effects on estimates of observed climate change", *Wiley Interdisciplinary Reviews: Climate Change*, **1**, 123–133.

Patterson, R.T. (2007), *Read the Sunspots*, 20 June (*www.canada.com/nationalpost/financialpost/comment/story.html?id = 597d0677-2a05-47b4-b34fb84068db11f4*).

Peacock, S., Lane, E., and Restrepo, J.M. (2006), "A possible sequence of events for the generalized glacial–interglacial cycle", *Global Biogeochemical Cycles*, **20**, GB2010– GB2026.

Pearson, P.N., and Palmer, M.R. (2000), "Atmospheric carbon dioxide concentrations over the past 60 million years", *Nature*, **406**, 695–699.

Pearson, P.N., Foster, G.L., and Wade, B.S. (2009), "Atmospheric carbon dioxide through the Eocene–Oligocene climate transition", *Nature*, **461**, 1110–1113.

Pederson, G.T., Fagre, D.B., Gray, S.T., and Graumlich, L.J. (2004), "Decadal-scale climate drivers for glacial dynamics in Glacier National Park, Montana", *Geophysical Research Letters*, **31**, L12203–L12207.

Peng, S., Piao, S., Ciais, P., Friedlingstein, P., Ottle, C., Breon, F.-M., Nan, H., Zhou, L., and Myneni, R.B. (2012), "Surface urban heat island across 419 global big cities", *Environmental Science and Technology*, **46**, 696-703.

Perovich, D.K., Richter-Menge, J.A., Jones, K.F., Light, B., Elder, B.C., Polashenski, C., Laroche, D., Markus, T., and Lindsay, R. (2011), "Arctic sea-ice melt in 2008 and the role of solar heating", *Annals of Glaciology*, **52**, 355–359.

Peters, S.E., Carlson, A.E., Kelly, D.C., and Gingerich, P.D. (2010), "Large-scale glaciation and deglaciation of Antarctica during the Late Eocene", *Geology*, **38**, 723–726.

Peterson, T.C. (2003), "Assessment of urban versus rural *in situ* surface temperatures in the contiguous United States: No difference found", *Journal of Climate*, **16**, 2941–2959.

Peterson, T.C., and Vose, R.S. (1997), "An overview of the global historical climatology network temperature database", *Bulletin of the American Meteorological Society*, **78**, 2837–2849.

Petit, J.R., Jouzel, J., Raynaud, D., Barkov, N.I., Barnola, J.-M., Basile, I., Bender, M., Chappellaz, J., Davis, M., Delaygue, G., *et al.* (1999), "Climate and atmospheric history of the past 420,000 years from the Vostok ice core, Antarctica", *Nature*, **399**, 429–436.

Philipona, R., Behrens, K., and Ruckstuhl, C. (2009), "How declining aerosols and rising greenhouse gases forced rapid warming in Europe since the 1980s", *Geophysical Research Letters*, **36**, L02806.

Pielke, R., Jr., Wigley, T., and Green, C. (2008a), "Dangerous assumptions: How big is the energy challenge of climate change?", *Nature Commentary*, **452**, 531–532.

Pielke, R., Sr. (2003), "Heat storage within the Earth system", *Bulletin of the American Meteorological Society (BAMS)*, March, 331–335.

Pielke, R., Sr., Nielsen-Gammon, J., Davey, C., Angel, J., Bliss, O., Doesken, N., Cai, M., Fall, S., Niyogi, D., Gallo, K., *et al.* (2007a), "Documentation of uncertainties and biases associated with surface temperature measurement sites for climate change assessment", *Bulletin of the American Meteorological Society*, **88**, 913–928.

Pielke, R.A., Jr., Gratz, J., Landsea, C.W., Collins, D., Saunders, M.A., and Musulin, R. (2008b), "Normalized hurricane damage in the United States: 1900–2005", *Natural Hazards Review*, February, 29–42.

Pielke, R.A., Jr., Landsea, C., Mayfield, M., Laver, J., and Pasch, R. (2005), "Hurricanes and global warming", *Bulletin of the American Meteorological Society*, **86**, 1571–1575.

Pielke, R.A., Sr. (2009), "Reply to comment by David E. Parker et al. on 'Unresolved issues with the assessment of multidecadal global land surface temperature trends'", *Journal of Geophysical Research*, **114**, D05105.

Pielke, R.A., Sr., Adegoke, J.O., Chase, T.N., Marshall, C.H., Matsui, T., and Niyogi, D. (2007b), "A new paradigm for assessing the role of agriculture in the climate system and in climate change", *Agricultural and Forest Meteorology*, **142**, 234–254.

Pielke, R.A., Sr., Davey, C.A., Niyogi, D., Fall, S., Steinweg-Woods, J., Hubbard, K., Lin, X., Cai, M., Lim, Y.-K., Li, H., *et al.* (2007c), "Unresolved issues with the assessment of multi-decadal global land-surface temperature trends", February (*http://climatesci.colorado.edu/publications/pdf/R-321.pdf*).

Pierrehumbert, R.T. (2009), "Principles of planetary climate" (*http://geosci.uchicago.edu/~rtp1/ClimateBook/ClimateBook.html*).

Pinker, R.T., Zhang, B., and Dutton, E.G. (2005), "Do satellites detect trends in surface solar radiation?", *Science*, **308**, 850–854.

Pokhrel, Y.N., Hanasaki, N., Yeh, P.J.-F., Yamada, T.J., Kanae, S., and Oki, T. (2012), "Model estimates of sea-level change due to anthropogenic impacts on terrestrial water storage", *Nature Geoscience*, 20 May (*www.nature.com/ngeo/journal/vaop/ncurrent/abs/ngeo1476.html*).

Polissar, P.J., Abbott, J.B., Wolfe, A.P., Bezada, M., Rull, V., and Bradley, R.S. (2006), "Solar modulation of *Little Ice Age* climate in the tropical Andes", *Proceedings of the National Academy of Sciences (PNAS)*, **103**, 8937–8942.

Pollack, H.N., and Huang, S. (2000), "Climate reconstruction from subsurface temperatures", *Annual Review of Earth and Planetary Sciences*, **28**, 339–365.

Polyak, L., Alley, R.B., Andrews, J.T., Brigham-Grette, J., Cronin, T.M., Darby, D.A., Dyke, A.S., Fitzpatrick, J.J., Funder, S., Holland, M., *et al.* (2010), "History of sea ice in the Arctic", *Quaternary Science Reviews*, **29**, 1757–1778.

Polyakov, I., Walsh, D., Dmitrenko, I., Colony, R.L., and Timokhov, L.A. (2003a), "Arctic Ocean variability derived from historical observations", *Geographic Research Letters*, **30**, 31–34.

Polyakov, I.V., and Johnson, M.A. (2000), "Arctic decadal and interdecadal variability", *Geophysical Research Letters*, **27**, 4097–4100.

Polyakov, I.V., Alekseev, G.V., Bekryaev, R.V., Bhatt, U., Colony, R., Johnson, M., Karklin, V.P., Walsh, D., and Yulin, A.V. (2003b), "Long-term ice variability in Arctic marginal seas", *Journal of Climate*, **16**, 2078–2085.

Polyakov, I.V., Bekryaev, R.V., Alekseev, G.V., Bhatt, U.S., Colony, R.L., Johnson, M.A., Maskshtas, A.P., and Walsh, D. (2003c), "Variability and trends of air temperature and pressure in the maritime Arctic, 1875–2000", *Journal of Climate*, **16**, 2067–2077.

Power, S.B., and Smith, I.N. (2007), "Weakening of the Walker Circulation and apparent dominance of El Nino both reach record levels, but has ENSO really changed?", *Geophysical Research Letters*, **34**, L18702–118705.

Prohom, M.J., Esteban, P., Martn-Vide, J., and Jones, P.D. (2003), "Surface atmospheric circulation over Europe following major tropical volcanic eruptions, 1780–1995", in *Volcanism and the Earth's Atmosphere*, Geophysical Monograph 139, American Geophysical Union.

Prothero, D. (2011), "The record tornado season of 2011", *The Skeptic Magazine* (*www.skeptic.com/eskeptic/11-06-01/#feature*).

Przybylak, R. (2002), "Changes in seasonal and annual high-frequency air temperature variability in the Arctic from 1951 to 1990", *International Journal of Climatology*, **22**, 1017–1032.

Quiggan, J. (2006), *Stern and the Critics Discounting* (*johnquiggin.com/wpcontent/uploads/2006/12/sternreviewed06121.pdf*).

Quirk, T. (2012), "Did the global temperature trend change at the end of the 1990s?", *Asia-Pacific Journal of Atmospheric Sciences*, **48**, 339–344.

Rabatel, A., Francou, B., Jomelli, V., Naveau, P., and Grancher, D. (2008), "A chronology of the Little Ice Age in the tropical Andes of Bolivia (16S) and its implications for climate reconstruction", *Quaternary Research*, **70**, 198–212.

Radic, V. (2008), "Modeling future sea level rise from melting glaciers", Ph.D. dissertation, University of Alaska at Fairbanks.

Radick, R.R., Lockwood, G.W., Henry, G.W., and Baliunas, S.L. (2004), "The variability of Sun-like stars on decadal timescales", in A.K. Dupree and A.O. Benz (eds), *Stars as Suns: Activity, Evolution and Planets*, IAU Symposium, Vol. 219.

Rahmstorf, S. (2007), "A semi-empirical approach to projecting future sea-level rise", *Science*, **315**, 368–370.

Ramanathan, V. (1981), "The role of ocean–atmosphere interactions in the CO_2 climate problem", *Journal of the Atmospheric Sciences*, **38**, 918–930.

Ramanathan, V. (1988), "The greenhouse theory of climate change: A test by an inadvertent global experiment", *Science*, **240**, 293–299.

Ramanathan, V., and Carmichael, G. (2008), "Global and regional climate changes due to black carbon", *Nature Geoscience*, **1**, 220–226.

Ramanathan, V., Cicerone, R.J., Singh, H.B., and Kiehl, J.T. (1985), "Trace gas trends and their potential role in climate change", *Journal of Geophysical Research*, **90**, 5547–5566.

Rapp, A.D., Kummerow, C., Berg, W., and Griffith, B. (2005), "An evaluation of the proposed mechanism of the adaptive infrared iris hypothesis using TRMM VIRS and PR measurements", *Journal of Climate*, **18**, 4185–4194.

Rapp, D. (2012), *Ice Ages: Measurements, Interpretation and Models*, 2nd edn, Springer/Praxis Publishing, New York, U.S./Chichester, U.K.

Ravelo, A.C., Andreasen, D.H., Lyle, M., Lyle, A.O., and Wara, M.W. (2004), "Regional climate shifts caused by gradual global cooling in the Pliocene epoch", *Nature*, **429**, 263–266.

Raymo, M.E. (1992), "Global climate change: A three million year perspective", in G.J. Kukla and E. Went (eds), *Start of a Glacial*, Vol. I3, Springer-Verlag, Heidelberg.

Reid, G.C. (1997), "Solar forcing of global climate change since the mid-17th century", *Climatic Change*, **37**, 391–405.

Reifen, C., and Toumi, R. (2009), "Climate projections: Past performance no guarantee of future skill?", *Geophysical Research Letters*, **36**, L13704.

Richey, J.N., Poore, R.Z., Hower, B.P., and Quinn, T.M. (2007), "1400 yr multi-proxy record of climate variability from the northern Gulf of Mexico", *Geology*, **35**, 423.

Ridgwell, A., Zondervan, I., Hargreaves, J.C., Bijma, J., and Lenton, T.M. (2007), "Assessing the potential long-term increase of oceanic fossil fuel CO_2 uptake due to CO_2-calcification feedback", *Biogeosciences*, **4**, 481–492.

Rignot, E., and Kanagaratnam, P. (2006), "Changes in the velocity structure of the Greenland Ice Sheet", *Science*, **311**, 986–990.

Rignot, E., Velicogna, I., van den Broeke, M.R., Monaghan, A., and Lenaerts, J. (2011), "Acceleration of the contribution of the Greenland and Antarctic ice sheets to sea level rise", *Geophysical Research Letters*, **38**, L05503.

Rind, D. (2008), "The consequences of not knowing low- and high-latitude climate sensitivity", *Bulletin of the American Meteorological Society*, **89**, 855–864.

Rind, D., Lean, J., and Healy, R. (1999), "Simulated time-dependent climate response to solar radiative forcing since 1600", *Journal of Geophysical Research*, **104**, 1973–1990.

Rind, D., Shindell, D., Perlwitz, J., Lerner, J., Lonergan, P., Lean, J., and McLinden, C. (2004), "The relative importance of solar and anthropogenic forcing of climate change between the Maunder Minimum and the Present", *Journal of Climate*, **17**, 906–929.

Robinson, A.R., Robinson, N.E., and Soon, W. (2007), "Environmental effects of increased atmospheric CO_2", *Journal of American Physicians and Surgeons*, **12**, 79–90.

Robinson, M.M., Dowsett, H.J., and Chandler, M.A. (2008), "Pliocene role in assessing future climate impacts", *EOS, Transactions American Geophysical Union*, **89**, 501–502.

Robock, A. (2000), "Volcanic eruptions and climate", *Reviews of Geophysics*, **38**, 191–219.

Robock, A. (2004), "Climatic impact of volcanic emissions", in S. Sparks and C. Hawkesworth (eds), *State of the Planet*, Monograph 150, American Geophysical Union.

Robock, A. (2005), "Comment on 'Climate forcing by the volcanic eruption of Mount Pinatubo' by David H. Douglass and Robert S. Knox", *Geophysical Research Letters*, **32**, L20711.

Robock, A., Ammann, C.M., Oman, L., Shindell, D., Levis, S., and Stenchikov, G. (2009), "Did the Toba volcanic eruption of <74 ka B.P. produce widespread glaciation?", *Journal of Geophysical Research*, **114**, D10107.

Roe, G.H. (2009), "Feedbacks, timescales and seeing red", *Annual Review of Earth and Planetary Sciences*, **37**, 93–115.

Roe, G.H. (2010), "What do glaciers tell us about climate variability and climate change?" (*earthweb.ess.washington.edu/roe/GerardWeb/.../Roe_GlaciersClimate.pdf*).

Roe, G.H., and Armour, K.C. (2011), "How sensitive is climate sensitivity?", *Geophysical Research Letters* (*http://onlinelibrary.wiley.com/doi/10.1029/2011GL047913/abstract*).

Roe, G.H., and Baker, M.B. (2007), "Why is climate sensitivity so unpredictable?", *Science*, **318**, 629–632.

Roe, G.H., and Bauman, Y. (2011), "Should the climate tail wag the policy dog?" (*earthweb.ess.washington.edu/roe/GerardWeb/.../RoeBauman_FatTail.pdf*).

Rogers, M.L., Richards, M.T., and St. P. Richards, D. (2006), "Long-term variability in the length of the solar cycle", *Astro-Ph*, 1–15 (*http://adsabs.harvard.edu/abs/2006astro.ph.6426R*).

Rogner, H.-H. (1997), "An assessment of world hydrocarbon resources", *Annual Review of Energy and the Environment*, **22**, 217–262.

Rohde, R., Curry, J., Groom, D., Jacobsen, R., Muller, R.A., Perlmutter, S., Rosenfeld, A., Wickham, C., and Wurtele, J. (2011), "Berkeley Earth temperature averaging process" (*www.berkeleyearth.org/*).

Rohde, R., Muller, R.A., Jacobsen, R., Muller, E., Perlmutter, S., Rosenfeld, A., Wurtele, J., Groom, D., and Wickham, C. (2012), "A new estimate of the average Earth surface land temperature spanning 1753 to 2011", submitted to JGR, The Third Santa Fe Conference on Global and Regional Climate Change, Manuscript # 2012JD018202.

Rohling, E.J., Grant, K., Hemleben, C., Siddall, M., Hoogakker, B.A.A., Bolshaw, M., and Kucera, M. (2008), "High rates of sea-level rise during the last interglacial period", *Nature Geoscience*, **1**, 38–42.

Roper, L.D. (2006), *Relationship of Antarctica Temperature to Atmospheric Carbon Dioxide and Methane* (*www.roperld.com/science/CO2_Temp.pdf*).

Rørvik, K.-L., Grøsfjeld, K., and Hald, M. (2009), "A late Holocene climate history from the Malangen fjord, North Norway, based on dinoflagellate cysts", *Norwegian Journal of Geology*, **89**, 135–147.

Rothman, D.H. (2002), "Atmospheric carbon dioxide levels for the last 500 million years", *Proceedings of the National Academy of Sciences (PNAS)*, **99**, 4167–4171.

Royer, D.L. (2006), "CO_2-forced climate thresholds during the Phanerozoic", *Geochimica et Cosmochimica Acta*, **70**, 5665–5675.

Royer, D.L. (2010), "Fossil soils constrain ancient climate sensitivity", *Proceedings of the National Academy of Sciences (PNAS)*, **107**, 517–518.

Royer, D.L., Berner, R.A., and Park, J. (2007), "Climate sensitivity constrained by CO_2 concentrations over the past 420 million years", *Nature*, **446**, 530–532.

Royer, D.L., Berner, R.A., Montañez, I.P., Tabor, N.J., and Beerling, D.J. (2004), "CO_2 as a primary driver of Phanerozoic climate", *GSA Today*, **14**, 4–10.

Royer, D.L., Pagani, M., and Beerling, D.J. (2011), "Geologic constraints on Earth system sensitivity to CO_2 during the Cretaceous and early Paleogene", *Earth System Dynamics Discussions*, **2**, 211–240.

Rozelot, J.P., Damiani, C., and Lefebvre, S. (2009), "Variability of the solar shape (before space dedicated missions)", *Journal of Atmospheric and Solar-Terrestrial Physics*, **71**, 1683–1694.

Ruckstuhl, C., Philipona, R., Behrens, K., Collaud Coen, M.C., Dürr, B., Heimo, A., Mätzler, C., Nyeki, S., Ohmura, A., Vuilleumier, L., *et al.* (2008), "Aerosol and cloud effects on solar brightening and the recent rapid warming", *Geophysical Research Letters*, **35**, L12708.

Ruddiman, W.F. (2010), "A paleoclimatic enigma?", *Science*, **328**, 838–839.

Rutherford, S., Mann, M.E., Osborn, T.J., Bradley, R.S., Briffa, K.R., Hughes, M.K., and Jones, P.D. (2005), "Proxy-based Northern Hemisphere surface temperature reconstructions: Sensitivity to method, predictor network, target season and target domain", *Journal of Climate*, **18**, 2308–2329.

Rutledge, D. (2007), *Hubbert's Peak and Climate Change*, California Institute of Technology, presentation at Jet Propulsion Laboratory, Pasadena, CA (May) (*www.rutledge.caltech.edu/*).

Sabine, C., and Gruber, N. (2005), "Response to comment on 'The ocean sink for anthropogenic CO_2'", *Science*, **308**, 1743.

Sabine, C.L., Feely, R.A., Gruber, N., Key, R.M., Lee, K., Bullister, J.L., Wanninkhof, R., Wong, C.S., Wallace, D.W.R., Tilbrook, B., *et al.* (2004), "The oceanic sink for anthropogenic CO_2", *Science*, **305**, 367–371.

Santer, B.D., Mears, C., Doutriaux, C., Caldwell, P., Gleckler, P.J., Wigley, T.M.L., Solomon, S., Gillett, N.P., Ivanova, D., Karl, T.R., *et al.* (2011), "Separating signal and noise in atmospheric temperature changes: The importance of timescale", submitted to *Journal of Geophysical Research (Atmospheres)*, **116**.

Santer, B.D., Mears, C., Wentz, F.J., Taylor, K.E., Gleckler, P.J., Wigley, T.M.L., Barnett, T.P., Boyle, J.S., Bruggemann, W., Gillett, N.P., *et al.* (2007), "Identification of human-induced changes in atmospheric moisture content", *Proceedings of the National Academy of Sciences (PNAS)*, **104**, 15248–15253.

Santer, B.D., Thorne, P.W., Haimberger, L., Taylor, K.E., Wigley, T.M.L., Lanzante, J.R., Solomon, S., Free, M., Gleckler, P.J., Jones, P.D., *et al.* (2008), "Consistency of modeled and observed temperature trends in the tropical troposphere", *International Journal of Climatology*, **28**, 1703–1722.

Santer, B.D., Wigley, T.M.L., Mears, C., Wentz, F.J., Klein, S.A., Seidel, D.J., Taylor, K.E., Thorne, P.W., Wehner, M.F., Gleckler, P.J., *et al.* (2005), "Amplification of surface temperature trends and variability in the tropical atmosphere", *Science*, **309**, 1551–1556.

Scafetta, N. (2009), "Total solar irradiance satellite composites and their phenomenological effect on climate", in D. Easterbrook (ed.), *Evidence-Based Climate Science*, Chapter 12, Elsevier.

Scafetta, N., and West, B.J. (2006), "Phenomenological solar contribution to the

1900–2000 global surface warming", *Geophysical Research Letters*, **33**, L05708.1–L05708.4.
Schatten, K., and Presnell, W.D. (2007), "Solar Cycle d24 and the solar dynamo" (*ntrs.nasa.gov/archive/nasa/casi.ntrs.nasa.gov/20070032658_2007033016.pdf*).
Schewe, J., Levermann, A., and Meinshausen, M. (2011), "Climate change under a scenario near 1.5C of global warming: monsoon intensification, ocean warming and steric sea level rise", *Earth System Dynamics*, **2**, 25–35.
Schimel, D.S., House, J.I., Hibbard, K.A., Bousquet, P., Ciais, P., Peylin, P., Braswell, B.H., Apps, M.J., Baker, D., Bondeau, A., *et al.* (2001), "Recent patterns and mechanisms of carbon exchange by terrestrial ecosystems", *Nature*, **414**, 169–172.
Schmitt, C., and Randall, D.A. (1991), "Effects of surface temperature and clouds on the CO_2 forcing", *Journal of Geophysical Research*, **96**, 9159–9168.
Schmidt, G., and Archer, D. (2009), "Too much of a bad thing", *Nature*, **458**, 1117–1118.
Schmidt, G.A., Ruedy, R., Miller, R.L., and Lacis, A.A. (2010), "The attribution of the present-day total greenhouse effect", *Journal of Geophysical Research*, **115**, D20106
Schmidt, M.W., Vautravers, M.J., and Spero, H.J. (2006), "Rapid subtropical North Atlantic salinity oscillations across Dansgaard–Oeschger cycles", *Nature*, **443**, 561–564.
Schmittner, A., Oschlies, A., Matthews, H.D., and Galbraith, E.D. (2008), "Future changes in climate, ocean circulation, ecosystems and biogeochemical cycling simulated for a business-as-usual CO_2 emission scenario until year 4000 AD", *Global Biogeochemical Cycles*, **22**, GB1013.
Schneider, B., and Schneider, R. (2010), "Global warmth with little extra CO_2", *Nature Geoscience*, **3**, 6–7.
Schrijver, C.J., Livingston, W.C., Woods, T.N., and Mewaldt, R.A. (2011), "The minimal solar activity in 2008–2009 and its implications for long-term climate modeling", *Geophysical Research Letters*, **38**, L06701.
Schulte, K.-M. (2008), "Scientific consensus on climate change?", *Energy and Environment*, **19**, 281–286.
Schuster, U., and Watson, A.J. (2007), "A variable and decreasing sink for atmospheric CO_2 in the North Atlantic", in A.J. Watson (2007), "A variable ... North Atlantic", *Journal of Geophysical Research: Oceans*, **112** (*http://onlinelibrary.wiley.com/doi/10.1029/2006JC003941/abstract*).
Schwartz, S.E. (2003), "Requirements for empirical determination of Earth's climate sensitivity", AAAS Annual Meeting, Denver, CO, 14–18 February (*www.ecd.bnl.gov/steve/abstracts/Empirical.html*).
Schwartz, S.E. (2004), "Uncertainty requirements in radiative forcing of climate change", *Journal of the Air and Waste Management Association*, **54**, 1351–1359.
Schwartz, S.E. (2007), "Heat capacity, time constant and sensitivity of Earth's climate system", *Journal of Geophysical Research*, **112**, D24S05 (*www.ecd.bnl.gov/ pubs/BNL-76939-2006-AB.pdf*).
Schwartz, S.E., Charlson, R.J., and Rodhe, H. (2007), "Quantifying climate change: Too rosy a picture?", *Nature Reports, Climate Change*, **2**, 23–24.

Schwartz, S.E., Charlson, R.J., Kahn, R.A., Ogren, J.A., and Rodhe, H. (2010), "Why hasn't Earth warmed as much as expected?", *Journal of Climate*, **23**, 2453–2464.

Scotese, C.R. (2002), "Paleomap Project Climate History" (*www.scotese.com/climate.htm*).

Screen, J.A., and Simmonds, I. (2010), "The central role of diminishing sea ice in recent Arctic temperature amplification", *Nature*, **464**, 1334–1337.

Seager, R., Kushnir, Y., Nakamura, J., Ting, M., and Naik, N. (2010), "Northern Hemisphere winter snow anomalies: ENSO, NAO and the winter of 2009/10", *Geophysical Research Letters*, **37**, L14703.

Sedlacek, J., and Mysak, L.A. (2008), "Sensitivity of sea ice to wind-stress and radiative forcing since 1500: A model study of the Little Ice Age and beyond", *Climate Dynamics*, **32**, 817–831.

Seki, O., Foster, G.L., Schmidt, D.N., Mackensen, A., Kawamura, K., and Pancost, R.D. (2010), "Alkenone and boron-based Pliocene pCO_2 records", *Earth and Planetary Science Letters*, **292**, 201–211.

Sellers, A., and Meadows, A.J. (1975), "Long term variations in the albedo and surface temperature of the Earth", *Nature*, **254**, 44.

Senior, C.A., and Mitchell, J.F.B. (2000), "The time dependence of climate sensitivity", *Geophysical Research Letters*, **27**, 2685–2688.

Serreze, M.C., and Barry, R.G. (2011), "Processes and impacts of Arctic amplification: A research synthesis", *Global and Planetary Change*, **77**, 85–96.

Serreze, M.C., Barrett, A.P., Stroeve, J.C., Kindig, D.N., and Holland, M.M. (2008), "The emergence of surface-based Arctic amplification", *The Cryosphere Discussions*, **2**, 601–622 (*www.the-cryosphere-discuss.net/2/601/2008/*).

Serreze, M.C., Holland, M.M., and Stroeve, J. (2007), "Perspectives on the Arctic's shrinking sea-ice cover", *Science*, **315**, 1533–1537.

Shakhova, N., Semiletov, I., Salyuk, A., Yusupov, V., Kosmach, D., and Gustafsson, Ö (2010), "Extensive methane venting to the atmosphere from sediments of the East Siberian Arctic Shelf", *Science*, **327**, 1246–1250.

Shakun, J.D., and Carlson, A.E. (2010), "A global perspective on Last Glacial Maximum to Holocene climate change", *Quaternary Science Reviews*, **29**, 1801–1816.

Shapiro, A.I., Schmutz, W., Rozanov, E., Schoell, M., Haberreiter, M., Shapiro, A.V., and Nyeki, S. (2011), "A new approach to the long-term reconstruction of the solar irradiance leads to large historical solar forcing", *Astronomy and Astrophysics*, **529**, A67–A74.

Shaviv, N., and Veizer, J. (2003), "Celestial driver of Phanerozoic climate?", *GSA Today*, **13**, 4–10.

Sheffield, J., Wood, E.F., and Roderick, M.L. (2012), "Little change in global drought over the past 60 years", *Nature*, **491**, 435–440.

Shepherd, A., and Wingham, D. (2007), "Recent sea-level contributions of the Antarctic and Greenland ice sheets", *Science*, **315**, 1529–1532.

Sherwood, K., and Idso, C. (2006), "Highly over-hyped: Greenland's and Antarctica's impacts on sea level CO_2", *Science*, **9**, 29 (*www.co2science.org/scripts/CO2ScienceB2C/articles/V9/N13/EDIT.jsp*).

Shindell, D. (2007), "Estimating the potential for twenty-first century sudden climate change", *Philosophical Transactions of the Royal Society A*, **365**, 2675–2694.

Shindell, D., and Faluvegi, G. (2009), "Climate response to regional radiative forcing during the twentieth century", *Nature Geoscience*, **2**, 294–300.

Shine, K.P. (2005), "Comments on 'Contrails, cirrus trends and climate'", *Journal of Climate*, **18**, 2781–2782, a reply to P. Minnis, *Journal of Climate*, **18**, 2783–2784.

Siegel, D.A., and Dickey, T.D. (1986), "Variability of net long wave radiation over the eastern north Pacific Ocean", *Journal of Geophysical Research*, **91**, 7657–7666.

Siegenthaler, U., Stocker, T.F., Monnin, E., Luthi, D., Schwander, J., Stauffer, B., Raynaud, D., Barnola, J.-M., Fischer, H., Mas-son-Delmotte, V., *et al.* (2005), "Stable carbon cycle–climate relationship during the Late Pleistocene", *Science*, **310**, 1313–1318.

Sigman, D.M., and Boyle, E.A. (2000), "Glacial/interglacial variations in atmospheric carbon dioxide", *Nature*, **407**, 859–869.

Sigman, D.M., Hain, M.P., and Haug, G.H. (2010), "The polar ocean and glacial cycles in atmospheric CO_2 concentration", *Nature*, **466**, 47–55.

Silverman, J., Lazar, B., Cao, L., Caldeira, K., and Erez, J. (2009), "Coral reefs may start dissolving when atmospheric CO_2 doubles", *Geophysical Research Letters*, **36**, L05606.

Sime, L.C., Wolff, E.W., Oliver, K.I.C., and Tindall, J.C. (2009), "Evidence for warmer interglacials in East Antarctic ice cores", *Nature*, **462**, 342–346.

Simmons, J.P., Nelson, L.D., and Simonsohn, I. (2011), "False-positive psychology: Undisclosed flexibility in data collection and analysis allows presenting anything as significant", *Psychological Science*, **22**, 1359–1366.

Simpson, I.J., Andersen, M.P.S., Meinardi, S., Bruhwiler, L., Blake, N.J., Helmig, D., Sherwood Rowland, F., and Blake, D.R. (2012), "Long-term decline of global atmospheric ethane concentrations and implications for methane", *Nature*, **488**, 490–494.

Singer, S.F., and Avery, D.T. (2007), *Unstoppable Global Warming—Every 1500 Years*, Rowman & Littlefield, Landham, MD, U.S.

Skinner, L.C. (2006), "Glacial–interglacial atmospheric CO_2 change: A simple 'hypsometric effect' on deep-ocean carbon sequestration?", *Climate of the Past Discussions*, **2**, 711–743 (*www.clim-past-discuss.net/2/711/2006/*).

Smerdon, J.E., and Kaplan, A. (2007), "Comment on 'Testing the fidelity of methods used in proxy-based reconstructions of past climate': The role of the standardization interval", submitted to *Journal of Climate* (*www.ldeo.columbia.edu/jsmerdon/papers/2006_agu_smerdonetal.pdf*).

Smith, A.G., and Pickering, K.T. (2003), "Oceanic gateways as a critical factor to initiate icehouse Earth", *Journal of the Geological Society, London*, **160**, 337–340.

Smith, D.M., Eade, R., Dunstone, N.J., Fereday, D., Murphy, J.M., Pohlmann, H., and Scaife, A.A. (2010), "Skilful multi-year predictions of Atlantic hurricane frequency", *Nature Geoscience*, **3**, 846–849.

Smith, R.L. (2011), "Understanding sensitivities in paleoclimatic reconstructions" (*www.stat.unc.edu/postscript/rs/Sensitivities-Submitted.pdf*).

Smith, S.J., van Aardenne, J., Klimont, Z., Andres, R.J., Volke, A., and Delgado Arias, S. (2011), "Anthropogenic sulfur dioxide emissions: 1850–2005", *Atmospheric Chemistry and Physics*, **11**, 1101–1116.

Soden, B.J., Jackson, D.L., Ramaswamy, V., Schwarzkopf, M.D., and Huang, X. (2005), "The radiative signature of upper tropospheric moistening", *Science*, **310**, 841–844.

Soden, B.J., Wetherald, R.T., Stenchikov, G.L., and Robock, A. (2002), "Global cooling after the eruption of Mount Pinatubo: A test of climate feedback by water vapor", *Science*, **296**, 727–730.

Sokolov, A.P., Stone, P.H., Forest, C.E., Prinn, R., Sarofim, M.C., Webster, M., Paltsev, S., Schlosser, C.A., Kicklighter, D., Dutkiewicz, S., *et al.* (2009), "Probabilistic forecast for 21st century climate based on uncertainties in emissions (without policy) and climate parameters", *MIT Joint Program on the Science and Policy of Global Change*, Report No. 169, January.

Solanki, S., Krivova, N., and Wenzler, T. (2005), "Irradiance models", *Advances in Space Research*, **35**, 376–383.

Solanki, S.K., and Fligge, M. (1998), "Solar irradiance since 1874 revisited", *Geophysical Research Letters*, **25**, 341–344.

Solanki, S.K., and Fligge, M. (1999), "A reconstruction of total solar irradiance since 1700", *Geophysical Research Letters*, **26**, 2465–2468.

Solanki, S.K., and Fligge, M. (2000), "Reconstruction of past solar irradiance", *Space Science Reviews*, **94**, 127–138.

Solanki, S.K., and Krivova, N.A. (2004), "Solar irradiance variations: From current measurements to long-term estimates", *Solar Physics*, **224**, 197–208.

Solanki, S.K., Usoskin, I.G., Kromer, B., Schussler, M., and Beer, J. (2004), "Unusual activity of the Sun during recent decades compared to the previous11,000 years", *Nature*, **431**, 1084–1087.

Solomon, S., Plattner, G.-K., Knutti, R., and Friedlingstein, P. (2009), "Irreversible climate change due to carbon dioxide emissions", *Proceedings of the National Academy of Sciences (PNAS)*, **106**, 1704–1709.

Solomon, S., Rosenlof, K.H., Portmann, R.W., Daniel, J.S., Davis, S.M., Sanford, T.J., and Plattner, G.-K. (2010), "Contributions of stratospheric water vapor to decadal changes in the rate of global warming", *Science*, **327**, 1219–1223.

Soon, W., and Baliunas, S. (2003a), "Lessons and limits of climate history: Was the 20th century climate unusual?", informal report, George C. Marshall Institute, Washington.

Soon, W., and Baliunas, S. (2003b), "Reconstructing climatic and environmental changes of the past 1000 years: A reappraisal", *Energy and Environment*, **14**, 233–299.

Soon, W., and Yaskell, S. (2007), *Year without a Summer* (*www.mitosyfraudes.org/Calen/Year1816.html*).

Sopkin, K.L. (2008), "Heat fluxes in Tampa Bay, Florida", Master of Science thesis, College of Marine Science, University of South Florida.

Spencer, R.W., and Braswell, W.D. (2007), "Potential biases in feedback diagnosis from observational data: A simple model demonstration", *Journal of Climate*, **21**, 5624–5628.

Spencer, R.W., and Braswell, W.D. (2010), "On the diagnosis of radiative feedback in the presence of unknown radiative forcing", *Journal of Geophysical Research*, **115**, D16109.

Spencer, R.W., and Braswell, W.D. (2011), "On the misdiagnosis of climate feedbacks from variations in Earth's radiant energy balance", *Remote Sensing*, **3**, 1603–1613.

Spencer, R.W., Braswell, W.D., Christy, J.R., and Hnilo, J. (2007), "Cloud and radiation budget changes associated with tropical intraseasonal oscillations", *Geophysical Research Letters*, **34**, L15707–L15711.

Spielhagen, R., Werner, F.K., Sørensen, S.A., Zamelczyk, K., Kandiano, E., Budeus, G., Husum, K., Marchitto, T.M., and Hald, M. (2011), "Enhanced modern heat transfer to the Arctic by warm Atlantic water", *Science*, **331**, 450–453.

Spörer, F.W.G. (1887/1889), "Über die Periodizität der Sonnenflecken seit dem Jahre 1618", *Vierteljahrssch. Astron. Ges. (Leipsig)*, **22**, 323 (1887), *Bull Astron.* **6**, 60 (1889).

Stahle, D.W., Cook, E.R., Cleaveland, M.K, Therrell, M.D., Meko, D.M., Grissino-Mayer, H.D., Watson, E., and Luckman, B.H. (2000), "Tree-ring data document 16th century megadrought over North America", *EOS Transactions, American Geophysical Union*, **81**, 121–125.

Stainforth, D.A., Allen, M.R., Tredger, E.R., and Smith, L.A. (2007), "Confidence, uncertainty and decision-support relevance in climate predictions", *Philosophical Transactions of the Royal Society*, **A365**, 2145–2161.

Stammerjohn, S.E., Martinson, D.G., Smith, R.C., Yuan, X., and Rind, D. (2008), "Trends in Antarctic annual sea ice retreat and advance and their relation to El Niño–Southern Oscillation and Southern Annular Mode variability", *Journal of Geophysical Research*, **113**, C03S90–C03S109.

Stauffer, B., Blunier, T., Dallenbach, A., Indermuhle, A., Schwander, J., Stocker, T.F., Tschumi, J., Chappellaz, J., Raynaud, D., Hammer, C.U., and Clausen, H.B. (1998), "Atmospheric CO_2 concentration and millennial-scale climate change during the last glacial period", *Nature*, **392**, 59–62.

Steig, E.J., Schneider, D.P., Rutherford, S.D., Mann, M.E., Comiso, J.C., and Shindell, D.T. (2009), "Warming of the Antarctic ice-sheet surface since the 1957 International Geophysical Year", *Nature*, **457**, 459–453.

Steinhilber, F., Beer, J., and Frohlich, C. (2009), "Total solar irradiance during the Holocene", *Geophysical Research Letters*, **36**, L19704.

Stephens, G.L., Li, J., Wild, M., Clayson, C.A., Loeb, N., Kato, S., L'Ecuyer, T., Stackhouse, P.W., Jr., Lebsock, M., and Andrews, T. (2012), "An update on Earth's energy balance in light of the latest global observations", *Nature Geoscience* (*www.aos.wisc.edu/˜tristan/publications/2012_EBupdate_stephens_ngeo1580.pdf*).

Sterl, A., Severijns, C., van Oldenborgh, G.J., Dijkstra, H., Hazeleger, W., van den Broeke, M., Burgers, G., van den Hurk, B., van Leeuwen, P.J., and van Velthoven, P. (2009), "The ESSENCE project: signal to noise ratio in climate projections", *Geophysical Research Letters* (*www.knmi.nl/˜sterl/Essence/essence_1_v2.2.pdf*).

Stern, D.I. (2005a), "Global sulfur emissions from 1850 to 2000", *Chemosphere*, **58**, 163–175.

Stern, D.I. (2005b), "Reversal in the trend of global anthropogenic sulfur emissions", Rensselaer working papers in economics, No. 0504 (May) (*www.rpi.edu/sternd/GEC2006.pdf*).

Stevenson, K., Alessa, L., Altaweel, M., Kliskey, A.D., and Krieger, K.E. (2012), "Minding our methods: How choice of time series, reference dates and statistical approach can influence the representation of temperature change", *Environmental Science and Technology*, **46**, 7435–7441.

Stone, D.A., and Weaver, A.J. (2003), "Factors contributing to diurnal temperature range trends in twentieth and twenty-first century simulations of the CCCma coupled model", *Climate Dynamics*, **20**, 435–445.

Stott, L., Cannariato, K.G., Thunell, R., Haug, G.H., Koutavas, A., and Lund, S. (2004), "Decline of surface temperature and salinity in the western tropical Pacific Ocean in the Holocene epoch", *Nature*, **431**, 6–59.

Stott, L., Timmermann, A., and Thunell, R. (2007), "Southern Hemisphere and deep-sea warming led deglacial atmospheric CO_2 rise and tropical warming", *Science*, **318**, 435–439.

Stott, P.A., Tett, S.F.B., Jones, G.S., Allen, M.R., Mitchell, J.F.B., and Jenkins, G.J. (2000), "External control of 20th century temperature by natural and anthropogenic forcings", *Science*, **290**, 2133–2137.

Streutker, D.R. (2003), "A study of the urban heat island of Houston, Texas", Ph.D. thesis, University of Houston.

Sundqvist, H.S., Zhang, Q., Moberg, A., Holmgren, K., Kornich, H., Nilsson, J., and Brattstrom, G. (2010), "Climate change between the mid and late Holocene in northern high latitudes. Part 1: Survey of temperature and precipitation proxy data", *Climate of the Past*, **6**, 591–608.

Svalgaard, L., Cliver, E.W., and Kamide, Y. (2005), "Sunspot cycle 24: Smallest cycle in 100 years?", *Geophysical Research Letters*, **32**, L01104.

Svensmark, H. (2000), "Cosmic rays and Earth's climate", *Space Science Reviews*, **93**, 155–166. *http://www.junkscience.com/Greenhouse/influence-of-cosmic-rays-on-the-earth.pdf*

Svensmark, H. (2007), "Cosmoclimatology: A new theory emerges", *Astronomy & Geophysics*, **48**, 1.18–1.24.

Svensmark, H., and Friis-Christensen, E. (1997), "Variation of cosmic ray flux and global cloud coverage: A missing link in solar–climate relationships", *Journal of Atmospheric and Solar-Terrestrial Physics*, **59**, 1225–1232.

Svensmark, H., and Friis-Christensen, E. (2007), "Reply to Lockwood and Fröhlich—The persistent role of the Sun in climate forcing, Danish National Space Center Scientific Report, March 2007" (*www.spacecenter.dk/publications/scientific-report-series/Scient_No._3.pdf/view*).

Sverdrup, H.U., Johnson, M.W., and Fleming, R.H. (1942), *The Oceans, Their Physics, Chemistry and General Biology*, Prentice-Hall, New York (*publishing.cdlib.org/ucpressebooks/data/.../kt167nb66r_ch04.pdf*).

Swinbank, W.C. (1963), "Long-wave radiation from clear skies", *Quarterly Journal of the Royal Meteorological Society*, **89**, 339–348.

Tapping, K.F., Boteler, D., Crouch, A., Charbonneau, P., Manson, A., and Paquette, H. (2006), "Modelling solar magnetic flux and irradiance during and since the Maunder Minimum" (*www.lps.umontreal.ca/paquetteh/Maunder_SP.pdf*).

Taylor, G. (2006), "Antarctic temperature and sea ice trends over the last century: An assessment of Antarctic climate data (March)" (*http://64.119.172.31/awpf.pdf, www.co2science.org/scripts/CO2ScienceB2C/articles/V9/N13/EDIT.jsp?Print=true*).

Taylor, K.E., Crucifix, M., Braconnot, P., Hewitt, C.D., Doutriaux, C., Broccoli, A.J., Mitchell, J.F.B., and Webb, M.J. (2000), "Estimating shortwave radiative forcing and response in climate models", *Journal of Climate*, **20**, 2530–2544.

Taylor, K.E., Hewitt, C.D, Braconnot, P., Broccoli, A.J., Doutriaux, C., and Mitchell, J.F.B. (2001), "Analysis of forcing response and feedbacks in a paleoclimate modeling experiment", Lawrence Livermore National Laboratory, Report UCRL-JC-143363.

Tedesco, M., and Monaghan, A.J. (2010), "Climate and melting variability in Antarctica", *EOS, Transactions, American Geophysical Union*, **91**, L18502.

Tedesco, M., Fettweis, X., Mote, T., Steiner, N., and Box, J.E. (2011a), "Year 2011 Greenland melting remains well above the (1979–2010) average, close-to-record mass loss" (*http://greenland2011.cryocity.org/*).

Tedesco M., Fettweis, X., van den Broeke, M.R., van de Wal, R.S.W., Smeets, C.J.P.P., van de Berg, W.J., Serreze, M.C., and Box, J.E. (2010), "Record summer melt in Greenland in 2010", *EOS AGU*, **92**, 126.

Tedesco, M., Fettweis, X., van den Broeke, M.R., van de Wal, R.S.W., Smeets, C.J.P.P., van de Berg, W.J., Serreze, M.C., and Box, J.E. (2011b), "The role of albedo and accumulation in the 2010 melting record in Greenland", *Environmental Research Letters*, **6**, 014005.

Thejll, P. (2009), "Update of the solar cycle length curve and the relationship to the global mean temperature", Danish Climate Centre Report 09-01 (*www.dmi.dk/dmi/dkc09-01.pdf*).

Thejll, P., and Lassen, K. (2000), "Solar forcing of the Northern Hemisphere land air temperature: New data", *Journal of Atmospheric and Solar-Terrestrial Physics*, **62**, 1207–1213.

Thompson, D.W.J., and Solomon, S. (2002), "Interpretation of recent Southern Hemisphere climate change", *Science*, **296**, 895–899.

Thompson, L.G., Davis, M.E., Mosley-Thompson, E., Sowers, T.A., Henderson, K.A., Zagorodnov, V.S., Lin, P.-N., Mikhalenko, V.N., Campen, R.K., Bolzan, J.F., *et al.* (1998), "A 25,000-year tropical climate history from Bolivian ice cores", *Science*, **282**, 1858–1864.

Thorne, P.W., Lanzante, J.R., Peterson, T.C., Seidel, D.J., and Shine, K.P. (2011), "Tropospheric temperature trends: History of an ongoing controversy", *WIRES Climate Change*, **2** (*http://wires.wiley.com/WileyCDA/WiresArticle/wisId-WCC80.html*).

Timmermann, A., Timm, O., Stott, L., and Menviel, L. (2008), "The roles of CO_2 and orbital forcing in driving southern hemispheric temperature variations during the last 21,000 years", *Journal of Climate*, **22**, 1626–1640.

Tisdale, B. (2009), (*http://bobtisdale.wordpress.com/*).

Titus, J.G. (1990), "Greenhouse effect, sea level rise and barrier islands: Case study of Long Beach Island, New Jersey" (*http://yosemite.epa.gov/oar/globalwarming.nsf/content/ResourceCenterPublicationsSLRBarrier_Islands.html*).

Trainer, T. (2011), "A critique of the 2011 IPCC Report on Renewable Energy" (*http://bravenewclimate.com/2011/08/09/ipcc-renewables-critique/*).

Trenberth, K. (2010), "The ocean is warming, isn't it?", *Nature*, **465**, 304.

Trenberth, K.E. (2004), "Comment on 'Impact of urbanization and land-use change on climate' by E. Kalnay and M. Cai in *Nature* **423**, 528–531 (2003)", *Nature*, **427**, 214.

Trenberth, K.E., and Hoar, T.J. (1997), "El Nino and climate change", *Geophysical Research Letters*, **24**, 3057–3060.

Trenberth, K.E., Fasullo, J.T., and Kiehl, J. (2009), "Earth's global energy budget", *Bulletin of the American Meteorological Society (BAMS)*, March, 311–324.

Trigo, R.M., Vaquero, J.M., Alcoforado, M.-J., Barriendos, M., Taborda, J., Garca-Herrerag, R., and Luterbacherh, J. (2009), "Iberia in 1816, the year without a summer", *International Journal of Climatology*, **29**, 99–115.

Tripati, A.K., Roberts, C.D., and Eagle, R.A. (2009), "Coupling of CO_2 and ice sheet stability over major climate transitions of the last 20 million years", *Science*, **326**, 1394–1397.

Tsurutani, B.T., Echer, E., and Gonzalez, W.D. (2011), "The solar and interplanetary causes of the recent minimum in geomagnetic activity (MGA23): A combination of mid-latitude small coronal holes, low IMF BZ variances, low solar wind speeds and low solar magnetic fields", *Annuls of Geophysics*, **29**, 839–849.

Turner, J., Colwell, S.R., Marshall, G.J., Lachlan-Cope, T.A., Carleton, A.M., Jones, P.D., Lagun, V., Reide, P.A., and Iagovkina, S. (2005), "Antarctic climate change during the last 50 years", *International Journal of Climatology*, **25**, 279–294.

Turner, J., Comiso, J., Marshall, G.J., Lachlan-Cope, T.A., Bracegirdle, T., Maksym, T., Meredith, M., Wang, Z., and Orr, A. (2009), "Non annular atmospheric circulation change induced by stratospheric ozone depletion and its role in the recent increase of Antarctic sea ice extent", *Geophysical Research Letters*, **36**, L08052.

University of Colorado (2010), "Sea level change" (*http://sealevel.colorado.edu/current/sl_ib_ns_global.pdf*).

Uriarte, A. (2009), *Earth's Climate History*, 2nd edn, Gobierno Vasco, Spain. *http://web.me.com/uriarte/Earths_Climate/Earths_Climate_History.html*

Usoskin, I.G., Solanki, S.K., Schussler, M., Mursula, K., and Alanko, K. (2003), "Millennium-scale sunspot number reconstruction: Evidence for an unusually active Sun since the 1940s", *Physical Review Letters*, **91**, 211101-1–211101-4.

Van Andel, I.E. (2008), "The new climate theory of Dr. Ferenc Miskolczi" (*ruby.fgcu.edu/courses/twimberley/EnviroPhilo/Ferenc.pdf*).

Van de Wal, R.S., de Boer, B., Lourens, L., Kohler, P., and Bintanja, R. (2011), "Continuous and self-consistent CO_2 and climate records over the past 20 Myrs", *Climate of the Past Discussions*, **7**, 437–461.

Van de Wal, R.S., Boot, W., van den Broeke, M.R., Smeets, C.J.P.P., Reijmer, C.H., Donker, J.J.A., and Oerlemans, J. (2008), "Large and rapid Greenland ice sheet", *Science*, **321**, 111–113.

Van der Schrier, G., Briffa, K.R., Osborn, T.J., and Cook, E.R. (2006), "Summer moisture availability across North America", *Journal of Geophysical Research*, **111**, D11102.

Van Dorland, R. (1998), "Radiation and climate from radiative transfer modelling to global temperature response", Ph.D. dissertation, University of Utrecht.

van Soelen, E.E., Wagner-Cremerc, F., Sinninghe Damsté, J.S., and Reicharta, G.J. (2013), "Reconstructing tropical cyclone frequency using hydrogen isotope ratios of sedimentary n-alkanes in northern Queensland, Australia", *Palaeogeography, Palaeoclimatology, Palaeoecology*, **376**, 66–72.

Vaquero, J.M. (2007), *Historical Sunspot Observations: A Review* (*http://arxiv.org/pdf/astroph/0702068*).

Vaquero, J.M., Gallego, M.C., Trigo, R.M., Sanchez-Bajo, F., Cancillo, M.L., and Garcia, J.A. (2006), "A new reconstruction of total solar irradiance since 1832", *Atmosfera*, **19**, 267– 274.

Vaughn, A.P.M. (2007), "Climate and geology: A phanerozoic perspective", in M. Williams, A.M. Haywood, and F.J. Gregory (eds), *Deep-Time Perspectives on Climate Change*, Geological Society.

Vecchi, G.A., Soden, B.J., Wittenberg, A.T., Held, I.M., Leetma, A., and Harrison, M.J. (2006), "Weakening of tropical Pacific atmospheric circulation due to anthropogenic forcing", *Nature*, **441**, 73–76.

Vecchi, G.A., Swanson, K.L., and Soden, B.J. (2008), "Whither hurricane activity?", *Science*, **322**, 687–689.

Veizer, J. (2005), "Celestial climate driver: A perspective from four billion years of the carbon cycle", *Geoscience Canada*, **32**, 13–28.

Veizer, J., Godderis, Y., and Franceois, L.M. (2000), "Evidence for decoupling of atmospheric CO_2 and global climate during the Phanerozoic eon", *Nature*, **408**, 698–701.

Velicogna, I. (2009), "Increasing rates of ice mass loss from the Greenland and Antarctic ice sheets revealed by GRACE", *Geophysical Research Letters*, **36**, L19503.

Velicogna, I., and Wahr, J. (2006), "Measurements of time-variable gravity show mass loss in Antarctica", *Science*, **311**, 1754–1756.

Vermeer, M., and Rahmsdorf, S. (2009), "Global sea level linked to global temperature", *Proceedings of the National Academy of Sciences (PNAS)*, **106**, 21527–21532.

Vieira, L.E.A., Solanki, S.K., Krivova, N.A., and Usoskin, I. (2011), "Evolution of the solar irradiance during the Holocene", *Astronomy and Astrophysics*, **531**, A6–A25.

Villarini, G., Vecchi, G.A., Knutson, T.R., and Smith, J.A. (2011a), "Is the recorded increase in short-duration North Atlantic tropical storms spurious?", *Journal of Geophysical Research*, **116**, D10114.

Villarini, G., Vecchi, G.A., Knutson, T.R., Zhao, M., and Smith, J.A. (2011b), "North Atlantic tropical storm frequency response to anthropogenic forcing: Projections and sources of uncertainty", *Journal of Climate*, **24**, 3224–3238.

Vinje, T. (2001), "Anomalies and trends of sea-ice extent and atmospheric circulation in the Nordic seas during the period 1864–1998", *Journal of Climate*, **14**, 255–267.

Vinther, B.M., Andersen, K.K., Jones, P.D., Briffa, K.R., and Cappelen, J. (2006), "Extending Greenland temperature records into the late eighteenth century", *Journal of Geophysical Research*, **111**, D11105–D11118.

Vinther, B.M., Jones, P.D., Briffa, K.R., Clausen, H.B., Andersen, K.K., Dahl-Jensen, D., and Johnsen, S.J. (2010), "Climatic signals in multiple highly resolved stable isotope records from Greenland", *Quaternary Science Reviews*, **29**, 522–538.

Visser, K., Thunell, R., and Stott, L. (2003), "Magnitude and timing of temperature change in the Indo-Pacific warm pool during deglaciation", *Nature*, **421**, 152–155.

Vollweiler, N., Scholz, D., Muhlinghaus, C., Mangini, A., and Spotl, C. (2006), "A precisely dated climate record for the last 9 kyr from three high alpine stalagmites, Spannagel Cave, Austria", *Geophysical Research Letters*, **33**, L20703.

Volz, H. (2006), "Sea water emissivity: A neglected climate forcing", *Second International Conference on Global Warming and the Next Ice Age*, Santa Fe, New Mexico, 17–19 July.

von Storch, H.E., Zorita, E., and Gonzalez-Rouco, F. (2007), "Assessment of three temperature reconstruction methods in the virtual reality of a climate simulation", *International Journal of Earth Sciences*, **98**, 67–82.

von Storch, H.E., Zorita, E., Jones, J., Dimitriev, Y., Gonzalez-Rouco, F., and Tett, S. (2004), "Reconstructing past climate from noisy data", *Science*, **306**, 679–682.

Vose, R.S., Easterling, D.R., and Gleason, B. (2005), "Maximum and minimum temperature trends for the globe: An update through 2004", *Geophysical Research Letters*, **32**, L23822–L23826.

Vose, R.S., Karl, T.R., Easterling, D.R., Williams, C.N., and Menne, M.J. (2004), "Comment on 'Impact of urbanization and land-use change on climate' by E. Kalnay and M. Cai in *Nature* **423**, 528–531 (2003)", *Nature*, **427**, 214–215.

Vuille, M., Francou, B., Wagnon, P., Juen, I., Kaser, G., Mark, B.G., and Bradley, R.S. (2008), "Climate change and tropical Andean glaciers: Past, present and future", *Earth-Science Reviews*, **89**, 79–96.

Wada, Y., van Beek, L.P.H., Sperna Weiland, F.C., Chao, B.F., Wu, Y.-H., and Bierkens, M.F.P. (2012), "Past and future contribution of global groundwater depletion to sea-level rise", *Geophysical Research Letters*, **39**, L09402.

Wada, Y., van Beek, L.P.H., van Kempen, C.M., Reckman, J.W.T.M., Vasak, S., and Bierkens, M.F.P. (2010), "Global depletion of groundwater resources", *Geophysical Research Letters*, **37**, L20402.

Wahl, E.R., and Ammann, C.M. (2007), "Robustness of the Mann, Bradley, Hughes reconstruction of Northern Hemisphere surface temperatures: Examination of criticisms based on the nature and processing of proxy climate evidence", *Climatic Change*, **85**, 33–69.

Wallmann, K. (2004), "Impact of atmospheric CO_2 and galactic cosmic radiation on Phanerozoic climate change and the marine ä^{18}O record", *Geochemistry Geophysics Geosystems*, **5**, Q06004.

Wang, C. (2002), "A modeling study on the climate impacts of black carbon aerosols", MIT Joint Program on the Science and Policy of Global Change, Report No. 84, March.

Wang, K., Dickinson, R.E., Wild, M., and Liang, S. (2012), "Atmospheric impacts on climatic variability of surface incident solar radiation", *Atmospheric Chemistry and Physics*, **12**, 14009–14042.

Wang, M., and Overland, J.E. (2009), "A sea ice free summer Arctic within 30 years?", *Geophysical Research Letters*, **36**, L07502.

Wang, Y.-M., Lean, J., and Sheeley, N.R., Jr. (2002), "Role of a variable meridional flow in the secular evolution of the Sun's polar fields and open flux", *Astrophysical Journal*, **577**, L53–L57.

Wang, Y.-M., Lean, J.L., and Sheeley, N.R., Jr. (2005), "Modeling the Sun's magnetic field and irradiance since 1713", *The Astrophysical Journal*, **625**, 522–538.

Ward, B. (2008), *Communicating on Climate Change: An Essential Resource for Journalists, Scientists and Educators*, Metcalf Institute for Marine & Environmental Reporting (*www.metcalfinstitute.org/Communicating_ClimateChange.htm*).

Watson, P.J. (2011), "Is there evidence yet of acceleration in mean sea level rise around mainland Australia?", *Journal of Coastal Research*, **27**, 36377.

Watts, R.G. (1980), "Discussion of questions concerning the possible influence of anthropogenic CO_2 on atmospheric temperature", *Journal of Applied Meteorology*, **19**, 494–495.

Watts, R.G. (1982), "Further discussion of questions concerning the possible influence of anthropogenic CO_2 on atmospheric temperature", *Journal of Applied Meteorology*, **21**, 243–247.

Webster, M., Paltsev, S., Parsons, J., Reilly, J., and Jacoby, H. (2008), "Uncertainty in greenhouse gas emissions and costs of atmospheric stabilization", MIT Joint Program on the Science and Policy of Global Change, Report No. 165, November.

Webster, M., Sokolov, A.P., Reilly, J.M., Forest, C.E., Paltsev, S., Schlosser, A., Wang, C., Kicklighter, D., Sarofim, M., Melillo, J., *et al.* (2009), "Analysis of climate policy targets under uncertainty", MIT Joint Program on the Science and Policy of Global Change, Report No. 180, September.

Webster, P.J., Holland, G.J., Curry, J.A., and Chang, H.-R. (2005), "Changes in tropical cyclone number, duration and intensity in a warming environment", *Science*, **309**, 1844–1846.

Weckström, J., Korhola, A., Erasto, P., and Holmstrom, L. (2006), "Temperature patterns over the past eight centuries in Northern Fennoscandia inferred from sedimentary diatoms", *Quaternary Research*, **66**, 78–86.

Wegman, E.J., Scott, D.W., and Said, Y.H. (2006), *Ad Hoc Committee Report on the Hockey Stick Global Climate Reconstruction*, Congressional Committee on Energy and Commerce, Washington, D.C., 14 July (*http://republicans.energy commerce.house.gov/108/home/07142006_Wegman_Report.pdf*).

Weiler, C.S., Keller, J.K., and Olex, C. (2011), "Personality type differences between Ph.D. climate researchers and the general public: Implications for effective communication", *Climatic Change* (*disccrs.org/files/WeilerEtAl_2011_Climatic Change_MBTI.pdf*).

Weinkle, J., Maue, R., and Pielke, R., Jr. (2012), "Historical global tropical cyclone landfalls", *Journal of Climate*, **25**, 4729–4735.

Weitzman, M. (2007), "A review of the Stern Review of the Economics of Climate Change", *Journal of Economic Literature*, **45**, 703–724.

Weitzman, M. (2009), "Additive damages, fat-tailed climate dynamics and uncertain discounting", *Economics: The Open-Access, Open-Assessment E-Journal* **3**, 2009-39 (*www.economics-ejournal.org/economics/journalarticles/2009-39*).

Weitzman, M. (2011), "GHG targets as insurance against catastrophic climate damages" (*citeseerx.ist.psu.edu/viewdoc/download?doi=10.1.1.168.1527...*).

Wenzler, T., Solanki, S.K., and Krivova, N.A. (2005), "Can surface magnetic fields reproduce solar irradiance variations in cycles 22 and 23?", *Astronomy and Astrophysics*, **432**, 1057–1061.

Wickham, C., Curry, J., Groom, D., Jacobsen, R., Muller, R., Perlmutter, S., Rohde, R., Rosenfeld, A., and Wurtele, J. (2011), "Influence of urban heating on the global temperature land average using rural sites identified from MODIS classifications" (*www.berkeleyearth.org/*).

Wielicki, B.A., Wong, T., Allan, R.P., Slingo, A., Kiehl, J.T., Soden, B.J., Gordon, C.T., Miller, A.J., Yang, S.-K., Randall, D.A., *et al.* (2002), "Evidence for large decadal variability in the tropical mean radiative energy budget", *Science*, **295**, 841–844.

Wigley, T.M.L., Ammann, C.M., and Santer, B.D. (2005), "Using the Mount Pinatubo volcanic eruption to determine climate sensitivity: Comments on 'Climate forcing by the volcanic eruption of Mount Pinatubo' by David H. Douglass and Robert S. Knox (*Geophysical Research Letters*, 32, L05710, doi:10.1029/2004GL022119, 2005)", *Geophysical Research Letters*, **110**, D09107.

Wigley, T.M.L., Richels, R., and Edmonds, J.A. (1996), "Economic and environmental choices in the stabilization of atmospheric CO_2 concentration", *Nature*, **379**, 240–243.

Wijffels, S.E., Willis, J., Domingues, C.M., Barker, P., White, N.J., Gronell, A., Ridgway, K., and Church, J.A. (2008), "Changing expendable bathythermograph fall rates and their impact on estimates of thermosteric sea level rise", *Journal of Climate*, **21**, 5657–5672.

Willson, R. (2006), "A critical review of the time series of total solar irradiance satellite observations", AGU Fall Meeting 2006, Paper SH41A-02 (*www.acrim.-com/senior%20review%202007/SH41A-02_fall%20AGU_final.pdf*).

Wilmking, M., and Singh, J. (2008), "Eliminating the 'divergence problem' at Alaska's northern treeline", *Climate of the Past Discussions*, **4**, 741–759.

Wingham, D.J., Shepherd, A., Muir, A., and Marshall, G.J. (2006), "Mass balance of the Antarctic ice sheet", *Transactions of the Royal Society*, **364**, 1627–1635.

Winsor, P. (2001), "Arctic sea ice thickness remained constant during the 1990s", *Geophysical Research Letters*, **28**, 1039–1041.

Winstanley, D., and Wendland, W.M. (2007), "Climate change and associated changes to the water budget", in *Climate Change and Variations: A Primer for Teachers*, Vol. 1, National Council for Geographic Education Special Publications (*www.sws.uiuc.edu/docs/books/ww/WinstanleyWendland07.pdf*).

Winton, M. (2006), "Amplified Arctic climate change: What does surface albedo feedback have to do with it?", *Geophysical Research Letters*, **33**, L03701.

Wood, K.R., and Overland, J.E. (2010), "Early 20th century Arctic warming in retrospect", *International Journal of Climatology*, **30**, 1269–1279.

Woods, T. (2008), "Solar irradiance variability observed during Solar Cycle 23" (*solar.physics.montana.edu/SVECSE2008/pdf/woods_svecse.pdf*).

Woodworth, P.L. (1990), "A search for accelerations in records of European mean sea level", *International Journal of Climatology*, **10**, 129–143.

Woodworth, P.L., White, N.J., Jevrejeva, S., Holgate, S.J., Church, J.A., and Gehrels, W.R. (2009), "Evidence for the accelerations of sea level rise on multi-decade and century timescales", *International Journal of Climatology*, **29**, 777–789.

Woppelmann, G., Miguez, B.M., Bouin, M.-N., and Altamimi, Z. (2007), "Geocentric sea level trend estimates from GPS analyses at relevant tide gauges world-wide", *Global and Planetary Change*, **57**, 396–406.

Wopplemann, G., Pouvreau, N., Coulomb, A., Simon, B., and Woodworth, P.L. (2008), "Tide gauge datum continuity at Brest since 1711: France's longest sea-level record", *Geophysical Research Letters*, **35**, L22605.

Wu, L., Lee, D.E., and Liu, Z. (2005), "The 1976/77 North Pacific climate regime shift: The role of subtropical ocean adjustment and coupled ocean–atmosphere feedbacks", *Journal of Climate*, **18**, 5125–5140.

Wuebbles, D.J., and Hayhoe, K. (2000), "Atmospheric methane: Trends and impacts", in J. van Ham, A.P.M. Baede, L.A. Meyer, and R. Ybema (eds), *Non-CO_2 Greenhouse Gases: Scientific Understanding, Control and Implementation*, Kluwer Academic Publishers, Netherlands, pp. 425–432.

Wunsch, C. (1999), "The interpretation of short climate records, with comments on the North Atlantic and Southern Oscillations", *Bulletin of the American Meteorological Society*, **80**, 245–255.

Wunsch C., Ponte, R.M., and Heimbach, P. (2007), "Decadal trends in sea level patterns: 1993–2004", *Journal of Climate*, **20**, 5889–5911.

Yang, B., Braeuning, A., Johnson, K.R., and Yafeng, S. (2002), "General characteristics of temperature variation in China during the last two millennia", *Geophysical Research Letters*, **29**, 1324–1327.

Yang, B., Wang, J., Brauning, A., Dong, Z., and Esper, J. (2009), "Late Holocene climatic and environmental changes in arid central Asia", *Quaternary International*, **194**, 68–78.

Yang, P., Hong, G., Dessler, A.E., Ou, S.S.C., Liou, K.-N., Minnis, P., and

Harshvardhan (2010), "Contrails and induced cirrus optics and radiation", *Bulletin of the American Meteorological Society (BAMS)*, April, 473–478.
Yang, X., Hou, Y., and Chen, B. (2011), "Observed surface warming induced by urbanization in east China", *Journal of Geophysical Research*, **116**, D14113.
Zachos, J., Pagani, M., Sloan, L., Thomas, E., and Billups, K. (2001), "Trends, rhythms and aberrations in global climate 65 Ma to present", *Science*, **292**, 686–693.
Zapadka, T., WoŸniak, B., and Dera, J. (2007), "A more accurate formula for calculating the net long wave radiation flux in the Baltic Sea", *Oceanologia*, **49**, 449–470.
Zeebe, R.E. (2011), "Where are you heading Earth?", *Nature Geoscience*, **4**, 416–417.
Zemp, M., Hoelzle, M., and Haeberli, W. (2009), "Six decades of glacier mass-balance observations: A review of the worldwide monitoring network", *Annals of Glaciology*, **50**, 101.
Zhang, P., Imhoff, M., Wolfe, R., and Bounou, L. (2010), "Potential drivers of urban heat islands in the northeast USA", American Geophysical Union, Fall Meeting 2010, abstract #B21E-0362 (*www.nasa.gov/pdf/505254main_zhang.pdf*).
Zhang, Q.-B., Cheng, G., Yao, T., Kang, X., and Huang, J. (2003), "A 2,326-year tree-ring record of climate variability on the northeastern Qinghai–Tibetan Plateau", *Geophysical Research Letters*, **30**, 1739–1742.
Zhang, X., and Church, J.A. (2012), "Sea level trends, interannual and decadal variability in the Pacific Ocean", *Geophysical Research Letters*, **39**, L21701.
Zhou, L., Dickinson, R.E., Tian, Y., Fang, J., Li, Q., Kaufmann, R.K., Tucker, C.J., and Myneni, R.B. (2004), "Evidence for a significant urbanization effect on climate in China", *Proceedings of the National Academy of Sciences (PNAS)*, **101**, 9540–9544.
Zorita, E., and von Storch, H. (2005), "Methodical aspects of reconstructing non-local historical temperatures", *Memorie della Società Astronomica Italiana*, **76**, 794–801.
Zorita, E., Gonzalez-Rouco, F.J., and von Storch, H. (2007), "Comments on 'Testing the fidelity of methods used in proxy-based reconstructions of past climate'", *Journal of Climate*, **20**, 3693–3698.
Zorita, E., von Storch, H., Gonzalez-Rouco, F.J., Cubasch, U., Legutke, S., Fischer-Bruns, I., and Schlese, U. (2003), *Simulation of the Climate of the Last Five Centuries*, GKSS Report, 2003/12 (*http://coast.gkss.de/staff/storch/pdf/gkss_2003_12.pdf*).
Zubrin, R. (2007), "The hydrogen hoax", *The New Atlantis*, winter.
Zumbuhl, H.J., Steiner, D., and Nussbaumer, S.U. (2008), "19th century glacier representations and fluctuations in the central and western European Alps: An interdisciplinary approach", *Global and Planetary Change*, **60**, 42–57.
Zwally, H.J., Abadalati, W., Herring, T., Larson, K., Saba, J., and Steffan, K. (2002), "Surface melt-induced acceleration of Greenland ice-sheet flow", *Science*, **297**, 218–222.
Zwally, H.J., Giovinetto, M.B., Li, J., Cornejo, H.G., Beckley, M.A., Brenner, A.C., Saba, J.L., and Yi, D. (2005), "Mass changes of the Greenland and Antarctic ice sheets and shelves and contributions to sea-level rise: 1992–2002", *Journal of Glaciology*, **51**, 509–527.

Index

Bold = include figures
Italics = include tables

D. Rapp, *Assessing Climate Change*, Springer Praxis Books,
DOI: 10.1007/978-3-319-00455-6

The manufacturer's authorised representative in the EU is Springer Nature Customer Service Centre GmbH, Europaplatz 3, 69115 Heidelberg, Germany. If you have any concerns regarding our products, please contact ProductSafety@springernature.com

Printed and bound by CPI Group (UK) Ltd, Croydon, CR0 4YY
30/09/2026
02225238-0019